AF732642

Lecture Notes in Electrical Engineering

Volume 703

Series Editors

Leopoldo Angrisani, Department of Electrical and Information Technologies Engineering, University of Napoli Federico II, Naples, Italy
Marco Arteaga, Departament de Control y Robótica, Universidad Nacional Autónoma de México, Coyoacán, Mexico
Bijaya Ketan Panigrahi, Electrical Engineering, Indian Institute of Technology Delhi, New Delhi, Delhi, India
Samarjit Chakraborty, Fakultät für Elektrotechnik und Informationstechnik, TU München, Munich, Germany
Jiming Chen, Zhejiang University, Hangzhou, Zhejiang, China
Shanben Chen, Materials Science and Engineering, Shanghai Jiao Tong University, Shanghai, China
Tan Kay Chen, Department of Electrical and Computer Engineering, National University of Singapore, Singapore, Singapore
Rüdiger Dillmann, Humanoids and Intelligent Systems Laboratory, Karlsruhe Institute for Technology, Karlsruhe, Germany
Haibin Duan, Beijing University of Aeronautics and Astronautics, Beijing, China
Gianluigi Ferrari, Università di Parma, Parma, Italy
Manuel Ferre, Centre for Automation and Robotics CAR (UPM-CSIC), Universidad Politécnica de Madrid, Madrid, Spain
Sandra Hirche, Department of Electrical Engineering and Information Science, Technische Universität München, Munich, Germany
Faryar Jabbari, Department of Mechanical and Aerospace Engineering, University of California, Irvine, CA, USA
Limin Jia, State Key Laboratory of Rail Traffic Control and Safety, Beijing Jiaotong University, Beijing, China
Janusz Kacprzyk, Systems Research Institute, Polish Academy of Sciences, Warsaw, Poland
Alaa Khamis, German University in Egypt El Tagamoa El Khames, New Cairo City, Egypt
Torsten Kroeger, Stanford University, Stanford, CA, USA
Qilian Liang, Department of Electrical Engineering, University of Texas at Arlington, Arlington, TX, USA
Ferran Martín, Departament d'Enginyeria Electrònica, Universitat Autònoma de Barcelona, Bellaterra, Barcelona, Spain
Tan Cher Ming, College of Engineering, Nanyang Technological University, Singapore, Singapore
Wolfgang Minker, Institute of Information Technology, University of Ulm, Ulm, Germany
Pradeep Misra, Department of Electrical Engineering, Wright State University, Dayton, OH, USA
Sebastian Möller, Quality and Usability Laboratory, TU Berlin, Berlin, Germany
Subhas Mukhopadhyay, School of Engineering & Advanced Technology, Massey University, Palmerston North, Manawatu-Wanganui, New Zealand
Cun-Zheng Ning, Electrical Engineering, Arizona State University, Tempe, AZ, USA
Toyoaki Nishida, Graduate School of Informatics, Kyoto University, Kyoto, Japan
Federica Pascucci, Dipartimento di Ingegneria, Università degli Studi "Roma Tre", Rome, Italy
Yong Qin, State Key Laboratory of Rail Traffic Control and Safety, Beijing Jiaotong University, Beijing, China
Gan Woon Seng, School of Electrical & Electronic Engineering, Nanyang Technological University, Singapore, Singapore
Joachim Speidel, Institute of Telecommunications, Universität Stuttgart, Stuttgart, Germany
Germano Veiga, Campus da FEUP, INESC Porto, Porto, Portugal
Haitao Wu, Academy of Opto-electronics, Chinese Academy of Sciences, Beijing, China
Junjie James Zhang, Charlotte, NC, USA

The book series *Lecture Notes in Electrical Engineering* (LNEE) publishes the latest developments in Electrical Engineering - quickly, informally and in high quality. While original research reported in proceedings and monographs has traditionally formed the core of LNEE, we also encourage authors to submit books devoted to supporting student education and professional training in the various fields and applications areas of electrical engineering. The series cover classical and emerging topics concerning:

- Communication Engineering, Information Theory and Networks
- Electronics Engineering and Microelectronics
- Signal, Image and Speech Processing
- Wireless and Mobile Communication
- Circuits and Systems
- Energy Systems, Power Electronics and Electrical Machines
- Electro-optical Engineering
- Instrumentation Engineering
- Avionics Engineering
- Control Systems
- Internet-of-Things and Cybersecurity
- Biomedical Devices, MEMS and NEMS

For general information about this book series, comments or suggestions, please contact leontina.dicecco@springer.com.

To submit a proposal or request further information, please contact the Publishing Editor in your country:

China

Jasmine Dou, Editor (jasmine.dou@springer.com)

India, Japan, Rest of Asia

Swati Meherishi, Editorial Director (Swati.Meherishi@springer.com)

Southeast Asia, Australia, New Zealand

Ramesh Nath Premnath, Editor (ramesh.premnath@springernature.com)

USA, Canada:

Michael Luby, Senior Editor (michael.luby@springer.com)

All other Countries:

Leontina Di Cecco, Senior Editor (leontina.dicecco@springer.com)

**** This series is indexed by EI Compendex and Scopus databases. ****

More information about this series at http://www.springer.com/series/7818

S. N. Merchant · Krishna Warhade ·
Debashis Adhikari
Editors

Advances in Signal and Data Processing

Select Proceedings of ICSDP 2019

Springer

Editors
S. N. Merchant
Department of Electrical Engineering
Indian Institute of Technology Bombay
Mumbai, Maharashtra, India

Krishna Warhade
Department of Electronics and
Communication Engineering
MIT World Peace University
Pune, Maharashtra, India

Debashis Adhikari
School of Electrical Engineering
MIT Academy of Engineering
Pune, Maharashtra, India

ISSN 1876-1100 ISSN 1876-1119 (electronic)
Lecture Notes in Electrical Engineering
ISBN 978-981-15-8390-2 ISBN 978-981-15-8391-9 (eBook)
https://doi.org/10.1007/978-981-15-8391-9

© Springer Nature Singapore Pte Ltd. 2021
This work is subject to copyright. All rights are reserved by the Publisher, whether the whole or part of the material is concerned, specifically the rights of translation, reprinting, reuse of illustrations, recitation, broadcasting, reproduction on microfilms or in any other physical way, and transmission or information storage and retrieval, electronic adaptation, computer software, or by similar or dissimilar methodology now known or hereafter developed.
The use of general descriptive names, registered names, trademarks, service marks, etc. in this publication does not imply, even in the absence of a specific statement, that such names are exempt from the relevant protective laws and regulations and therefore free for general use.
The publisher, the authors and the editors are safe to assume that the advice and information in this book are believed to be true and accurate at the date of publication. Neither the publisher nor the authors or the editors give a warranty, expressed or implied, with respect to the material contained herein or for any errors or omissions that may have been made. The publisher remains neutral with regard to jurisdictional claims in published maps and institutional affiliations.

This Springer imprint is published by the registered company Springer Nature Singapore Pte Ltd.
The registered company address is: 152 Beach Road, #21-01/04 Gateway East, Singapore 189721, Singapore

Preface

The present era of technological advancements is moving at a very fast pace. It is therefore essential to keep oneself abreast with all the recent developments in various engineering domains. The book discusses various state–of–the-art developments in the diverse area of communications, data processing and signal processing, and the endeavor has been to bring together some of these developments in a concise platform that would benefit students, researchers, academicians and industry people. The chapters presented in the book have been selected on the basis of relevance and mathematical deliberations on the topics. Apart from the above-listed domains, this book has additionally included topics on social issues providing advanced technological solutions.

Chapter "Deep Semantic Segmentation for Self-driving Cars" introduces the technique of semantic segmentation of urban scene for a self-driving car that comprises three sub-systems in navigation, viz. lane finding, urban scene understanding and geo-positioning. In Chapter "Shot Boundary Detection Using Artificial Neural Network," hybrid video shot boundary detection process using feature extraction by mean log difference is discussed in combination with artificial neural network techniques. A system for leaf parameter analysis is proposed in Chapter "Custard Apple Leaf Parameter Analysis, Leaf Diseases, and Nutritional Deficiencies Detection Using Machine Learning," where detection of N, P and K deficiencies and leaf diseases is accomplished using K-nearest neighbors (k-NN) and support vector machine (SVM) algorithms. A typical problem of recognizing and removing the rain streaks on photographs by an improved convolutional neural network (CNN) architecture is discussed in Chapter "Single Image Rain Removal Using Convolutional Neural Network." A study of voice samples for two disorders—hypo and hyper—along with normal voice samples is considered in Chapter "A Robust Approach of Estimating Voice Disorder Due to Thyroid Disease" to create a databank for three classes—normal, hypo and hyper. A combined classifier, i.e., SVM and HMM (hidden Markov model), was utilized.

In Chapter "Face Recognition Using Golden Ratio for Door Access Control System," a combination of Viola–Jones face detection method with characteristics of extracting golden ratio is discussed to improve the security of sensitive places through facial recognition. This book has included chapters on topics such as underwater detection of objects, adaptive background subtraction models for shot detection, acoustic classification and evaluation of bird species using support vector machine and artificial neural networks. In Chapter "Feature-Based Model for Landslide Prediction Using Remote Sensing and Digital Elevation Data," a study to generate landslide susceptible maps and landslide hazard zonation maps is presented using the digital elevation model for the prediction of future landslides. Chapter "Emotion Recognition using Gamma Correction Technique Applied to HOG and LBP Features" discusses techniques on emotion recognition using gamma correction when applied to histogram of oriented gradient (HOG) and LBP features. The book also includes diverse areas on images and speech signal processing, besides the above-mentioned topics such as analysis of vocal tract parameters of speech, 3D reconstruction of plant features with non-destructive plant growth monitoring systems, digital image watermarking by fusion of wavelet and curvelet transform, and content-based image retrieval (CBIR) techniques.

This book also features many recent advancements on machine learning algorithms. Chapter "Automatic Gear Sorting Using Wireless PLC Based on Computer Vision" discusses conversion of wired PLC into wireless PLC by interfacing the PLC with the Wi-Fi module to enable real-time surveillance and control of the system of equipment sorting via Wi-Fi module interfacing with PLC. Chapter "Machine Learning Feature Selection in Archery Performance" discusses machine learning feature selection in Archery performance using Boruta algorithm, Chapter "Skin Lesion Classification Using Deep Learning" deliberates on skin lesion classification using deep learning, Chapter "Deep Learning-Based Paperless Attendance Monitoring System" deals with deep learning-based paperless attendance monitoring system, and Chapter "Image Analytics to Detect Cigarette in an Image Using Deep Learning" deals with image analytics to detect cigarette in an image using deep learning.

In the domain of antenna design and communication, Chapter "Frequency and Pattern Reconfigurable Antenna for WLAN and WiMAX Application" proposes an innovative bow tie frequency and pattern reconfigurable antenna for WLAN and WiMAX applications. Chapter "Design of a Power Efficient Multiband Patch Antenna" gives a design consideration of multiband patch antenna. Chapter "A Frequency Reconfigurable Antenna for Sub-GHz and TV White Space Applications" discusses a frequency reconfigurable antenna for sub-GHz and TV white space. Chapter "Comparative Analysis of Least Squares Method and Extended Kalman Filter for Position Estimation in GPS Receiver" provides a comparative analysis of position estimation techniques in a GPS receiver by using the least squares (LS) method and extended Kalman filter method (EKF). Fair scheduling non-orthogonal random access for 5G networks is presented in Chapter "Fair Scheduling Non-orthogonal Random Access for 5G Networks." Chapter "An Improved Carrier Frequency Offset Estimation Under Narrowband Interference in OFDM Cognitive Radio" surveys various techniques to estimate carrier frequency offset (CFO) for OFDM cognitive radio.

This book has also included few interesting power-related chapters. Chapter "Trends in Energy Management System for Smart Microgrid—An Overview" reviews several energy management systems developed based on different strategic approaches available for microgrid on demand-side management. Chapter "Discontinuous PWM Techniques to Eliminate Over-Charging Effects in Four-Level Five-Phase Induction Machine Drives" presents equivalent circuit modeling of a Li-ion battery cell and its state of charge estimation using the Kalman filter algorithm in MATLAB Simulink. A Transition Based Odd/Full Invert (TBO/FI) coding scheme, which focuses on crosstalk avoidance and low dynamic power consumption in NoC links, is also discussed. Chapter "Efficient Design of Drone Flight Control Using Delay Tolerant Algorithm" presents a study to reduce the human error parameter in the probable causes for drone crashes.

Various social issues are also presented by authors with their technical solutions. In Chapter "IRIS: An Application for the Visually Impaired Using Google Cloud API," the authors present the design considerations of a cost-effective and efficient visual aid which proposes a smart stick (IRIS) to help the user in obstacle detection and navigation. Chapter "Implementation of Hand Gesture Recognition System to Aid Deaf-Dumb People" considers a time system for hand gesture recognition that acknowledges hand gestures and then converts them into text and voice. Statistical validity of pre-smoking and post-smoking impact on heart rate variability among middle-age men is presented in Chapter "Statistical Validity of Presmoking and Postsmoking Impact on Heart Rate Variability Among Middle Age Men." Chapter "Analysis of Chronic Joint Pain Using Soft Computing Techniques" analyzes the chronic joint pain remedies using soft computing techniques.

A critical evaluation of each submitted chapter by at least two expert reviewers was carried out. The authors re-submitted with all suggested alterations given by the expert panel. The book would definitely be of immense help to passionate researchers, students and industry persons.

Mumbai, India S. N. Merchant
Pune, India Debashis Adhikari
Pune, India Krishna Warhade

Contents

About the Editors

Dr. S. N. Merchant is currently a Professor (Emeritus Fellow) in the Department of Electrical Engineering at IIT, Mumbai. He obtained his BTech, MTech and PhD degrees from the department of Electrical Engineering, IIT Mumbai. His broad area of research interest are wireless communication, wireless sensor network, signal processing, multimedia communication and image processing. His noteworthy contribution has been in solving state of the art signal and image processing problems faced by the Indian Defense forces. He serves on the editorial board of two international journals: International Journal of Ultra-Wideband Communication & Systems and International Journal of Distributed Sensor Networks. He is a fellow of Institution of Electronics and Telecommunication Engineers, IETE and is a recipient of 10th IETE Prof SVC Aiya Memorial Award for his contribution in the field of detection and tracking. He is also the recipient of the 9th IETE Prof SVC Aiya Memorial Award for "Excellence in Telecom Education". He is the winner of 2013 VASVIK Award in the category of Electrical & Electronic Sciences and Technology.

Dr. Krishna Warhade is currently the Dean Research, Development, Innovation and Consultancy and Professor and Head of the Department, department of Information Technology at MIT College of Engineering and Dr. Vishwanath Karad MIT World Peace University, Pune, India. He has obtained his BE (Electronics) and ME (Instrumentation) from Shri Guru Gobind Singhji Institute of Engg. & Tech., Nanded, under Swami Ramanand Teerth Marathwada University, Nanded. He obtained his PhD from IIT, Mumbai. His research interests include healthcare, video segmentation, wavelets, digital signal processing, digital Image processing, digital filter design and ICT in agriculture. He has worked on various collaborative projects with Bhaba Atomic Research Center (BARC), Blackberry and Konkan Railway Corporation Limited (KRCL).

Dr. Debashis Adhikari is currently the Professor and Dean, School of Electrical Engineering at MIT Academy of Engineering, Pune. He has obtained his BTech from the Institute of Radio Physics & Electronics, University of Calcutta, ME and

PhD from the Defense Institute of Advanced Technology (DRDO) (DU), Pune. His broad area of research interest is wireless communication and antenna systems. He has vast experience on surface to air missile systems, troposcatter communication, SATCOM systems of the Indian Air Force. He has been actively involved in many modification projects on SAM systems of the IAF. He is a fellow of IETE and is a regular reviewer of many reputed international journals.

Deep Semantic Segmentation for Self-driving Cars

Abhilash Sachin Kulkarni, Jyothi S. Nayak, Aditi Desai, Jahnavi Singh, and Shraddha Murali

Abstract A self-driving car comprises three subsystems in the navigation, viz. lane finding, urban scene understanding and geopositioning. This paper introduces the technique of semantic segmentation for urban scene understanding with various implementations in recent years and proposes a novel approach to fast and accurate semantic segmentation. The architecture of the model is designed using VGG16 as encoder, adopting hierarchical feature fusion technique to perform downsampling followed by a lightweight decoder to perform upsampling. Skip connections are introduced between the encoder and the decoder to improve the information flow. This new approach outperforms the previous implementations by giving 94% accuracy and 8.7 frames per second (FPS) simultaneously.

Keywords Semantic segmentation · FCN · CRFs · KITTI · VGG16

A. S. Kulkarni (✉) · J. S. Nayak · A. Desai · J. Singh · S. Murali
Department of Computer Science and Engineering, BMS College of Engineering, Basavangudi, Bengaluru, Karnataka, India
e-mail: abhilashkulkarni340@gmail.com

J. S. Nayak
e-mail: jyothinayak.cse@bmsce.ac.in

A. Desai
e-mail: aditi.desai298@gmail.com

J. Singh
e-mail: sjahnavi70@gmail.com

S. Murali
e-mail: shraddha_murali@yahoo.co.in

© Springer Nature Singapore Pte Ltd. 2021

S. N. Merchant et al. (eds.), *Advances in Signal and Data Processing*,
Lecture Notes in Electrical Engineering 703,
https://doi.org/10.1007/978-981-15-8391-9_1

1 Introduction

Semantic segmentation is a method of scene understanding in which classification is performed on every single pixel of an image. Semantic segmentation is used in autonomous vehicles to locate frontal objects such as roads, dividers, vehicles, pavements, etc. It is a vital subsystem of the vehicle's navigation system.

In the early days, deep neural networks were used for semantic segmentation. This is termed as the classical method [1], and approaches based on neural networks were used. Some of the networks include random forest classifier, condition random fields (CRFs) and texton forest. CRFs, a sequential modelling algorithm, are used for structured prediction problems. CRFs can identify and recognize patterns after they are trained.

Another approach uses fully convolution networks [2] such as pretrained VGG or ImageNet. FCNs have convolution layers that use kernels and perform computations across all regions. The computation is divided into multiple regions and generates a feature map. This feature map is then upsampled using deconvolution process. These networks are used for segmentation, and another network is employed to perform upsampling or deconvolution operation.

The aforesaid methods do not exploit the structure of the data. These methods tend to establish a structure in the data and learn from them. Condition random fields or CRFs [3] incorporate structure in the model. CRFs can be used to exploit the structure in segmentation of automotive scenes and for post processing after segmentation.

The most widely used architecture is the use of dilated convolution after the condition random fields network for considering contextual information. Another popular approach for using structure in data is the spatio-temporal approach. In this approach, the spatial and temporal information of the data is used to train the model. This is most widely used in video segmentation. For example, the use of long short-term memory (LSTMs) in fully convolution networks allow different layers to learn at different rates controlled by clockworks involving small clock signals or clock pulses [4]. The clockworks method is depicted in Fig. 1.

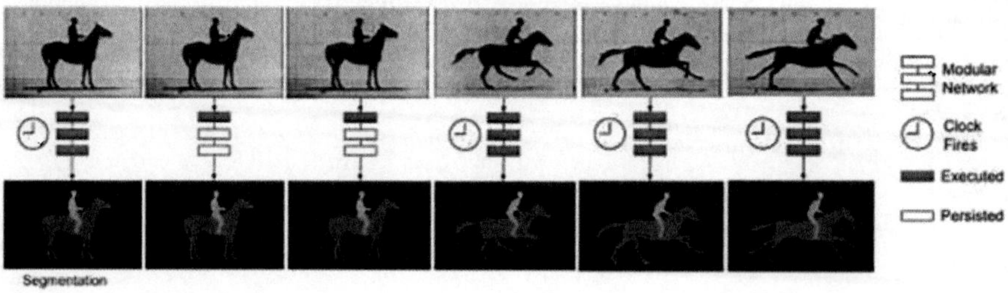

Fig. 1 Clockworks method

2 Related Work

This section discusses a few models of semantic segmentation:

2.1 SegNet

SegNet is a neural network that comprises an encoder, a decoder and a classification layer that performs pixel-wise classification. The encoder has thirteen convolution layers. Pretrained VGG16 is generally modified by replacing the fully connected layers with fully convolution layers. The feature maps generated by the layers are retained. For training, the model is initialized with the pretrained weights since VGG16 is trained for classification problems on ImageNet data.

The encoder network is followed by a decoder network. The decoder network has thirteen layers, and each layer corresponds to a layer in the encoder network. Pixel-wise class probabilities are produced by feeding the decoder output directly into a Softmax classifier. This produces a segmented image [5]. This is depicted in Fig. 2.

2.2 DeepLab

There are various methods of performing semantic segmentation using DeepLab [6] such as atrous convolution and atrous spatial pyramid pooling. Atrous convolution is used in deep convolution neural networks (DCNNs) for computation of undecimated wavelet transforms. Atrous convolution is used for its efficiency in computation. The response of layers can be computed at any resolution and can be seamlessly integrated with training and applied thereafter once trained.

Atrous spatial pyramid pooling is based on multiscale image representation technique and DCNNs. It has two approaches, viz. multiscale processing and spatial pyramid pooling method which is widely used in recurrent-CNNs. Multiscale image

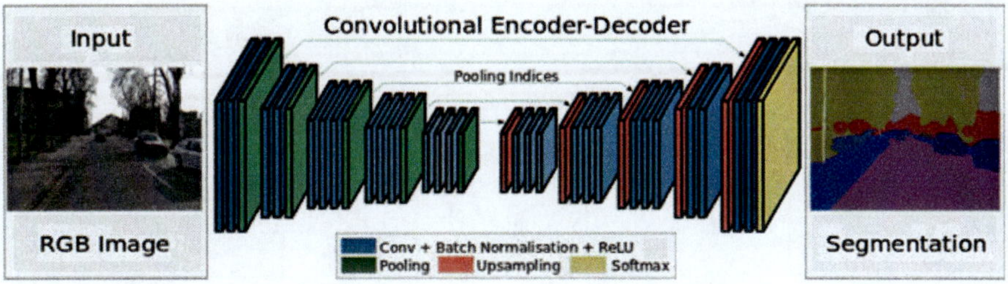

Fig. 2 SegNet architecture

representation can handle scale variability in semantics segmentation. R-CNN can efficiently and accurately classify regions of arbitrary scale. The model can be trained on datasets that contain objects of varying size.

2.3 RefineNet

Dilated convolution is a technique employed for faster convolution. Dilated convolution introduces zeros or holes in the convolution kernel between each pixel, thereby significantly increasing the effective receptive field of the kernel. However, dilated convolution is computationally expensive as it has to be applied on a huge number of feature maps that are of high resolution involving volumes of memory. Comparatively, DeepLab performs the same computation with predictions one-eighth the size of input image.

This approach uses ResNet-101 blocks for the encoder, while the decoder comprises RefineNet blocks. The decoder concatenates high-resolution feature maps generated by encoder and preceding RefineNet blocks [7]. The architecture of RefineNet is depicted in Fig. 3.

2.4 Fast Semantic Segmentation

Real-time semantic segmentation mandates execution speed to be traded against accuracy of segmentation. Fast semantic segmentation features a network that is

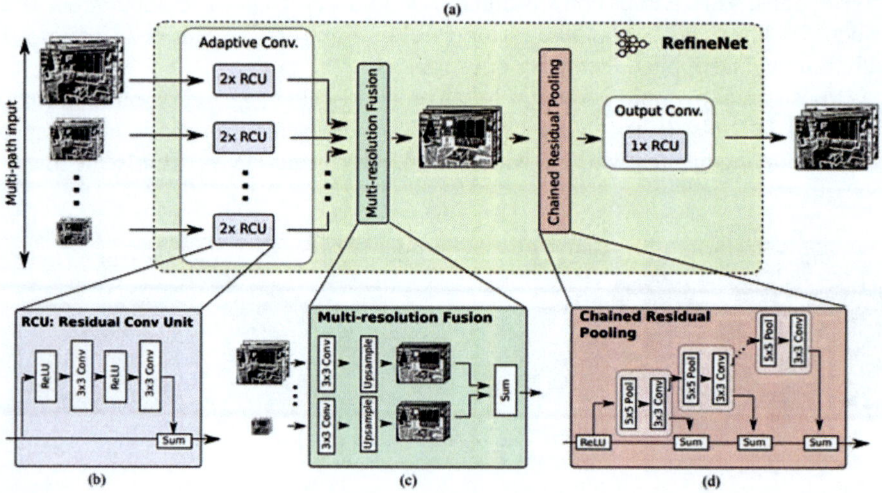

Fig. 3 RefineNet network and its components

structured as an encoder-decoder pair where the encoder has CNN to detect high-level objects in the input image and the decoder takes this information from the output of encoder and performs prediction for each pixel in the original input. Fire module and dilated convolution [8] is featured in the model which helps in improving the accuracy of the model. RefineNet module with ELU is also included in the architecture.

2.5 U-Net for Biomedical Imaging

U-Net is widely used in the field of biomedical image segmentation and was developed by Olaf Ronnieberger et al. U-Net is based on architecture by Ciresan et al. [9] that uses sliding window setup for training the model. This setup allows for the receptive fields to obtain the context of a region around the pixels by providing local regions around the pixel as inputs. This redundancy results in a trade-off between accuracy and context.

U-Net consists of an encoder and a decoder. The encoder contains four convolution layers and a point-wise convolution layer, and the decoder is a standard decoder network symmetric to the encoder network. Skip connections are introduced in the network between the encoder and the decoder for improving the information flow between them.

2.6 PSPNet

PSPNet aggregates different scales of global information by adding multi-scale pooling on top of the backend model [10]. Bilinear interpolation is used for upsampling, and 3×3 convolution fuses different levels of feature after concatenation. Bilinear interpolation in PSPNet learns less number of parameters during training as compared to deconvolution operation reducing the amount of data needed to train the network.

3 Proposed Architecture

The proposed model will have an encoder-decoder architecture similar to U-Net. The image is fed to the encoder block that performs downsampling or convolution operation producing a feature map. This feature map is then fed into the decoder block that performs upsampling or deconvolution operation and produces a pixel-wise classification or a segmented image. The high-level architecture of the proposed semantic segmentation network is depicted in Fig. 4.

The subsystems involved in the encoder and decoder blocks are as follows:

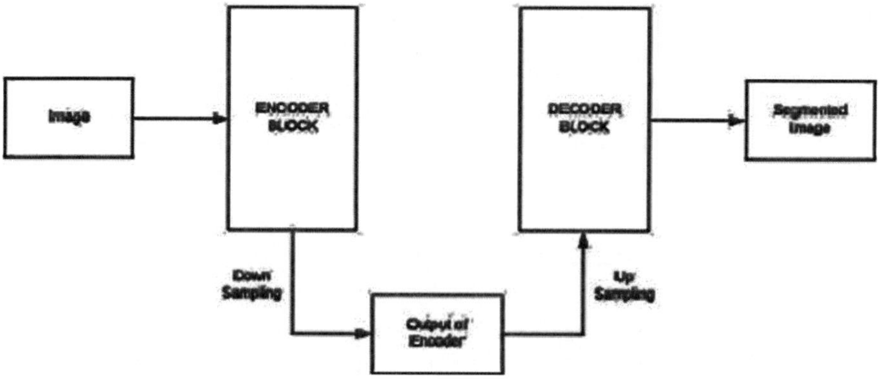

Fig. 4 High-level architecture

3.1 VGG16 (Encoder)

It is envisaged that the VGG16 network will form the backbone of the encoder block [11] since it offers very high accuracy with low loss. In the proposed architecture, the fully connected layers of the network are converted to fully convolutional network (FCN) in VGG16. The third, fourth and seventh fully connected layers are replaced with convolution layers with the convolution factorization principle resulting in the split of the computation into multiple steps applying the hierarchical feature fusion technique. These layers are rescaled for compatibility as a skip layer that introduces skip connection. Point-wise convolution is then applied to the rescaled layers to reduce the number of classes. The final layer of the VGG16 is a point-wise (1×1) convolution, and the number of filters is equal to the number of masks, i.e., the number of classes.

3.2 Hierarchical Feature Fusion

The proposed model will incorporate hierarchical feature fusion or HFF [12] technique, recently adopted in efficient spatial pyramid (ESP) block, to identify the dominant features without increasing the computation cost and intensity. ESP block uses reduce-split-transform method to handle downsampling. The reduce phase uses point-wise convolution to reduce the number of features, hence reducing the computation cost of the next phase. The convolution operation is then split into multiple operations where dilated convolution is applied with varying kernel sizes and holes. This produces multiple feature maps. HFF is applied in the transform phase. HFF concatenates the feature maps, generated by multiple dilated convolution operations in the split phase, hierarchically to obtain a feature map with the dominant feature(s). The use of hierarchical feature fusion in ESP block is depicted in Fig. 4.

3.3 Lightweight Decoder

The decoder employs a lightweight decoder network such as lightweight RefineNet [13]. This network performs the deconvolution operation at a faster rate and provides the desired speed in semantic segmentation.

3.4 Experimental Dataset

The following open-source datasets are available for training semantic segmentation networks:

1. Cityscapes dataset for semantic urban scene understanding [14]
2. Berkeley DeepDrive Video dataset (BDDV)
3. KITTI dataset [15]
4. ImageNet dataset
5. PASCAL VOC dataset
6. MNIST.

The dataset used for the project is KITTI pixel-level semantic segmentation dataset from KITTI Vision Benchmark Suite. The dataset contains training set and testing set. The training set contains original images, segmented images and calibration files. There are 289 original images and corresponding segmented images. Example images are shown in Fig. 5.

3.5 Implementation

3.5.1 Modification to VGG16

VGG16 as Encoder
Layers 3, 4 and 7 (fully connected layers) are replaced with fully convolution layers. L2 regularization (ridge regression) is used as a kernel regularizer, and the kernel is initialized with normal initialization. The kernel size and stride is 1, and the number of classes is 2. Padding remains the same.

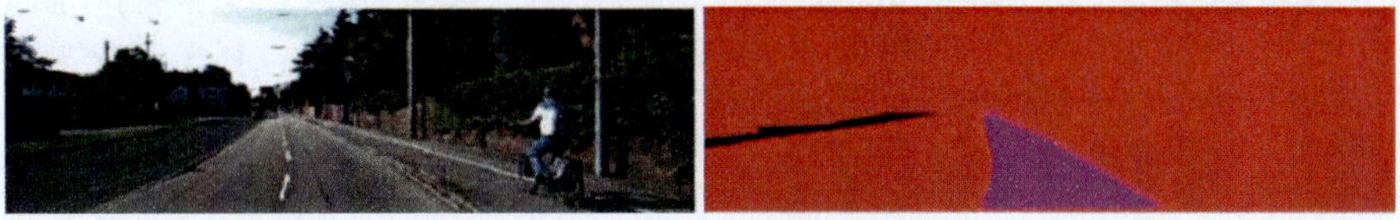

Fig. 5 Original image and corresponding segmented image

Decoder
Upsampling layer (transposed_conv) is placed corresponding to the 3rd, 4th and 7th layers of encoder. L2 regularization is used as kernel regularizer, and the kernel is initialized with zeros. The kernel size is 4, and strides is 2. The number of classes is 2, and padding remains the same.

Skip Connections
The output of the upsampling layers is rescaled, and skip connections are added between the encoder and decoder. The VGG layers are resized for compatibility as skip layers. The skip connections are added between layer 3 of encoder and the corresponding layer 3 of decoder and between layer 4 of encoder and the corresponding layer 4 of decoder.

Generator Function
Generator function is used to supply the images to the model. The function gets the path of the files in the dataset and loads the images randomly using cv2 package. The function generates batches of the images and resizes the images to same dimensions. The function returns an array with the images.

Optimization Function
The optimization function uses regularization loss and Softmax cross entropy loss to calculate loss and Adam optimizer.

3.5.2 Incorporation of Hierarchical Feature Fusion in the Model

The fully connected layers in third, fourth and seventh layers of VGG16 are replaced with fully convolution layers with dilation rate for dilated convolution. The convolution operation is split into multiple steps, and dilated convolution is performed with dilation rate from 2 to$n$. The feature maps generated is then concatenated hierarchically, and another convolution operation produces the final feature map which is fed into the next layer. The flowchart of the proposed architecture is shown in Fig. 6.

4 Experimental Setup and Results

After a critical assessment of the networks, U-Net has been chosen for this project owing to its accuracy in biomedical image segmentation and the same has been modified for our data. Semantic segmentation training was performed using KITTI dataset. Salient outcomes are appended below.

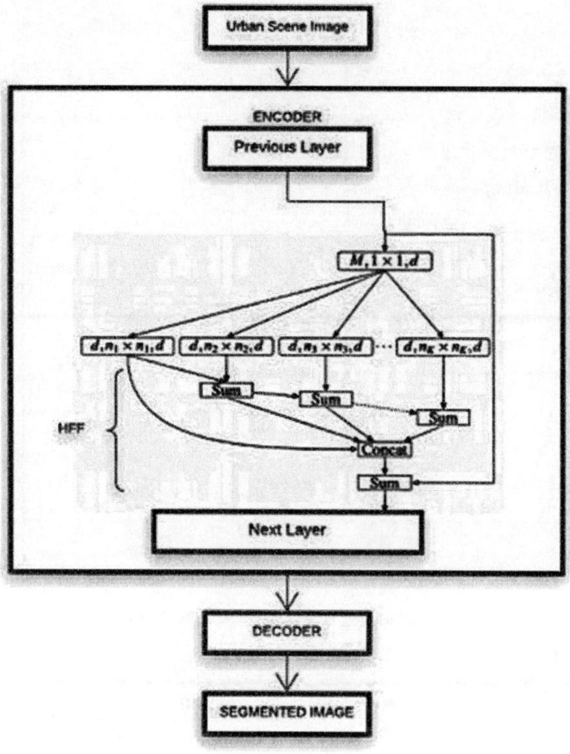

Fig. 6 Flowchart of the architecture

4.1 Architecture

The encoder has the standard U-Net architecture with multiple repeating convolution layers with rectified linear unit (RLUs) and max-pooling. The image is downsampled through these layers, and a feature map is generated.

The decoder uses the feature map generated by the encoder and performs upsampling or deconvolution operation to obtain a feature mask. This feature mask is the segmented image. The concatenation layer of the decoder only concatenates images with identical length and width. Hence, concatenation layer is modified to handle images with rectangular dimensions. The decoder architecture is depicted in Fig. 7.

4.2 Results

Training the model and its fine tuning resulted in the following: Hyperparameters for training are summarized in Table 1.

Accuracy and loss at the end of training are summarized in Table 2.

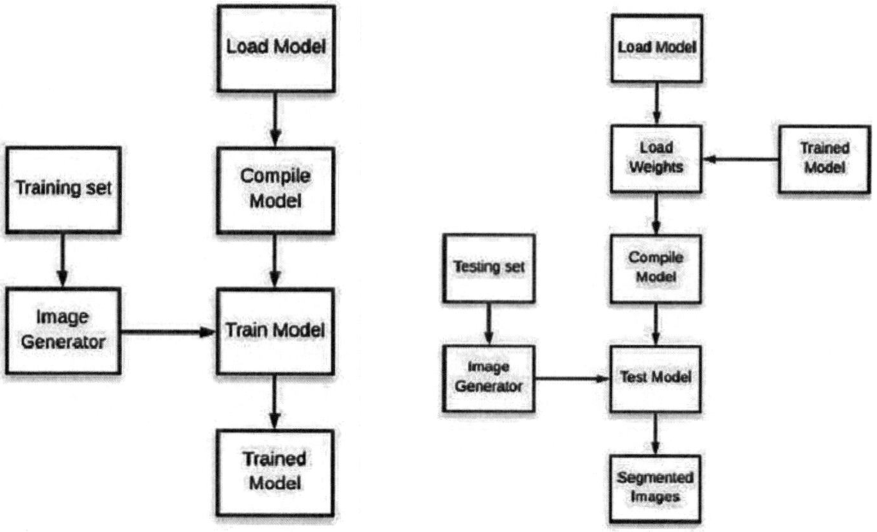

Fig. 7 Training process and testing process

Table 1 Hyperparameter for training U-Net

Hyperparameter	Value
Validation split	10%
Number of epochs	20
Batch size	32
Optimization function	ADAM

Table 2 Accuracies and losses at the end of training

	Training	Validation
Accuracy	88.79%	86.02%
Loss	0.83	0.85

4.2.1 System Testing

Training System

The training system uses the VGG16 architecture as encoder and a lightweight decoder and runs training for 25 epochs. The process of training is shown in Fig. 7.

In the training system, the model is loaded using the predefined configurations, and model is compiled with the optimization function. Image generator is used to get batches of images from the training dataset. These images (original and segmented) are fed into the model in every epoch. The model is trained on the images for 25 epochs using the same images. After the final epoch, a trained model is obtained with minimum loss, and the model and the weights are saved for testing.

The hyperparameters for training are summarized in Table 3.

Table 3 Hyperparameters for training

Hyperparameter	Value
Learning rate	0.009
Batch size	32
Epochs	25
Optimization function	ADAM

Table 4 Results of training against the hardware/platforms used

Platform/hardware	Training time (min)	Training accuracy	Training losses
Amazon Sagemaker	9	0.96	0.15
GCP	5	0.94	0.14
Lenovo Thinkpad	200	0.94	0.15

Each epoch runs training on 289 original images and 289 segmented images. The training time, training accuracy and training losses are shown against the hardware/platform used as shown in Table 4.

Testing System

The testing system uses the VGG16 trained model generated by the training system. The testing is performed on a test dataset containing 290 non-segmented images. The process of testing is shown in Fig. 7. In the testing system, the saved model is loaded into the instance. Then, the saved weights are loaded from the trained model, and the model is compiled. The image generator is used to make batches of images from the testing dataset. These images are then fed into the model, and the output is a set of segmented images.

4.2.2 Evaluation Metrics

The following evaluation metrics are used to evaluate the model during **training**:

1. Training Accuracy (TA): Accuracy obtained on the training set after every epoch.
2. Training Loss (TL): Loss calculated after every epoch on the training set.
3. Validation Accuracy (VA): Accuracy obtained on the validation set.
4. Validation Loss (VL): Loss (MSL) calculated on the validation set.
5. Training Time Per Epoch (TTPE): The time taken for training in one epoch.
6. Total Training Time (TTT): The total time taken for training. Table 5 shows the results of evaluation of the metric on the model.

The following evaluation metrics are used to evaluate the model during **testing**:

1. Total Testing Time: Total time taken to run segmentation on images

Table 5 Result of metric evaluation on the model

Epoch	TA	TL	VA	VL	TTPE (s)
1	0.61	1.049	0.53	0.943	508
5	0.73	0.677	0.69	0.775	489
10	0.79	0.420	0.77	0.501	482
15	0.85	0.218	0.79	0.230	486
20	0.91	0.165	0.89	0.181	481
25	0.95	0.140	0.92	0.149	482

$$L_i = -\log\left(\frac{e^{f_{y_i}}}{\sum_j e^{f_j}}\right) \text{ or equivalently } L_i = -f_{y_i} + \log \sum_j e^{f_j} \qquad L = \underbrace{\frac{1}{N}\sum_i L_i}_{\text{data loss}} + \underbrace{\lambda R(W)}_{\text{regularization loss}}$$

Fig. 8 Cross-entropy loss function and loss function with the regularization function

2. Frames Per Second (FPS): The ratio of the total number of test images and the total testing time. The higher the FPS, the faster the model.

The mean total testing time obtained was 33 s after running tests 15 times. Hence, the speed obtained was approximately 9 FPS.

The improvement in training and validation accuracy was achieved by customizing the optimization function and choosing an optimum batch size. The optimization function calculates the total loss and uses ADAM optimizer to minimize the loss in each step. This helps in increasing the training and validation accuracies at each step.

Loss Calculation

The loss calculation is performed by summing cross-entropy loss (Softmax loss), and L2 norm (regularization) that are calculated based on the difference between the actual probabilities and the predicted probabilities. The cross-entropy loss function is shown in Fig. 8.

The loss function along with the regularization function is shown in Fig. 8.

4.3 Performance Analysis

For analyzing performance, accuracy, total training time, total testing time, FPS and size of the model are considered. The results are shown in Table 6.

Modified VGG gives 94% accuracy compared to 86% of U-Net and 95% of PSPNet, but it gives 8.7 FPS compared to 3.7 of U-Net and 4.4 of PSPNet. The modified VGG16 outperforms U-Net and PSPNet in terms of FPS, but there’s a

Table 6 Results of three semantic segmentation networks

Model	Accuracy (approx. %)	Total training time (25 epochs, min)	Total testing time (290 images, s)	FPS	Size (MB)
Mod. VGG16	94	29	33	8.7	956
U-Net	86	21	78	3.7	128
PSPNet	95	24	66	4.4	398

heavy price to pay in terms of its size. The model performs accurately on the test images and provides accurate mask for the same.

5 Conclusion

The application of semantic segmentation systems that can perform at higher speeds without compromising accuracy is an imperative in autonomous vehicles. The encoder-decoder network as implemented in this paper utilizing VGG16 with HFF and lightweight decoder is expected to provides the requisite speed and accuracy for real-time deployment in autonomous vehicles.

References

1. Mennatullah S, Sara E, Martin J, Senthil Y (2017) Deep semantic segmentation for automated driving: taxonomy, roadmap and challenges
2. Jonathan L, Evan S, Trevor D (2015) Fully convolutional networks for semantic segmentation
3. Guosheng L, Chunhua S, Anton VDH, Ian R (2016) Efficient piecewise training of deep structured models for semantic segmentation
4. Mohsen F, Mohammad HS, Mohammad S, Mahmood F, Reinhard K, Fay H (2016) STFCN: spatio-temporal FCN for semantic video segmentation
5. Vijay B, Alex K, Roberto C (2017) SegNet: a deep convolutional encoder-decoder architecture for image segmentation
6. Liang-Chieh C, George P, Florian S, Hartwig A (2017) Rethinking atrous convolution for semantic image segmentation
7. Guosheng L, Anton M, Chunhua S, Ian R (2017) RefineNet: multi-path refinement networks for high-resolution semantics segmentation
8. Simon J, Michal D, David V, Adriana R, Yoshua B (2017) The one hundred layers tiramisu: fully convolutional densenets for semantic segmentation
9. Ciresan DC, Gambardella LM, Giusti A, Schmidhuber J (2012) Deep neural networks segment neuronal membranes in electron microscopy images
10. Hengshuang Z, Jianping S, Xiaojuan Q, Xiaogang W, Jiaya J (2017) Pyramid scene parsing network
11. Karen S, Andrew Z (2014) Very deep convolution networks for large-scale image recognition
12. Sachin M, Mohammad R, Anat C, Linda S, Hannaneh H (2018) ESPNet: efficient spatial pyramid of dilated convolution for semantic segmentation
13. Vladimir N, Chunhua S, Ian R (2018) Light-weight refinenet for real-time semantic segmentation

14. Marius C, Mohamed O, Sebastian R, Timo R, Markus E, Rodrigo B, Uwe F, Stefan R, Bernt S (2016) The cityscapes dataset for semantic urban scene understanding
15. Andreas G, Philip L, Christoph S, Raquel U (2013) Vision meets robotics: the KITTI dataset

Trends in Energy Management System for Smart Microgrid—An Overview

Harini Vaikund and S. G. Srivani

Abstract Smart Grid (SG) is an integrated grid to improve efficiency, consistency, and security with the incorporation of conventional and renewable energy sources, through automated control and modern communication technology. The development of Microgrid (MG) is a great solution for the incorporation of sustainable energy resources inside the smart grid environment. The emergence of Microgrid (MG) by decomposition of the grid is a combination of Distributed Energy Resources (DERs), Energy Storage System (ESS), loads and Control devices. This makes MG a single and controllable power supply system that can enhance versatility, dependability and furthermore incorporate the benefits of distributed generation. In MG, an EMS is fundamental for the ideal utilization of distributed energy sources in smart, secure, reliable and synchronized ways. The need for understanding the energy utilized is increasing since effective Energy Management is more challenging in Microgrid (MG). This paper reviews several Energy management systems developed based on different strategic approaches available for Microgrid on Demand-Side Management.

Keywords Microgrid · Energy management system · Demand-side management

1 Introduction

A Microgrid is a cutting edge distributed power system utilizing local sustainable energy sources designed through different smart grid initiatives. Renewable power resources like wind, solar, microturbines, latest generation technologies like combined heat and power (CHP) technology and fuel cell technologies become part of a Microgrid. Renewable energy resources normally have intermittency problems. To solve the intermittency problem electric vehicle technology, flywheel storage

H. Vaikund (✉)
Department of Electrical and Electronics, Dr. Ambedkar Institute of Technology, Bengaluru, India
e-mail: harini.vaikund@yahoo.com

S. G. Srivani
Department of Electrical and Electronics, R V College of Engineering, Bengaluru, India
e-mail: srivanisg@rvce.edu.in

© Springer Nature Singapore Pte Ltd. 2021

S. N. Merchant et al. (eds.), *Advances in Signal and Data Processing*,
Lecture Notes in Electrical Engineering 703,
https://doi.org/10.1007/978-981-15-8391-9_2

systems and Storage devices like battery energy storage systems can be used [1]. Due to their better coordination and control microgrid is preferred compared to the distributed generation sources. MG can be operated without the presence of the main grid (called as an islanded mode) so providing security for a local community. If it is connected to the grid it is grid-tied mode.

Three significant objectives of MG is

- Reliability—Cyber, physical
- Sustainability—environmental considerations
- Economics—efficiency, cost optimizing [2]. It is capable of operating with grid and/or without the grid (island mode). Figure 1 shows a MG architecture consisting of various DERs, critical and responsive loads. The main grid and the MG are linked through Point of Common Coupling (PCC). MG's advantage is, during grid-connected mode—power trading with the main grid is done and system stability shifts to the islanded mode when there are disturbances in the main grid. Microgrid Central Controller (MGCC) and Local Controllers (LCs) controls and coordinates the whole MG operation.

There are different classifications of MGs based on

1. Power type—AC and DC
2. Supervisory Control—Centralized and Decentralized
3. Operation Mode—Grid-connected and islanded
4. Phase—Single and Three Phases
5. Application—Residential/Commercial/Industrial and Utility/Municipality/Military.

A few focal points of MGs are a decrease in Green House Gases (GHG) outflows, voltage level improvement, power supply decentralization, Demand Response (DR) and incorporation of cogeneration. It additionally lessens losses in line and blackouts in transmission and distribution (T&D) systems [3, 4]. On different operating modes, the MGs to protect the grid, managing the load connected to the system and the

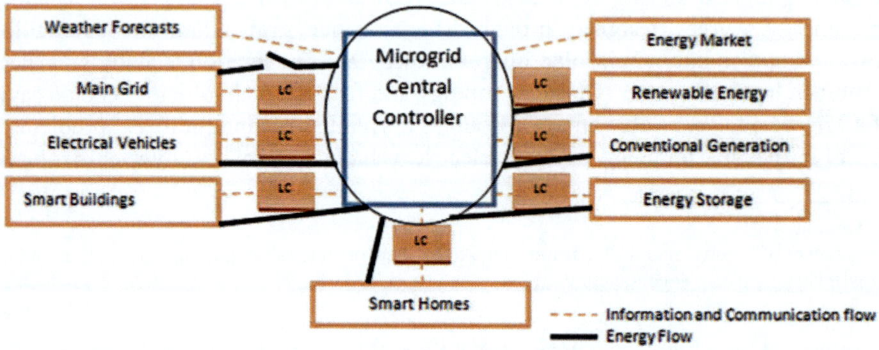

Fig. 1 Microgrid architecture

voltage and frequency adjustment is a requirement so resynchronization process and facilitation of the generation-side and load-side management is necessary.

Some restrictions of MGs are a high upfront cost of RERs, efficient use of power sources, problems due to control of MGs and non-availability of protected and administrative standards, and client security. Solving EMS problems is the key focus for researchers because of the non-availability of continuous RERs and increased integration of controllable loads [3].

In this paper, the Energy Management System (EMS) and the need for it is presented in Sect. 2. Demand-Side Management (DSM), necessity and different optimization techniques based on many mathematical optimization techniques which are used to solve the strategies are reviewed and discussed in Sect. 3. Finally, the paper is concluded in Sect. 4.

2 Energy Management System (EMS)—Introduction, Architecture, and Strategies

An energy management system (EMS) monitors, controls, optimizes the performance of the transmission and generation system and also saves overall operating cost in electric utility grids [5]. EMS is defined as "a computer system comprising a software platform providing basic support services and a set of applications providing the functionality needed for the effective operation of electrical generation and transmission facilities to assure adequate security of energy supply at minimum cost" [3]. Energy management is a venture to control all kinds of energy.

A basic illustration of MG EMS is shown in Fig. 2. The microgrid EMS consists of some components. Modules for efficient execution of decision-making strategies are carried out by Human Machine Interfaces (HMI), DERs/load forecasting and Supervisory Control and Data Acquisition (SCADA) modules by transferring optimal decisions to the storage, generation and load units. Optimizer to determine the optimal power flow trajectories and for electricity demand use of forecasting modules and

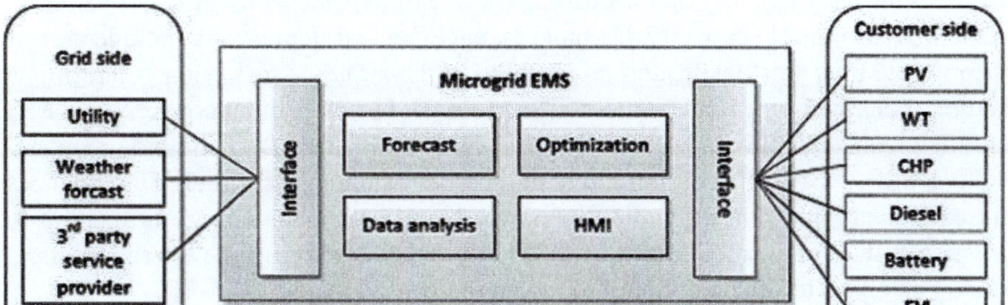

Fig. 2 An illustration of a microgrid energy management system

renewable energy sources power production are the main components of MG [6]. The local controller has the following functions—supervising the renewable energy sources (RES), energy storage (ES) and based on trajectories set by the optimizer collect the measurement data by activating the actuators.

In MG the main challenge is the reduction of energy balances caused due to the dynamic nature of electricity consumption and uncertainties in energy supply from renewable-based Distributed Generators (DGs). Small size imbalances can be handled by droop control or frequency control. In large supply-demand imbalances, these methods fail and which necessitates the need for the development of energy management strategies for microgrids [7]. These procedures give numerous advantages to live energy-saving, frequency regulation, low cost reduction, GHG emission reduction, and client privacy.

EMS of a MG incorporates both Demand-side and Supply-side Management, while fulfilling system constraints, to realize a reliable, cost-effective and sustainable working of MG. Scheduling DERs and loads, losses and system outages minimization, control of irregularity and unpredictability of RERs are the strategies of MG.

To achieve proficient and optimal operation of MG based on mathematical techniques various approaches [3, 8, 9] many researchers have been using to solve energy management strategies. An overview of these solution approaches and control strategies is described in the following sections. Table 1 gives some of the EMS based on strategic approaches used to solve the strategies.

3 Demand-Side Management

The expression "demand-side management" (DSM), also known as "energy demand management," stands for a various energy consumption related work. It incorporates the controls and alteration of energy use (e.g., energy preservation, efficiency, and storage) but also the practices that are engaged with the procedures [10].

Demand management refers to any strategy that reduces energy consumption or remodels the use of energy on the customer's side of the electricity grid as an alternative to increase supply capacity. Demand management strategies may include preservation programs, shift tasks, and an increase in the strategic task. Load control is a feature of demand-side management. The process of transforming conventional MG into the green system is facilitated by a powerful tool called Demand-side management (DSM). DSM is a cost-effective choice for balancing the energy imbalances caused by unreasonable utilization of power by adjusting the energy utilization to the supply. The main focal point of DSM strategies is moving the energy demand from peak hours to off-peak hours to lessen the peak demand [7].

A vital role is played by DSM for the development of industrial power, planning of energy and protection of the environment. In DSM there is a concept of Demand Response (DR), which refers to "changes in electric usage by end-use customers from

Table 1 EMS based on different strategic approaches

	Different methods	Approaches
Energy management system based on	Classical method	• Linear programming • Mixed integer linear programming • Nonlinear programming • Mixed nonlinear programming • Dynamic programming and rule-based methods
	Meta-heuristic approach	• Genetic algorithm • Particle swarm optimization • Differential evolution • Ant colony optimization • Gravitational search algorithm • Modified bacterial foraging • Artificial bee colony • Tabu search • Imperialist competition algorithm
	Artificial intelligence methods	• Fuzzy logic • Neural networks • Multi-agent system • Game theory • Markov decision process • Adaptive intelligence technique
	Stochastic and robust programming approaches	• Stochastic optimization • Robust optimization • Chance constrained programming • Scenario-based optimization
	Model predictive control	• Model predictive control
	Other methods	• Rolling horizon optimization • Hierarchical control • Homeostatic control • Predictor corrector proximal multiplier algorithm • Mesh adaptive direct search

their normal consumption patterns in response to changes in the price of electricity over time" [8].

Electricity Market can have the following advantages by the introduction of DSM [10]:

- Supply and demand information trading can be done instantly, transparent transactions can be facilitated and rapidly improve the development of an electricity price mechanism.
- The efficient market operation and good control of power of market is possible.

- Improve power system reliability by addressing peak hours demand congestion problems.
- Lessen the cost of power generation, transmission, and distribution.
- A new diagnosis for Energy management and emission reduction can be facilitated.

For improving energy efficiency by balancing supply and demand the combined work of advanced DSM and distributed energy resources (DERs) in a MG along with a good communication plan can play a vital role. The DSM by using developed optimal methods and information technologies can achieve a reduction in energy consumption, energy handling efficiency enhancement and cost reduction.

Due to the heterogeneous nature of the MG network which comprises of ESS, diesel generators, solar power, wind turbines, and electric vehicles, the specific objective function need to be defined, unlike the classical power grid where optimization is based on the centralized objective function. To control and enhance SG systems that include a MG distribution network it is natural to implement distributed analytical methods. In this regard, a few open issues in microgrids can be dealt with utilizing various strategies [11].

3.1 Classical Methods

An optimization method was proposed in [12] for the DSM EMS based on Linear Programming (LP) approach. EMS is for a given customer's hourly electricity prices. The proposed EMS which is developed using optimization algorithm based in LP, the customer can use their own plan to

- Control load and prices in the energy distribution system
- Forecast future energy use
- Efficiency improvement
- Losses reduction.

During the demand for energy excess energy which is stored in the battery can be utilized. In [13] focus is on both DSM and SSM load scheduling problem for users in SG using LP models to

- Minimize energy cost in DSM
- Maximize the load factor in SSM.

For the operation of intended appliances, the LP model offers various flexibilities to preset if the appliance has to operate without interruption and to define multiple time intervals. Uncertainty aspect of the MG system with RERs is addressed by designing and experimentally testing an adaptive online MG EMS in [14]. Interfacing with controller and incorporation with other modules and communication between them is tested using the designed EMS architecture. Optimization module is based on Mixed Integer Linear Programming (MILP).

3.2 Meta-Heuristic Approach

To solve the objective function the interior point method and for demand response, particle swarm optimization and artificial immune systems are applied in [15]. For both customer and utility in the MG environment the DR management proposed, a service provider who carries out the optimization decides a common solution. It has the following features

- For energy balance problems application of DR techniques
- Decrease the revelation of the supplier to the market instability by improving the cost of both supplier and consumer
- Utilized newest and extensively used realistic tools
- Easy adaptation in the energy industry to develop real-world applications.

The main focus of [16] is to develop a microgrid generation scheduling model using the intelligent meta-heuristic algorithm. Here a modified cuckoo search algorithm is developed and utilized for EMS in microgrids. The results of the studies conducted for

- Reducing operating costs with and without DR participation.
- Wind and PV resources uncertainties.

The results show that the operating costs are reduced by adaptive demand response programs.

Various heuristic techniques like Particle Swarm Optimization (PSO), Firefly Algorithm (FA), and Artificial Bee Colony (ABC) are applied in [17] for optimal consumption of renewable energy resources (RES). To incorporate increased energy demand at reduced cost batteries are used. Comparing these techniques ABC turns out to be very effective.

In paper [18], a combined solution using a genetic algorithm for residential loads for both economic dispatch and DSM in a MG is presented. The evaluations show that the given approach can

- Reduce the cost of operation for both suppliers and consumers in a single and multiple-facility microgrid.
- Cost generation reduction.
- Reduction in shifting of loads inconvenience.

The objective function of [19] containing discontinuous functions uses a Genetic Algorithm (GA) based solution for savings for the customer with DSM. For Real-time Microgrid Power Management problem two computational intelligence methods, particle swarm optimization (PSO) and Ant Colony Optimization (ACO) were introduced in [20]. A mathematical framework or multi-objective and multi-constraint optimization was presented and the advantages of intelligent methods over traditional computational techniques for optimization were discussed.

3.3 Artificial Intelligence Methods

A home automation economic model that has integration with the residential DERs, dynamic tariffs and enables actions to DR based on the Internet of Things was proposed in [21]. Decision making is based on the application of Artificial Intelligence (Fuzzy Logic) for automatic load management.

In paper [22] two Genetic Algorithm (GA) executed alternately, one for determining the microgrid scheduling and the fuzzy rules and others to tune the membership functions for optimizing microgrid operation. By using a hybridized Fuzzy and GA algorithm a MG generation schedule, day ahead wind generation electricity prices and based on load demand allocation of storage power is developed.

In [23, 24] a novel DSM method using dynamic game theory is proposed which can aid in

- Effective use of DERs by analyzing and coordinating the interactions among users [23].
- Reduce the total energy cost [23].
- Be modeled for variations in wind power using the Markov chain [23].
- Saves the cost of generation using DSM [24].
- Reduces the Peak-to-Average ratio [24].

A DSM framework integrating smart fuzzy load controller and DR in MG is proposed in [25]. The Fuzzy load controller for load shifting and load usage uses the data procured by the smart load monitoring system to make decisions. This method can

- lessen peak demand
- to minimize energy loss
- increase the efficiency
- effective cost saving.

In [26] a load management system for highly stochastic loads (treated as Markov models) is proposed. Markov decision process is used to

- reduce the overall cost linked with DR control action.
- increase the certainty of fulfilling the DR.
- reduce the load by modeling the load using each household as Markov chains.
- cost saving compared with the industry used model.

The key focus of an agent-based EMS proposed in [27] is to

- aid trading of power among microgrids with distributed storage and DR.
- make use of energy accessibility from the DERs and diversity in load usage patterns of the customers.
- DR in reducing the peak demand.
- to reduce electricity costs.

In paper [28] based on the Lagrangian multiplier method, Lagrange programming neural network is introduced for

- economic dispatch optimization
- reduction in the objective function
- functional cost reduction and
- increase in RERs power generation.

The optimal solutions for power production resources and ESS can be determined using this approach. Neural network helps in achieving prediction of renewable resources and load demand a day ahead.

3.4 Stochastic and Robust Programming Approaches

Stochastic model predictive control scheme energy scheduling for optimal EM, the supply and demand-side uncertainties are taken into consideration in [29]. In this solution is found efficiently because

- Uncertainties can be handled.
- The energy trading between MG and main grid is found to be in a assigned trajectory.
- References tracking including uncertainties have considerable improvement than the traditional scheduling scheme.

In [30] microgrid EMS framework based on agent-based modeling by introducing Robust Optimization (RO) is proposed. Uncertainties can be handled using the extension of the framework. Evaluation of each uncertainty impact on the cost accuracy and revenue function is made possible in this proposed optimization framework. An increase in the reliability indicators and reduction in energy shortage is evident from the reliability analysis.

3.5 Model Predictive Control

For MG optimization neural network load forecasting and model predictive control (MPC) are implemented in the paper [31]. The algorithm updates the optimal course at each stage for

- energy balance
- operational cost reduction
- reducing load forecasting errors.

A 2 stage Real-time demand-side management (RDSM) method using MPC based optimization model for Response Executers (RE) under an environment with uncertainties consist of the ESS and the Plug-in electric vehicles (PEV) is proposed in [32]. With the proposed framework MG can achieve

- economic benefits
- better dealing with uncertainty
- improve net load characters
- enhancing energy balance.

The analysis of MG DSMs based on different approaches, their contributions and limitation are given in Table 2. Further, this review will pave the way for developing a Demand-Side Management System considering the issues caused due to conventional generators, the privacy of consumers, incorporation of demand response, reliability, battery status and losses in the system.

4 Conclusions and Scope

Due to various challenges and opportunities, Microgrid penetration is presently rising across the globe. This paper attempted to provide a review of the various microgrid demand-side energy management approaches. The review consists of an introduction to microgrids, their components and associated benefits, limitations, which further followed by a discussion on Energy Management systems, its architecture, advantages, and constraints. The process of transforming conventional MG into the green system is facilitated by a powerful tool called Demand-side management (DSM). These MG DSMs mainly focuses to reduce the energy cost, utilization of renewable energy resources, reduction of MG system losses and energy trading with the main grid. Even though many efforts are taken for developing DSM algorithms for Intelligent Energy Management, the following challenges have to be met in the microgrid system: integration of demand response, the robustness of power management, losses cost of MGs and system reliability. The choice of approaches which is discussed in this paper to be implemented for optimization depends on various factors like uncertainties caused due to renewable energy resources, integration of demand response, computational time complexity, operational cost of battery and depth of discharge of battery of in energy storage systems and many more. Artificial Intelligence promises to be more effective and standalone approach to address the issues related to demand-side management while considering the different factors mentioned above. The benefits of Artificial intelligence on Microgrid are worth for future efforts.

Table 2 Contributions and limitation of MG DSM based on different approaches

References	Approach	Contributions	Limitations
[12]	Artificial neural network based linear programming	Losses reduction, optimal use of resources on demands, energy balance flexibility	Demand response was not considered which required external control techniques
[13]	Linear programming (LP)	Reduction in demand-side scheduling cost and improve the load factor while supply-side scheduling	Integration of 2 LP problems into one model not considered
[14]	Mixed integer linear programming	Operational cost reduction, able to operate on different modes	Advanced forecast, testing and optimization not implemented
[33]	Artificial immune systems and particle swarm optimization	Cost improvement by using the DR technique for both utility and consumer	The integration of renewable sources is ignored
[16]	Modified cuckoo search algorithm	Wind and PV resources uncertainty are considered and the objective function is analyzed. The operating costs reduced by adaptive demand response programs	Small scale power network considered
[17]	Artificial bee colony (ABC), particle swarm optimization (PSO), and firefly algorithm (FA)	Optimal use of RERs, reducing the cost of production and market-clearing price	Solar power not considered
[18]	Genetic algorithm	Reduction in energy cost	Constraint-capturing mechanisms and the larger MG level not considered
[19]	Genetic algorithm	Generation cost is more without DSM compared to with DSM	MG for a large area not considered
[21]	Fuzzy logic	Residential DER operation cost-benefit ratio improvement	Computational complexity not discussed
[22]	Fuzzy logic	Management of power in the battery and scheduling of MG	Non-consideration of loss of power and time complexity for higher computation
[23]	Game theory	Entire energy cost reduction in the isolated microgrid system	Uncertainties not considered

(continued)

Table 2 (continued)

References	Approach	Contributions	Limitations
[24]	Game theory	Benefitting the grid by reducing the ratio of peak to average and reduction in load profile are smoothened which are caused by supply constraints	Hybrid micro-grid is not considered for the study
[25]	Fuzzy logic	Reduction in cost of energy, loss of power, peak demand and inefficiency of the grid	More renewable energy sources not considered
[26]	Markovian model	Reaching demand response targets as well as reduction in cost of operation	Computational complexity not discussed
[27]	Multi-agent system	This EMS increases levels of thermal comfort and consumer's electrical and reduces the operational cost of MG	Microturbine emission cost is ignored. Non-consideration of the operational cost of the battery
[28]	Neural network	Minimization of the overall cost of MG, a neural network used for forecasting and the results are compared with PSO	Non-consideration of battery operation cost and complexity of the time of computation
[29]	Stochastic programming	Minimization of CG battery cost of operation, trading of energy cost and cost of degradation with main grid	Discussion of complexity due to time in computation is not covered
[30]	Robust programming	Realization of performance of MG (imbalance cost), reliability of MG (loss of expected energy) and load expectation loss	DR not included in a residential district. Complex
[31]	Model predictive control	Static control issues addressing and elimination of load forecast errors and reduction in operations costs	Compensating residual forecast errors not discusses
[32]	Model predictive control	Power balance maintenance and the operation cost reduction	Computational complexity not discussed

References

1. Khan B, Singh P (2017) Selecting a meta-heuristic technique for smart micro-grid optimization problem: a comprehensive analysis. IEEE Access 5
2. Bayindir R et al (2014) A comprehensive study on microgrid technology. Int J Renew Energy Res 4(4)
3. Zia MF, Elbouchikhi E, Benbouzid M (2018) Microgrids energy management systems: a critical review on methods, solutions and prospects. Appl Energy
4. Parhizi S et al (2015) State of the art in research on microgrids: a review. IEEE Access 3
5. Borase PB, Akolkar SM (2017) Energy management system for microgrid with power quality improvement. In: International conference on microelectronic devices, circuits and systems (ICMDCS). IEEE
6. Kowalczyk A et al (2016) Microgrid energy management system. In: 21st international conference on methods and models in automation and robotics (MMAR). IEEE
7. Kumar Nunna HSVS et al (2016) Multi-agent based demand response management system for combined operation of smart microgrids. Sustain Energy Grids Netw 6:25–34
8. Khan AA et al (2016) A compendium of optimization objectives, constraints, tools and algorithms for energy management in microgrids. Renew Sustain Energy Rev 58:1664–1683
9. Ma J, Ma X (2018) A review of forecasting algorithms and energy management strategies for microgrids. Syst Sci Control Eng 6(1):237–248
10. Li D, Chiu WY, Sun H (2017) Demand side management in microgrid control systems. In: Microgrid advanced control methods and renewable energy system integration, pp 203–230
11. Saad W et al (2012) Game theoretic methods for the smart grid. IEEE Signal Process Mag
12. Loganathan N, Lakshmi K (2014) Demand side energy management system using ANN based linear programming approach. In: IEEE international conference on computational intelligence and computing research
13. Tan W-N et al (2016) Optimization models for demand-side and supply-side scheduling in smart grids. In: IEEE 16th international conference on environment and electrical engineering (EEEIC). IEEE
14. Luna AC et al (2016) Online energy management systems for microgrids: experimental validation and assessment framework. IEEE Trans Power Electron
15. Rat C et al (2018) Microgrid supply management. Procedia Soc Behav Sci 238:46–52
16. De M et al (2018) A reliable energy management and generation scheduling model in microgrids using modified cuckoo search algorithm. In: Emerging trends in electronic devices and computational techniques (EDCT). IEEE
17. Paliwal NK et al (2016) Demand side energy management in hybrid microgrid system using heuristic techniques. In: IEEE international conference on industrial technology (ICIT). IEEE
18. Arif A et al (2014) Integrating renewable economic dispatch with demand side management in micro-grids: a genetic algorithm-based approach. Energy Effic 7:271–284
19. Jayadev V, Shanti Swarup K (2013) Optimization of microgrid with demand side management using genetic algorithm. In: IET conference on power in unity: a whole system approach. IET
20. Nehrir MH et al (2010) Towards real-time microgrid power management using computational intelligence methods. In: IEEE PES general meeting
21. Souza AT et al (2018) Development of a demand response system integrated to photovoltaic microgeneration and energy storage using IoT and artificial intelligence. In: 15th international conference on the European energy market (EEM). IEEE
22. Fossati JP et al (2015) Optimal scheduling of a microgrid with a fuzzy logic controlled storage system. Electr Power Energy Syst 68:61–70
23. Wu C et al (2011) Demand side management for wind power integration in microgrid using dynamic potential game theory. In: IEEE international workshop on smart grid communications and networks
24. Noor S et al (2018) Energy demand side management with supply constraints: game theoretic approach. Energy Procedia 145:368–373

25. Zunnurain I et al (2018) Implementation of advanced demand side management for microgrid incorporating demand response and home energy management system. Infrastructures 3:50
26. Thomas E et al (2019) Towards demand side management control using household specific Markovian models. Automatica 101:450–457
27. Kumar Nunna HSVS et al (2013) Energy management in microgrids using demand response and distributed storage—a multiagent approach. IEEE Trans Power Deliv
28. Wang T et al (2017) Neural networks for power management optimal strategy in hybrid microgrid. Neural Comput Appl
29. Kou P et al (2016) Stochastic energy scheduling in microgrids considering the uncertainties in both supply and demand. IEEE Syst J
30. Kuznetsova E et al (2014) An integrated framework of agent-based modeling and robust optimization for microgrid energy management. Appl Energy 129:70–88
31. Jaboulay P-A et al (2017) Real-time energy management optimization using model predictive control on a microgrid demonstrator. In: IEEE international conference on energy internet (ICEI). IEEE
32. Yang X et al (2017) Real-time demand side management for a microgrid considering uncertainties. IEEE Trans Smart Grid
33. Herath P et al (2018) Computational intelligence based demand response management in a microgrid. IEEE Trans Ind Appl Comput

IRIS: An Application for the Visually Impaired Using Google Cloud API

Kajal Naik, Nikita Sawant, Gauravi Kamat, Siddhi Kandolkar, and Niyan Marchon

Abstract The visually impaired struggle to walk safely without having any human guidance. They face several problems in their lives which include identifying the hindrances when they are walking. Traveling from one place to another is a tedious task for them requiring the help of other people. White canes and guide dogs are usable but are not fully reliable. Due to this disability, they have to go through a lot of hardships to conquer over the problems in navigation. Thus there is a need for a smart solution which would overcome the challenges faced. Focus is on designing one kind of a visual aid which is cost-effective and efficient. In this context, we propose a smart stick (IRIS) to help the user in obstacle detection and navigation. A conventional PVC pipe forms the stick on which ultrasonic sensors, camera, GPS module, GSM Module which are all interfaced with Raspberry Pi are mounted at different positions to detect and identify obstacles in the path of the user. Google Cloud Vision API recognizes the objects in the surroundings and delivers an audio signal about the same via earphones. The sense of sight of the user is now achieved by the sense of hearing. It also uses a GSM Module and a GPS module to contact the user's relatives in emergency situation by sending them the user's location. The whole device's design ensures independence to the user.

Keywords Ultrasonic sensors · GPS module · GSM module · Raspberry Pi · Google Cloud

---

K. Naik · N. Sawant (✉) · G. Kamat · S. Kandolkar · N. Marchon
Padre Conceicao College of Engineering, Verna, Goa, India
e-mail: nikita.sawant.248@gmail.com

K. Naik
e-mail: kajalbbnaik@gmail.com

G. Kamat
e-mail: gauravihkamat@gmail.com

S. Kandolkar
e-mail: siddhikandolkar21@gmail.com

N. Marchon
e-mail: niyanmarchon@gmail.com

© Springer Nature Singapore Pte Ltd. 2021

S. N. Merchant et al. (eds.), *Advances in Signal and Data Processing*,
Lecture Notes in Electrical Engineering 703,
https://doi.org/10.1007/978-981-15-8391-9_3

1 Introduction

Blindness is the inadequacy of a person to perceive the things around his/her surrounding with the eyes. According to the World Health Organization, it is globally estimated that about 285 million people live with partial or complete blindness [9]. Currently the visually impaired use a white cane or a guide dog to find out if any obstacles are present in front of them. But these are inefficient in various aspects and the person using it has to face several problems. In an unknown environment, the visually impaired might get confused and this will restrict their movement. It becomes difficult for them to do their routine activities. The proposed smart stick can help them in moving freely and allowing them to carry out their work with ease. Using existing technologies we deliver a better result to overcome the disadvantages of the earlier system. A cost-effective system is necessary which can be used by them to work smoothly and comfortably. Technology progress should be reflected in such cases, providing a smart solution solving cost and efficiency issues. Billions of visually impaired across the World suffer from daily problems that could risk their lives. This smart stick is a perfect and secure solution. Our focus is to bring out the best of technology and to help the visually impaired to live independently. In this fast growing technology globe, where people strive to live independently, this project proposes a smart stick (IRIS) to support them and help them to navigate with ease. Efficiency is the main priority, hence one can make use of it very easily.

2 Literature Review

This section is a study on the researches that have been carried out in the assistive technologies sector. The paper [1] puts forth a system which consists of two cameras. These cameras are fixed on the glasses of the visually impaired. The proposed project uses a blind stick onto which the sensors are attached. Infrared sensors are used which use infrared waves to detect objects in front of the user and an audio signal is sent to the user. The information about the obstacles must be trained into the system. This paper [2] proposes a system which consists of Raspberry Pi, camera and ultrasonic sensors to detect and identify the obstacles. It uses digital image processing for obstacle identification. This system [3] gives the data of different types of objects along with their size and distance from the user. The signal is processed using MATLAB. Apart from audio signal, it also gives out a vibration when an object gets detected. This is done using ultrasonic sensor. Whenever an obstacle is detected by ultrasonic sensor, the motor vibrates. This paper [4] proposes a smart stick using machine learning techniques, Google Assistant and ultrasonic sensor to detect the obstacles and alert the user via vibration. Also the user can find out his/her location through an audio output. The project [5] proposes a system using Android Smartphone and image processing libraries such as OpenCV and Google Cloud Vision API to detect an image using the camera of the Smartphone. An audio signal is given out which alerts the user of an obstacle that is present in front.

There are various other systems that are present in the market. These are Laser Cane, Sonic Torch, Sonic Path Finder, Mowat Sensor, Meldog, Navbelt, etc. These systems are called Electronic Travel Aid (ETA) which are developed for the visually impaired aiding in their mobility. Majority of these systems just detect the obstacle but does not identify it. The visually impaired do not know what kind of obstacle is present. Also, these are either bulky or quite expensive and not very accurate. According to the National Center for Biotechnology Information [6], most of these commercially available ETAs cost between 300$ and 3000$. Although there are some cheaper options available in India at a price range of 40$–50$, it only uses sensors such as ultrasonic sensor or laser to detect an obstacle but does not identify it.

3 Proposed Device and Method

3.1 Construction

The skeleton of the device (IRIS) is made of PVC material on which different components are mounted. Raspberry Pi is the vital part of the device. All the components are interfaced with it for the efficient working of the device. Some of the components like GSM module (USB dongle), GPS module, power bank (power source for the Raspberry Pi), and wireless adapter (for Internet connectivity) along with the Pi are encased in a box. The box is then positioned at a convenient distance near the upper end. The rest of the components lie outside the box and are positioned at various levels on the stick. Two ultrasonic sensors are placed at different levels on the stick to provide obstacle detection above and below the knee level. The first Ultrasonic sensor is implemented on the lower side of the stick to avoid small obstacles below the knee level. A buzzer is used which buzzes when an object is discovered by this sensor. The second ultrasonic sensor is implemented to detect the obstacle above the knee level. This sensor is set to a particular range to trigger the camera mounted on the stick to capture the image of the obstacle detected by it for the identification of the obstacle. The information regarding the type of the obstacle and its distance from the user is conveyed via headphones or earphones connected to the audio jack of the Raspberry Pi using text-to-speech convertor, espeak. Two push buttons are implemented on the stick. One of the push buttons is used for shutting the system off when not in use and is called the power button. A switch is placed on the box which is used to cut off the power supply to the Raspberry Pi as soon as the power button is pressed. The second push button can be pressed by the visually impaired when they feel they have lost their path or reached an unknown place and need some help and this button is called the SOS button. The GPS and the GSM module come into picture as soon as this button is pressed. The location of the user is traced by using the GPS Module. The device also sends an emergency message (I am in trouble)

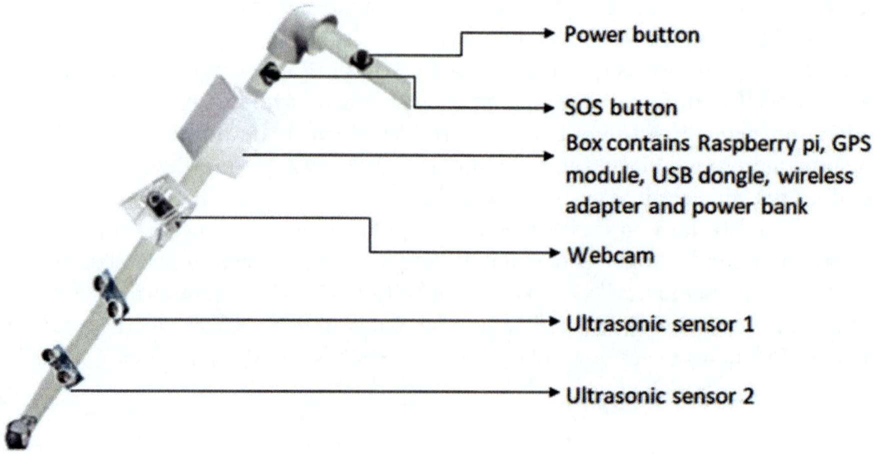

Fig. 1 Smart walking stick prototype (IRIS)

together with the user's location on a saved mobile number using GPS and GSM modules. Additionally, the user gets to know about his/her location via headphones using espeak.

Figure 1 demonstrates IRIS: Smart Walking Stick prototype.

3.2 Module Description

The device consists of two major units: The obstacle detection and recognition unit, and the navigation and alert unit for sending emergency messages. The block diagram of these two units is shown in Fig. 2.

For obstacle detection and recognition, a camera, ultrasonic sensors and a speaker (earphones) is used. When a visually impaired encounters an obstacle, the Ultrasonic sensor 1 present on the stick detects the obstacle. This sensor can be programmed to recognize the obstacle within a range of 3 m from the user. When the obstacle is detected by this sensor, the information is received by the Raspberry Pi which triggers the camera to capture a picture of the obstacle. The captured image is converted into base64 format and sent to the Google Cloud Vision API for object recognition. The labels that are detected from the image are sent back to the Raspberry Pi which communicates the data to the client via audio output using espeak. The ultrasonic sensor 2 is used to recognize the obstacle at a lower height than the ultrasonic sensor 1. This sensor can also be programmed and set to a particular range. Once an obstacle crosses this range, buzzer will start buzzing and the obstacle distance can also be conveyed to the user via headphones using espeak. In addition an IR sensor can also be implemented for level detection.

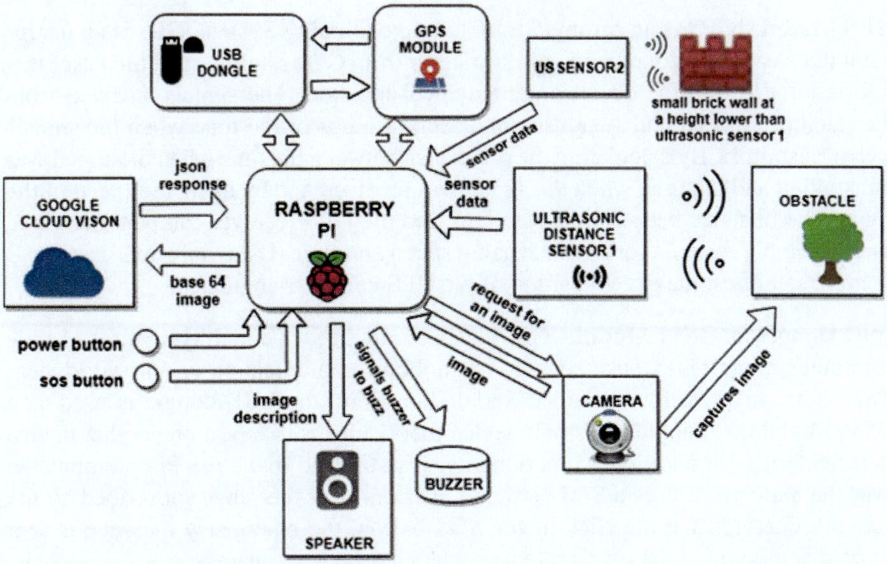

Fig. 2 Block diagram of the smart stick

For navigation and alert System, GPS and GSM modules are used. The current position of the user is extracted by the GPS Module. The SOS button can be pressed by the user when he/she is in danger. As soon as this button is pressed, the user's location is sent to the saved mobile numbers as a text along with an emergency message via the GSM module. In addition, the user also gets to know his/her current location on pressing this button and also the status of the emergency message (delivered/undelivered). The power button is used to switch off the system when not in use. The specification and working of the individual modules are given below.

Ultrasonic Sensor An ultrasonic sensor is a device that calculates the distance to a target by using ultrasonic sound waves. It uses two transducers to send and receive ultrasonic pulses that transmits information about an object's proximity. High-frequency sound waves reflect from boundaries of the target to produce distinct echo patterns. The sensor determines the range to a target by calculating time intervals between the sending and receiving of the ultrasonic pulse.

Raspberry Pi The Raspberry Pi is a compact high-performance computer which is cost-effective and can be connected to a monitor or TV. It utilizes a mouse and keyboard. It can be programmed in languages like Python and Scratch. It is well suited for doing everything from using the Internet and running HD videos, to creating documents, and engaging in playing games.

Camera A camera is mounted on the stick which is used to capture the image of the obstacle present in front of the user for its identification. The image captured is converted into base64 format before sending it to the Google Cloud for object recognition.

GPS Module GPS is an acronym for Global Positioning System. GPS is an instrument that is effective in receiving information from GPS satellites and then uses this information to find the device's geographical location. The signals received from the satellites and ground stations contain time stamps of the time when the signals were transmitted. By calculating the difference between the time when the signal was transmitted and the time when the signal was received and by using the speed of the signal, the distance between the satellites and the GPS receiver can be determined using a simple formula for distance using speed and time. Using information from 3 or more satellites, the exact position of the GPS can be triangulated.

USB Dongle as GSM Module GSM is an acronym for Global System for Mobile communication. It is extensively used for mobile communication system worldwide. USB is an acronym for Universal Serial Bus. Here, the USB dongle is used as a GSM Module. A dongle is a small device that connects to a port on another device to provide it with additional functionality. It is used to make the Pi communicate over the network. It uses a SIM card and operates over the range subscribed by the network operator. On the click of the SOS button, the emergency message is sent using this module to the user's relatives when he/she is in distress.

Audio Module This module is used for giving instructions to the visually impaired user about the obstacles via a speaker or headphones. When the object present in front of the user is detected and identified, it is conveyed to the user by using the python library, espeak which is a text-to-speech converter.

Switch A switch is an electrical element that is used for connecting or disconnecting an electrical circuit by disrupting the current or rerouting it from one conductor to another.

Buzzer A buzzer, also known as beeper is used as a sound signaling device. In this project piezoelectric buzzer is used. It is a sound producing device which works on the principle that as soon as an electric potential is applied across a piezoelectric substance, a variation in pressure is generated which in turn gives out a loud sound.

Push Button Push buttons are generally open physical switches. These buttons enable us to energize the circuit or make any connection only when the button is pressed. In simple words, it connects the circuit when pressed and disconnects when released.

Google Cloud Vision Google Cloud Vision Application Programming Interface (API) consists of pre-trained machine learning models that enable developers to understand the content of an image [7]. It quickly classifies the objects present in an image with various labels. When the camera clicks an image, the image is converted into base64 format and is sent to the Google Cloud via the Internet in the form of JSON Request. In the Google Cloud Vision API, this image is parsed, and the detected labels are sent back as JSON Response. Since the JSON Response consists of many labels, the label having the highest score is taken into consideration and this label is given out as audio output to the user.

Python Python is an object-oriented, interpreted programming language. It has various features such as high level built-in data structures. It supports dynamic binding and dynamic typing, making it very attractive for Rapid Application Development, as well as for use as a scripting language [8]. Python's simple syntax makes it easy for the person to read and hence reduces the cost of maintaining the program.

3.3 Working

Whenever the user encounters an obstacle in front, the ultrasonic sensor 1 will detect the obstacle and send the data to the Raspberry Pi. The camera clicks a picture whenever an object less than 3 m away from the user is detected. The picture is stored in the Raspberry Pi and converted into base64 format before sending it to the Google Cloud Vision API. The captured images are run against the Google Cloud Vision API to find out objects in the image. The HTTP request is built using Google Credentials and an encoded image (image content). The JSON for the body of the request is shown in Fig. 3.

The response is also JSON and contains a label attribute which is a string containing the name of the most prominent object in the scene. The requested image is shown in Fig. 4 and the returned result is shown in Fig. 5.

Similarly ultrasonic sensor 2 detects the obstacles at a lower height than ultrasonic sensor 1 and triggers the buzzer to buzz.

The GPS receiver module uses UART communication to communicate with controller or PC terminal. Before using UART on Raspberry Pi, it should be configured and enabled. The GPS Module interfaced with the Raspberry Pi extracts the GPS information. The extracted GPS information consists of NMEA GPGGA string which contains the latitude, longitude and time information. By using these latitude


```
{
  "requests":[
    {
      "image":{
        "content":"/9j/7QBEUGhvdG9...image contents...eYxxxzj/Coa6Bax//Z"
      },
      "features":[
        {
          "type":"LABEL_DETECTION",
          "maxResults":1
        }
      ]
    }
  ]
}
```


Fig. 3 JSON request

Fig. 4 Image sent as request to Google Cloud Vision API

and longitude values, the current position of the user is located on the Google Map. After programming the GPS Module, the output is shown in Fig. 6.

To plot the location on Google map, URL link for the Google map is called. The following link "http://maps.google.com/?q=<latitude>,<longitude>" is used for opening Google map with the extracted longitude and latitude values.

The output on the Google Map is shown in Fig. 7.

An alert message is sent to the user's chosen contacts when he/she is in danger using the GSM Module. These contacts are stored in the python program. The location obtained using the GPS Module is also sent along with a message. When the SOS button is pressed, this Python program gets executed (as shown in Fig. 8) and the received message is shown in Fig. 9.

The flow of our project is depicted in Fig. 10 for obstacle detection and identification and Fig. 11 for SOS system.

4 Results

The final implementation of IRIS is shown in Fig. 12. The Pi is powered by a power bank placed inside the box and is provided with wireless Internet connectivity via the wireless adapter. The implemented programs are stored in auto start function of the Raspberry Pi so that when the user powers on the device, all the programs start running automatically without manually executing them.

A picture of an object was clicked as shown in Fig. 13.

The labels detected are shown in Fig. 14.

The detected labels are in the form of a string which are transformed into audio by using text-to-speech conversion library, espeak and the audio output is heard using headphones.

```
{
  "responses": [
    {
      "labelAnnotations": [
        {
          "mid": "/m/0bt9lr",
          "description": "dog",
          "score": 0.97346616
        },
        {
          "mid": "/m/09686",
          "description": "vertebrate",
          "score": 0.85700572
        },
        {
          "mid": "/m/01pm38",
          "description": "clumber spaniel",
          "score": 0.84881884
        },
        {
          "mid": "/m/04rky",
          "description": "mammal",
          "score": 0.847575
        },
        {
          "mid": "/m/02wbgd",
          "description": "english cocker spaniel",
          "score": 0.75829375
        }
      ]
    }
  ]
}
```


Fig. 5 JSON response

On the click of the SOS button, current location is heard by the user and an SMS is sent to the user's stored contacts along with an emergency message and the user's current location, providing safety and security to the user in navigation.

5 Conclusion

This paper outlines a system to help the visually impaired in mobility. The proposed system design demonstrates how Google Cloud Vision API can be used to develop a technology based visual aid. IRIS will help them to gain personal independence by detecting and identifying obstacles around them and providing safety and security to

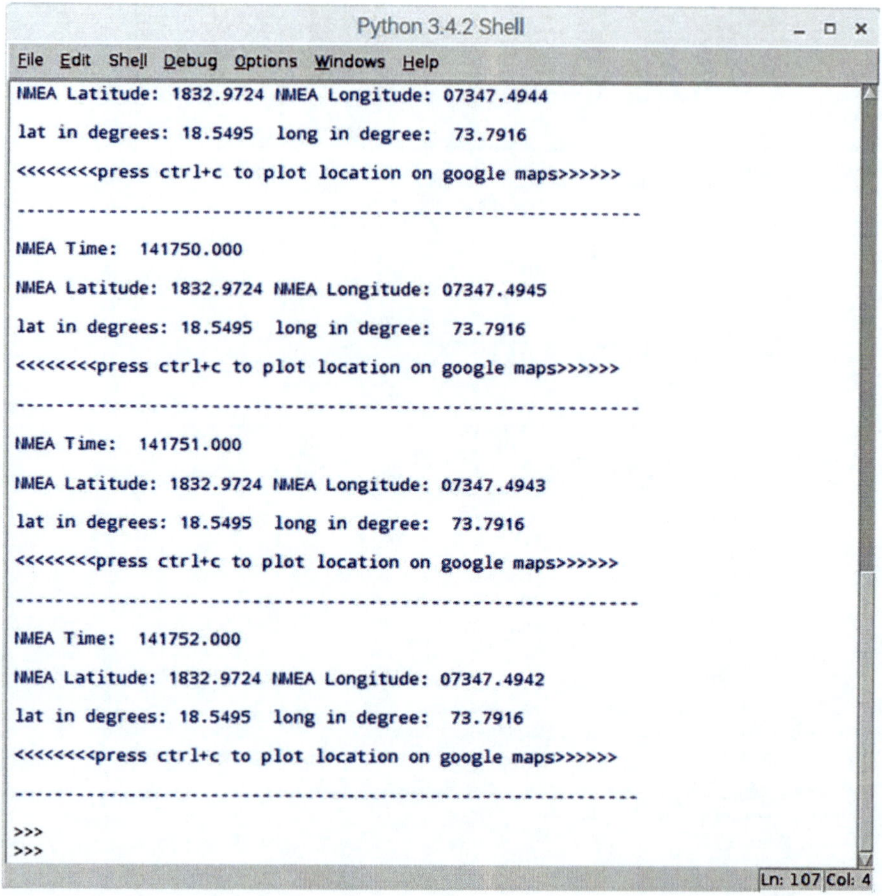

Fig. 6 Output of GPS module

the user. The project began with a drive to find a solution to the problems of visually impaired and IRIS (smart stick) is considered to be an efficient and secure solution to the problems faced.

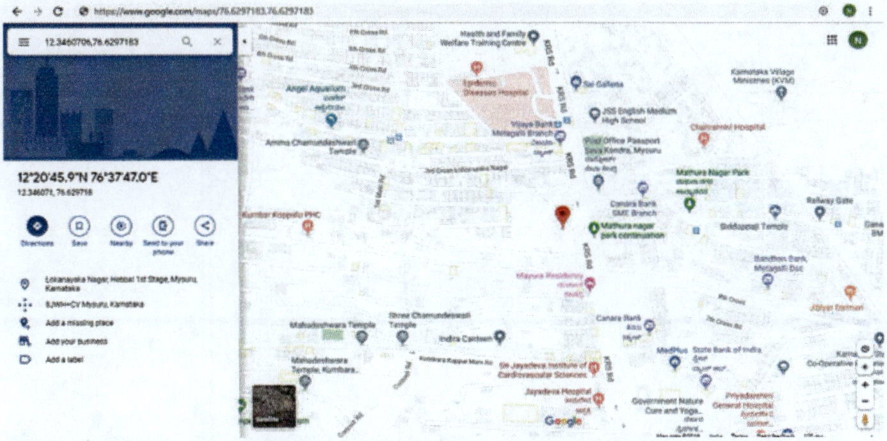

Fig. 7 Output of Google map

Fig. 8 Execution of python program for sending message

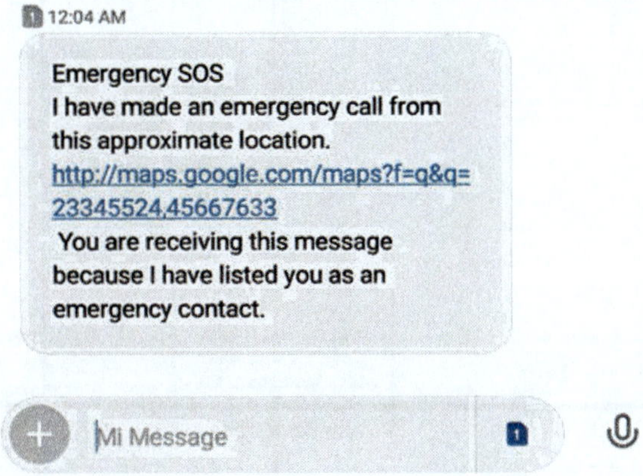

Fig. 9 SOS message

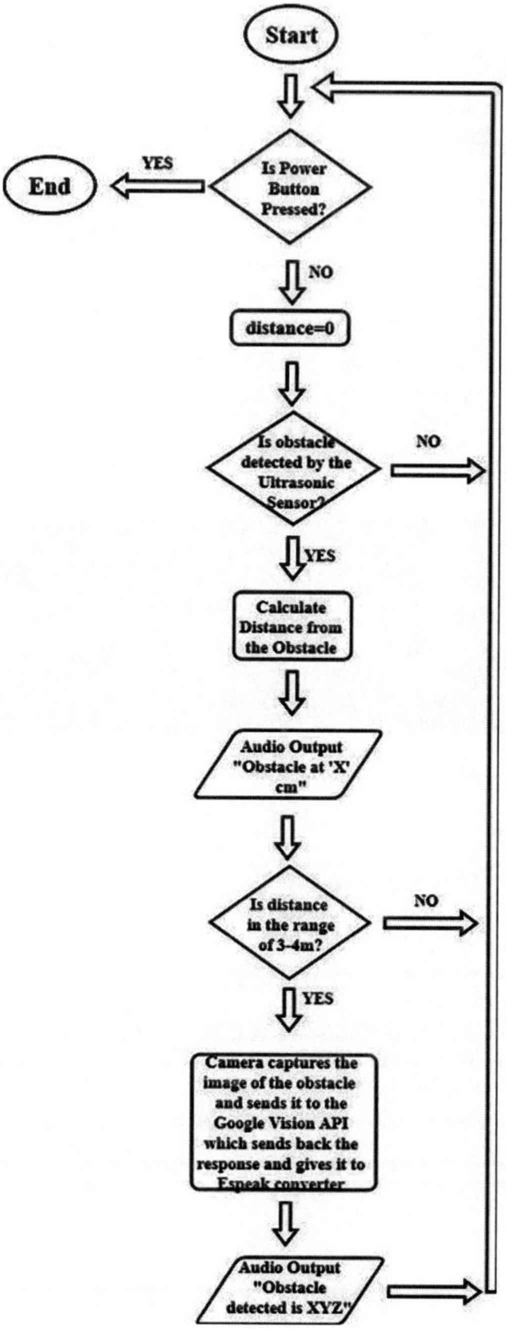

Fig. 10 Flowchart for obstacle detection and identification

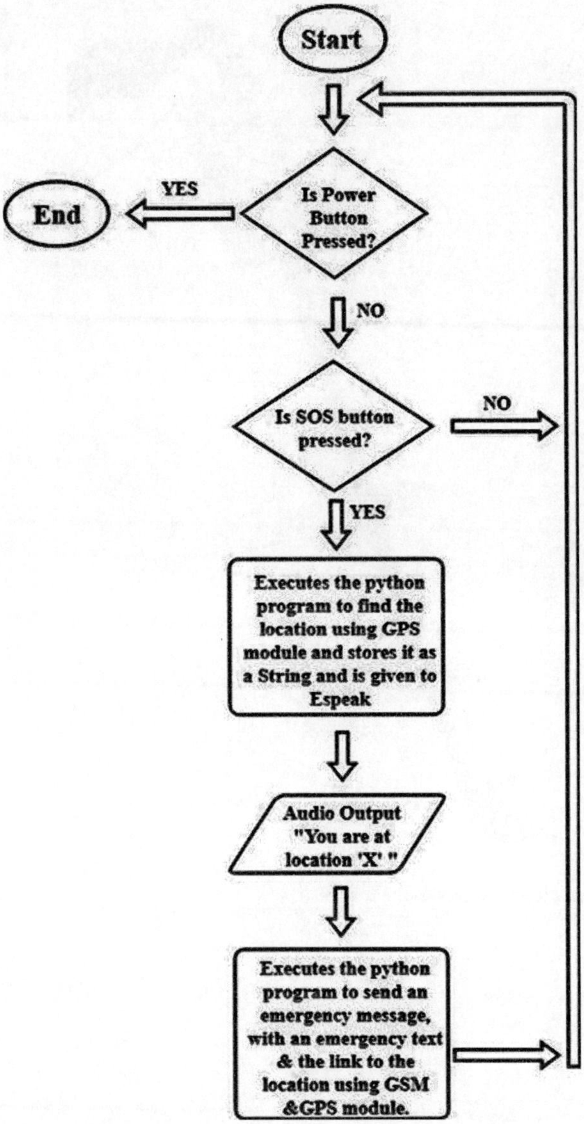

Fig. 11 Flowchart for SOS system

Fig. 12 Implementation of smart stick

Fig. 13 Image captured by the camera

Fig. 14 Detected labels

References

1. Seema U, Shivaji S, Sujit G, Dinesh K, Suraj C (2006) Object detection system for blind people. Int J Innov Res Comput Commun Eng
2. Vanitha M, Rajiv A, Elangovan K, Vinoth Kumar S (2018) A smart walking stick for visually impaired using raspberry pi. Int J Pure Appl Math
3. Daniyal R, Faheem A, Habib A, Zakir Ahmed S, Aamir S (2014) Smart obstacle detector for blind person. J Biomed Eng Med Imaging
4. Mohamedarif R, Khazi SBA, Sushmita S, Mrityunjay CK, Sharda KS (2019) Smart stick for blind using machine learning. Int J Innov Sci Res Technol
5. Roshan R, Dinesh P, Paresh K, Deesha R, Indu D (2018) Proposed system on object detection for visually impaired people. Int J Inf Technol (IJIT)
6. Electronic Travel Aids for the Blind. https://www.tsbvi.edu/orientation-and-mobility-items/1974-electronic-travel-aids-for-the-blind
7. Cloud Vision Documentation. https://cloud.google.com/vision/docs/
8. What is Python? https://www.python.org/doc/essays/blurb/
9. Global Data on Visual Impairments 2010. https://www.who.int/blindness/GLOBALDATAFINALforweb.pdf

Shot Boundary Detection Using Artificial Neural Network

Neelam Labhade-Kumar, Yogeshkumar Sharma, and Parul S. Arora

Abstract As of late video is the most utilized information type on the Internet. Content, sound, and pictures are consolidated to establish a video, so recordings are enormous in size. The human mind can process visual media quicker than it can process message. This expansion in information has required the investigation of powerful strategies to process and store information content. In this paper, we have suggested a hybrid video shot boundary detection process using feature extraction by mean log difference which is combined with artificial neural network. We devised two-step method for automatic shot boundary detection. Firstly, features are extracted using H, V, S procedure along with histogram distribution technique, and then, this mean log difference array is applied as an input to ANN which identifies video shots based on probability function. We have incorporated feed-forward network structure which processes nonlinear factual information to calculate shot boundary detection considering probability function. Finally, we have evaluated the results using precision, recall, and $F1$ measure. An experimental result indicates that ANN along with mean log difference, it offers efficient representation of shot boundaries and the results are satisfactory. Comparing the proposed method with improved block color feature method, there is a sort of trade-off relation between the two algorithms, and it is observed that for fast characteristic variations, ANN performs moderately better while for complex videos improved block color feature method is suited in better way.

Keywords Artificial neural network (ANN) · Content-based image retrieval (CBIR) · Shot boundary detection (SBD) · Mean log difference

N. Labhade-Kumar (✉) · Y. Sharma
Shri J.J.T. University, Churella, Jhunjhunu, Rajasthan, India
e-mail: neelam.labhade@gmail.com

N. Labhade-Kumar · P. S. Arora
JSPM's ICOER, Wagholi, Maharashtra, India

© Springer Nature Singapore Pte Ltd. 2021

S. N. Merchant et al. (eds.), *Advances in Signal and Data Processing*,
Lecture Notes in Electrical Engineering 703,
https://doi.org/10.1007/978-981-15-8391-9_4

1 Introduction

1.1 Data Mining

Data mining is a practice of extracting earlier unknown information and sensing the exciting patterns from a huge set of data. Its models and algorithms were also brought up for image classification and clustering [1]. Video mining is used to realize information, configurations, patterns ,and events of interest in the video data, and its advantages range from efficient surfing and summarization of video content to facilitating video access and retrieval in a large database or online multimedia repository. The aim of video data mining is to recognize and define stimulating patterns from the enormous quantity of video records as it is one of the essential problem areas of the data mining. There are many inspiring investigation problems in effect with video mining.

1.2 Shot Boundary Detection (SBD)

Shot boundary detection (SBD) otherwise called transient video division is the principal procedure in CBVIR, and its yield altogether influences the consequent procedures. The point of SBD is to parcel a video into its essential units (shots) to be sent to the remainder of the CBVIR modules for further investigation [2]. Structure of video is as presented in Fig. 1. A video shot is characterized as a constant grouping of edges taken from a same camera and speaking to an activity after some time [3].

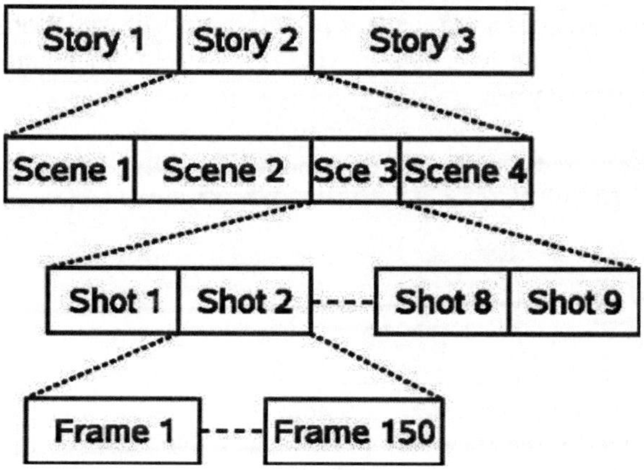

Fig. 1 Structure of video and shots

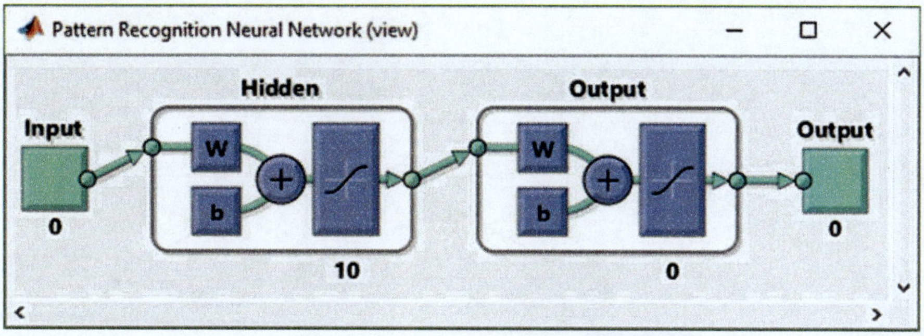

Fig. 2 Structure of ANN

1.3 Artificial Neural Networks (ANN)

A neural network is a progression of calculations that try to perceive basic connections in a lot of information through a procedure that copies the manner in which the human mind works [4]. Artificial neural networks (ANNs) are generally unrefined electronic models dependent on the neural structure of the mind. The cerebrum gains as a matter of fact. In this method, we used feed-forward network which consists of a series of layers. This paper gives a presentation into ANN and the manner in which it is utilized.

Structure of ANN is shown in Fig. 2. ANN consists of neurons. These neurons comprise data sources and output. At the point when a signal arrives, it gets increased by weight (W) esteem. Weight of neurons decides if any change in input then how much impact it has on the results. A weight close to zero indicates that fluctuating input will not alter the output. Negative weights mean increasing this input will decrease the output. Neuron biases (b) are supplementary to feed-forward neural networks, and it helps to absorb patterns. Bias nodes function like an input node that always produces constant value 1 or other constant. Because of this property, they are not connected to the preceding layer. The bias node in a neural network is a node which is always 'on,' and its value is set to 1. A sigmoid transfer function is used as an activation function squashes values between ranges 0 and 1. Mathematically, sigmoid activation function is represented by Eq. (1)

$$f(x) = \tan^{-1}(x) \tag{1}$$

1.4 Feed-Forward Neural Network

First and simplest type of artificial neural network devised was the feed-forward neural network. It contains multiple neurons/nodes arranged in layers. Neurons/nodes

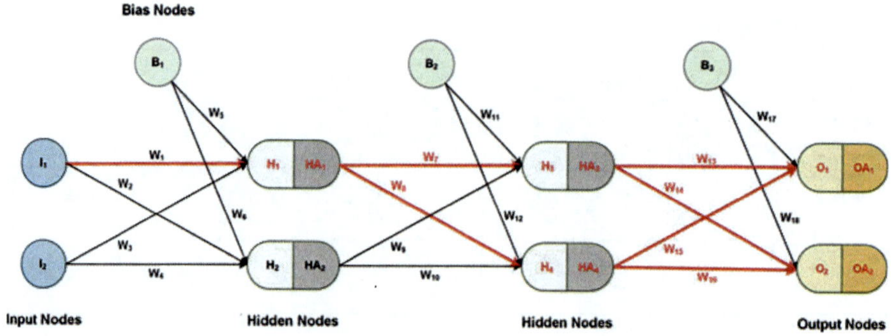

Fig. 3 Structure of feed-forward ANN

from adjacent layers have connections between them. All these connections have weights associated with them. An example of a feed-forward neural network is shown below. Two examples of feed-forward networks are given below:

1. **Single Layer Perceptron**—This is the simplest feed-forward neural network and does not contain any hidden layer.
2. **Multi-Layer Perceptron**—A Multi-layer perceptron has one or more hidden layers. Figure 3 shows a multi-layer perceptron with a single hidden layer.

Input Layer: This is the first layer in the neural network. It takes input signals (values) and passes them on to the next layer. It does not apply any operations on the input signals (values) and has no weights and biases values associated.

Hidden Layers: Hidden layers have neurons which apply various changes to the info information. One hidden layer is a gathering of neurons stacked vertically. Keep going, as hidden layer passes on qualities to the output layer. Every one of the neurons in a hidden layer is associated with every single neuron in the next layer.

Output Layer: This layer is the last layer in the system and gets contribution from the last hidden layer. With this layer, we can get wanted number of qualities and in an ideal range, In a feed-forward network, the information moves in only forward direction—from the input nodes, through the hidden nodes, and to the output nodes. There are no cycles or loops in the network.

1.5 Proposed Algorithm

Our projected algorithm starts with frame extraction from the given video, and then, we have incorporated RGB-HVS converter to identify the histogram feature of H, V, S values. We have plotted graphically H plane histogram, S plane histogram, and V plane histogram. In the further step, we have used mean log difference method for the feature extraction of the corresponding frame. By averaging, the pixel in rows and

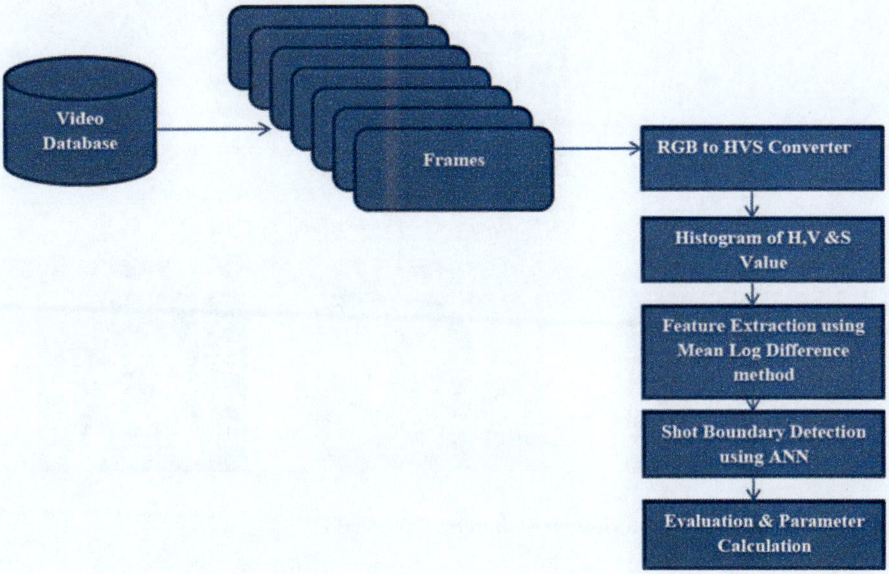

Fig. 4 Flowchart of proposed algorithm

columns and taking the corresponding difference between successive frames gives mean log difference. In the final step, this statistical array generated by mean log difference is given as input to feed-forward network of ANN. ANN with its inherent characteristics calculates shot boundaries with the help of features extracted by means of log difference method (Fig. 4).

2 Results and Discussions

In order to evaluate our method and compare with other methods, we employed five video clips like sport, movie, cartoon, wildlife, and news and extracted frame sequence of each video. Then, we manually marked the shot boundary indexes of transitions of each video and compared same with detected shots by ANN.

2.1 RGB-HVS Conversion

In RGB feature extraction method, R, G, B are co-related to the color luminance which is similar to intensity, and therefore, separating color information from luminance is difficult. RGB defines color in terms of a combination of primary colors. In circumstances where shading portrayal assumes an essential job, the HVS shading

Fig. 5 **a** Original frame from video. **b** RGB to HVS conversion

model is frequently favored over the RGB model. To beat this restriction, here, we use HVS for highlight extraction which is only hue, value, and saturation. The HVS model depicts hues also to how the human eye will in general observe shading. In circumstances where shading portrayal assumes a vital job, the HVS shading model is regularly favored over the RGB model. Mathematical equation for RGB-HVS conversion is given by Eqs. (2)–(4) (Fig. 5)

$$\text{Max} = \max(R, G, B)$$

$$\text{Min} = \min(R, G, B)$$

$$\Delta = \text{Max} - \text{Min}$$

$$H = \frac{G - B}{\Delta} \quad \text{Max} = R \tag{2a}$$

$$H = 2 + \frac{B - R}{\Delta} \quad \text{Max} = G \tag{2b}$$

$$H = 4 + \frac{R - G}{\Delta} \quad \text{Max} = B \tag{2c}$$

$$V = \text{Max} \tag{3}$$

$$S = \frac{\Delta}{\text{Max}} \quad \text{Max} > 0 \tag{4a}$$

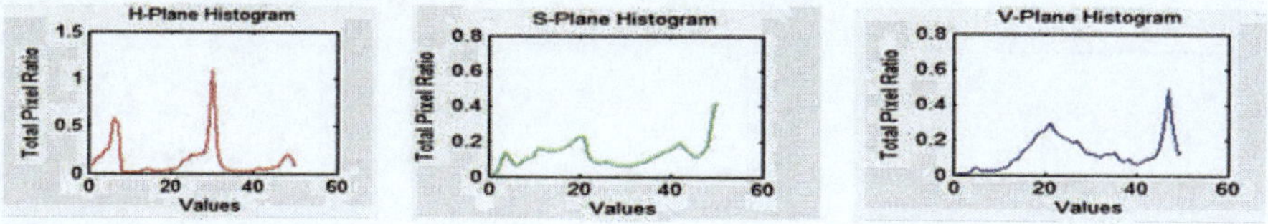

Fig. 6 Histogram of H, V, S frames

$$S = 0 \qquad \text{Max} = 0 \tag{4b}$$

2.2 *Histogram*

A picture histogram is graphical portrayal of the color dispersion in a computerized picture. It plots the quantity of pixels for each color worth [5]. This histogram is a representation appearing number of pixels in a picture at each extraordinary strength worth found in that picture (Fig. 6).

A histogram is a precise portrayal of the dissemination of numerical information. It is a gage of the likelihood circulation of a constant variable.

Histogram Calculation

$$H(i) = \sum_{i=1}^{255} \text{SUM}(I\text{pixelValues} == i) \tag{5}$$

2.3 *Means Log Difference Method*

We have incorporated differentiation between current frame ($I1$) and the previous frame ($I2$) with the function for finding mean. Mathematical expressions for calculating mean log difference is given by Eqs. (6)–(9). In the next step, we have converted to double precision floating point of the value corresponding to the current and previous frame. Further this value is divided with the mean value and assigned to the respective variables. In final step, difference between these values and its absolute value in the integer form is calculated. This is depicted in corresponding Fig. 7a–e.

$$\text{Mean}\ (\mu 1) = \frac{\sum_{i=1}^{R} \sum_{j=1}^{C} I1(i, i)}{R * C} \tag{6}$$

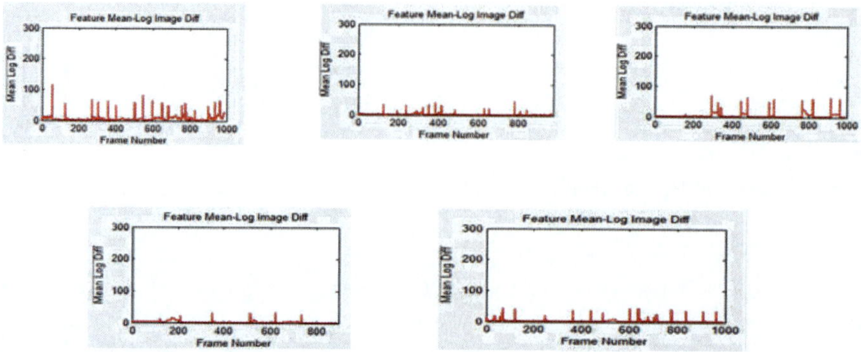

Fig. 7 **a** Mean log difference of "movie." **b** Mean log difference of "cartoon." **c** Mean log difference of "sports." **d** Mean log difference of "wildlife." **e** Mean log difference of "news"

$$\text{Mean}\ (\mu 2) = \frac{\sum_{i=1}^{R} \sum_{j=1}^{C} I2(i, j)}{R * C} \tag{7}$$

$$\text{Mean Difference Image} = \text{Abs}\left(\frac{I1}{\mu 1} - \frac{I2}{\mu 2}\right) \tag{8}$$

$$\text{Log Difference Image} = |\text{Log}(I1) - \text{Log}(I2)| \tag{9}$$

2.4 Shot Boundary Detection Using ANN

The mean log difference values are given as a input for calculating ANN decision considering probability function. Incorporating handle function and displaying legend access a graph between frame number on X-axis and ANN decision with probability on Y-axis is represented. The peak represents not only ANN shot decision value but also shot probability function values which are calculated during feature extraction process. In GUI represented, function also calculates count of total shot boundaries detected by algorithm. While this graph represents the same values in the form of shots, that means the area between two peaks represents one shot. In this way, the total count calculated and the shots detected by ANN shot detection shot probability function are analogous to each other (Fig. 8).

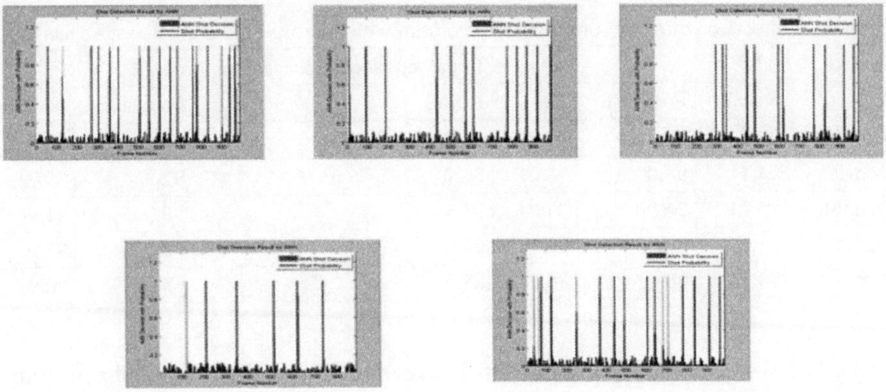

Fig. 8 **a** ANN decision with probability of "movie." **b** ANN decision with probability of "cartoon." **c** ANN decision with probability of "sports." **d** ANN decision with probability of "wildlife." **e** ANN decision with probability of "news"

2.5 *Evaluation and Parameter Calculation*

For each video, an array of shot boundary indexes was developed, comparing detection boundary results for each video to the shot boundary reference data constructed manually [6].

Following parameters are used to evaluate the shot boundary detection method. It is observed that experimental results depicts that identification of video shot boundary is effective.

But performance of suggested algorithm for videos with camera motion and random changes in characteristics is not up to the mark. Table 1 shows experimental results of proposed algorithm.

Table 1 Experimental results of proposed algorithm

S. No.	Video	Recall (%)	Precision (%)	$F1$ (%)
1	Cartoon_1	92.30	90.46	91.37
2	Cartoon_2	92.30	75.38	82.99
3	Movie_1 (Chennai Express)	96.87	85.75	90.97
4	Movie_2 (Bable)	94.11	98.00	96.01
5	Movie_3 (Transformer)	72.72	82.66	77.37
6	Wildlife	85.71	84.00	84.84
7	Nature	70.00	86.22	77.26
8	News	94.11	80.47	86.76
9	Sports_1 (Tennis)	91.66	89.83	90.74
10	Advertise (Bornvita)	95.45	93.54	94.49

Table 2 Comparison of results of proposed algorithm with improved block color feature method

S. No.	ANN			Improved block color feature		
	Recall	Precision	$F1$	Recall	Precision	$F1$
Cartoon	92.30	90.46	91.37	89.7	94.5	92.0
Movie	94.11	98.00	96.01	86.7	92.9	89.7
Wildlife	85.71	84.00	84.84	93.8	90.0	91.9
Sports	91.66	89.83	90.74	92.0	92.0	92.0
News	94.11	80.47	86.76	88.9	93.0	90.9

Table 2 shows comparison of results between proposed algorithm and algorithm based on improved block color feature method. Initially, if we consider carton video proposed algorithm results are satisfactory in terms of recall, precision, and $F1$ measures which comes out to be more than 90% while improved block color method depicts slightly less values of recall and $F1$ measures. For the movie video, again recall and $F1$ values are substantially high as compare to improved BCF, but the precision value of the proposed algorithm is comparatively less which shows that as the complexity of video varies the precision value decreases.

In wildlife video, the evaluative parameters are moderately less which is below 85% while improved BCM performs well, and this shows that ANN method needs more training in the memory system. Sports video particularly tennis video shows moderately high-performance parameters with the proposed algorithm while improved block code feature method projects slightly less values. This indicates that drastic change in pixels is well managed by AN method finally news video gives equal values of all performance parameters for both the algorithms.

3 Conclusion

In this paper, we have projected novel approach based on ANN along with mean log difference to the detection of video shot boundary. We have first incorporated mean log difference to calculate features like H, V, S and histogram distribution. In the second part, ANN is implemented for SBD. In this work, particularly we have used feed-forward network for input to output mapping incorporating series of layers. The first layer is networked with input succeeding toward hidden layer. Further, final layer produces networks output. ANN has the skill to learn and model nonlinear complex relationship which is essential in applications of video segmentation. Experimental results depicts that the suggested method very effectively identifies video shot boundary consistently.

Our future work will be planned for identifying VSB using one more novel approach focusing on eliminating disturbances caused by various aspects like camera motion and random changes in video characteristics which is considered to be measure challenges in video shot boundary detection.

References

1. Sharma YK, Sharif GM (2018) Framework for privacy preserving classification in data mining. J Emerg Technol Innov Res 9(5):178–183
2. Bhalotra PA, Patil BD (2014) Video shot boundary using random projection method and ridgelet transform method: a comparison. In: Proceedings of "The 2nd international conference on applied information and communications technology" ICAICT 2014. ISBN 9789351072850
3. Bhalotra PA, Patil BD (2013) Shot boundary detection using random projection transform. Int J Signal Image Process (IJSIP) 4(3):60–64, 2152–5048
4. Labhade NS, Arora PS, Sharma Y (2019) Study of neural networks in video processing. J Emerg Technol Innov Res (JETIR) 6(3):330–335. ISSN 2349-5162
5. Liu H, Li T (2016) An adaptive threshold shot detection algorithm based on improved block color features. In: 2016 12th international conference on natural computation, fuzzy systems and knowledge discovery (ICNC-FSKD), 978-1-5090-4093-3/16/$31.00 ©2016. IEEE, pp 1852–1857
6. Bi C, Yuan Y, Zhang J, Shi Y, Xiang Y, Wang Y, Zhang R (2018) Dynamic mode decomposition based video shot detection. https://doi.org/10.1109/ACCESS.2018.2825106
7. Bhalotra PA, Patil BD (2013) Video shot boundary detection using Ridgelet transform. In: Springer AISC series, vol 249, pp 163–171. ISBN 978-3-319-03094-4
8. Yang J, Zhong W (2016) On the image enhancement histogram processing. In: 2016 3rd international conference on informative and cybernetics for computational social systems (ICCSS), Jinzhou, 10 Oct 2016. Electronic ISBN: 978-1-5090-3367-6, INSPEC Accession Number: 16359679

Custard Apple Leaf Parameter Analysis, Leaf Diseases, and Nutritional Deficiencies Detection Using Machine Learning

Appasaheb Gargade and Shridhar Khandekar

Abstract Custard apple (Annona Squamosa L.) is the oldest fruit plant in the dry land. It is begun from a tropical area of America and widely disseminated all through the tropics and subtropics. The custard apple fruits are cultivated in many states in India on a commercial scale. Disease detection and health monitoring in a plant are essential for sustainable agriculture. Nutrients play a crucial role in influencing tree growth, fruit production, and fruit quality. It is arduous for human vision to identify the particular leaf disease and nutrient deficiency by naked eyes. In this paper, an attempt is made to propose a system for leaf parameter analysis, detection of N, P, K deficiencies, and leaf diseases. K-nearest neighbors (k-NN), and Support vector machine (SVM) algorithms are applied for the classification of leaf deficiencies and leaf diseases. Database of 125 and 80 Custard apple leaf images are used for leaf diseases and deficiencies, respectively. Experimental results showed that the proposed leaf parameter measurement system had attained 99.5% accuracy. This paper exercise a supervised machine learning approach using image processing.

Keywords Leaf parameter · Leaf disease · Nutrient deficiency · Image processing · Machine learning · Custard apple

1 Introduction

The custard apple is called as sweets-sop, sugar-apple, and as sitafal. The custard apple provides many health benefits due to the presence of an anti-oxidant like vitamin C, vitamin A, vitamin B6, copper, and magnesium. The plant growth and yield depend on different environmental conditions and soil quality [1]. Leaf parameters (maximum length, maximum width, area, and perimeter of a leaf) are a critical

A. Gargade (✉) · S. Khandekar
Department of Electronics, School of Electrical Engineering, MIT Academy of Engineering, Alandi, Pune, India
e-mail: appagargade@gmail.com

S. Khandekar
e-mail: sakhandekar@etx.maepune.ac.in

© Springer Nature Singapore Pte Ltd. 2021

S. N. Merchant et al. (eds.), *Advances in Signal and Data Processing*,
Lecture Notes in Electrical Engineering 703,
https://doi.org/10.1007/978-981-15-8391-9_5

crux in scientific research. Traditionally, different methods like paper weighing, grid counting, regression equation are used for leaf parameter measurements. These are simple methods but have the drawbacks like low precision and consume more time [2].

The disease is a freak state of a plant. The plant leaves get infected due to various bacteria, viruses, and fungi [3, 4]. Many times, farmer or expert is not able to identify leaf illnesses by bare eyes. Early recognition of various diseases diminishes the consumption of pesticides on the farm. As an analytic apparatus, visual perception can be constrained by different variables, including pseudo deficiencies and hidden hunger in plant [5–7]. Various techniques are used for leaf diseases, defects classification, and parameter measurements [8–10]. The custard apple leaves get affected by different diseases such as Alternaria alternata, Cercospora leaf spots, leaf holes, anthracnose, and bacterial blight.

Nutrient inadequacy occurs when fundamental nutrients are not available in sufficient quantity to fulfill the requirements of growing plants. Common nutrient deficiencies observed in agriculture fields are primary nutrient deficiencies (such as nitrogen, phosphorus, and potassium (NPK)) [11]. The conventional methods used to identify nutrient deficiencies in plants are by using plant assessment, visual inspection, and soil analysis. Visual inspection is a prejudicial analysis and depends on signs and symptoms on different parts of plants such as yellowing of leaves, undersized growth due to nutrient stress. Signs and symptoms of phosphorus deficiencies are dark green leaf and often developing purple or red color. Due to nitrogen deficiencies, leaves are light green or yellow and due to potassium deficiencies, burning of leaf margins, and sometimes spotting along the leaf margins [7].

A review of previous work is summarized in Sect. 2. The projected system methodology and implementation are described in Sect. 3. The results are discussed in Sect. 4, and an overall conclusion is highlighted in Sect. 5.

2 Review of Previous Work

Many research scholars have made significant contributions allied to leaf parameter analysis, leaf diseases, and deficiencies detection using various techniques. Detail literature survey of related work is summarized in the review paper [12].

2.1 Leaf Parameter Analysis

In work linked to leaf parameter measurements, authors of paper [13] proposed the algorithm to calculate the leaf surface area of healthy and unhealthy regions using threshold segmentation techniques. The accuracy in separating sick, and hygienic leaf portions are approximately 96%, and 98% respectively. Authors in paper [2] recommended an algorithm to measures the area of black gram, green gram, and pigeon pea plant leaf. In this algorithm, the coin is taken as a reference object, and digital camera (Make: Nikon, Model: D5300, Resolution: 24 megapixels) is utilized to acquire a leaf image. The average accuracy of 97% is obtained using this method. In paper [14], leaf parameters such as width, height, perimeter, and area of clove plant leaf are measured. Different paper labels are used as a reference object. The accuracy of 97% is achieved in width, height measurements, and 93% in the area, perimeter measurement of a leaf.

Authors in paper [15] uses an image scanner (Make: Gemstar, Model: JT-DBG00) for leaf area measurement of the various plant. OTSU's thresholding, morphological filtering, blob analysis technique are used. The leaf area measurement accuracy of 99% is obtained using this technique. Researchers in paper [1] uses OTSU's thresholding and region filling techniques. The CCD camera (Make: Soney, Resolution: 16.1 megapixels) is used to grab a peepal plant leaf image along with a square reference object. The leaf area measurement accuracy using this method is 99%. Authors of paper [16] uses a flatbed scanner to measure the area, width, and length of various plant leaves with 99% accuracy. Writers of paper [17] uses eight megapixels canon make, digital camera to capture cucumber leaf image with A3 paper, circle as a reference object. In this work area, length, and width of cucumber leaf is measured with 97% accuracy. The algorithm results are correlated with different measurement techniques like leaf area meter, paper weighing, and grid counting method (millimeter graph method).

2.2 *Leaf Diseases, Nutritional Deficiencies Classification*

In leaf disease detection, authors of paper [8] worked on tomato, potato, and pepper leaf diseases suchlike leaf spot, and late blight. In this work, Otsu's segmentation, color co-occurrence method, and genetic algorithm is applied. Authors in paper [10] worked on the classification of different diseased leaves of a citrus plant. In this work, the classification accuracy of 90% is achieved. In paper [18], authors proposed a system for mango plant leaf diseases, like algal spot, and anthracnose. In this work, 89.9% accuracy is obtained using a decision tree model. Authors in paper [3] done invention on segregation of five tomato leaf diseases. Accuracy of 97.3% is attained. The main intention of paper [19] is to classify infected and healthy tomato leaves using GLCM feature extraction and SVM classifier.

Authors in paper [7] represented a system to detect nitrogen (N), potassium (K), and phosphorus (P) deficiencies in paddy tree leaf using pattern and color analysis. The authors managed to meet 90% accuracy. The review paper [4] gives enlightenment on cotton leaf diseases, various segregation, and segmentation approaches. Authors in paper [6] put forward a system for cotton leaf disease categorization using shape and color based feature extraction. The paper [5] suggested an algorithm to detect black spot and sun scorch leaf diseases in the orchid plant with an accuracy of 81.8% and 90.1% in recognition, respectively. In paper [9], the machine learning technique is proposed to classify five different plant diseases using the SVM classifier.

The papers [20, 21] gives a review of advanced methods for identifying and sorting various plant leaf diseases. Authors in paper [22] presents an algorithm for sorting maize leaf diseases viz flyspeck, gray spot, curvularia spot, brown spot, and rust using binary tree search, and machine vision with 80% accuracy. Authors in paper [23] proposed a system to segregate sugar beet leaf diseases, namely powdery mildew, leaf rust, and Cercospora leaf spot using multiclass SVM. Accuracy of 86% is attained. In this paper [24], authors suggested a methodology for detecting five different plant leaf diseases using Gabor wavelet transform (GWT) and SVM with 89% accuracy. An advanced algorithm is necessary to enhance sorting accuracy.

3 Proposed Methodology

3.1 Proposed System Implementation

The proposed system implementation is divided into leaf parameters measurement, leaf diseases, and leaf nutritional deficiencies classification. The system is implemented in MATLAB 9.2 (R2017b) software. The proposed workflow sequence of a system for leaf diseases and leaf nutritional deficiencies classification is as displayed in Fig. 1.

The significant stages of custard apple leaf diseases and leaf nutritional deficiencies classification using machine learning technique [25] are training the predictive model with feature vectors, labels, and then testing with query images. Different steps used for system training and testing are standard image database collection, image enhancement, segmentation (K-means), statistical feature extraction (GLCM), and classification technique (k-NN, and SVM) [26].

3.2 Leaf Parameters Measurement

The image acquisition setup is as shown in Fig. 2. Camera (Logitech C 922 pro) and the white platform are used. First, Custard apple leaf is placed on the white platform along with a reference object (IC P89v51rd2). The captured image is stored in the laptop connected through USB cable. Custard apple leaf image with a reference object is processed for analysis.

Leaf parameters measurement GUI is shown in Fig. 3. It shows the original leaf image, total area of leaf, affected area of the leaf, extent, orientation, leaf major, and leaf minor axis length. The grid counting approach for custard apple leaf parameter measurement is as shown in Fig. 4.

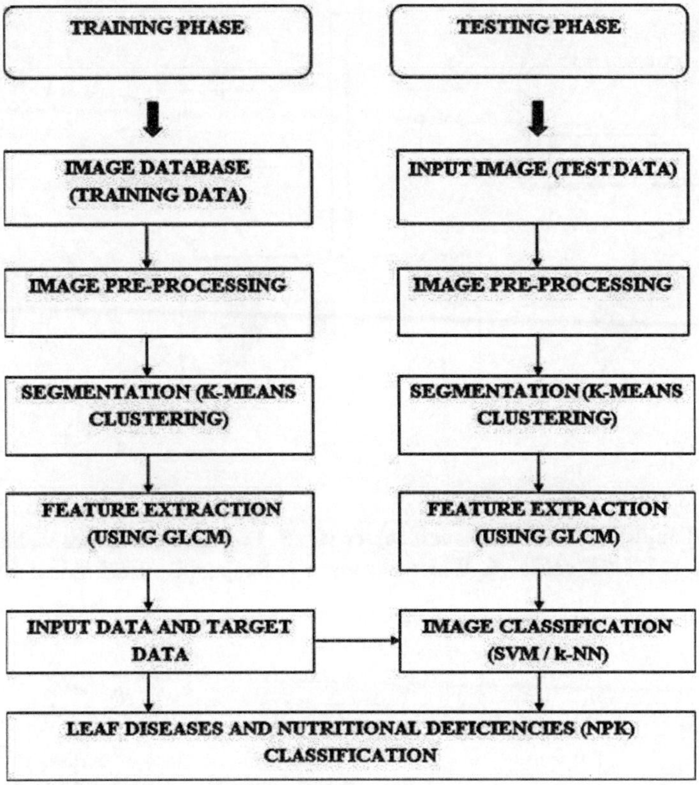

Fig. 1 Proposed system block diagram

For leaf parameter calculation, regionprops (L, properties) function in MATLAB is used. After preprocessing, the captured image is separated into leaf image and reference object image. The total pixels in each image are measured. Let $P1$ and $P2$ are the total pixel count in leaf image and a reference image. The entire leaf area (A) is calculated by comparing and taking proportion with the reference object's known area as given below [2, 12],

$$A = \frac{R * P1}{P2}\ (\text{mm}^2) \tag{1}$$

where

A: Leaf surface area (mm^2)

R: Reference object surface area (mm^2)

$P1$: Total pixel count in leaf surface area

$P2$: Total pixel count in reference object surface area.

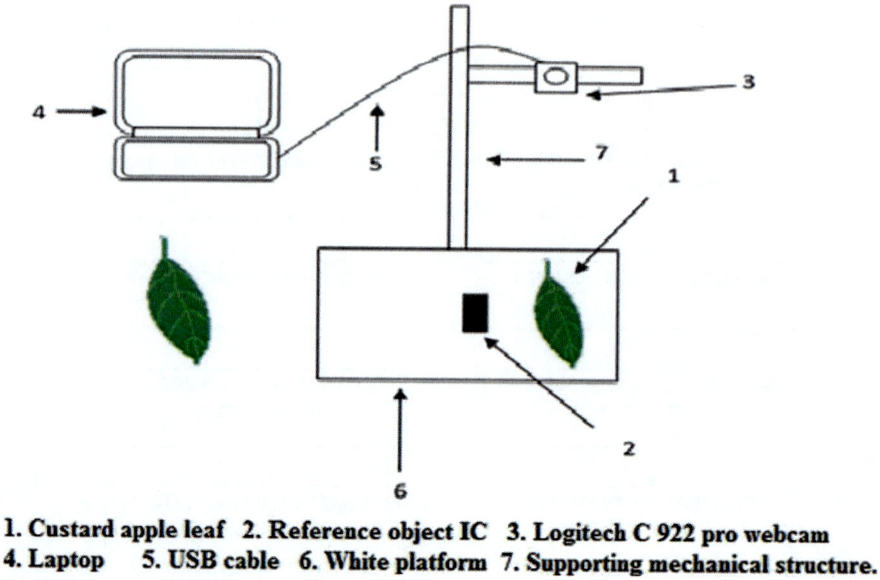

Fig. 2 Image acquisition

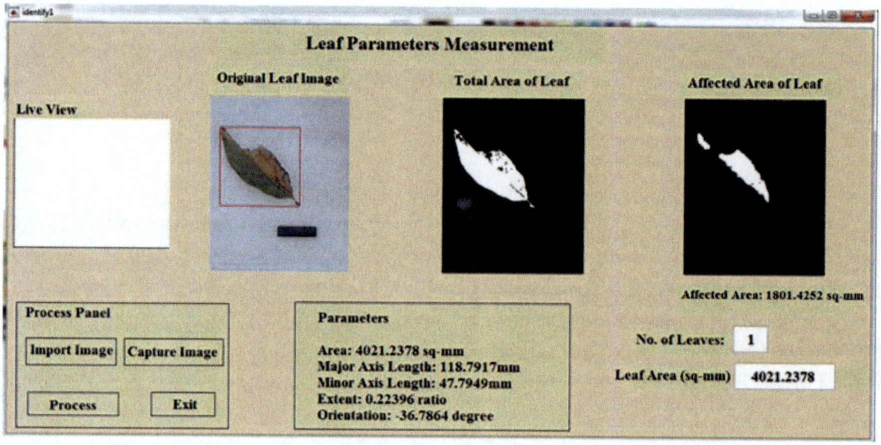

Fig. 3 Leaf parameter measurement GUI

Leaf area computed using this algorithm is compared with grid counting technique.

$$\text{Calibration Factor (CF)} = \frac{52\,\text{mm}}{L} \tag{2}$$

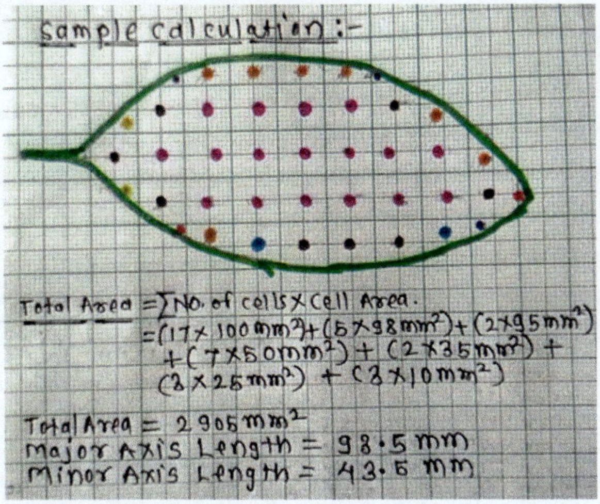

Fig. 4 Area calculation by using grid counting method

where 52 mm = Reference object length.

L = Total pixel count in reference object length (1 mm length equivalent number of pixels).

$$\text{Leaf Area}\left(\text{mm}^2\right) = \text{No. of pixels in Area} * \text{CF} \quad (3)$$

$$\text{Major Axis Length (mm)} = \text{No. of pixels} * \text{CF} \quad (4)$$

$$\text{Minor Axis Length (mm)} = \text{No. of pixels} * \text{CF} \quad (5)$$

$$\text{Percentage Error Rate} = \frac{A2 - A1}{A1} * 100 \quad (6)$$

where $A1$ is the result obtained using graph paper grid counting, and $A2$ is the result obtained using MATLAB. The affected or diseased area of the leaf is calculated by using the HSV color threshold value. The leaf parameters and affected area are shown using GUI. Leaf parameter measurement correctness using this algorithm depends on the resolution of a digital camera used to capture images.

3.3 Leaf Diseases and Nutritional Deficiencies Detection

The details of supervised machine learning algorithm implementation for custard apple leaf diseases (Alternaria alternata, Cercospora leaf spots, anthracnose, bacterial blight), and nutritional deficiencies (nitrogen, phosphorus, and potassium (NPK)) detection and classification are as given below.

Image Enhancement For image enhancement Imadjust and stretchlim function in MATLAB are used [26]. The result of image contrast enhancement is as shown in Fig. 5.

Image Segmentation The input query image is transformed to CIE 1976 $L * a * b$ color format and then segmented into three clusters using the K-means algorithm. A partition-based cluster analysis method is used in K-means. For image segmentation, kmeans(ab,nColors,distance,sqEuclidean,Replicates,3) MATLAB function is used. Image segmentation helps to identify the affected region from the input leaf image [26]. These segmented image features are extracted and used during training and testing. The results of image segmentation are shown in Fig. 6.

Image Feature Extraction Image feature extraction is an essential step for training and testing the machine learning algorithm. In this system, different features viz homogeneity, correlation, contrast, and energy are obtained from the gray-level co-occurrence matrix. These are the statistical characteristics of the segmented image. GLCM matrix measures the spatial relationship in pixels [19]. Other features extracted from segmented ROI image are mean, entropy, root mean square (RMS), smoothness, standard deviation (SD), kurtosis, skewness, variance, and inverse difference moment (IDM). All these features are enforced as inputs to classifiers. Detail calculation equations of all these features are as explained below [10].

Features:

- Contrast: Contrast provides information about the structural variations in the image. It figures out the intensity contrast between nearby two pixels over the entire image.

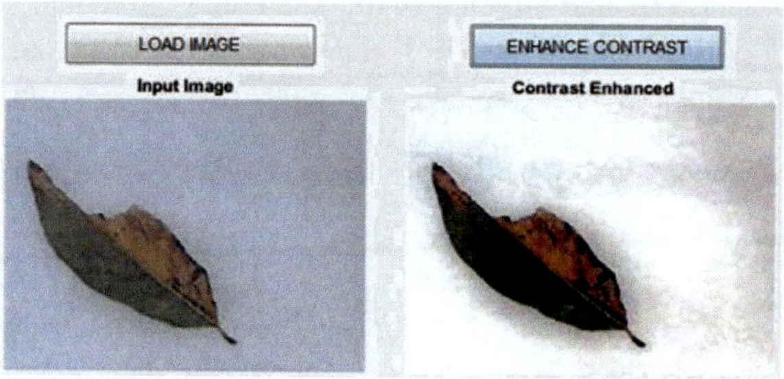

Fig. 5 Image enhancement

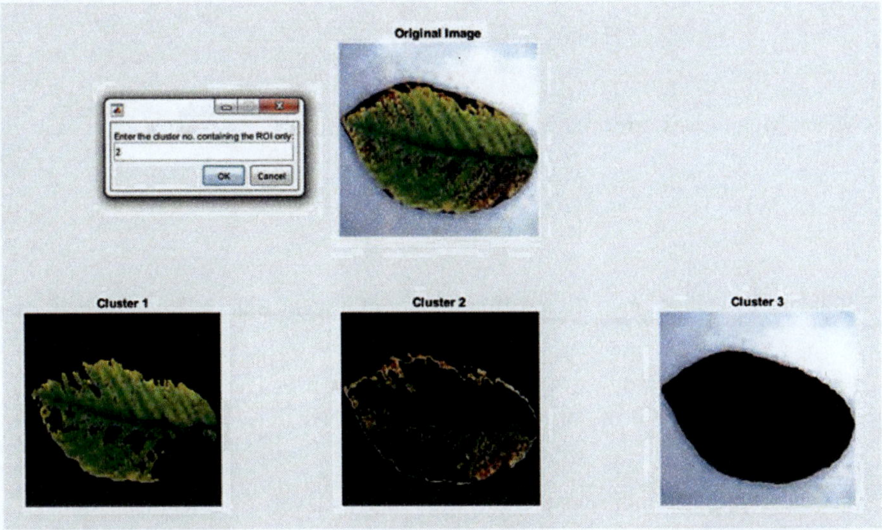

Fig. 6 Image segmentation

$$\text{Contrast} = \sum_{x,y} |x - y|^2 p(x, y) \tag{7}$$

where $P_{x,y}$ is pixel at location (x, y).
P is the matrix of primitive texture cell corresponding to the ROI segmented image.

- Correlation: This feature finds how nearby pixels are correlated to each other in the complete image.

$$\text{Correlation} = \frac{\sum_x \sum_y (x, y) P_{x,y} - \mu^2}{\sigma^2} \tag{8}$$

where σ is standard deviation (SD) of the GLCM.

- Energy: Energy is calculated using the summation of all squared elements in GLCM. Energy value varies in between 0 and 1. Energy calculation equation is given as,

$$\text{Energy} = \sum_{x,y} p(x, y)^2 \tag{9}$$

- Homogeneity: Homogeneity is related to the adjacency of the dissemination of points in the GLCM to it's diagonal. It's value varies in between 0 and 1. It is defined as,

$$\text{Homogenity} = \sum_x \sum_y \frac{P_{x,y}}{1 + |x - y|} \tag{10}$$

- Mean: Mean gives information about image brightness.

$$\mu_P = \frac{1}{n^2} \sum_{x=0}^{n-1} \sum_{y=0}^{n-1} P_{x,y} \tag{11}$$

- Standard Deviation: σ gives information about image contrast. It is given as,

$$\sigma_P = \left[\frac{1}{n^2} \sum_{x=0}^{n-1} \sum_{y=0}^{n-1} \left[P_{x,y} - \mu_P \right]^2 \right]^{\frac{1}{2}} \tag{12}$$

- Variance: Variance indicates the deviation of gray levels from its mean values.

$$\text{Variance} = \sum_x \sum_y (x, y) P_{x,y} - \mu^2 \tag{13}$$

  where μ: Mean of the GLCM.
- Entropy: Entropy gives information about the randomness that depicts the texture of the given input image.

$$E = -\sum_x p_x \log_2 p_x \tag{14}$$

  where p is the histogram counts.
- Kurtosis: Kurtosis gives information about the tailedness of the probability distribution. K = KURTOSIS(A) function in MATLAB returns the kurtosis of the values in A.
- Skewness: Skewness tells about the (lack of) symmetry. If the skewness is negative, the histogram is negatively skewed. The positive skewness is the opposite. S = SKEWNESS(A) function in MATLAB returns the skewness of the values in A.
- Inverse Difference Movement: Inverse difference moment gives information about local homogeneity, and it is given as,

$$\text{IDM} = \sum_x \sum_y \frac{1}{1 + (x - y)^2} P_{xy} \tag{15}$$

Image Classification The leaf images of different diseases and nutritional deficiencies are classified using SVM and k-NN Classifier.

- Support Vector Machine: For data analysis, pattern recognition, and classification SVM is used. It is a supervised machine learning method [27] (Fig. 7).

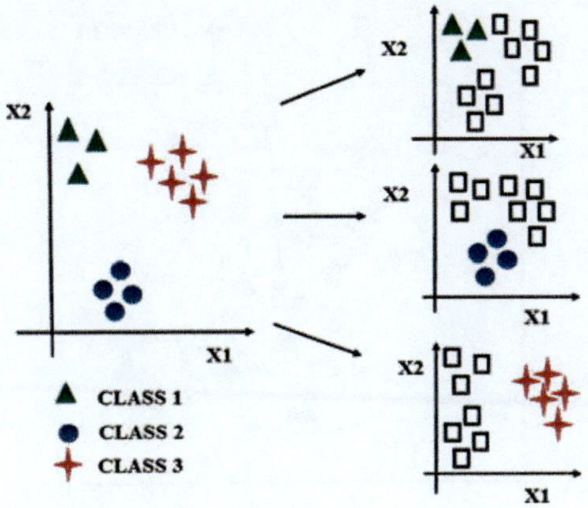

Fig. 7 Multiclass support vector machine classifier

Using different kernel equations, SVM reforms data from input feature scope into a high-dimensional feature scope. SVM is memory efficient and versatile. SVM is used for binary or multiclass classification. Multiclass classification is done using one versus rest or one versus one method [28]. In this application, the classification learner app in MATLAB is used to train a system using a multiclass model, which uses error-correcting output codes (ECOC) having SVM binary learner. A multiclass ECOC model is trained using fitcecoc function. SVM is more effectual when the quantity of dimensions is higher than the quantity of training data in a high dimensional space.

- k-Nearest Neighbors: k-NN classifies test images on similarity measures. It is a distribution-free approach adopted for classification. In k-NN, the test point is grouped by a maximum vote of its neighbors. The test point is allotted to the group, which is most familiar between its k nearest neighbors [29]. k-Nearest Neighbors model is trained using fitcknn function in MATLAB (Fig. 8).
  The classification learner app in MATLAB, function returns a trained SVM and k-NN classifier model and it's validation accuracy. The classifier make predictions of test data using the function yfit on test data. The output of yfit function gives a class prediction for each test data point. The image feature dataset acts as input to the different classifiers. These image feature datasets are divided into testing and training dataset. The classifier is trained using a training dataset and then used to classify the testing dataset. The different classifier's performance is checked by correlating the actual labels and anticipated labels.

$$[\text{tClass}, \text{vAcc}] = \text{trainclassifier}(\text{traindata}) \tag{16}$$

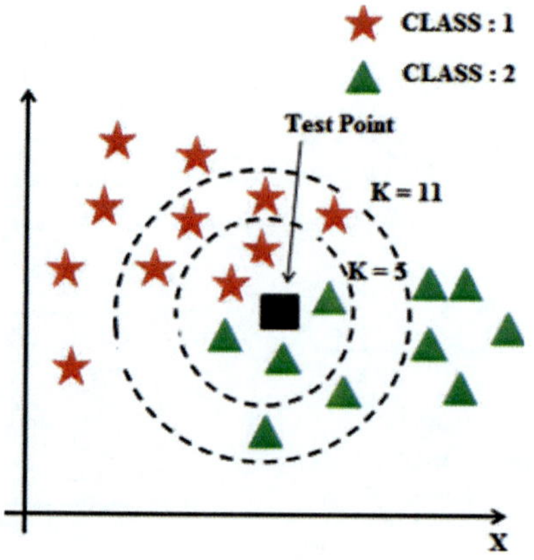

Fig. 8 k-nearest neighbors classifier

3.4 Algorithm

Step-wise algorithm for leaf parameters calculation.

Algorithm 1 For Leaf Parameters Calculation

1: Import image from folder location. OR Capture image through an external camera.
2: Check image is rgb or gray image.
3: If rgb then convert into gray image.
4: Convert gray image into binary image using threshold value.
5: Calculate width and length of the reference object (Reference object is IC P89v51rd2, $L =$ 52 mm, $W = 14$ mm).
6: Calculate calibration factor value, i.e., no equivalent pixel for 1 mm length.
7: Calculate leaf image parameters using regionprops function in MATLAB.
8: Calculate leaf image extent, orientation, area, minor axis length, and major axis length by applying calibration factor.
9: Convert original rgb leaf image into hsv image.
10: Apply hsv color threshold values to hsv image.
11: Calculate the affected area of leaf.
12: Display leaf parameters and affected leaf area.

Step-wise algorithm for leaf disease detection and leaf nutritional deficiencies (NPK) detection.

Algorithm 2 For Leaf Diseases and Nutritional Deficiencies Detection

1: Import leaf image from database.
2: Apply imresize function for image re-size.
3: Apply image contrast enhancement using stretchlim function in MATLAB.
4: Create color transformation structure using $C = makecform(type)$ function.
5: Convert image from RGB to CIE 1976 $L * a * b *$ color space.
6: Apply image segmentation function $varargout = kmeans(X, k, varargin)$.
7: Select a segmented image based on ROI. Convert segmented image into grayscale image.
8: Extract image features using GLCM.
9: **if** Training Phase **then**
10: Models (Leaf Diseases, Leaf Nutritional Deficiencies Detection) are first trained by using training dataset feature vectors, labels of different diseases and nutritional deficiencies separately.
11: Training Phase use above steps 1–8 followed by class labels.
12: **else if** Testing Phase **then**
13: The trained models predict disease and deficiency of Custard Apple leaf using SVM and k-NN classifiers.
14: Testing Phase use above step 1 to 8 followed by classifiers.
15: **end if**

4 Results and Discussion

4.1 Custard Apple Leaf Parameter Analysis

The results for leaf parameter measurements using MATLAB and grid counting method for ten samples are given in Table 1. The table contains the leaf parameter result for leaf major axis length (in mm), leaf minor axis length (in mm), and leaf area (in mm^2).

4.2 Leaf Diseases Detection

The result for the leaf disease detection method is as shown in Fig. 9. The custard apple leaf diseases such as Alternaria alternata, Cercospora leaf spots, anthracnose, and bacterial blight are detected and classified using SVM and k-NN classifier. First, the training database is prepared using 125 images of different leaves, with 100 diseased leaves (25 leaves of Alternaria alternata disease, 25 leaves of anthracnose disease, 25 leaves of bacterial blight disease, 25 leaves of Cercospora leaf spots), and 25 healthy leaves. The trained model is used for detection and classification (using SVM and k-NN classifier) of different custard apple leaves. The performance per class in leaf diseases detection in term of confusion matrix plot for SVM and K-NN

Table 1 Custard apple leaf parameters measurement

S. No.	Sample	Area	Major axis length	Minor axis length	Area	Major axis length	Minor axis length
		Using MATLAB			Grid counting method		
1	S1	6268.98	135.01	61.73	6240	135	60
2	S2	3736.93	111.55	43.37	3700	110	42
3	S3	4021.23	118.79	47.79	4055	120	47
4	S4	4147.87	107.66	50.3	4150	109	50
5	S5	2935.05	100.32	38.17	2940	100	37
6	S6	3137.9	91.39	45.11	3130	90	45
7	S7	4693.76	128.43	48.59	4650	126.5	48
8	S8	6048.61	137.06	57.72	6100	139.5	55.5
9	S9	3411.02	104.48	45.94	3465	107	44
10	S10	2966.57	92.1	41.74	2905	93.5	43.5

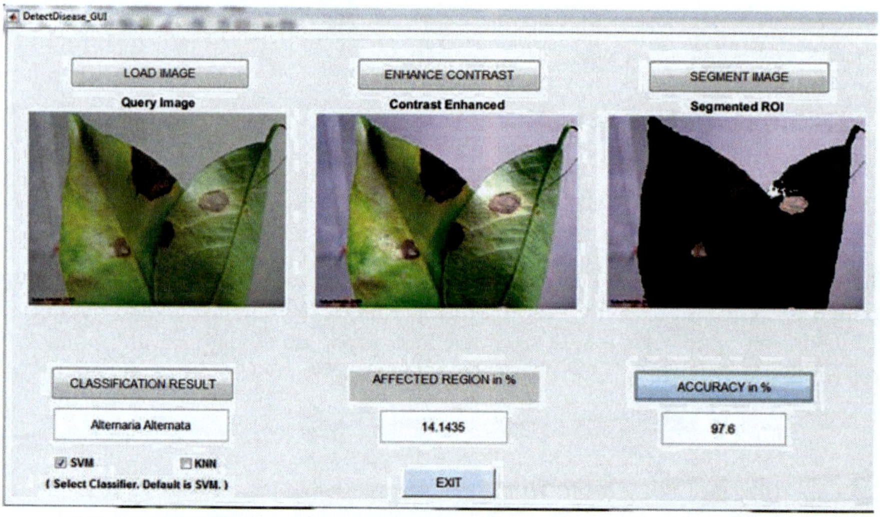

Fig. 9 Leaf diseases detection

classifier are as shown in Fig. 10. In this confusion matrix, classes are mapped as 0: Alternaria Alternata, 1: Cercospora Leaf Spot, 2: Anthracnose, 3: Bacterial Blight, 4: Healthy Leaf. The accuracy for leaf disease detection using SVM and k-NN classifier are 97.6% and 99% respectively.

Database The database of 125 leaf images is used for leaf disease classification, and a database of 80 leaf images is used for leaf deficiency classification. Different pictures of custard apple leaf are collected from Krushi Vigyan Kendra, Narayangaon (Pune), and nearby farms in Pune District in Maharashtra.

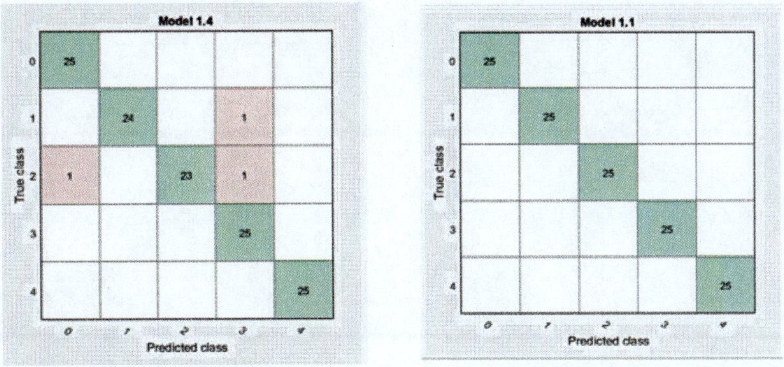

(a) SVM Classifier (b) k-NN Classifier

Fig. 10 Confusion matrix for leaf diseases

4.3 Leaf Nutritional Deficiencies Detection

The result for leaf nutritional deficiencies detection method is as shown in Fig. 11. The custard apple leaf nutritional deficiencies such as nitrogen, phosphorus, and potassium (NPK) are detected and classified using SVM and k-NN classifier.

First, the training leaf database of 80 images is created, with 20 nitrogen-deficient leaves, 20 phosphorus-deficient leaves, 20 potassium deficient leaves, and 20 healthy leaves. The trained model is used to identify nutritional deficiencies in custard apple leaves. The performance per class in leaf deficiencies detection in term of confusion

Fig. 11 Leaf nutritional deficiencies detection

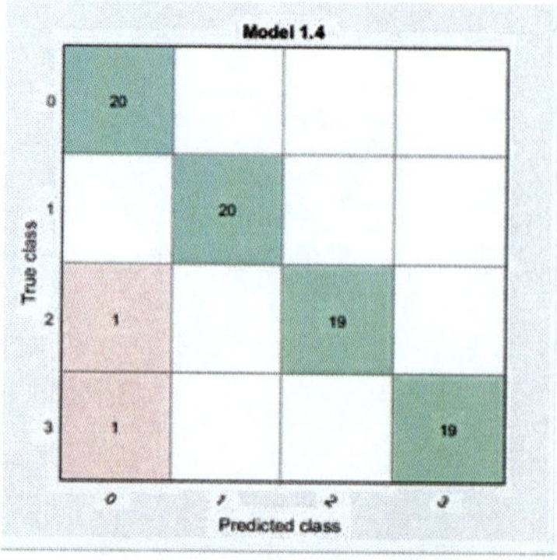

(a) SVM Classifier

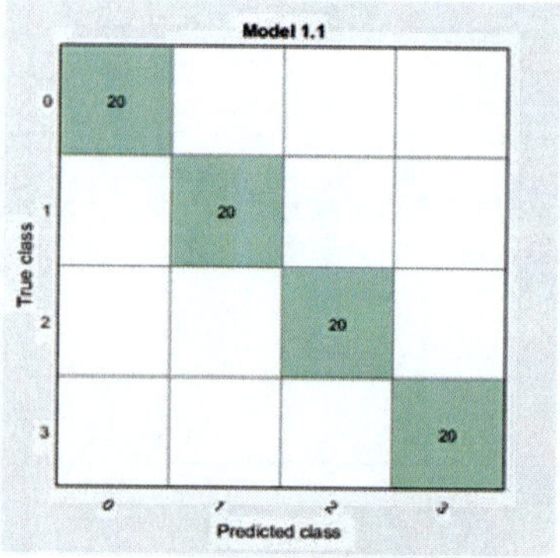

(b) k-NN Classifier

Fig. 12 Confusion matrix for leaf deficiencies

matrix plot for SVM and K-NN classifier are as shown in Fig. 12. In this confusion matrix, classes are mapped as 0: Nitrogen Deficiency, 1: Phosphorus Deficiency, 2: Potassium Deficiency, 3: Healthy Leaf. The accuracy for leaf deficiency detection using SVM and k-NN classifier are 97.5% and 99% respectively.

5 Conclusion

Custard apple leaf parameter measurement system is implemented in MATLAB, and measurement results are correlated with the grid counting technique. The average accuracy of approximately 99.5% is achieved in measurements. The accurate detection and classification of the custard apple leaf nutritional deficiencies and leaf diseases are vital for successful cultivation, and this is implemented using a machine learning method.

This paper discussed leaf parameter analysis, feature extraction technique to extract the features of infected custard apple leaf, classification of diseases, and NPK deficiencies using SVM and k-NN classifier. The k-NN classifier accuracy is found better than the SVM classifier for this application. The assessment of visual signs and symptoms in agriculture is an economical and fast approach for identifying potential threats such as diseases and nutrient deficiencies.

References

1. Chaudhary P, Godara S, Cheeran A, Chaudhari AK et al (2012) Fast and accurate method for leaf area measurement. Int J Comput Appl 49(9):22–25
2. Jadon M, Agarwal R, Singh R (2016) An easy method for leaf area estimation based on digital images. In: International conference on computational techniques in information and

communication technologies (ICCTICT). IEEE, pp 307–310
3. Sabrol H, Satish K (2016) Tomato plant disease classification in digital images using classification tree. In: International conference on communication and signal processing (ICCSP). IEEE, pp 1242–1246
4. Prajapati BS, Dabhi VK, Prajapati HB (2016) A survey on detection and classification of cotton leaf diseases. In: International conference on electrical, electronics, and optimization techniques (ICEEOT). IEEE, pp 2499–2506
5. Fadzil WNWM, Rizam MBS, Jailani R, Nooritawati M (2014) Orchid leaf disease detection using border segmentation techniques. In: Conference on systems, process and control. IEEE, pp 168–173
6. Rothe P, Kshirsagar R (2014) Automated extraction of digital images features of three kinds of cotton leaf diseases. In: International conference on electronics, communication and computational engineering (ICECCE). IEEE, pp 67–71
7. Latte M, Shidnal S (2016) Multiple nutrient deficiency detection in paddy leaf images using color and pattern analysis. In: International conference on communication and signal processing (ICCSP). IEEE, pp 1247–1250
8. Arya M, Anjali K, Unni D (2018) Detection of unhealthy plant leaves using image processing and genetic algorithm with Arduino. In: International conference on power, signals, control and computation (EPSCICON). IEEE, pp 1–5
9. Pooja V, Das R, Kanchana V (2017) Identification of plant leaf diseases using image processing techniques. In: Technological innovations in ICT for agriculture and rural development (TIAR). IEEE, pp 130–133
10. Prakash RM, Saraswathy G, Ramalakshmi G, Mangaleswari K, Kaviya T (2017) Detection of leaf diseases and classification using digital image processing. In: International conference on innovations in information, embedded and communication systems (ICIIECS). IEEE, pp 1–4
11. Dobermann A, Cassman K (2002) Plant nutrient management for enhanced productivity in intensive grain production systems of the United States and Asia. Int J Plant-Soil Relat 247(1):153–175
12. Gargade A, Khandekar SA (2019) A review: custard apple leaf parameter analysis and leaf disease detection using digital image processing. In: 3rd international conference on computing methodologies and communication (ICCMC). IEEE, pp 267–271
13. Chumuang N, Thaiparnit S, Ketcham M (2016) Algorithm design in leaf surface separation by degree in hsv color model and estimation of leaf area by linear regression. In: 12th international conference on signal-image technology & internet-based systems (SITIS). IEEE, pp 628–631
14. Demin Z, Wei C, Lijing L (2012) An image processing based system for leaf parameter measuring. In: Symposium on photonics and optoelectronics. IEEE, pp 1–4
15. Kaiyan L, Junhui W, Jie C, Huiping S (2014) Measurement of plant leaf area based on computer vision. In: Sixth international conference on measuring technology and mechatronics automation. IEEE, pp 401–405
16. Shivling V, Singla A, Ghanshyam C, Kapur P, Gupta S (2011) Plant leaf imaging technique for agronomy. In: International conference on image information processing. IEEE, pp 1–5
17. Tian YW, Wang XJ (2009) Analysis of leaf parameters measurement of cucumber based on image processing. In: WRI world congress on software engineering, vol 3. IEEE, pp 34–37
18. Trongtorkid C, Pramokchon P (2018) Expert system for diagnosis mango diseases using leaf symptoms analysis. In: International conference on digital arts, media and technology (ICDAMT). IEEE, pp 59–64
19. Mokhtar U, El Bendary N, Hassenian AE, Emary E, Mahmoud MA, Hefny H, Tolba MF (2015) SVM-based detection of tomato leaves diseases. In: Intelligent systems'2014. Springer International Publishing, Cham, pp 641–652
20. Arnal Barbedo JG (2013) Digital image processing techniques for detecting, quantifying and classifying plant diseases. SpringerPlus 2(1):660
21. Sankaran S, Mishra A, Ehsani R, Davis C (2010) A review of advanced techniques for detecting plant diseases. Comput Electron Agric 72(1):1–13. https://www.sciencedirect.com/science/article/pii/S0168169910000438

22. Zhao YX, Wang KR, Bai ZY, Li SK, Xie RZ, Gao SJ (2009) Research of maize leaf disease identifying models based image recognition. In: Crop modeling and decision support. Springer Berlin Heidelberg, Berlin, Heidelberg, pp 317–324
23. Rumpf T, Mahlein AK, Steiner U, Oerke E, Dehne H, Plumer L (2010) Early detection and classification of plant diseases with support vector machines based on hyperspectral reflectance. Comput Electron Agric 74(1):91–99
24. Prasad S, Kumar P, Hazra R, Kumar A (2012) Plant leaf disease detection using gabor wavelet transform. In: Panigrahi BK, Das S, Suganthan PN, Nanda PK (eds) Swarm, evolutionary, and memetic computing. Springer Berlin Heidelberg, Berlin, Heidelberg, pp 372–379
25. Alpaydin E (2010) Introduction to machine learning, 2nd edn. The MIT Press
26. Gonzalez RC, Woods RE (2009) Digital image processing. Pearson Publications
27. Vapnik V (2013) The nature of statistical learning theory. Springer Science & Business Media
28. Dorj UO, Lee KK, Choi JY, Lee M (2018) The skin cancer classification using deep convolutional neural network. Multimed Tools Appl 77(8):9909–9924
29. Hu LY, Huang MW, Ke SW, Tsai CF (2016) The distance function effect on k-nearest neighbor classification for medical datasets. SpringerPlus 5(1):1304

Discontinuous PWM Techniques to Eliminate Over-Charging Effects in Four-Level Five-Phase Induction Machine Drives

J. Balakrishna, Teegala Bramhananda Reddy, and Marapu Vijaya Kumar

Abstract Four-level dual inverter configuration fed five-phase induction motor drive utilizing unequal DC voltages for the dual two-level inverters on both sides of the open-ended stator is discussed in this paper. The DC voltages of the two inverters are maintained in the ratio of 2:1. By properly adding zero-sequence signals to the reference signals, continuous and discontinuous pulse width modulation (DPWM3) can be generated. But in continuous PWM technique, extensive speed control is not possible since over charging of capacitors and the DC voltage ratio of 2:1 fail to maintain. To solve this problem, an off-set value addition to the reference modulating signal is implemented. Compared to continuous PWM technique, in DPWM3 techniques, a better improvement in the quality of output voltage and current is observed and is discussed in terms of harmonic order; number of levels in common mode voltage (CMV), total harmonic distortion (THD). The proposed techniques are designed using a simple, easier, scalar-based approach which depends upon instantaneous values of the reference voltages and it does not need complex mathematical calculations. Simulation and experimental studies have been carried to show the efficient performance of the drive.

Keywords Discontinuous pulse width modulation (DPWM) · Five-phase motor drives · Open-end winding · Over-charging effects · Common mode voltage

1 Introduction

In many literatures, multiphase machines offer numerous benefits compared to three-phase machines. Multiphase machines predominantly controlled by different voltage source inverters. The conventional space vector-based PWM techniques have been

J. Balakrishna (✉) · M. V. Kumar
Department of EEE, JNT University, Anantapur, Andhra Pradesh, India
e-mail: balakrishna.caid@gmail.com

T. B. Reddy
Department of EEE, G Pulla Reddy Engineering College, Kurnool, Andhra Pradesh, India

© Springer Nature Singapore Pte Ltd. 2021

S. N. Merchant et al. (eds.), *Advances in Signal and Data Processing*,
Lecture Notes in Electrical Engineering 703,
https://doi.org/10.1007/978-981-15-8391-9_6

developed for five-phase machines in [1–3]. Dual inverter fed open-end winding (OeW) is developed which allows the machine to run under two inverters on both sides. OeW topologies of two-level inverter with equal and unequal DC voltages or a combination of multi-level and two-level inverts have been extensively studied in [4, 5]. For these space vector PWM scheme has been investigated majorly to improve quality of output voltage, converter losses, CMV reduction, etc. Basic conventional space vector PWM technique involves complex calculations for vector state identification, dwell time, sector selection, etc., since there are a greater number of voltage vectors present compared to three-phase system. OeW five-phase drives discussed in [6–8] illustrate the space vector PWM schemes.

The developments of PWM techniques for five-phase machine are more complicated and are different from the three-phase as the space vectors are mapped into α–β and x–y plane. Generation of sinusoidal output voltages using two large and two medium and carrier-based PWM for the five-phase voltage source inverters is investigated and comparative studies say that in both the techniques, switching patterns are similar for all sectors in output voltage generation. Investigations made in [9–11] say that carrier-based PWM techniques are easy to execute and it does not require complex calibrations that are made in space vector modulation methods. A comparative analysis of continuous and discontinuous PWM schemes for two-level voltage source inverter fed five-phase induction motor drive was made and study of switching characteristics in five-phase motors with SVPWM and discontinuous PWM are extensively studied in these papers. But all these papers are implemented with complex mathematical calculations and are limited to two-level five-phase systems. Discontinuous PWM techniques for three-phase inverters have studied in [12–16]. A decoupled modulation schemes for the four-level OeW five-phase drives have implemented experimentally for high performance drive applications. OeW five-phase machine drive maintaining the inverters in 2:1 ratio has proposed with two carrier-based modulation strategies PD and APOD PWM techniques with capacitor over-charging elimination methods and also indirect field orientation control methods are proposed to study the performance of the drive. First technical paper on decoupled PWM-based controlling strategies for four-level five-phase drives has been proposed in [17]. The over-charging of DC capacitors has been solved by proposing decoupled carrier-based modulation methods.

Literature with space vector approach is studied in [18, 19]. Both these articles adapted zero vector placements using complex space vector approach and compared the output variables in terms of voltage and current THDs. Isolated and non-isolated DC source with different control methods for CMV reduction along with output voltage quality improvement. A simplified approach of discontinuous PWM schemes for the three-phase drives much discussed in literature [20, 21] show a better improvement in quality of output voltage, reduced zero-sequence voltages, and current ripples. All these papers proposed a simple and easiest way of generating different continuous and discontinuous modulating signals. And also, these PWM schemes do not require complex mathematical calculations.

This paper focuses on a simple scalar-based PWM techniques for four-level dual inverter configuration fed five-phase induction machine. This PWM technique implementation does not involve complex calibrations like sector and its respective vector identification. Comparative studies between continuous and DPWM3 methods are studied in terms of CMV, number of levels in effective phase voltage, and its harmonic spectrum. Over-charging effect is eliminated by implementing an offset addition subtraction method. To validate the results, MATLAB/SIMULINK-based simulation and experimental results are presented.

This paper is organized in five sections. Section 2 provides description of conventional space vector-based modulation strategy for the four-level OeW five-phase topology. The proposed scalar-based approach for continuous and discontinuous modulating signals is explained in Sect. 3. Over-charging effects of DC-link capacitors are addressed in Sect. 4. Finally, simulation and experimental results are performed in Sects. 5 and 6, respectively.

2 Conventional SVPWM Technique for the Dual Inverter Fed OeW Five-Phase Motor Drive

Figure 1 indicates four-level topology of dual inverter fed OeW five-phase drives, with the Inverters I and II are maintained at a ratio of 2:1. A two-level inverter generates 32 space vectors. The number of switching states of the dual inverter supplied machine with five-phase is given by $n^5 \times n^5$ where n indicates the number of levels of the inverter. When two 2-level five-phase inverters are applying a combination of $32 \times 32 = 1024$ switching states with 30 distinct active vectors and 34 zero vector states and 211 space vector positions which is plotted in Fig. 2. Figure 3 indicates space vector distribution for the four-level topology with unequal voltages maintaining in 2:1 ratio. As the DC-link voltage changes, the number of space vectors also changes. This topology produces 1024 switching states with 728 voltage space

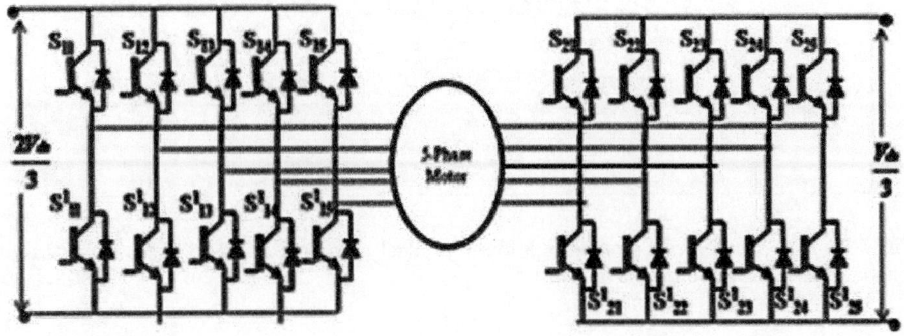

Fig. 1 Four-level dual inverter topology of a five-phase open-end winding motor drives

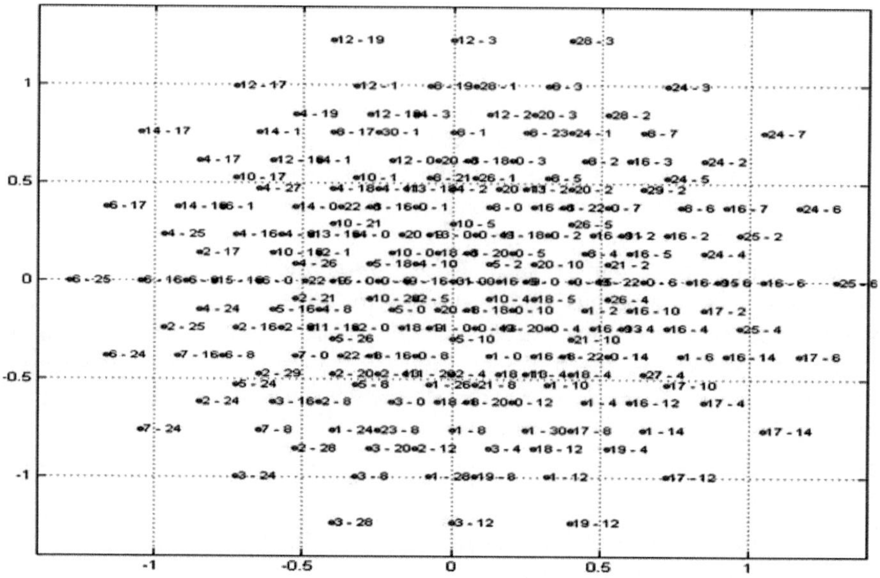

Fig. 2 Voltage space vector plotting of dual inverter fed OeW five-phase drive with equal DC voltages

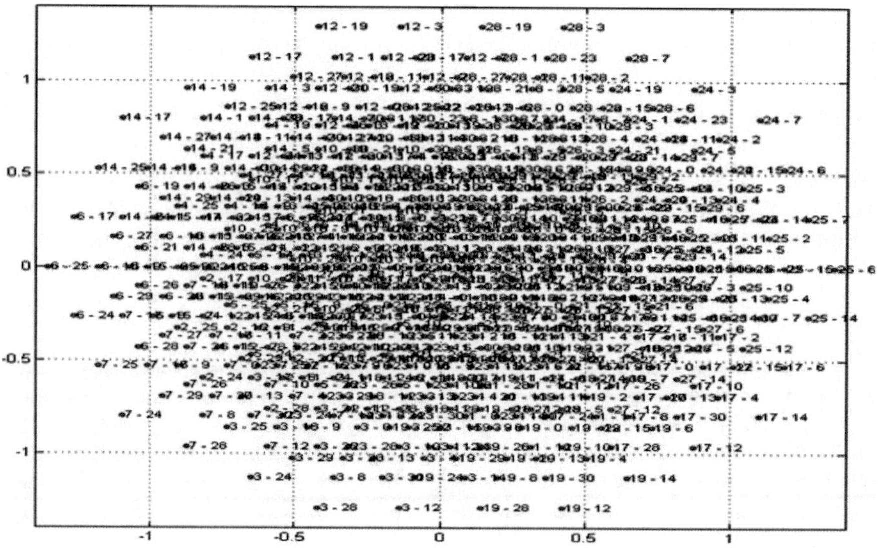

Fig. 3 Voltage space vector plotting of a four-level dual inverter fed OeW 5-phase motor drive with unequal DC-link voltage ratio of 2:1

vectors generated. In this topology with $V_{dc1} = 2V_{dc}/3$ and $V_{dc2} = V_{dc}/3$, results in four equivalent phase voltages are possible, namely $-V_{dc}/3, +V_{dc}/3, 0$ and $+2V_{dc}/3$.

SVPWM technique implementation for a dual inverter fed five-phase motor is more complicated than for the three-phase counterpart. The additional complexity appears not only because of the higher number of voltage vectors but also due to the existence of the additional subspace since the harmonic content is mapped into different planes. Developing PWM techniques for this topology is very difficult since it involves complex calculations for vector identification, sector identification, dwell time calculations, and switching state selection. The best solution is to split the reference voltages and then applying the high frequency triangular carriers. Based on the intersections made, gate pulses are generated. The proposed PWM method is a simplified scalar-based algorithm which generates different continuous and discontinuous modulating signals with a good quality output voltage with reduced CMV and switching losses. This method is easy to implement since it does not do complex calculations. A comparative result shows significant changes in THD of phase voltages and currents. Firstly, simulations are made in MATLAB/SIMULINK environment and the performance of conventional continuous PWM and proposed discontinuous PWM methods are compared.

Unequal DC-link voltage is applied to the two 2-level inverters in the ratio of 2:1. An input voltage of $2V_{dc}/3$ is applied to the Inverter I and $V_{ic}/3$ for the Inverter II. In this case, V_{ao}, V_{bo}, V_{co}, V_{do}, V_{eo} indicates pole voltages of Inverter I, $V_{a^1o^1}$, $V_{b^1o^1}$, $V_{c^1o^1}$, $V_{d^1o^1}$, $V_{e^1o^1}$ indicates pole voltages of Inverter II. V_{aa^1}, V_{bb^1}, V_{cc^1}, V_{dd^1}, V_{ee^1} indicates effective phase voltages; V_{ab}, V_{bc}, V_{cd}, V_{de}, V_{ea} indicates line voltages. Table 1 shows four-level pole voltage by utilizing the switching patterns of Inverter I and Inverter II.

$V_{oo'}$ is the voltage drop between the two neutrals of the inverters, which indicates the CMV. The expression for the CMV derived from pole voltages as in (1)

Table 1 Effective pole voltage calculation

ON state switches in Inverter I	Pole voltage of Inverter I	ON state switches in Inverter II	Pole voltage of Inverter II	Effective pole voltage
S_{12} or S_{14} or S_{16} or S_{18} or S_{10}	0	S_{21} or S_{23} or S_{25} or S_{27} or S_{29}	$+\frac{V_{dc}}{3}$	$-\frac{V_{dc}}{3}$
S_{12} or S_{14} or S_{16} or S_{18} or S_{10}	0	S_{22} or S_{24} or S_{26} or S_{28} or S_{20}	0	0
S_{11} or S_{13} or S_{15} or S_{17} or S_{19}	$+\frac{2V_{dc}}{3}$	S_{21} or S_{23} or S_{25} or S_{27} or S_{29}	$+\frac{V_{dc}}{3}$	$+\frac{V_{dc}}{3}$
S_{11} or S_{13} or S_{15} or S_{17} or S_{19}	$+\frac{2V_{dc}}{3}$	S_{22} or S_{24} or S_{26} or S_{28} or S_{20}	0	$+\frac{2V_{dc}}{3}$

$$V_{\mathrm{CMV}} = V_{00^1} = \frac{1}{5}((V_{a0} - V_{a'0'}) + (V_{bo} - V_{b'0'}) + (V_{co} - V_{c'0'}) \\ +(V_{do} - V_{d'0'}) + (V_{e0} - V_{e'0'})) \quad (1)$$

3 Scalar-Based PWM Algorithm for Continuous and Discontinuous Modulating Signals

The mathematical expression for the five reference signals is given in (2)

$$V_i = V_{\max}\cos\left(\omega t - \frac{2a\pi}{5}\right) \quad (2)$$

where $i = a, b, c, d, e$ and $a = 0, 1, 2, 3, 4$. The five sinusoidal reference signals each with phase displacement of 72° with a triangular carrier signal V_t. Figure 3 shows the comparison of the reference modulating signal with the triangular carrier signal and the intersection points define the control signals. For the proposed four-level configuration, three level-shifting triangular signals (with magnitudes of 0–1/3; 1/3–2/3 and 2/3–1) are required to generate gate pulses. Three different carrier triangular signals can be generated by using XOR logical gates as shown in Fig. 4.

The zero-sequence signal is calibrated using Eq. (3)

$$V_{\mathrm{zs}} = \frac{V_{\mathrm{dc}}}{2}(2a_0 - 1) - a_0 V_{\max} + (a_0 - 1)V_{\min} \quad (3)$$

where $V_{\min} = \mathrm{Min}(V_a, V_b, V_c, V_d, V_e)$ and $V_{\max} = \mathrm{Max}(V_a, V_b, V_c, V_d, V_e)$. The new set of reference signals obtained are given by $V_i^* = V_i + V_{\mathrm{ZS}}$ where $i = a, b, c, d, e$.

The isolation of the terminals O, O^1 eliminates the tripled harmonic current path in dual inverter configuration. The lower-order harmonic signals are injected to the

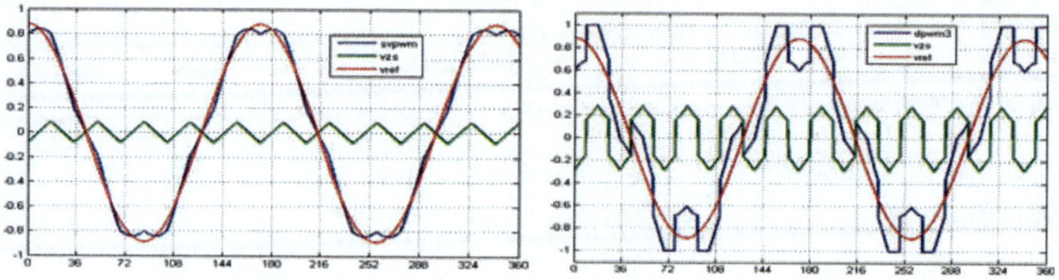

(a) (b)

Fig. 4 Off-set addition to the reference modulating signal at $M = 0.88$, fs = 1000 Hz for **a** continuous PWM, **b** DPWM3 reference signals using min-max method. The constant a_o value lies between 0 and 1. The modulating signals can be differentiated by the selection of a_o value. For a_o, value of 0.5 continuous modulating signals which is similar to SVPWM can be obtained. If the condition given in (4) is satisfied, then DPWM3 is generated. Even though several discontinuous modulating signals like DPWMMIN, DPWMMAX, DPWM0, DPWM1 and DPWM2 are present, the one which has efficient performance for the four-level topology of the drive is selected. Using (4), addition of zero-sequence signals to the reference signals linear modulation region can be extended. This crosses the level-shifting carrier set; this does not change the effective voltage but improves DC bus utilization.

$$\begin{aligned} V_{\min} + V_{\max} &< 0 \quad \text{then}\, a_0 = 0 \\ V_{\min} + V_{\max} &\geq 0 \quad \text{then}\, a_0 = 1 \end{aligned} \tag{4}$$

The off-set addition to the reference modulating signals is shown in Fig. 4. The diagram of generating the gate pulse pattern is shown in Fig. 5. The obtained continuous and DPWM3 modulating signals are plotted in Fig. 5a, b for $M = 0.55$ and Fig. 6a, b for $M = 0.88$. The generated gate pulses of the two inverters are shown for continuous and DPWM3. For explanation purpose, the pattern is differentiated into four different regions. For Fig. 5a, for continuous modulating signals in region 1, modulating signal interacts with carrier triangle signal V_{t3} then Inverter I is clamped to negative DC bus which means upper switches of Inverter I are in OFF state and Inverter II is in ON state. In region 2, modulating signal magnitudes are greater than the carrier triangular signals V_{t1} or V_{t2} but less than V_{t3} then lower switches of inverter 1 and inverter 2 are in complimentary state. A quite opposite phenomenon takes for the regions 3 and 4 as in 1 and 2, respectively. For DPWM3 in Fig. 6b, the peak level of modulating signal is clamped either to +DC or −DC supply with gate

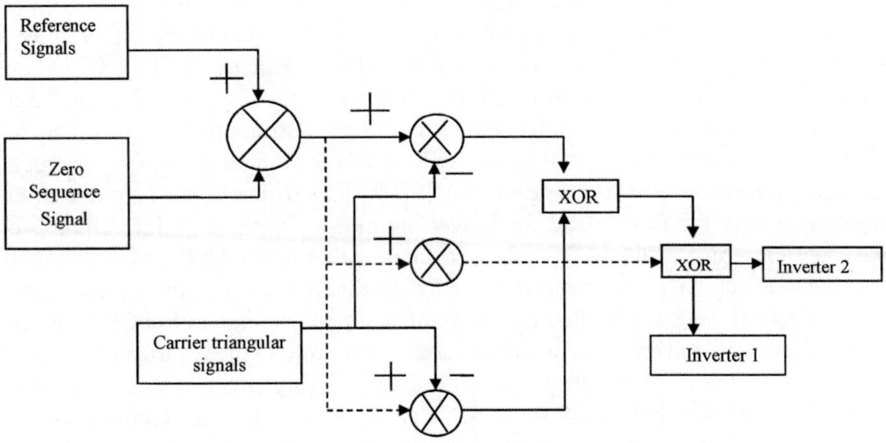

Fig. 5 Block diagram showing gate pulse generation for four levels OeW drive

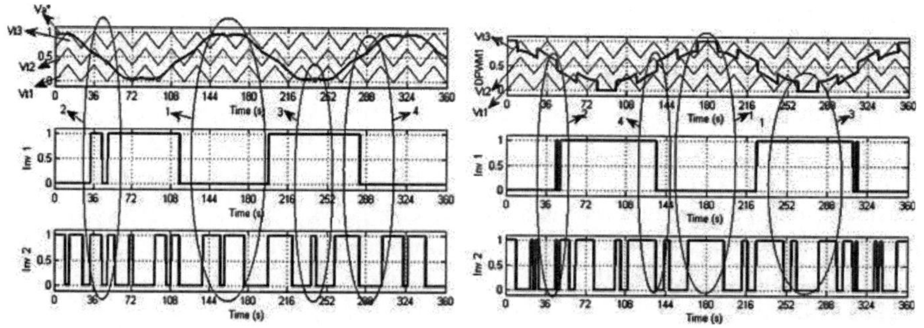

Fig. 6 Pulse patterns of Inverter I and Inverter II for scalar-based **a** continuous, **b** discontinuous PWM technique for $m = 0.65$; Fs = 100 Hz

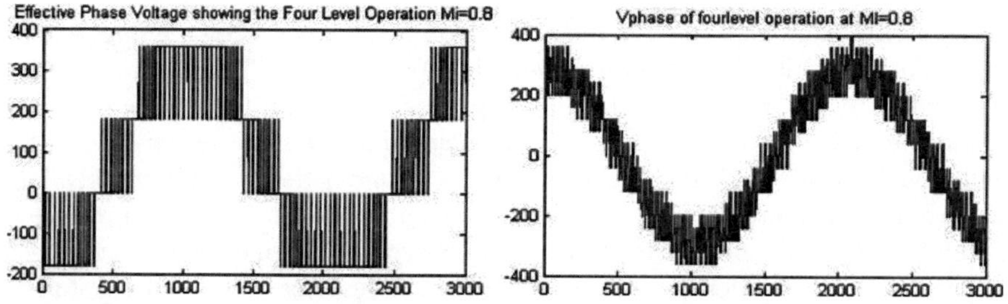

Fig. 7 Calibrated four-level voltage and effective phase voltage at $M = 0.88$; fs = 1 kHz

pulses of Inverter 1 are in OFF state and Inverter 2 in ON state (region 1) and vice versa in region 4. Whenever the modulating signal cuts V_{t1} and V_{t2} Inverter 1 gate pulses clamps to continuously ON state till the modulating signal until less than V_{t3} (region 3). A quite opposite phenomenon takes place in region 4 (Fig. 7).

The continuous modulating signal is not covered in whole region of three-triangular carriers V_{t1}, V_{t2}, V_{t3} for lesser modulation indices of $M \leq 0.8$. Gate pulses of Inverter I are completely OFF or ON states in the regions 1, 2, and 3. But for DPWM3, modulating signal has covered all the tree-triangular carriers. Due to which the gate pulses are not completely OFF or ON for longer durations. The high DC input voltage inverter is made to turn ON for longer durations. This makes to supply power to the motor load for longer durations. Continuous flow of current makes the capacitor with low DC voltage to charge. During the OFF state of Inverter I as shown in region 1, the charged capacitor discharges and the current flows into the windings of the machine and it fails to maintain DC voltage ratio of 2:1. But in DPWM3, the gate pulses are switched in faster durations and the Inverter I is continuous turned ON or OFF. Because of this, the over-charging effects does not exist for DPWM3 at low modulation indexes of $M = 0.55$. For $M = 0.88$, the continuous and DPWM3 modulating signals intersect with all the tree-triangular carriers as shown in Fig. 6a, b shown better pulse pattern compared to continuous PWM.

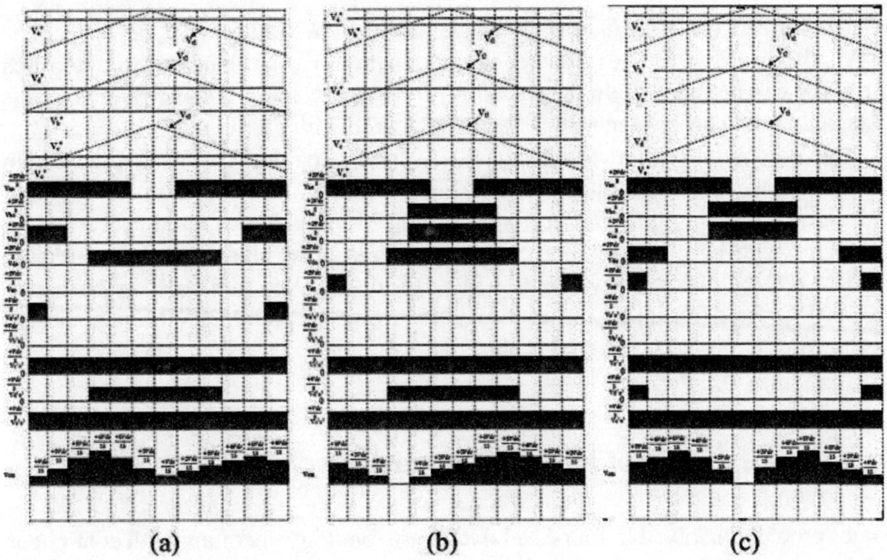

Fig. 8 Scalar-based continuous PWM-based CMV measurments using pole voltages of Inverter I and II in different regions

Mathematical calculations of CMV are shown in Fig. 8. Out of five modulating signals, two modulating signals cut at lower- and upper-level-shifting carrier signals other three modulating signals cut in intermediate values. The entire region is divided into ten slots with 36° each which is shown in Fig. 8a. This figure indicates five modulating signals and three carrier signals at the top of the trace and pole voltages of Inverter I and II in the bottom trace. The CMV is measured by comparing the pole voltages of Inverter I with the pole voltages of the Inverter II and it is shown in the lower trace of Fig. 8. In the entire region, $0° \leq \omega t \leq 36°$, the modulating signals v_a^*, v_b^*, v_c^*, v_d^*, v_e^* are compared with the three-level-shifting carrier signals v_{t1}, v_{t2}, v_{t3} which generate the pulse pattern. Among these five modulating signals, v_a^* is having maximum instantaneous value and is compared with v_{t3}; v_b^* is having an intermediate value (less than that of v_a^*) which is compared with upper half of the carrier signal v_{t2}; v_c^* is also having an intermediate value (slightly greater than v_d^*) which is compared with lower half of the carrier signal v_{t2}; Contrarily, v_d^* is compared with upper half of the carrier signal v_{t1} and v_e^* is having a minimum instantaneous value and is compared with the lower half of the carrier signal v_{t1}. Based on this analysis, the four-level dual inverter configuration of the five-phase induction motor drive is partitioned into three different regions ($A1$, $A2$, and $A3$) for the entire region $0° \leq \omega t \leq 36°$. Assumptions are made for the entire 2π region is symmetrical and hence the discussion is limited the pole voltages and CMV is discussed in the following section for the three different regions. The CMV is calibrated using (2). By this method, the CMV varies from a minimum value of $-V_{dc}/15$ to a maximum value of $+4V_{dc}/15$ since the pole voltages of Inverter I varies from $+2V_{dc}/3$ or 0 and Inverter II pole voltage varies from $+V_{dc}/3$ or 0. By conventional CPWM, the CMV varies between $+4V_{dc}/15$ to $-V_{dc}/15$. The reference signals positions can be changed through which the CMV can be reduced. In the region A1 ($0° \le \omega t \le 36°$), reference signal V_a* is compared with carrier signal V_{t1}; V_d* is compared with carrier signal V_{t2} and V_b*, V_c* are compared with a carrier signal V_{t3}. In this region, CMV has a maximum value of $+2V_{dc}/3$ with a multiple jump of $+V_{dc}/3$ and a minimum value of $-V_{dc}/3$. In the region A2 ($36° \le \omega t \le 72°$) reference signals V_a*, V_b* compared with V_{t1}; V_d* compared with V_{t2}; V_c*, V_d* are compared with a carrier signal V_{t2}. In this region, the CMV has a maximum value of $+5V_{dc}/15$ to $+V_{dc}/15$. Similarly, in the region A3, CMV varies from $+V_{dc}/15$ to $+4V_{dc}/15$. However, using DPWM3 signals, the positions of the reference signal is changed.

4 Over-Charging of DC-Link Capacitors

Major problem involved in four-level five-phase topology operating with continuous PWM techniques is over-charging of DC-link capacitors. The effects of this damage the performance of the drive and it does not able to control the voltage and speed. Among the two rectifiers, the one which has lower DC-link voltage leads to capacitor overcharging. Hence, a bidirectional current flow and it does not maintain 2:1 ratio of voltages. This problem can be solved by an offset value addition or subtraction to the reference modulating signal. Figure 9 shows addition and subtraction of offset value of a single modulating signal for modulation indexes of 0.33 (Figs. 10 and 11). It is observed that for $M = 0.33$, the modulating signal interacts with only lower high frequency carrier triangles V_{t1} and with which only the gate pulses of Inverter I are produces. But the gate pulses of Inverter II are constant and are clamped to either positive DC-link or negative DC-link terminals. This is indicated by using equivalent circuit in Fig. 12a, b. Figure 12b clearly indicates that the capacitor with voltage magnitude of $2V_{dc}/3$ clamps to either positive or negative terminals or current stops flowing through it. The current flowing from Inverter I charges the other capacitor

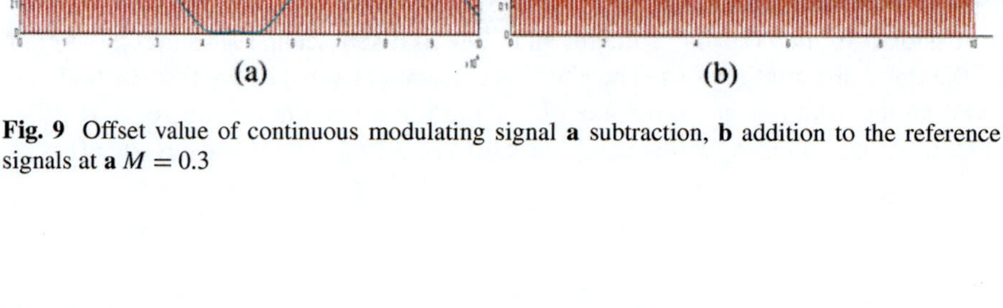

(a) (b)

Fig. 9 Offset value of continuous modulating signal **a** subtraction, **b** addition to the reference signals at **a** $M = 0.3$

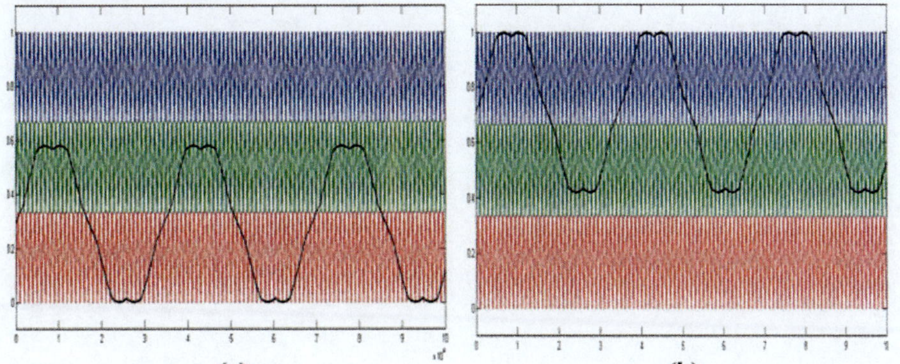
(a) (b)

Fig. 10 Offset value of continuous modulating signal **a** subtraction, **b** addition to the reference signals at a $M = 0.55$

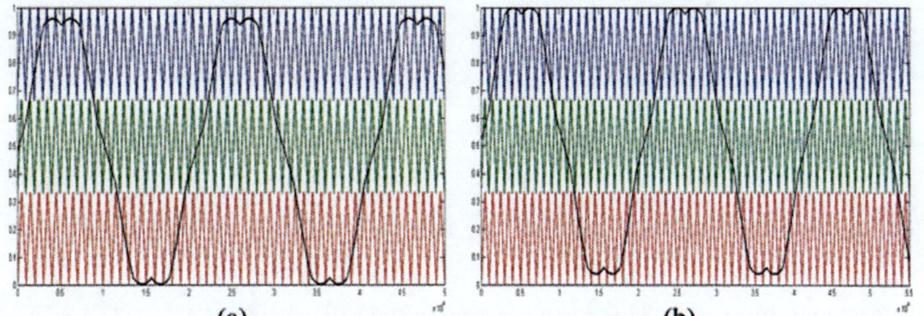
(a) (b)

Fig. 11 Offset value of continuous modulating signal **a** subtraction, **b** addition to the reference signals at a $M = 1.10$

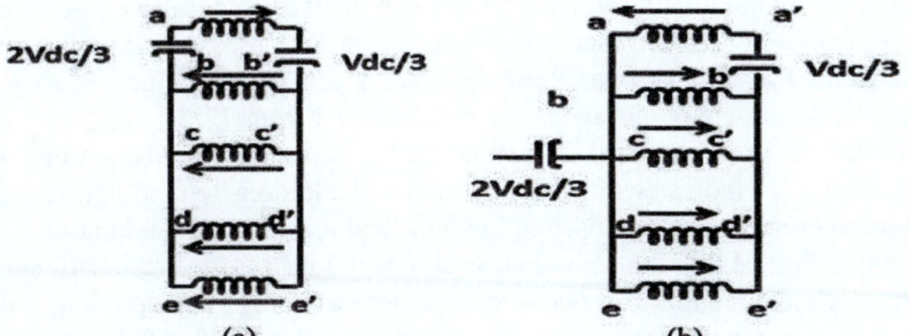

(a) (b)

Fig. 12 Switching state combination circuits for four-level five-phase drives at $M = 0.33$

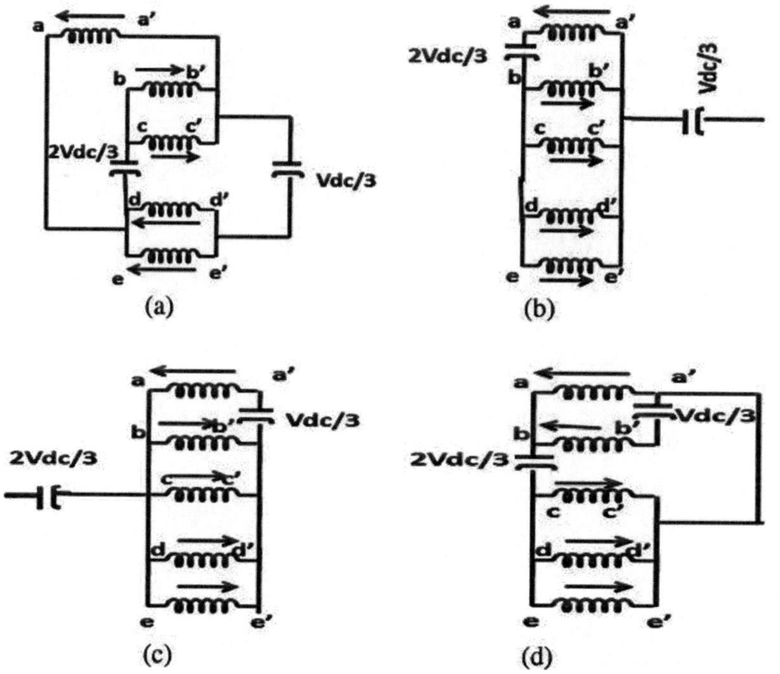

Fig. 13 Switching state combination circuits for four-level five-phase drives at $M = 0.55$

connected to Inverter II. This capacitor charges up to a voltage of $V_{dc}/3$. Hence, an over-charging of capacitor takes place. But in DPWM3, for $M = 0.33$, the DPWM3 modulating signal covers the entire region of all three triangular carrier signals V_{t1}, V_{t2} and V_{t3} hence, there is no possibility of over-charging of capacitors. For $M = 0.55$, offset addition of modulating signal is shown in Fig. 10a, b and the five combinations of equivalent circuits are formed and are shown in Fig. 13. Among which, in two combinations (Fig. 13b, c), over-charging effect occurs for the range $0 \leq \omega t \leq 72°$. In Fig. 13a, d over-charging effects are not observed from $72° \leq \omega t \leq 360°$ since the current flows throughout the circuit and the DC capacitors are in operational. For $M = 1.10$, Fig. 11 indicates offset subtraction and additions to modulating signals. The capacitor chagrining and discharging indicated for different modulation indices are drawn in Fig. 14. For the combination shown in Fig. 14a, in Inverter II, currents flow in opposite direction. But due to the shorter switching state duration, it might have very less chance of the capacitor to overcharge. Remaining other combination circuits in Fig. 14, the currents are flowing from one inverter to the other and are free from over-charging effects.

This concluded that continuous PWM technique has over-charging effects for the linear modulation range (0–1.10) and these effects are not seen for over modulation region (1.10–1.154). Due to these effects, the extensive speed control is not possible and badly impacts on the performance of the drive. Also, the over-charging effects

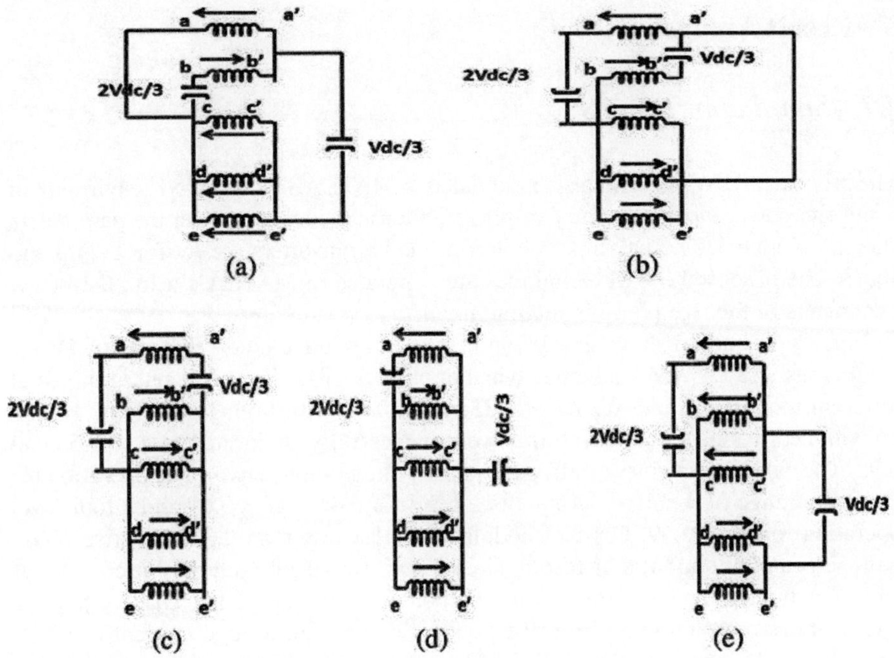

Fig. 14 Switching state combination circuits for four-level five-phase drives at $M = 1.10$

are only in the regions $0 \leq \omega t \leq 72°$. With the application of DPWM3 techniques for the four-level five-phase OeW drive, the DC-link voltages are perfectly maintained in the ratio of 2:1 and hence, there is no effect of over-charging. A comparison on existence of over-charging for different modulation indices on continuous and DPWM3 is tabulated in Table 2.

Table 2 Over-charging effect

Modulation index	Continuous PWM	DPWM3
0–0.33	Yes	No
0.33–0.77	Yes	No
0.77–1.10	Yes	No
1.10–1.15	Yes	No

5 Result Analysis

5.1 Simulation Results

The proposed DPWM3 scheme is modulated in MATLAB/SIMULINK environment using sim power block set. The five-phase motor is modulated as per the parameters shown in Table 3. The DC-link voltages of the two inverters are set for $2V_{dc}/3$ and $V_{dc}/3$. The proposed PWM techniques are simulated in MATLAB using the below parameters of the five-phase induction motor.

Figures 15 and 16 show simulation results of effective phase voltage; CMV for continuous and DPWM3 schemes when applied for four-level five-phase drives at different modulation indexes $M = 0.33$; 0.55; 0.88; 1.0. Voltage levels of 19 and 15 with nearly sinusoidal waveforms were successfully achieved for $M = 0.65$ and 0.95. The calculated values of effective phase voltage show a two-level operation for the entire range of $0 < M < 0.46$; a three-level for $0.46 < M < 0.69$ and a four-level operation for $M > 0.69$. For the modulation index less than 0.38, effective phase voltage of continuous PWM results similar to a two-level voltage source inverter and this effective phase voltage is not sufficient to run the motor. Due to this, the motor draws more currents from the power supplies and a huge amount of CMV (0 to $+5V_{dc}/15$) is formed. For DPWM3 at $M = 0.33$, effective phase voltage also forms a two-level VSI but the number of levels in CMV ($-4V_{dc}/15$ and $+9V_{dc}/15$) are less. For $M = 0.55$, the effective phase voltage forms similar to a three-level VSI and CMV variations are observed. In CPWM, the CMV levels of $-V_{dc}/15$ to $+6V_{dc}/15$ and in DPWM3, voltage level $-3V_{dc}/15$ to $+8V_{dc}/15$ levels are observed. For $M \geq 0.8$, the four-level pole voltage is perfectly observed. In both CPWM and DPWM3, the number of levels in effective phase and CMV are same. In both techniques' voltage has the total of 19 levels with each level has a magnitude of $V_{dc}/20$ (10 V) for $1.23 \leq M \leq 0.8$. For the modulation range $0.79 \leq M \leq 0.66$, total of 17 voltage levels are observed in effective phase voltage. These voltage levels are quite

Table 3 Parameters of five-phase motor

Parameter	Value
Stator resistance Rs	0.8 Ω
Rotor resistance Rr	0.6 Ω
Stator leakage inductance Lls	2.6 MH
Rotor leakage inductance Llr	2.6 MH
Mutual inductance Lm	151 MH
Rotor inertia J	0.047 kg mts2
Pole pair P	2
Switching frequency Fs	1000 Hz
Rated motor current I	3.4 A
Rated speed Ns	1450 rpm

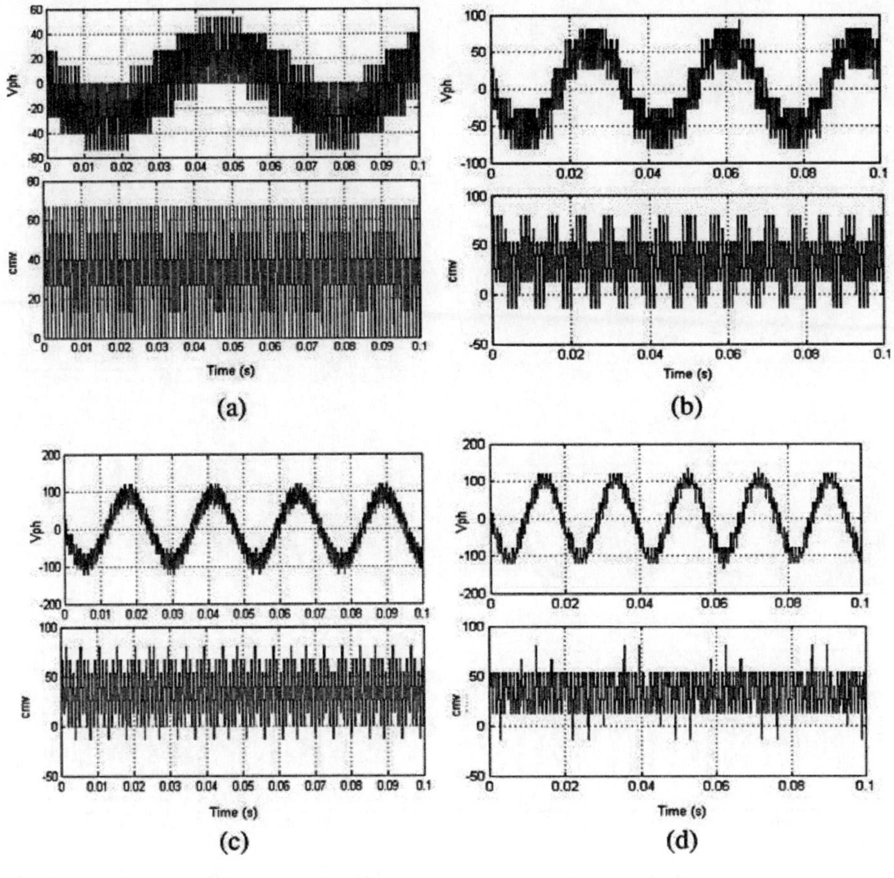

Fig. 15 Simulation results of effective phase voltage and CMV of continuous PWM-based four-level five-phase drive **a** $M = 0.33$, **b** $M = 0.55$, **c** $M = 0.88$, **d** $M = 1.0$

similar to the effective phase voltages obtained in three-level open-end five-phase. For the modulation region $0.69 \leq M \leq 0.35$ total of 12 voltage levels are observed. Similarly, for $M \leq 0.33$, total of nine voltage levels are observe which is similar to the voltage levels formed under a two-level five-phase VSI. The number of voltage levels in effective phase voltage remains the same for all modulating signals at same modulation indices M and changes are observed only by varying the modulation indices M. The total harmonic distortion (THD) depends upon employed frequency, voltage-levels, number of steps, and magnitude of each step voltage.

Hence, variation in THD is observed in continuous PWM, and DPWM3. But due to the change in pulse position, DPWM3 has shown better THD reduction when compared to continuous PWM. These voltage levels for effective phase voltage and CMV are tabulated for different modulation indexes in Tables 4 and 5. Results conclude that in over modulation region, the DPWM3 reduces the harmonic content

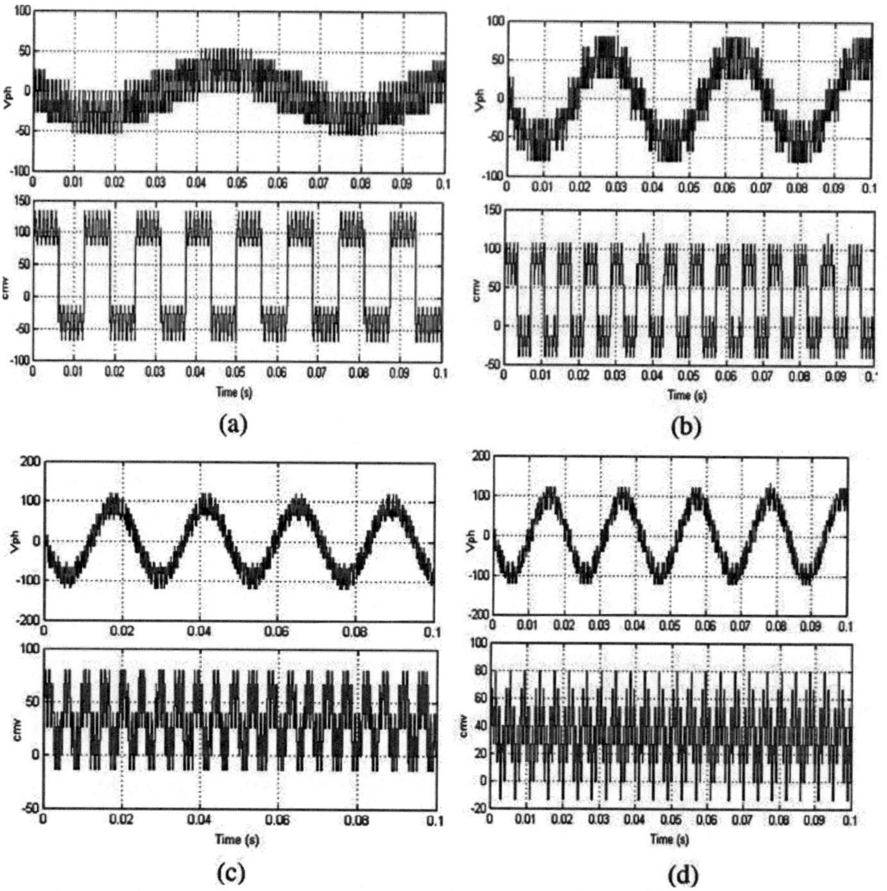

Fig. 16 Simulation results of effective phase voltage and CMV of DPWM3-based four-level five-phase drive **a** $M = 0.33$, **b** $M = 0.55$, **c** $M = 0.88$, **d** $M = 1.0$

compared to continuous PWM through which a better improvement in the performance of the five-phase drive system operating at over modulation region can be observed. In the over modulation region (1.05–1.23), the generated CMV with DPWM3 has CMV lesser number of levels ($+V_{dc}/15$ and $+4V_{dc}/15$) and reduced harmonic content (3rd, 5th, and 7th) compared to continuous PWM. Peak-to-peak voltage levels of CMV for DPWM3 and continuous PWM are observed. The differentiation is made in selecting the ranges in modulation indices by considering the number of levels generated in CMV. The highlighted indicates lesser number of levels in CMV. On observation, in over modulation region, DPWM3 has lesser number of levels compared to the continuous PWM.

Table 4 Number of levels in effective phase voltage

Modulation index range	Voltage levels	Number of voltage levels
0.8–1.23	$+13V_{dc}/20$; $+3V_{dc}/5$; $+V_{dc}/2$; $+9V_{dc}/20$; $+2V_{dc}/5$; $+8V_{dc}/25$; $+13V_{dc}/50$; $+V_{dc}/5$; $+3V_{dc}/25$; $+3V_{dc}/50$; 0; $-V_{dc}/5$; $-3V_{dc}/25$; $-3V_{dc}/50$; $-8V_{dc}/25$; $-13V_{dc}/50$; $-9V_{dc}/20$; $-13V_{dc}/25$; $-3V_{dc}/5$	19
0.79–0.66	$+13V_{dc}/20$; $+23V_{dc}/50$; $+2V_{dc}/5$; $+8V_{dc}/24$; $+13V_{dc}/49$; $+2V_{dc}/5$; $+13V_{dc}/100$; $+3V_{dc}/46$; 0; $-3V_{dc}/46$; $-13V_{dc}/100$; $-2V_{dc}/5$; $-13V_{dc}/49$; $-8V_{dc}/24$; $-2V_{dc}/5$; $-23V_{dc}/50$; $-13V_{dc}/20$	17
0.69–0.35	$+4V_{dc}/50$; $+V_{d}/20$; $+3V_{d}/20$; $+V_{dc}/5$; $+V_{dc}/4$; $+7V_{dc}/20$; 0; $-7V_{dc}/20$; $-V_{dc}/4$; $-V_{dc}/5$; $-3V_{dc}/20$; $-V_{dc}/20$	12

Table 5 Number of levels in CMV

M	CPWM	DPWM3
1.05–1.23	0 to $+5V_{dc}/15$	$+V_{dc}/15$ to $+4V_{dc}/15$
0.92–1.05	$-V_{dc}/15$ to $+6V_{dc}/15$	$-V_{dc}/15$ to $+6V_{dc}/15$
0.69–0.92	$-V_{dc}/15$ to $+6V_{dc}/15$	$-V_{dc}/15$ to $+6V_{dc}/15$
0.52–0.69	$-V_{dc}/15$ to $+6V_{dc}/15$	$-3V_{dc}/15$ to $+8V_{dc}/15$
Less than 0.52	0 to $+5V_{dc}/15$	$-4V_{dc}/15$ to $+9V_{dc}/15$

5.2 Experimental Results

The experimental setup shown in Fig. 17 has two two-level Inverter setups supplied with a three-phase diode rectifier. The isolation transformer is used between the rectifiers (for VS1 and VS2). Two isolation transformers are used to adjust the DC voltages of VS1 and VS2 at 66.6 and 133.3 V to set a total of 200 V. The DC rectifiers are connected to a three-phase, 415 V, 9.2 kVA, 13 A DC-link converters. The inverters are designed externally by parallel connection of five semicron IGBT sets. Each semicron IGBT set consists of one-leg (two semicron IGBTs connected in series) having a rating of 2000 V/100 A. It has provision to provide control signals for the upper and lower gate terminals; DC supply terminals; external triggering terminals. The stator of the five-phase motor is opened and the terminals are connected to the two inverters on both sides of the motor. dSPACE 1104 is used to generate the control

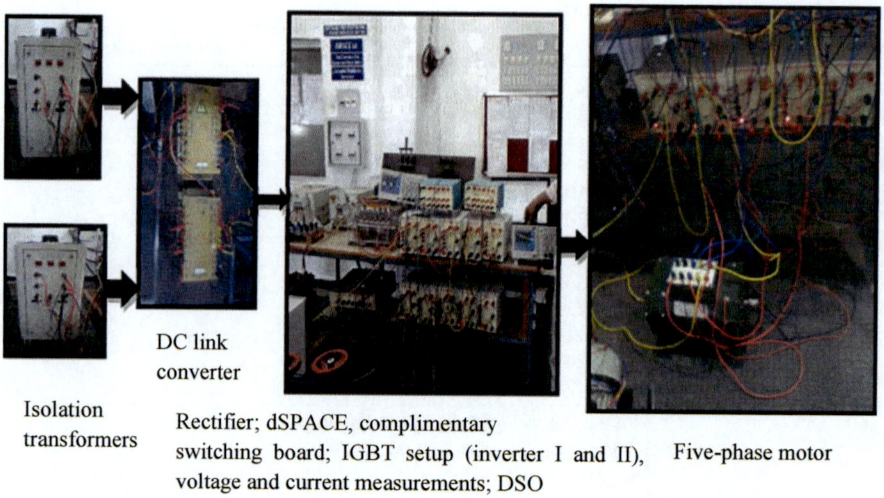

Fig. 17 Experimental setup of open-end winding four-level five-phase motor drives

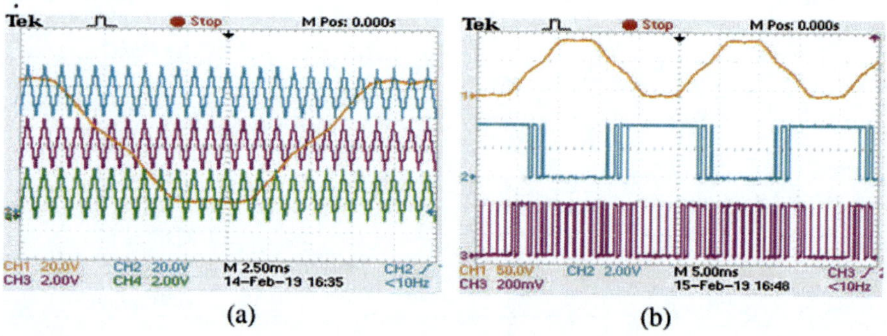

Fig. 18 **a** Continuous modulating signal with three carriers, **b** pulse pattern of Inverter I and II at $M = 0.88$; Fs = 1000 Hz

signals with a switching frequency of 1 kHz. LV20-P (500–3.3 V regulator) and LA-55P (50 A to 3.3 V with three-turns) are used for voltage and current measurements. For this experimental setup, dSPACE 1104 generates five control signals for each inverter. Each control signal is connected to a complimentary switching board. The function of this board is to generate pulses for the gate terminals of upper and lower IGBTs with a dead time of 20 μs. The complete block diagram with photograph of the decoupled PWM-based dual inverter fed OeW five-phase motor is shown in Fig. 17.

During the experimental analysis, the digital voltmeters that are used to measure the DC voltage across the input terminals of both the inverters are checked weather the voltages are in the ratio of 2:1. It is clearly shown the voltages around 133 and 67 V. All the results shown in Figs. 18, 19, 20, 21 and 22 are taken for the cases

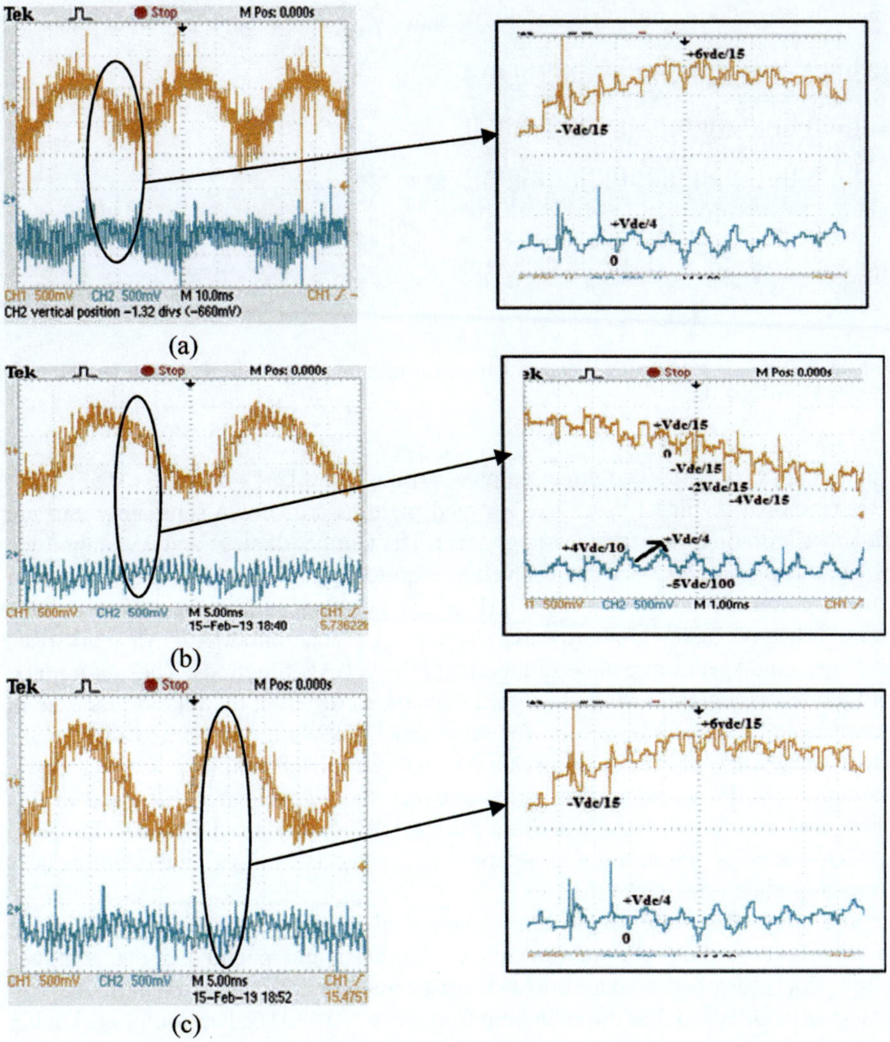

Fig. 19 Experimental results of output effective phase voltage and CMV for continuous PWM technique-based four-level five-phase drives at **a** $M = 0.55$; Fs = 1000 Hz, **b** $M = 0.88$; Fs = 1000 Hz, **c** $M = 1.0$; Fs = 1000 Hz

where the proposed method of offset value addition that is discussed in Sect. 4. The results shown in Fig. 22 are for DPWM3 at lower modulation indexes of $M = 0.33$ and the motor is rotating with low speed of 610 rpm. These results are similar to hardware results that discussed in previous section. During the lower modulation, i.e., $M \leq 0.45$ effective phase voltage resembles similar to a two-level five-phase voltage and the performance of the voltage is poor. Figure 18a, b shows the continuous modulating signal with three high frequency carriers and pulse pattern of Inverter

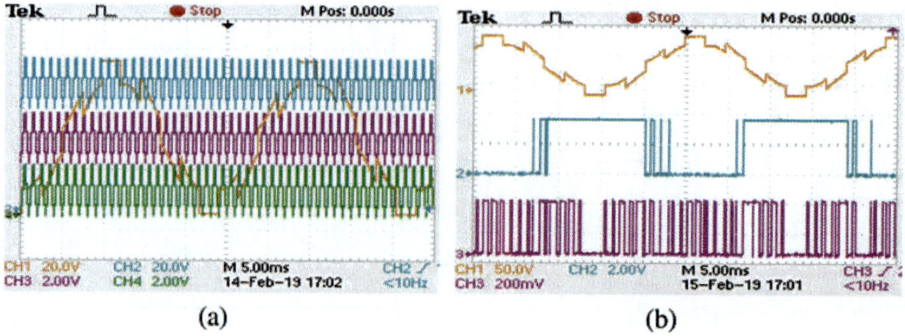

Fig. 20 **a** DPWM3 modulating signal with three carriers, **b** pulse pattern of Inverter I and II at $M = 0.88$; Fs = 1000 Hz

I and II at $M = 0.88$. The effective phase voltage and CMV for $M = 0.55$; 0.88; 1.0 are shown in Fig. 19a–c. The zoomed pictures are shown separately and are indicated with their respective voltage level. The number of steps and the magnitude of each step in terms of voltage levels are quite similar to that of the simulation studies presented in previous section (Fig. 23). Harmonic spectrum of the effective phase voltages obtained in hardware results can be taken into MATLAB workspace and harmonic spectrum can be plotted using MATLAB functions. These are noted and plotted as a graph in Fig. 24 for all considered different modulation indexes in experimental results. The results for proposed DPWM3 are presented in Figs. 19 and 20. The DPWM3 modulating pulse with three carriers and its pulse pattern are shown in Fig. 19. A clear observation between the two schemes discussed in this paper and it is found that throughout the cycle (0–360°), the DPWM3 clamps to either positive or negative DC links of the inverter and pulses generated are less compared to continuous PWM.

The number of steps and voltage levels dual inverter fed five-phase induction motor drive are similar to simulation results. The harmonic spectra of both techniques when observed a better spectrum with lesser magnitudes and fewer harmonic are observed in DPWM3. For the switching frequency of 1000 Hz, harmonics are having greater magnitude in continuous PWM. Even multiples of switching frequency are also having fewer magnitudes in DPWM3. In over modulation, the effective phase voltage and CMVs are improved better in DPWM3 compared to continuous PWM. This indicates that in over modulation region, DPWM3 techniques performance is better compared to CPWM. Less than 0.52 modulation is not suitable for DPWM3 compared to CPWM as it does not produce fifth-order harmonic component. THD comparison of effective phase voltage and stator currents at study state for linear modulation and over modulation is plotted in Fig. 26. In over-modulation region, THD is affected due to the variations in pulse position and there is no change in number of steps. Effective phase voltage THD is similar for all continuous and DPWM3 in linear modulation region. But as modulation index increased to over modulation region DPWM3 shows lower THD (for $M = 1.05$–1.2) compared to

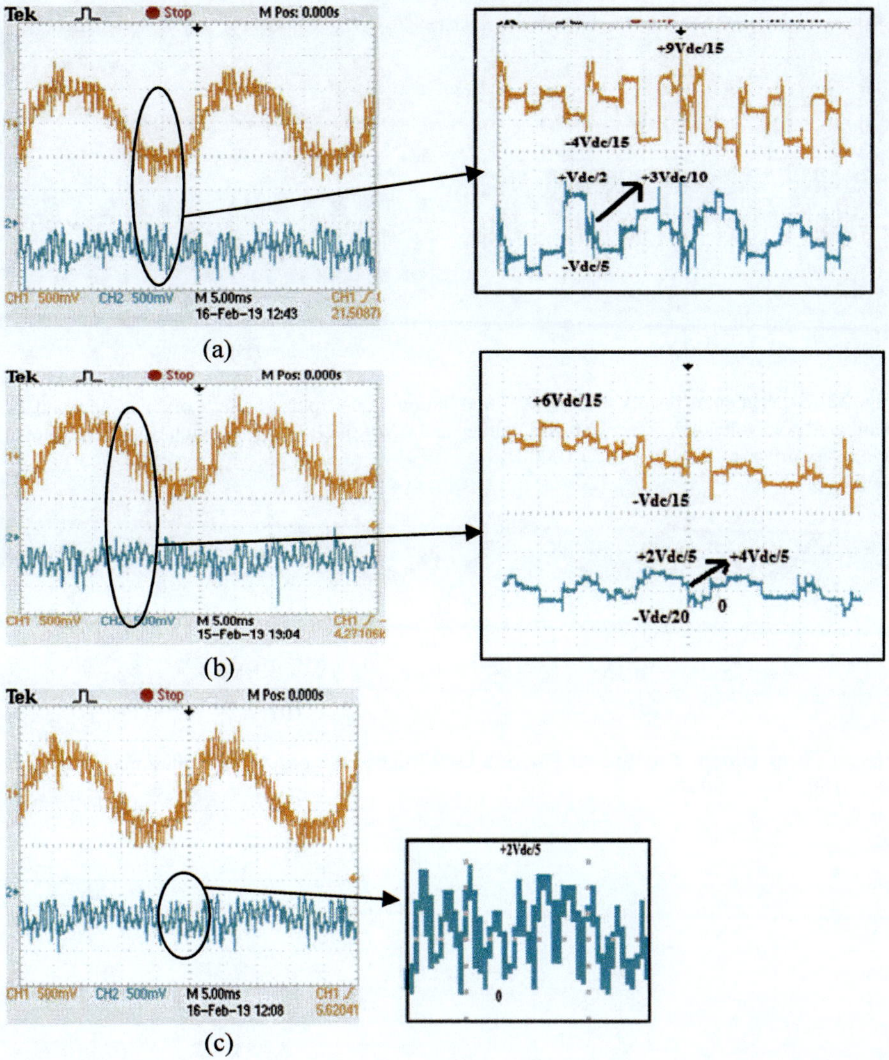

Fig. 21 Experimental results of output effective phase voltage and CMV for DPWM3 technique-based four-level five-phase drives at **a** $M = 0.55$; Fs = 1000 Hz, **b** $M = 0.88$; Fs = 1000 Hz, **c** $M = 1.0$; Fs = 1000 Hz

continuous PWM techniques. This concludes comparisons between DPWM3 and continuous PWM. It shows that DPWM3 shows better performance in linear and over-modulation region.

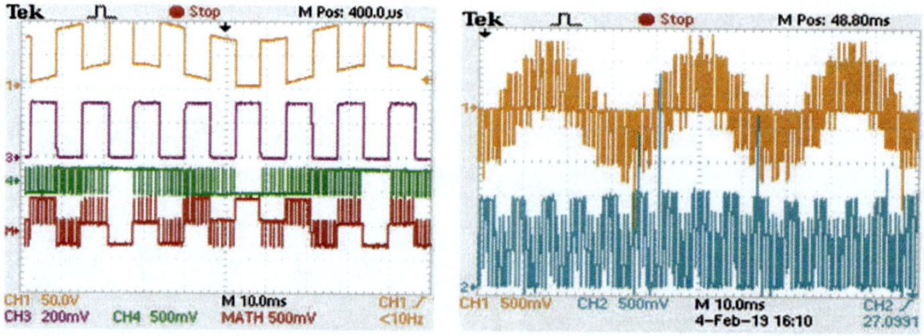

(a) (b)

Fig. 22 Experimental results **a** DPWM3 modulating signal; pulse pattern of Inverter I and II; calibrated pole voltage, **b** effective phase voltage and CMV of DPWM3 technique-based four-level five-phase drives at $M = 0.3$; Fs = 1000 Hz

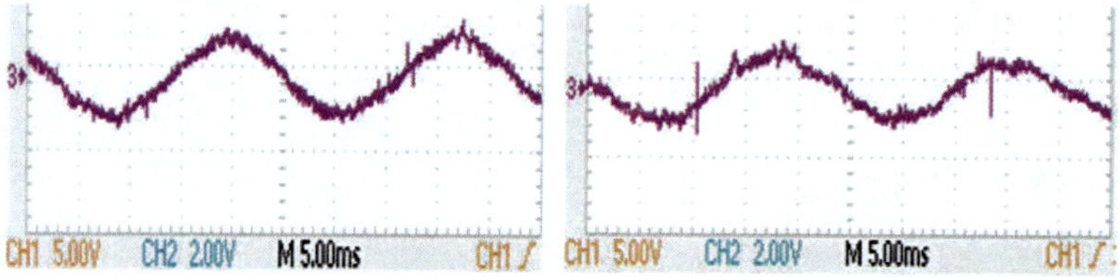

(a) (b)

Fig. 23 Stator current **a** continuous PWM, **b** DPWM3 based four-level five-phase motor drive at $M = 0.88$; Fs = 1000 Hz

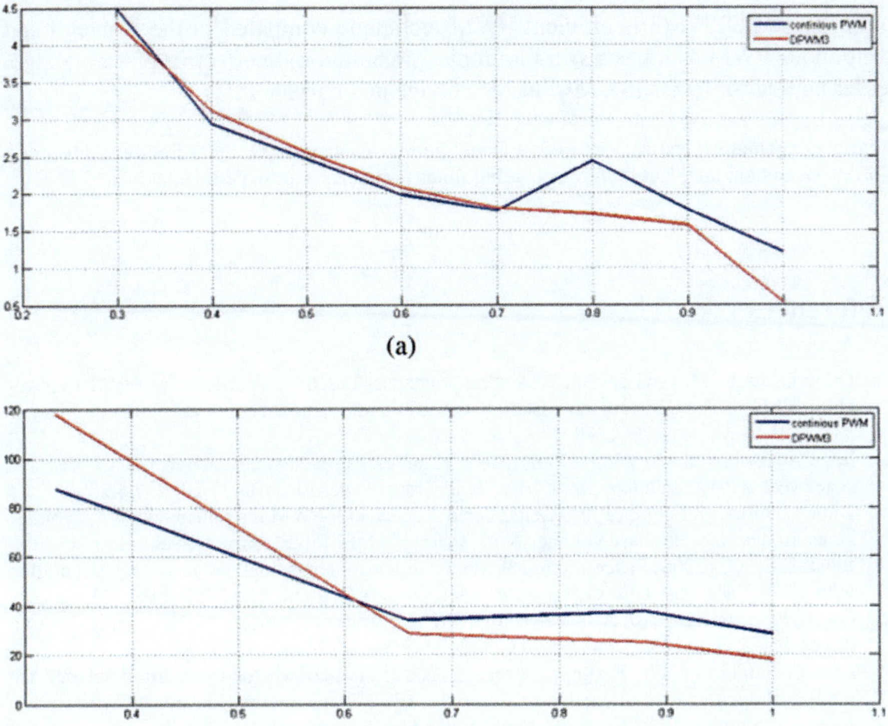

(b)

Fig. 24 Comparative graph for hardware results of **a** effective phase voltage, **b** currents at different modulation indexes

6 Conclusion

In this paper, a simplified carrier-based continuous and discontinuous PWM (DPWM3) techniques were discussed for the four-level dual inverter fed open-end winding five-phase induction motor drives. Mathematical calculations for reference voltage vector, sector identification, switching state selection are not essential to implement these PWM techniques as in case of conventional space vector-based PWM techniques. Through this scalar-based approach, different continuous and discontinuous PWM schemes are implemented. Two methods are proposed to eliminate the over chargining effects that occur in four-level five-phase topology. The major problem involved in over-charging of capacitor is solved by implementing an off-set value addition and subtraction method. Another method is using discontinuous DPWM3 modulating signals. Experimental results are presented for the proposed four-level five-phase drives that maintain 2:1 ratio of DC-link voltages. The proposed DPWM3 scheme reduces number of switching's up to 1/5th of the time period compared to continuous PWM and conduction losses can be reduced by 20%. It is observed from the results that in both linear and over modulation region, the proposed DPWM3 is efficient PWM technique compared to the conventional continuous PWM as it has lesser harmonics (magnitude) in effective phase voltage; lesser number of levels in CMV; lesser voltage and current THDs.

Acknowledgements The authors like to thank the management of G. Pulla Reddy Engineering College (Autonomous), Kurnool for providing funds to purchase the experimental kits.

References

1. Dujic D, Jones M, Levi E (2007) Continuous carrier-based vs. space vector PWM for five-phase VSI. In: The IEEE region 8 EUROCON 2007: international conference on 'computer as a tool', Warsaw, pp 1772–1779
2. Iqbal A, Moinuddin S (2009) Comprehensive relationship between carrier-based PWM and space vector PWM in a five-phase VSI. IEEE Trans Power Electron 24(10):2379–2390
3. Dujic D, Jones M, Levi E (2009) Generalized space vector PWM for sinusoidal output voltage generation with multiphase voltage source inverters. Int J Ind Electron Drives 1(1):1–13
4. Iqbal A, Levi E (2006) Space vector PWM techniques for sinusoidal output voltage generation with a five-phase voltage source inverter. Electr Power Compon Syst 34(2):119–140
5. Gao L, Fletcher JE (2010) A space vector switching strategy for three-level five-phase inverter drives. IEEE Trans Ind Electron 57(7):2332–2343
6. Levi E, Satiawan INW, Bodo N, Jones M (2012) A space-vector modulation scheme for multilevel open-end windings five-phase drives. IEEE Trans Energy Convers 27(1):1–10
7. Jones M, Satiawan IN (2012) A simple multi-level space vector modulation algorithm for five-phase open-end winding drives. Trans Int Assoc Math Comput Simul (IMACS)
8. Jones M, Vukosavic SN, Levi E, Iqbal A (2005) A six-phase series-connected two-motor drive with decoupled dynamic control. IEEE Trans Ind Appl 41(4):1056–1066
9. Jones M, Satiawan IN, Bodo N, Levi E (2012) A dual five-phase space-vector modulation algorithm based on the decomposition method. IEEE Trans Ind Appl 48(6):2110–2120 (2012)
10. Hava AM, Kerkman RJ, Lipo TA (1997) A high performance generalized discontinuous PWM algorithms. IEEE Trans Ind Appl 34(5):1059–1071
11. Krishna Murty H, Narayanan G, Narayanan VT, Ayyanar R (2003) Design of space vector based hybrid PWM techniques for reduced current ripple. In: Eighteenth annual IEEE applied power electronics conference and exposition, Miami Beach, pp 583–588
12. Hava AM, Onur Cetin N (2011) A generalized scalar PWM approach with easy implementation features for three-phase three wire voltage source inverters. IEEE Trans Power Electron 26(5):1385–1395
13. Park H-J, Youn M-J (2017) A new time-domain discontinuous space-vector PWM technique in over modulation region. IEEE Trans Ind Electron 58(2)
14. Darijevic M, Jones M, Dordevic O, Levi E (2017) Decoupled PWM control of a dual-inverter four-level five-phase drive. IEEE Trans Power Electron 32(5)
15. Baiju MR, Mohapatra KK, Kanchan RS, Gopakumar K (2004) A dual two-level inverter scheme with common mode voltage elimination for an induction motor drive. IEEE Trans Power Electron 19(3):794–805
16. Shivakumar EG, Gopakumar K, Sinha SK, Pittet A, Ranganathan VT (2002) Space vector PWM control of dual inverter fed open-end winding induction motor drive. Eur Power Electron Drives J 12(1):9–18
17. Somasekhar VT, Shivakumar EG, Gopakumar K, Pittet A (2002) Multilevel voltage space phasor generation for an open-end winding induction motor drive using a dual inverter scheme with asymmetrical DC-link voltages. Eur Power Electron Drives J 12(3):21–29

18. Somasekhar VT, Baiju MR, Gopakumar K (2004) Dual two-level inverter scheme for an open-end winding induction motor drive with a single DC power supply an improved DC bus utilization. Proc IEE Electr Power Appl 151(2):230–238
19. Reddy BV, Somasekhar VT, Kalyan Y (2011) Decoupled space-vector PWM strategies for a four-level asymmetrical open-end winding induction motor drive with waveform symmetries. IEEE Trans Ind Electron 58(11):5130–5141
20. Reddy MHV, Reddy TB, Reddy BR, Surya Kalavathi M (2015) Reduction of CMV in asymmetrical dual inverter configuration using discontinuous modulating signal based PWM technique. J Power Electron 15(6):1524–1532
21. Reddy MHV, Reddy TB, Reddy BR, Surya Kalavathi M (2017) Discontinuous PWM technique for the asymmetrical dual inverter configuration to eliminate the over-charging of DC-link capacitor. IEEE Trans Ind Electron 41(4):1056–1066

State of Charge Estimation Using Extended Kalman Filter

Kshitija A. Gaikwad and Vijaykumar Bhanuse

Abstract As the Automobile Industry is currently focusing on the development of Electric vehicles, battery engineering has taken a boon. State of Charge is the ratio of the amount of charge available in the battery to the rated charge capacity of the battery. This paper presents equivalent circuit modeling of a Li-ion battery cell and its state of charge estimation using the Kalman Filter algorithm in MATLAB Simulink. The simulation result indicates an error of about 1% in estimation.

Keywords Battery management system · Extended Kalman filter · Kalman filter · State of charge

Abbreviations

BMS	Battery Management System
CC	Coulomb Counting
EKF	Extended Kalman filter
EVs	Electric Vehicles
HEVs	Hybrid Electric Vehicles
KF	Kalman Filter
OCV	Open Circuit Voltage
SOC	State of Charge

K. A. Gaikwad (✉) · V. Bhanuse
Instrumentation and Control Department, Vishwakarma Institute of Technology, Pune, India
e-mail: kshitijagaikwad23@gmail.com

V. Bhanuse
e-mail: vijaykumar.bhanuse@vit.edu

© Springer Nature Singapore Pte Ltd. 2021

S. N. Merchant et al. (eds.), *Advances in Signal and Data Processing*,
Lecture Notes in Electrical Engineering 703,
https://doi.org/10.1007/978-981-15-8391-9_7

1 Introduction

The automobile industry is on the path of development of Electric (EVs) and Hybrid Electric vehicles (HEVs). A high voltage rechargeable battery is the heart of any EV/HEV. An intensive sensing and control unit known as the Battery Management System (BMS) is necessary for the safe operation of the battery. Sensors, electronic control units with communication capabilities together form the BMS.

State of Charge (SOC) at a given instant is defined as the ratio of the charge available at that point to the maximum charge capacity. SOC is expressed in percent, from 100% which indicates a fully charged battery to 0% which indicates a completely discharged battery. SOC estimation helps in operating the battery within the limits of maximum charge and minimum discharge. Now let us see why do we say SOC 'Estimation' and not 'measurement'? Since SOC is the internal state of the battery, there is no sensor or an instrument available to measure SOC directly. We have to estimate SOC from indirect measurements of battery current and battery voltage.

Murnane and Ghazel in [1] proposes SOC and SOH estimation based on the coulomb counting method, the voltage method, and the Kalman filter method. In [2] Youngryul et al. introduced the design of BMS simply and, as an estimation method of SOC, dual extended Kalman filter is proposed and proven through simulation based on the real data of current and voltage of a battery.

2 Methodology

2.1 *Kalman Filter*

Kalman Filter (KF) is an optimal state estimator algorithm. A system state cannot be measured directly. KF is used to estimate such a state. Main feature of KF is that it works in presence of process and sensor noises from different sources [3]. Below figure is the Block diagram representation of KF (Fig. 1).

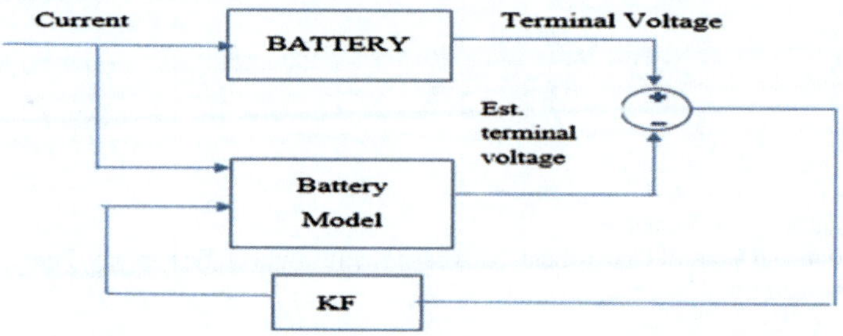

Fig. 1 Block diagram

$$x_k = Ax_{k-1} + Bu_k \tag{1}$$

$$y_k = Cx_k + Du_k \tag{2}$$

KF is built on a state-space representation of a linear system. As Yuan et al. presents in [4], a mathematical model of the real system is built. This model is fed with a known input and corresponding estimated output is obtained. But, the model is only a close approximation of your real system, and hence the real system output and the estimated output do not match. Our goal is to minimize the difference between them. KF acts as a controller to control its error difference. In the case of SOC estimation, Li-ion battery is the real system with SOC as one of its internal states. As SOC cannot be measured directly we use KF to find its best estimate from indirect measurements of battery terminal voltage and current. KF combines the measurement from the sensor and an estimated output predicted from the model. KF is the gain adjusting the error difference. (Figure 2) shows the 2 step KF algorithm.

Initialization:

$$\hat{x}_0^+ = E[x_0]$$

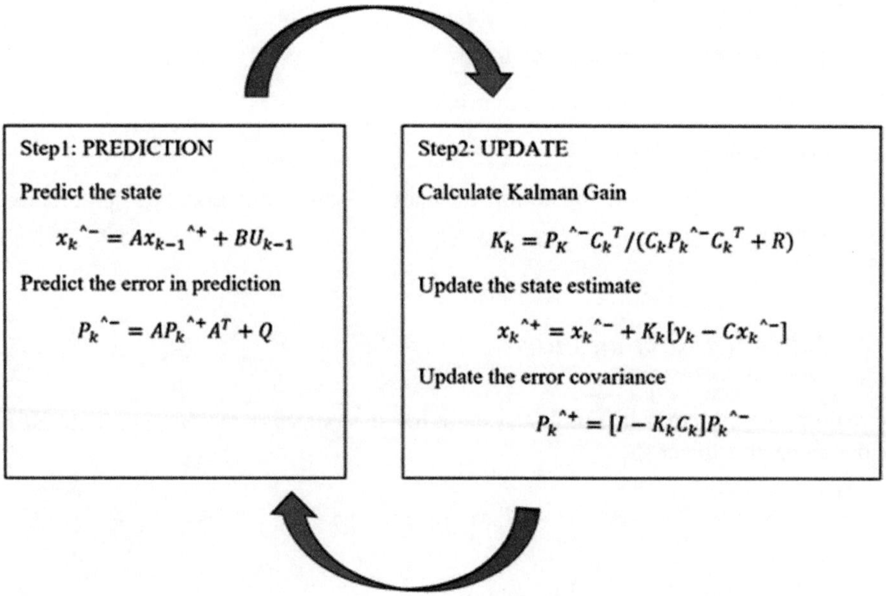

Fig. 2 Kalman filter algorithm

$$\hat{P_0^+} = E\left[(x - x_0)(x - x_0)^T\right]$$

Prediction:

Here priori state estimate $\hat{x_0^+}$ and error covariance matrix $\hat{P_0^+}$ is calculated.

Update:

Uses $\hat{x_0^+}$ and $\hat{P_0^+}$ along with current measurement to and a more accurate Posterior estimate is obtained. KG is calculated such that the difference between the real output and predicted output minimizes

where

$\hat{x_k^-}$ = Priori state estimate matrix

$\hat{P_k^-}$ = priori error covariance matrix

A = state matrix

B = state input matrix

C = state observation matrix

Q = process error covariance matrix

$\hat{x_k^+}$ = Posterior state estimate matrix

K_k = Kalman Gain

$\hat{P_k^+}$ = posterior error covariance matrix

R = measurement error covariance matrix

y_k = measurement at instant k.

KG indicates whether the posterior estimate depends on model or prediction measurement.

2.2 *Extended Kalman Filter*

In real life, no system is linear. In a system, the state states or the output function or both can be non-linear [5].

$$x_k = f(x_{k-1}, u_k) + w_k \tag{3}$$

$$y_k = g(x_k, u_k) + v_k \tag{4}$$

$$w_k \sim N(0, Q), \quad v_k \sim N(0, R)$$

$$A = \frac{\partial f(x_{k-1}, u_k)}{\partial x_{k-1}} \qquad B = \frac{\partial f(x_{k-1}, u_k)}{\partial u_k}$$
$$C = \frac{\partial g(x_k, u_k)}{\partial x_k} \qquad D = \frac{\partial g(x_k, u_k)}{\partial u_k}$$

For this, a non-linear state estimate is required. In the case of non-linearity, an Extended Kalman Filter (EKF) is used. EKF linearizes around the mean of the current state estimate. Linearization at each step is performed locally using Taylor series expansion. Only the first-order derivatives are considered and Jacobian matrices are calculated.

3 Implementation of EKF on Battery Model

For Implementation of EKF for SOC Estimation of A lithium-Ion cell following are the steps to be followed:

3.1 Battery Model

Batteries are highly nonlinear electrical systems. When a current step is applied as input, we observe a dynamic response. The battery's cell voltage continues to evolve. Voltage does not immediately return to its OCV but gradually relaxes. It is necessary to include this phenomenon while modeling the battery. This phenomenon is caused by slow diffusion processes in the cell, so it is known as diffusion voltage. Diffusion voltages can be closely approximated in a circuit using one or more parallel resistor-capacitor sub-circuits [6]. Battery response to a step discharging current is shown in (Fig. 3).

Thevenin model is a reasonable description of cell dynamics. The above circuit has a dependent voltage source which is a function of the state of charge and temperature. It has one resistor-capacitor branch to model diffusion and it has an equivalent series resistor to model the internal resistance of the battery [6, 7] as shown in (Fig. 4).

Thevenin Model Equations The Thevenin model equations are as follows.

$$\mathrm{SOC}(k) = \mathrm{SOC}(k-1) + \frac{\Delta t}{\mathrm{Cn}} \tag{5}$$

Diffusion Voltage State equation:

$$\mathrm{Vd}_k = \mathrm{e}^{\frac{-\Delta t}{\mathrm{Cp*Rp}}} * \mathrm{Vd}_{k-1} + R_p\left(1 - \mathrm{e}^{\frac{-\Delta t}{\mathrm{Cp*Rp}}}\right) * I \tag{6}$$

Battery Terminal Voltage:

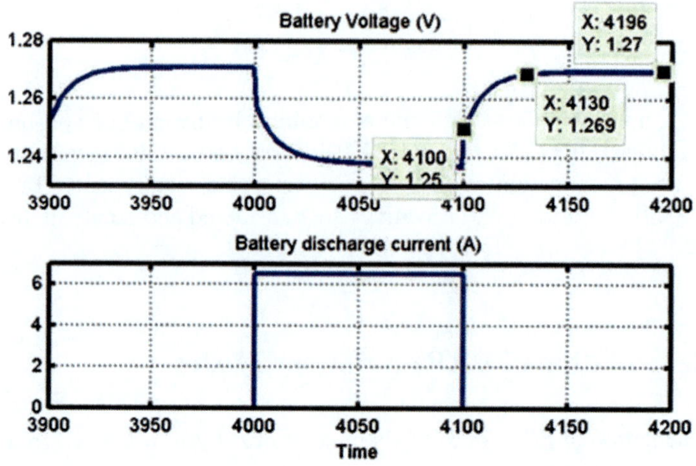

Fig. 3 Transient response of battery cell

Fig. 4 Thevenin battery model

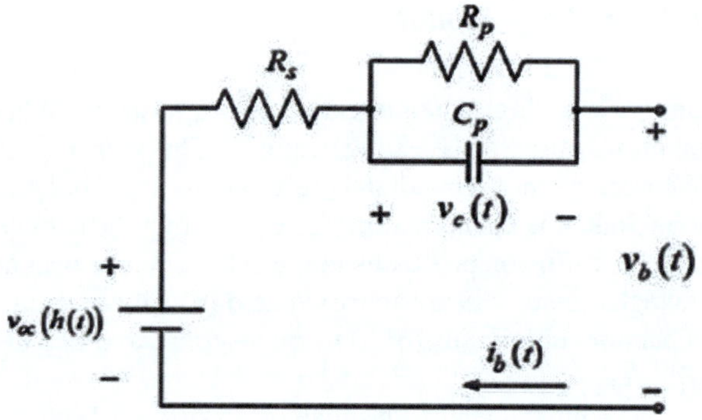

$$\mathrm{Vt} = \mathrm{OCV(SOC)} - \mathrm{Rs} * I - \mathrm{Vd}_k \tag{7}$$

Parameters Rs, Cp, and Rp vary with SOC and Temperature.

State Space Representation of Battery Cell. In Thevenin battery model, we have two system states, i.e., SOC and diffusion voltage as shown in Eqs. 8 and 9. Input is battery current.

$$X_k = AX_{k-1} + Bu_k \tag{8}$$

Battery output, i.e., terminal voltage is non-linear function of its state SOC. A Matlab function is built for this model. Therefore, it is represented as following in (Fig. 5)

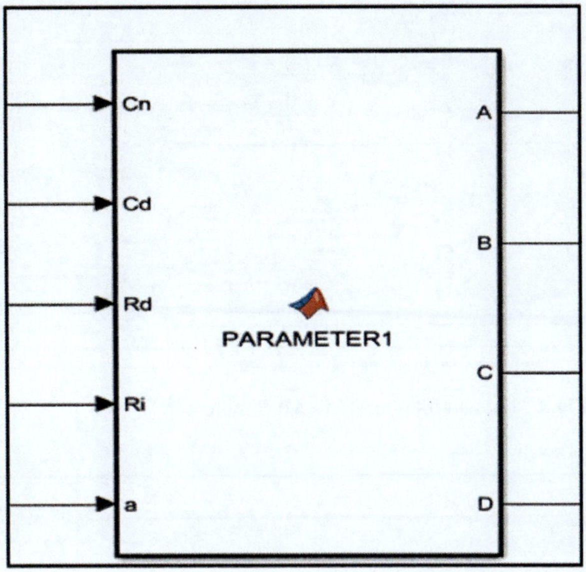

Fig. 5 State space representation function

$$y_k = g(X_k, u_k) \tag{9}$$

$$A = \begin{matrix} 1 & 0 \\ 0 & e^{\frac{-\Delta t}{Cp*Rp}} \end{matrix} \qquad B = \begin{matrix} \frac{\Delta t}{Cn} \\ R_p\left(1 - e^{\frac{-\Delta t}{Cp*Rp}}\right) \end{matrix}$$

$$C = \frac{OCV}{SOC} \; -1 \qquad D = Rs$$

3.2 *Implementation of EKF*

KF is built on its basic six equations (see Fig. 6).

3.3 *Integration of the Battery Model and EKF*

The battery model is integrated with KF block (see Fig. 7) and the battery model is been developed on the basis of Thevenin equivalent circuit discussed previously. Parameters Cp, Rp, Rs are input to the battery model. Output of this model is given as input to the EKF. EKF estimates battery internal state SOC.

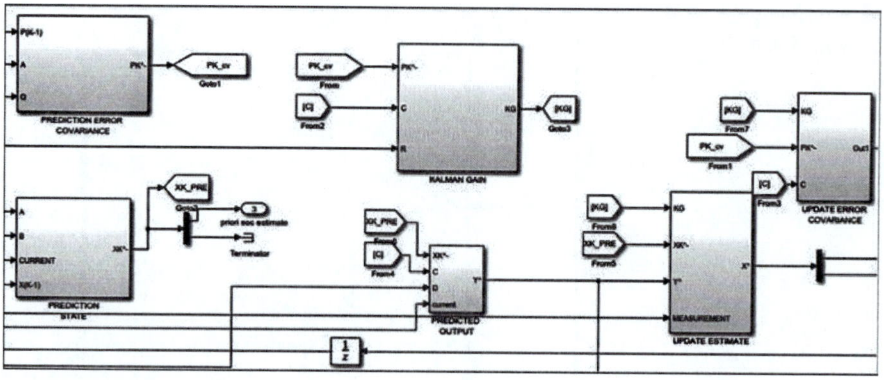

Fig. 6 Kalman filter in MATLAB Simulink

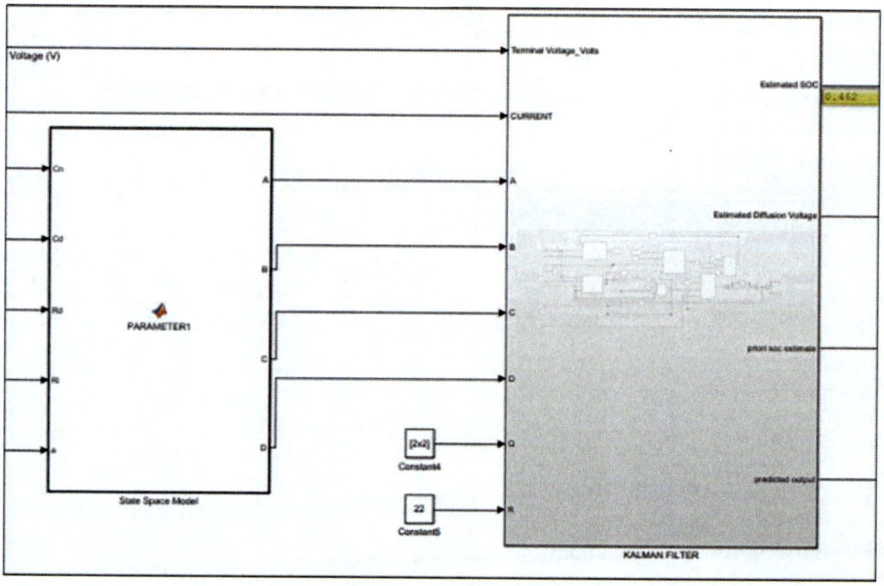

Fig. 7 Integration of battery model and EKF

4 Results

Battery current (charging or discharging) is input to the battery model. Temperature is kept constant for simplicity (see Fig. 8).

In the output plot (see Fig. 9), the yellow line indicates true SOC of the battery whereas the blue line indicates SOC estimation using EKF. As observed, the EKF estimated SOC follows the true SOC with an error of about 2 units.

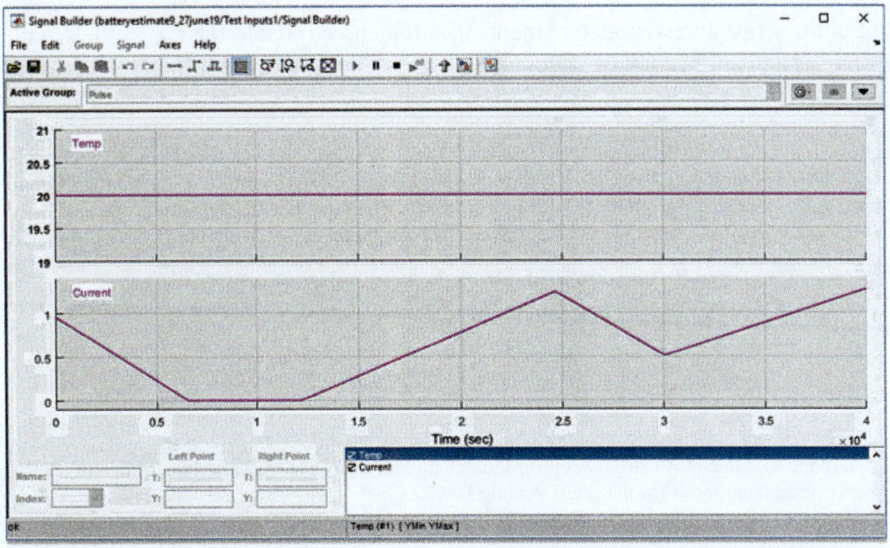

Fig. 8 Current and temperature input

Fig. 9 Comparison of SOC

5 Conclusion

SOC estimation is a complex task. This paper presented application of a non-linear technique, i.e., the EKF algorithm for SOC estimation of a Lithium-ion battery cell. Along with a 1-RC equivalent circuit battery model, EKF can be used with an accuracy of 1%.

Future scope includes deployment of sufficiently reliable battery models with hysteresis and application of Unscented Kalman Filter (UKF) and Cubature Kalman Filter (CKF).

Acknowledgements I, thank Mr. M. R. Saraf (H.O.D-SDL, AML and TG, ARAI) and Mrs. Ujjwala Karle (GM, Technology Group, ARAI) for allowing me to work as an intern at Technology Group Dept. I also extended my gratitude towards my guide Mr. Ravindra. G. Shah for his constant guidance and my mentor Mr. M. Siva for his insights and all the members of Technology Group for their valuable contribution.

References

1. Murnane M, Ghazel A (2017) A closer look at state of charge (SOC) and state of health (SOH) estimation techniques for batteries. Analog Devices
2. Youngryul K, Sunghyun Y, Junho L (2017) SOC estimation and BMS design of Li-ion battery pack for driving. In: 14th international conference on ubiquitous robots and ambient intelligence (URAI). Maison Glad Jeju, Jeju
3. Yuan S, Wu H, Yin Y (2013) State of charge estimation using the extended Kalman filter for battery management systems based on the ARX battery model. Energies 6:444–470. https://doi.org/10.3390/en6010444
4. Farag M, Attari M, Gadsden A, Habibi S (2017) Lithium-ion battery state of charge estimation using one state hysteresis model with nonlinear estimation strategies. Int J Mater Metall Eng 11(3)
5. He H, Xiong R, Fan J (2011) Evaluation of lithium-ion battery equivalent circuit models for state of charge estimation by an experimental approach. Energies 4:582–598. https://doi.org/10.3390/en4040582
6. Yu Z, Huai R, Xiao L (2015) State-of-charge estimation for lithium-ion batteries using a Kalman filter based on local linearization. Energies 8:7854–7873. https://doi.org/10.3390/en8087854
7. Nejad S, Gladwin D, Stone D (2015) Enhanced state-of-charge estimation for lithium-ion iron phosphate cells with flat open-circuit voltage curves. In: IECON2015-Yokohama, 9–12 Nov 2015

Frequency and Pattern Reconfigurable Antenna for WLAN and WiMAX Application

Anuradha A. Palsokar and Swapnil L. Lahudkar

Abstract An innovative bow tie frequency and pattern reconfigurable antenna is proposed in this paper which can be used for WLAN and WiMAX applications. The proposed antenna is a modified bow tie antenna with an additional branch on patch side. Two p-i-n junction (PIN) diodes are used for obtaining the required frequency and radiation pattern. One diode is placed on the patch side and another diode is placed on the ground side. The antenna operates at three different frequency values 3.35, 3.6 and 5.2 GHz with changing radiation patterns for each operating frequency. During simulation maximum efficiency of 80% along with a maximum gain of 1.2 dBi is obtained for on switch condition and an efficiency of 91% and gain of 2.1 dBi is obtained for off switch condition.

Keywords Frequency reconfigurable antenna · Pattern reconfigurable antenna · PIN diode · WLAN · WiMAX

1 Introduction

Today's smart handheld communication devices support operation over various bands such as Bluetooth, LTE, WLAN, WiMAX, GSM etc. and reconfigurable antennas are replacing multiband antennas due to their inherent advantages in all such applications. In literature, we observe discussion and design of various reconfigurable antennas. A rectangular patch with two longitudinal slits is presented in [1]. This antenna is equipped with a pair of PIN diodes along the slits to achieve frequency and pattern agility. Simulation of frequency and radiation pattern reconfigurable ring-type patch antenna is given in [2] that shows operation over 2.7, 2.8 and 3.6 GHz. An antenna with two quarter-wavelength radiators with a truncated quarter-wavelength ground

A. A. Palsokar (✉)
Rajarshi Shahu College of Engineering, Pune 411033, India
e-mail: panuradha2010@gmail.com

S. L. Lahudkar
Imperial College of Engineering and Research, Pune 412207, India

© Springer Nature Singapore Pte Ltd. 2021

S. N. Merchant et al. (eds.), *Advances in Signal and Data Processing*,
Lecture Notes in Electrical Engineering 703,
https://doi.org/10.1007/978-981-15-8391-9_8

plane and two quarter-wavelength parallel strips at the substrates edges capable to steer the radiation pattern in four directions using four PIN diodes is designed in [3]. A frequency reconfigurable antenna designed using a bowtie antenna and slots is explored in [4, 5], where band switching is achieved using switches between the slots and bowtie antenna. V-shaped slot antenna with T and C-shaped resonators is designed in [6]. This antenna controls its operation using four switches, at various bands over the frequency range from 970 MHz to 3.76 GHz with a stable pattern.

A microstrip patch antenna with two slots is designed to obtain three frequency reconfigurable bands over the range 2.4–5.8 GHz. Two PIN diodes are added along each slot to alter the electrical length of the antenna [7]. Rectangular patch antenna equipped with T-shaped slotted feed line, operating at 1.8, 2.3, and 2.4 GHz is designed in [8]. A bow tie antenna operating over Bluetooth, WiMAX and WLAN is presented in [9]. PIN diodes are used over bow tie arms for tuning operating bands.

The design of antenna presented here is a simple frequency and pattern reconfigurable antenna for WLAN and WiMAX applications.

2 Antenna Design and Geometry

The antenna proposed in this work is a bow tie antenna, derived from the design given in [10]. The structure of antenna is shown in Fig. 1.

Basic design of proposed antenna is a bow tie antenna printed on both side of substrate with an operating frequency of 5.5 GHz, the design equations are followed from [9, 10]. Optimized dimensions of antenna are Ws = 30 mm, Ls = 26 mm, Lf = 3 mm, a = 13.5 mm, b = 10 mm, e = 7.5 mm, d = 16 mm, g = 10 mm and h = 10.8 mm. Antenna makes use of two switching diodes each one placed on patch and ground structure. First diode is placed on the patch side in between the two branches

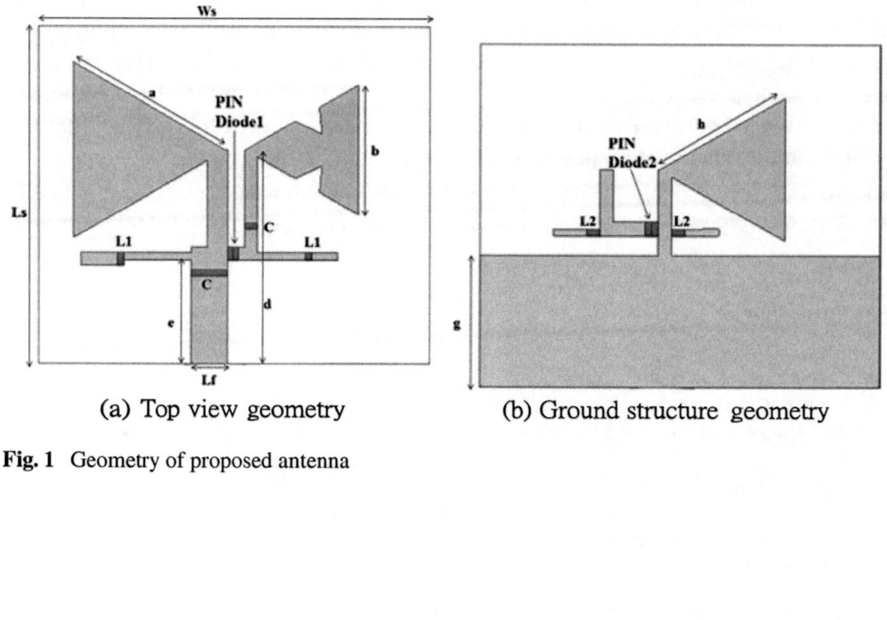

(a) Top view geometry (b) Ground structure geometry

Fig. 1 Geometry of proposed antenna

of patch, second diode is placed on the ground side between the bow-shaped element and a stub. Biasing and DC blocking elements are added on antenna, which consists of two inductors (L1) of 33 nH and two capacitors (C) of 10 pF each. Two inductors (L2) of 1.8 nH are used for biasing purpose on ground side also. PIN diodes are basically used to alter the electrical length of the antenna so as to achieve frequency agility.

3 Simulation Results

ANSYS High-Frequency Structural Simulator (HFSS), an electromagnetic software for designing high-frequency electronic products is used for simulation of proposed antenna. The performance of antenna is observed in two modes. In the first mode when both the diodes are off the antenna supports WiMAX application with a frequency of operation 5.2 GHz. Gain of 2.1 dBi and efficiency of 91.05% is obtained for this mode during simulation. When both the diodes are on the antenna supports WLAN application with resonant frequency 3.35 GHz and 3.6 GHz. For 3.35 GHz gain of 1 dBi and efficiency of 74% and for 3.6 GHz a gain of 1.2 dBi with an efficiency of 80% is obtained during simulation. Return loss for the simulated antenna design is shown in Fig. 2. The performance of antenna is summarized in Table 1.

The antenna is simulated considering low loss FR4 substrate, whose dimensions are 30 mm * 26 mm * 1.6 mm. While obtaining the simulation results BAR 64 SC79 PIN diode is considered as switch. For on switch case, each PIN diode is replaced with

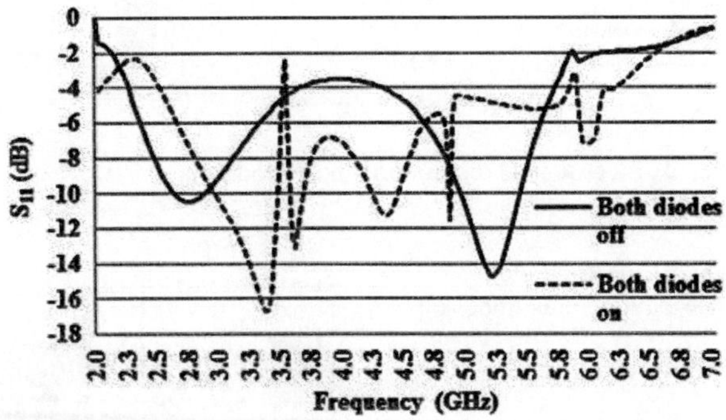

Fig. 2 Simulated return loss for proposed antenna

Table 1 Performance of proposed antenna

Mode	Switch position	Frequency (GHz)	Return loss (dB)	VSWR	Gain (dBi)	Efficiency (%)	% Bandwidth
I	Both on	3.35	−16.74	1.34	1	74	14
		3.36	−13.1	1.44	2.1	80	2
II	Both off	5.2	−14.76	1.44	2.1	91.05	9.6

a resistance of 2 Ω and for off switch case, it is replaced with a parallel combination of a resistance of 3 kΩ and a capacitor of 0.17 pF. In both the modes, a series inductance of 1.8 nH is also considered in addition to the above stated equivalent circuit.

4 Reconfigurable Operation

Reconfigurable operation is obtained with two PIN diodes. As the state of diodes is changed the electric length of the antenna is changed and it results in changing frequency of operation. The electric field distribution for the first mode when both the switches are on is shown in Fig. 3. Figure 4 shows electric field distribution for both switches in off condition. As the switch configuration is changed the electric dimensions of the antenna are changed which leads to change in electric field distribution and surface current distribution on the antenna. From these figures, we note that for each case the electric field distribution is different, the result of which is

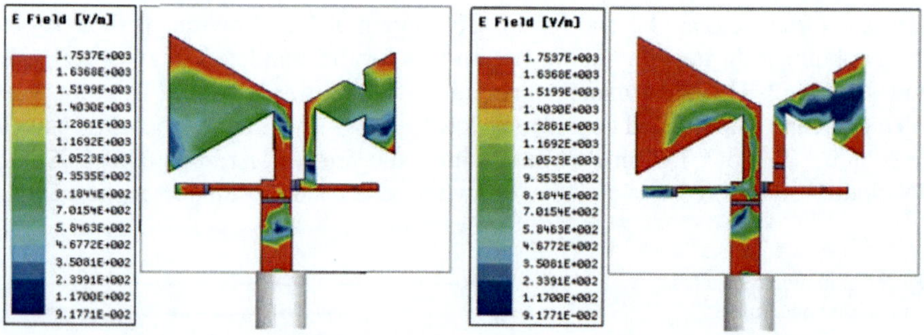

(a) At 3.35 GHz (b) At 3.6 GHz

Fig. 3 Electric field distribution: both diodes on

Fig. 4 Electric field distribution: both diodes off

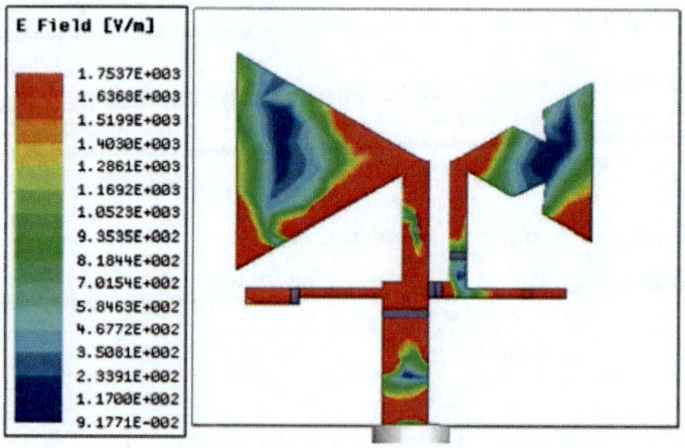

reflected in varying radiation patterns for each condition. Thus pattern reconfiguration is achieved along with frequency reconfigurability. The radiation pattern for the proposed antenna is shown in Figs. 5, 6, and 7.

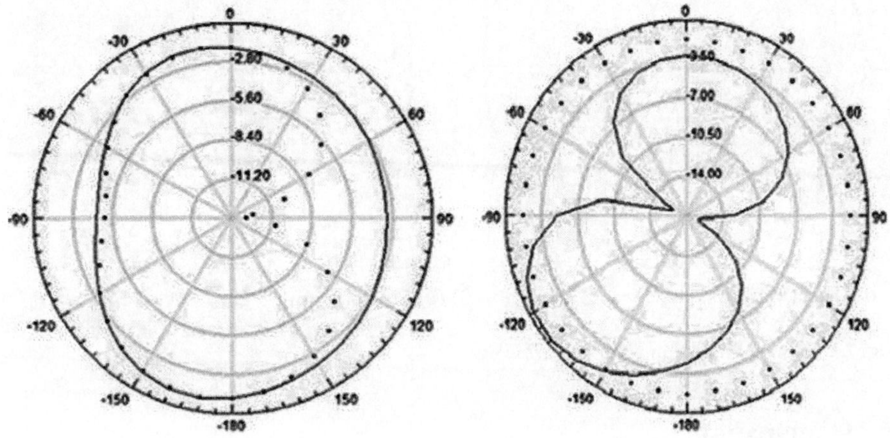

(a) Theta variation (– $E_\varphi = 90$; . . . $E_\varphi = 0$) (b) Phi variation (– $E_\theta = 90$; . . . $E_\theta = 0$)

Fig. 5 Simulated radiation pattern at 3.35 GHz

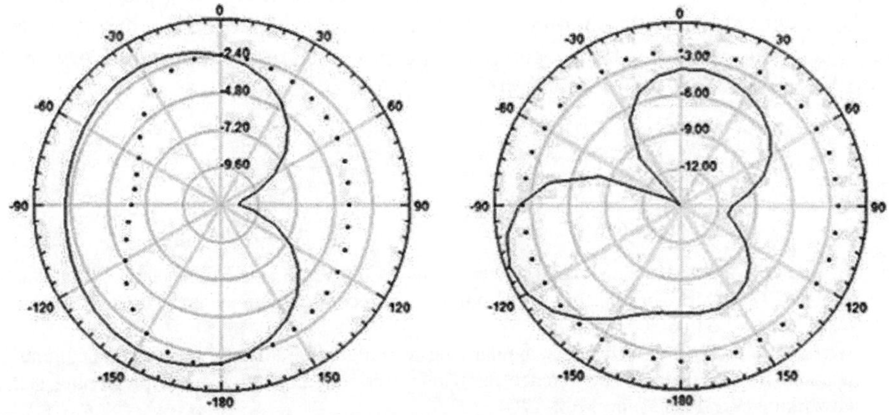

(a) Theta variation (– $E_\varphi = 90$; . . . $E_\varphi = 0$) (b) Phi variation (– $E_\theta = 90$; . . . $E_\theta = 0$)

Fig. 6 Simulated radiation pattern at 3.6 GHz

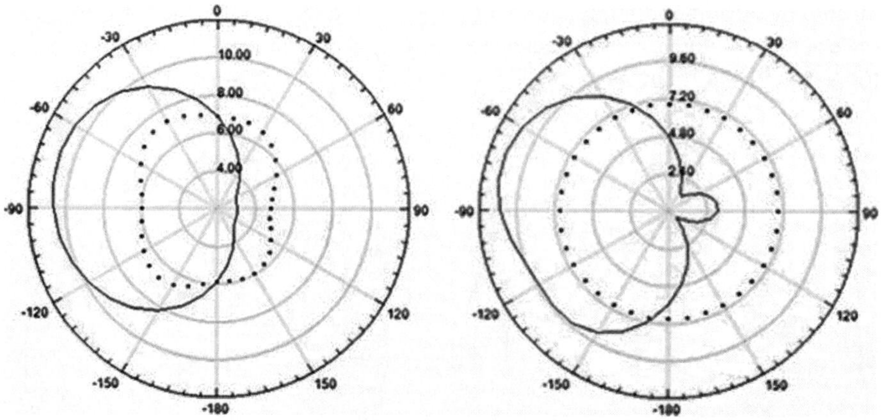

(a) Theta variation (- $E_\varphi = 90$; . . . $E_\varphi = 0$) (b) Phi variation (- $E_\theta = 90$; . . . $E_\theta = 0$)

Fig. 7 Simulated radiation pattern at 5.2 GHz

5 Conclusion

A new design of a compound reconfigurable antenna using two switching diodes is proposed in this paper. The antenna works in two modes, in one mode operating frequency is 5.2 GHz and in another mode the frequency of operation is 3.35 GHz and 3.6 GHz, with a varying radiation pattern for each mode; hence the antenna exhibits frequency and radiation pattern agility. The proposed antenna supports WLAN and WiMAX applications with a small profile.

References

1. Selvam YP, Malathi K, Gulam Nabi Alsath M, Sangeetha V, Saffrine K, Sangeetha S, Ramana Rao V (2017) A low-profile frequency- and pattern-reconfigurable antenna. IEEE Antennas Wirel Propag Lett 16:3047–3050
2. Sweta SS, Ashish KS (2017) Rectangular ring reconfigurable antenna for wireless communication. In: International conference on wireless communications, signal processing and networking (WiSPNET), pp 1702–1704
3. Alam MS, Abbosh AM (2017) Wideband pattern-reconfigurable antenna using pair of radial radiators on truncated ground with switchable director and reflector. IEEE Antennas Wirel Propag Lett 16:24–28
4. Mansoul A, Seddiki ML (2018) Multiband reconfigurable bowtie slot antenna using switchable slot extensions for WiFi, WiMAX, and WLAN applications. Microw Opt Technol Lett 60:413418
5. Woosung L, Hyungrak K, Young JY (2008) A frequency reconfigurable bow-tie slot antenna with wide bandwidth. Microw Opt Technol Lett 50:404–406
6. Sahar C, Mohamad RH, Muhammad RK, Farid G (2015) Reconfigurable multiband tapered slot antenna. Microw Opt Technol Lett 57:2182–2186

7. Han L, Wang C, Chen X, Zhang W (2016) Compact frequency-reconfigurable slot antenna for wireless applications. IEEE Antennas Wirel Propag Lett 15:1795–1798
8. Sabapathy T, Bashah MA, Jusoh M, Soh PJ (2016) Frequency reconfigurable rectangular antenna with T-slotted feed line. In: International conference on radar, antenna, microwave, electronics, and telecommunications, pp 81–84
9. Li T, Zhai H, Wang X, Li L, Liang C (2015) Frequency-reconfigurable bow-tie antenna for Bluetooth, WiMAX, and WLAN applications. IEEE Antennas Wirel Propag Lett 14:171–174
10. Rahim MKA, Abdul Aziz MZA, Goh CS (2005) Bow-tie microstrip antenna design. In: 13th IEEE international conference on networks jointly held with the 2005 IEEE 7th Malaysia international conference on communications, pp 17–20. https://doi.org/10.1109/ICON.2005.1635425

Implementation and Analysis of Low Power Consumption Full Swing GDI Full Adders

Deepgandha Shete and Anuja Askhedkar

Abstract Power efficiency of any design can be obtained in terms of PDP. The approach used to design any system defines the performance of system. Here, Gate Diffusion Input (GDI) design techniques as well as CMOS design techniques are used for designing full adder circuits. Full swing GDI technique is utilized to reduce power consumption and delay. Full swing GDI technique gives better speed of operation as compared to CMOS technique. While keeping these parameters at best GDI maintains low complexity of design. Full adder circuits are designed using 180 nm technology node in Cadence Virtuoso with supply voltage of 1.8 V.

Keywords Cadence Virtuoso · CMOS · Full swing GDI · Power-delay product

1 Introduction

There is tremendous growth in IC computing power that brings huge improvement in digital world. According to Moore's law, the numbers of transistors are getting doubled every 1.5 years [1]. Hence on same chip area, transistor count increases which requires more power and speed downs the devices. There are several ways to reduce power consumption but along with that the requirements of high speed of operation of design should also be fulfilled that indicates that the power delay product (PDP) of design should be less.

The power optimization can applicable at various levels like circuit level, system level and architecture level [2]. Circuit level power optimization reduces switching activity of transistors in combinational and sequential circuits. To design full adder circuits different logic styles are used where each one is having own merits and bottle-necks [3].

D. Shete (✉) · A. Askhedkar
MIT WPU, Pune, India
e-mail: deepgandhashete1995@gmail.com

A. Askhedkar
e-mail: anuja.askhedkar@mitwpu.edu.in

© Springer Nature Singapore Pte Ltd. 2021

S. N. Merchant et al. (eds.), *Advances in Signal and Data Processing*,
Lecture Notes in Electrical Engineering 703,
https://doi.org/10.1007/978-981-15-8391-9_9

Any digital circuit's performance is dependent on the performance of arithmetic circuit. Hence design of full adder circuit with better performance is necessary to enhance the total performance of the system. But in some cases, full adder circuits dissipates large amount of power and increases delay.

To design full adder circuit using basic gates, it requires 28 transistors to perform addition operation. As the transistor count increases power consumption, area and delay also increases, hence decrement in transistor count is necessary to save total power required [4]. The CAD tool uses standard cells for designing chip, hence stand-are cells play an important role in chip design. Standard cells are predesigned and verified which can be used in many designs [5].

In this paper, two techniques are compared CMOS (Complementary Metal Oxide Semiconductor) technique and full swing GDI technique. To reduce power consumption and delay full swing GDI technique is used [6]. Full swing GDI is the most power-saving technique used in any digital circuits.

Organization of paper is as follows: Sect. 2 describes analog design flow. 1 bit full adder circuits using CMOS technique is explained in Sect. 3. The proposed full swing GDI technique is described in Sect. 4. Section 5 describes design of one more combinational circuit i.e. multiplexer using CMOS and GDI technique. Section 6 gives the details about comparison of different technologies based on analysis of GDI and CMOS full adder circuits. Lastly, conclusion and future scope are drawn in Sect. 7.

2 Design Steps for Analog Design Flow

The steps involved for full adder circuit design using Cadence tool are schematic, layout, Design Rule Check (DRC), Layout Versus Schematic (LVS) and parasitic extraction. For design of 1-bit full adder circuits, different platforms of cadence are used. Cadence Spectre Simulation platform is used to design schematic of full adder circuits. For layout design Cadence Virtuoso Layout Suite XL is used. DRC and LVS are carried out by using Assura [7]. The designs in this work are carried out at 180 nm technology using 1.8 V supply. The W/L ratio defines how much power is required for design; hence for 180 nm technology length is fixed, i.e., 0.18 μm and width can vary according to designer requirement.

To design full adder circuits, first step is to draw schematic design using spectre. After correct design of schematic, layout is generated. To verify layout DRC and LVS are used. After verification of DRC and LVS, parasitic extraction is carried out.

3 Design of CMOS Full Adder Circuits

Conventional CMOS technique uses 28 numbers of transistors for designing full adder circuits; delay and power requirement of this circuit is more [8]. The switching activity of transistor defines the power consumption in the design. Static power dissipation and dynamic power dissipation are two types of power dissipation occurred in any design [9]. Design 1 and design 2 are designed using CMOS technique at 180 nm technology node with supply voltage 1.8 V. The width of nMos and pMos are 540 nm and length is 180 nm, hence ratio of width and length is 3.

Design 1

Design 1 circuit is designed with smaller area and low power utilization using CMOS technique. But due to threshold voltage defeat problem [10] the outputs of design 1 is not full swing. The decrement in voltage swing leads to slow switching problem. Design 1 gives longer delay and poor driving capability due to threshold voltage drops.

The outputs sum and carry are not full swing hence to eliminate this problem additional buffers are added at the output sides. Figures 1 and 2 shows the schematic and layout of this design. Hence in design 1, total 18 transistors are used to obtain full swing output.

Design 2

In this design 2, Low Power Full Adder (LPFA) CMOS technique is used to design full adder circuit [11]. The design of XOR gate is done by using XNOR gate followed by an inverter, XOR and XNOR gate gives better driving capability [12]. But the outputs of full adder circuit using LPFA technique are not full swing hence additional buffers are required to overcome this problem. The schematic and layout of this design is shown in Figs. 3 and 4 respectively. To obtain full swing output this technique requires 22 transistors.

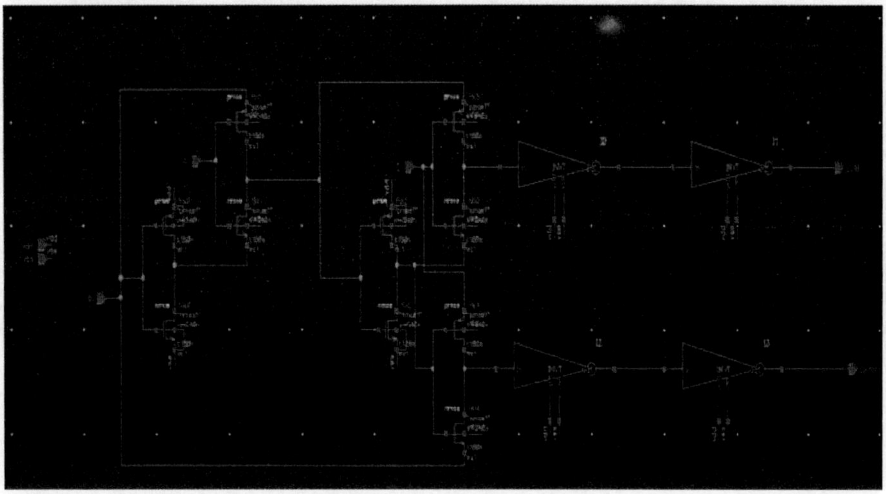

Fig. 1 Schematic of design 1 full adder

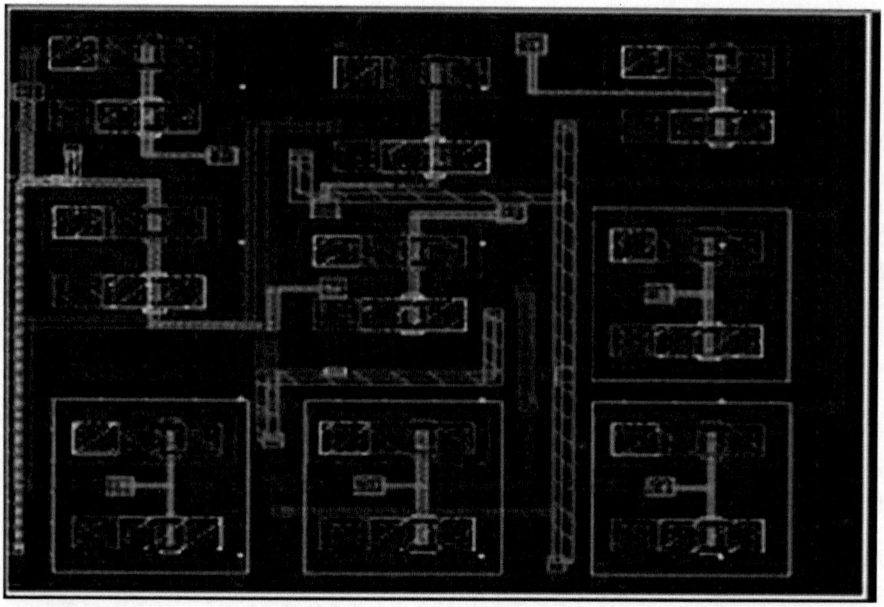

Fig. 2 Layout of design 1 full adder

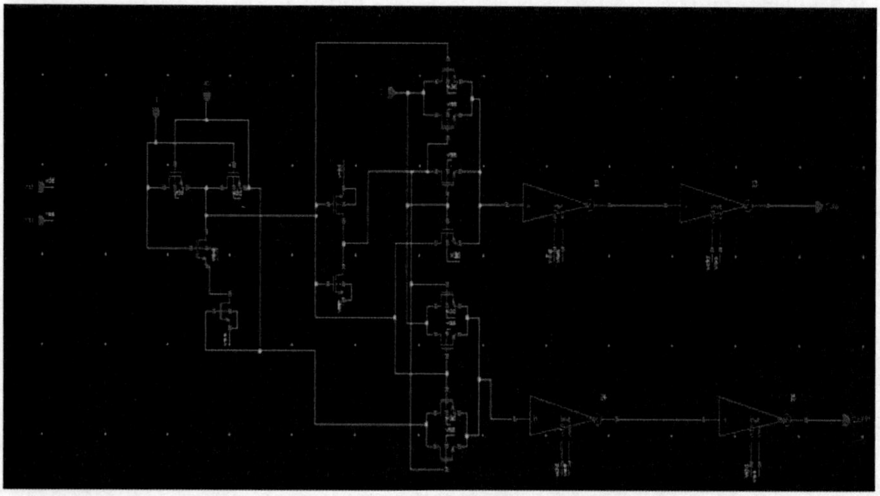

Fig. 3 Schematic of design 2 full adder

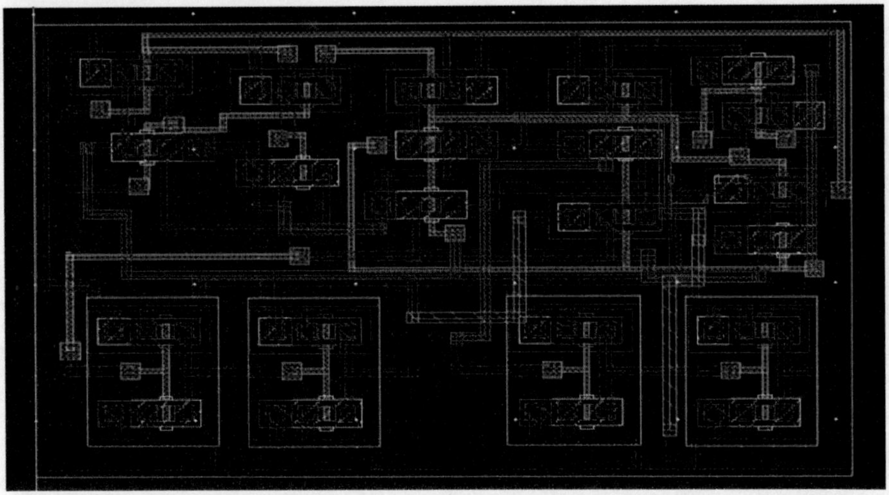

Fig. 4 Layout of design 2 full adder

4 Proposed Full Swing GDI Full Adder

In the design 1 and design 2, the outputs sum and carry of full adder circuits are not full swing, to make it full swing further buffers are inserted at outputs which increases power consumption and delay. To overcome this problem Gate Diffusion Inputs (GDI) technique is used. In design 1 and design 2 full adders are designed using CMOS technique in proposed design full adders are designed using GDI technique.

Gate Diffusion Input (GDI) Technique

Morgenshtein et al. have proposed basic GDI cell [13]. GDI technique gives reduction in power and delay in VLSI digital circuits. Basically, GDI cell has three inputs, these are G is the common gate input of pMOS and nMOS, P is the input to source/drain of pMOS and N is the input to source/drain of nMOS [14]. Figure 5 shows schematic of basic GDI cell. Unlike in CMOS source terminal of pMOS is not connected to the VDD; it is used as input terminal. Similarly, for nMOS instead of connecting source to ground, it is used as input terminal [15].

To design full adder circuit using GDI technique the following basic expressions are used:

$$\text{SUM} = A \,\text{xor}\, B \,\text{xor}\, \text{Cin} \tag{1}$$

$$\text{CARRY} = (A \,\text{xor}\, B)\,\text{Cin} + (A \quad B) \tag{2}$$

Design of 1-Bit Full Adders Using Full Swing GDI

The design used for implementing full adder is shown in Fig. 6.

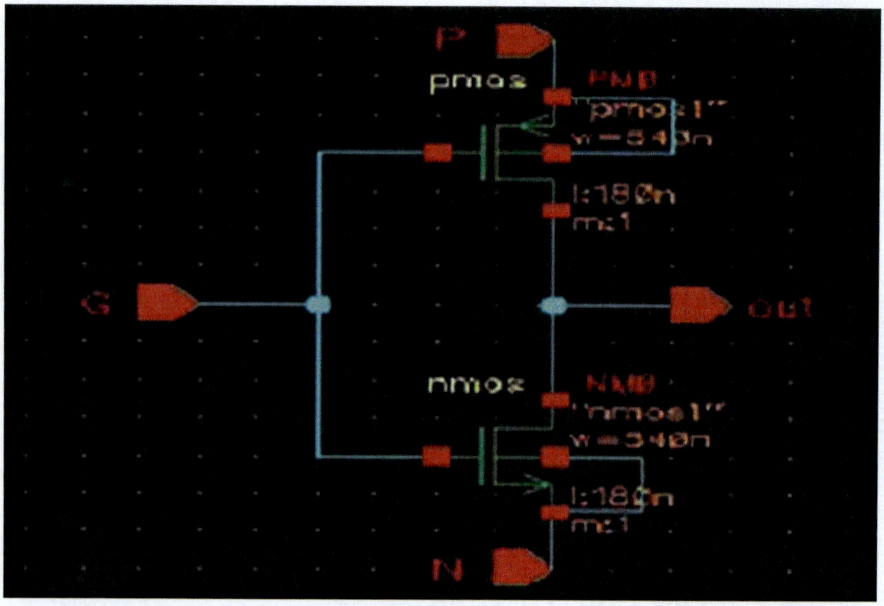

Fig. 5 Basic GDI cell

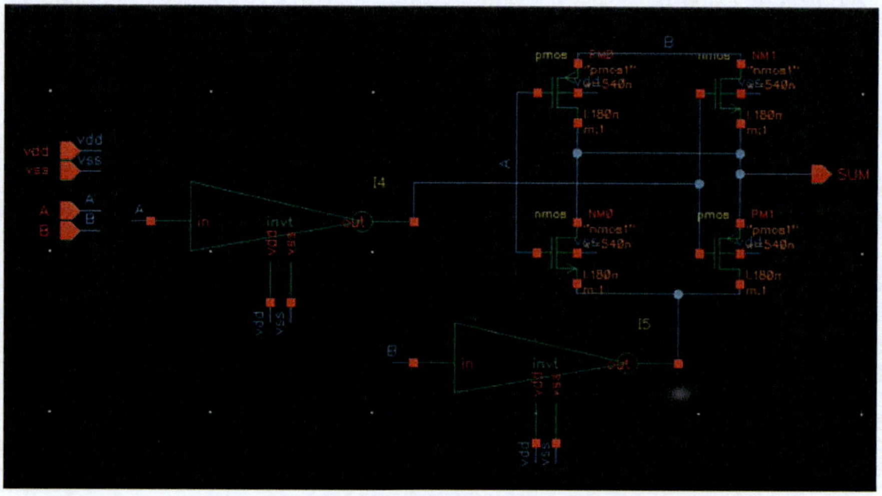

Fig. 6 Schematic of full swing GDI XOR

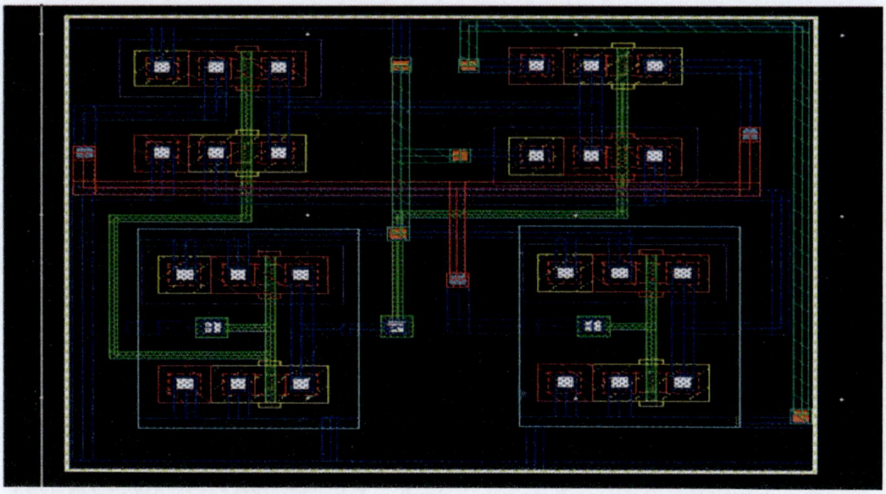

Fig. 7 Layout of full swing GDI XOR

The design basically consists of the following modules:

1. Full Swing GDI AND Gate
2. Full Swing GDI XOR Gate
3. Full Swing GDI OR Gate.

The schematic and layout of full swing XOR gate using GDI technique are shown in Figs. 6 and 7 respectively. The input and output waveforms of the full swing GDI XOR gate are shown in Fig. 8. The 1-bit full adder circuits are designed using full swing GDI technique at 180 nm technology with 1.8 V power supply and keeping width is 540 nm and length is 180 nm. Figures 9 and 10 show the schematic and layout of GDI based full adder. For all the three designs the input, output waveforms are shown in Fig. 11.

5 Design of Combinational Circuit

The combinational circuit like multiplexer is designed using CMOS technique and full swing GDI technique. n:1 MUX has 2^n input lines, one output line, and n select lines. 'S' is the select input line which selects one of the input line and forwarded towards the output line [16]. In this paper, 2:1 MUX is designed using 2 input lines and one select line. To design 2:1 MUX using CMOS technique we required 12 transistors and for same design using full swing GDI technique only 6 transistors are used. Hence area, power consumption and delay for full swing GDI is less as compared to CMOS technique.

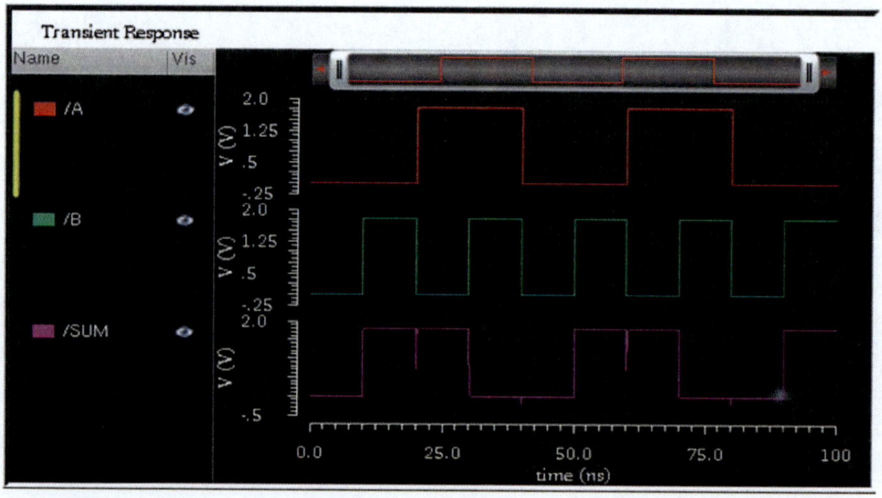

Fig. 8 Input, output waveforms of full swing GDI XOR

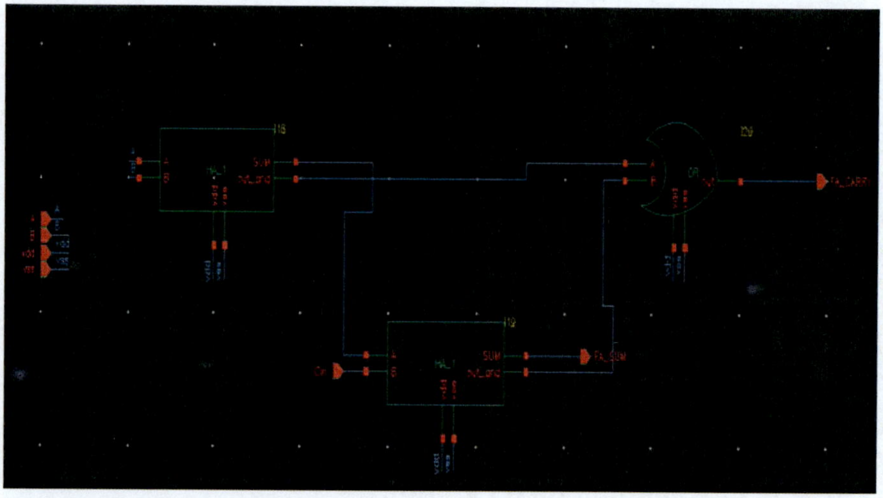

Fig. 9 Schematic of full swing GDI full adder

The output expression for Fig. 12 basic block of 2:1 MUX is as follows where $I0$ and $I1$ are input lines and S is the select input line.

$$Y = \overline{S}.I0 + S.I1 \tag{3}$$

2:1 MUX Using CMOS

See Figs. 13, 14 and 15.

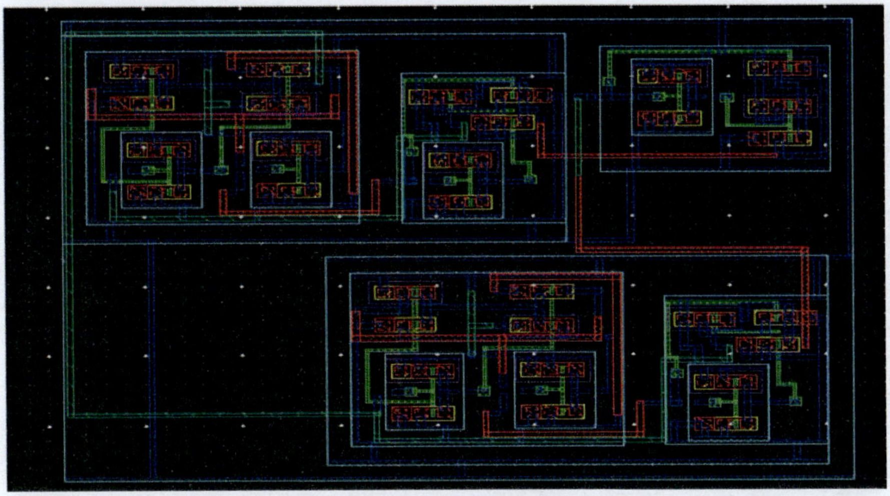

Fig. 10 Layout of full swing GDI full adder

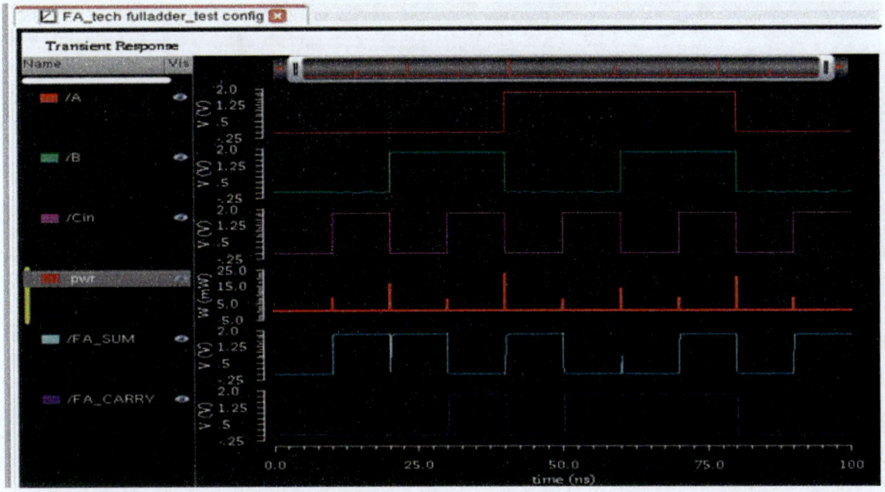

Fig. 11 Input output waveforms of all full adder circuits

Fig. 12 Basic block of 2:1 MUX

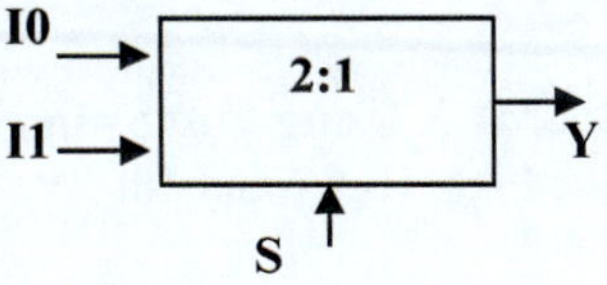

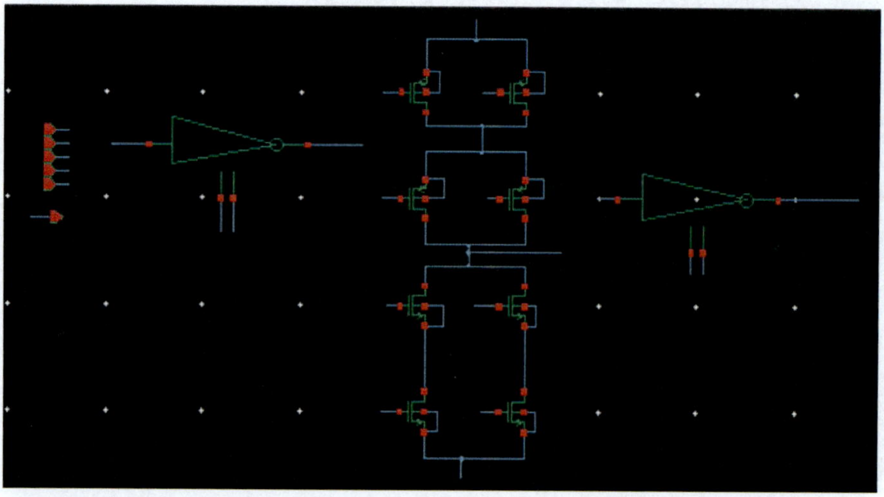

Fig. 13 Schematic of CMOS 2:1 MUX

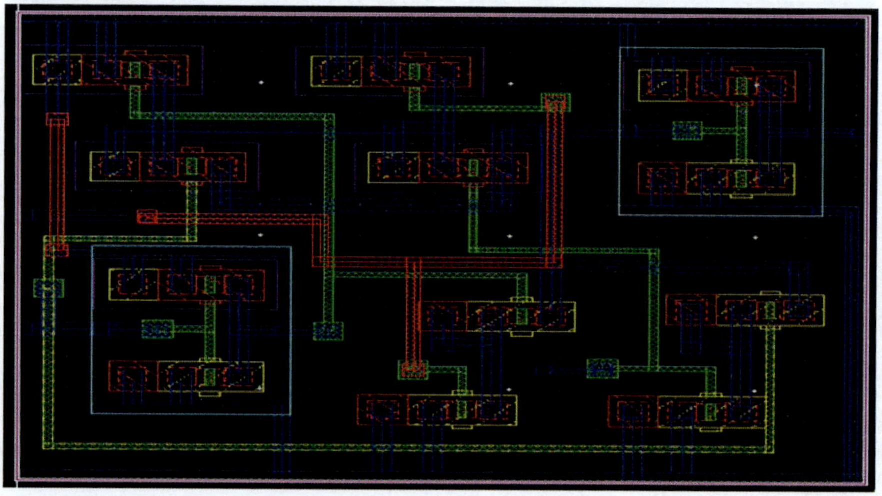

Fig. 14 Layout of CMOS 2:1 MUX

2:1 MUX Using Full Swing GDI

See Figs. 16, 17 and 18.

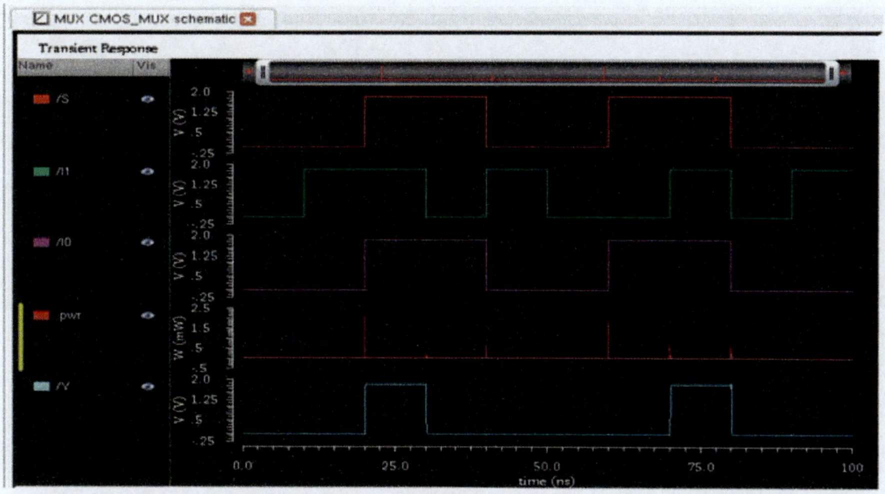

Fig. 15 Input output waveforms CMOS 2:1 MUX

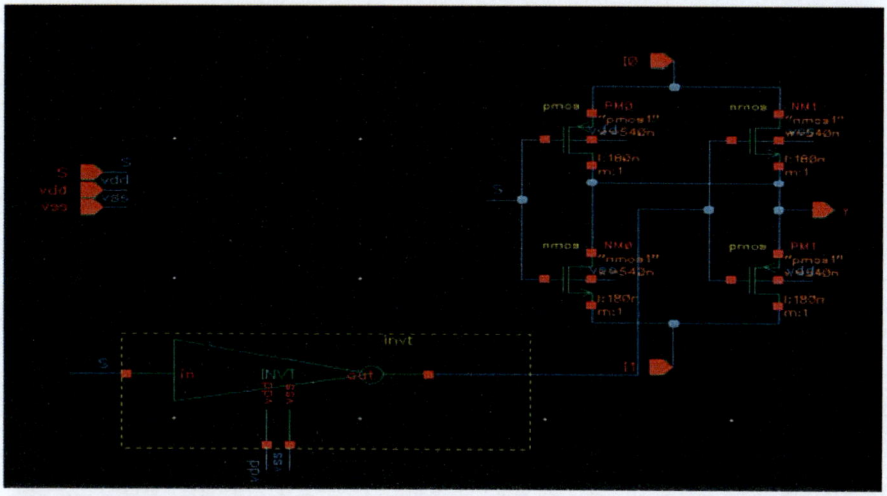

Fig. 16 Schematic of full swing GDI 2:1 MUX

6 Results

The full adder circuits and multiplexers are designed using Cadence Virtuoso ADE L schematic editor environment at 180 nm technology using 1.8 V as supply voltage and 0.9 V as threshold voltage. Delay, power consumption, PDP and area obtained for all the three designs are shown in tables.

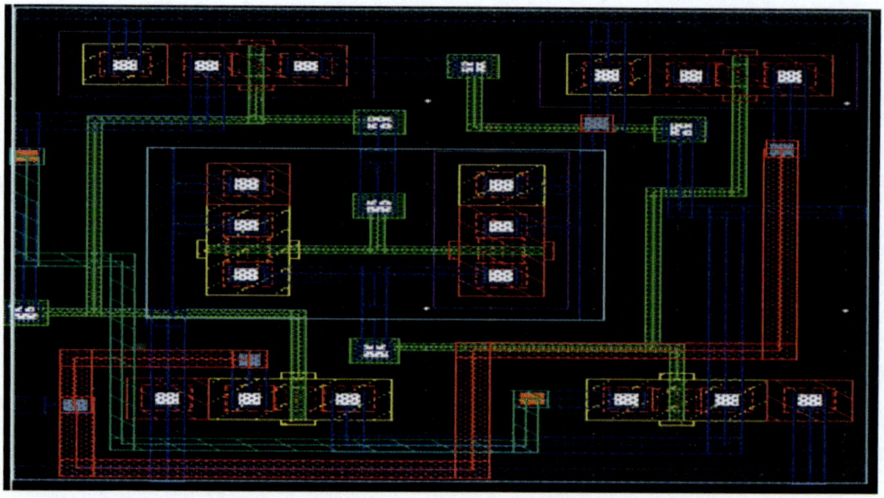

Fig. 17 Layout of full swing GDI 2:1 MUX

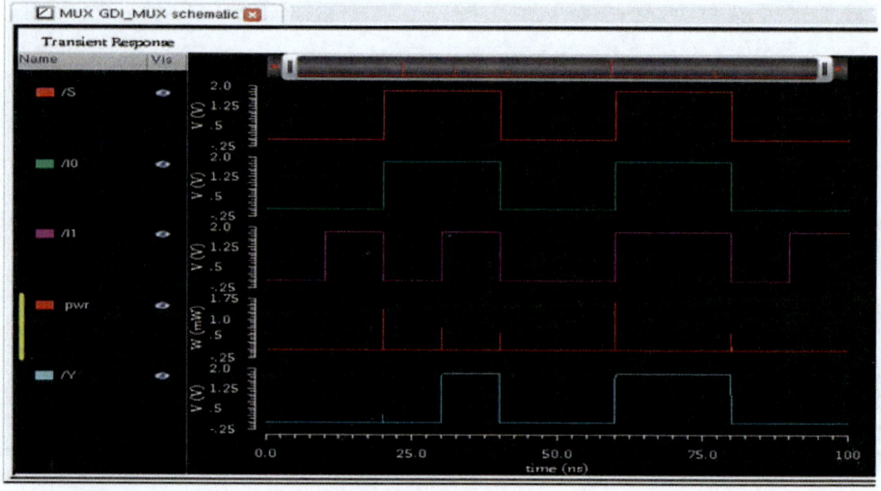

Fig. 18 Input output waveforms full swing GDI 2:1 MUX

1. **Delay**

The propagation delay calculated for each design is shown in Tables 1 and 2. GDI based circuit has lesser delay as compared to CMOS design. The percentage improvement in delay for GDI based full adder circuit is over 22% for design 2 and over 25% for design 1.

The percentage improvement in delay for GDI based 2:1 MUX circuit is over 86.26% for CMOS based MUX circuit.

Table 1 Maximum delay for full adder circuits

Types of full adder	Delay (ps)	% Improvement
Design 1	326.3	25.65
Design 2	311.0	22.1
Full swing GDI full adder	242.6	–

Table 2 Maximum delay for 2:1 MUX

Types of MUX	Delay (ps)	% Improvement
CMOS MUX	114.2	86.26
Full swing GDI MUX	15.68	–

2. **Power Consumption**

Power consumption calculated for full swing GDI circuits and CMOS circuits are shown in Tables 3 and 4. The percentage improvement for GDI based full adder is 70% for design 1 and over 26% for design 2.

The percentage improvement for GDI based 2:1 MUX is 44.21% for CMOS MUX circuits.

3. **Power Delay Product (PDP)**

The PDP obtained for all the designs is shown in Tables 5 and 6. The percentage improvement in PDP for GDI based full adder is over 77% for design 1 and over 42% for design 2.

The percentage improvement in PDP for GDI based 2:1 MUX is over 92.61% for CMOS MUX design.

Table 3 Power consumption of each full adder circuits

Types of full adder	Total average power (μW)	% Improvement
Design 1	28.93	70
Design 2	11.74	26.07
Full swing GDI full adder	8.679	–

Table 4 Power consumption of 2:1 MUX

Types of MUX	Total average power (μW)	% Improvement
CMOS MUX	1.547	44.21
Full swing GDI MUX	0.863	–

Table 5 Power-delay-product of each full adder circuits

Types of full adder	PDP (pJ)	% Improvement
Design 1	0.0094	77.77
Design 2	0.0036	41.66
Full swing GDI full adder	0.0210	–

Table 6 Power-delay-product of 2:1 MUX

Types of MUX	PDP (fJ)	% Improvement
CMOS MUX	0.176	92.61
Full swing GDI MUX	0.013	–

Table 7 Area of each full adder circuit

Types of full adder	Area (μm^2)
Design 1	337.77
Design 2	426.78
Full swing GDI	1346

Table 8 Area of 2:1 MUX

Types of MUX	Area (μm^2)
CMOS MUX	263.85
Full swing GDI MUX	117.15

4. **Area**

The area of full adders and 2:1 MUX using CMOS and GDI techniques are listed in Tables 7 and 8 respectively. The design 1 requires lesser area as compared to design 2 and GDI based full adder circuits.

The full swing GDI MUX requires less area as compared with CMOS design which is listed in Table 8.

7 Conclusion

The proposed full swing GDI circuit requires low power consumption, less delay and hence less PDP, which helps to improve the efficiency of the combinational circuit design. The percentage improvement for PDP is very high of GDI based full adder is over 77% with respect to design 1 and over 41% for design 2. Similarly for GDI based MUX design also has low PDP and less area requirement. Although the area required for GDI based full adder design is quite large as compared to CMOS based designs, the improvements in delay, power and PDP make it a suitable choice for many high-performance applications.

References

1. Navi K, Maeen M, Foroutan V, Timarchi S, Kavehei O (2009) A novel low-power full-adder cell for low voltage. Integration 42:457–467
2. Devadas S, Malik S (1995) A survey of optimization techniques targeting low power VLSI circuits. In: 32nd ACM/IEEE design automation conference
3. Bhattacharyya P, Kundu B, Ghosh S, Kumar V, Dandapat A (2014) Performance analysis of a low-power high speed hybrid 1-bit full adder circuit. IEEE Trans VLSI Syst 1–8
4. Vardharajan SK, Nallasamy V (2017) Low power VLSI circuits design strategies and methodologies: a literature review. In: IEEE conference on emerging devices and smart systems, pp 245–251
5. Farid N (1994) A survey of power estimation techniques in VLSI circuits. IEEE Trans VLSI Syst 2(4):446–455
6. Goyal R, Sharma S (2017) Single bit hybrid full adder cell by gate diffusion input and pass transistor logic technique. In: IEEE international conference on advances in electrical technology for green energy, pp 37–42
7. Hiremath S, Mathad A, Hosur A, Koppad D (2017) Design of low power standard cells using full swing gate diffusion input. IEEE, pp 940–945
8. Manjunath KM, Abdul Lateef Haroon PS, Pagi A, Ulaganathan J (2015) Analysis of various full-adder circuits in Cadence. IEEE J 90–97
9. Kumar A, Shrivatava B, Dardoriya A (2017) Low power high speed 1-bit full adder circuit design at 45 nm CMOS technology. In: International conference on information, communication, instrumentation and control (ICICIC), Paper Id: 262. IEEE, pp 427–431
10. Gupta R, Pandey SP, Akashe S, Vidyarthi A (2013) Analysis and optimization of active power and delay of 10T full adder using power gating technique at 45 nm technology. IOSR J VLSI Signal Process (IOSR-JVSP) 2(1):51–57
11. Kumar P, Mishra S, Singh A (2013) Study of existing full adders and to design a LPFA (low power full adder). Int J Eng Res Appl (IJERA) 3(3):509–513. ISSN: 2248-9622
12. Musala S, Reddy BR (2013) Implementation of a full adder circuit with new full swing EX-OR/EX-NOR gate. In: IEEE Asia Pacific conference on postgraduate research in microelectronics and electronics, pp 29–33
13. Morgenshtein A, Fish A, Wagner IA (2002) Gate-diffusion input (GDI)—a technique for low power design of digital circuits: analysis and characterization. IEEE Trans VLSI Syst 477–480
14. Morgenshtein A, Shwartz I, Fish A (2010) Gate diffusion input (GDI) logic in standard CMOS nanoscale process. IEEE Trans VLSI Syst 776–780
15. Thenmozhi V, Muthaiah R (2017) Optimized low power full adder design. In: International conference on networks & advances in computational technologies, July 2017, pp 86–89
16. Anugraha Rose V, Durga Devi S, Avudaiammam R (2017) Design and performance analysis of 2:1 multiplexer using multiple logic families at 180 nm technology. In: 2017 2nd IEEE international conference on recent trends in electronics information & communication technology (RTEICT), 19–20 May 2017, pp 1849–1853
17. Ahmed MA, Abdelghany MA (2018) Low power 4-bit arithmetic logic unit using full-swing GDI technique. In: 2018 international conference on innovative trends in computer engineering (ITCE 2018), pp 193–196

Single Image Rain Removal Using Convolutional Neural Network

P. Musafira and K. S. Shanthini

Abstract The visibility of the image is highly affected by the rain streaks. This can influence the performance of numerous visual tasks, for example, image enhancement, object tracking, recognition, surveillance and autonomous navigation. Process of recognition and removal of rain streaks is a quite complex and difficult task since there is no spatial-temporal content of rain streaks in a single image as compared to the dynamic video. This paper proposes an improved convolutional neural network (CNN) architecture to recognize and remove the rain streaks. Linear additive composite model is used for making rainy image model. Network is trained on the pre-processed image, which helps to enhance the learning of the network weights and training without huge increase in training data or computational resources. The experimental work shows that the CNN architecture successfully performed on both synthesized and real-world rainy images.

Keywords Rain removal · Image decomposition · CNN architecture

1 Introduction

In most of the computer vision research and applications, like image enhancement, object tracking, recognition, etc. Under rainy weather conditions, quality of the image will be reduced by rain streaks. This will reduce the visibility of an image. This paper focuses on effect of rain in image processing. Area of applications of this project includes intelligent transport systems, driver assistance systems, etc. This project can also be applied to research area like rain density measurement using single image.

The method based on single image performs two processes, rain streak recognition rain removal. Normally rain drop will affect certain areas of an image. So process of detection and removal is quite complex and difficult. Existing methods distinguish the presence of rain drop from a single image using image characteristics such as pixel

P. Musafira (✉) · K. S. Shanthini
Department of Electronics and Communication Engineering, Government College of Engineering Kannur, Kannur, India
e-mail: musafiramustafa@gmail.com

© Springer Nature Singapore Pte Ltd. 2021

S. N. Merchant et al. (eds.), *Advances in Signal and Data Processing*,
Lecture Notes in Electrical Engineering 703,
https://doi.org/10.1007/978-981-15-8391-9_10

(a) Rainy image [1] (b) Our result

Fig. 1 An example of proposed result

intensity, pixel gradient orientation and color. These methods cannot be succeeded if some characteristics of the image are matching with the characteristics of the rain streaks. To solve the failure of the existing techniques, we introduce the new technique based on convolutional neural network (CNN). CNN method can obtain the best visual results which effectively performed on both synthesized and realistic rain images.

This paper describes the current methods of related topics in Sect. 2. Section 3, discuss the proposed model of the network architecture and training details. Section 4 deals with the experiment setup and analysis and comparison of the results obtained from our network architecture. Finally Sect. 5 concludes the work with some applications (Fig. 1).

2 Literature Review

There are several approaches for removing the rain from images. They are broadly classified as single image based approaches and video based approaches. In [2, 3] they focus on removal of rain drops from video sequences. Rain removal from an image is best field research in recent years. This is different and more challenging than video based approaches. Here we concentrate only on single image based rain removal tasks.

Kang et al. introduced a morphological component analysis related approach [4]. This technique decomposes input rainy image into the low frequency (LF) and high frequency (HF) layer by applying a bilateral filter. The LF layer contains the basic image details and HF layer contains rain streaks, edges or texture information. The dictionary learning and sparse coding are used for separating HF part into rain and non-rain parts. In dictionary learning, a group of overlapping patches is extracted from HF part. Then separate atoms in the dictionary into 2 sub-dictionaries to point out rain component and textural component of HF part. Then apply K-means algorithm for separating all these atoms into two sub-dictionaries. Sparse coding is performed on HF part and then gets the rain-free output of the input rainy image by merging LF and non-rain image of HF part.

Chen and Hsu introduced a globalized low-rank appearance model for rain removal [5]. This method does not need dictionary learning. This method used similar repeating rain streak patterns and developed the tensor structure.

Chen et al. proposed a structure by accurately designing rain removal based on sparse representation [6]. This method effectively worked in color images. Initially, a rainy image is partitioned into a LF part and a HF part by applying the guided filtering [7]. HF part contains rain streaks. This method is an improved version of method [4].

Kim et al. proposed an adaptive algorithm [8]. They notice that a specific rain streak has an elongated elliptical shape with a vertical direction. So, by using this algorithm they first need to recognize an area of rain streak by examining the aspect ratio and rotation angle of the kernel at every pixel position. After this, they carry out the non-local filter on the recognized rain streak regions by choosing the weights of non-local neighbor pixels.

Luo et al. proposed an algorithm for visual effects of rain removal [9]. Rainy image is partitioned into two layers. They used non-linear generative model for rainy image modelling. This method is built on dictionary learning with mutual exclusive property.

Li et al. solve the issue of rain streak removal as a layer decomposition problem [10]. This method uses simple patch based layer priors, which is a Gaussian mixture model. This method is quantitatively and qualitatively better than previously discussed methods. But output contains some artifacts in rain streak regions.

Fu et al. solve the rain streaks by deep learning method [1]. The CNN network had a three layer networks.

This network uses image enhancement step. This CNN network performs better than other current methods.

3 Methodology

Figure 2 shows the proposed framework for removal of rain streaks. Complete steps are briefly discussed in this section.

The following are the steps for removal of rain streaks:

1. Create synthetic rainy image using photoshop [11].
2. Each input rainy image is decomposed into a low-frequency base layer and a high-frequency detail layer.
3. Build a CNN architecture.
4. The detail layer is applied to the first layer of CNN.
5. Train the high pass detail layer.
6. Rain streaks estimated from final layer of network.
7. Rain streak subtracted from detail layer.
8. Derained image is acquired by adding derained detail layer to the base layer.

The linear additive composite model is used for making rainy image model.

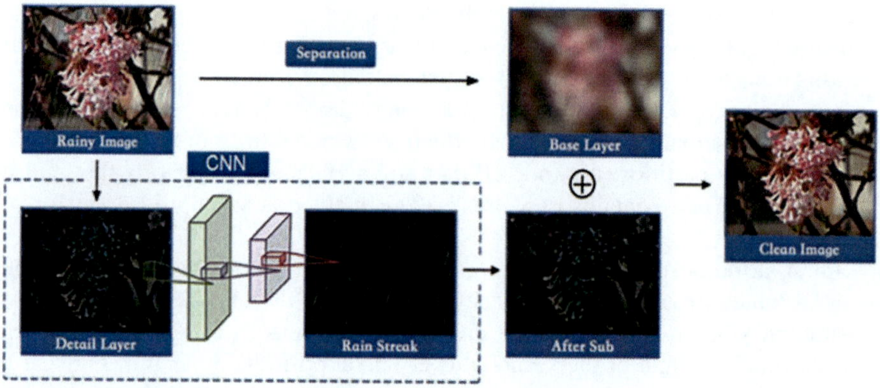

Fig. 2 The proposed framework

$$I = B + R \tag{1}$$

where I, B, and R denote the rainy image, clean background region and rain streak region respectively.

3.1 Training Data Collection

We require the dataset of rainy images and corresponding ground truth images for network training. We do not have a rainy image with corresponding ground truth images in real-world. So, create a synthetic rainy image using photoshop [11]. First, we collected 1000 clean outdoor images from the Google. These images are used to create synthetic rainy images. Fourteen different versions of rain streaks are created from each outdoor image. Fourteen rainy images have various rain streak orientations and intensity. Finally, dataset contains 1000 × 14 = 14,000 rainy images and 1000 ground truth clean images.

3.2 Image Preprocessing

Input rainy image is first passed through a low pass filter to separate the low-frequency and high-frequency components. Apply guided filter [12] on input image, we get the low-frequency base layer (I_{base}). Then subtract the base layer from input image to obtain the high-frequency detail layer (I_{detail}) which is the input of CNN.

$$I = I_{\text{base}} + I_{\text{detail}} \tag{2}$$

$$I_{\text{detail}} = I - I_{\text{base}} \tag{3}$$

The advantages of training using detail layer are that it can enhance the learning of the network weights and training without large training data or computational resources. Also, mapping range is significantly decreased. Another advantage is that time required for convergence of training is much less as compared to that of whole image. So we used detail layer as the input of CNN instead of rainy image.

3.3 Network Design

The aim of the network design is to minimizes the objective function for training, If R represents the rainy image, and I represents ground truth image, then function is given by,

$$L = \frac{1}{M} \sum_{m=1}^{M} \| f_w(R^m) - I^m \|_F^2, \tag{4}$$

where W represents the network parameter, F is Frobenius norm, m is image index and M is the number of training images.

3.4 Network Architecture

Simply cascaded blocks cannot obtain better results [1]. Here we use the residual block [13] structure to recognize the rain streaks from image details effectively. The shortcut connection can even directly transverse lossless data through all parameter layers. This shortcut connection help in handling the vanishing gradient problem in a deep network.

The proposed network is composed of repeated blocks. Each block contains many convolutional layers and a skip connection. This network transverse rain streak data from one block to next block. Architecture of this network is shown in Fig. 3. First, we apply the input detail layer to the first layers. Each parameter layers contain convolutional layers with Rectified Linear Unit (ReLU) and batch normalization (BN). Batch normalization is used to solve covariate shift [15] thereby increase the speed of the network, Rectified Linear Unit [15] is used for non-linear activation function. Pooling layer removes spatial feature information, so pooling layer is not used in our network. Rain streak is estimated from the final parameter layer. Rain streak is

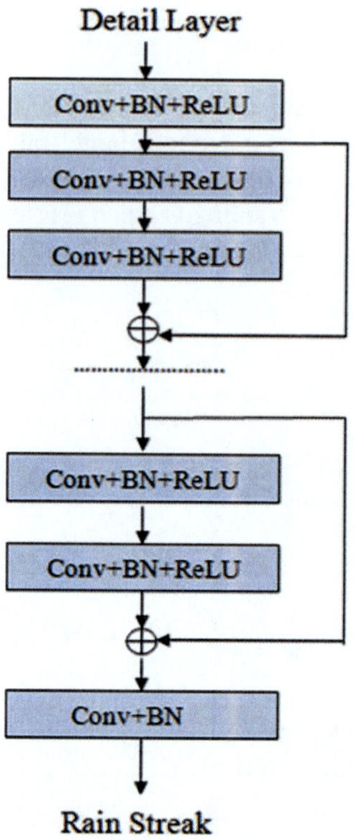

Fig. 3 The rain removal architecture of CNN network

defined to be the difference between ground truth image (I) and corresponding rainy image (R). Predicting the rain streaks is easy for learning the network. Rain streak subtracted from the detail layer, we get derained detail layer. The output image is acquired by adding this derained detail layer to the base layer (Fig. 4).

Fig. 4 The pre-processed images

3.5 Training

We optimize the network with Stochastic gradient descent (SGD). We do not have a rainy image with corresponding ground truth images in real-world. So create a synthetic rainy image using photoshop [11]. Thousand numbers of the clean outdoor images collected from Google are used to create synthetic rainy images. The fourteen thousand synthetic rainy images thus created will have various rain streak orientations and intensity. Then, use one thousand rainy-clean image patches of size 64×64 for training.

4 Experiment

We use Keras for software implementation written in python language. But Keras can not work by itself, it needs a backend for low-level operations. Thus we installed a dedicated software library TensorFlow. As a development environment, PyCharm is used. For local training, we use NVDIA GTX 1060 GPU with 6 GB graphics memory and 8 GB RAM is used. We also use google colab to train the model on Tesla K80 24 GB Graphics Processing Unit (GPU) and 12 GB RAM.

4.1 Parameters Setting

The network contains 26 layers (Network depth = 26). Use optimizer result with weight decay of 10^{-10}, a mini-batch size of 20 and momentum of 0.9. Network learning rate starts with 0.1, the number of iteration is 210 K. Filter sizes are set as 3 and filter numbers as 16. This method is applied to colour images.

4.2 Result on Synthesized Data

Figure 5 shows the complete stage result of synthesized rainy images. Figure 6 shows the result analysis of three synthetic rainy images used for testing. Synthetic rainy images are evaluated by the structure similarity index (SSIM) value. Since the ground truth as a reference image is available, we measure SSIM [17] for quantitative analysis. A higher SSIM value means that derained image is similar to the ground truth based on structural similarities. (SSIM value of ground truth equals one.) Table 1 represents the quantitative performance of synthesized test images.

Fig. 5 Result on synthetic rainy images

Fig. 6 Result analysis on synthetic rainy images

Table 1 Quantitative result analysis using SSIM

Images	Ground truth	Rainy image	Method [9]	Method [10]	Method [1]	Our results
Umbrella	1	0.75	0.81	0.86	0.88	0.88
Rabbit	1	0.72	0.77	0.79	0.85	0.91
Bird	1	0.57	0.62	0.75	0.82	0.92

(a) Rainy images [1] (b) Our results

Fig. 7 Results obtained on rainy images in real-world

4.3 Result on Real-World Data

Figure 7 shows two qualitative results on rainy images in real-world. These images can not be evaluated by structure similarity index (SSIM) value. In the case of real-world rainy images, there is no reference image for calculating SSIM value. It is clear from Fig. 7 that our method shows better visual result in real-world rainy images. So this network works effectively on real-world applications.

4.4 Comparison with Another Rain Removal Methods

In Table 1, we show quantitative result analysis using SSIM on three synthetic images. In our method, we use a deeper structure. That is we increase depth by adding more hidden layers and increase the network breadth by using more filters in each hidden layer. As compared to the method [1], there are only three layers. Our network performance is better than the three layer network. In direct network [1], there is a chance of gradient vanishing problem. So we developed a network that contains short cut connection to skip over some layers, which helps to remove the vanishing gradient problem by using activation from a preceding layer up to the next layer learn its weights. So this can make the learning process much easier. Our network generates better results without having any image enhancement techniques. Training on the detail layer rather than the rainy image can enhance the learning of the network weights and training without large training data or computational resources. Therefore we obtained the result with less training data instead of using large amount of data. Using skip connection and preprocessed images can help deraining and better output quality.

5 Conclusion

In this work, an improved CNN architecture for single image rain removal is proposed. This technique learns the rain streaks feature from detailed layer of rainy images. Though the network is trained on synthetic rainy/non-rainy image patches, it works well on both synthetic images and realistic rainy images. Also, the results indicate that our technique outperforms the existing techniques.

In future, we plan to use realistic image dataset instead of synthetically generated dataset for single image rain detection. The developed system can be very well be extended for more complex rain and fog conditions. We also plan to find the possibility of using this network in real-world problems. For example, network modification to perform rain density measurement.

References

1. Fu X, Huang J, Ding X, Liao Y, Paisley J (2017) Clearing the skies: a deep network architecture for single-image rain removal. IEEE Trans Image Process 26(6):2944–2956
2. Garg K, Nayar SK (2004) Detection and removal of rain from videos. In: International conference on computer vision and pattern recognition (CVPR)
3. Barnum PC, Narasimhan S, Kanade T (2010) Analysis of rain and snow in frequency space. Int J Comput Vis 86(2–3):256–274
4. Kang LW, Lin CW, Fu YH (2012) Automatic single-image-based rain streaks removal via image decomposition. IEEE Trans Image Process 21(4):1742–1755
5. Chen Y-L, Hsu C-T (2013) A generalized low-rank appearance model for spatio-temporally correlated rain streaks. In: Proceedings of IEEE international conference on computer vision, pp 1968–1975
6. Chen D-Y, Chen C-C, Kang L-W (2014) Visual depth guided color image rain streaks removal using sparse coding. IEEE Trans Circuits Syst Video Technol 24:1430–1455
7. He K, Sun J, Tang X (2013) Guided image filtering. IEEE Trans Pattern Anal Mach Intell 35(6):1397–1409

8. Kim JH, Lee C, Sim JY, Kim CS (2013) Single-image deraining using an adaptive nonlocal means filter. In: IEEE ICIP, pp 914–917
9. Luo Y, Xu Y, Ji H (2015) Removing rain from a single image via discriminative sparse coding. In: International conference on computer vision (ICCV)
10. Li Y, Tan RT, Guo X, Lu J, Brown MS (2016) Single image rain streak decomposition using layer priors. IEEE Trans Image Process 26(8)
11. https://www.photoshopessentials.com/photo-effects/rain/
12. Zhang Q, Shen X, Xu L, Jia J (2014) Rolling guidance filter. In: European conference on computer vision (ECCV)
13. He K, Zhang X, Ren S, Sun J (2016) Deep residual learning for image recognition. In: CVPR
14. Ioffe S, Szegedy C (2015) Batch normalization: accelerating deep network training by reducing internal covariate shift. In: ICML
15. Wang Z, Bovik AC, Sheikh HR, Simoncelli EP (2004) Image quality assessment: from error visibility to structural similarity. IEEE Trans Image Process 13(4):600–612
16. Kingma DP, Ba J (2014) Adam: a method for stochastic optimization. Preprint arXiv:1412.6980
17. Krizhevsky A, Sutskever I, Hinton G (2012) ImageNet classification with deep convolutional neural networks. In: NIPS

Ring Oscillator-Based Physical Unclonable Functions

Shruti Sakhare and Dipti Sakhare

Abstract Physical unclonable function is playing an important and efficient role in system security. Ring oscillator is basically a delay-based PUF, and during fabrication process variations, the delay introduced is used for detecting secrecy of the PUF design. A framework of ring-oscillator PUF is built to check the unpredictability of the response based on the challenge created by a 4 bit-LFSRs. Simulation results of ring-oscillator PUF show that the response bits generated are unique for every challenge. Experimental results of National Institute of Standard technology Test (NIST) Suite demonstrate that the PUF's secrecy generated by the ring-oscillator PUF is random, and it varies from different FPGA platforms. The ring-oscillator PUF is evaluated according to the metrics namely security, uniqueness, and randomness of the response bits generated. The RO-PUF uniqueness and randomness calculated are more efficient in comparison with any other RO-PUF implemented.

Keywords Physical unclonable function · Linear feedback shift register · Field programmable gate arrays

1 Introduction

Hardware is playing very important role in terms system security. First, the concept of PUF was introduced by Pappu et al. [1] and was introduced to enhance the system security. Delavar et al.'s [2] physical unclonable functions (PUFs) are advance security primitives mainly used for security applications. These security-related applications are secret key generation and key extraction, integrated circuits (IC) obfuscation, IP protection, and reconfiguration binding. Physical unclonable function (PUF) has

S. Sakhare (✉)
VLSI Design and Embedded System, MIT Academy of Engineering, Pune, India
e-mail: svsakhare@mitaoe.ac.in

D. Sakhare
School of Electrical Engineering, MIT Academy of Engineering, Formerly MAE Alandi, Pune 412105, India
e-mail: dysakhare@etx.maepune.ac.in

© Springer Nature Singapore Pte Ltd. 2021

S. N. Merchant et al. (eds.), *Advances in Signal and Data Processing*,
Lecture Notes in Electrical Engineering 703,
https://doi.org/10.1007/978-981-15-8391-9_11

certain properties such as uniqueness, unclonability, unpredictability, and robustness. These properties make PUF very different. The reason for the fact that two identical integrated circuits (ICs) are not the same due to random variations in silicon-based PUFs is that differences in fabrication process are difficult to predict, and complexity makes fabrication process more advanced. The overall motive behind selecting a physical unclonable function (PUF) is that it enables security, generates a unique response, and can be used for low-cost authentication with respect to system security. With the increasing system security, these PUF circuits will be in much demand in the near future.

2 Physical Unclonable Function and Its Types

Gao et al.'s [3] physical unclonable functions (PUFs) are the tiny piece of circuit embedded in the design which removes the fabrication variations of a silicon chip and uses this kind of intra-physical features for safety applications. The PUFs of silicon are further classified into different types namely ring oscillator [4], configurable RO-PUF [5], arbiter PUF [6], butterfly PUF [7], flip-flop PUF [8], SRAM PUF [9], and bistable PUF [10]. These PUFs are used for security purposes like security key generation, circuit identifiers, and cryptography. This framework of RO-PUF generates bit streams based on the delay introduced at the input of the ring oscillators. Delay introduced is the variations created by the manufacturers. There are many proposals to improve system security and randomness. Proposed work includes the implementation of RO-PUF on Nexsys 4™ DDR board in order to improve the system security and to test the randomness of the output response bits. This approach is more efficient due to the unique combinations of the bits.

The following is a summary of contributions to this paper: (1) Simulation and synthesis of ring oscillator PUF. (2) Generation of response bits to achieve security of the circuit. (3) Experimental results of the National Institute of Standard Technology Test (NIST) Suite indicate the proposed RO-PUF that generates more random bits than existing PUFs.

3 Proposed Work

Ravishankar's [11] basic ring oscillator is the combination of an odd n number of NOT and a NAND gate as shown in Fig. 1. The frequencies of operation of two identical ring oscillator varies due to the process variations and are unique for each ring oscillator. This property of ring oscillator is used to implement a ring-oscillator physical unclonable function.

Rahaman et al.'s [12] RO-PUF consists of N identical ROs, multiplexers, comparator, and two counters. Variations in the manufacturing process cause each RO to oscillate with a unique frequency. The input challenge is further forwarded

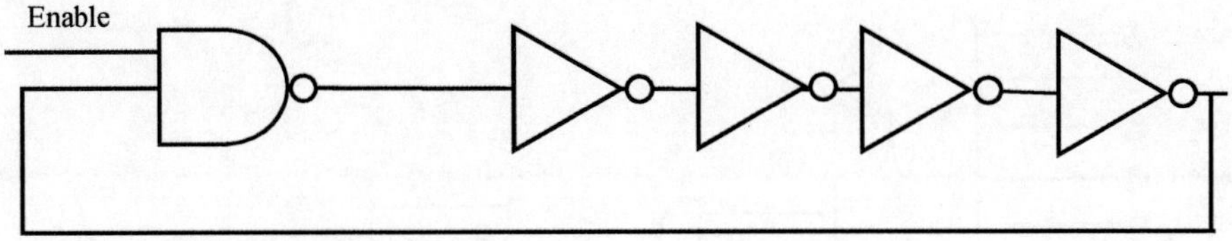

Fig. 1 Basic ring oscillator circuit [11]

to the multiplexer in order to select one pair of RO. Counter counts the number of oscillations for a specified period of time. The counter's output is compared to the output generated.

The comparator output is set to '1' and '0' depending on which oscillator is faster from the selected RO pair. The proposed block diagram of RO-PUF is shown in Fig. 2, which consists of 16-pair of ring oscillator, two multiplexer, two 4-bit LFSRs (for creating challenges), combinational circuits (for reset), and two counters. At the output, the response is captured in '1' and '0' with the overflow bit.

Figure 3 shows implementation of RO-PUF, 16 ring oscillators in combination with other components to generate the output. Multiplexer is used to select which ring oscillators feed in the counter. The ring oscillator used here has a large number of inverters added to the loop, which in turn results in oscillator in the circuit. The Frequencies of operation of two ROs vary due to process variations and are unique to each other. This property of ROs is being used to implement ring-oscillator PUF. Multiplexers are used to select which ring oscillator is to be selected. At the select input of the multiplexer, 4-bit LFSR is used in order to create challenges response

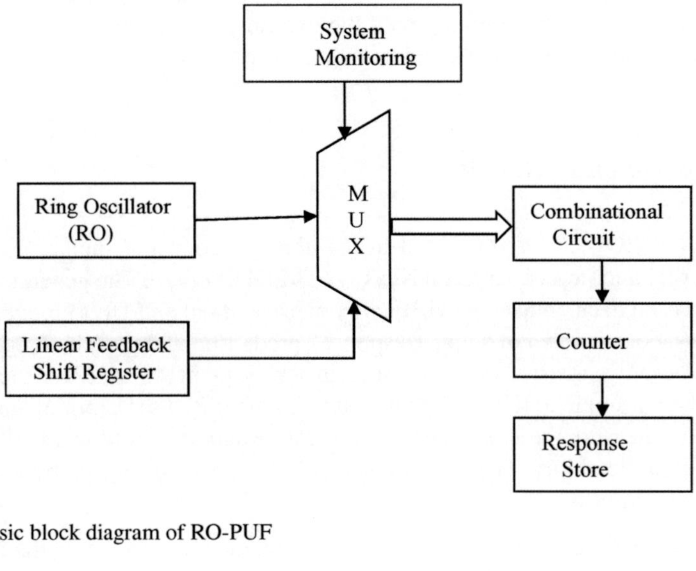

Fig. 2 Basic block diagram of RO-PUF

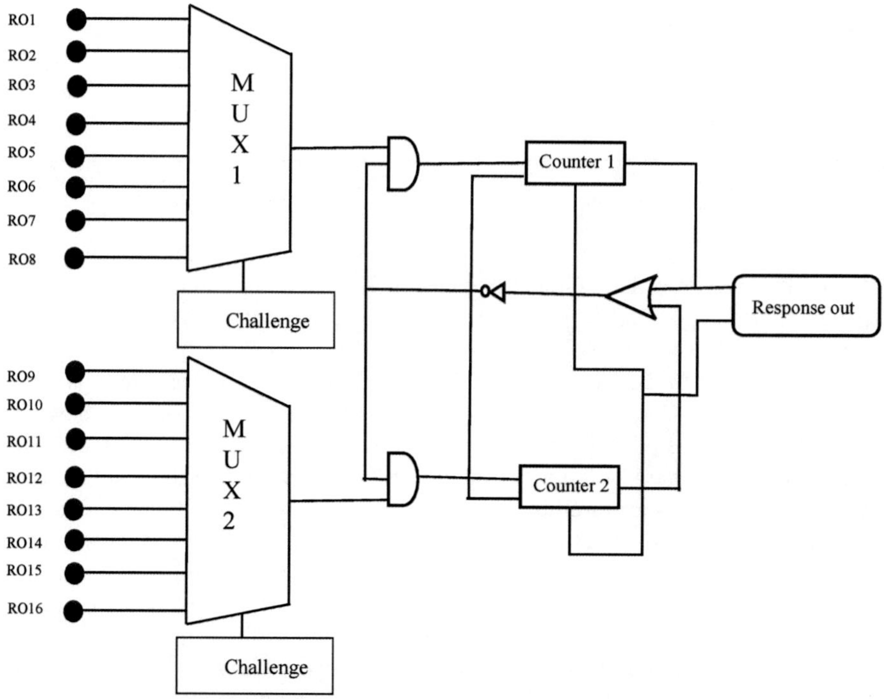

Fig. 3 Ring-oscillator physical unclonable function

pairs (CRPs). The 4-bit linear feedback shift registers (LFSRs) start creating challenges. The counter starts counting with the overflow bit. As soon as the counter-1 overflows, response stored is '1', and counter-2 response is '0'.

3.1 *Experimental Result*

The proposed RO-PUF as illustrated in the above section is simulated on Xilinx Vivado 2014.4 and implemented on Nexsys 4TM DDR board. The generated output response is used to calculate the PUF quality metrics based on PUF's uniqueness and randomness. For a number of output (response), prediction of input (challenges) is not possible. These two metrics are mostly used to assess quality of the PUF. Uniqueness measures how a single PUF is differentiated from other PUFs. Hamming distance is calculated to measure the uniqueness of the chip. Randomness tests a PUFs output response unpredictability. The randomness of PUF can be measured by a practical approach such as NIST.

(A) Simulation Results The simulation results obtained on simulating the RO-PUF on Xilinx Vivado 2014.4 is shown in Fig. 4. The simulated RO-PUF design is able

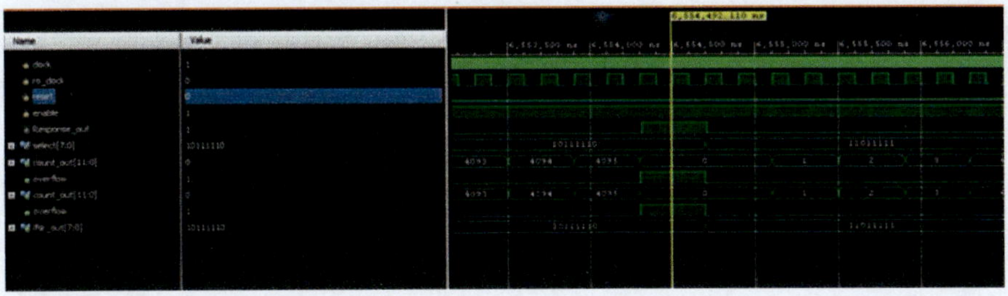

Fig. 4 Simulation result of RO-PUF

to output the response bits. The operating clock frequency is 50 MHz. The ring-oscillator clock frequency is used to select which RO is selected at the output of the multiplexer. The ring-oscillators clock frequency depends on the delay applied. The RO clock frequency ranges from 10 to 20 MHz. The Select [7:0] signifies that the challenges of 4-bit LFSR are applied to the MUX-1 and MUX-2. The select lines are used as challenges with some seed value '1111' and '0000' (initial value) as LFSR starts its shifting operation on its own. Two counters, Counter-1, and Counter-2, of 12-bit work on the overflow bit generation operation. Whichever counter overflows first, the response will be captured. This response bits generated helps to achieve a higher level of secrecy as it is very difficult to predict the challenges applied at select lines of multiplexers. This helps to stop counterfeiting of the chip.

Figure 5 shows sample response bit for a challenge applied at the input of the multiplexers. The challenge bits are calculated manually because 10 MHz clock frequency was getting divided so as to catch the response bits as '1' and '0' on every overflow of the counter. Table 1 shows the samples of challenges created and the response bit generated with the overflow of the counters

(B) Uniqueness of the PUF—Response Uniqueness is a metric of how clearly PUF can be identified within a group of FPGAs. For many security applications, the PUF output information is used secret key generation, and for cryptography, this shows that different chip shows different PUF output. The RO-PUF is implemented on 2 FPGAs. Uniqueness is defined by U. Maiti and Schaumont [13] that the Hamming distance between the two n-bit responses, $R1$, and $R2$ is generated by a PUF from a

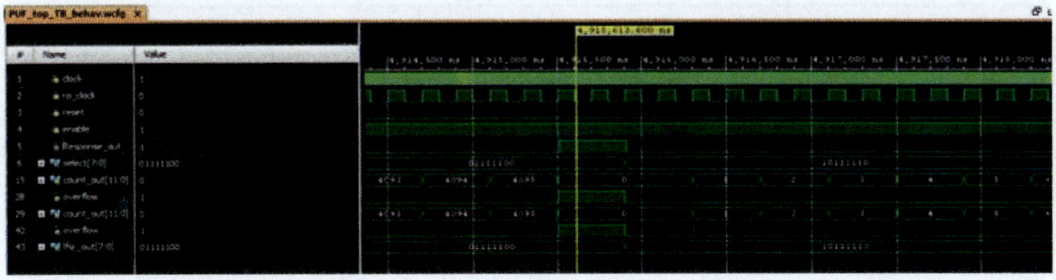

Fig. 5 Response bit generation by ILA

Table 1 Sample value of challenge response pairs (CRPs)

Sr. no.	Challenges	(Response) Counter-1	(Response) Counter-2
1	1 0 0 0 0 0 0 0	1 (Overflow)	0
2	1 1 1 1 0 0 0 0	0	1 (Overflow)
3	1 1 1 1 1 1 1 1	1 (Overflow)	0
4	0 1 0 1 0 1 0 1	1 (Overflow)	0
5	0 0 0 1 0 1 1 0... (*n* number of challenges)	0	1 (Overflow)

pair of FPGAs, $F1$ and $F2$, and this is a good approximation of the PUF's uniqueness.

$$U = \frac{2}{k(K-1)} \sum_{i=1}^{i=K-1} \sum_{J=i+1}^{j=k} \frac{hij}{n} * 100\% \tag{1}$$

hij = Hamming distance between two *n*-bit response, and *i* and *j* are, respectively, for a challenge *C*. Zhang et al. [14] uniqueness demonstrates how unique the PUF output can be, deciding the response value of the PUF. If the different PUFs produce the same response when fed with the same challenge, this type of PUFs is not acceptable. Tables 2 and 3 show the calculated Hamming distance of a number of response bits generated and comparison of pairwise Hamming distance (HD) on two different platforms. Figure 6 is the histogram of the HD outputs of the implemented RO-PUF, where the bell-shaped curve is obtained. This shows that the PUF outputs are unique, and two platforms will not output the same response.

The percentage of bits in Tables 2 and 3, it is observed that as the Hamming distance increases, the uniqueness varies.

(C) National Institute of Standard Technology Test Suite (Randomness) Andrew Rukhin et al. [15] the NIST believes that these procedures are useful in detecting deviations of binary sequences from randomness. Randomness is a probabilistic property, and it can be described in terms of probability. Ritter [16] randomness can be calculated using a variety of practical methods, depending on the statistical complexity and transformation. Depending on the length of the input for ensuring

Table 2 Hamming distance; % of uniqueness

HD	1	2	3	4	5
%	0.01	0.20	0.31	0.41	0.50

Table 3 Hamming distance; % of uniqueness

HD	6	7	8	9	10
%	0.6	0.7	0.80	0.9	0.10

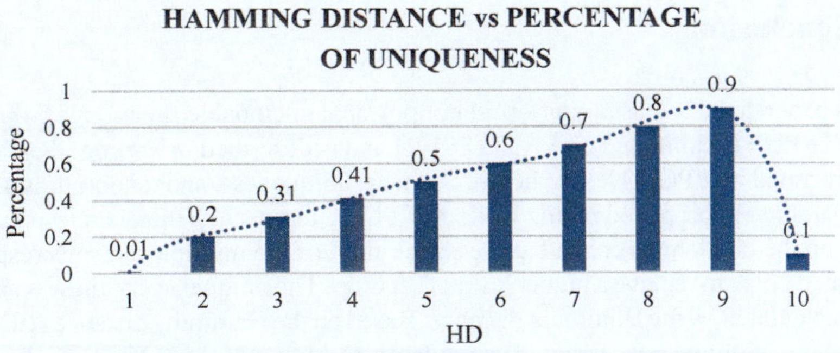

Fig. 6 Hamming distance and percentage of uniqueness

accuracy for the NIST, statistical test is performed on different bits streams obtained. For the test sequence "*P*-value," the criteria for threshold value is 0.001, and the test results obtained are much higher than the threshold value.

$$\mathbf{P-Value > 0.001\ (Random\ Sequence)}$$

$$\mathbf{P-Value < 0.001(Non-RandomSequence)}$$

The NIST test results for the bit stream obtained are given in Table 4. If the *P*-value obtained is >0.001, the sequence is random, otherwise the sequence is non-random.

The NIST tests result consist of different parameters. This randomness of bits justifies the occurrence and repetition of events, the occurrence of '1′ and '0′ occurrence, the sum of partial sequences occurring in the sequence, the number of sequences and zeros of different lengths from a random sequence of bits, and the number of overlapping patterns expected from a normal random sequence.

Table 4 NIST tests results of 5-bit streams

	1	2	3	4	5
Frequency	0.984573	0.981268	0.983471	0.979065	0.976262
Block frequency	0.524011	0.524206	0.52406	0.523429	0.52428
Cumulative sums	0.003991	0.004409	0.132472	0.023671	0.101678
Runs	0.977963	0.974659	0.986776	0.974658	0.979064
Serial	1.000000	1.000000	1.000000	1.000000	1.000000

4 Conclusion

In this paper, a delay-based ring-oscillator physical unclonable function is proposed. This RO-PUF is simulated on Vivado 2014.4 and is described in Verilog. RO-PUF is demonstrated on FPGA Nexsys board. Security, uniqueness, and randomness are the PUF parameters proposed in this work. RO-PUF attempts to generate a response bit based on the challenges created at the select input of the multiplexer. This response generated helps to receive a higher level of security. The uniqueness of these response bits is calculated by the Hamming distance. Based on the Hamming distance (HD), the percentage of uniqueness varies. The randomness of the response bits are calculated on the NIST test suite, and the results obtained are much random as compared to the threshold value. The proposed RO-PUF limits to simulation and implementation of RO-PUF on FPGA platform.

References

1. Pappu R, Recht B, Taylor J, Gershenfeld N (2002) Physical one-way functions. Science 297(5589):2026–2030
2. Delavar M, Mirzakuchaki S, Mohajeri J (2016) A ring oscillator-based PUF with enhanced challenge response pairs. Can J Electr Comput Eng 39(2)
3. Gao M, Lai K, Qu G (2014) A highly flexible ring oscillator PUF. In: Proceedings of the 51st design, automation conference. ACM/IEEE, pp 1–6
4. Yin C-E, Qu G (2009) Temperature-aware cooperative ring oscillator PUF. In: International Workshop on Hardware-Oriented Security and Trust, HOST/IEEE
5. Xin X, Kaps J-P, Gaj K (2011) A configurable ring-oscillator-based PUF for Xilinx FPGAs
6. Edward Suh G, Devadas S (2007) Physical unclonable functions for device authentication and secret key generation. ACMDAC
7. Kumar SS, Guajardo J, Maesyz R, Schrijen G-J, Tuyls P (2008) Extended abstract: the butterfly PUF protecting IP on every FPGA. International Workshop on Hardware-Oriented Security and Trust, HOST/IEEE, pp 67–70
8. Maes R, Tuyls P, Verbauwhede I (20058) Intrinsic PUFs from flip-flops on reconfigurable devices. Workshop on Information Workshop on System Security WISSec
9. Holcomb DE, Burleson WP (2009) Power-up SRAM state as an identifying fingerprint and source of true random numbers. IEEE Trans Comput 58(9):1198–1210
10. Chen Q, Csaba G, Lugli P, Schlichtmann U, Rührmair U (2011) The bistable ring PUF: a new architecture for strong physical unclonable functions. International Workshop on Hardware-Oriented Security and Trust, HOST/IEEE
11. Shankar YR (2010) PUFs—an extensive survey. Master Thesis. George Manson University, Fairfax
12. Rahman T, Forte D, Fahrny J, Tehranipoor M (2014) ARO-PUF: an aging resistant ring oscillator PUF design. In: Design automation, and test in Europe conference and exhibition, pp 1–6
13. Maiti A, Schaumont P (2010) Improving the quality of a physical unclonable function using configurable ring oscillator. In: International Conference on FPGA, pp 703–707
14. Zhang JL, Qu G, Lv Y-Q, Zhou Q (2018) A survey on silicon PUFs and recent advances in Ring oscillator PUFs. J Comput Sci Technol 29, ver. 4:664–678
15. Rukhin A, Soto J et al (2010) A statistical test suite for random and pseudorandom number generators for cryptographic applications. NIST Special Publication 800-22 Revision La

16. Ritter T, [Online]. https://www.ciphersbyritte.com/RES/RANDTEST.HTM. Randomness tests a literature survey

A Robust Approach of Estimating Voice Disorder Due to Thyroid Disease

Namrata V. Kanase, Satyajit A. Pangoankar, and Ashish R. Panat

Abstract Thyroid is butterfly-shaped gland present in the lower anterior of the neck. The main root of thyroid disease is the improper working of thyroid gland. Thyroid disease is mainly categorized into two types, i.e., hypothyroidism and hyperthyroidism. In this study, voice samples for two disorders—hypo and hyper along with normal voice samples are considered. A databank is created for three classes—normal, hypo, and hyper. The structure of the robust approach of diagnosing thyroid disease contains four stages. In the first stage, preprocessing is performed by considering framing, windowing, and filtering. In the second stage, feature extraction is performed by using mel-frequency cepstral coefficient (MFCC) method. In the third stage, classification is achieved by using combined classifier, i.e., support vector machine (SVM), and hidden Markov model (HMM). In the fourth stage, performance evaluation for diagnosing thyroid disease is achieved by estimating accuracy, confusion matrix, and precision. The classification accuracy of a robust approach for diagnosing thyroid disease is obtained about 97.28%.

Keywords Thyroid gland · MFCC · SVM · HMM · Confusion matrix

1 Introduction

Thyroid gland is one of the largest endocrine glands in human body. The main job of thyroid gland is to create and produce thyroid hormone to fulfill the need of human body as well as controls the metabolism of the body. The normal weight

N. V. Kanase (✉) · S. A. Pangoankar
Department of Electronics Engineering, MIT Academy of Engineering, Alandi, Pune, India
e-mail: nvkanase@mitaoe.ac.in

S. A. Pangoankar
e-mail: sapangoankar@entc.maepune.ac.in

A. R. Panat
Department of Electronics and Communication Engineering, MITADT University, Rajbaug, Pune, India
e-mail: ashish.panat@mituniversity.edu.in

© Springer Nature Singapore Pte Ltd. 2021

S. N. Merchant et al. (eds.), *Advances in Signal and Data Processing*,
Lecture Notes in Electrical Engineering 703,
https://doi.org/10.1007/978-981-15-8391-9_12

of thyroid gland is about 30 g [1]. Thyroid gland has main two active hormones, i.e., thyroxine and triiodothyronine. These two thyroid hormone effects on energy production, energy regulation, protein production, and temperature regulation of human body [2]. The most common root of thyroid disease is improper working of thyroid gland and iodine deficiency, eating habits, existing life styles, and heredity problem. Due to dysfunction of thyroid gland, various thyroid-related diseases occur [3]. According to survey held in India, 15–30% of peoples have thyroid disease [4]. In India, survey represents women are four times more susceptible to thyroid disease than men. Mainly thyroid disease is divided into two categories, i.e., hypothyroidism and hyperthyroidism. In hypothyroidism, thyroid gland does not create and produce sufficient amount of hormone to human body. Due to this hypothyroidism condition, many thyroid-related problems occur like hoarseness in voice, weight gain, joint pain, and infertility. In hypothyroidism, thyroid glands produce excess quantity of thyroid hormone which is not required to human body. Due to that excess creation of thyroid, hormone causes many thyroid-related disease like Plummer's graves and thyroiditis disease [5]. There are many diagnosis tests that are available to check the categories of the thyroid disease (hypo or hyper). In medical practice, various types of approaches are used to detect thyroid disease like blood examination, clinical evaluation, imaging examination, tissue biopsy, and self-assessment by speech therapist [2].

This area has enormous potential for conducting research and building new system. There are many systems developed for detecting the thyroid disease. This work came across task done by different persons using various methods to differentiate the type of thyroid disease. After careful study of pros and cons for each method and their outcomes, we choose the best technique and features from their work.

Gour et al. [4] have discussed various methods that are available to diagnose the thyroid disease. Methods discussed are acoustic voice analysis and their parameters, statistical method, biospectral analysis, and anomalous frequency analysis using music. By considering the various parameters like jitter, shimmer, fundamental frequency, standard deviation, harmonic-to-noise ratio, voice turbulence index, etc. According to this parameter analysis, it is observed that vocal cord is affected and the voice parameters are changes in thyroid patients.

Dogantekin et al. [6] have developed the diagnosis system that depends on thyroid gland (DSTG) for diagnosis of thyroid disease. In this study, principle component analysis (PCA) is used for feature extraction and least square support vector machine (LS-SVM) classifier is used. Further, performance evaluation is performed by using K-fold cross-validation method and confusion matrix. By using this technique, system got 97.67% of accuracy.

Zabidi et al. [7] have discussed the analysis of hypothyroidism in newborn infant cry. In this system, preprocessing a signal processing techniques is used. In preprocessing, segregation of unvoiced and voiced signal is carried out. In signal processing, MFCC is used for feature extraction from newborn cry signal and fisher ration is used for feature selection. By using this system, differentiate between infants suffering from hypothyroidism and healthy infants.

In this study, the robust approach of voice disorder system on thyroid gland is used for diagnosing thyroid disease. This system contains four levels, which are preprocessing, feature extraction, classification, and performance evaluation. In preprocessing, framing, windowing, and filtering are carried out. Feature extraction is parametrical representation of original signal. In proposed system, mel-frequency cepstral coefficient (MFCC) technique is used for feature extraction. Further classification is carried out. Classification is the procedure that is based on the feature vectors which are made by the combinations of two or more algorithms. By using one or more classifier, system improves the accuracy of the signal. In this work, combined classifier is used, i.e., support vector machine (SVM) and hidden Markov model (HMM). Further, performance evaluation is performed by considering confusion matrix, precision, and accuracy.

The paper is organized as follows: In Sect. 2, the proposed methodology of diagnosing thyroid disease is discussed, in that the thyroid gland database used in this study is explained also MFCC for feature extraction and HMM + SVM classifier are explained, respectively. In Sect. 3, implementation and results of thyroid disease diagnosis are presented. In Sect. 4, the conclusion is presented.

2 Proposed Methodology

2.1 Block Diagram

See Fig. 1.

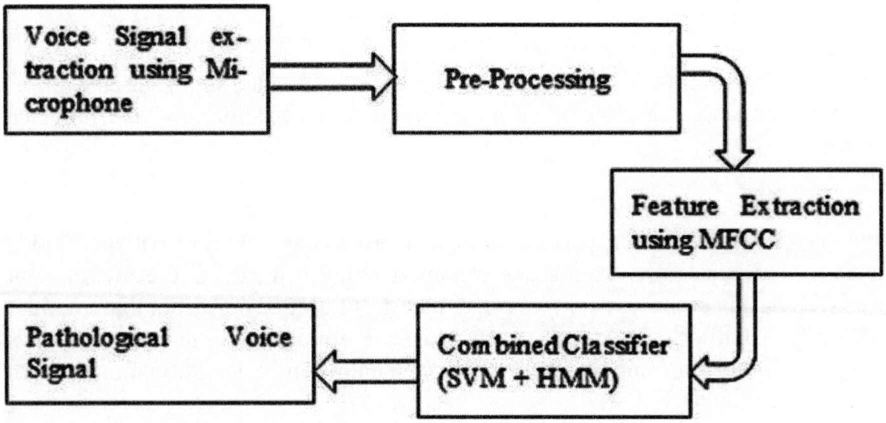

Fig. 1 Block diagram of the proposed system

Table 1 Number of testing and training data for each class

Data bank	Number of training data	Number of testing data
1. The normal function class (Class-1)	60	15
2. The hypo function class (Class-2)	90	30
The hyper function class (Class-3)	10	3

2.2 *Database*

The data of voice samples is collected from Dr. Milind Patwardhan at Dr. Patwardhan's Endocrine and Diabetes Research Centre and Pathology Laboratory, Miraj, Maharashtra, India. 150 voice samples of male and female with and without thyroid disease have been collected. In this database, there are three classes and 150 samples. These three classes are normal, hypothyroidism (hypo), and hyperthyroidism (hyper). The 50 samples of 150 belong to normal function class namely as Class-1. The 90 samples of 150 belong to hypo function class namely as Class-2. The 10 samples of 150 belong to hyper function class namely as Class-3. The voice samples are recorded in Wav format. A sampling frequency is chose for this study is 44,100 Hz (Table 1).

2.3 *Algorithm*

The main algorithm for diagnosing thyroid disease is divided into four categories, i.e., preprocessing, feature extraction, combined classification, and performance evaluation.

Preprocessing

Preprocessing is important part of the signal processing. Thyroid patients' voice signal is continuous time varying and responsive to noise, mostly as the environment is noisy. By using preprocessing, we can reduce the unwanted noise. For that framing, windowing and filtering processes are carried out. Framing defines the short windows of signal. Windowing function is used for calculating finite length or finite length weight of the signal (Fig. 2).

$$Y(n) = X(n) \cdot W(n) \tag{1}$$

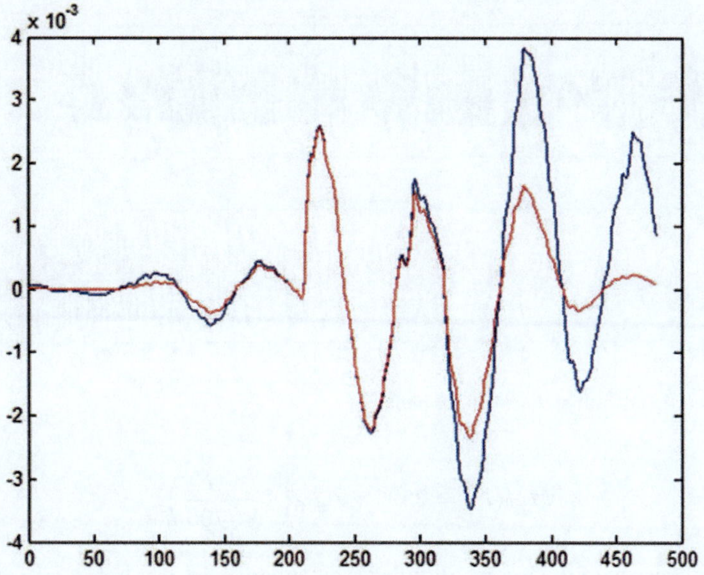

Fig. 2 Thyroid patients voice signal with windowing

$$W(n) = \begin{cases} 0.54 - 0.46\cos\left(\frac{2\pi n}{N-1}\right), & 0 \le n \le N-1 \\ 0 & \text{otherwise} \end{cases} \tag{2}$$

Filtering is used to discard the unwanted noise content present at the voice signal. This is the overall preprocessing process.

Feature extraction

Mel-frequency cepstral coefficient (MFCC) is most commonly used feature extraction technique in speech signal processing applications. The MFCC mimics the human perception of loudness and pitch of human auditory system [8]. Basically MFCC is perception-based feature extraction technique. This includes plotting of acoustic frequency data to linear perceptual scale is called as the "mel-scale."

After preprocessing, taking the fast Fourier transform (FFT) of the processed signal converts time domain signal into frequency domain.

$$Y(w) = H(w) * X(w) \tag{3}$$

The sampling frequency of the signal is 44100 Hz, and the length of FFT signal is 2048 points to optimize the frequency and time resolution of the voice signal. The result of log spectrum for each FFT bin is in the mel-scale. Due to that, triangular filter equal to 300 mels is used. So the first filter will start from 300–600 mels. And here 50% of overlapping is used. This spectral vector is passing through 28 mel-scale logarithmic filter bank (Fig. 3).

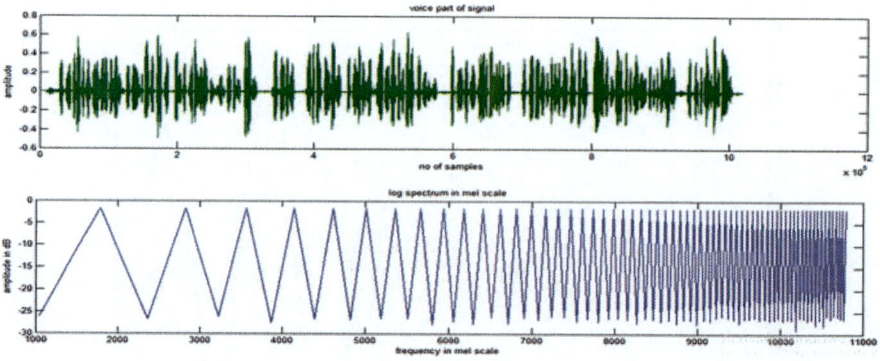

Fig. 3 Voice signal with mel-scale filter bank

$$Mel(F) = 2595 * \log_{10}\left(1 + \frac{F}{700}\right) \tag{4}$$

Spectral energy of the log magnitudes is calculated by considering output of mel-filter. Then, mel-frequency cepstral coefficients are calculated by performing DCT of log magnitude of spectral energy (Fig. 4).

$$C_n = \sum_{k-1}^{k} \log(Sk) \cos\left[n\left(k - \frac{1}{2}\right)\frac{\pi}{k}\right] \tag{5}$$

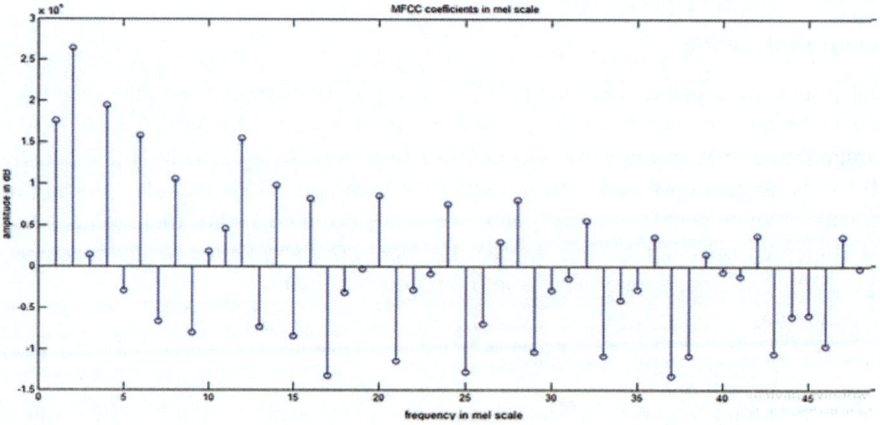

Fig. 4 MFCC coefficients in mel-scale

Classification

Classification is the most significant part in speech signal processing. In this work, combined classification technique is used. Hidden Markov model (HMM) and support vector machine (SVM) algorithms are combined. HMM is used for selecting features which is extracted by using MFCC method and SVM is used for training and testing data.

Hidden Markov model (HMM)—A stochastic procedure has the Markov property if the conditional probability distribution of the future state of the process (conditioned on both the previous and present state) depends only on the present state, not on the order of events that proceeded it. A process with this property is called a Markov process.

Generally, the term "state" is used to refer the hidden states and "observations" are used to refer observed state. HMM emits the observations based on the emission probability distribution from a given state and state transition occurs in a sequence as per the transition probability.

HMM to be generating the observation sequence and the state sequence that has the maximum probability of generating the observed sequence is considered the output of HMM of those problems where we are required to determine the hidden state sequence (Fig. 5).

Support vector machine (SVM)—SVM is basically used for classification and regression. The main aim of SVM algorithm is fine the hyperplane in *N* the number of features that distinctly classifies the data features [9]. To calculate the maximum margin between the data features and hyperplane, the loss function is used. The loss function is calculated by using below function;

$$C(x, y, f(x)) = \begin{cases} 0 & \text{if } y * f(x) \geq 1 \\ 1 - y * f(x) & \text{else} \end{cases} \tag{6}$$

Hyperplane is used as decision boundaries that support to classify the data features. In this work, three classes are used so that hyperplane is two-directional. The

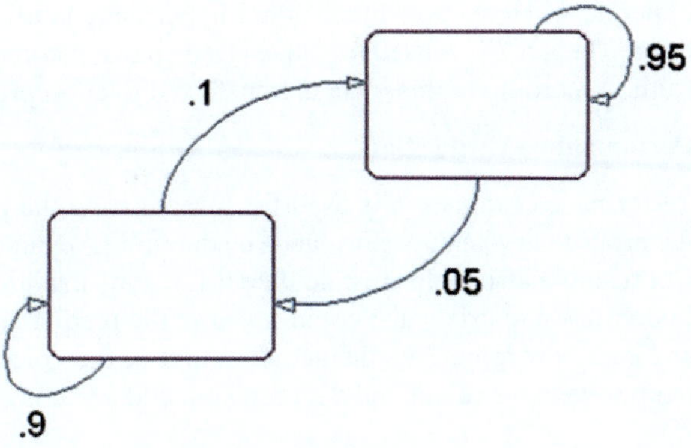

Fig. 5 State diagram of HMM classifier

following equation is used for calculating the accuracy of SVM.

$$Y(x) = \text{sign}\left[\sum_{i=1}^{L} \alpha_i \gamma_i \Psi(x, x_i) + b\right] \quad (7)$$

where
ψ = kernel function
α_i = Lagrange multiplier
b = real constant.
x_i = support vectors.
If the data is separable,

$$Y_i\left[W^{\mathrm{T}}\varphi(X_i) + b\right] \geq 1, \quad i = 1, \ldots, L \quad (8)$$

If data is non-separable, it is used as a slack variable ξ_{i}

$$Y_i\left[W^{\mathrm{T}}\varphi(X_K) + b\right] \geq 1 - \xi_i \quad \xi_i \geq 0 \;\; i = 1, \ldots, L. \quad (9)$$

The SVM cracks the optimization problem by using following equation.

$$\text{Min}_{w,b,e} J_{L,S}(w, b, e) = \left(\frac{1}{2}\right) W^{\mathrm{T}} W + C \sum_{i=1}^{1} -\xi_i \quad (10)$$

Combined approach—In this proposed system, HMM is used for the feature selection. HMM is the selected feature by considering the "state" changes and emitted output. The state represents hidden state, and emission output shows output emitted by that state. Considering both outputs, observational and conditional probability is calculated. And which output having highest probability is considered as an output of HMM classifier. For this system, selected probabilities are 0.95: 0.05; 0.9: 0.1. According to these probabilities, HMM selects the features. Basically, SVM supervises algorithm so that it selects the maximum margins between two-dimensional hyperplane. Here, two-dimensional hyperplane is used because the class length is three. Then, SVM is used for supervised so that it compares the data with the ground truth. According to this, data is transferred to appropriate category.

Performance Evaluation

Performance evaluation is used for investigating the roughness of the classifier. In the proposed system, performance evaluation is obtained by using confusion matrix. Confusion matrix helps to calculate test results that are more accurate and valuable. In confusion matrix, each column shows the predicted outputs of the voice samples and each row represents the actual output of the voice sample. By using confusion matrix, we have calculated the precision and accuracy of classifier.

Table 2 Confusion matrix for 70 voice samples

Confusion matrix	Normal	Hypo	Hyper
Normal	22	0	1
Hypo	2	42	1
Hyper	0	0	2

Based on classification, result of the system is accurate and whether the system gives true positive (TP), false negative (FN), false positive (FP), and true negative (TN) for each class can be determined.

$$\text{Precision} = \frac{\text{TP}}{\text{TP} + \text{FP}} \tag{11}$$

$$\text{Accuracy} = \frac{\text{TP} + \text{TN}}{\text{TP} + \text{TN} + \text{FP} + \text{FN}} = \frac{\text{TP} + \text{TN}}{\text{N}} \tag{12}$$

Precision is the ratio of overall number of correctly predicted samples to the overall number of all predicted samples in that category.

Accuracy is the ratio of correctly predicted samples to the total number of samples (Table 2).

3 Results and Discussion

This proposed system includes the results of feature extraction technique, designing and execution of combined classifier, and outcomes are obtained from performance evaluation technique. These described results of all techniques can be obtained by using MATLAB. In feature extraction part, appropriate features may change due to disordered voices that are considered. The features can be MFCC—log energy, spectral energy, pitch, DCT, shimmer, jitter, etc. The dimension of feature matrix is 48. Further, SVM and HMM classifiers are used to classify the data. SVM is supervised algorithm so that it requires ground truth or actual dataset which is mentioned in first row in Table 3. And second row shows the predicted output of the proposed system. By considering actual and predicted output, confusion matrix is formed, and by using confusion matrix, precision and accuracy are calculated.

Table 3 Output of 19 voice samples

Ground truth	2	2	2	2	2	2	1	1	3	2
	1	1	2	2	2	2	2	2	2	
Predicted output	2	2	2	2	2	2	1	1	3	2
	1	1	2	2	2	2	2	2	2	

$$\text{Precision} = \frac{\text{TP}}{\text{TP} + \text{FP}} = 1.000000$$

$$\text{Accuracy} == \frac{\text{TP} + \text{TN}}{\text{TP} + \text{TN} + \text{FP} + \text{FN}} = \frac{\text{TP} + \text{TN}}{\text{N}} = 1.00000$$

In Table 3, 19 voice samples are analyzed from that dataset: 4 samples are from Class-1 (normal), 14 samples are from Class-2 (hypo), and 1 sample is from Class-3 (hyper).

$$\text{Precision} = \frac{\text{TP}}{\text{TP} + \text{FP}} = 0.975000000$$

$$\text{Accuracy} == \frac{\text{TP} + \text{TN}}{\text{TP} + \text{TN} + \text{FP} + \text{FN}} = \frac{\text{TP} + \text{TN}}{\text{N}} = 0.975000000$$

In Table 4, 40 voice samples are analyzed from that dataset: 10 samples are from Class-1 (normal), 28 samples is from Class-2 (hypo), and 2 samples are from Class-3 (hyper).

$$\text{Precision} = \frac{\text{TP}}{\text{TP} + \text{FP}} = 0.8055000$$

$$\text{Accuracy} == \frac{\text{TP} + \text{TN}}{\text{TP} + \text{TN} + \text{FP} + \text{FN}} = \frac{\text{TP} + \text{TN}}{\text{N}} = 0.9428000000$$

In Table 5, 70 voice samples are analyzed from that dataset: 22 samples are from Class-1 (normal), 42 samples are from Class-2 (hypo), and 2 samples are from Class-3 (hyper). There are 2 samples detected hyper but actually it is normal and hypo, respectively. There are 2 samples detected normal when actually it is from hypo category.

Competitive Analysis

See Table 6.

Table 4 Output of 40 voice samples

Ground truth	2	2	2	2	2	2	1	1	3	2
	1	1	2	2	2	2	2	2	2	2
	2	2	2	2	2	2	1	1	1	1
	1	1	3	2	2	2	2	2	2	2
Predicted output	2	2	2	2	2	2	1	1	3	2
	1	1	2	2	2	2	2	2	2	2
	2	2	2	2	2	2	1	1	1	1
	1	1	3	2	2	2	2	2	2	1

Table 5 Output of 70 voice samples

Ground truth	2	2	2	2	2	2	1	1	3	2
	1	1	2	2	2	2	2	2	2	2
	2	2	1	2	2	2	1	1	1	1
	1	1	3	2	2	2	2	2	2	2
	2	2	2	2	2	1	1	1	1	2
	2	1	1	1	1	1	1	1	1	2
	2	2	2	2	2	2	2	2	2	2
Predicted output	2	2	2	2	2	2	1	1	3	2
	1	1	2	2	2	2	2	2	2	2
	2	2	3	2	2	2	1	1	1	1
	1	1	3	2	2	2	2	2	2	2
	2	2	2	2	2	1	1	1	1	2
	2	1	1	1	1	1	1	1	1	2
	2	2	2	2	2	2	1	1	3	2

Table 6 Competitive analysis of different methods for diagnosing thyroid disease

Sr. no.	Author	Methods		Precision (%)	Accuracy (%)
		Feature extraction	Classifiers		
1	Dogantekin et al. [6]	LPC	LS-SVM	84.77	97.67
2	Zabidi et al. [7]	MFCC	*F*-ratio method	71.94	90.71
3	Proposed work	MFCC	HMM + SVM	80.55	97.28

4 Conclusion

This proposed system represents the pathological results of thyroid patient's voice signal. This system has been developed to help pathologist, speech therapist, and doctors to give an accurate diagnosis of thyroid disease.

It is concluded that the proposed system for diagnosing thyroid disease gives very promising results in distinguishing disease of thyroid patients. The performance evaluation of this system for diagnosing thyroid disease was estimated by using confusion matrix, precision, and accuracy. The results obtained by using combined classifier are best as compared with literature. Combined classifier technique gives higher accuracy to diagnose thyroid disease.

By considering these approaches, further research can be smoothened. This paper assists the researchers for a robust approach of feature extraction and classifier algorithm for detecting and analyzing the thyroid disease and its accurate diagnosis.

Acknowledgements We express our sincere gratitude toward authorities of our institute, MIT Academy of Engineering, Alandi (D), Pune, for providing us encouragement and supporting environment, as well as Dr. Milind Patwardhan, at Dr. Patwardhan's Endocrine and Diabetes Research Centre and Pathology Laboratory, Miraj, Maharashtra, India., and their staff for providing innumerous support for collecting voice samples.

References

1. Gour GB, Udayshankara (2015) Voice disorder analysis in thyroid patients. Int J Comput Sci Mobile Comput 4(5), 720–727
2. Dogantekin E, Dogantekin A, Avsi D (2016) An automatic diagnosis system on thyroid gland: ADSTG. J Exp Syst Appl 37, 6368–6372
3. Meek P, Carding PN, Howard DH, Leannard TW (2008) Voice change following thyroid and parathyroid surgery. J Voice 22(6):765–772
4. Zabidi A, Mansoor W, Khaun LY, Shahak R, Yasmin F, Rahman A (2009) Mel-frequency cepstrum coeefficient analysis of infant cry with hypothyroidism. In: International colloquium on signal processing and its applications, pp 204–208
5. Xiu M, Fauth C, Vaxilaire B,Rodier J-F, Volkmar PP, Sock R (2016) A post-thyroidectomy voice quality in patients suffering or not from recurrent laryngeal paralysis
6. Naikare K., Nirmal JH, Lad N (2018) Classification of voice disorder using i-vector analysis. In: International conference on communication, information and computing technology
7. Carson M (2009) Assessment and management of patients with hypothyroidism. Nursing Standard, vol 23, No. 18
8. Ma L, Ma C, Liu Y, Wang X (2019) Thyroid diagnosis from SPECT images using convolutional neural networks with optimization. J Comput Intell Neurosci 1–12
9. Firdos S, Umrani K (2016) Disordered voice classification using svm and feature selection using GA. In: International conference on cognitive computing and information processing, pp 1–6
10. Hammami I, Salhi L, Labidi S (2016) Pathological voice detection using support vector machine. In: International conference on advance technologies for signal and image processing, pp 662–666

Smart Glasses: Digital Assistance in Industry

Trupti Sutar and Savita Pawar

Abstract New media developments have revolutionized the behavior of people in an unprecedented technique in the latest decades. Mobile phones have created an always online mentality. However, what is next? Recent developments underline the increase of a technology known as "Wearable devices." Augmented reality smart glasses (ARSG) such as Microsoft HoloLens and Google Glass are very good examples of these technology. It provides huge potential for innovation for firms and manufacturing industries. ARSG is becoming very common and important technology that promotes shop floor operators to fulfill industry 4.0 requirements. Augmented reality is currently an interesting and hot research topic in manufacturing industries. The main goal of this research paper is to improve the use of smart glasses for operator training with augmented reality. It will assist to increase effectiveness and shorter learning times for the individual operator. ARSG products available in the market are very expensive. It will help to find an affordable solution for the industries. It provides new methods for reducing the efforts of the operators working online. It mainly focuses on minimizing disadvantages of the existing products.

Keywords Augmented reality smart glasses (ARSG) · Industrial operator support · Smart factory

1 Introduction

The growth of new communication and IT has had a huge effect on how individuals communicate with each other and how companies interact with clients. Because of mobile phones such as smartphones and tablets, consumers are always and everywhere online [1]. But what is next? A review of the recent innovations by leading

T. Sutar (✉)
VLSI and Embedded Systems, MIT AOE, Alandi, Pune, India
e-mail: trupti.sutar23@gmail.com

S. Pawar
School of Electrical Engineering, MIT AOE, Alandi, Pune, India
e-mail: srpawar@etx.maepune.ac.in

© Springer Nature Singapore Pte Ltd. 2021
S. N. Merchant et al. (eds.), *Advances in Signal and Data Processing*, Lecture Notes in Electrical Engineering 703,
https://doi.org/10.1007/978-981-15-8391-9_13

technology firms like Microsoft, Google, and Facebook shows that what we call ARSG could be the next development in media technology. These smart glasses are wearable devices that meld in the field of view of the customers real and virtual data. Recent marketing and academic projections indicate that intelligent glasses in media evolution are likely to be the next "major thing" [2].

Microsoft HoloLens, Google Glass, and Everywhere are three examples of commercially announced intelligent glasses. We define smart glasses as "Wearable Augmented Reality (AR) Systems" that are worn like regular glasses and combine the user's field of perspective with virtual data with physical information smart glasses are distinguished from other smartphone and wearable devices in different ways. Most significantly, they merge digital and physical data while being carried, rather than merely providing a virtual reality. Existing wearables, such as smart watches or virtual reality glasses, are not built from any parts of augmented reality. For example, many AR techniques are stationary, mirrors or AR screens. Furthermore, AR applications that can be used on smartphones or tablets must be managed proactively by the user, while intelligent glasses function separately and can be regulated by speech instructions [3].

The fourth industrial revolution is here, requiring a basic shift in attitude to intelligent factories using fresh techniques and manufacturing philosophies to notice brief product life cycles and extreme mass customization in a cost-effective manner [4, 5]. The smart factory concept is designed to enable flexible and self-conforming manufacturing processes with products and machines that function smartly and autonomously using concepts such as the Internet of things and cyber-physical systems [6]. This fundamental change in policy and a new manufacturing form will change the conditions for the operators working in the workshop, as the work activities of the operators will no longer be static and predetermined, but will be dynamic and constantly changing [7, 8]. This will position the operator's flexibility and adaptability with heavy demands. Operators need to be equipped with effective technology to promote optimal decision making and intervention to satisfy these requirements effectively.

Augmented reality smart glasses (ARSG) have been recognized in the latest years as a strong technology that supports plant operators performing multiple duties such as installation, maintenance, quality control, and handling of materials. While working on a line operator is bounded by timeline, he has to complete the task within time. If he misses any required task or steps the entire work he has done will come for rework and this rework costs much higher. To avoid such kind of instances, operator must be equipped with the device that is continuously guiding him about the next step. This task is done by a device called augmented reality smart glasses (ARSG).

ARSG is fundamentally a hand-to-hand transparent screen that integrates a miniature portable computer that adds virtual data to what the user sees [9]. ARSGs are hands-free devices with eye-level data correct where it is required, making them a perfect user interface for an industrial operator. Using ARSG for operators enables the precise data required to be automatically provided at the correct location and moment to manage a particular situation or job assignment in an ideal manner. In the latest years, ARSG's clear advantages have made technology interest grow quickly, and its growth is presently driven by several industries, such as games, sports, and tourism. ARSG shipments are expected to explode in the years to come.

There are many ARSG producers and a broad variety of products available on the market, but very few manufacturing businesses have embraced ARSG [4] despite this general availability. This may seem surprising at first glance given the obvious benefits of using ARSG on store floors, but at least two main reasons can be identified for this lack of adoption. First, the products of today are mainly sold either as goods of the general customer or as professional office products. Second, with big variations in design, technology, and functionality, the products are extremely varied. This wide variety of products makes it both complex and time-consuming for a manufacturing business to define the ideal product for its distinctive shop floor context, creating a threshold for ARSG adoption. The aim of this research is to remove this limit and promote the industry's inexpensive and easy-to-access solution, which clearly costs less than the products already on the market. The potential of new business models for innovative apps occurs in relation to enhancing the efficiency of current functions.

Knowledge on the motivations of prospective customers is required to create effective intelligent glasses and apps. Consequently, the study findings are discussed and expanded in Sect. 2 and thus give readers an overview of promising approaches. Section 3 explains the methodology used to design intelligent lenses. Section 4 provides results and discussion. By incorporating intelligent glasses, the objective is to make readers think about fresh and innovative business ideas. We finish the paper in Sect. 5. The discussion given also stimulates thoughts for studies on leadership that is academically essential.

2 Related Work

Smart glass performance can encourage ongoing research into past research and technology adoption models. A study was carried out to analyze the smart glass design factor [10].

2.1 Smart Glasses

Smart glass users can see the display irrespective of the user's position. They provide technology and data opportunities for users. Smart glasses allow individuals to do things like connecting to the Internet, sending messages, taking photos and videos, discovering locations, and running mobile apps. Companies continue to operate on virtual reality and enhanced reality to alter visual information. The main fields of use of smart glasses are medical, schooling, leisure, sports, and commerce [11].

This article provides guidance on smart glass assessment parameters, which will help the sector to choose from among the products accessible. Authors comment only on the product comparison they do not provide any technical data. But the nice thing

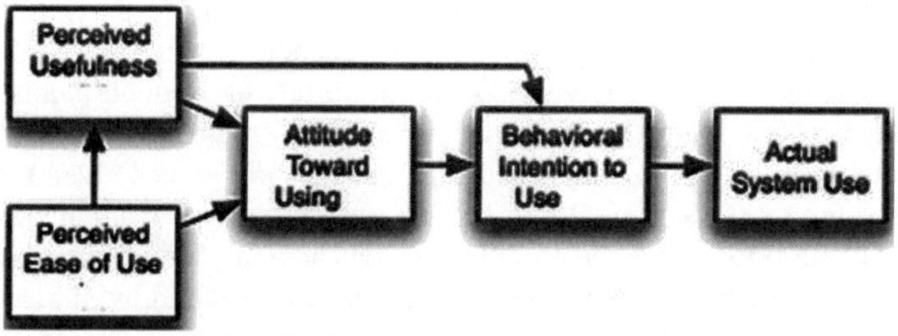

Fig. 1 Technology acceptance model (TAM)

about the document is that writers are exploring the future smart glass expansions [4].

2.2 Technology Acceptance Model

The technology acceptance model (TAM), with its roots in data systems, is a commonly cited and extended framework to explain the intentions and behaviors of potential customers in adopting particular technologies. Research on technology acceptance, particularly TAM, appears to be a helpful starting point for exploring the basic mechanisms of smart glass adoption. In particular, different researchers have implemented modified TAM models on wearable techniques. For example, for studying accepting smart watches, Kim and Shin used a revised TAM [12]. The technology acceptance model (TAM) provides an effective solution that predicts the implementation of new technology [13] (Fig. 1).

2.3 Usability

Usability is the capacity with fulfillment to communicate efficiently with data systems for par-specific purposes. The goal of interaction issues with human computers is to create more usable systems. Volker Paelke proposes a fresh strategy to the use of vibrant information content that automatically adapts to the individual operator and their learning progress in order to improve effectiveness and shorter learning times. In this paper, Buti Al Delail and Chan Yeob Yeun introduce the original experience with this scheme, which has already been used effectively by several hundred customers without prior installation assignment experience [14].

2.4 Interaction Methods

There are various elements, such as vision-based and not vision-based approaches, gesture-based and not gesture-based approaches. An alternative element divides approaches to communication into three classes that are less handheld, touch and touch.

From the perspective of interaction objectives, interaction methods, including spectacular smart glass frames, rings, wristbands, body surfaces, body movements, gloves and cameras, LIK-HANG, LEE and PAN, HUI, are evaluated. Their input capabilities are discussed based on the suggested touch and touchless input classification scheme [15]. In conclusion, touch input and freehand interaction are the most popular research topics in smart glass communication.

2.5 The Lying Challenges

This type of technology has many difficulties. One of the primary problems is that in which product category has intelligent glasses [16]. There are physical, mental, social, psychological, and technological issues related to the category of product difficulties. These are as illustrated in Fig. 2 and further discussed.

While companies find excellent alternatives to workflow through eyewear technology, the general public will still have to wait a little longer to reap the benefits

Fig. 2 Lying challenges for smart glasses

of mass accessibility and use. It happens that the challenge of achieving a harmonious balance between functionality and wearability at an inexpensive cost must first be overcome by companies in order to eventually attain mass market usage. High-performance smart glasses currently tend to be bulky and stand at a cost range that is not yet sufficiently (nor fashionable) convincing for daily social use. On the other hand, fashionable products need to sacrifice effectiveness to get a sleeker look and still tend to keep a higher range of prices. Furthermore, in either functionality, wearability or both, affordable parts will be lacking. In addition to this use of smart lenses, some eye illnesses will also tend to occur. Continuous use of eye-sensitive technology and finding that the brain and eye can quickly adapt and thus be affected by ongoing use [17].

More usually, there is a possibility of computer vision syndrome [18], resulting from looking close to the eye for a long time. It may possibly generate various types of cognitive dissonance and pain [19]. How the technology affects the eye is therefore an unresolved issue.

The main regulatory problem is mainly whether and how it is legal to film without knowing and accepting it, and partially how the information will be stored and who has access to it.

3 Methodology

3.1 Motivation

Wearable technology is considered as the next big thing and seems to be the most promising hi-tech in our lives. Our growing technological dependence desires further innovation. Innovation, communication channels, time, and social structures will determine the future. But if we look at Gartner's hype cycle (2018) shown in Fig. 3, wearable user interfaces are at the top of the hype but have a timeline of up to 10 years until they are widely used. According to the cycle logic, the hype around [20].

The products will be disillusioned before the products achieve wide execution.

However, most commentators think it will take less than 10 years for growth. In the next 3–5 years, there will probably be a powerful market selling some kind of intelligent glasses—probably not the Google Glass design we understand today.

3.2 System Design

The primary processor choice is guided by the device type and characteristics. As glasses should be wearable, processor should be lightweight and should have capability to process the images. Here, in this design, Raspberry Pi is used as an application processor because it works on its own operating system which is open source.

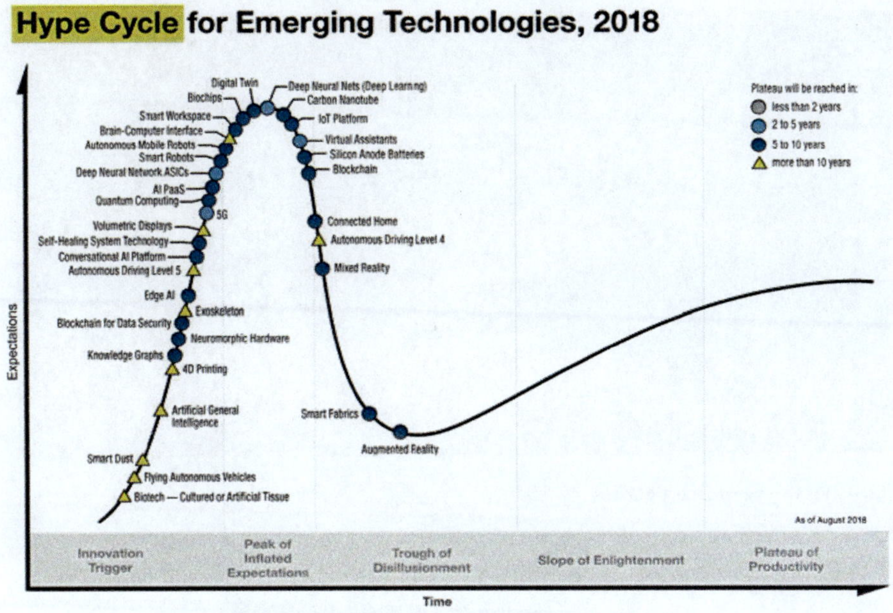

Fig. 3 Gartner's hype cycle 2018

To display the images, LCD used is of smallest size. To make the images viewable by eyes, later the images are magnified by using the convex lens and are projected by using the pentagonal prism which is semitransparent. So, it will not affect the real field of view of the operator. By assembling all these component altogether, the requirements for the wearable devices are fullfilled. How this device will look is shown in Fig. 4. This prototype can be used at left as well as right side by rotating the image and can also be useful for the prescribed eyeware user.

3.3 Working

Figure 5 shows the system block diagram. The main objective of this design is to provide assistance to the workers in the industry, i.e., it will display the work instructions at the operator's eye level. In this system, Raspberry Pi works as application processor which will drive the LCD module for display of images containing the work instructions of operator.

In this system, the interaction method for smart glasses used is push button. When user will press the key/button, the next image in the line will be displayed onto the LCD module.

The view field is a key parameter that directly affects how much information can be displayed and where it can be put. The horizontal field of view is particularly

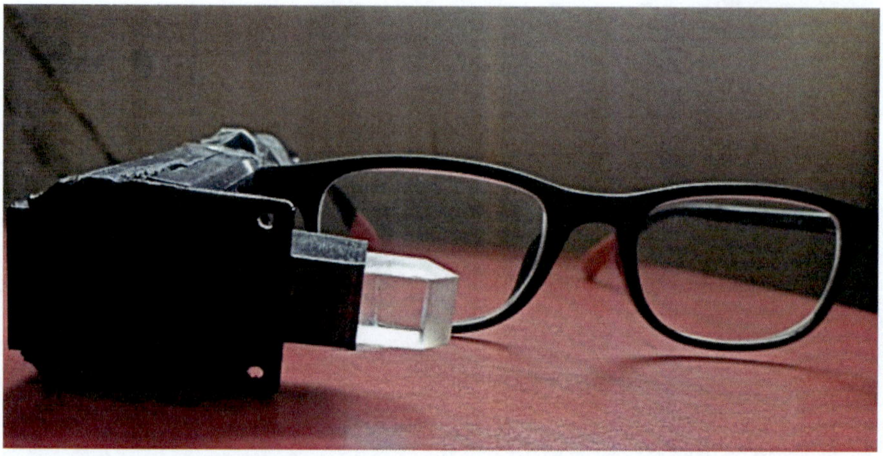

Fig. 4 Prototype of smart glasses

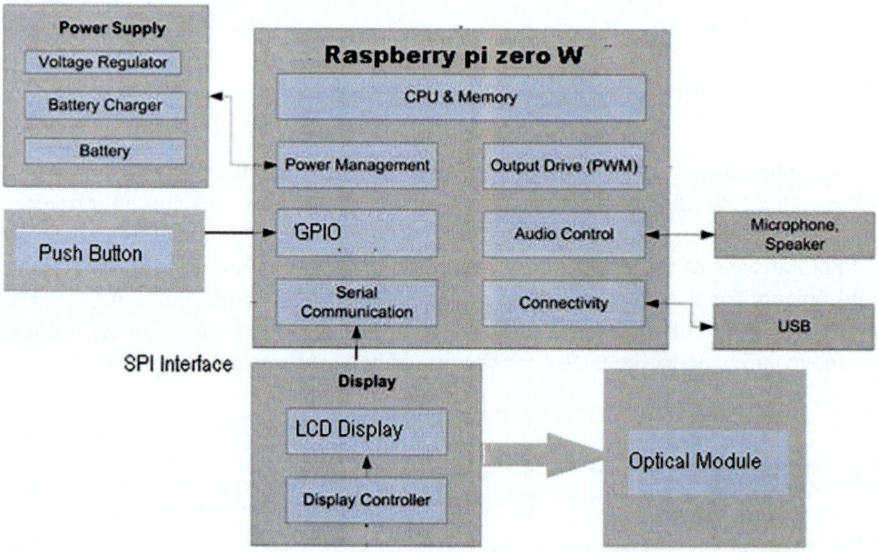

Fig. 5 Smart glasses block diagram

important because it is a wide horizontal field from the point of view, the edge of information can be displayed, keeping the sight clear to see the real world [4].

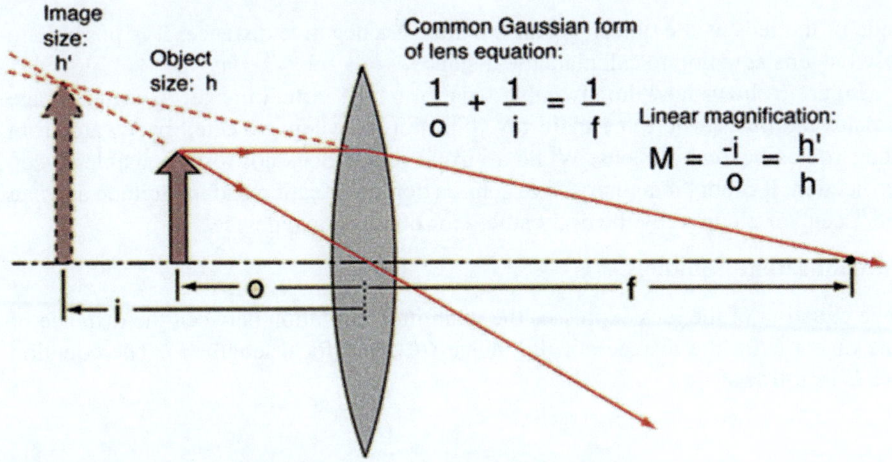

Fig. 6 Virtual image formation using positive lens

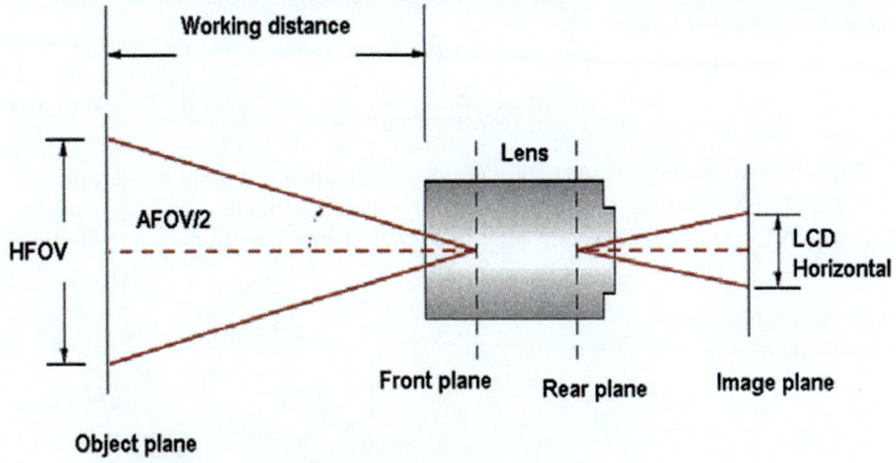

Fig. 7 Relationship between HFOV, working distance (WD), and display size

3.4 *Block Diagram of the Smart Glasses*

Human vision field is nearly 180° horizontal, but the present ARSG is far from that. We think that the ARSG is at least 30° (landscape) in a realistic, acceptable field of perspective.

We can calculate the object distance by using the lens equation to create a virtual image with convex lens. In order to match the illustration, if an object distance smaller than the focal length is entered, the image is an enlarged virtual image on the same side of the lens as the object, giving the image a negative distance. It is possible to use the lens equation to calculate the distance.

Figure 6 shows how the favorable lens forms the virtual image. A virtual image creates the place where the main ray routes cross when projected backward from their routes beyond the lens. While a virtual image does not form a visible screen projection, it is not "imaginary," i.e., it has a definite location and magnitude and can be "seen" or pictured by the eye, camera, or other optical device

Magnification Equation

The equation of the lens expresses the quantitative relation between the distance of the object (*d*0), the distance of the image (*di*), and focal length (*f*). The equation reads as follows:

$$\frac{1}{f} = \frac{1}{d0} + \frac{1}{di} \tag{1}$$

The equation of magnification relates the percentage of image range and object size to the ratio of image height (*hi*) and object height (*h*0). The equation for magnification is as follows:

$$M = \frac{hi}{h0} = -\frac{di}{d0} \tag{2}$$

These two equations are used to produce data about image range and image height.

Figure 7 shows the relationship between horizontal field of view (HFOV), working distance (WD), and display size. Using this relationship, we can calculate the field

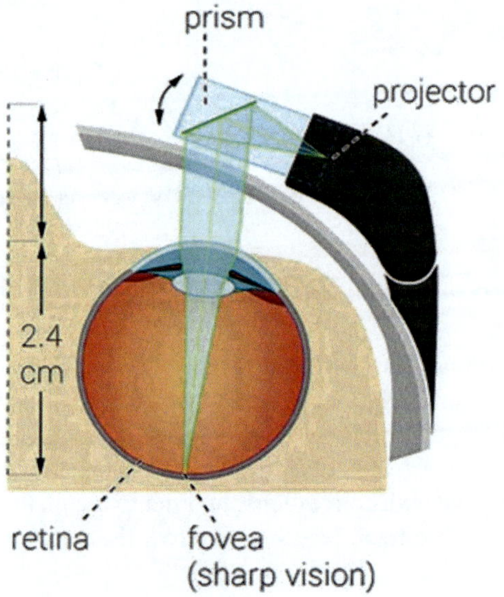

Fig. 8 Smart glasses prism and retina (eye)

of view of the above design of smart glasses.

$$f = \left(\frac{h * \mathrm{WD}}{\mathrm{HFOV}}\right) \tag{3}$$

In Eq. 3, h is the horizontal dimensions (horizontal pixels multiplied by the pixel size) and f is the focal length of the lens.

The above discussion is all about lens calculation for the magnification of image and field of view. Now, the question is how the magnified image is presented at the eye level using prism? It is summarized below.

Here, Fig. 8 demonstrates how the retina picture is focused by the prism (Fovea = point of sharp visual picture). The layer appears in the bottom correct corner or in the center of the visual field depending on how you wear the smart glasses. When the smart glasses are high on the nose, you need to turn your eye up to see the picture sharply so that you can virtually see through underneath. You can also position it straight in front of the pupil because the prism is semitransparent. You have the sharp layer right in front of your eyes in that case.

4 Results and Discussion

Smart glass design features consist of five variables, stand-alone device, interaction, field of perspective, cost and screen resolution. Various intelligent glasses are prepared for launch and some are already on the market. Here, we can make a comparison table consisting of the above five factors for the betterment of the smart glasses design explained in this paper (Table 1).

Discussions:

Table 1 Comparison with the available products in the market

Alternative	Stand-alone device	FOV (in degrees)	Interaction	Display resolution	Price (rupees)
1	Yes	65	Voice recognition and touchpad	640 × 360	42,000
2	Yes	12	Hand gestures	1024 × 768	42,000
3	No	65	Voice recognition and touchpad	1024 × 768	42,000
4	No	65	Hand gestures	1024 × 768	42,000
5	Yes	65	Hand gestures	1024 × 768	69,000
Smart glass design	Yes	28	Key	160 × 80	10,000

Fig. 9 Experiences of participants in terms of points out of 10

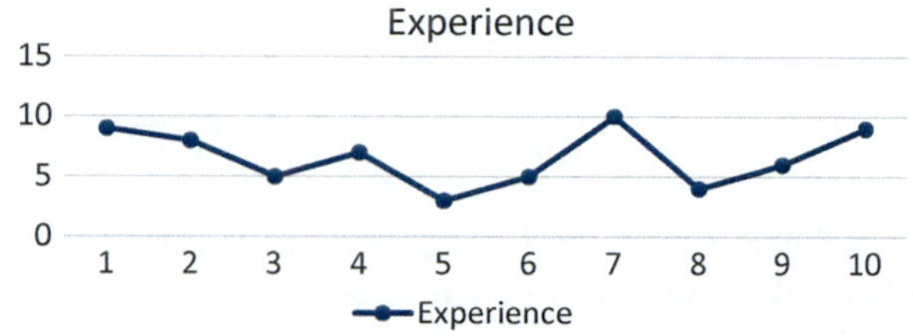

Some of the key observations are also noted with the help of few participants. A series of task is carried out to access their benefits both with and without use of smart glasses. In the design of the smart glasses, display is small and rectangle within central point of field of view, expectations from people is that they get more benefit from the use of smart glasses. Two people with same characteristics, e.g., age, gender, profession, visual field, sight loss condition may have different experience when using a smart glass.

Here, Fig. 9 is plot drawn from the feedback given by participants. There is no single predictor as to why a person may or may not benefit from the smart glasses. There are so many factors (age, sight loss condition, vision level, visual field, environmental light levels, and various smart glass methods) that it is hard to identify the future profile of a prospective glass recipient.

5 Conclusion

This study aims to take the manufacturing industry one step closer to the adoption of ARSG on the shop floor by providing them with a low cost and easily available solution. It is very much obvious that these smart glasses are very cost effective and applications-oriented. By using these smart glasses for the operator working on a production line or in a maintenance section, mistakes can be minimized rather I would say eliminated. As there are some pros and cons of the design, it has very much scope for the future development. Some future scope points for this design are listed down.

A. **Extending the field of view**: Without a doubt, the field of perspective is one of ARSG's most difficult problems. The natural human field of perspective is almost horizontally 18°, but this design's field of perspective is 28°.
B. **Making the glasses wearable**: Current ARSG is actually wearable, but for extended periods it cannot be worn. There is still a scope to reduce the size and weight.
C. **Improving interaction method**: In industrial shop floor applications, communication with ARSG must be hands-free, as the user must use his/her hands to perform job duties. Using speech commands is the most common way of implementing hands-free communication. However, there are significant noises

from machinery and transportation in the industrial shop floor. In the presence of sound, voice recognition becomes an important task and with high certainty the program needs to implement distinctive features to reduce noise and accept the correct instructions. Apparently, in the context of ARSG, this challenge is not being considered and this is an important topic for further research.

References

1. Hennig-Thurau T, Malthouse EC, Friege C et al (2010) The impact of new media on customer relationships. J Serv Res 13(3):311–330
2. Rauschnabel PA, Ro YK (2016) Augmented reality smart glasses: an investigation of technology acceptance drivers. Int J Technol Market 11(2):123–148
3. Rauschnabel PA, Brem A, Ro Y (2015) Augmented reality smart glasses: definition, conceptual insights, and managerial importance. Working paper, The University of Michigan-Dearborn
4. Syberfeldt A, Danielsson O, Gustavsson P (2017) ARSG in the smart factory: product evaluation and guideline in IEEE. Access 5:9118–9130
5. Veza I, Mladineo M, Gjeldum N (2015) Managing innovative production network of smart factories. IFACPapersOnLine 48(3) (2015)
6. Gorecky D, Schmitt M, Loskyll M, Zühlke D (2014) Human-machine interaction in the industry 4.0 era. In: Proceedings of 12th IEEE international conference industrial information
7. Syberfeldt A, Ayani M, Holm M, Wang L, Lindgren-Brewster R (2016) Localizing operators in the smart factory: A review of existing techniques and systems. In: International symposium on flexible automation (ISFA)
8. Syberfeldt A, Danielsson O, Holm M, Wang L (2016) Dynamic operator instructions based on augmented reality and rule-based expert systems
9. Delail BA, Yeun CY (2015) Recent advances in smart glass application security and privacy. ICITST
10. Due BL (2014) The future of smart glasses. An essay about challenges and possibilities with smart glasses. Working papers on interaction and communication
11. Ok AE, Basoglu NA, Daim T(2015) Exploring the design factors of smart glasses. In: Proceedings of PICMET
12. Kim KJ, Shin D (2015) An acceptance model for smart watches: Implications for the adoption of future wearable technology. Internet Res Electr Netw Appl Policy 25. https://doi.org/10.1108/IntR-05-2014-0126
13. Li J, Deng L, Gong Y, Haeb-Umbach R (2014) An overview of noise robust automatic speech recognition. IEEE/ACM Trans Audio Speech Lang Process 22(4)
14. Paelke V (2014) Augmented reality in the smart factory: Supporting workers in an industry 4.0 environment. In: Proceedings of IEEE Emerging Technologies and Factory Automation
15. Lee L, Hui P (2017) Interaction methods for smart glasses. ACM Comput Surv 1, Article 1
16. Dvorak JL (2010) Moving wearables into the mainstream: taming the Borg (Softcover reprint of hardcover 1st 2008 edition. Springer, New York
17. Mann S (2013) Vision 2.0. IEEE Spectr 50(3):42–47
18. Blehm C, Vishnu S, Khattak A, Mitra S, Yee RW (2005) Computer vision syndrome: a review. Surv Ophthalmol 50(3):253–262
19. Patterson R, Winterbottom MD, Pierce BJ (2006) Perceptual Issues in the Use of Head-Mounted Visual Displays. Hum Fact J Hum Factors Ergon Soc 48(3):555–573
20. Hype cycle for emerging tchnologies. https://www.gartner.com/en/research/methodologies/gartner-hype-cycle
21. Azuma R, Baillot Y, Behringer R, Feiner S, Julier S, MacIntyre B (2001) Recent advances in augmented reality. IEEE Comput Graph Appl 21(6)

22. Rauschnabel PA, Brem A (2015) Augmented reality smart glasses: definition, conceptual insights, and managerial importance. Research Gate Publications
23. Downes L, Mui C (2000) Unleashing the killer app: digital strategies for market dominance. Harvard Business Press
24. Starner T (2013) Project glass: an extension of the self. IEEE Pervasive Comput 12(2):14–16
25. Due BL (2014) Don't be a glasshole. Kommunikationsforum
26. Pascual-Leone A, Amedi A, Fregni F, Merabet LB (2005) The plastic human brain cortex. Annu Rev Neurosci 28(1):377–401
27. Small G, Vorgan G (2009) iBrain: surviving the technological alteration of the modern mind, 1st edn. William Morrow Paperbacks, New York
28. Stone L (2014) Continuous partial attention. Retrieved from https://lindastone.net/qa/continuouspartial-attention/
29. Tractica, an Informa business. https://www.tractica.com/newsroom,05-07-2018, 12:44pm
30. Hardware development community https://hackaday.io/project, 26–06–2018, 2:15pm

Implementation of Hand Gesture Recognition System to Aid Deaf-Dumb People

Supriya Ghule and Mrunalini Chavaan

Abstract In recent years, the population of deaf-dumb victims has increased because of birth defects and other issues. Since a deaf and mute person cannot talk with an ordinary person in order that they ought to rely on some kind of communication system. The gesture shows some physical movements of the hand that convey a piece of information. Gesture recognition is the analytical interpretation of the movement of an individual through an information processing system. Linguistic communication provides the most effective conversation platform for the mute person to speak with an ordinary person. The aim of this paper is to build up a time system for hand gesture recognition that acknowledges hand gestures and then converts them into text and voice. In this paper, efforts have been done to detect 8 different gestures. Each gesture has assigned unique sound and text output. In experimental results, 800 samples were taken into the consideration out of which 760 samples were detected correctly and 40 samples were detected wrongly. Hence, the proposed system gives accuracy of 95%.

Keywords Raspberry Pi · Python · OpenCV · Feature extraction · Contours

1 Introduction

In our day-to-day routine, we will speak with one another by using speech. Gestures are a lot of preferred and common to interact with computers for humans, so it assembles a scaffold among machines and people. For several hard of hearing person, linguistic communication is their fundamental language making a powerful sense of social and cultural character. The arranged framework is predicated on vision-based

S. Ghule (✉)
VLSI Design and Embedded System, School of Electrical Engineering, MIT Academy of Engineering, Alandi, Pune 412105, India
e-mail: ssghule@mitaoe.ac.in

M. Chavaan
School of Electrical Engineering, MIT Academy of Engineering, Alandi, Pune 412105, India
e-mail: mhchavaan@etx.maepune.ac.in

© Springer Nature Singapore Pte Ltd. 2021

S. N. Merchant et al. (eds.), *Advances in Signal and Data Processing*,
Lecture Notes in Electrical Engineering 703,
https://doi.org/10.1007/978-981-15-8391-9_14

hand acknowledgment approach that is a ton of common and needn't bother with any information to detect the real signal. The hand signals ought to be known under variable brightening conditions. There are several extraction ways and classification strategies are offered and therefore the call on which of them to utilize could be a difficult task. The planned methodology carries out background separation of the hand from the information. It comprises of feature extraction ways to calculate angle between two fingers of hand, then at last the gestures are detected and converted into text and voice. The planned system relies on Raspberry Pi, Pi camera module, 5 inches Resistive HDMI Touch Screen Display, Open Source Computer Vision (OpenCV) library, Python Programming Language. The Pi camera has 5-megapixel resolution to capture the images of a person's hand. In this paper, efforts have been done to detect eight different gestures. Each gesture has assigned unique sound and text output. In experimental results, 800 (100 samples for each gesture) samples were taken into the consideration out of which 760 samples were detected correctly and 40 samples were detected wrongly.

Rathi et al. [1] framework for perceiving a dynamic hand words motion of Indian signs and change of perceived signal into the text to voice and vice versa. Eigenvectors and Eigen esteem method has utilized for highlight extraction. Eigenvalue worth weighted Euclidean Distance based for the most part classifier has been utilized. Nath and Anu [2] ARM CORTEX A8 Processor is used. For image characterization, Haar classifier is used though 1-D HMM is utilized for Speech modification. Marking acknowledgment has created its significance in very few zones like Human–computer Interactions (HCIs), mechanical controls, and home computerization. Quiapo et al. [3], it was ready to fulfill the needs of a Sign Language Translator. The task was ready to boost the fluctuate of the flex detection parts although as well as new types of detector states for included sifting. The procedure GUI conveyed the bigger a part of the capacities that were needed within the two-way interpretation technique. Tapadar et al. [4] includes training with the acquired alternatives that square measure near particular for different hand motions. In this way, we will be prepared to set up gesture-based communications and thus assemble crippled individuals socially satisfactorily. Use the distinctive feature extraction.

Jalab and Omer [5] hand motion interface for prevailing media player misuse neural system. The anticipated standard recognizes a gathering of four explicit hand signals, to be specific: Reverse, Forward, Play, and Stop. Our standard is predicated on four stages, Image procurement, Hand division, alternatives extraction, and Classification. Nath and Arun [6] implemented framework for marking recognition for not too sharp people in ARM CORTEX A8 processor board misuse convex sunken body standard and model coordinating principle. The framework is utilized to oversee gadgets like an instrument, car Audio Systems, and home apparatuses. Shweta et al. [7] build up a genuine time framework for hand motion acknowledgment that recognizes hand signals, alternatives of hands like pinnacle figuring and edge computation thus convert motion pictures into voice and contrariwise using image processing. Abed and Rahman [8] the versatile instrument is developed and tried to demonstrate the viability of the anticipated guideline. The instrument movement and route happy with very surprising headings: Right, Left, Forward, Backward, and Stop. The ubiquity rate of the automated framework reached about ninety-eight using Raspberry Pi, camera module, and Python Language. Javed et al. [9] Digital Dactylology Converser (DOC) that could be a gadget that changes over a sign language into voice sign and instant message. The anticipated gadget will function admirably and translates the letters, letters to content, and sound. Nandy et al. [10] give efficient acknowledgment exactness to a restricted arrangement of dynamic ISL motions. It incorporates the amazing outcomes for Euclidian separation and K-Nearest neighbor measurements.

The paper is organized as follows: Sect. 2 introduces the basic system models, including system architecture and methodology in Sect. 3 implementation and result analysis including algorithm, flowchart, and result and analysis part, in Sect. 4 we conclude the paper.

2 System Models

2.1 Methodology

The camera coupled with the Raspberry Pi initially captures the image of the hand to be processed and identified. The input image must be converted to a specific format before processing and identifying the gesture. After the process, the gesture is identified and the text is generated. This text is for ordinary people to read, and text-to-speech messages are available if you cannot see them (Fig. 1).

- **Image Capture**:
  The input information is a picture or series of pictures taken by one camera and directed at the user's hand. The gesture pictures are real images of various sizes taken with a camera.

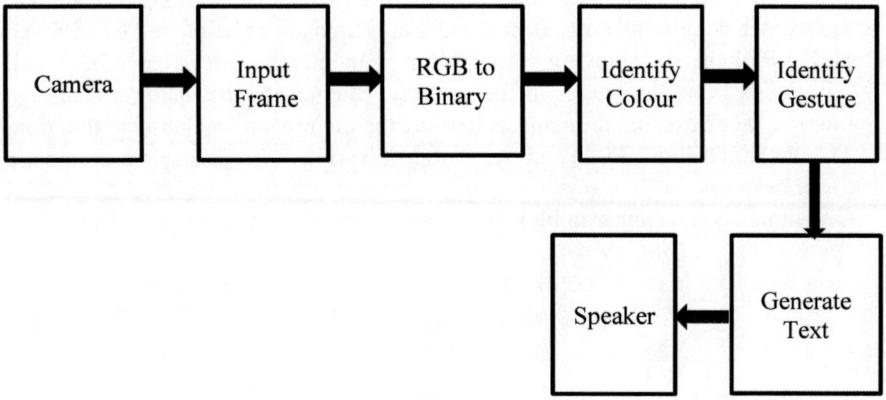

Fig. 1 Block diagram of working module

- **RGB to Binary**:
  Two skin color areas are defined, one is the top area and one is the bottom area with respect to HSV, where the input image is captured by the camera and then converted to a grayscale image.
- **Identify Color**:
  In this step, the skin color is extracted from the object frame. Next, the image was cropped to get rid of the unwanted parts of the initial pictures. Finally, we get clear results images with uniform size and consistent background.
- **Identify Gesture**:
  Identify the contour and convex hull in the green box around the hand. In this green box, the gap between two fingers is called as defects. If the angle between two fingers is less than 90° and greater than 30°, it is considered defects. These defects are marked with a blue dot. Based on the hand area and the defects we can identify the gesture.
- **Generate Text and Sound**:
  The corresponding text will be shown on the LCD screen according to the gesture. This text is then converted to sound in the output.

2.2 System Architecture

1. **Frame Capture**:
   The input file is an image, taken by a Pi camera pointed toward the user's hand. A 5MP Pi camera that is suited for recording of 1080p video and captures still image. The picture was clicked with proper brightening conditions and plain background. Finally, captured image is converted to gray scale image.
2. **Blur Image**:
   In image processing, a Gaussian blur is the result of blurring a picture by using a Gaussian function. Blur frame is important to a method for picture improvement and for obtaining smart outcomes. Blurring is utilized for smoothing pictures and scale down details and noise from the pictures. Image can be filtered by LPF and HPF. LPF helps in removing noises, blurring the picture and on the other hand, HPF helps in finding edges in the pictures. Mathematically, using a Gaussian blur to a picture is that the same as looping the picture with a Gaussian function. The one dimension (Eq. 1) and two dimensions (Eq. 2) of Gaussian function are given below.
   One-dimensional Gaussian blur function:

$$G(x) = \frac{1}{\sqrt{2\pi\sigma^2}} \mathrm{e}^{-\frac{x^2}{2\sigma^2}}. \tag{1}$$

   Two-dimensional Gaussian blur function:

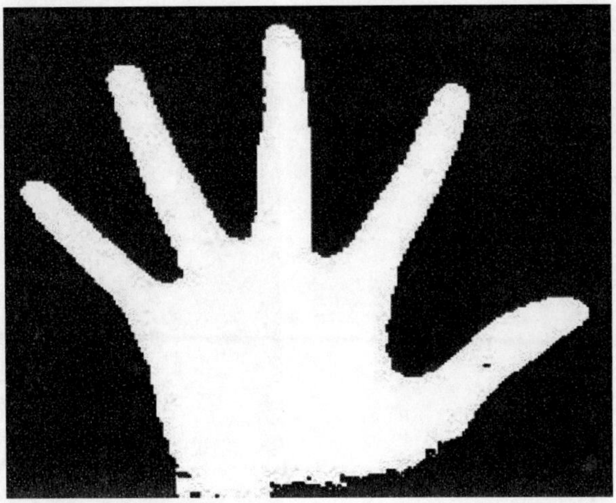

Fig. 2 Thresholding process

$$G(x, y) = \frac{1}{2\pi\sigma^2} e^{-\frac{x^2+y^2}{2\sigma^2}} \quad (2)$$

3. **Fame Segmentation**:
   Frame segmentation is that the beginning of any frame recognition method. The goal of hand segmentation is to distinguish hand area from the background within the picture. To realize this, totally different image segmentation algorithms are used like thresholding method, this result is shown in (Fig. 2).
4. **Find Convex Hull and Contours:**
   The green line around the hand (Fig. 3) is termed a convex hull that is utilized to get the fingertips, is the arched set encasing the hand space. Convex hull can look the same as contour approximation however it's not. (Both could offer an equivalent result in the same case). Contour is a curve joining all the continuous points. It is a fundamental instrument for shape examination, identification, and acknowledgment. In OpenCV Software, finding contours is like finding a white color object from the black color background.
5. **Find Convexity Defects and Area Ratio**:
   Any deviation of the object from this hull can be considered as convexity defects denoted as blue dots (Fig. 3). There are 3 points start point, far point (defect point), and end point between the two fingers. If the angle between two fingers is greater than 30° and less than 90°, then the cavity formed will be termed as a defect. Any two fingers. The defects will be most likely of triangular shape, which will have three corner points, i.e., start point, far point, and end point (Fig. 4). After joining these three points, triangle gets formed having sides a, b, c. Formulas of these sides are given below (Eqs. 3, 4 and 5). From these sides, we calculate the area between two fingers (Eq. 7) and angle (Eq. 9) between two fingers. Finally, we will calculate the area ratio based on convex hull area and contours area (Eq. 8).

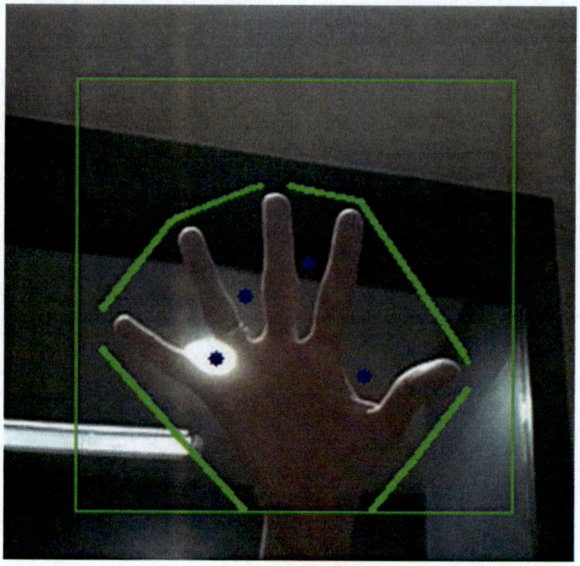

Fig. 3 Convex hull and defects

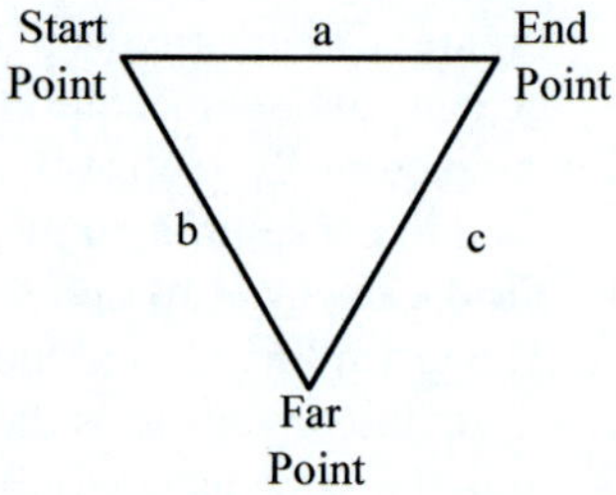

Fig. 4 Angle between two fingers

Formulas for finding defects due to fingers (Eqs. 3–6):

$$a = \sqrt{(\text{end}(0) - \text{start}(0))^2 + (\text{end}(1) - \text{start}(1))^2} \tag{3}$$

$$b = \sqrt{(\text{far}(0) - \text{start}(0))^2 + (\text{far}(1) - \text{start}(1))^2} \tag{4}$$

$$c = \sqrt{(\text{end}(0) - \text{far}(0))^2 + (\text{end}(1) - \text{far}(1))^2} \tag{5}$$

$$s = \frac{a + b + c}{2} \tag{6}$$

Calculate the area ratio between two fingers (Eqs. 7, 8)

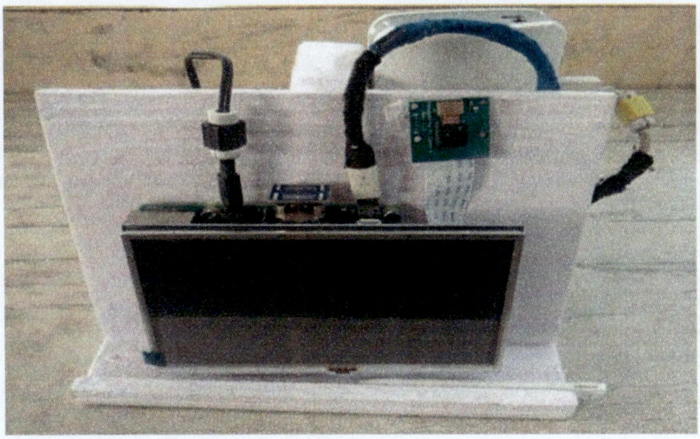

Fig. 5 Hardware module

$$\text{ar} = \sqrt{s(s-a)(s-b)(s-c)} \tag{7}$$

$$\text{area ratio} = \frac{\text{area of convex hull} - \text{area of contours}}{\text{area of contours}} * 100 \tag{8}$$

Calculate the angle between two fingers (Eq. 9)

$$\text{Angle} = \frac{a\cos\left(b^2 + c^2 - a^2\right)}{2bc} * 57 \tag{9}$$

3 Implementation and Result Analysis

3.1 Hardware Implementation

This system was implemented using Raspberry Pi 3 b+ , Pi camera module, 5-inch HDMI touch screen LCD display, intex power bank, 3.5 mm jack headphone, or speaker. These all components mounted on wooden block (Fig. 5).

3.2 Flow Chart

See Fig. 6.

Fig. 6 Flowchart

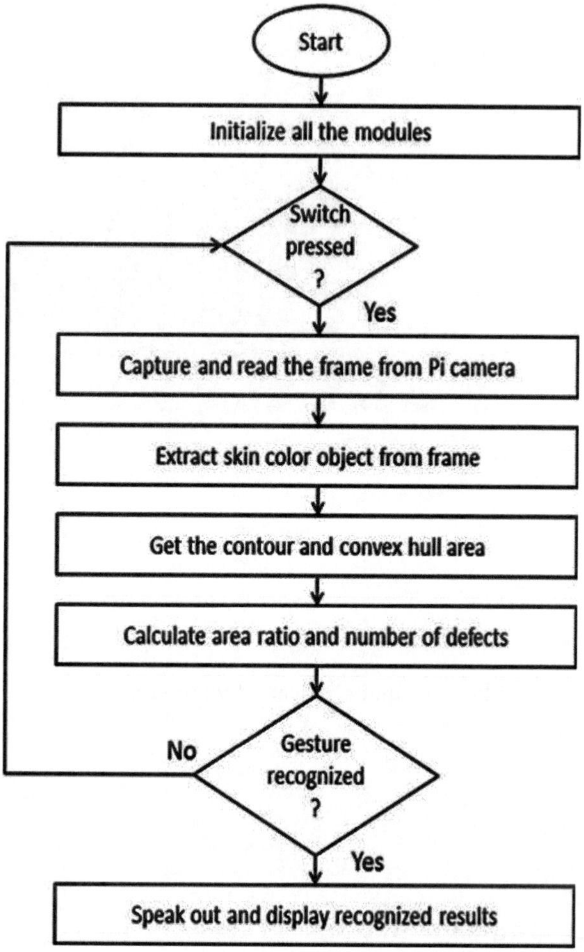

3.3 Result

Table 1 represents how input image gets converted into sound and display output. Each and every gesture has unique sound and display output. We will discuss this below.

- Gesture 1: In the first gesture there are no defects. Area ratio of this gesture is less than 12. This gesture is assigned to display "NO" and also speaks out "NO".
- Gesture 2: In the second gesture there are no defects. Area ratio of this gesture is less than 17.5. This gesture is assigned to display "HELLO" and also speaks out "HELLO".

Table 1 Image to sound conversion

Gestures	Input image	Threshold image	Identified text image output	Identified audio file output	Wrongly detected images
1	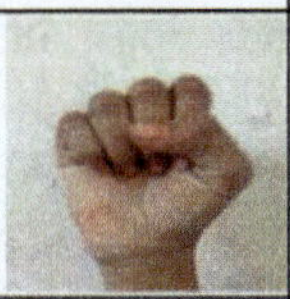				
2	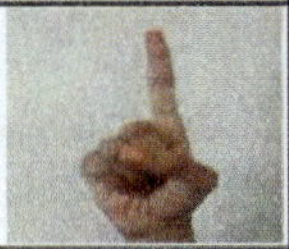				
3					
4					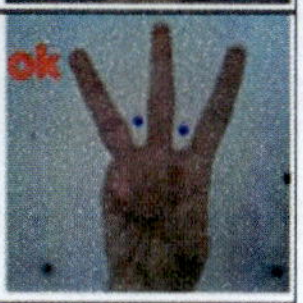
5	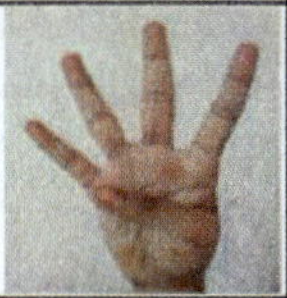		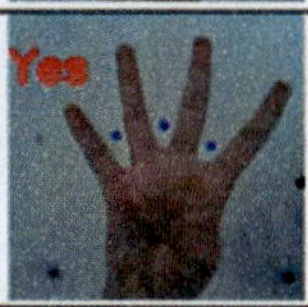		
6	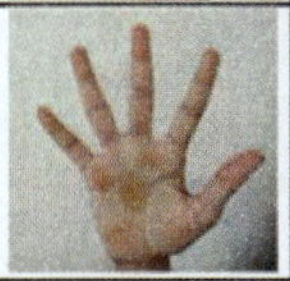			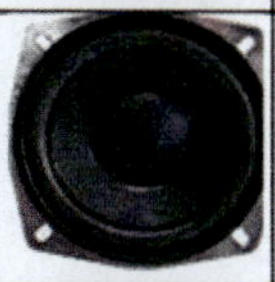	
7					
8	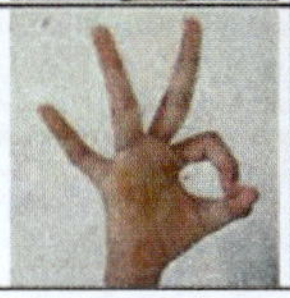		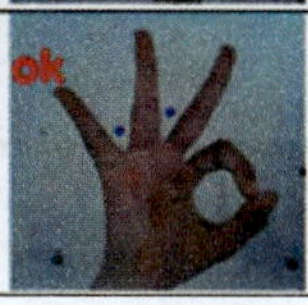		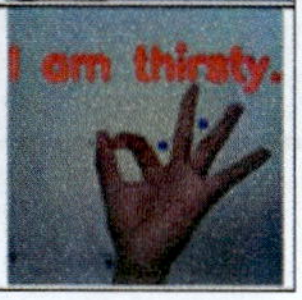

- Gesture 3: In the third gesture there is one defect. (Area ratio of this gesture is less than 17.5. This gesture is assigned to display "PLEASE HELP" and also speaks out "PLEASE HELP".
- Gesture 4: In the fourth gesture there are two defects. Area ratio of this gesture is less than 27. This gesture is assigned to display "I AM THIRSTY" and also speaks out "I AM THIRSTY".
- Gesture 5: In the fifth gesture there are three defects. Area ratio of this gesture is less than 27. This gesture is assigned to display "YES" and also speaks out "YES".
- Gesture 6: In the sixth gesture there are four defects. Area ratio of this gesture is less than 27. This gesture is assigned to display "THANK YOU" and also speaks out "THANK YOU".
- Gesture 7: In the seventh gesture there are no defects. Area ratio of this gesture is less than 17.5. This gesture is assigned to display "ALL THE BEST" and also speaks out "ALL THE BEST".
- Gesture 8: In the eighth gesture there are two defects. Area ratio of this gesture is less than 27. This gesture is assigned to display "OK" and also speaks out "OK".

3.4 Analysis Part of Result

For hand gesture detection we have a tendency to take eight different gestures. Every gesture repeated hundred times, so the total range of tested pictures was 800 among that pictures correct recognition was 760 and wrong recognition was 40 with a mean detection accuracy of gestures is 95% (Table 2).

$$\text{Accuracy} = \frac{\text{No. of correctly detected gestures}}{\text{Total no. of tested gestures}} * 100\% \tag{10}$$

Table 2 Confusion matrix

Actual output	Predicted output							
	1	2	3	4	5	6	7	8
1	100	0	0	0	0	0	0	0
2	0	100	0	0	0	0	0	0
3	0	4	95	0	0	0	1	0
4	0	0	3	90	2	0	0	5
5	0	0	0	0	100	0	0	0
6	0	0	0	0	0	100	0	0
7	7	6	2	0	0	0	85	0
8	0	0	3	5	2	0	0	90

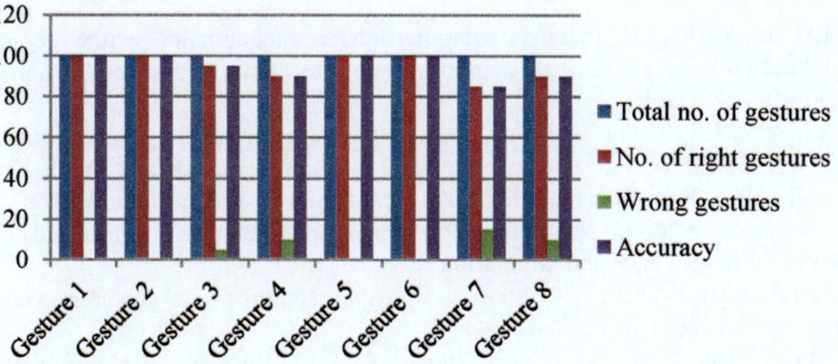

Fig. 7 Graph for average recognition rate

The chart for average recognition rate of various gestures is demonstrated as follows (Fig. 7).

4 Conclusion

The planned system is simple to implement as there's no complicated feature calculation. This system was implemented using Raspberry Pi, Pi camera module, 5-inch HDMI touch screen LCD display, Python Programming Language, and Open Source Computer Vision (OpenCV) library. The system was used to acknowledge sign language utilized by deaf and dumb person. This system is used to overcome the communication gap between mute person and ordinary person. There's a necessity for research within the area feature extraction and illumination therefore the system becomes more reliable. The system was used to acknowledge sign language utilized by deaf and dumb persons. The deaf and dumb person will use the hand gestures to do linguistic communication and it'll be converted into voice and text with accuracy 95%.

References

1. Rathi S, Gawande U (2017) Development of full duplex intelligent communication system for deaf and dumb people. In: 7th International Conference on Cloud Computing, Data Science & Engineering—Confluence https://doi.org/10.1109/confluence.2017.7943247 (2017)
2. Nath GG, Anu VS (2017) Embedded sign language interpreter system for deaf and dumb people. In: International Conference on Innovations in information Embedded and Communication Systems (ICIIECS) (2017)
3. Quiapo CEA, Ramos KNM (2016) Development of a sign language translator using simplified Tilt, Flex and contact sensor modules. In: IEEE Region 10 Conference (TENCON) (2016)
4. Tapadar S, Chatterjee SK, Saha HN, Ray S, Saha S (2018) A machine learning based approach for hand gesture recognition using distinctive feature extraction. In: 8th annual computing and communication workshop and conference (CCWC), IEEE

5. Jalab HA, Omer HK (2015) Human computer interface using hand gesture recognition based on neural network. In: 5th national symposium on information technology towards new smart world, 17–19 Feb, IEEE
6. Nath GG, Arun CS (2017) Real time sign language interpreter. In: International Conference on Electrical, Instrumentation and Communication Engineering (ICEICE), (2017)
7. Shweta SS, Autee RM, Bhosale VK (2016) Real time two way communication approach for hearing impaired and dumb person based on image processing. In: International conference on computational intelligence and computing research, IEEE
8. Abed AA, Rahman SA (2017) Python-based Raspberry Pi for hand gesture recognition. Int. J. Comput,. Appl. 173(4): 0975–8887
9. Javed MY, Gulzar MM, Rizvi STH, Junaid Asif M, Iqbal Z (2016) Implementation of image processing based digital dactylology converser for deaf-mute persons. In: International conference on intelligent system engineering (ICISE), Jan 2016, IEEE
10. Nandy A, Prasad JS, Mondal S, Chakraborty P, Nandi GC (2010) Recognition of isolated indian sign language gesture in real time. In: BAIP 2010, CCIS 70, pp 102–107. Springer, Berlin
11. Yadav K, Bhattacharya J (2016) Real-time hand gesture detection and recognition for human computer interaction. In: Intelligent systems technologies and applications, advances in intelligent systems and computing, vol 384. Springer International Publishing Switzerland
12. Farulla GA, Russo LO, Pintor C, Pianu D, Micotti G, Salgarella AR, Camboni D, Controzzi M, Cipriani C, Oddo CM, Rosa S, Indaco M (2014) Real-time single camera hand gesture recognition system for remote deaf-blind communication. In: AVR 2014, LNCS 8853, pp 35–52. Springer International Publishing Switzerland
13. Gonzalez R, Goods R (14) Digital image processing. Pearson Education Inc., New Jersey
14. Umbaugh S (1998) Computer vision and image processing. A practical approach using Cviptools with Cdrom,. Prentice Hall PTR

Robust Underwater Animal Detection Adopting CNN with LSTM

Harishchandra Jagtap and Mrunalini Chavaan

Abstract Underwater detection of objects valuable problem for many civil and military applications such as hydrographic surveys for the purpose of ensuring navigation. The objective of this work is to ensure flexibility, speed, and precise recognition of object underwater system for use in a variety of low-level underwater images captured. These imaging systems are used in separate occasions and under distinct weather and bathymetric circumstances from underwater imaging systems. This paper discusses the use of deep learning in the latest past to analyze underwater imagery. The methods to analyze are classified according to the object of detection, highlighting the characteristics and architectures used for deep learning. In the evaluation of digital sea bed imagery using deep neural networks, it is found that there is excellent scope for automation, particularly for the detection and tracking of detected object is harmful or non-harmful.

Keywords CNN · Underwater object detection · LSTM · HOG

1 Introduction

In most recent years, various investigation fields have achieved great results with the quick development of profound learning, and computer vision has arrived at another stage, trailed by relentless improvement of neural systems in convolution. Convolution neural network architecture is continuously improving. In addition, the return of the neural network convolution also considerably enhances computer vision application such as face recognition, object detection, object tracking, semantic segmentation, etc.

H. Jagtap (✉)
VLSI Design and Embedded System, MIT Academy of Engineering, Alandi, Pune 412105, India
e-mail: hjagtap01@gmail.com

M. Chavaan
School of Electrical Engineering, MIT Academy of Engineering, Alandi, Pune 412105, India
e-mail: mhchavaan@etx.maepune.ac.in

© Springer Nature Singapore Pte Ltd. 2021

S. N. Merchant et al. (eds.), *Advances in Signal and Data Processing*,
Lecture Notes in Electrical Engineering 703,
https://doi.org/10.1007/978-981-15-8391-9_15

Research has focused on object detection as one of the major Computer-related vision applications, and the neural network convolution has made excellent strides in object detection. Detection of objects grows from recognition of a single object to recognition of multi-objects. The meaning of the first is to identify a single object only from an image, it can be said that it is a classification problem, and it is not only the meaning of the later that can identify all the objects in an image. Deep learning has created a CNN-based, LSTM-based mainstream object identification algorithm, refreshing the greater precision in a number of renowned datasets.

The method in [1] showed the sonar change detection system is intended on the grounds of feedforward neural network and the network of pattern recognition. To recognize items, the Adaptive-Neuro Fuzzy System used submarine sonar pictures. Under supervised classification techniques such as feedforward and pattern recognition network, algorithm detection is introduced. This paper describes the development of fuzzy C-mean clustering, proposed by Chang et al. [2]. Shadow-Removal system for benthic article location. Since the acoustic shadow is joined by the gathered side output sonar picture, the precision of submerged article discovery will in general debase extensively. section the thing to finish the ID task by means of the Otsu calculation. Simulation studies showed that this paper's present strategy clearly And of course covers the quality of the underwater image degradation issue satisfactorily [2]. A novel method using the surveillance algorithm used by Faster R-CNN and Kernelized Correlation Filter (KCF) for real-time detection of seafood products like sea cucumber, sea urchin, and scallop, etc. [3].

Presented a full algorithmic pipeline for the identification and evaluation of underwater objects and, in particular, a novel calculation for the discovery of multifunctional items to discover ancient rarities made by individuals [4]. Li et al. [5] Encouraged by the exceptional detection precision and velocity property of Fast R-CNN, we use this promising network to automatically define fish to assist marine biologists to estimate fish presence and amount and efficiently comprehend geographical and biological ocean environments. Galceran et al. [6] states a novel algorithm is suggested in forward-looking underwater imagery to detect man-made objects underwater. The proposed technique does not require information from training as it considers the attributes of the detected sonar data. Suggested to use detectors of monocular vision to detect underwater items. We utilize light transmission information in connection to every now and again utilized tactile highlights, for example, shading and power to investigate the potential outcomes of identification of articles submerged. We recommend using monocular vision detectors to detect underwater items [7].

Xinhua et al. [8] another, convolutionary engineering for encoding-unraveling object acknowledgment submerged. It uses the significant encoding system to obtain the discriminative qualities from the boisterous low-light submerged picture. This proposed framework intended to give visual data, for example, route, review, mapping, and submerged assignments mediation. Our plan, called Fugu-f (Fugu adaptable), has the essential advantages of being vigorous in both the mechanical and programming parts, its adaptability since it is introduced as an inward module and can be custom fitted to various vehicles and exercises, and its ability to work in genuine time [9]. Sun et al. [10] A CNN learning move structure for submerged article acknowledgment and addresses the issue of removing discriminative highlights from generally small complexity pictures [10]. In this paper, first, we outline some profound learning calculations for particle identification, and after that apply one of the calculations to a new dataset to check its wide relevance.

The paper is organized as follows: Sect. 2 includes the basic system models, introduce system architecture of CNN with LSTM and flowchart. In Sect. 3 experiments and Sect. 4 results analysis including graphs and true table. In Sect. 5 we conclude the paper.

2 Proposed System

2.1 Architecture Selection

Choosing the correct architecture can be a challenging issue for a proposed framework built upon CNN and LSTM, since the process requires a great deal of tinkering for fine-tuning the hyperparameters. It can be quite difficult to know how many layers to use, what should be filter sizes, correct values for stride, etc. These are not trivial questions and no fixed standard set by researchers. This is on the grounds that, a system will generally rely upon the kind of information that may thus fluctuate by size of the picture, unpredictability of given information, accessible equipment assets, and considerably more. Through looking at the above-mentioned dataset and experimenting between several hyperparameters, I have chosen the best-performing model using right combination that created abstractions of image at a proper scale and exhibited promising outcomes. The CNN LSTM architecture, as proposed for the problem in this research, was comprised of a pattern, in which Convolutional, Dropout, Convolutional, and Max Pooling layers were placed sequentially (Fig. 1).

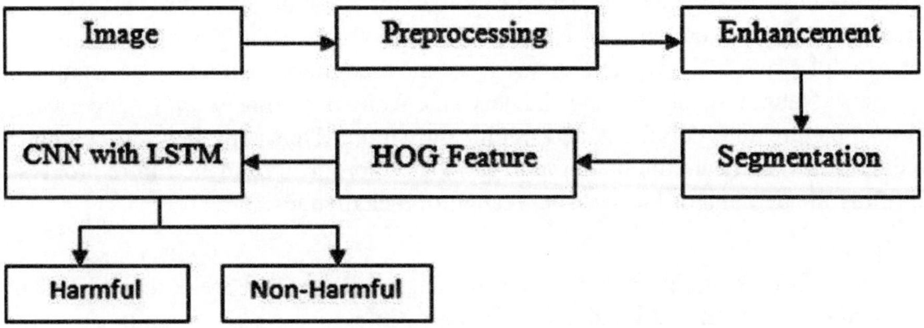

Fig. 1 Architecture diagram of underwater object detection using CNN with LSTM

2.2 Histogram of Gradients (HOG)

The significant concept histogram of gradients focused is that the local form and appearance object within a picture can be defined by the allocation of gradients of intensity or edge directions. The picture is split into tiny linked areas called cells, and a histogram of gradient directions is plotted for the pixels within each cell. These histograms are linked by the descriptor.

In a nutshell, The HOG descriptor is determined by figuring picture inclinations catching shape, and grayscale picture outline information. Gradient data is pooled into a 1-D orientation histogram, transforming a 2-D picture into a much narrower 1-D vector that forms the input for machine learning algorithms such as random forests, vector supporting machines, or logistical regression classifiers.

2.3 Convolutional Neural Network (CNN)

Deep neural networks were observed to be the most influential among all innovations in the field of computer vision, generating remarkable performance on Underwater Animal image classification. CNNs this class of artificial vision systems is particularly fascinating as an instrument for studying biological vision exhibits visual recognition capabilities that are comparable to those of human observers. As these models enhance their efficiency of recognition, they also appear to become more efficient in anticipating neural reactions in human ventral cortex. Recent benchmarks have shown that deep CNNs are excellent approaches for underwater object recognition [11].

Innovative improvements have even permitted the utilization of top of the line designs processor units gpust accelerate numerical issue settling utilizing this methodology. They resort not exclusively to bring down computational time, yet additionally permit thinking about a lot bigger systems. Thus, PCs are currently ready to drive further, more extensive and all the more dominant models. Cutting edge CNNs have accomplished human-like execution in a few acknowledgment assignments, for example, manually written character acknowledgment, Human face acknowledgment, Natural scene naming, Underwater Animal picture search, Underwater Animal picture auto-comment, and significantly more. Thus applying this technique to empirical data bears the potential to show great promise to make future progress fulfilling the demand of latest visual recognition challenges.

1. Information Layer
   The decision features from the pre-taken care of sensor data are portrayed as data vector of the Convolutional neural framework:

$$X = [x_1, x_2, \ldots x_k] \quad (1)$$

where k is the amount of features per window in the wake of figuring. To quicken the association of the model, we use Min–Max Normalization proposed in [12], by which the characteristics in every segment of data are straightly changed and institutionalized to [0, 1] region:

$$x = \left[\frac{x - \min}{\max - \min}\right] \tag{2}$$

min is the base of each area and max is the farthest point of each portion. By then, we make 2D picture like data by reshaping the 1 by 169 shallow features into a 13 by 13 square system before convolution movement.

2. Convolution Layer
The yield of the jth highlight map on the ith unit of the l convolution layer is:

$$x_i^{l,j} = \sigma\left[b_j + \sum_{a=1}^{m} w_a^j x_{i+a-1}^{l-1,j}\right] \tag{3}$$

b_j is the inclination term for jth feature map, m is the bit size, j a w is the largeness of jth feature guide and ath channel record and σ is the establishment work. We use ReLu as sanctioning work, which has been checked to have snappier getting ready speed and adequacy [13], closer to the human restless synaptic effect.

3. Max-Pooling Layer
The pooling layer determines the total insights for adjacent yield of l, j, I, x, which can lessen the measurement and affectability of the yield and accomplish scale-invariant element protection. The pooling activity utilized in this paper is max pooling. The maximum pooling layer separates the yield highlights of the convolution layer into a few segments and finding the greatest in each parcel [14].

The output of the max-pooling layer is:

$$x_i^{l,j} = \max_{n=1}^{r}\left(x_{(i-1)*Tn}^{l-1,j}\right) \tag{4}$$

where n is pooling size and T is pooling stride.

Intermittent neural systems (RNNs) model worldly elements by mapping shrouded state information arrangements and concealed yield states through the accompanying repeat conditions

$$h_t = g(W_{xh^{x_t}} + W_{hh^{h_{t-1}}} + b_h) \tag{5}$$

$$z_t = g(W_{hz^{h_t}} + b_z) \tag{6}$$

where g is a component insightful non-linearity, for example, sigmoid or hyperbolic digression, x_t is the info $h_t \in R^N$ is the shrouded state with N concealed units, and

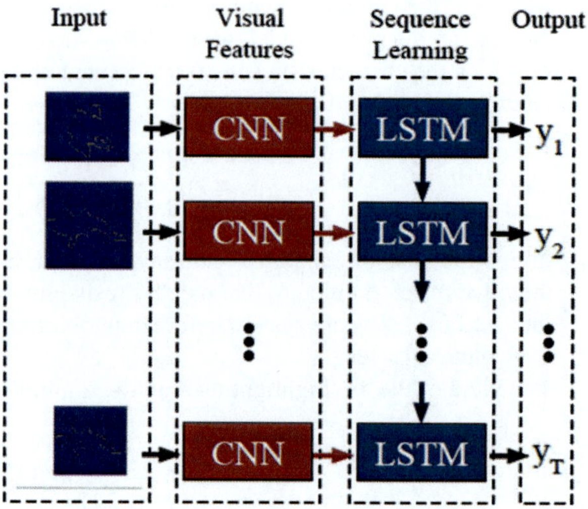

Fig. 2 Working diagram of CNN with LSTM

Z_t is the yield at time t. For a length T info grouping $(x_1, x_2, \ldots, x_T)$, the updates above are processed sequentially as $h1$ (letting $h_0 = 0$), $z_1, h_2, z_2, \ldots, h_t, z_T$ (Fig. 2).

Time Complexity of the Network The time unpredictability of all convolutional layers is:

$$O\left(\sum_{l=1}^{d} n_{l-1} \cdot s_1^2 \cdot n_1 \cdot m_1^2\right) \tag{7}$$

here l is the list of the convolutional layer, and d is the quantity of convolutional layers. n_l is the quantity of channels in the lth layer, and n_{l-1} is otherwise called the quantity of information channels of the lth layer. s_l is the spatial size of the channel and m_l is the spatial size of yield highlight map. The time cost of completely associated layers and pooling layers regularly takes 5–10% computational time, which isn't engaged with the above detailing.

2.4 Long Short Term Memory (LSTM)

Long Short-Term Memory (LSTM) [15] model is a variation of Convolutional Neural Network that has been proposed as an answer for fathom angle blast or abatement inferable from long time slacks in the Convolutional Neural Network (CNN) model learning process during back proliferated blunder.

LSTM might be viewed as a LSTM unit organize. Each LSTM unit is fitted with three doors to control the progression of information: (1) input entryways to decide when the info is adequately imperative to recollect; (2) overlook doors to decide when the unit ought to recall or overlook the worth; and (3) yield door to decide when the unit should show the worth.

In the previous decade, LSTM models have been recognized as solid models that show grouping data learning abilities. LSTM's capacity lies in its ability to catch long-go conditions and gain from variable groupings of span proficiently. A few examinations have uncovered that LSTMs have been effective in fathoming the accompanying issues: grouping of edge savvy phonemes, order of scene pictures, age of pictures. Likewise, to recognize fake card exchanges, LSTM models were examined.

LSTM engineering is a variation of RNN, which is intended to go up against the disappearing inclination issue. It utilizes memory squares to store and access data over extensive stretches of time. So LSTM is truly reasonable for consistent acknowledgment undertakings, which interest for utilizing long haul relevant data [15].

The formulas to update LSTM at time t are described as follow,

$$i_t = \sigma_i(W_{xi^{x_t}} + W_{hi^{h_{t-1}}} + W_{ci^{c_{t-1}}}) \tag{8}$$

$$o_t = \sigma_o(W_{xo^{x_t}} + W_{ho^{h_{t-1}}} + W_{co^{c_{t-1}}}) \tag{9}$$

$$f_t = \sigma_f(W_{xf^{x_t}} + W_{hf^{h_{t-1}}} + W_{cf^{c_{t-1}}}) \tag{10}$$

$$c_t = f_t c_{t-1} + i_t \sigma_c(W_{xc^{x_t}} + W_{hc^{h_{t-1}}}) \tag{11}$$

$$h_t = o_t \sigma_t(c_t) \tag{12}$$

where σ, is the non-linear function, W is the weight between two connected units, x_t is the input vector, and i_t, o_t, f_t, c_t, h_t represent outputs of input gate, output gate, forget gate, cell, and hidden state vector respectively. In our model, the weights are initialized with uniform random numbers of scale 0.01. The activation function tanh(h) is used in the concealed layer. The dropout is used between the input vector and the input gate, with dropout speed 0.5 to prevent the network overflow.

Softmax function is used to estimate distribution with prior layers output, and our model's cost function is the probability loss function. To train the LSTM model, RMSProp technique 16 is used to minimize the loss function, which is the optimization of gradient descent. The weights are updated using the backpropagation algorithm.

Our proposed model joins the convolutional neural system (CNN) for pictures portrayal and the LSTM for successive label rundown demonstrating. This model is roused by late work of image inscribing utilizing RNN. The differentiation is that in picture inscribing, the image includes just legitimately adjusts the beginning condition of the RNN for subtitle age. Once the first expression of the inscription is produced, the rest of the words are created simply dependent on past shrouded

Fig. 3 Flowchart

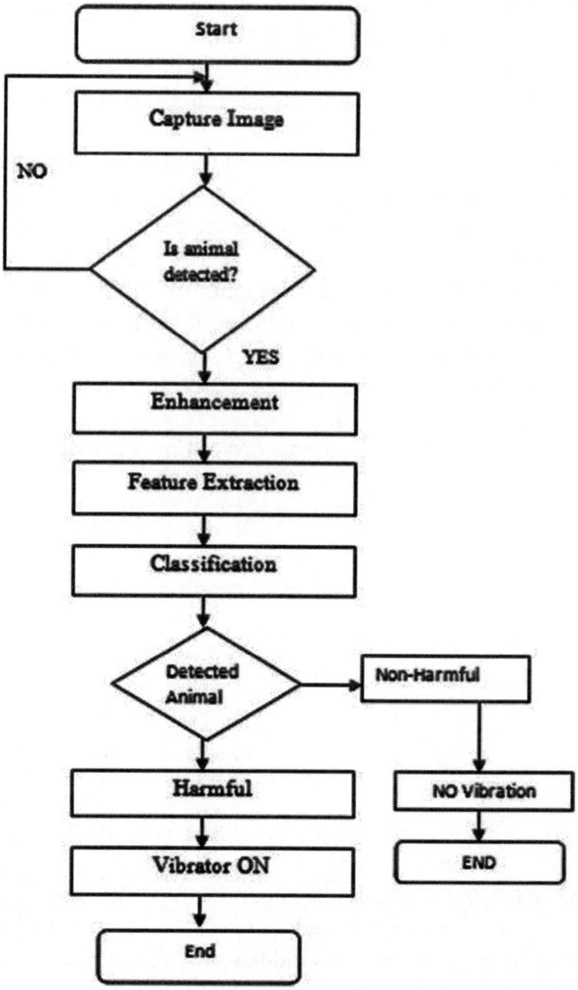

conditions of the RNN. For sentence age, this bodes well since the words in a sentence will in general have solid conditions, so it is sensible to encourage the picture include just to the underlying state [15] (Fig. 3).

3 Experiments

For experimenting with the proposed CNN & LSTM framework, it was important to choose a well-adjusted dataset, a feasible architectural design built upon suitable hyperparameters, and required hardware resources to fit the training process. The detailed experimental procedure, which was followed based on these factors, is given in the sections below.

3.1 Training

The suggested model was trained using a logarithmic loss function with a stochastic gradient downward optimization algorithm configured with a big decline in momentum and weight that began with a learning rate of 0.01. It was equipped with 10 epochs and a big batch size of 128, which was discovered through a small experiment. Normally the number of epochs for this problem would be one or two orders of magnitude larger. Each epoch took an average computational time of 123 s, causing the entire training phase to finish in approximately 13 min. These epochs acted like smaller training sessions that ran over all of the data given in training set. During that run, values or weights of filters were adjusted through a process called backpropagation. It was the part where value of the layers were correctly tuned with respect to the loss function while carrying out forward and backward passes. The ultimate objective was to obtain a set of parameters that could generalize towards fresh information. And the accuracy of validation expressed that capacity.

Running my proposed model delivered precision and loss of classification function values on the training dataset through each epoch, where the best-achieved classification accuracy without data augmentation was 97.17% along with a loss function value of 0.1146. However, a light augmentation resulted in having classification accuracy up to 91.03% followed by a loss function value of 0.3404. This training process was performed on a personal computer featuring a dual-core Intel Core i7-CPU, 8 GB system memory.

4 Results

This section reports the best accuracy achieved by the proposed model for the VOR task and provides an analysis as well as highlights a list of top recognition rates achieved by professional researchers, followed by a discussion showing some scope for improvement.

4.1 Testing

The trained model built in accordance to specified configuration, as described in the previous section, was tested on the validation dataset to estimate its ability to generalize toward new data, because the validation set contains only the model's information has never seen before and therefore cannot just memorize. Although the model initially achieved a classification accuracy of 91.03% along with a loss function value of 0.3404, data augmentation pushed the classification to reach a baseline accuracy of 93.46% followed by a loss function value of 0.4512.

4.2 Analysis

Look at it closer, the difference between the classification accuracy and loss function values obtained from training and validation datasets before and after performing image data.

There is plenty decisions to make when designing and configuring a deep learning model such as the one which I have proposed. Most of these decisions must be resolved empirically through by trial and mistake, and evaluating the model on problem data. As it is therefore of critical importance to have a robust way to evaluate the performance of the deep neural networks.

Below are a few ways that could have been followed to elevate the proposed model's performance:

Train for more epochs: For a few epochs, the current model has been trained, 100 to be precise. It is prevalent for a few hundreds or thousands of epochs to train big convolutionary neural networks to achieve required outcomes. State-of the-art performance can be accomplished by increasing the amount of training periods considerably.

Non-Harmful Class:

See Fig. 4.

Harmful Class:

See Figs. 5 and 6.

Table 1 shows that predicted result and true label. Our proposed algorithm gives better accuracy with minimal false recognition (Fig. 7).

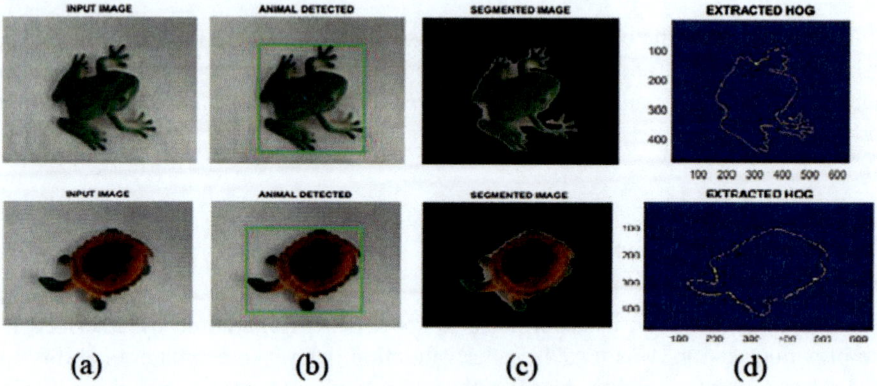

Fig. 4 **a** Input image, **b** animal detected, **c** segmented image, **d** expected HOG

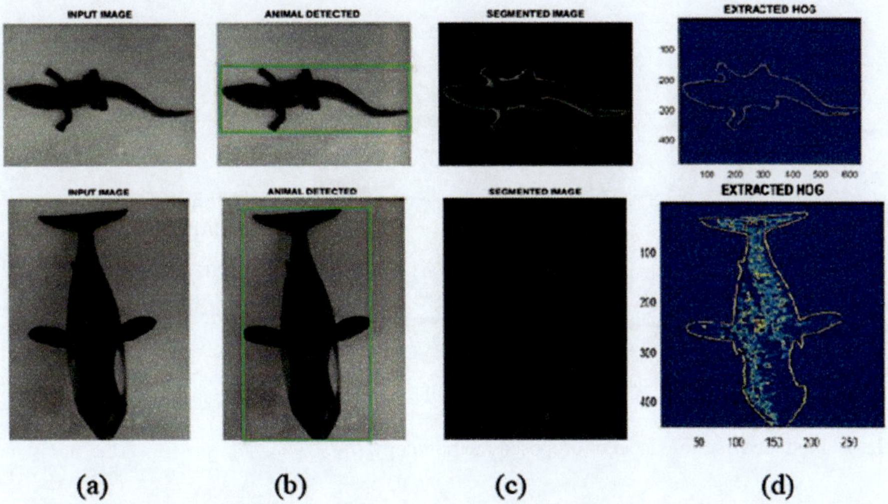

Fig. 5 **a** Input image, **b** animal detected, **c** segmented image, **d** expected HOG

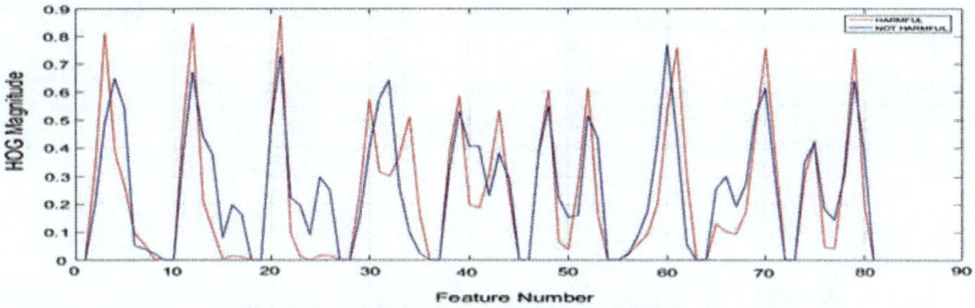

Fig. 6 Combine feature vector of harmful and non-harmful animal

Table 1 Proposed methodology predicted and true label:

Animal name	Total no. of image captured	Predicted as harmful	Predicted nonharmful
Frog	100	1	99
Crocodile	100	100	0
Tortoise	100	2	98
Shark	100	99	1

Table 2 shows that predicted result and true label. Our proposed CNN + LSTM algorithm gives better performance than rest of the methods (Fig. 8).

Table 3 shows that comparison result of LBP features and HOG features extraction from experiment taken, it is clear that HOG performs better than LBP feature (Fig. 9).

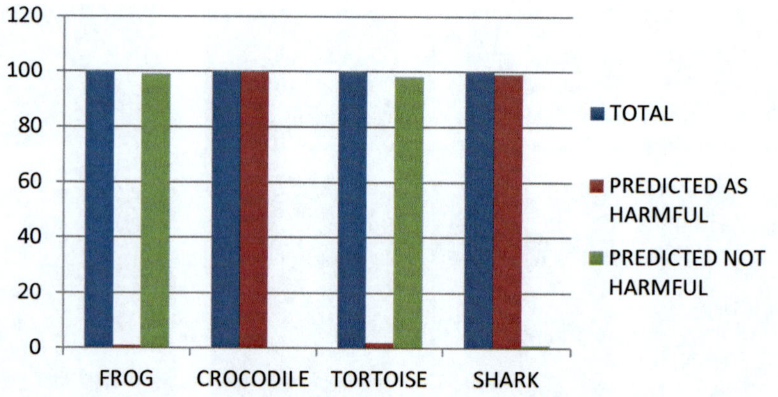

Fig. 7 Predicted and true label of proposed methodology

Table 2 Comparison of traditional classifier and proposed CNN + LSTM:

Animal name	Total no. of image captured	KNN	ANN	SVM	CNN	Proposed CNN + LSTM
Frog	100	84	93	95	96	99
Crocodile	100	80	91	93	96	100
Tortoise	100	86	88	93	89	98
Shark	100	97	96	94	97	99

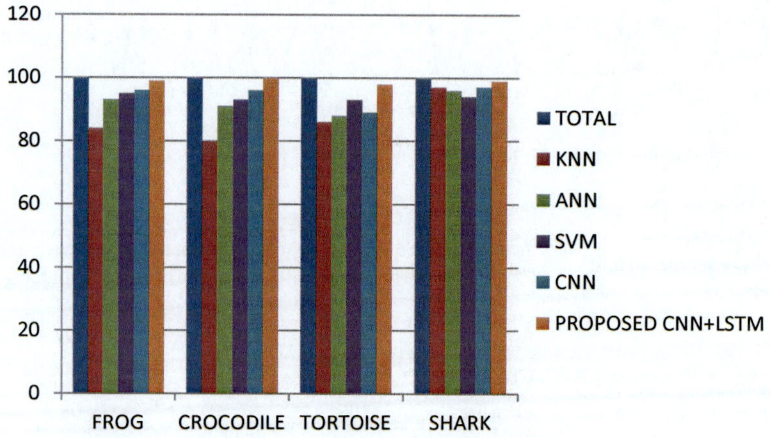

Fig. 8 Comparison of traditional classifier and proposed CNN + LSTM

Table 3 Comparison of LBP features and HOG features

Animal name	Total no. of image captured	LBP	HOG
Frog	100	93	99
Crocodile	100	98	100
Tortoise	100	78	98
Shark	100	97	99

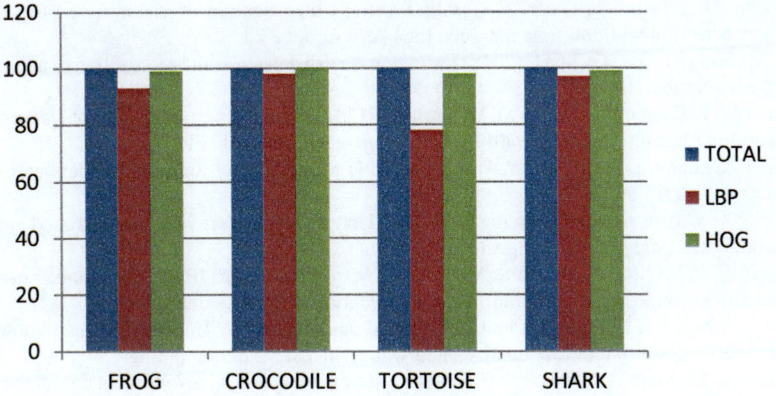

Fig. 9 Comparison of LBP features and HOG features

5 Conclusion

In this paper, recent methods are discussed to detect and classify different underwater animals using deep learning. Used image features and architecture of profound learning are summarized. All approaches to data analysis had to be highlighted in a single article to make it simple to concentrate on the opportunities of future work based on profound neural network techniques. It has been discovered that more research has been done to detect and classify corals using profound learning, but no work has been done to detect objects that are harmful or non-harmful, which is equally essential to oceanic ecosystems. When combining color and texture-based characteristics, the efficiency, precision, and robustness of any identification and classification algorithm can be considerably improved. For object detection and classification, the accumulation of hand-crafted characteristics and neural networks can yield better outcomes. Therefore, there is a chance to create efficient and effective underwater imagery deep learning that will be the focus of our future work.

References

1. Anitha U, Malarkan S (2018) Underwater object identification and recognition with sonar images using soft computing techniques. Indian J Mar Sci **47**(03)
2. Chang R, Wang Y, Hou J, Qiu S, Nian R, He B, Lendasse A (2016) Underwater object detection with efficient shadow-removal for side scan sonar images. IEEE
3. Xu F, Ding X, Peng J, Yuan G, Wang Y, Zhang J, Fu X (2018) Real-time detecting method of marine small object with underwater robot vision. IEEE
4. Rizzini DL, Kallasi F, Oleari F, Caselli S (2015) Investigation of vision-based underwater object detection with multiple datasets. Int J Adv Robot Syst
5. Li X, Shang M, Qin H, Chen L (2015) Fast accurate fish detection and recognition of underwater images with fast R-CNN, MTS
6. Galceran E, Djapic V, Carreras M, Williams DP (2011) A real-time underwater object detection algorithm for multi-beam forward looking sonar. TRIDENT EU FP7
7. Chen Z, Zhang Z, Dai F, Bu Y, Wang H (2017) Monocular vision-based underwater object detection. MDPI Sens
8. Wang X, Ouyang J, Li D, Zhang G (2019) Underwater object recognition based on deep encoding-decoding network. Springer, Berlin
9. Bonin-Font F, Oliver G, Wirth S, Massot M, Negre PL, Beltran JP (2015) Visual sensing for autonomous underwater exploration and intervention tasks. Ocean Eng
10. Sun X, Shi J, Liu L, Dong J, Plant C, Wang X, Zhou H (2017) Transferring deep knowledge for object recognition in low-quality underwater videos. Neurocomputing
11. Tobias L, Ducournau A, Rosseau F, Fablet R, Mercier G (2016) Convolutional neural networks for object recognition on mobile devices: a case study. In: International conference on pattern recognition. Research Gate, Cancun, pp 2–7
12. Sola J, Sevilla J (1997) Importance of input data normalization for the application of neural networks to complex industrial problems. IEEE Trans Nucl Sci 44(3): 1464–1468
13. Glorot X, Bordes A, Bengio Y (2011) Deep sparse rectifier neural networks. In: International Conference on Artificial Intelligence and Statistics, pp 315–323
14. Nagi J, Ducatelle F, Di Caro GA et al (2011) Max-pooling convolutional neural networks for vision-based hand gesture recognition. In: 2011 IEEE international conference on signal and image processing applications (ICSIPA), IEEE, pp 342–347
15. Yang S, Zhu Q (2017) Continuous Chinese sign language recognition with CNN-LSTM. In: Ninth international conference on digital image processing
16. Gauthier I, Tarr MJ (2016) Visual object recognition: do we (finally) know more now than we did? Ann Rev Vis Sci 2:377–396
17. Graves A (2012) Supervised sequence labelling with recurrent neural networks. Springer, Berlin

Face Recognition Using Golden Ratio for Door Access Control System

Prajakta S. Gaikwad and Vinayak B. Kulkarni

Abstract The paper introduces a method-based correspondence scheme that integrates a permutation of Viola–Jones face detection method with characteristics of extracting golden ratio. The purpose of this article is to help users improve the security of sensitive places through facial recognition. We come to resolve the issue of low precision in the suggested technique. Here, we propose a new scheme for aligning the face using Viola–Jones and support vector machine face detection method followed by extraction technique of golden ratio function. The suggested technique is very effective, more realistic and accurate compared to other face detection techniques. The module contains a secure face identifier. The scheme is designed to satisfy the requirements of classification face to face in real scenario.

Keywords Viola–Jones · Golden ratio · Support vector machine · Face detection

1 Introduction

The face represents a significant role in transmitting identity in our social interactions. It is noteworthy the human capacity to recognize faces. Even after years of separation, we can acknowledge thousands of faces learned in our lives and realize acquainted faces concisely. In spite of the great changes in the visual style due to the circumstances of visualization, the expressions, the aging and the distractions, like glasses or the modifications of the hairstyle, the capacity is quite robust. But it is quite troublesome to develop a process model of face classification, as faces are often complicated, multidimensional and altered over time. Typical applications of face organization are human–robot interaction, human–computer interaction, driver‘s license, sensible cards, national ID, passports, elector registration, security

P. S. Gaikwad (✉) · V. B. Kulkarni
School of Electrical Engineering, MIT Academy of Engineering, Alandi, Pune, India
e-mail: prajktagaikwad53@gmail.com

V. B. Kulkarni
e-mail: vbkulkarni@entc.maepune.ac.in

© Springer Nature Singapore Pte Ltd. 2021

S. N. Merchant et al. (eds.), *Advances in Signal and Data Processing*,
Lecture Notes in Electrical Engineering 703,
https://doi.org/10.1007/978-981-15-8391-9_16

system, criminal identification, personal device logon, desktop logon, information security, database safety, computer network security, Web access, medical records video observation, CCTV and suspicious pursuit and investigation.

Face recognition involves capturing the facial image from a web camera. This is the visitor image capture and image comparison with the stored database. A face detector must say if there is a human face in an image of arbitrary size and if so, wherever it is. A natural framework to address this issue is binary classification, wherever a classifier is constructed to reduce with a sensible computational budget. Techniques applied on integral image and additional cascade classifier create the Viola–Jones algorithmic rule extremely effective supplied with a picture flow produced in real time from a conventional webcam. It works well in face classification and extracts golden ratio characteristics. Using SVM classifier, it is one of the most effective and significant image analysis and classification applications. Since there can be no objective of allocation a specified image's real prior probability of having a face, an formula should minimize each false negative and false positive rates to get a suitable output. This work wants a certain numerical description of what constitutes human faces.

Pawar et al. [1] this document contains a technique for evaluating correspondence based on a novel method combining permutation of the Viola–Jones face detection method with the Eigen face classification (identification) technique. The human face is a complex multidimensional visual model, and the development of a computational model is therefore very hard to recognize. The paper pre-sends a methodology based on the characteristics acquired from the image to identify the human face. The first phase uses a Viola–Jones algorithm to define in a image the human face. The first stage utilizes an algorithm of Viola–Jones to identify the human face in a picture. In the next phase, the principle component analysis and another feed forward neural network fusion acknowledge the picture face recognized. Dabhi and Pancholi [2], in this document stated that for applications (like eye tracking, face recognition, face expression recognition, face monitoring and lip reading) locating facial features is an important step in face identification. They present a method of identification of the face of a live image. The face is recognized from the full picture using algorithm Viola–Jones. The hair-based AdaBoost algorithm is used to extract the face region from the image.

Hassaballah and Murakami [3], they propose another assessment measure for face identification calculations by abusing an organic property called golden ratio of the ideal human face. Contrasted with the current measure, the new assessment measure is increasingly sensible and exact. Li et al. [4] proposed a system for assessing most extreme probability of correspondence (MLCE) to learn customized correspondence dependent on the reason of greatest probability of frontal face. We can orchestrate virtual frontal photos of the profile faces for ensuring acknowledgment in the wake of procuring the semantic correspondence encoded in the educated dislodging. Sekhon and Agarwal [5] present a fake neural system-based back spread learning calculation to perceive human appearances. We have been taking a shot at eight recognizable proof attributes.

Ding and Tao [6] were addressed the issues inborn in PIFR and present an intensive assessment of the strategies created. Existing PIFR systems can be arranged into four groupings, for example, present powerful extraction ways to deal with highlights, multi-see ways to deal with subspace learning, face amalgamation approaches and half and half approaches. Multi-see dependability infer a speculation blunder connected to multi-see solidness and Rademacher intricacy, demonstrating that multi-see integrally is valuable to steadiness and speculation. Wisarutchantara [7] presents a strong method for identifying items dependent on fast coordinating layouts. To accurately isolate the objective picture, we utilize a segment search. We can adequately make each model in the sub-pictures during this stage contingent upon twisting measures. Narasimha–Shenoi [8], while some may debate the importance of the golden ratio, it is clear that through our history there has been an interest with it. Many will guess on the legitimacy of it in nature, just as in our history.

The document is disciplined as follows: Sect. 1 includes introduction and related work. Section 2 is consisting of proposed system in which architecture is shown. In this, techniques used for face detection, eyes detection, nose detection, mouth detection are included. Flowchart and hardware module are shown in this section. Section 3 is consisting of results and charts of the results. We concluded in the Sect. 5.

2 Proposed System

In the proposed architecture, the inputs are taken from the camera. There are three work modules. This is the registration module, the authentication module and the application module. There is a database that has stored images. For tests, each time the user arrives in front of the camera, the latter captures the image and sends it to the authentication module. The authentication module verifies the authenticity of the user by checking the data of the registered database. The result will go to the application module. The proposed system architecture is illustrated in Fig. 1.

In Fig. 2, block diagram of hardware module is shown. Raspberry Pi is itself a mini computer. It performs all operations like computer. Camera has connected to the Raspberry Pi. Camera will capture the images and will send to controller. It will train images in registration module. In authentication module, it will check authenticity of person. And at a time, result will go to the display as well as motor for further operations.

2.1 Image Preprocessing

The image preprocessing is an important stage in many face recognition methods. It makes easy way to detect the facial features by revising face images to the same size.

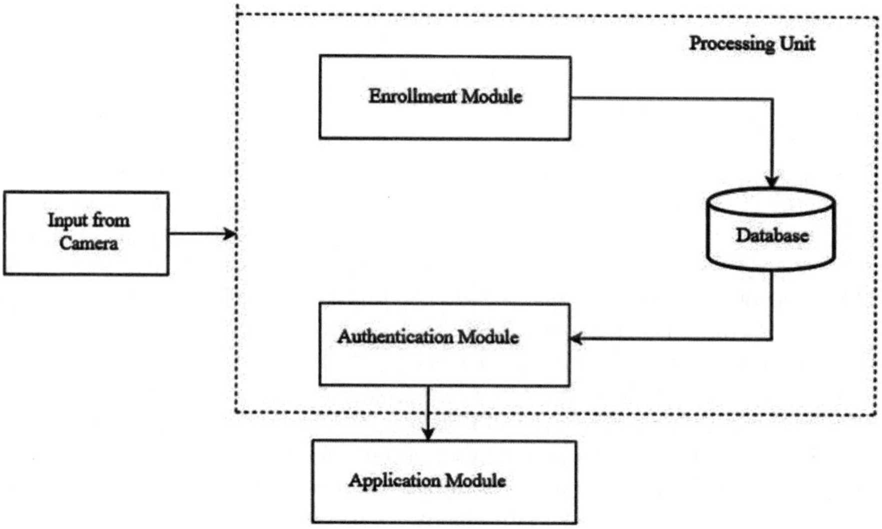

Fig. 1 Proposed system architecture

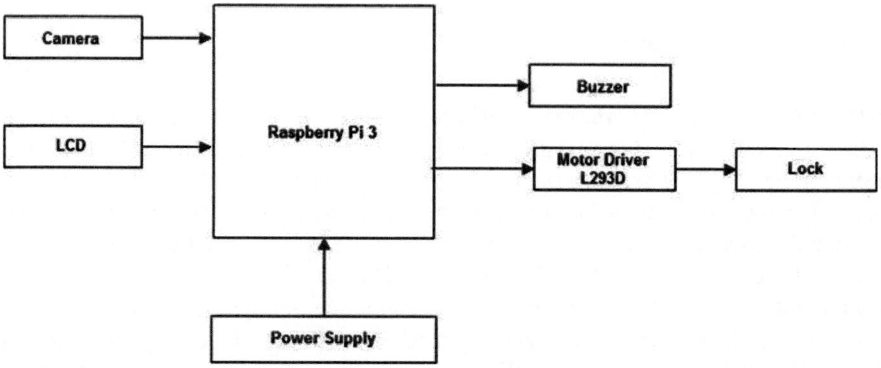

Fig. 2 Block diagram for hardware module

2.2 Noise Removing

After segmentation, there may be some small regions, which are not facial features are served as noise. We remove noise by opening operation in morphology.

2.3 Face Detection

After the process of noise removing, we can get candidate face region. In our research, there is only one face for detection. Therefore, the biggest region is just the face region. We can easily get face region by an eight-connected-component labeling. Therefore, we can choose the max region to be a face region. In Fig. 3, we have shown proposed processing steps applied. In Fig. 4, we have shown the detected face from whole image.

Viola–Jones

For instance, there are a total of $M = 162{,}336$ feasible characteristics in a normal 24×24 pels sub-window. The object detection framework employs a variant of the AdaBoost learning algorithmic program to both choose the most effective features and to train classifiers that use them. This algorithmic program constructs a "strong" classifier as a linear combination of weighted straightforward "weak" classifiers [9].

$$h(x) = \operatorname{sgn}\left(\sum_{j=1}^{m} \alpha_j \, h_j(x)\right) \tag{1}$$

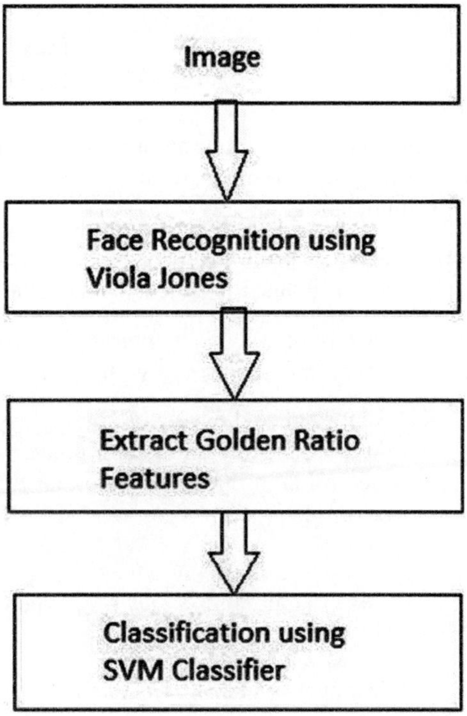

Fig. 3 Proposed processing steps

Fig. 4 Face detection

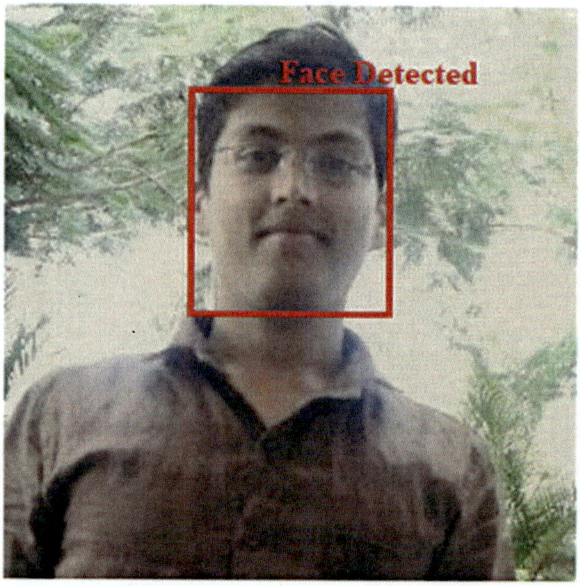

Each weak classifier is a feature-based limit.

$$h_j(x) = \begin{cases} -8_j \text{ if } f_j < \theta_j \\ 8_j \quad \text{otherwise} \end{cases} \tag{2}$$

The amount of limitation and the pleasure of pollis snoops in training, as well as campaigners. Here is what is reported a streamlined version of the learning algorithmic program:

Input: Set of N positive and negative trained images on their labels.
(x^i, y^i). If image i is a face $y^i = 1$, otherwise $y^i = -1$.

1. Initialization: assign a weight $w_1^i = \frac{1}{N}$ to every image i.
2. For every characteristic $f_i$with$j = 1, \ldots, M$

   a. Renormalize the weights so that they total to one.
   Apply the function to each image within the learning set, and then, notice the optimum threshold and polarity minimizing the weighted classification error. Then, it is

$$\theta_j, s_j = \arg\min_{\theta, s} \sum_{i=1}^{N} w_j^i \in_j^i \quad \text{where} \quad \in_j^i = \begin{cases} 0 \text{ if } y^i = h_j(x^i, \theta_j, s_j) \\ 1 \text{ otherwise} \end{cases} \tag{3}$$

   b. Assign weight $\propto_j$ to h_j that is inversely proportional to the error rate. In this way, best classifier is c.

c. The weights for the successive iteration, i.e. w^i_{j+1}, are reduced for correctly classified images.

3. Set the ultimate classifier to

$$h(x) = \text{sgn}\left(\sum_{j=1}^{M} \alpha_j h_j(x)\right) \tag{4}$$

2.4 Eye Detection

The eyes are distinguished on the basis of the hypothesis that they are darker than alternative parts of the face. A pair of potential ocular regions is considered an eye if it satisfies certain constraints based on the anthropological characteristics of human eyes [9].

We use biometrics to detect facial features. Based on the ratio of size between eyes and the face, we can locate eyes from different size of pictures. Also, we use the human face golden ratio to detect the region of eyes. By the human face golden ratio and experimental results, the size of eyes is varied between 2 and 4% of the size of face region. Moreover, the locations of eyes are also important information to verify the eyes. At first, we use a box to locate the face region that has length (Disx) and wide (Disy). Suppose that the upper-left corner of this box is the coordinate center, the vertical location of the center of eyes will be located in between 33 and 66% of the width of face. In Fig. 5, we have shown the detection of eyes.

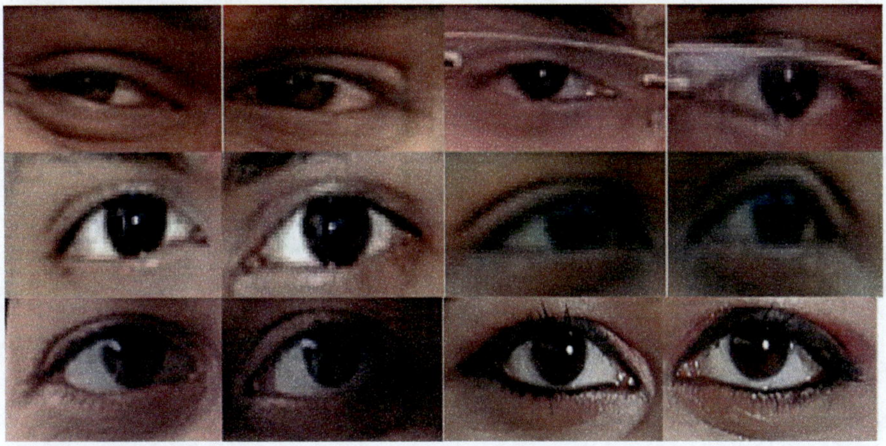

Fig. 5 Eye detection

2.5 Nose Detection

The pixel values of nose are more close to those of face color. Thus, the methods based on pixel values will have poor performance for the segmentation of nose. In our approach, we combine human face golden ratio to estimate the location of nose region which is shown in Fig. 6.

The nose has three completely different local (native) characteristics used as follows:

1. Similarity on both sides: The left and right sides of the nose are comparable in a front face.
2. Dark–White–Dark (DWD) Property: The lower part of the nose area is described by two blunt nostrils and a clear sub-place due to the light impression on the nose.
3. The variety in the property of the lower/upper parts: When the face is turned a few degrees, these expectations are lost, and the most obvious property is the variety between the lower part and the upper part. This variety can be estimated by the change in each part. Based on this survey, a quest is performed for a specific district among the ten highest recognized localities due to the impression of light at the end of the nose.

On both sides of the nose, the same double image is sought for the edges of the nose, and the separation between these edges is considered as the width of the nose [9].

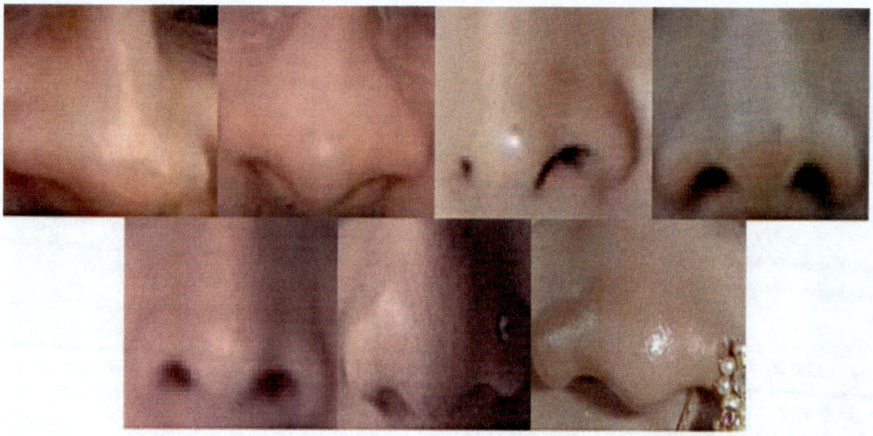

Fig. 6 Nose detection

2.6 Mouth Detection

By the human face golden ratio and experimental results, the size of mouth is varied between 3 and 7% of the size of face region. Moreover, the vertical location of mouth denoted as Mouth(y) is located in the lower 33% of the face. Also, the horizontal location of mouth denoted as Mouth(x) is located from 30 to 70% of the width of face.

Location and extraction highlights from the quarter of the mouth; this model is composed of fragile classifiers, for a choice of stump, which uses Haar's strengths to encode the subtleties of the mouth [9].

From the point of view of human visual recognition, the lips are anything but difficult to separate from the face skins for all races of individuals because of their various complexities in shading. In Fig. 7, we have shown the detection of mouth. Similarly, the conversion of the RGB color space to YCbCr is performed to determine the skin area following the equation. In order to detect skin color, the following rules were found to be more accurate than other models.

$$Y = 0.257R + 0.504G + 0.098B + 16 \tag{5}$$

$$C_b = -0.148R - 0.291G + 0.439B + 128 \tag{6}$$

$$C_r = 0.439R - 0.368G - 0.071B + 128 \tag{7}$$

Golden Ratio

$$\frac{h}{w} = 1.618, \tag{8}$$

and

$$\frac{d_{\text{eye}}}{m} = 1.618 \tag{9}$$

where h, w, d_{eye}, and m are: height of the face, width of the face, distance between the center of eyes, and the length of the mouth, respectively. Fig. 8 is showing detection

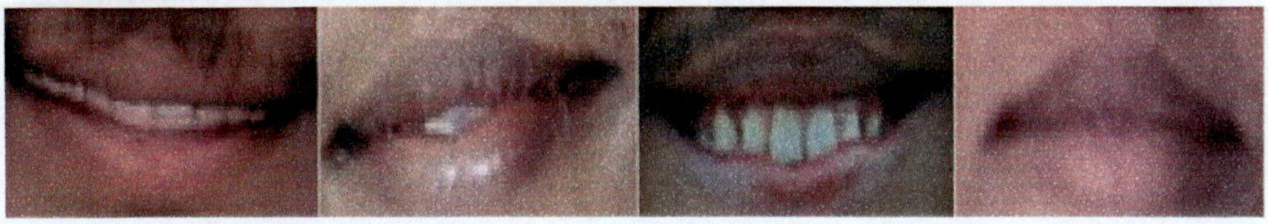

Fig. 7 Mouth detection

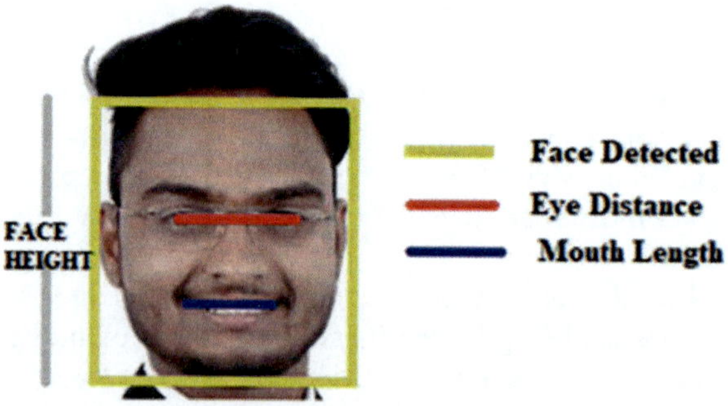

Fig. 8 Figure showing face detected and following parameters: facial height, facial width, pupillary distance and the length of the mouth, respectively

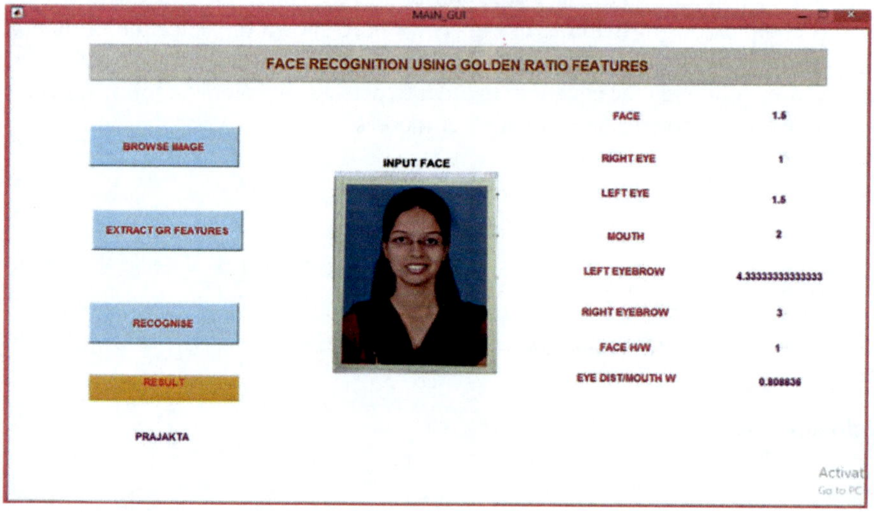

Fig. 9 MATLAB GUI output

of face and following parameters: facial height, facial width, pupillary distance and the length of the mouth, respectively.

In Fig. 9, output of the offline MATLAB system is shown. In this, after clicking on the browse image button, it will take an input image, which we have to recognize. In Table 1, we have made list of different facial ratios called as golden ratios. After that clicking on feature extraction button, feature points are extracted, and its golden ratio values are shown in list (Table 2). After that, person will be get recognized.

2.7 SVM Classifier

According to the nonlinear classifiers proposed by Vapnik (1992), it happens regularly that our golden proportion homes are not directly detachable in a *p*-dimensional

Table 1 Different ratios used called as golden ratios

Ratio No.	Description
1	Length of the ear to the interocular distance
2	Length of the ears to the width of the nose
3	Eye distance to interocular distance
4	Distance between the eyes and the width of the nose
5	Width of the mouth at the interocular distance
6	Lip–chin distance at the interocular distance
7	Distance lip–chin to the width of the nose
8	Interocular distance to the width of the eye cleft
9	Interocular distance at lip height
10	Width of the nose to the width of the eye slot
11	Width of the nose at the height of the lips
12	Distance from eye crack to nose-to-mouth distance
13	Height of the lip at the distance nose–mouth
14	Face length to the width of the face
15	Distance from chin to chin, distance from lips to chin
16	Distance from the nose to the nose–mouth distance
17	Width of the mouth to the width of the nose

Table 2 Tested real-time golden ratio values

Feature	Harish	Umesh	Prajakta
Face	1.5185	1.5172	1.5
Right eye	1.03	1.34	1
Left eye	1.5	1.4838	1.5
Mouth	1.9722	2.0158	2
Left eyebrow	4.0909	3.8333	4.3333
Right eyebrow	2.9285	2.9333	3
Face h/w	1	1	1.21
Eye dist/mouth w	0.0199	0.0078	0.8088

(finite) space [10]. When $k(x_i, x_j)$ is a partial work, x_i and x_j are bright proportion a vector highlight the space and d is the polynomial capacity level.

$$k\left(\overrightarrow{x_i}, \overrightarrow{x_j}\right) = \left(\overrightarrow{x_i} \cdot \overrightarrow{x_j}\right)^d \tag{10}$$

Selecting the SVM Hyperplanes for classifying Golden ratio features

Linear Separation: For brilliant proportion, include which can be isolated directly; we select two parallel hyperplanes that differentiate the two classes of highlight information, the goal being that the separation between the two lines is extreme. The regional parameters b/w of these two hyperplanes are called “edge”, and the most extreme of the hyperedges is the one that surrounds them.

$$\vec{\omega}x_i - b \geq 1 \quad \text{if} \quad \theta_i = 1 \tag{11}$$

$$\vec{\omega}x_i - b \leq 1 \quad \text{if} \quad \theta_i = -1 \tag{12}$$

where $||\vec{\omega}||$ is normal vector to the hyperplane, θ_i represents classes and x_i represents features. The distance between two hyperplanes is $\frac{2}{||\vec{\omega}||}$ to maximize this distance denominator value; you must minimize it, that is $\|\vec{\omega}\|$ should be minimized.

For a correct classification, we can construct a combined equation:

$$|\vec{\omega}||_{\min} \text{ for } \theta_i(\vec{\omega}x_i - b) \geq 1 \quad \forall i = 1, 2, \ldots, n \tag{13}$$

Nonlinear Separation: To create classifier for nonlinear golden ratio data, we try to minimize

$$\left[\frac{1}{n}\sum_{i=1}^{n} \max\left(0, 1 - y_i\left(\vec{\omega} \cdot \vec{x_i} - b\right)\right) + \lambda||\vec{\omega}||^2\right] \tag{14}$$

Here, the max () method will be zero (0), if x_i is on the right side of the margin. For data located on the opposite side of the margin, the value of the function is proportional to the distance of the margin, where λ determines a compromise between the increase in the size of the margin and the fact that $\vec{x_i}$ is on the correct side of the margin.

Hardware Details

In this modern world, crime has become ultra modern too! In this current time, a lot of incident occurs like robbery, stealing unwanted entrance happens abruptly. So, the security does matters in this daily life. People always remain busy in their day to day work also want to ensure their safety of their beloved things. Sometimes, they forget to look after their necessary things like keys, wallet, credit cards, etc. Without these, they are unable to access their home or any place they want. To prevent such incidents, many scholars came into action and tried to prohibit them.

Hardware implementation is the most important part of the project in a precise way. The project’s electronics components are easily understood, and it includes connection to the Raspberry Pi microcontroller of various hardware devices and various other available ports without the need to solder the wires for connection. The system proposed includes Raspberry Pi 3b+, buzzer, LCD, camera, etc.

Fig. 10 Hardware module

Camera will start capturing the image if person is authentic, the door will get automatically opened. If the captured image is unauthentic, then buzzer will turn automatically on, and result will display on LCD. Hardware module has shown in Fig. 10 and flowchart of the working of system in Fig. 11.

Methodology for combining hardware and software:

Firstly, we have done offline training and testing of data on MATLAB 2016a software. We added and installed hardware support packages required for the work that is Raspberry Pi. Then accordingly, we installed operating system on memory card which is inserted in Raspberry Pi. We uploaded the code. For that, mex code has been generated. Then, from vncviewer (used for Raspberry Pi), we can connect Raspberry Pi on computer. Then, no need to open MATLAB. Testing will directly be done on hardware.

3 Results

In the result section, recognized faces have shown in Fig. 12. In Fig. 12a, there is a snapshot of a corresponding author 1 itself. In Table 3, result of expected users detected and users recognized quantity has shown. 100 tests are done for every person

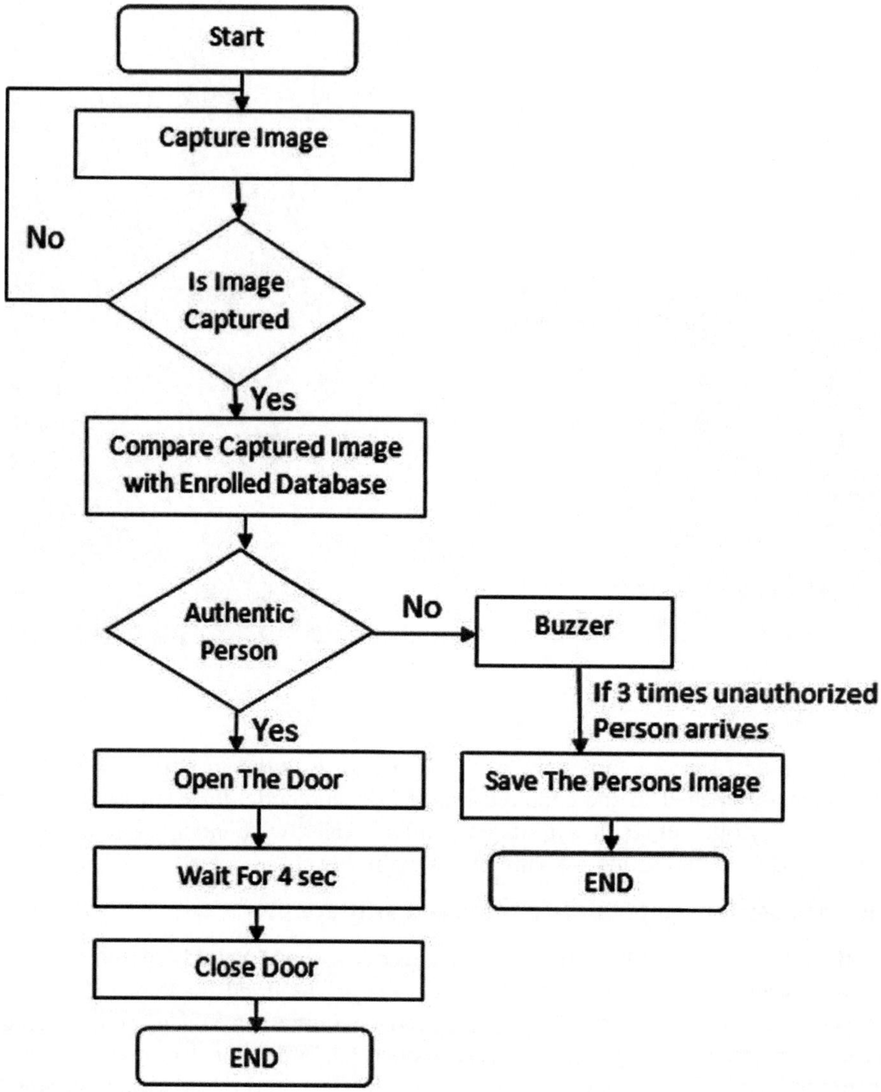

Fig. 11 Flowchart

as shown in table. Figure 11 shows graph of proposed methodology tested results—expected and recognized values. Users Preshit and Swapnil having good a result that means 100 times they came in front of system and they get recognized. Vice versa Snehal gets bad results as compared to the Preshit and Swapnil. The same analysis we can make from the graph.

Fig. 12 Faces recognized successfully

Table 3 Proposed methodology tested results—expected and recognized

User name	No. of expected result	No. of recognized result
Preshit	100	100
Harish	100	98
Prajakta	100	97
Snehal	100	89
Smruti	100	92
Umesh	100	98
Swapnil	100	100
Urvi	100	98
Supriya	100	93

Database details

We have made our new database by collecting data of faces of nearby peoples of around 100. From that, we have tested each user for 200 times and checked its features and values to get its optimum. We have shown result comparison in graph (Fig. 13).

Table 4 shows comparative values of traditional classifier with proposed SVM classifier tested on images. The NMD classifier [11] and ANN [12] and SVM have compared. Each user tested for 200 times for each classifier. As compared with other two techniques, SVM has highest classification rate in the proposed system. ANN also good but for face recognition SVM is preferred because of its linear and nonlinear separability. The same analysis we can make from Fig. 14.

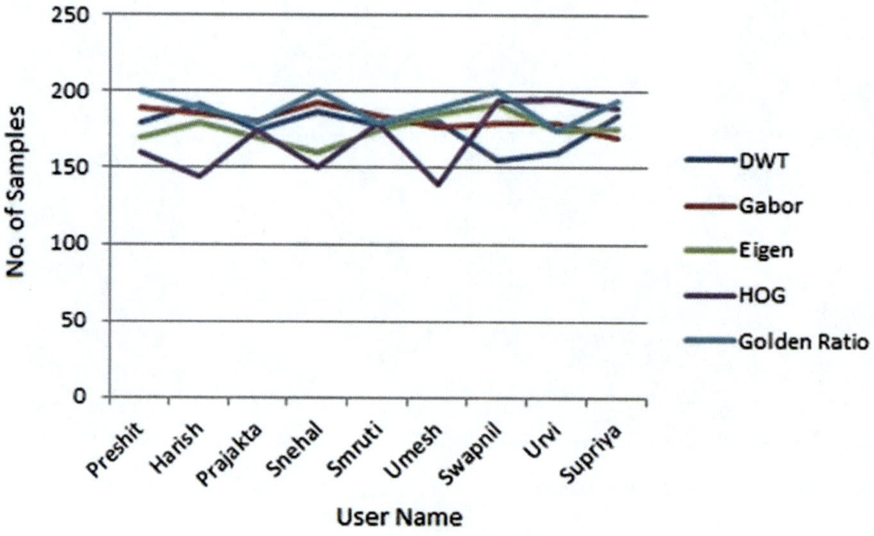

Fig. 13 Graph of proposed methodology tested results—expected and recognized

Table 4 Comparative result of traditional classifiers with proposed SVM classifier

Name of person/classifier	NMD	ANN	SVM
Preshit	160	180	200
Harish	180	190	190
Prajakta	170	160	180
Snehal	180	190	200
Smruti	180	180	180
Umesh	160	180	190
Swapnil	175	190	190
Urvi	160	185	180
Supriya	190	180	195

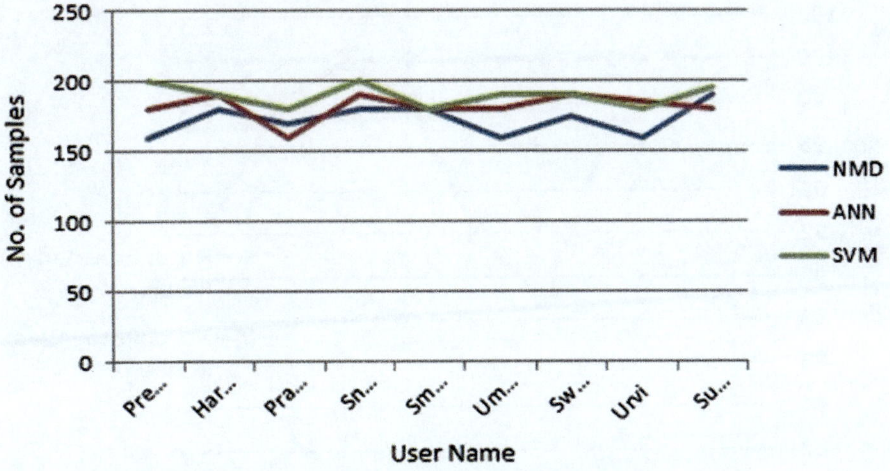

Fig. 14 Comparative chart of traditional classifier and proposed SVM

Table 5 shows comparative results of golden ratio features and other classifier features. DWT [13], Gabor [14], Eigen [8], HOG [15] techniques are compared with our proposed classifier feature extraction technique. For these techniques, citations have provided. From that, we can find out its advantages and drawbacks. Each user tested for 200 times for each classifier. As compared with other techniques, golden ratio extracts face features rapidly. The same analysis we can make from graph as shown in Fig. 15.

In Table 6, confusion matrix has shown. Confusion matrix is also known as error matrix. Total samples are taken 900. For each user, 100 samples are taken. Number of times the user came in front of camera and person gets recognized or not, the number has displayed in the table. Correct recognized samples are 865, and wrong recognized samples are 35 with mean accuracy rate of samples of 96.11%. When

Table 5 Comparative results of golden ratio features and other classifier features

Name of person/Classifier	DWT	Gabor	Eigen	HOG	Golden ratio
Preshit	180	190	170	160	200
Harish	192	186	180	145	190
Prajakta	175	181	170	175	180
Snehal	187	193	160	150	200
Smruti	178	185	175	180	180
Umesh	181	177	186	140	190
Swapnil	155	180	192	195	200
Urvi	160	180	175	196	175
Supriya	185	170	176	190	195

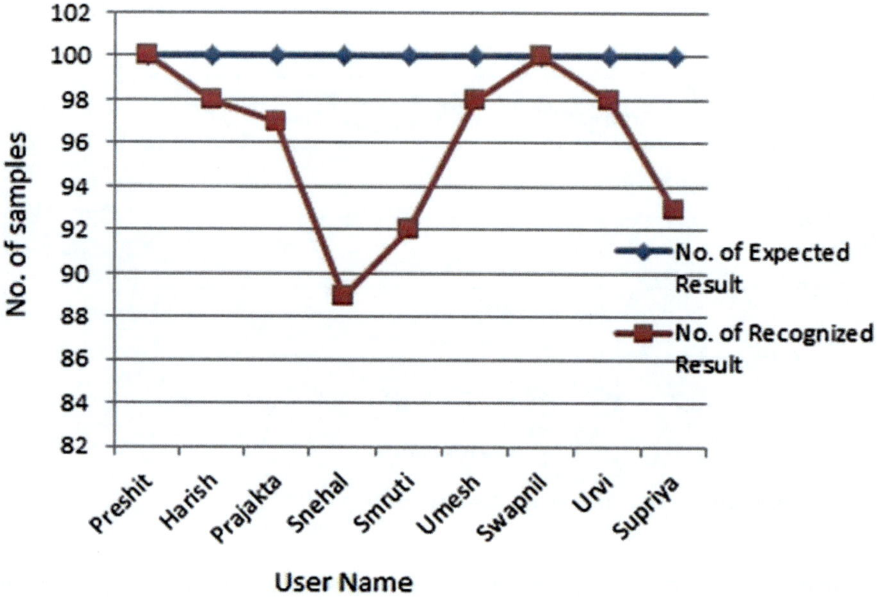

Fig. 15 Comparative results of golden ratio features and other classifier features

Table 6 Confusion matrix

Actual	Expected								
	Preshit	Harish	Prajakta	Snehal	Smruti	Umesh	Swapnil	Urvi	Supriya
Preshit	100	0	0	0	0	0	0	0	0
Harish	1	98	0	0	0	1	0	0	0
Prajakta	0	0	97	3	0	0	0	0	0
Snehal	0	0	5	89	3	0	0	0	3
Smruti	0	0	0	4	92	0	0	0	4
Umesh	1	1	0	0	0	98	0	0	0
Swapnil	0	0	0	0	0	0	100	0	0
Urvi	0	0	1	0	1	0	0	98	0
Supriya	0	0	1	4	2	0	0	0	93

two persons are twins, in extremely rare conditions, system may not recognize the same person. Golden ratio is almost same for that users. In that condition, accuracy degraded. Accuracy can be calculated as

$$\text{Accuracy} = \frac{\text{No. of recognized images}}{\text{Total No. Tested Images}} \times 100\% \tag{15}$$

Table 7 shows percentage accuracy for various users, Fig. 16 shows chart for average recognition rate, and it is upto 96%. Table 8 verifies identification accuracy of various techniques used for face recognition.

However, with an increase in the speed of displaying objects, an increase in the number of noises and objects in the image, the recognition rate and its quality of our brain and the brain of primates decrease dramatically. Even the simplest convolutional neural network recognizes objects better. So in our work, we extract feature points of the face. Using artificial intelligence and deep learning, we can also campare the

Table 7 Percentage accuracy for various users

User name	Total no. of tested images	No. of recognized images	No. of unrecognized Images	Accuracy in percentage
Preshit	100	100	0	100
Harish	100	98	2	98
Prajakta	100	97	3	97
Snehal	100	89	11	89
Smruti	100	92	8	92
Umesh	100	98	2	98
Swapnil	100	100	0	100
Urvi	100	98	2	98
Supriya	100	93	7	93
Total	900	865	35	96.11

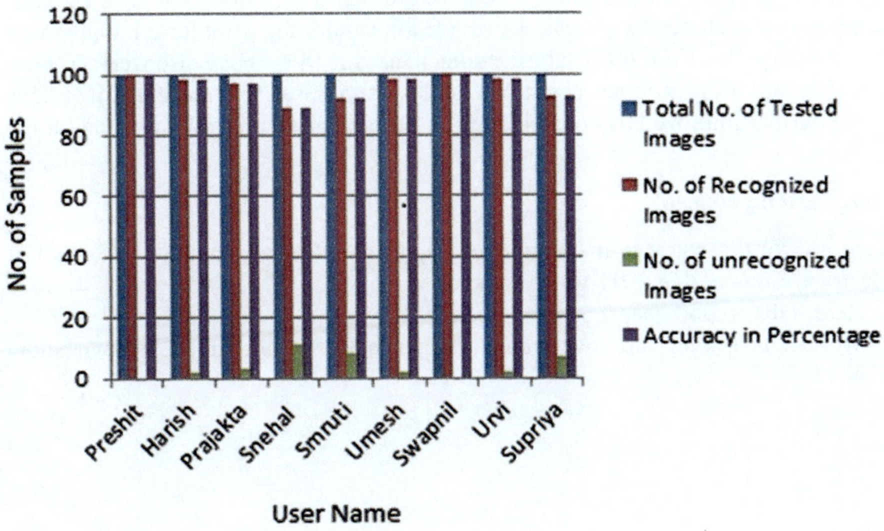

Fig. 16 Chart for average recognition rate

Table 8 Various techniques and its identification accuracy

Techniques	Identification accuracy (%)
PCA [8]	72
ANN [12]	92
(a) Nonsegmented (b) Segmented [10]	84 90
Viola–Jones algorithm and artificial neural networks [16]	88.64
Viola–Jones algorithm, neural networks [17]	90.31
Viola–Jones algorithm, ANN and PCA [1]	94
Proposed method [Viola–Jones algorithm, golden ratio features and SVM]	96.11

results. But it requires high resolution images for testing. But in our case, we can apply it on any image.

4 Conclusion

This paper proposed a procedure for locating and arranging the face (identification) by joining two of the most important strategies that is golden ratios used for face and combination of SVM and Viola–Jones for this path. Viola–Jones strategy is legitimately integrated into virtually every enterprise and programming system that manages the PC vision. In addition, the extraction of the highlighting system of brilliant proportions has existed for a long time. In this work, we have consolidated these two strategies in order to provide a special framework for grouping faces continuously. Precision for database countenances is 99%. Face discovery productivity is about 98%, accuracy about 96% and is nonpartisan to force and points. This can be further improved by consolidating systems to make up for lighting and force varieties.

Statement on consent

Approval for this study from committee: Department Review committee, School of Electrical Engineering, MITAOE, Alandi.

Refer DRC report snapshot shown below (Fig. 17).

We also have ethical approval from participants. Refer consent letter shown below (Fig. 18).

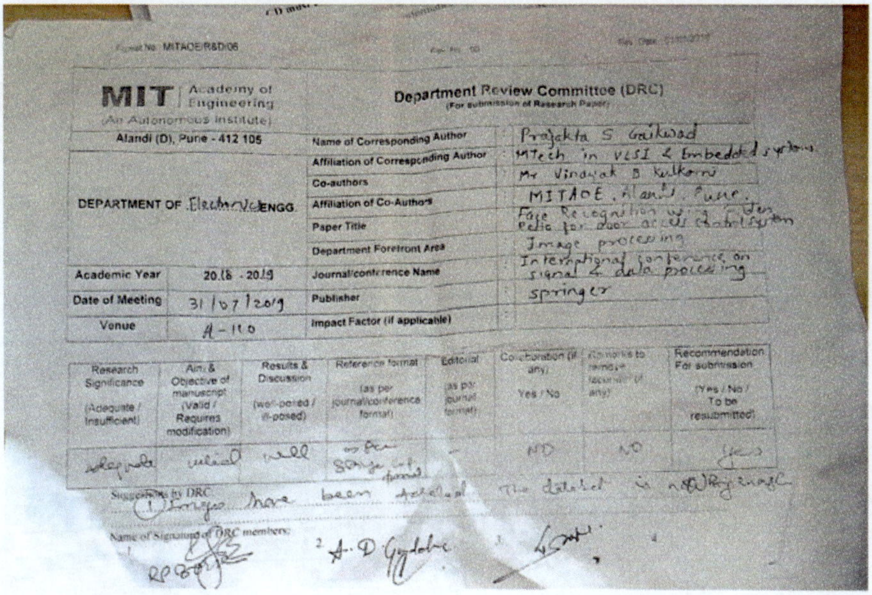
Format No. MITAOE/R&D/06

MIT Academy of Engineering (An Autonomous Institute) Alandi (D), Pune - 412 105

Department Review Committee (DRC) (For submission of Research Paper)

DEPARTMENT OF Electronic ENGG		Name of Corresponding Author	Prajakta S. Gaikwad
		Affiliation of Corresponding Author	MTech in VLSI & Embedded system
		Co-authors	Mr Vinayak B Kulkarni
		Affiliation of Co-Authors	MITAOE, Alandi, Pune.
		Paper Title	Face Recognition using golden ratio for door access control system
		Department Forefront Area	Image processing
Academic Year	20.18 - 2019	Journal/conference Name	International conference on signal & data processing
Date of Meeting	31/07/2019	Publisher	springer
Venue	A-110	Impact Factor (if applicable)	

Research Significance (Adequate / Insufficient)	Aim & Objective of manuscript (Valid / Requires modification)	Results & Discussion (well-posed / ill-posed)	Reference format (as per journal/conference format)	Editorial (as per journal format)	Collaboration (if any) Yes / No	Remarks to remove (if any)	Recommendation For submission (Yes / No / To be resubmitted)
adequate	valid	well	as per Springer format	–	NO	NO	yes

Suggestions by DRC: 1) Images have been added. The detailed ...

Name of Signature of DRC members: 1. RP Borkar 2. A. D. Gujdale 3. 4.

Fig. 17 DRC report

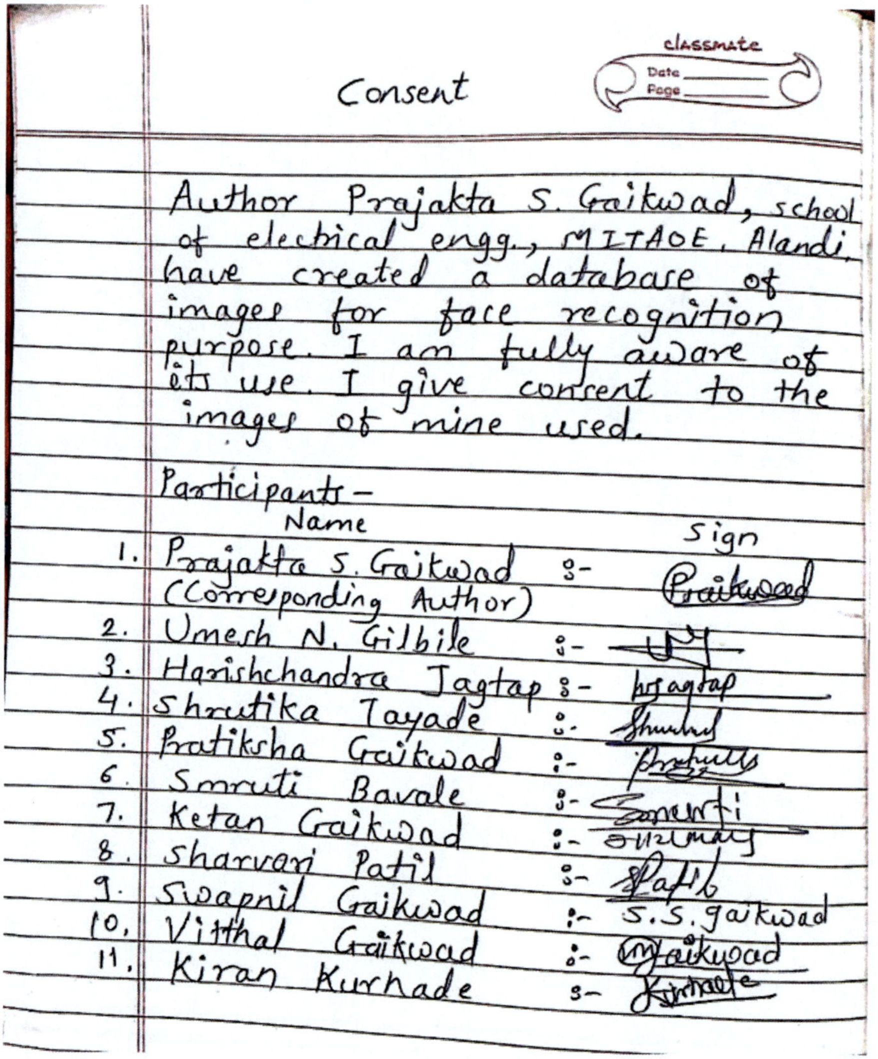

classmate
Date
Page

Consent

Author Prajakta S. Gaikwad, school of electrical engg., MITAOE, Alandi, have created a database of images for face recognition purpose. I am fully aware of its use. I give consent to the images of mine used.

Participants –

	Name		Sign
1.	Prajakta S. Gaikwad (Corresponding Author)	:-	Pgaikwad
2.	Umesh N. Gilbile	:-	[illegible]
3.	Harishchandra Jagtap	:-	hsjagtap
4.	Shrutika Tayade	:-	Shrutika
5.	Pratiksha Gaikwad	:-	Pratiksha
6.	Smruti Bavale	:-	Smruti
7.	Ketan Gaikwad	:-	[illegible]
8.	Sharvari Patil	:-	Patil
9.	Swapnil Gaikwad	:-	S.S. gaikwad
10.	Vitthal Gaikwad	:-	Vgaikwad
11.	Kiran Kurhade	:-	Kurhade

Fig. 18 Consent letter

References

1. Pawar K.B., Biradar V, Mirajkar F, Fatima R (2017) A novel practice for face classification. In: International Conference on Current Trends in Computer, Electrical, Electronics and Communication, pp 822–825
2. Dabhi MK, Pancholi BK (2013) Face detection system based on Viola–Jones algorithm. Int J Sci Res
3. Hassaballah M, Murakami K (2013) Face detection evaluation: a new approach based on the golden ratio. Signal Image Video Process

4. Li S, Liu X, Chai X, Zhang H, Lao S, Shan S (2014) Maximal likelihood correspondence estimation for face recognition across pose. IEEE Trans Image Process 23(10)
5. Sekhon A, Agarwal P (2015) Face recognition using back propagation neural network technique. In: IEEE International Conference on Advances in Computer Engineering and Applications. pp 226–230
6. Ding C, Tao D (2016) A comprehensive survey on pose-invariant face recognition. ACM Trans Intell Syst Technol 7(3), Article 37
7. Wisarutchantara Y (2015) Object detection based on fast template matching through adaptive partition search. In: Proceedings in 12th International Joint Conference on Computer Science and Software Engineering (JCSSE), pp 1–6
8. Prasanth G. Narasimha-Shenoi(2012) Golden ratio in human-anatomy. https://www.researchgate.net/publication/234054763, researchgate
9. Support vector machine. https://dataaspirant.com/2017/01/13/support-vector-machine-algorithm
10. Green CD (1995) All that glitters: a review of psychological research on the aesthetics of the golden section. Perception 24:937–968
11. Abhishree TM, Latha J, Manikantan K, Ramachandran S (2015) Face recognition using Gabor filter based feature extraction with anisotropic diffusion as a pre-processing technique. Procedia Comput Sci 45:312–321
12. Chakraborty D, Saha SK, Bhuiyan MAA (2012) Face recognition using eigenvector and principle component analysis. Int J Comput Appl 50(10)
13. Dadi HS, Pillutla GKM (2016) Improved face recognition rate using HOG features and SVM classifier. IOSR J Electron Commun Eng 11(4):34–44
14. Hassaballah M, Murakami K, Ido S (2013) Face detection evaluation: a new approach based on the golden ratio Φ. SIViP 7(2):307–316
15. LNCS Homepage. https://www.springer.com/lncs, last accessed 2016/11/21
16. Hsu M-K, Lin H-Y, Chen F-C (2017) NMD classifier: a reliable and systematic classification tool for nonsense-mediated decay events. PLOS ONE https://doi.org/10.1371/journal.pone.0174798
17. Yadav AR, Dewal ML, Anand RS, Gupta S (2013) Classification of hardwood species using ANN classifier. In: 2013 fourth national conference on computer vision, pattern recognition, image processing and graphics (NCVPRIPG), pp 1–5

Efficient Design of Drone Flight Control Using Delay Tolerant Algorithm

Priyanka Dange and Bhairavi Savant

Abstract As the activities of drones which are specifically concerns to flying, racing are growing, drone crashes are also growing. Many of these crashes are linked with operational problems, and hence, improving drone controls is urgently needed. The main objective of this entire study was to reduce the accident of drone due to mistakes made by a human. The approach given in the paper describes a composable pipe model for task scheduling. The primary objective of this system is to enhance the efficiency of task scheduling; for this, we have used two proportional integration derivative (PID), controllers. Two end-to-end terms are analyzed using the pipe model: reaction time and freshness time. We have used Cleanflight control firmware with Real-Time Operating System (RTOS). The experimental results convey that the latency time and delay time of task are getting reduced.

Keywords End-to-end timing analysis · Flight controller · PID controller · RTOS · VCPUs

1 Introduction

Due to the widespread use of uninhabited aerial vehicles or civil drones in various areas such as merchandise delivery film making and hobbies, drone accidents, which damage people and property, are also growing. Basically to establish end-to-end time constraints and to establish it on systems, it is very crucial to analyze make or break their correctness: [1]. In some systems, very accurate results are required. This is not only significant for the individual processes to be carried out within the deadlines, as well as for the start-to-end lag specifications of the cause-effect chains should meet to the necessary demands: [2]. Drone crashes have raised worldwide safety as well as

P. Dange (✉) · B. Savant
School of Electrical Engineering, MIT Academy of Engineering, Alandi, Pune, India
e-mail: priyankadange27@gmail.com

B. Savant
e-mail: bnsavant@entc.maepune.ac.in

© Springer Nature Singapore Pte Ltd. 2021

S. N. Merchant et al. (eds.), *Advances in Signal and Data Processing*,
Lecture Notes in Electrical Engineering 703,
https://doi.org/10.1007/978-981-15-8391-9_17

security concerns including infringement of civil aviation legislation and seclusion: [3]. Drone accidents are increasingly reported because of operational mistakes. For example, the 11-month-old child was badly injured by a control error in the USA in September 2015: [4]. Accidents caused by drones having the equal amount of flight time which in hours were fifty times the amount of actual accidents regarding flight, as well as crashes triggered by user errors, this about the action of controlling at which drone make up 32% of all accidents concerned: [5]. Without safety improvements, it is an actually challenging job to avoid rigid regulations or a ban on civil drones: [6]. Therefore, a solution must be found to reduce the risk of accidents caused because of the drones users and to maximize the potential for industrial development: [7, 8]. Mainly, we are focusing on the accomplishment or in other word implementation of the flight controller which should be real time based. Our first stage is just to refactor a popular flight control firmware for use with real-time information called Cleanflight. Cleanflight is used on human-operated racing drones through control of the radio. By configuring Cleanflight as an implementation in real time, we can incorporate it with extra features needed for a completely autonomous flight or aircraft management system. A goal is to substitute control which should be radio along with onboard functions that perform configurable aircraft or flight missions. For example, aerial photography, packages delivery then search and rescue operations.

The challenge, for instance, is to determine the time limitations required for end-to-end latency checks using a flight controller having multirotor. Sensor data should be processed as well as sampled at a lower rate to guarantee that the drone operates with the latest figures of its attitude, speed, direction, and altitude. Likewise, the parameter mainly speeds of the rotor as it has a crucial role in performing controlling action must be modified to reflect current sensor measurements within certain time limits. Therefore, in aspects reaction time of the sensor information actuator and freshness, we specified two end-to-end constraints which are related to timing limitations that limit the issue of composing a sustainable task pipeline. The problem with a pipeline set of duties, each assigned to a separate distinct virtual central processing units (VCPU), is to decide a relevant set of budget and time spans to guarantee end of analysis of timing of freshness and waiting times at the time of all virtual CPUs are designed.

Our job is to implement a pipe model which is composable that ensures limited end of analysis of timing execution or processing and inter-processing that is communication delays among a combination of all activated tasks which can be one set which are independently mapped to servers that maintain bandwidth. Secondly, a method for deriving task budget and periods from the limitations of the end of analysis of timing spans in the stage of application design . Our main contribution in this system is parallelism. In this, we will take two controllers that will perform two tasks at a time, which will improve the performance in term of speed and reduce the delay and latency.

This paper demonstrates the methodology we are described in Sect. 2. In Sect. 3, we have described simulation model while in Sect. 4, we have demonstrated simulation experiments, and lastly in Sect. 5, we have given conclusion from the simulation experiments and graphs obtained from that.

2 Methodology

In this paper, our target is to introduce an independent multirotor drone flight management system. We have to use the Cleanflight firmware with real-time operating system. Cleanflight's key software components are a proportional integration derivative (PID) controller then sensor and also actuator drivers; the algorithm used in this system is priority scheduling algorithm for task scheduling, and many communication stacks, as well as the logging system. To analyze different models, it is necessary to categorize in three subsections as mentioned in: [9].

2.1 *Task Model*

We design the flight controller code as a set as a collection of regular real time $\{\tau_1, \tau_2, \ldots, \tau_n\}$ assignments that are arbitrary when the flight controller is initially released.

Each task τ_j is defined by the worst execution moment e_j, period T_j, and a time limit which is equivalent to period. In addition, to exchange and process data, each and every job communicates with no that is zero or it can be possible to more successors as well as predecessors. All the regular tasks are carried out using threads at the user level.

2.2 *Scheduling Model*

Threads are organized in any environment using a scheduling hierarchy along with the threads having two-level linked to virtual central processing units (VCPUs) assigned to actual processors. A processor capability reserve is specified for each VCPU [4] composed of a budget capability, C and time, T. The C and T value is decided by the task e and T mapped to the VCPU. At least C must be given to a VCPU time units for each T time unit when it can be run. As far as concern, a test of schedulability is get over when it is created new VCPUs: [5]. This is the way, in which customization of the subsystem having scheduling which ensures temporary isolation in the environment between threads of runtime system platform like used in this proposed approach.

2.3 *Communication Model*

Control flow inside the controller of the flight is affected by the information route originating from sensory information that is from its inputs and actuation end. Inputs include different parameters such as inertial sensors, GPS devices, as well as optional cameras while actuators include core part of the system like engines that

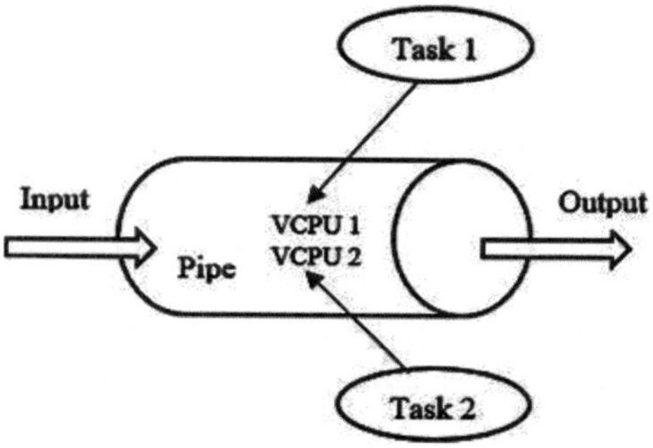

Fig. 1 Illustration of pipe model [9]

influence rotor speeds and drone behavior such as its attitude. The flow of the data includes a pipeline of communication duties, leading in a design of communication distinguished by:

(1) The inter-arrival times of tasks or threads in the concurrent structure means of the pipeline,
(2) Buffering of the inter-jobs or tasks, and
(3) Tasks access the pattern to special registers or simply called as buffers which are concerns to communication.

2.4 Pipe Modeling

We sorted out the factors that affect delay corresponding to end to end regarding the communication process. Among them, basically the absolute time regarding the processing and the data size which has to be transferred are resolved by this structure. To capture remaining timing characteristics too, we designed the composable model also known as the pipe using the described scheduling approach. A job and pipe have a relationship which is one to one, as specified in Fig. 1.

A pipe mechanism has single terminal of the pipe as well as pipe ends which are total two, for input and output, respectively. A pipe terminal is depicted by a virtual CPU, which guarantees units C of processing or execution span after every T time described in the unit when it is in run condition. Here, as illustrated about the contribution part, we have added two virtual CPU with the pipe. Terminals of the pipe are correlated with traditional tasks linked to two virtual CPUs and paths of the kernel control connected to I/O VCPUs.

2.5 Different Phases of Timing Analysis

The performance of any task is basically categorized into three phases, including.

1. Reading phase,

2. Processing phase: Nothing but operation, and
3. Writing data or information.
   To scrutinize the analysis of timing, we consider that tasks can complete the stages of reading and can be write within single moment of the pipe end that is terminal is mapped to the task. This is realistic for application, especially for a flight controller. The reasons behind this are:

   1. The information to be transmitted is generally low
   2. All three stages or phases can typically be finished within one period of time.

However, we do not enforce any restrictions on the duration of the processing phase in order to maintain generality.

Generally, in real-time application-oriented systems, time analysis which is end to end can be performing using techniques. To prevent repetition, it is needed to abbreviate the end of analysis of timing assessment of freshness timing only by concentrating on the specific instances outlined. Basically, the freshness time is described as the time span or interval between an input's arrival time and its last corresponding output's departure. We are therefore investigating the interval between a particular input data instance.

Flowchart

See Fig. 2.

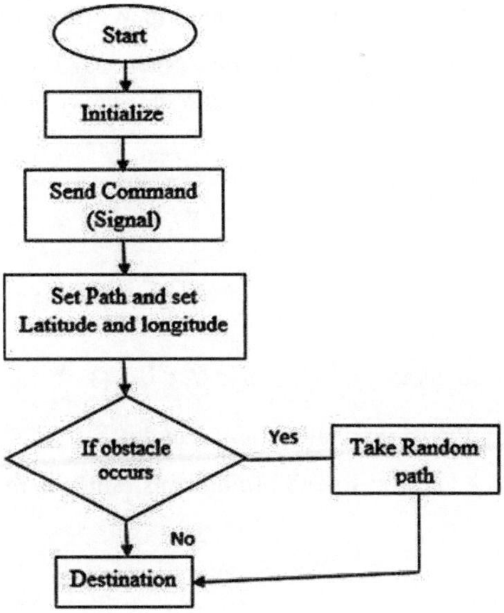

Fig. 2 Flowchart of proposed system

3 End-To-End Design

3.1 Problem Definition

By combining the pipe model's timing analysis with linear optimization methods, we generate task based from the end-to-end limitations of timings. In order to utilize the simulation as per given in this paper, it is possible to generalize for a large scope of real-time applications with the help of the following considerations and assumptions, which are given in: [9].

Consider a set of tasks.

$$\Gamma = \left\{\tau', \tau^2, \ldots, \tau^n\right\} \tag{1}$$

$$\Pi = \{\pi_1, \pi_2, \ldots, \pi_n\}, \tag{2}$$

where $\tau_j = \left(d_i^j, p^j, d_o^j\right)$.

$$\pi_j = \left(\left(W_i^j, \delta_i^j\right), \left(C^j, T^j\right)\left(W_o^j, \delta_o^j\right)\right) \tag{3}$$

While the second one gives the pipes set.

Including these with the help of following equations problem statement is defined.

1. The mapping between Γ and Π. For ease of indication, we assume tasks are mapped to the pipe with the subscript, which is same and henceforth

$$\forall j \in \{1, 2, \ldots, n\}, \quad \tau_j \rightarrow \pi_j \tag{4}$$

$$\forall j \in \{1, 2, \ldots, n\} \tag{5}$$

the value of d_i^j, p^j and $d^j_o \forall j \in \{1, 2, \ldots, n\}$.

the value of W_i^j, δ_i^j, W_o^j and d_o^j;

2. The end-to-end timing constraints, namely the value of

$$\underset{i \rightarrow_i \tau_j \rightarrow \pi_j}{E_{\tau\pi|}} \quad |\ldots|\underset{k \rightarrow_k}{\tau\pi}| \quad \text{and /o r } F_{\underset{p \rightarrow p\tau_q \rightarrow \pi_q}{\tau\pi|}} |\ldots| \underset{r \rightarrow_r}{\tau\pi} | \tag{6}$$

where $i, j, k, p, q, r \in \{1, 2, \ldots n)$.

The intention is to find set of $\left\{\left(C_j, T_j\right)\right\}$ pairs for $j \in \{1, 2, \ldots, n\}$ which are a feasible set that:

1. Should meet all possible specified timing constraints,
2. Necessary to perform schedulability test,
3. Normally but not inherently minimizes the CPU use.

3.2 Solution

To get the solution over this constraint, some assumptions are important with the block diagram which has taken from: [9]. Consider the topology graph of the pipe given in Fig. 2, in which total six tasks are shown and these tasks are assigned to six corresponding pipes. First, three tasks that are task number 1, 2, 3 scan sensor inputs, while tasks 4 as well as 6 send respective outputs to the actuators. Finally, task 5 is a mediator or an intermediary accountable for complex handlings, for example, PID control or it can be the fusion of sensor data. The most significant thing, i.e., characteristics timing of the model which is pipe structure and task are presented in Table 1. Basically, the timings of execution are supposed to be same for all assigned tasks (Fig. 3).

In practice, it is not compulsory to be the case but it does not create any impact on the approach generality.

- **Reaction time**
  It is nothing but the time span it takes for input data samples to flow between the overall system, which is affected by the time span or period of an each and every consumer in the structure of the pipeline. Basically, a reaction time constraint limits the interval of time in between input of the sensor and the very first corresponding actuator output.
- **Freshness time**
  It is nothing but the time span or interval in which a different sampled data inputs are there. This affects the process and mostly the time span of each pipeline producer. A fresh timing constraint limits the interval in between both the input of the sensor as well as the last correlating actuator output.

This basic is given in tabular analysis for the above-given problem statement and its solution which helps in stabilization graphs.

Table 1 Application timing characteristics with the time span of reaction and freshness time period: [9]

Reaction	Freshness
$E_{\underset{1\to 4}{\tau\pi}\mid\underset{4\to 4}{\tau\pi}} \le 10,$	$F_{\underset{1\to 1}{\tau\pi}\mid\underset{4\to 4}{\tau\pi}} \le 20,$
$E_{\underset{2\to 4}{\tau\pi 2}\mid\underset{\to 4}{\tau\pi}} \le 15,$	$F_{\underset{2\to}{\tau\ \pi\ 2}\mid\underset{4\to 4}{\tau\ \pi}} \le 30,$
$E_{\underset{2\to 2}{\tau\pi}\mid\underset{5\to 5}{\tau\pi}\mid\underset{6\to 6}{\tau\pi}} \le 25,$	$F_{\underset{2\to 2}{\tau\pi}\mid\underset{5\to 5}{\tau\pi}\mid\underset{6\to 6}{\tau\pi}} \le 50,$
$E_{\underset{3\to 3}{\tau\pi}\mid\underset{6\to 6}{\tau\pi}} \le 15;$	$F_{\underset{3\to 3}{\tau\pi}\mid\underset{6\to 6}{\tau\pi}} \le 20;$

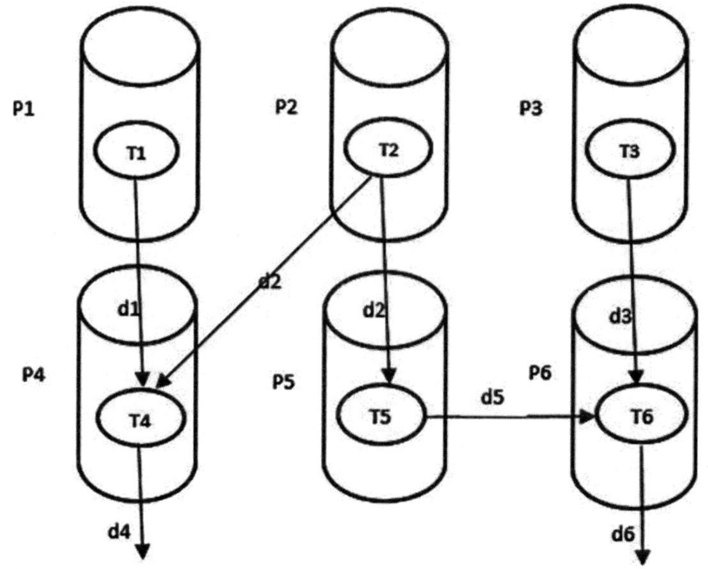

Fig. 3 Basic diagram of the workflow: [9]

4 Experiments of the Simulation

We have developed the model simulation which is compatible for both Windows and for other operating system to predict the end-to-end time constraints with the MATLAB simulation for customization. We have to use a piping structure for scheduling. By using FIFO structure, we have scheduled the task. We have minimized the delay time. We used three terminologies like reaction time, freshness time, schedulability and execution time.

Figure 4 shows the source and destination set to the drone. The drone will fly from source to destination based on the geographical latitude and longitude. Ti will turn left right up and down according to the instruction. It has a delay in communication which we will remove (Fig. 5).

In Fig. 6, we have given some obstacle which will disturb the drone. When the drone is disturbed by a particular obstacle, it will first find the nearest track where he can join the particular scheduler and travels further (Fig. 7).

The top model includes two areas of blocks:

4.1 Model 1: Functional Components

It includes two closed-loop systems. Each one has a proportional controller running a plant.

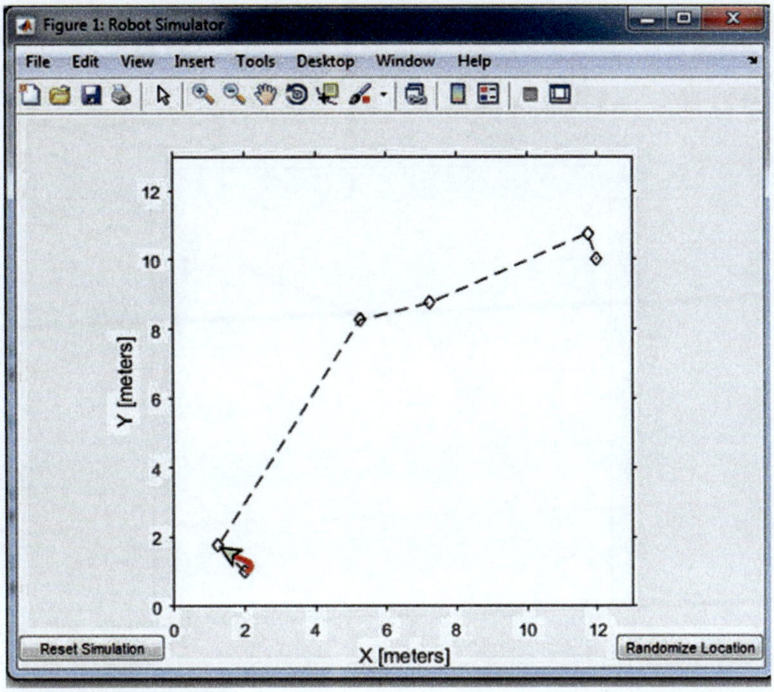

Fig. 4 Starting of drone flight

4.2 Model 2: Architectural Components

It involves Sim Events blocks that are modeling the tasks and scheduler of this control system.

Figures 8 and 9 specify task scheduling model architecture.

This is the scheduler model where we will see the input and output. This controller will manage task scheduling.

4.3 Modeling Tasks and Scheduler

This instance prototype a controller is a function model imported by Simulink. It maps server implementation to a software job or a task that is periodically scheduled and executed by a working scheme. It can be possible to fragment a task into several sections. These sections must be performed in sequential order due to data dependencies.

A task with the following parameters is specified:

ID: Unique task identifier.

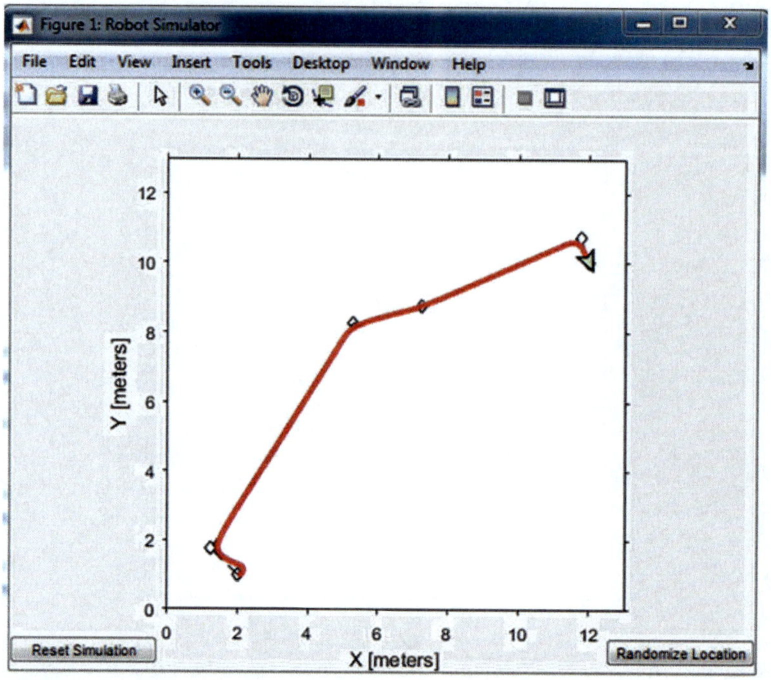

Fig. 5 Ending of drone flight

Period: How often is the job installed for execution?

Priority: The task's priority.

List of sections that can be run: Executable aligned with each segment of tasks. Executable associated with each job segment.

Segment execution duration Time: To finish a task section if it is performed on a processor without disruption.

Needs resource of disk I/O for each segment: Whether a task section needs the use of a resource which is generally shared resource protected by a mutex.

For instance, block task 2 indicates a second controller task. The task involves two sections, "t2_run" and "t2_write," both described in model seExampleSwc Controller 2 as Simulink features. "t2 write" needs the use of the shared resource protected by mutex in these segments.

The operating system scheduler is formed with the components specified as follows: Task creation: Block Task 1 and similarly to Task 2 generate tasks and handle tasks.

A*Sim Events entity is nothing but a simple task example. The properties of a task are modeled as object characteristics.*

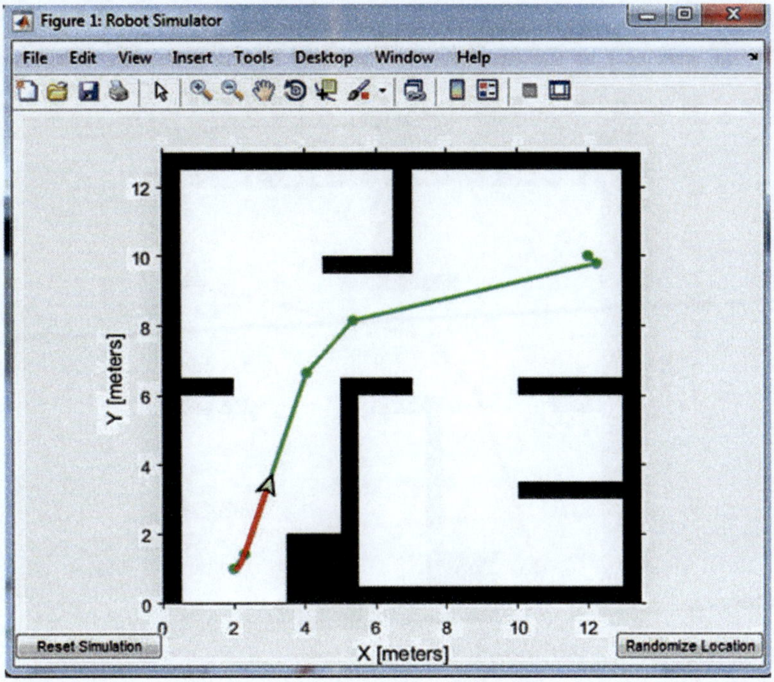

Fig. 6 Drone flight starting with obstacle

Task queue: Upon installation, a task get connects to a ready task queue, which is represented by the block known as Entity Queue OS task queue. The block of the queue is designed to type tasks by task priority characteristic to represent scheduling policy concern to a non-preemptive priority based.

CPU: The system processor is modeled as the Entity Server block CPU. It receives entities from either OS task queue and processes the entity for the duration specified by the segment execution duration parameter of the task. After this delay is over, the correlating Simulink function of such a task segment can be called as part of the complete action of the block's service.

1. **Lock/unlock Mutex**: Before a task segment, it enters the CPU block and it must obtain the necessary asset in the previous Lock Mutex block. Once the segment of task finishes and it exits the CPU block, the resource is getting released in the block which is known as an Unlock Mutex block.
2. **Managing the all task states**: Blocks under task 1 and task 2 it is very important to manage the state of tasks at the run time. Upon closure of a section nothing but a segment of the task, the task is redirected to the task of the OS queue if the task has sections which are subsequent to perform. Otherwise, this task example will be finished and dismissed.

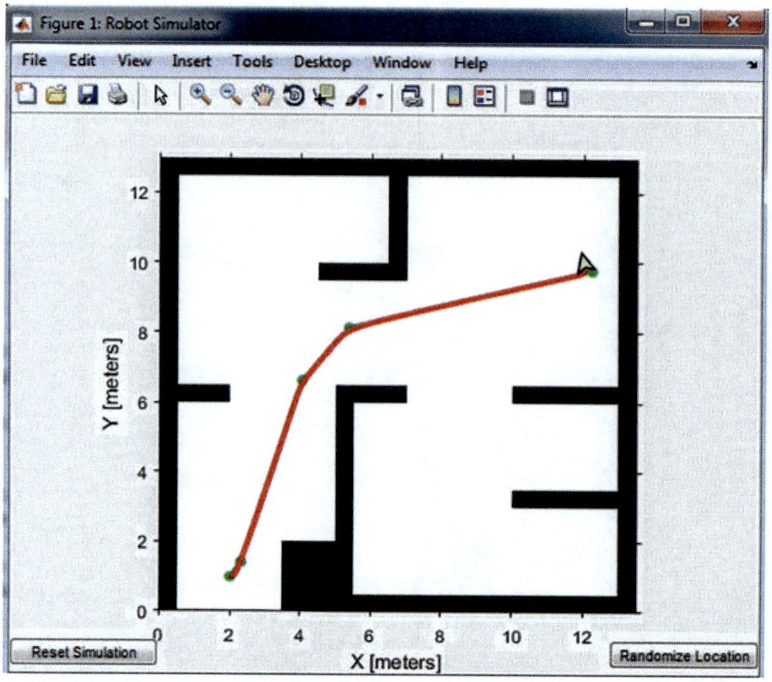

Fig. 7 Drone flight ending with obstacle

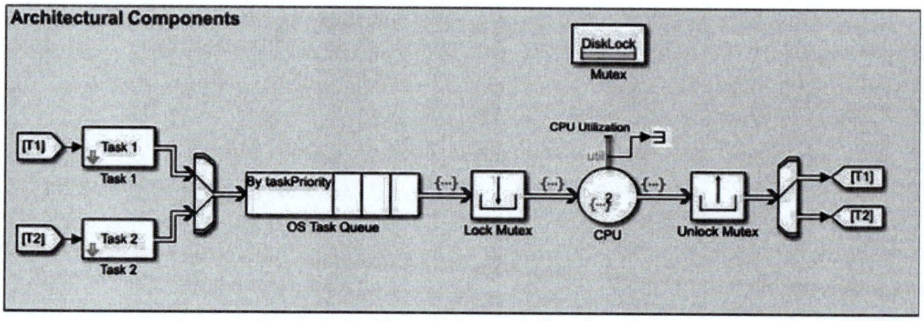

Fig. 8 Architectural component of task scheduling model

4.4 *Summary of the Results*

The two-core block CPU is installed. The simulation of the model produces the previous Gantt chart.

- Task 2 (shown by red bars) of higher priority is assigned to core 1, i.e., ($y = 1$).
- Task 1 (shown by blue bars) of lower priority task is assigned to core 2, i.e., ($y = 2$)

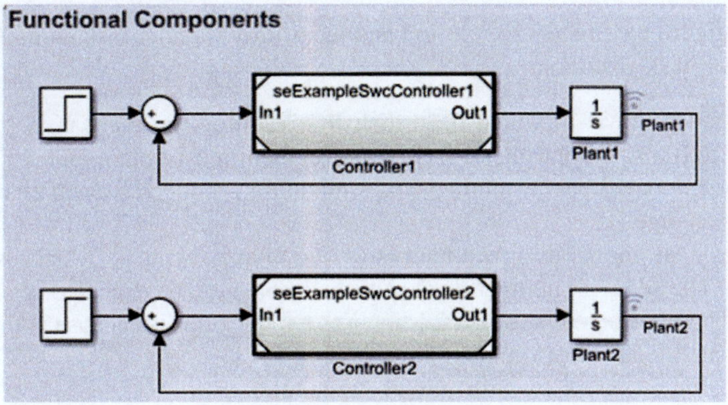

Fig. 9 Functional component of task scheduling model

- Task 2 second number segment utilizes the DiskLock mutex. Green bars show the utilization ($y = 3$) (Fig. 10).

  Change the following parameters and explore how task schedules and controller performance changes with reconfigured architectural parameters.

  - Configure the parameter capability to alter the number of cores in the block CPU.

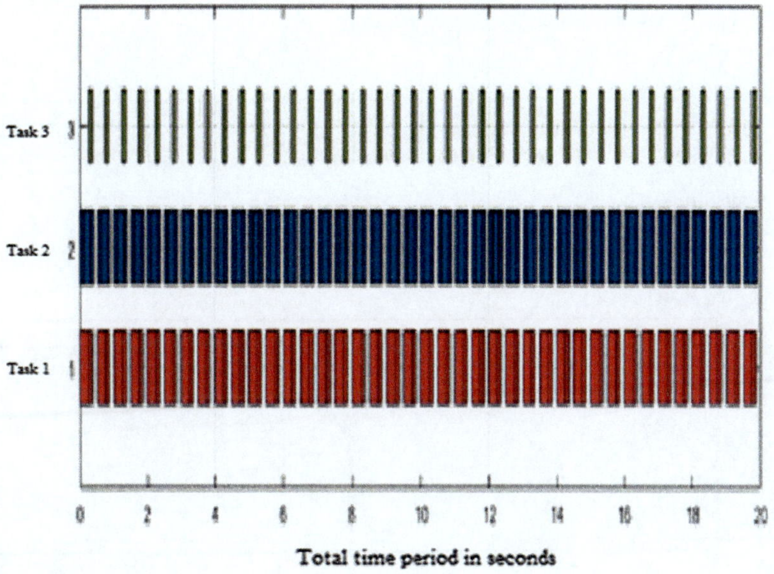

Fig. 10 Execution time

- Configure parameters like period and the priority to alter task requirements in task 1, task 2 blocks.

For instance, if we modified need disk I/O resource for every task 1 segment parameter to [0 0 1], Controller 1's t1_write segment must obtain mutex DiskLock before that can occur begin running. The simulation produces a Gantt chart, which shows this change.

Both tasks as shown by green bars at ($y = 3$) have segments which use mutex DiskLock. The segment number third of task 1 must now be executed serially with the second number segment of task 2 (shown as a $y = 1$); this is because both sections communicate the DiskLock mutex.

Figure 11 shows the CPU utilization of schedule.

Plant 1 shows the delay taken is higher. In-plant 2 delay is consumed by any instruction, which will be reduced by using pipeline structure (Figs. 12, 13 and 14).

Cleanflight Simulation: It is basically used to observed time constraints and achieves compatibility of the algorithm to reduce latencies.

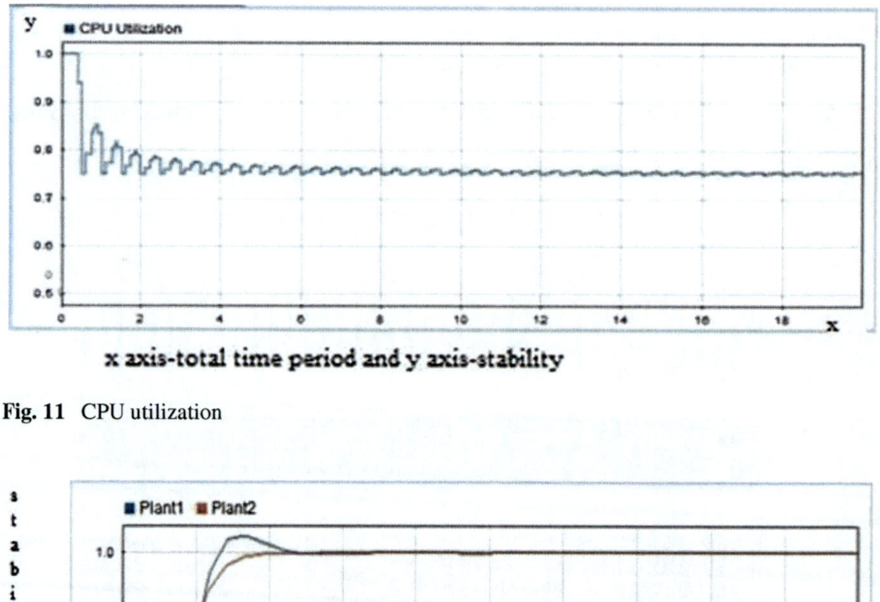

Fig. 11 CPU utilization

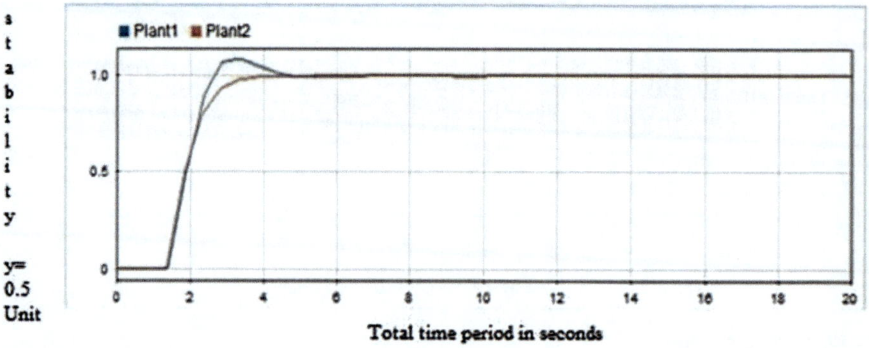

Fig. 12 Delay taken by instruction

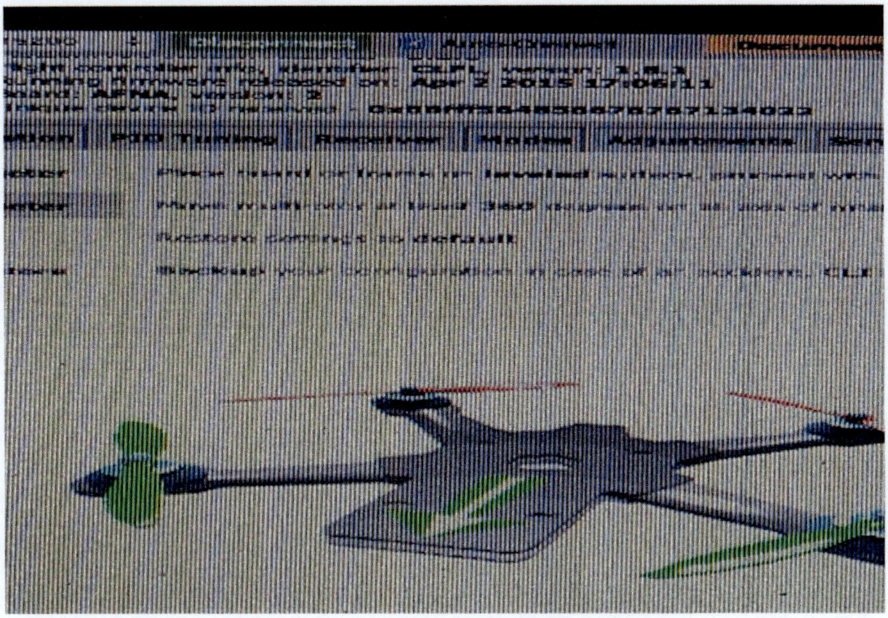

Fig. 13 Cleanflight simulation

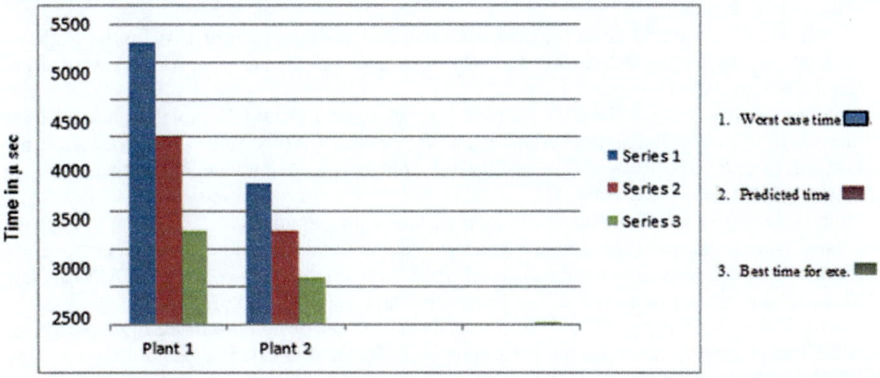

Fig. 14 Cleanflight times analysis has plotted through observations

5 Conclusion

This paper describes how an autonomous flight management system for multirotor drones is implemented. We have used pipelining terminology for task scheduling. In the previous system, they had used one controller for managing the task which resulted in increased latency and delay time. To reduce latency time and delay time, we have used two PID controllers by which the performance of the system gets increased. For task scheduling, we used reaction time, freshness time, scheduling and execution time. In the result, we have shown the first model of drone flight control, and in the second model, we have shown the scheduling of the task. For scheduling, we have used a FIFO structure on a priority basis. Our system gives better performance than the previous system.

Future scope of this system involves advanced techniques such as ad-hoc networks with the fuzzy logic which further improves the performance of the system.

References

1. Gerber R, Hong S, Saksena M (1995) Guaranteeing real-time requirements with resource-based calibration of periodic processes. IEEE Trans Softw Eng 21(7):579–592. https://doi.org/10.1109/32.392979
2. Becker M, Dasari D, Mubeen S, Behnam M, Nolte T (2016) Synthesizing job- level dependencies for automotive multi-rate effect chains. In: The 22nd IEEE International conference on embedded and real-time computing systems and applications (RTCSA). https://doi.org/10.1109/RTCSA.2016.41
3. drone-crash-injures-baby-highlighting-faa concerns
4. https://www.nytimes.com/2015/09/23/business/drone-crash-injures-baby-highlighting-faa-concerns.html
5. Shallwani P, Armaghan S, Gold D (2013) Remote controlled helicopter kills man in Brooklyn. Wall Street J. https://blogs.wsj.com/metropolis/2013/09/05/remote-control-helicopterkills-man-in-brooklyn
6. Weibel RE, Hansman RJ (2005) Safety considerations for operation of UAV in the National Airspace System (No. ICAT2005-1). MIT International Center for Air Transportation, Cambridge, pp 105–109
7. Manning SD, Rash CE, LeDuc PA, Noback RK, McKeon J (2004) The role of human causal factors in U.S. Army Unmanned Aerial Vehicle Accidents. T1-Army Aeromedical Research Lab, Fort Rucker, AL. The Role of Human Causal Fators in U.S. Army Unmanned Aerial Vehicle Accidents (No. USAARL-200411)
8. Hing JT, Oh PY (2008) Integrating motion platforms with unmanned aerial vehicles to improve control, train pilots and minimize accidents. In: ASME 2008 international design engineering technical conference and computers and information in engineering conference. ASME, New York. 32nd mechanisms and robotics conference, Parts A and B, pp 867–875
9. Cheng Z, West R, Einstein C (2018) End-to-end analysis and design of a drone flight controller. IEEE Trans Comput Aided Des Integr Circuits Syst 37(11):2404–2415. https://doi.org/10.1109/TCAD.2018.2857399

Adaptive Background Subtraction Models for Shot Detection

Dattatraya A. Jadhav, Yogeshkumar Sharma, and Parul S. Arora

Abstract The paper projected a unique prospective on background subtraction for object identification which is in motion as a structural block for many multimedia application beaning the primary applicable stage for successive detection, classification and analysis of actively of task. Since color information is not sufficient for addressing issue like sadden changes in elimination or visibilities in foreground object and color conflict. We have projected this work in which subtraction of background for detection of shot boundary in video is based on adaptive technique. This method depends on detecting the difference of mean gray value of current frame and previous frame is incorporated the result as arranging. We have calculated the result for five video inputs and evaluated outcomes interims of precision, recall and F1 measure.

Keywords Background subtraction · Shot boundary detection (SBD)

1 Introduction

Video shot boundary detection is the first step toward various video processing techniques like video abstraction, video indexing, video segmentation, video analysis and contain base video retrieval. Video segmentation involves segmentation video according to many factor based on pixel difference, feature extraction, background forward pixel difference, etc. Much work is reported for segmentation video based on diverse technique for various applications. The most commonly technique in scene capturing in by static camera in background subtraction. It is based on modeling of background of the scene and extraction of moving object which are foreground by subtraction model background from the current frame background subtraction is one

D. A. Jadhav (✉) · Y. Sharma
Shri J.J.T.University, Churela, Jhunjhunu, Rajasthan, India
e-mail: lncs@springer.com

D. A. Jadhav · P. S. Arora
ICOER, Pune, India

© Springer Nature Singapore Pte Ltd. 2021

S. N. Merchant et al. (eds.), *Advances in Signal and Data Processing*,
Lecture Notes in Electrical Engineering 703,
https://doi.org/10.1007/978-981-15-8391-9_18

Fig. 1 Simple approach image at time

of the original technique for analysis of video which works on the bases of identifying the difference between the changes in current and previous frame. It is quite challenging to develop an accurate background subtraction method. In this contest, several algorithms have been put forward by researcher. Background subtraction method instasize on frame difference considering the parameter color texture, edge improve the detection. This method identifies the object by associating two frames and finds the difference to create the distance matrix [1–10] (Fig. 1).

1.1 Simple Approach Steps

- Estimate the background for time *t*.
- Subtract the estimated background from the input frame.
- Apply a threshold, Th, to the absolute difference to get the foreground mask.

Background is estimated between previous and current frame. Background subtraction equation then becomes:

$$B(x,\ y, t) = I(x,\ y, t\ -\ 1) \tag{1}$$

$$|I(x,\ y, t) - I(x, y, t - 1)| > \mathrm{Th} \tag{2}$$

Wren et al. [6] projected a very simple method of subtraction by structuring the stable background at each pixel level with 3D Gaussian distribution [7, 8]. In order to modal the foreground pixel, deviation between each pixel of the input frame and the model frame is calculated after following the modeling of background. Dynamic background complexity is not handling satisfactory by Gaussian function which is single.

Grimson and Stauffer [9, 10] reported there research on model of Gaussian mixture the effect of background texture caused by the some external condition [8]. In the model of Gaussian mixture, each pixel is structured with a combination of K number of function of Gaussian. In these techniques, the existing frame is linked with background structure related with Gaussian model until the match found.

When the similarity is detected, the two parameters, i.e., variance and mean of the compare Gaussian model updated.

Spagnolo et al. [8] proposed a reliable history subtraction method that integrates temporal photoanalysis with reference heritage model. Similar case are address many video processing technique like analysis of video, investigation technique I video traffic, flow checking, game activity.

Zivkovic [10] In this technique, recurrent-based method incorporates to apiaries parameter effectively as well as to select the appropriate number of components for each pixel simultaneously. In this work, the conventional pixel aliened method is used.

Lee et al. [9] proposed an powerful approach to improve the convergence charge without converting the balance of Gaussian mixture model [10]. Associating this technique with card lance frame work subtraction of background will lead toward improving the performance of video segmentation.

2 Modeling Framework for Adaptive Subtraction:

In adaptive difference modeling framework for background subtraction each position x in video stream V, we denote as $V(x) = \{\ldots, v_{t-1(x)}, v(x), v_{t+1(x)}, \ldots\}$, where $v_t(x)$ denotes the pixel value on position x. we define four difference status, S, D, B, F, with two difference discriminators as

$$d_t^v(x) = \begin{cases} S \text{ if } |v_t(x) - v_{t-1}(\text{x})| < K \\ D, \text{ otherwise} \end{cases} \tag{3}$$

$$d_t^v(x) = \begin{cases} B, \text{ if } |v_t(x) - b(x)| < K \\ F, \text{ otherwise} \end{cases} \tag{4.}$$

where K is a decision threshold and $b(x)$ denotes the corresponding background value. Discriminator $d_t^v(x)$ models the difference process of adjacent frames, while discriminator $d_t^v(x)$ models the process of the difference between current value $v_t(x)$ and background value $b(x)$. Corresponding to $V(x)$, there are two histories of states [1].

$D^v(x) = \{\ldots, d_{t-1}^v(x), d_t^v(x), d_{t+1}^v(x), \ldots\}$ and $D^b(x) = \{\ldots\ldots, d_{t-1}^b(x), d_t^b(x), d_{t+1}^b(x), \ldots..\}$

The evaluation of each pixel as an alternation process of foreground and background segments. The steadiness duration, l denotes the length of the S segment immediately preceding the current frames (Fig. 2),

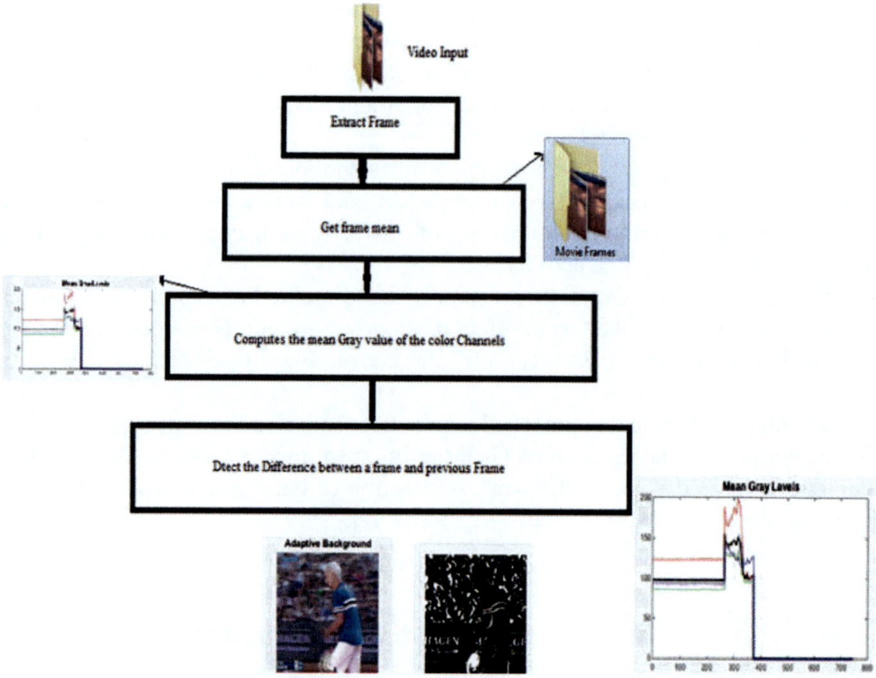

Fig. 2 Flowchart of background subtraction method

$$l = \begin{cases} l+1, \text{ if } d_t^v(x) = S \\ 0, \quad\quad \text{otherwise} \end{cases} \tag{5}$$

as an example of $D^v(x)$ and its corresponding l sequence is

$$D^v(x) = \{\ldots, S, S, D, S, S, S, S, S, D, \ldots\}, \tag{6}$$

$$l = \{\ldots, 1, 2, 0, 1, 2, 3, 4, 5, 0, \ldots\}.$$

The results obtain according to the equation

$$R_t(x) = \begin{cases} \text{foreground}, & \text{if } d_t^v(x) = D \\ \text{foreground}, & \text{else if } d_t^b(x) = F \\ \text{background}, & \text{otherwise} \end{cases}. \tag{7}$$

In this method, two steps are very important.

1. Mean gray level
2. Adaptive background.

3 Results and Discussion

We have consider video data base of more than 150 videos. Among this video in this paper, we have projected results of five videos with diver side parameters to disclose with we initially take the video "Gods of Egypt" having 6933 frames. The total numbers of shots detected by adaptive shot boundary detection are 18. Manually, total numbers of shot in this video are 20, so almost the evaluation parameter precision, recall, F1 measure comes out to be 90% which is consistently satisfactory.

Secondly, we have taken in to consideration the video frames "tennis sports" having 746 frames. In an analogs way, conscientiously the evaluation parameter is again consistent, i.e., 88%. Also, we have under taken "bulletin information" video has 275 frames. In these case, numbers shot detected are less is 22 as compare to actual number of shot 28. The complexity of these video varies as frame parameters are changing in subsequently frame, so precession is 73%, recall is much better, i.e., 78% and F1 measure is 74%.ken.

Next, we have under "traffic playback site visitors" with 1545 frames. For these video, numbers of shot detected are 12 and actual shot are 10 which is again given conscientiously with value of precision, recall and $F1$ measure which comes out to be 80, 85, 82, respectively. Finally "Krishna carton" is taken with 440 frames, maintaining shot detection and total shots. In the video giving satisfactory value of precision, recall and F1 measure . We take in to consideration diversified video with variance in complexity, type, size of video. Overall the proposed algorithm performs well except video with changes in inherent semantics property. In next algorithm, we take care of this shots coming including illumination effects, camera motion and object motion.

3.1 Input and There Result

See Figs. 3, 4, 5 and 6; Tables 1 and 2.

Fig. 3 Movie input

Fig. 4 Adaptive background

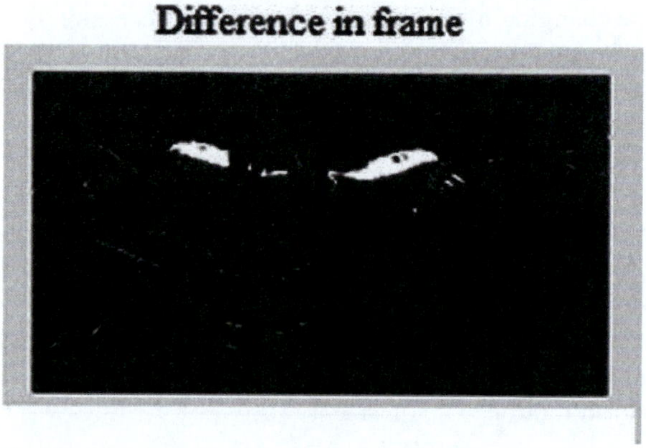

Fig. 5 Difference between frames

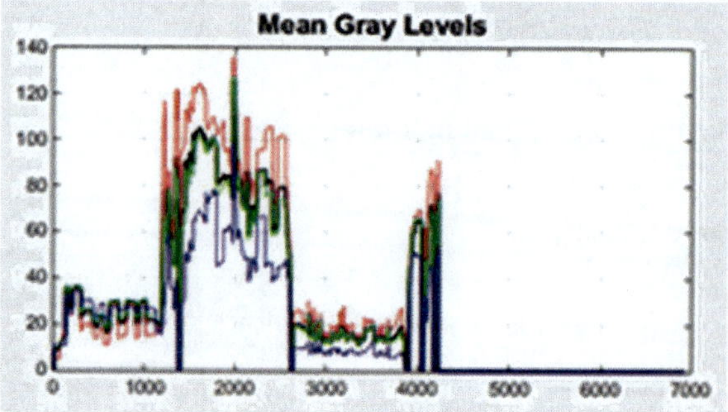

Fig. 6 Mean gray levels gods of Egypt

Sport Video Inputs

See Figs. 7, 8, 9 and 10.

Comparative analysis of histogram disparity and proposed method indicates trade-off between proposed method and histogram disparity method. News video shows comparatively good results in terms of precision, recall, $F1$ for proposed algorithm.

Table 1 Number of shot detection and evaluation parameters

Video inputs	Total no. of frames	Total shot	Shot detected	Precision (%)	Recall (%)	$F1$ (%)	Average % of detection
Movie Gods of Egypt	6933	20	18	90	90	90	93.87
Tennis sports	746	18	16	88	88	88	
Bulletin news	2751	28	22	73	78	74	
Traffic Videoplayback	1545	10	12	80	85	82	
Krishna carton	440	22	24	91	92	90	

Table 2 The performance comparison of two method [11]

video input	Histogram disparity			Proposed method		
	Precision	Recall	$F1$	Precision	Recall	$F1$
News	80.5	93.77	86.63	73	78	74
Movie	93.3	100	96.53	90	90	90
Carton	59.1	100	75.05	91	92	90

Fig. 7 Movie input

Fig. 8 Adaptive background

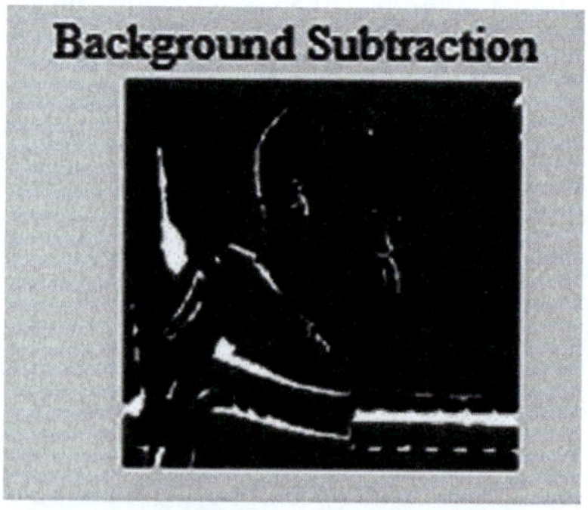

Fig. 9 Difference between frames

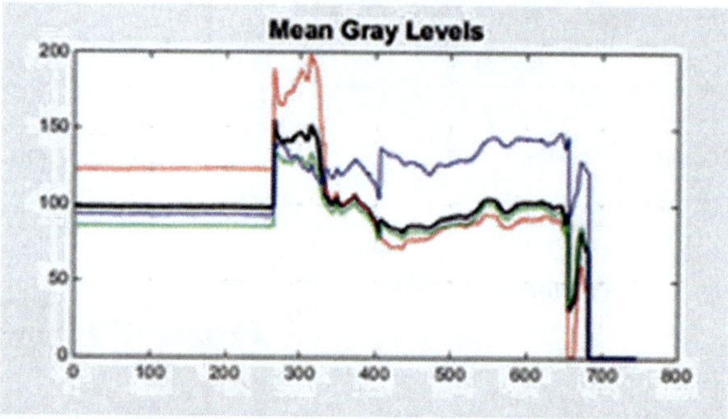

Fig. 10 Mean gray levels of tennis sport

Which histogram disparity given out decrees value of precision and $F1$. For the movie video, histogram disparity method performance well as compare to proposed method as precision and recall values of histogram disparity method are much method. For the carton video, proposed method performs well. We have higher value of evaluation parameter compare to histogram disparity method (Figs. 11 and 12).

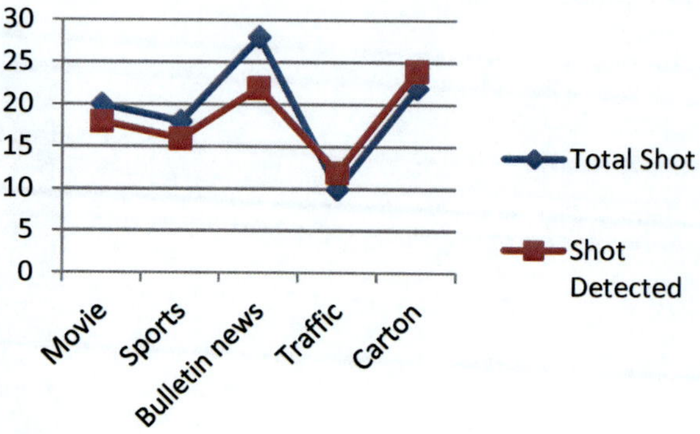

Fig. 11 Number of shot detection

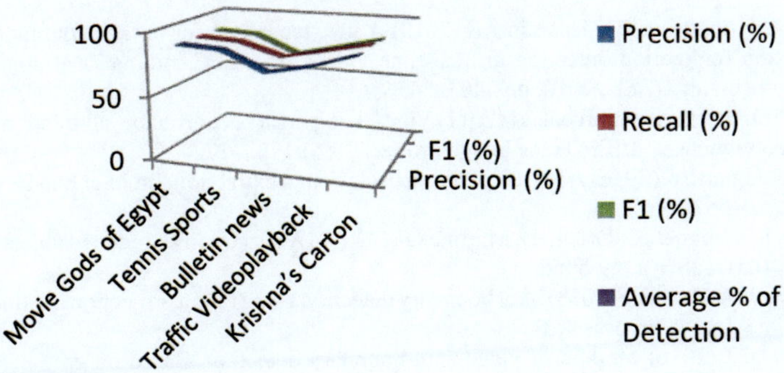

Fig. 12 Evaluation parameters

4 Conclusion

Video shot detection is one of the fundamenta steps for video shot analysis, video segmentation, video indexing, etc. In this paper, we have projected a technique for video shot detection using adaptive background subtraction model. This method is based background modeling of the scene and extracting the foreground object which is moving by subtracting the background frame with current frame. The projected method tacks in account several feature of video frames linked, texture, color and edges information to enhance the shot boundary detection. The results are satisfactory for movie video input of Gods of Egypt, Krishna carton, bulletin news, tennis sports, videoplayback traffic. We have observed satisfactory shot detection with 93.87%. We have evaluated our method using precession, recall and F1 measure which comes out to be quite satisfactory I the feature work we will focus on more novel techniques like motion activity, elimination effects, etc.

References

1. Chiu S-Y, Chiu CC, Xu SS-D (2018) A background subtraction. Algorithm in complex environments based on category entropy analysis. Appl Sci 8:885. https://doi.org/10.3390/app8060885
2. Singh S, Prasad A, Srivastava K, Bhattacharya S (2018) A novel method to improve basic background subtraction methods for object detection in video surveillance system. Int J Appl Eng Res 13(4):1866–1873. ISSN 0973-4562
3. Yao G, Lei T, Zhong J, Jiang P, Jia W (2017) Comparative evaluation of background subtraction algorithms in remote scene videos captured by MWIR sensors. Sensors 17:1945. https://doi.org/10.3390/s17091945
4. Sahasri M, Gireesh C (2017) Object motion detection and tracking for video surveillance. IJETT 161–164
5. Devi RB, Chanu YJ, Singh KM (2016) A survey on different background subtraction method for moving object detection. IJRIEST 3(10)

6. Azab MM, Shedeed HA, Hussein AS (2010) A new technique for background modeling and subtraction for motion detection in. real-time videos. In: IEEE international conference on image processing. IEEE, 2010, pp 3453–3456
7. Barnich O, Van Droogenbroeck M (2011) Vibe: A universal background.. subtraction algorithm for video sequences. IEEE Trans Image Process 20(6):1709–1724
8. Danve J, Jagtap SK (2016) A review paper on background extraction. Int J Sci Eng Res 7(1):373. ISSN 2229–5518
9. Luna E, San Miguel JC, Ortego D, Martínez JM (2017) A survey on abandoned object detection. JETIR 4(05). ISSN-2349-5162
10. Bhalotra PSA, Patil BD (2013) Shot boundary detection using radon projection method. ACEEE Int J Signal Image Process 4(3)
11. Liu X, Dai J (2016) Method of video-shot-boundary detection. Based on grey modeling for histogram sequence. Int J Signal Process Image Process Pattern Recogn 9(4):265–280

Automatic Gear Sorting Using Wireless PLC Based on Computer Vision

Yogesh Darekar and Smita Kulkarni

Abstract Gears are the most important components of the device and are usually used in the design of transmission of cars and other pivoting devices. In this paper, computer vision is suggested to sort out the defective equipment based on image processing depending on their amount of teeth image processing instrument and sensory circuitry used to solve the gear sorting issue. Through less human participation, a programmable logic controller (PLC) is used in sectors to automatically execute the entire manufacturing cycle to prevent human errors. A low-cost automation (LCA) has emerged in sophisticated technology that is used to prevent wiring composition and to effectively acquire control over the process. This paper involves converting wired PLC into wireless PLC by interfacing the PLC with the Wi-Fi module. To enable real-time surveillance and control of the system of equipment sorting via Wi-Fi module interfacing with PLC.

Keywords Wi-fi module · Firebase server · Wireless PLC · Centroid · Mobile app

1 Introduction

In gear manufacturing industry gear is an extensively used mechanical part. In mechanical industry power transfer from one shaft to other shaft by using gears. There are of various types gear are available namely spur gear, worm gears, rack gear, helical gears, etc. Gear are driven by various kinds of machines like automobiles, metal cutting tools, rolling mills, material handling equipment, marine power plants, etc. In the power transmission equipment, the friction and other losses are comparatively very low. In this work, we proposed image processing to determine

Y. Darekar (✉)
Department of Electrical Engineering, VLSI Design and Embedded System, MIT Academy of Engineering, Alandi, Pune 412105, India
e-mail: yndarekar@mitaoe.ac.in

S. Kulkarni
Department of Electrical Engineering, MIT Academy of Engineering, Alandi, Pune 412105, India
e-mail: sskulkarni@entc.maepune.ac.in

© Springer Nature Singapore Pte Ltd. 2021

S. N. Merchant et al. (eds.), *Advances in Signal and Data Processing*,
Lecture Notes in Electrical Engineering 703,
https://doi.org/10.1007/978-981-15-8391-9_19

gear parameters and sort them to faulty or non-faulty ones. Captured image is pre-processed to de-noise it and then it is segmented and enhanced. Finally, image is classified into faulty or non-faulty one. If image is faulty then rod placed at the end of conveyor rotates to 45° to let it fall in box of faulty gears. Classification algorithm classifies given image of gear as faulty or non-faulty according to standard specifications fed to it like 24 tooth gear is correct according to standard then less or more tooth gear are rejected or are classified as faulty one. For pushing the object servo motor is used to sort objects from conveyor to sorted bin whereas DC motor 12 V is used to run conveyor belt continuously. In this work we used soapbox PLC is used for designing ladder diagram for the batch process. As to make system automatically wireless PLC is used. For transmitting and receiving signals from PLC to the process we are using Node MCU Wi-Fi module as the communication interface which is used for transmitting and receiving signals from the PLC to Android App and vice-versa. A Central Processing Unit (CPU) is like the brain of PLC which is used to perform most of the calculations. As central CPU intimate to the processor, it consists of complicated electronic circuitry, which is used to perform the task of executing stored program instructions. CPU consists of two units: (1) The Control Unit (CU) and (2) The Arithmetic Logic Unit (ALU) [1] which is used to perform some arithmetic's operation. The control unit which is the brain of PLC does not execute the instructions but it guides to other parts of the systems in accomplishing this, whereas ALU is used for executing the logical and arithmetic instructions. There is some step that performed by CPU such as fetching the code, then decoding it and last is execute and store. Figure 1 shows the architecture of PLC it contains CPU with program memory, data memory, and input/output module, and isolation barrier. Memory which is used by CPU gives changeless capacity to the working framework

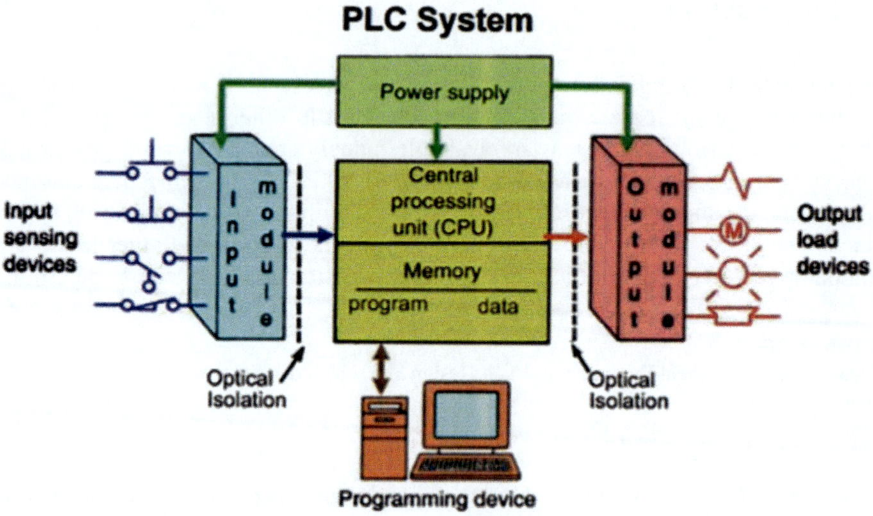

Fig. 1 PLC system [1]

for information. There are three types of PLCs are used, i.e., fixe type, modular type, and rack type. We can choose any of these as per our applications. In this paper, we choose Open Soap-box PLC in which microcontroller Atmega328 is used.

In wireless technology with up-gradation, the arena devices can also be controlled and monitor wirelessly. In industry using wireless technology, we can diminish human efforts by reducing the wiring or cabling, capital, have fewer nuisances with connectors. This technology is very helpful where it is difficult to cable and troubleshoot as it saves time and also provides a backup communication connection in ordinary network failure. Since ZigBee and Bluetooth protocol is one of the WPANs used for transmitting information in the 10 m range, the data transmission rate is very small. Due to this feature, the scan time of the whole process increases. It is, therefore, difficult to apply it in fields or procedures where it is necessary to accomplish a particular job with the least interval pulses of the encoder, whereas it is suggested to use Wi-Fi technology which is low-cost and of high transmission rate.

2 Literature Survey

Exploration of Gear Measurement utilizing MATLAB that has been carried out by centering two highlights of apparatus picture object [2]. Determined the area of the device and included the teeth by using an image processing algorithm. This paper consisted of the five rigging picture objects prepared from the created MATLAB code; all device picture items discovered having different zone estimates and shifting teeth with each other. The estimate is a basic endeavor to explicitly restrict the apparatus. The main research has been finished by using image preparation to quantify two stuff that is its most important highlights (a) To quantify the apparatus picture object and (b) To include the number of teeth in the rigging picture object.

The study exploits MATLAB picture preparing for rigging deformity recognition [3]. Rigging shows unique constructions of imperfection during the process of use. The most important kinds are destructive setting and constant loss.

Investigated the approach consists of capturing the image of the tooth profile and the contact spot by means of a computerized camera, recreating the contact spot of the hypoid gear by means of PC illustration processes, ensuring the comparative position on the edges between the tooth surface and the spot, and assessing the size, position and bearing of the contact spot in order to provide the institution with the necessary information [4].

The research consists of developing image processing code by using image preparation, reading the picture's distinctive apparatus object and transforming it into a dark-scale image over a distinctive rigging object, and then determining the edge assessment of the dim-scale image and using limit esteem transformed it into a paired image over the dim-scale image. The study has estimated the properties of the picture object locales, after that they have curved the polygon which are in areas, at last, it has changed over into districts important to the areas veil through which it has been featured the district with lines which demonstrates the teeth area of a rigging object. Clearly, through this procedure, it has estimated the rigging object zone and tallied the teeth by utilizing the MATLAB instrument.

Prof. Nilima Bargal et al. [5] attempted to make an arrangement that will diminish human exertion and prevailing to a degree by utilizing the minimal effort robotization framework (LAC) to dodge hazard, improve exactness, increment speed of generation, and decrease the duration of the process. There will be Confinements because of the handy troubles in designing of programming of the undertaking concurring the accessibility of the materials and their parts. This can be accomplished by utilizing the different sensors, different items, apparatuses, with a high level of precision and quality with a robotization.

The improvement of an LCA (Low-Cost Automation) [6] framework to sort articles as indicated by their tallness has been carried out by Rohit et al. [7]. This LCA framework is constrained by (PLC). This venture comprises two sections, first comprising of programming which contains stepping stool rationale programming which is utilized to program PLC which controls the entire procedure of the task well ordered by information grouping. Second is the equipment part which comprises of transports used to move the items, for detecting the tallness sensors is used (for example, laser sensors) of the items, electronic framework used to sort the articles, and engines to drive the transports. To sort things, one should most likely look at them, i.e., to decide if the article getting through the transport is littler, more prominent, or equivalent to the regarded stature.

The study completed a basic item arranging framework utilizing the (PLC) [8]. The framework comprises of a transport model in which the arranging procedure happens. The item will get through the transport framework to the detecting part, in light of the stature of the article the item will be dealt with the transport framework by utilizing the Programmable Logic System (PLC). The programmed transport framework is a mechanized control strategy for controlling and dealing with the arranging instrument simultaneously keeping up the effectiveness of the business and nature of the products. HMI for the programmed transport framework is viewed as the essential method for controlling every task. Content showcases are accessible just as graphical touch screens. It is utilized in contact boards and neighborhood observing of machines. This paper manages the productive utilization of PLC in the programmed transport framework and furthermore constructing the exactness in it [9]. Paper [10] proposes a pre-preparing strategy for the first deviations of tooth profiles as indicated by the qualities of the anomalies from the tooth profiles and the highlights of the involute apparatus profile mistake indicator model and builds up a model of exception identification and revision for the deviations of tooth profiles. The primary point of research work [11, 12] is to structure and build up a minimal effort machine vision framework and transport line arrangement for programmed acknowledgment and arranging of mechanical parts. In this task, the calculation for highlight extraction and item recognition has been created and tried in MATLAB programming conditions. From the outcome, it is demonstrated that the proposed framework can ready to distinguish, perceive, and arranging of parts in like manner as required for the applications.

3 Proposed System

The suggested system is split into five parts (1) Fault gear identification by image processing, (2) Conveyor belt assembly, (3) Atmega328 Open PLC for automation process control and monitoring. (4) Wi-Fi Module, and (5) Android App (Fig. 2).

3.1 Image Processing

The camera captures gear image when a proximity sensor detects gear on the conveyor belt. For noise removal, the gear picture is pre-processed to improve the picture information of the equipment that suppresses unwanted distortions or enhances gear picture characteristics that are essential for further processing.

The image of the gear becomes a gray image; it measures the threshold of the gray image and transforms the gray image into a binary image using the threshold value. After this, the small objects from the binary image had been removed, the object hole filled and the binary image center measured. The image center is calculated accordingly:

$$X_C = \frac{1}{M}\sum_{i=1}^{n} x_i m_i \quad (1)$$

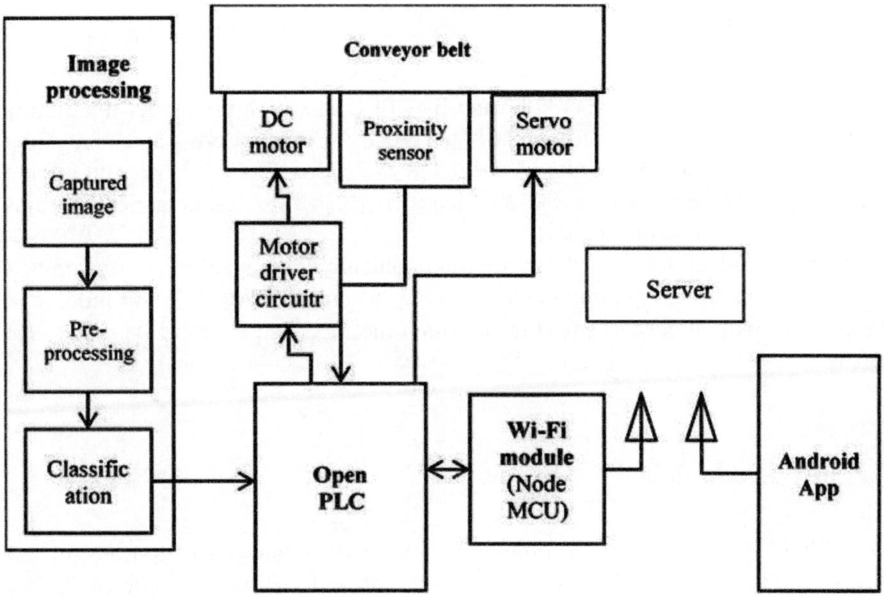

Fig. 2 Block diagram of proposed system

where 'M' is the sum of intensity is the pixel intense value, x_i and y_i are pixels on the image, here n is the complete pixel count. Image is then masked using adaptive masking. In image processing, adaptive filters are most commonly used to enhance or re-store data by removing unwanted noise without significantly blurring the structures in the image. Masking involves setting certain of the values pixel in an image to zero, or "background" value. The mask image covers the region of interest which contains gear position. Mask image is multiplied with Black and white images to find no. of gears. If no. of white blocks is equal to no. of teeth in gear then the gear is accepted otherwise it is rejected (Fig. 3).

3.2 Conveyor Belt

To detect gear on the conveyor belt, the proximity sensor is connected to the conveyor belt. The conveyor belt is powered by the use of a DC motor regulated by the motor driver circuitry. Motor Driver circuits are current amplifiers. The input to the motor driver IC or motor driver circuit is a low current signal. The role of the motor driver circuit is to transform the low-current signal to a high-current signal when the IR detector detects the equipment PLC can stop the motor. The motor then receives this high-current signal. Servo motor is used to distinguish defective gears from non-fault gears.

3.3 Open PLC

In this paper, we used Atmega328 Soap Box PLC in which the ladder diagram is designed for the implementation of the batch process. For monitoring and controlling the entire process automatically. Atmega328 Soap Box PLC. It is an automation platform based on open-source PC We used soapbox PLC to design ladder diagrams in this work as shown in Fig. 4.

Use wireless PLC to create the system automatically. The signal can be transmitted or received in wireless communication via the system node MCU Wi-Fi module to the defined communication interface between the PLC and Android App and vice versa.

3.4 Wi-Fi Module

Node MCU Wi-Fi module to a communication interface that is used to transmit and receive signals from the PLC to the Android App and vice versa. Node MCU is a platform based on open source IoT. Node MCU Wi-Fi module consists of firmware running from Express-if on the ESP8266 Wi-Fi SoC and the configuration is based

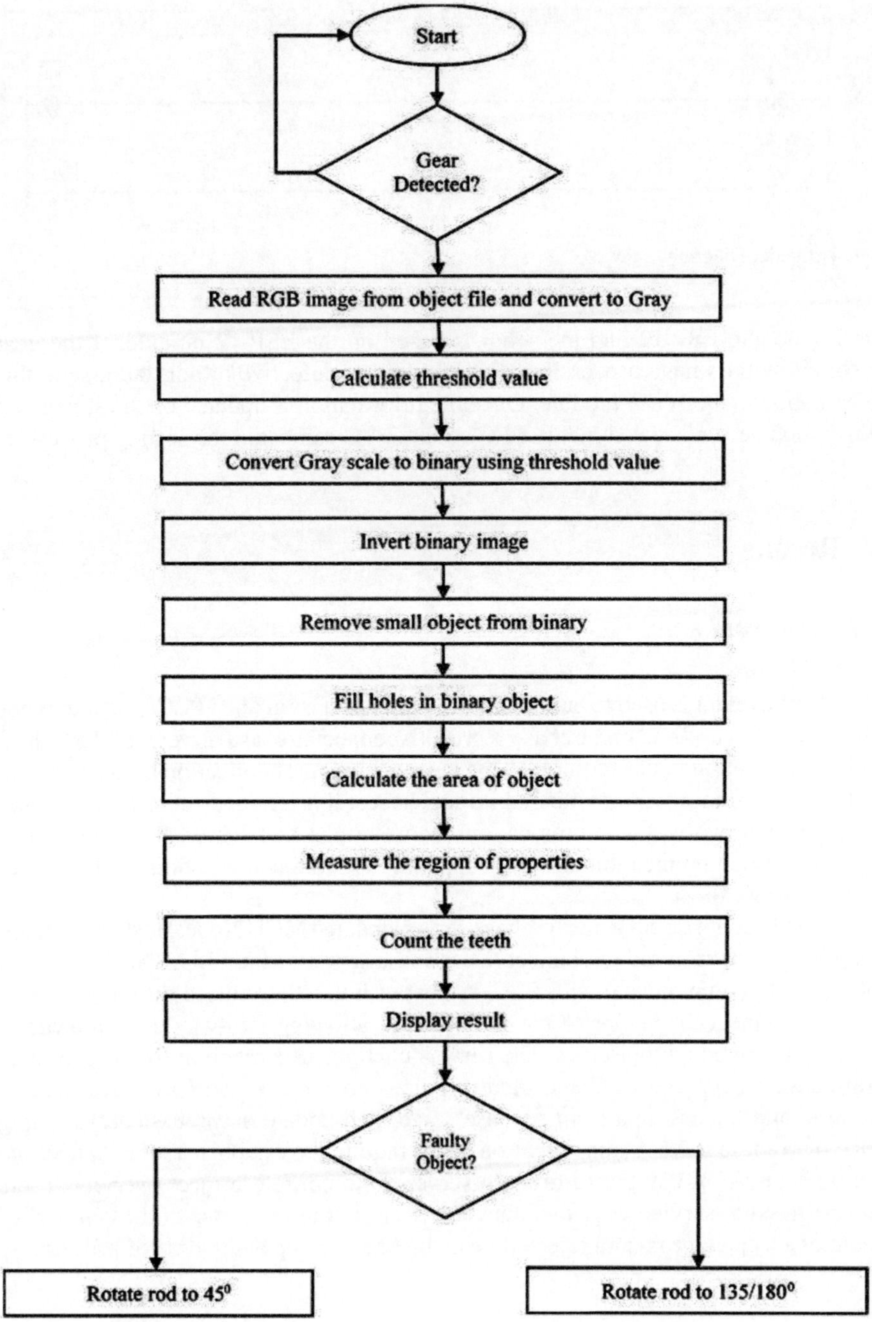

Fig. 3 Flow chart of proposed system

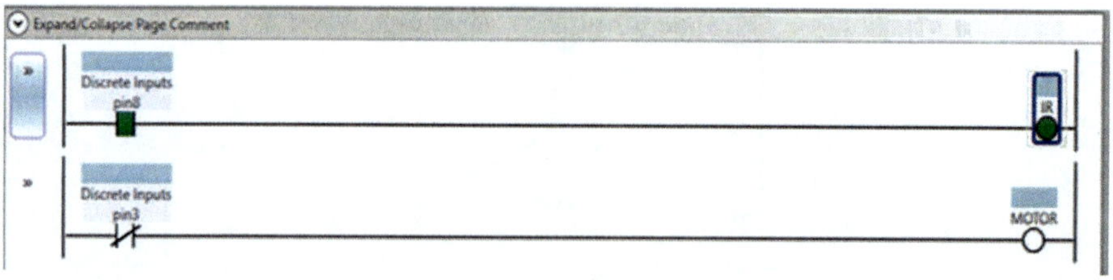

Fig. 4 Ladder Diagram

on the module ESP-12 and the setup is based on the ESP-12 module. If the gear is sorted by the image processing algorithm either defective or non-fault, send this information to the Wi-Fi module. Once the information is updated on the server, the Wi-Fi module sends the signal to PLC so that PLC can begin the sorting process.

4 Results

4.1 Android App

MIT App Inventor is used to build Android App. Data from open PLC is sent to Node MCU, which is updated on Firebase server. To connect firebase server with PLC, host and authentication key is required. This key enables communication between server and PLC. To connect server with mobile app, host or link is added while programming which allows two-way communications between app and server. Server acts as a medium for communication between App and PLC and hence coding or data is not required (Fig. 5).

Open PLC data is sent to the Firebase server-updated MCU Node. Host and authentication key are required to connect the firebase server to the PLC. This key allows the server to communicate with PLC. A host or link is introduced during programming to connect the server to the mobile app, allowing two-way communication between the app and the server. The server functions as a medium for communication between App and PLC and therefore does not require coding or information. Android app fetches data from firebase server and update on firebase server using URL of firebase server. Authentication key is required to establish a safe connection. For authentication key got to firebase server-go to console-project overview-users and permission-service accounts-database secrets. Figure 6 shows the number of count of accepted gears and rejected gears in Android app at the time of processing.

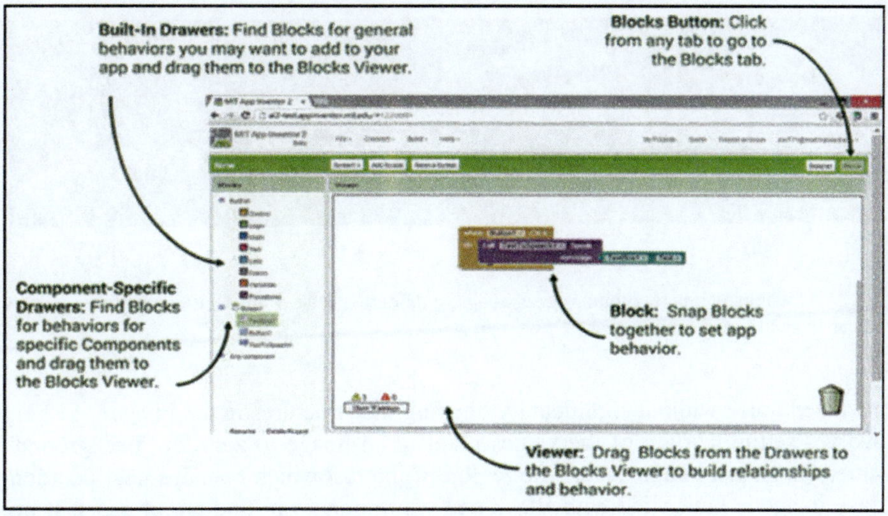

Fig. 5 Connection to android phone

Fig. 6 Android App

4.2 *Image Analysis*

Captured color image is first converted into a grayscaled image and then by using morphological operations noise is removed to suppress unwilling distortions. De-noised image is shown in figure Fig. 7a. After de-noising the image, it is converted to black and white image or grey level and then measures the threshold of the gray image and transforms the gray image into a binary image using the threshold value as shown in Fig. 7b. Image is then masked using adaptive masking. In image processing, adaptive filters are most commonly used to enhance or re-store data by removing

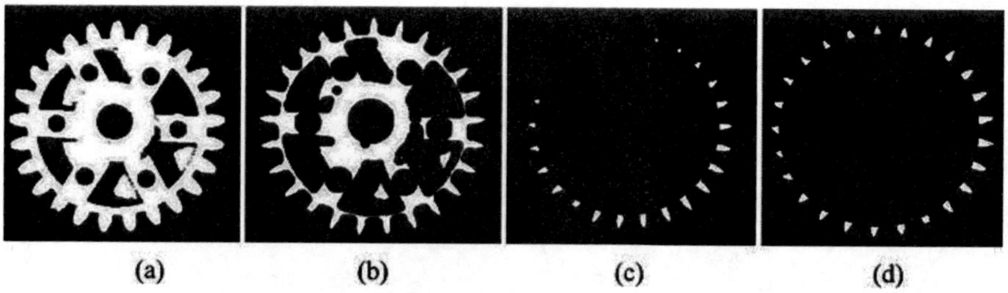

Fig. 7 **a** De-noised image, **b** binary gear image, **c** defective gear, **d** image after multiplying mask to gear image (non-defective gear)

unwanted noise without significantly blurring the structures in the image. Masking involves setting certain of the values pixel in an image to zero, or "background" value. This mask image covers the region of interest which contains gear position. Mask image is multiplied with Black and white images to find no. of gears. If no. of white blocks is equal to no. of teeth in gear then the gear is accepted as shown in Fig. 7c otherwise it is rejected shown in Fig. 7d.

Table Visualize the final result with confusion matrix. The numbers on the bottom side of the matrix give the overall accuracy. We achieved the final accuracy of 89% as shown in Table 1.

No of Data Set Frames: 200.

Positive (P): 178.

Negative (N): 22.

The manual calculations of the process are given below:

$$\text{Accuracy} = \frac{\text{TP} + \text{FN}}{\text{P} + \text{N}} = \frac{86 + 92}{200} = 89\% \tag{2}$$

$$\text{Sensitivity} = \frac{\text{TP}}{\text{TP} + \text{FN}} = \frac{86}{86 + 14} = 95.5\% \tag{3}$$

$$\text{Specificity} = \frac{\text{TN}}{\text{TN} + \text{FP}} = \frac{92}{92 + 8} = 92\% \tag{4}$$

Table 1 Graph showing detected and non-detected images

	Detected	Non-detected
Detected	86 [TP]	92 [TN]
Non-detected	8 [FP]	14 [FN]

5 Conclusion

In the manufactured industry there are different vibration-based signal processing sorting techniques are used but this technique requires more parameters and they are very complex in their process the result of loss of time and production cost because of machine and human error. To sort out the faulty gear, we have implemented image processing which gives us correct result of faulty gear and it has been implemented successfully. For continuous monitoring and controlling the automatic gear sorting system, we used PLC which give us accurate result. We have successfully done the wireless communication between PLC and firebase server so that the system can monitor and control processes wirelessly by using Android app. From the experimental result we have developed low-cost automatic gear sorting system which improves efficiency, reduces the cost of production as well as cycle time the accuracy of sorting the faulty gear is 89%.

References

1. Ghildiyal S, Bhimani K, Manimozhi M (2018) Design to convert a wired plc into wireless PLC. Int J Res Appl Sci Eng Technol (IJRASET). ISSN: 2321–9653. IC Value: 45.98; SJ Impact Factor: 6.887 6(I):1–2
2. Nawaz H, Ali H (2014) Gear measurement using image processing in MATLAB. Int J Innov Technol Explor Eng (IJITEE) 3:43–47
3. Pengfei C, Henan FC (2013) Characteristic: value extraction of gear defect based on image processing. J Multimedia 8:198–205
4. Jing Z (2011) The research on the application to digital image technology in hypoid gear contact zone detection. J Comput 6:1519–1525
5. Bargal N, Deshpande A, Kulkarni R, Rucha M (2016) PLC based object sorting automation. Int Res J Eng Technol (IRJET) 03(07):103. e-ISSN: 2395-0056
6. Bhuiya K, Anish K, Parekh D, Sahiti KL (2013) Low cost wireless control and monitoring using PLC and R. In: Gonzalez, RG (ed) Digital image processing. Pearson Education, New Jersey. Inc. SCADA. Int J Sci Res Publ 3(9):1–4
7. Yadav RR, More SS, Mutekar ON, Mogal AL, Mahale RS (2019) Multilevel object sorting system using PLC controller. Int Res J Eng Technol (IRJET) 06(05):1910
8. Thirumurugan P, Aravind R, Arun Kumar M, Dharshan Manjunath S, Kalaiselvan R (2018) Automatic sorting in process industries using PLC. In: GRD J Global Res Dev J Eng 3(3):8
9. Aruna YV, Beena S (2015) Automatic convey or system with in–process sorting mechanism using PLC and HMI system. Int J Eng Res Appl 5(11):37–42
10. Dhavalikar MN, Gawande V, Bhise S, Khandelwal P, Bhat A (2018) geometric verification of gears using image processing technique. Int J Adv Res Innov Ideas Educat IJARIIE 4(2):1644
11. Senthilkumar K, Aswinkumar R, Gunaseelan V (2017) Low cost machine vision system for sorting of mechanical parts using MATLAB & ARDUINO embedded hardware. Int J Recent Trends Eng Res (IJRTER). Conference on Electronics, Information and Communication Systems (CELICS'17) Special Issue, pp 1–2 2017, 11
12. Kale DS, Maknikar AV (2016) Automatic material handling and sorting of defective gear using image processing. Int J Eng Techn 2(6):201

Statistical Validity of Presmoking and Postsmoking Impact on Heart Rate Variability Among Middle Age Men

S. R. Rathod and C. Y. Patil

Abstract *Background:* Cigarette smoking is associated with various forms of an acute heart attack such as myocardial infarction, arrhythmia, and atrial and ventricular fibrillation. Increased sympathetic activity triggered by cigarette smoking is one of the major risk factors for cardiovascular extinction. *Objective:* To analyze the acute effects of smoking on the control of neuro-cardiovascular by evaluating the time domain, frequency domain, and nonlinear HRV indices in middle-aged smokers. *Method:* Thirty-six male participants of age between 40 and 60 were evaluated and divided into two groups, i.e., control and smokers. The ECG recorded for 15 min from control and smokers male participants. In the case of smokers, data recorded before smoking and 10 min after smoking. The heart rate variability (HRV) indices in the time domain (mean HR, mean RR, SDNN, and RM SSD), frequency domain (LF, HF, and LF/HF ratio), and nonlinear parameter (SD1, SD2, SD1/SD2, Poincare plot, detrended fractal analysis, approximate entropy, and sample entropy) were evaluated. The electrocardiogram (ECG) was recorded using our own designed ECG module, we used lead II data for experimentation. Statistical analysis performed between control versus presmoking using independent student '*t*' test and presmoking versus postsmoking using dependent student '*t*' test. The presmoking values are considered as baseline values. *Result:* When control group compared with the smoking group SBP, DBP, mean HR, LF (ms^2), LF(nu), and LF/HF ratio were significantly increased and also mean RR, SDNN, RM SSD, TP (ms^2), HF (ms^2), HF (nu), SD1, SD1/SD2, AppEN, and SampEN significantly decreased after smoking. *Conclusion:* The results of this study show that smoking has an acute effect on the autonomy of the brain, causing impaired vagal activity and an overbalance of sympathetic function.

S. R. Rathod (✉) · C. Y. Patil
Department of Instrumentation and Control Engineering, College of Engineering Pune, Pune, Maharashtra, India
e-mail: rsr18.instru@coep.ac.in

C. Y. Patil
e-mail: cyp.instru@coep.ac.in

© Springer Nature Singapore Pte Ltd. 2021

S. N. Merchant et al. (eds.), *Advances in Signal and Data Processing*,
Lecture Notes in Electrical Engineering 703,
https://doi.org/10.1007/978-981-15-8391-9_20

Keywords Neuro-cardiovascular regulation · Heart rate variability · Standard deviation (SD) · Detrended fractal analysis · Approximate entropy (AppEN) · Sample entropy (SampEN)

1 Introduction

Smoking leads to increased acute heart disease such as cardiac arrhythmia and ventricular fibrillation centered on long-term smoking of cigarettes [1, 2]. Various researchers have found that due to smoking there are changes in autonomous cardiac nerve function occurs. It has been observed among smokers that cardiac sympathetic activity has a prominent parameter which shows the link between smoking and acute heart diseases. The variability of heart rate is the most fragile and abstract measurement of specific markers of sympathetic and parasympathetic activity in the heart rate variability. In clinical practice, HRV is a noninvasive tool for assessing cardiovascular function autonomously. The time domain, frequency domain, and nonlinear parameter (Poincare plot) are recommended with short-term recording for HRV analysis [2, 3]. Although a few reviews examined the acute effect of cigarette smoking by observing the HRV postsmoking parameter, no investigation contrasted the acute effect of smoking on HRV in a single sitting for control and smokers (presmoking and postsmoking). This study was, therefore, intended to observe the effect of smoking in cigarette smokers on HRV parameters [2, 4, 5].

2 Method

For this study, the data are analyzed from 36 middle-aged people of the male gender of the same age (control—48.96 ± 4.88, smoker—50.59 ± 3.53). The people were divided into two groups: control group ($n = 36$, male = 36) and smoker group ($n = 36$, male = 36). Predominantly, we have taken the ECG recording of a male smoker because the percentage of smoking in the male is more as compared to females. Before recording a complete detail, the history of the control and smoker group is received. The participants in the control group with the post-history of hypertension, heart disease, diabetics, habits of drugs, and alcohol consumption are excluded from the study. Before the ECG recording, the control and smoker group participant had asked to take rest in the supine position for 10 min, and the hemodynamic variables, like systolic blood pressure (SBP) and diastolic blood pressure (DBP), were noted. ECG recording of a control group and smokers group is taken for 15 min. ECG recording of the smokers group was observed at baseline (presmoking), and then the subject was allowed to smoke one cigarette. HRV recording of the same subject was taken 10 min after smoking (postsmoking).

The ECG recorded using our own designed module, for experimentation purpose lead II data were used. The quality of the ECG signal is analyzed [6]. HRV analysis was carried out using the time domain, frequency domain, and non-linear indices in heart rate variability analyzer (Biomedical Workbench-Lab VIEW). The HRV parameter is measured using time domain methods such as mean HR, mean RR, SDNN, and RMSSD. The SDNN represents the standard deviation from beat to beat RR interval and represents the parasympathetic activity. The RMSSD has defined the root mean square difference of successive NN interval, which interprets the cardio autonomic control through the vagal nerve. Both SDNN and RMSSD represented in milliseconds [2–4]. HRV analysis is performed in the autoregressive frequency domain, because the value of total power (TP), low-frequency predominance. Poincare plot is a visual way to understand the variability of RR interval. In the Poincare plot, the length of each RR interval (RR$i + 1$) is plotted against the preceding RR interval (Ri), which is natural and information method for visual presentation of HR variability changes and assessment of autonomic control [7].

Detrended fractal analysis (DFA) has been gaining importance for analyzing heart rate variability based on self-similarity features. This feature represented by α, i.e., scaling exponent indicates correlation properties of the signal, i.e., the correlation between different timescales. The measuring complexity and recurrence with a time series is a mathematical algorithm used called entropy quantifier. Approximate entropy (AppEN) indicates the complexity of the signal where sample entropy (SampEN) represents repeatability in time series signal [7].

All data obtained in the control and smokers group are expressed as mean ± standard deviation. The parameter obtained between control and smoker (presmoking) participant is compared using independent student 't' test while parameter received in the smoker before smoking (presmoking) and after smoking (postsmoking) is analyzed using dependent 't' test. P-value $p < 0.05$ is accepted as a level of significance.

3 Results

In this study, all smokers had significantly higher mean values of systolic and diastolic blood pressure ($P < 0.001$) than the control group (Table 1).

In this study, time domain parameters of HRV such as mean RR, SDNN, and RMSSD significantly decreased while mean HR significantly increases in smokers. The smokers group shows significant reduction in TP values (ms^2), HF values (ms^2), and HF (nu); whereas, there is an increase in values of LF (nu) and LF/HF ratio in the analysis of frequency domain parameter of HRV. Analysis of HRV using nonlinear parameter shows a significant decrease in SD1, SD1/SD2 ratio, AppEN, and SampEN. In contrast, another parameter shows a reduction in the value but not significant when compared control with smokers group.

Table 1 Hemodynamic variable of control and smokers group

	Control group (male = 36)	Smokers group (male = 36)	*P*-value
Age (year)	48.96 ± 4.88	50.59 ± 3.53	0.1032
BMI (kg/m^2)	25.66 ± 2.74	30.59 ± 3.53	0.6011
SBP (mmHg)	118.01 ± 4.64	125.27 ± 4.03	<0.001
DBP (mmHg)	79.71 ± 4.46	84.65 ± 4.94	<0.001

Mean ± standard deviation, statistical difference compared to the control group (independent student '*t*' test, $p < 0.05$)
Abbreviation: *m* meter, *kg* kilograms, *BMI* body mass Index, *kg/m*^{2} kilogram per meter square, *SBP* systolic blood pressure, *mmHg* millimeter of mercury, *DBP* diastolic blood pressure

Table 2 Time domain parameter of HRV in control and smokers group

Control group		Smokers group	
		Presmoking	Postsmoking
Mean HR (ms)	71.43 ± 6.79	90.29 ± 10.42**	96.57 ± 13.67####
Mean RR (ms)	758.52 ± 91.96	713.47 ± 92.12**	662.38 ± 101.45##
SDNN (ms)	133.90 ± 26.48	70.11 ± 27.03**	57.99 ± 24.70####
RMSSD (ms)	64.15 ± 22.96	51.49 ± 21.18***	37.54 ± 21.74###

Mean ± standard deviation, statistical difference compared with control group
*Control group versus presmoking – independent student '*t*' test (**=$P < 0.05$, ***=$P < 0.01$, ****=$P < 0.001$)
#Presmoking versus postsmoking—dependent '*t*' test (##=$P < 0.05$, ###=$P < 0.01$, ####=$P < 0.001$).
Abbreviations: *Mean HR* heart rate, *Mean RR* RR interval, *SDNN* standard deviation of normal to normal interval, *RMSSD* root mean square of standard deviation

Table 3 Frequency domain parameter of HRV in control and smokers group

	Control group	Smokers group	
		Presmoking	Postsmoking
TP (ms^2)	3443.70 ± 109.72	2701.10 ± 157.26****	2439.64 ± 152.52####
LF (ms^2)	665.88 ± 35.49	880.33 ± 53.17****	1019.63 ± 86.68####
HF (ms^2)	596.61 ± 24.06	463.78 ± 33.93****	380.26 ± 36.45####
LF (nu)	53.47 ± 1.82	73.03 ± 1.34 ****	75.17 ± 1.05####
HF (nu)	35.11 ± 0.22	28.58 ± 1.34****	24.06 ± 1.23####
LF/HF	1.50 ± 0.02	2.29 ± 0.10****	2.93 ± 0.13####

Mean ± standard deviation, statistical difference compared with control group
*Control group versus presmoking – independent student '*t*' test (**=$P < 0.05$, ***=$P < 0.01$, ****=$P < 0.001$)
#Presmoking versus postsmoking—dependent '*t*' test (##=$P < 0.05$, ###=$P < 0.01$, ####=$P < 0.001$)
Abbreviations: *ms*^{2} millisecond square, *nu* normalized unit, *TP* total power, *LF* low frequency, *HF* high frequency

Table 4 Nonlinear parameter of HRV in control and smokers group

	Control group	Smokers group	
		Presmoking	Postsmoking
SD1	28.60 ± 5.89	20.31 ± 7.44 **	17.32 ± 5.29****
SD2	47.72 ± 9.21	47.23 ± 7.99	45.82 ± 7.49
SD1/SD2	0.40 ± 0.09	0.35 ± 0.09 **	0.28 ± 0.11###
DFA—α1	1.31 ± 0.21	1.31 ± 0.22	1.26 ± 0.17
DFA—α2	1.00 ± 0.10	0.95 ± 0.10	0.94 ± 0.10
AppEN	1.48 ± 0.34	1.20 ± 0.40***	0.92 ± 0.30###
SampEN	1.58 ± 0.22	1.35 ± 0.26***	1.22 ± 0.09###

Mean ± standard deviation, statistical difference compared with control group
*Control group versus presmoking – independent student '*t*' test (** = $P < 0.05$, *** = $P < 0.01$. **** = $P < 0.001$)
Presmoking versus postsmoking – dependent '*t*' test (##=$P < 0.05$, ###=$P < 0.01$, ####=$P < 0.001$)
Abbreviations: *ms* milliseconds, *SD1* standard deviation 1, *SD2* standard deviation 2, *DFA* detrended fractal analysis, *AppEN* approximate entropy, *SampEN* sample entropy

4 Discussion

In the present study, we have investigated the smoking effect on cardiovascular health by examining the cardiac autonomic nervous system in middle-aged male. In addition, the acute impact of smoking on autonomous cardiac nerve function is measured by analyzing information in cigarette smokers before and after smoking. In all cases, the autonomous operation of the cardiac nerve is evaluated by time domain features, frequency domain features, and nonlinear HRV features.

Factors, such as age and gender, also affect HRV; in this study, we have included males of the middle-aged group belongs from 40 to 60 years. In the case of the hemodynamic variable, the blood pressure was found significantly higher in the smokers group compared with the control group. This can be indicated by using [(control—118.01 ± 4.64) versus (smokers—125.27 ± 4.03)] and [(Control—79.71 ± 4.46) versus (84.65 ± 4.94)] systolic blood pressure. The blood pressure values of smoker subjects are already in the range of cardiovascular risk. Higher values of blood pressure indicate the predominance of sympathetic activity [5].

In all cases, time domain measures such as mean RR interval and mean heart rate reflect the vagal activity. Measures of SDNN and RMSSD are for general autonomous equalization. When control group is compared with smokers, the significant decrease in RR interval [(control-758.52 ± 91.96, presmoking—713.47 ± 92.12, postsmoking—662.38 ± 101.45)] indicates significant increase in HR value [(control—71.43 ± 6.79, presmoking—90.29 ± 10.42, postsmoking—96.57 ± 13.67)]. The decline in parasympathetic activity is verified by reduction in RMSSD [(control—64.15 ± 22.96, presmoking—51.49 ± 21.18, postsmoking—37.54 ± 21.74)], whereas overall change in variability is represented by reduction in SDNN [(control—133.90 ± 26.48, presmoking—70.11 ± 27.03, postsmoking—57.99 ±

24.70)]. HRV analysis in the frequency domain shows significantly increase in the value of LF (ms^2) [(control—665.88 ± 35.49, presmoking—880.33 ± 53.17, postsmoking—1019.63 ± 86.68)], LU (nu) [(control—53.47 ± 1.82, presmoking—73.03 ± 1.34, postsmoking—75.17 ± 1.05)], and LF/HF ratio [(control—1.50 ± 0.02, presmoking—2.29 ± 0.10, postsmoking—2.93 ± 0.13)] and significantly decrease in the value of total power TP(ms^2) [(control—3443.70 ± 109.72, presmoking—2701.10 ± 157.26, postsmoking—2439.64 ± 152.52)], HF (ms2) [(control—596.61 ± 24.06, presmoking—463.78 ± 33.93, postsmoking—380.26 ± 36.45)], and HF(nu) [(control—35.11 ± 0.22, presmoking—28.58 ± 1.34, postsmoking—24.06 ± 1.23)].

The HF (ms^2) and LF (ms^2) reflect the parasympathetic, i.e., vagal activity and sympathetic activity, respectively [2, 8]. A significant decrease in HF power indicates the inhibition of parasympathetic activity, and a substantial increase in LF power indicates the sympathetic predominance. The normalized unit of LF and HF represents the change in sympathetic and parasympathetic regulation. The LF/HF ratio indicates an overall balance between the sympathetic and parasympathetic systems. In the present case, the LF/HF ratio reflects the dominance of the sympathetic system. Total power measures overall autonomic activity, and a significant decrease in total power indicates a reduction in sympathetic activity.

In the analysis of nonlinear indices of HRV, SD1, SD1/SD2, AppEN, and SampEN values significantly decreased. Rest all the parameter shows declination from their average values but not significant. The Poincare plot was analyzed using SD1, SD2, and SD1/SD, which indicate a reduction in parasympathetic activity in the smokers group. Poincare plot is presented in Fig. 1.

Detrended fractal analysis is based on the self-similarity of the time series signal. In a delicate subject, DFA-α1 and DFA-α2 show the lower value, which is irregular

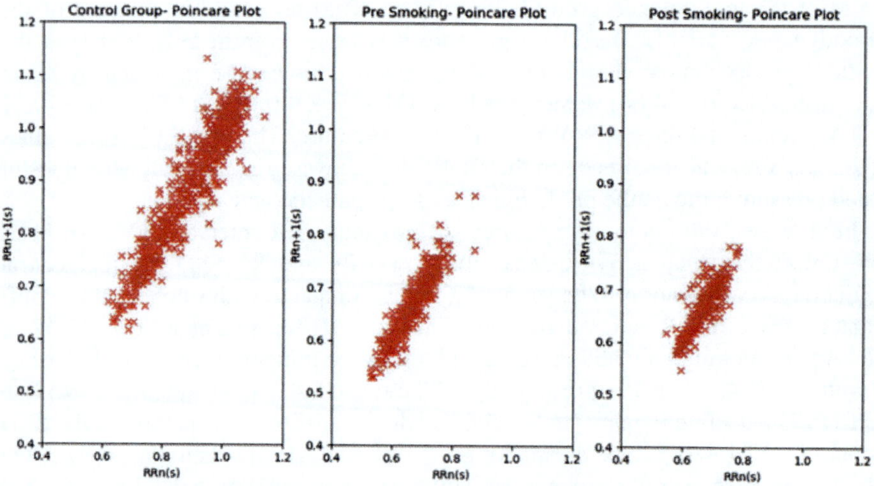

Fig. 1 Graphical representation of Poincare plot of the control and smokers group

and less complicated. In our study, we have found that DFA values are degraded, which majorly predict that subject may suffer from cardiac arrest. AppEN and SampEN measure the complexity of the time series signal. Higher values of AppEN and SampEN represent a more complex tachogram, which is a good indication. It is observed that AppEN and SampEn significantly decreased, which indicates that the signal loses its complexity, which is a strong predictor of future diseases related to heart [7].

5 Conclusion

Smoking cigarettes altered the autonomic functions of the nerve measurement by HRV in the smokers of middle age. The findings of this study show that smoking corresponds to a rapid and intermittent decline in heart parasympathetic activity and enhance sympathetic activity. Sympathovagal balance may shift to strong sympathetic dominance as a result of cigarette smoking. Therefore, the smoker is more susceptible to an increased adverse cardiovascular event.

Conflict of Interest This study was conducted at institute level for research purpose only and approved by Research and Development committee of College of Engineering Pune. There is a voluntary agreement before recruitment of volunteer to participate in research study.

References

1. Hallstrom AP, Cobb LA, Ray R (1986) Smoking as a risk factor for recurrence of sudden cardiac arrest. N Engl J Med 314(5):271–275
2. Ferdous M, Ferdousi S (2014) Acute effect of cigarette smoking on HRV in current cigarette smokers. J Bangladesh Soc Physiol 9(2):59–64
3. Hayano J, Yamada M, Sakakibara Y, Fujinami T, Yokoyama K, Watanabe Y, Takata K (1990) Short and long term effects of cigarette smoking on heart rate variability. Am J Cardiol 65(1):84–8
4. Barutu I, Esen AM, Kaya D, Turkmen M, Karakaya O, Melek M, Esen OB, Basaran Y (2005) Cigarette smoking and heart rate variability: the powerful influence of parasympathetic and sympathetic maneuvers. Ann Noninvas Electrocardiol 10(3):324–329
5. Ferdous M, Ferdousi S (2018) Acute impact of cigarette smoking on power spectral measures of HRV. J Bangladesh Soc Physiol 13(1):8–12
6. Shashikant CP (2020) Assessment of ECG signal quality. Advances in decision sciences, ımage processing, security, and computer vision. In: ICETE 2019. Learning and analytics in ıntelligent systems, vol 3. Springer, Cham
7. Jindal GD, Deepak KK, Jain RK (2010) Advanced application of physiological variability. Physiol Variability
8. Ferdousi S, Ferdous M, Islam MS (2014) Impact of smoking status on autonomic functions assessed by spectral analysis of heart rate variability. Int J Clin Exp Physiol 1(1):57

Transition Based Odd/Full Invert Coding Scheme for Crosstalk Avoidance and Low Power Consumption in NoC Links

M. Vinodhini and N. S. Murty

Abstract In this paper, a Transition Based Odd/Full Invert (TBO/FI) coding scheme, which focuses on crosstalk avoidance and low dynamic power consumption in NoC links is proposed. This scheme is designed and implemented at both architectural and logic level and is evaluated using synthetic traffic scenarios for both 4- and 8-wire links. All the evaluations are performed for the worst-case, the best-case, and the typical-case scenarios. TBO/FI coding scheme has the maximum reduction percentage for both the switching activities in all the cases for 4-wire link and same is true for 8-wire link except for the worst-case scenario. TBO/FI coding scheme allows NoC power savings of up to 25.5% and 40.4% for 4- and 8-wire links with worst-case scenario and with other scenarios, NoC power consumption increases. However, this increase is lower than that of other existing schemes. These results are achieved despite the NoC router area and power overheads of 117% and 50% for 4-wire and 52% and 26% for 8-wire link, respectively. NoC router area and power reduces by 16% and 30%, respectively, with increase in link width and this reduction is more compared to all other schemes.

Keywords NoC interconnects · Crosstalk effects · Lowpower links · Coding techniques · Coupling capacitance · Switching capacitance

1 Introduction

Marching towards Ultra Deep SubMicron (UDSM) technology has empowered design engineers to integrate many number of processing elements into a System on Chip (SoC) [1–3]. Even though this benefited in performance, bus based on chip communication constraints forced them to move toward Network on Chip (NoC). Scalable NoC transfers messages from one IP core to another IP core in packetized form through routers and links.

M. Vinodhini (✉) · N. S. Murty
Department of Electronics and Communication Engineering, Amrita School of Engineering, Amrita Vishwa Vidyapeetham, Bengaluru, India
e-mail: m_vinodhini@blr.amirta.edu

© Springer Nature Singapore Pte Ltd. 2021

S. N. Merchant et al. (eds.), *Advances in Signal and Data Processing*, Lecture Notes in Electrical Engineering 703,
https://doi.org/10.1007/978-981-15-8391-9_21

In SoC, significant proportion of the overall power budget is used up by NoC. Links, buffers, and crossbar switches contribute in different proportions to total power budget of NoC. Buffers, crossbar switch, routing computation block, and arbiter form the NoC router. In NoC router, all the components get involved in routing packets and hence, contribute to total power budget of NoC.

For example, in the three mesh-based NoCs, TeraFLOPs, TRIPS, and RAW on an average, 29% of the total power of NoC is absorbed by links and buffers each individually, 26% is absorbed by crossbar switches and 16% is absorbed by arbiter and routing computation blocks [4]. This shows clearly that one of the major contributions to the total power budget is by the NoC links. In fact, this communication architecture also has major impact on various design goals such as area, performance, and reliability. As technology shrinks, the wires which form the NoC links, become more aggressive in power consumption and performance [5]. Further, cumulative effect of various noise sources such as leakage noise, crosstalk noise, power supply noise, and so on increases with technology scaling [6]. These noises lead to effects like glitches, delays, faults, and signal distortion. Hence, propagation delay and dynamic power consumption of NoC links increase with technology scaling and this is mainly due to coupling and load capacitance of the link [7]. Coupling capacitance depends on technology and various other structural factors like wire width, wire spacing, wire material, wire length, and so on. Coupling capacitance is also larger in magnitude than the load capacitance and is dependent on the volume and nature of the data transitions [6]. Therefore, in NoC-based communication system, which supports multi-cores, accomplishing low link power consumption [8] and crosstalk avoidance [9] are the critical issues.

NoC link power consumption is directly proportional to self and coupling switching activities and also to voltage swing of the signal. Approaches used to reduce the voltage swing of the signal rely on reducing the supply voltage and adapting low swing signaling which makes router design crucial [4]. Another common approach to deal with switching activity is to adapt coding techniques. Apart from achieving low link power consumption in NoC, the other critical issue is crosstalk avoidance. Coupling capacitance existing between adjacent wires is a potential cause of crosstalk [10]. Various factors influencing coupling capacitance are spacing between metal wires, transition of data on the wires, and the switching activity between adjacent wires. International Technology Roadmap for Semiconductors (ITRS) has predicted that by 2020, the total wire length will extend to 7000 m/cm^2 [11–13]. Therefore, the coupling and self switching capacitances in lengthy and adjacent wires become more predominant and lead to several crosstalk effects. These effects result in incorrect transmission of data, link speed variations, and high link power consumption in NoC.

At different layers of abstraction, several mechanisms are proposed to handle crosstalk effects. However, these mechanisms solve crosstalk effects only partially with significant area overhead [9]. Despite this, mechanisms like repeater insertion, shielding, wire spacing, and timing skewing are employed at the physical and switch level in NoCs [13, 14]. Several coding mechanisms are commonly used for low-power link and crosstalk avoidance [4].

The work proposed in this paper is a Transition Based Odd/Full Invert (TBO/FI) coding scheme that encodes the transition patterns to minimize crosstalk and self switching activities during transmission on the NoC links. Impact of implementing codec in each router of NoC is assessed by performing various simulations to compute the percentage of power savings and area and power overheads. Further, improvements in NoC performance are estimated in terms of latency and throughput with different data streams.

The remainder of this paper is presented in this manner. Brief discussion on related work is provided in Sect. 2. Section 3 explains the proposed coding scheme. Section 4 presents the discussion on area, power, and delay overhead characteristics of the proposed scheme compared to other schemes and the work is concluded in Sect. 5.

2 Related Works and Contributions

Average number of transitions like 0 to 1 or 1 to 0 in the data bits communicated through single wire over a period of time is commonly known as self switching activity. This self switching activity on single wire causes transitions on the adjacent wires due to the coupling capacitance. This activity is normally known as coupling switching activity. Effects of this coupling switching activity are glitches and excess delay in the data signal transmitted through the adjacent wires. Crosstalk effects due to coupling capacitance between adjacent wires are classified as 0C, 1C, 2C, and 3C for 2-wire link [15] as given in Table 1. As per the literature, several coding schemes have been suggested to prevent these activities in the data signal communicated to avoid crosstalk and to attain low power in NoC links. Mainly, self switching activity occurring in a wire and coupling switching activity occurring in the adjacent or victim wires are focused on by these coding schemes.

As per the survey, in earlier years, large number of research works were focusing only on reducing the power consumption due to self switching activity by ignoring coupling switching activity [4, 8, 15]. Bus Invert (BI) [16], Fast INC-XOR [17], Gray code [18], T0 [19], Working zone encoding [20], T0-XOR [21], Quadro coding [22], Multi-coding [23], and Data flipping coding [24] are the diverse coding schemes which fall under the above-mentioned category. Later, when technology shrank, it was recognized that in the overall power consumption of the NoC links, significant contribution is by the coupling switching activity. Coding schemes proposed to

Table 1 Crosstalk effect classification for 2-wire link transition patterns

Crosstalk effect level	Transition pattern
0C	$--$
1C	$\uparrow\uparrow$, $\downarrow\downarrow$
2C	$-\uparrow$, $-\downarrow$, $\uparrow-$, $\downarrow-$
3C	$\uparrow\downarrow$, $\downarrow\uparrow$

address coupling switching activity mainly concentrated on removing the transition patterns like 010 and 101. Forbidden Transition Code (FTC) [25], Forbidden Pattern free Code (FPC) [25], Forbidden Overlapped Code (FOC) [25], and One Lambda Codes [26] are the distinctive coding schemes proposed to address this activity.

In the recent years, significant number of coding schemes concentrated on both self switching and coupling switching activities. Enhanced Crosstalk Detector Bus Invert (ECDBI) [7], Odd Even Full invert (OEF) [8], SC [15], Calculated Odd/Even Bus Invert (COEBI) [27], efficient RC low-power [28], and Coupled driven Bus Invert (CBI) [29] are the different coding schemes focused on both the switching activities.

Now, the coding schemes with which our proposed coding scheme is compared are discussed in detail. In hop-to-hop-based BI coding scheme [16], number of self transitions present between the current and the next data bits of a flit is computed and if the number of transitions is greater than half of the number of wires in a link, then the next data bits are inverted. This scheme purely concentrates on the self switching activity by ignoring the effects of coupling switching activity which is said to be the dominating factor in USDM technology [16]. Further to ensure decoding, an extra bit is appended with each data pattern with increased link width and switching activities. Another hop-to-hop coding scheme proposed in [29] concentrates on coupling capacitance effects. In this CBI scheme, the encoder counts the number of transitions present between the current and the next data bits with specifically one and two weight coefficients. If this count is higher than half of the number of wires in a link then the next data bits are inverted. Issue faced with this scheme is for few data bit patterns, full inversion using CBI encoding results in higher switching activity. Extra wire issue persists in this scheme too. In FPC scheme [25], the occurrence of patterns like 010 and 101 are avoided by using shielding wire and coding four bit pattern to five bit pattern. This way of coding effectively reduces the coupling and self switching activities, however, increases the communication bandwidth, traffic, and time.

In [7], ECDBI encoder uses crosstalk and switching detector which estimates the self and coupling switching activities in parallel manner. Based on the amount of activities present, the next data bits are inverted and one extra wire line is added to transmit one control bit to indicate about the inversion to the decoder. This scheme effectively reduces the transition patterns which generate high level of crosstalk but increases the power dissipation. This scheme also increases the number of transitions which generate low level of crosstalk effect considerably. In [27], COEBI encoder, by estimating the number of coupling transitions, performs any one of the following changes in the data pattern transmitted. These changes are (1) all bus line are inverted (2) all bus lines are kept unaltered (3) even bus lines are inverted and (4) odd bus lines are inverted. To support efficient decoding with each data pattern, two bits are appended which adds on to the link bandwidth and traffic load.

SC [15] and OEF [8] coding schemes group all the possible 2 bit communication patterns as Type I, II, III, and IV transition pattern. In SC scheme, the number of 01 to 10 and 10 to 01 (i.e., Type II) and 01 to 01 and 10 to 10 transitions (i.e., Type IV) present between adjacent data patterns are calculated and if 01 to 10 and 10 to 01 transitions are more in number then the next data bits are inverted. To indicate this inversion to the decoder, one extra bit is added with each data bit pattern. This added bit increases the flit size and the self and coupling switching activities. In OEF scheme, odd, even, full, and no inversions are performed to reduce the effect of crosstalk and power dissipation. This scheme uses a complex encoder which counts for specific type of transitions present in the data pattern and decides to perform any one of the inversions. Two extra bits are added for each pattern in the encoder and corresponding operation in reverse way is performed in the decoder. This scheme increases the hardware complexity of encoder and decoder with increased link bandwidth. In all these schemes discussed, notable amounts of reduction in self and coupling switching activities are achieved either with complex codec design and/or with additional link bandwidth. Another issue to be noted in all these schemes is considerable increase in the switching activity due to the appended extra bits used for correct decoding. In NoC, bits belonging to a single flit are communicated at a same time through the available parallel wires between the routers. For example, 32 bits of a single input flit are transmitted through 4-wire link as shown in Fig. 1. Commonly used methodology in all coding schemes as per literature is to compute the number of self and coupling switching activities present between the current (e.g. 1, 2, 3, 4 bits) and next (e.g. 5, 6, 7, 8 bits) transmitted data bits.

Depending upon the level of switching activity present between them any one of the modification like invert, shift, odd invert, even invert ,and so on is performed on the next data bits. These newly generated data bits are compared with next data bits and this process continues till the last set of data bits of a flit are processed. Instead of comparing and analyzing each and every pair of parallelly transmitted data bits for computing switching activities i.e., column-wise, our proposed TBO/FI coding scheme analyzes all consecutive bits transmitted through each and every wire, i.e., row-wise. This is the first novel idea in our coding scheme. Total number of bits compared in all the schemes remains the same at the end. However, TBO/FI scheme has higher switching activity reduction. Second novelty in our scheme is that in order to indicate the encoding process to the decoder, one extra bit is appended with each set of data bits transmitted through a wire and transfer the encoded flit bits along with taking extra bit transmission time. The benefit of doing so is there is no need of extra wire to transmit the appended extra bit which is not the case in other coding schemes.

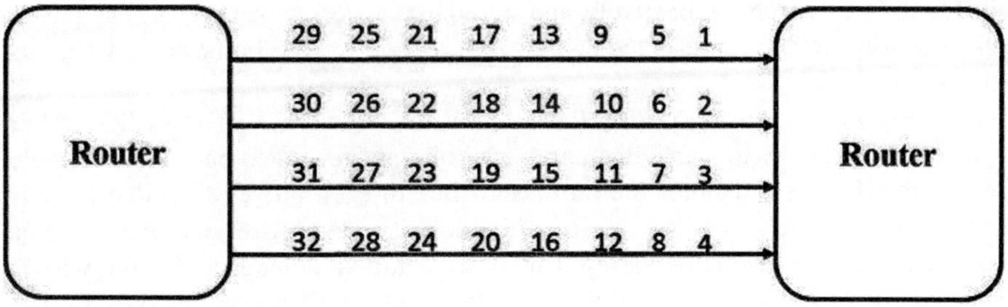

Fig. 1 Parallel data transmission

In the existing schemes, depending upon the encoding logic minimum, one wire and maximum three wires are required for sending the extra bits which are transmitted parallely with original data bits. Third novelty in our proposed scheme is that the appended extra bits are dependent on the previous bits transmitted. This is done so to retain the same number of switching activities if it is already less.

3 Proposed TBO/FI Coding Scheme

TBO/FI coding scheme reduces the total number of 0 to 1 and 1 to 0 transitions occurring in the message transmitted through the wires as well as avoids the strongest transition patterns occurring in the adjacent wires. This coding scheme is implemented at switch-to-switch level of NoC. Figure 2 shows the 4 × 4 NoC architecture and the internal architecture of the router. Each router has five ports and codec is placed at north, east, west, and south ports except, at the local port. Message in the form of flits during its traversal on a chip is decoded and encoded at each router's entry and exit, respectively. TBO/FI encoder and decoder are placed at each output and input port, respectively. TBO/FI codec architecture is suitable for different sizes of link width in wires (i.e., 2-wire, 3-wire, 4-wire, 5-wire, and so on). Number of parallel bits communicated at a time in NoC depends on the link width. Further, generic encoder and decoder architectures are proposed in this work which are applicable for different sizes of link width.

3.1 TBO/FI Encoder

In TBO/FI encoder, the group of input flit bits transmitted across each wire is named as bit-set. For example, if link width is four and input flit size is 32, then input bits are transmitted in four bit-sets of 8 bits each as shown in Fig. 1. Each bit-set transmitted across a wire is separately analyzed to decide whether to odd invert or full invert the bits in that bit-set as shown in Fig. 3. In the codec architecture shown in Fig. 3, B_1, B_2, …, B_N denote the bit- sets of one flit which are received across parallel wires wire1, wire2,…, wireN, respectively, and stored in registers present across each wire. Bits in bit-set1 B_1 are represented as B_{11}, B_{12}, B_{13}, ..., B_{1n}. In the same way, bits in bit-setN.

BN are represented as $BN_1, BN_2, BN_3, \ldots, BN_n$. In our work, N denotes the number of available parallel wires and also number of bit-sets generated based on available link width whereas n denotes the number of bits in each bit-set. Encoded bit-sets are represented as C_1, C_2, C_3, ... C_N. To each encoded bit-set one extra bit r is appended to ensure proper decoding. Bit-set transmitted along each parallel wire is either odd or full inverted depending upon the number self transitions present in it i.e., number of 0 to 1 and 1 to 0 transitions (self) present between consecutive bits.

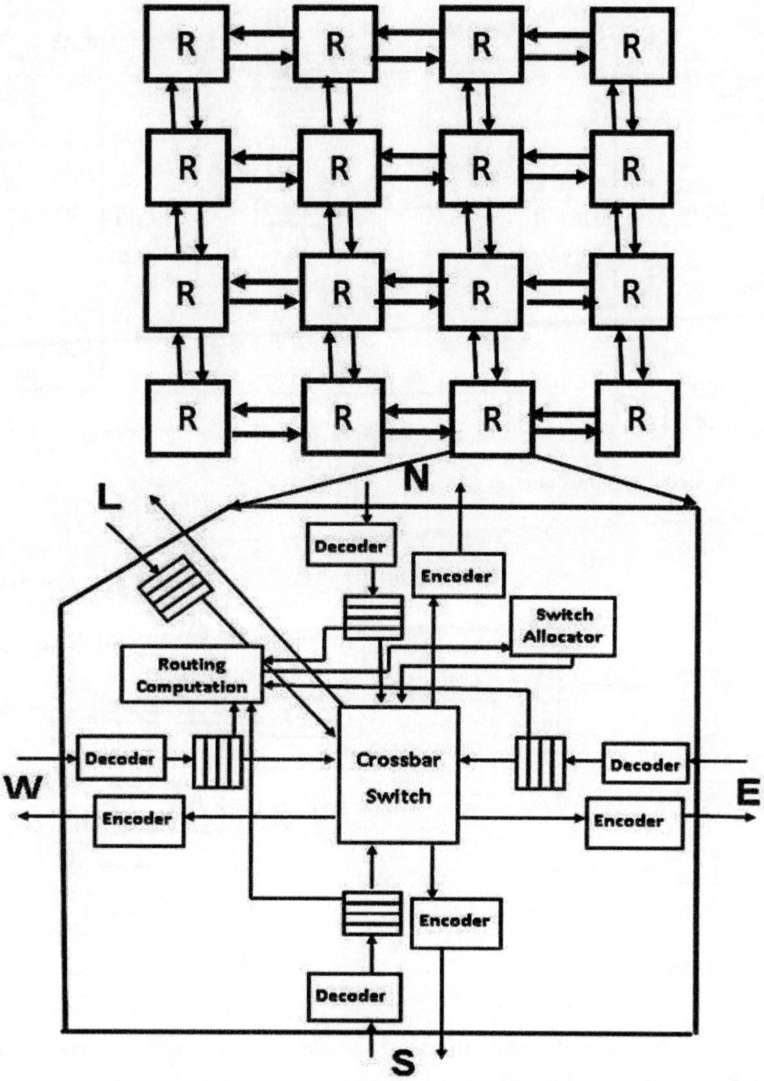

Fig. 2. 4 × 4 NoC architecture and router internal architecture

If the number of transitions are greater than half of the number of bits in bit-set then those bits are odd inverted otherwise full inverted as shown in Fig. 4.

Consecutive bits in bit-set B_1 are XORed using 2 input XOR gates to compute the number of self transitions (T_s) present in it as shown in Fig. 4. T_s is compared with T in comparator. T is equal to $n/2$. If T_s is greater than T then odd bits in set B_1 are inverted. Otherwise, all the bits in set B_1 are inverted. If all the bits are not inverted when T_s is lesser than T then total number of switching activity increases.

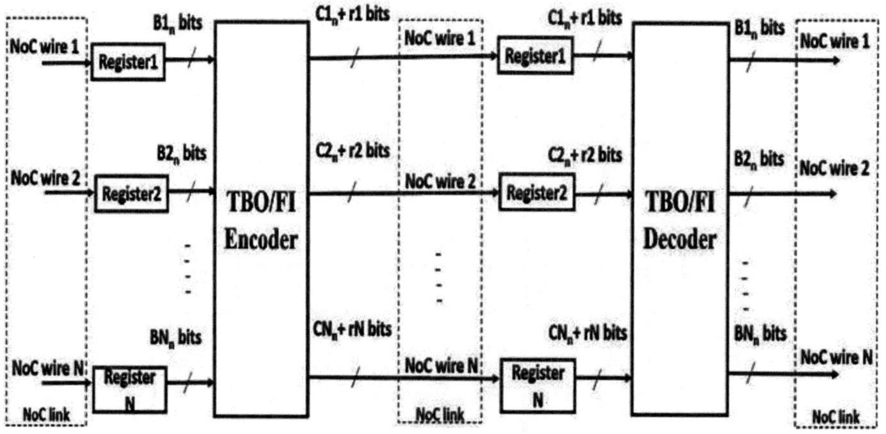

Fig. 3 TBO/FI codec architecture

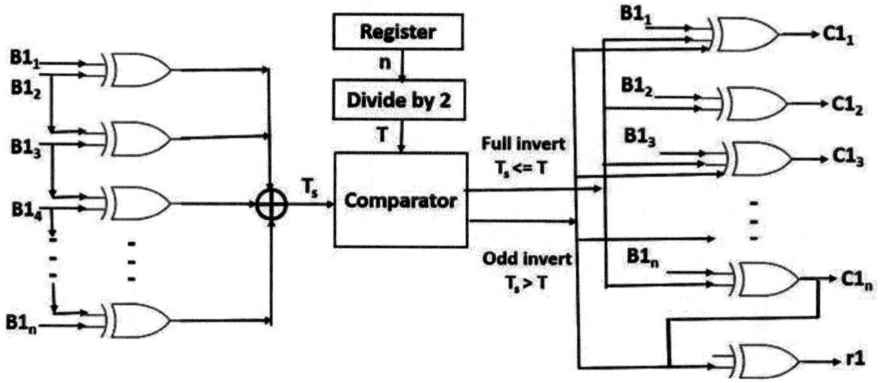

Fig. 4 TBO/FI encoder

To indicate the full or odd inversion to the decoder, with each bit-set one extra bit r is appended. In Fig. 4, encoded bits of set.

B_1 are denoted as C_{11}, C_{12}, …, C_{1n} with extra bit $r1$. $r1$ bit is the same as C_{1n} bit for full inversion otherwise $r1$ is inverted C_{1n} bit for odd inversion of the bit-sets. Similar encoding process is performed with other bit-sets transmitted across wire2, wire3… wireN.

3.2 TBO/FI Decoder

In TBO/FI decoder, each encoded bit-set is received and stored in registers and decoded separately based on the extra bit r. $r1$ is compared with C_{1n} using 1 bit

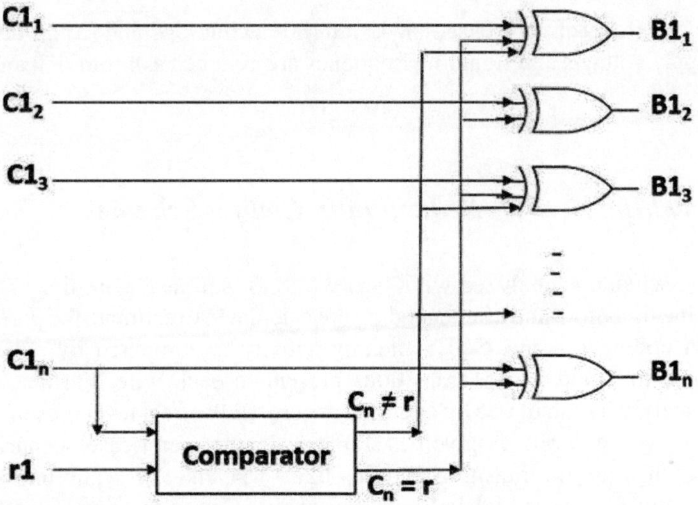

Fig. 5 TBO/FI decoder

comparator as shown in Fig. 5. For the encoded bit-set1, if $r1$ is equal to C_{1n} then received C_{11}, C_{12}, ..., C_{1n} bits are full inverted. Otherwise, the received encoded bit-set is odd inverted. Similar decoding process is performed with the bits received across other wires.

4 Results and Discussions

In our work, TBO/FI coding scheme is designed using Verilog HDL and synthesized in Cadence RTL compiler at 45-nm technology library. Further, input flit size of 32 bits transmitted through 4-wire and 8-wire links are considered for the evaluation of the proposed TBO/FI coding scheme. TBO/FI coding scheme is compared against BI, CBI, COEBI, ECDBI, FPC, OEF, and SCS coding schemes in all the evaluation processes and discussions. Aim of our TBO/FI scheme is to lessen the NoC link power consumption and crosstalk effects by reducing the coupling and self switching activities on the NoC links. The dynamic power consumption of NoC link (P_{LINK}) is [8]

$$P_{\text{LINK}} = [T_s C_W + T_C C_C][V_{DD}]^2 F_{\text{clk}} \quad (1)$$

where T_s is the number of transitions present between each set of two successive bits transmitted, C_W is the total wire capacitance, T_C is the number of transitions present between adjacent wires, and C_C is the coupling capacitance. In (1), total wire capacitance and coupling capacitance are considered as 257.22 fF/mm and

205.77 fF/mm, respectively, based on 45 nm [30] technology and [31]. Further, link length, supply voltage, and signaling frequency are assumed as 1 mm, 1 V and 1 GHz, respectively.

4.1 Switching Activity Reduction in Coding Schemes

Our coding scheme is analyzed with 4-wire link as well as 8-wire link. T_s and T_C present in the flit before and after encoding are calculated to estimate the performance of TBO/FI coding scheme. Self switching activity is computed by counting the number of 0 to 1 and 1 to 0 transitions present in each wire. Further, coupling switching activity is calculated by counting the crosstalk effect level present between the adjacent set of 2 bits as given in Table 1. In the worst-case scenario, if the flit consists of only the transition patterns like "↑↓" and "↓↑" alternatively, then switching activities are at higher levels. For analysis purpose, we considered best-case, worst-case, and typical-case scenarios. Our proposed scheme reduces the levels of switching activities, if they are high and retains the same level if they are already minimal.

For the three different cases of bit patterns and data types, reduction in number of transition patterns with 3C crosstalk effect level and increase in number of transition patterns with 0C crosstalk effect level before and after encoding is computed for 4- and 8-wire links and provided in Figs.6 and 7. It is clearly noticed that for all the different data types, number of transition pattern with 3C crosstalk effect is either reduced or retained and number of transition pattern with 0C crosstalk effect is increased with coded flits when comparing the uncoded flits.

Comparison of our coding scheme with other schemes for percentage reduction in switching activities for worst-case, best-case, and typical-case scenarios is given in Table 2. In 4-wire link, our scheme has the lower self and coupling switching activities for all the three scenarios. In 8-wire link, for the worst-case, our coding

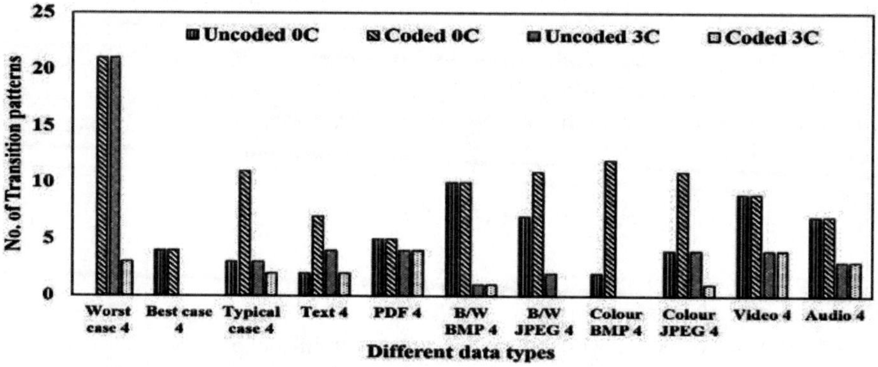

Fig. 6 Analysis of 0C and 3C transition patterns for 4-wire link

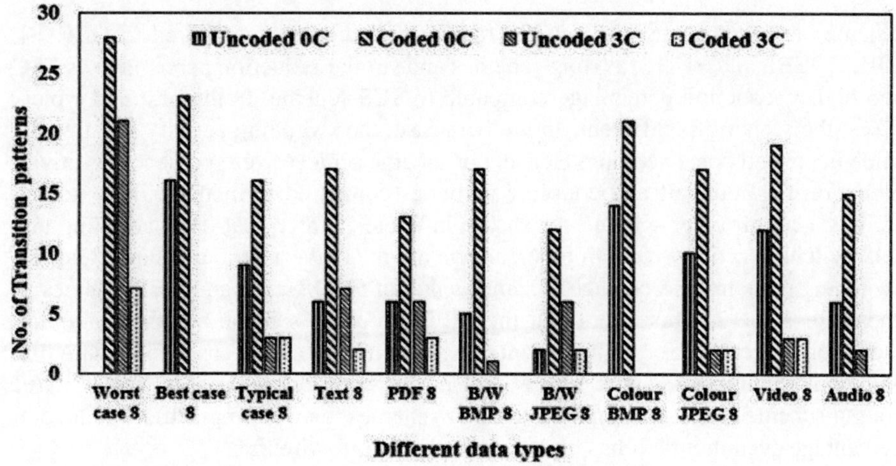

Fig. 7 Analysis of 0C and 3C transition patterns for 8-wire link

Table 2 Reduction percentage of switching activity for different coding schemes for best, worst, and typical-case scenarios

Schemes	Self switching activity reduction (%)			Coupling switching activity reduction (%)		
	Best	Worst	Typ.	Best	Worst	Typ.
4 wire link						
FPC	−100	21	−17	-25	3	3
BI	0	75	23	−33	73	3
CBI	0	75	29	−33	73	3
ECDBI	0	75	29	−33	73	8
COEBI	−75	50	6	−83	60	−22
OEF	0	14	−6	−33	49	−27
SCS	−100	32	−35	−100	9	−47
TBO/FI	0	85	35	0	83	25
8 wire link						
FPC	−133	66	10	−10	61	0
BI	0	87.5	20	−20	90	12
CBI	0	87.5	20	−20	90	12
ECDBI	0	87.5	20	−20	90	12
COEBI	−100	75	10	−80	85	−8
OEF	−333	79	−10	−320	80	−28
SCS	0	58	−40	−20	58	−60
TBO/FI	0	66	20	0	66	16

scheme has lesser switching activity reduction percentage compared to OEF, BI, CBI, COEBI, and ECDBI coding schemes and similar reduction percentage as FPC and higher reduction percentage compared to SCS scheme. In the best and typical cases, the scenario is different. In the best-case, the switching activity remains the same in pre and post encoding. Hence, our scheme achieves zero percentage activity reduction for both self and coupling activities compared to increase in switching activity with all other schemes as shown in Table 2. In typical-case, coupling and self switching activity reduction percentage are the highest for our scheme and self switching reduction percentage is same as that of ECDBI, BI, and CBI schemes as shown in Table 2. This infers that the TBO/FI coding scheme has the maximum reduction percentage for both the switching activities in all the cases for 4-wire link and same applies for 8-wire link except in the worst-case scenario. The TBO/FI coding scheme is better compared to other schemes in switching activity reduction percentage even though it has high bit overhead for 8-wire link.

4.2 Power Consumption, Area Occupancy, Delay, and Bit Overhead of Different Codecs

Proposed TBO/FI coding scheme is implemented at switch-to-switch level of NoC. Wormhole-based baseline router operating at a higher frequency of 1.55 GHZ is designed and synthesized in 45 nm technology. By incorporating the TBO/FI coding scheme in the wormhole router further evaluation is performed. The power consumption and area occupancy of router are 0.068 mW and 3635 μm^2, respectively, based on which the power, area, and delay overheads are evaluated. Power consumption, area occupancy, and delay of the different codecs for 4- and 8-wire links are specified in Tables 3 and 4.

Table 3 Power consumption, area occupancy, delay, and bit overhead of different codec for 4-wire link

Codec	Area (μm^2)	Power (mW)	Delay (ns)	Bit Overhead (%)
BI	489	0.029	0.74	25
CBI	655	0.032	0.79	25
COEBI	394	0.018	0.59	50
ECDBI	554	0.032	0.7	25
FPC	558	0.021	0.58	62.5
OEF	1725	0.091	0.67	50
SCS	763	0.047	0.61	25
TBO/FI	450	0.019	0.57	12.5

Table 4 Power consumption, area occupancy, delay, and bit overhead of different codec for 8-wire link

Codec	Area (μm^2)	Power (mW)	Delay (ns)	Bit Overhead (%)
BI	475	0.036	0.84	12.5
CBI	1108	0.047	1.11	12.5
COEBI	302	0.014	0.59	12.5
ECDBI	509	0.030	0.82	12.5
FPC	434	0.018	0.58	34.3
OEF	1959	0.111	0.75	25
SCS	859	0.045	0.66	12.5
TBO/FI	240	0.008	0.56	25

It is noted from Table 3, that the area and power of TBO/FI codec are only marginally higher than COEBI which has the lowest value. Even though area occupancy and power consumption of COEBI scheme are lower than TBO/FI scheme, COEBI has more bit overhead as in this scheme for each parallel bits transmitted two extra wires are required for proper decoding. This increases the link bandwidth and traffic. TBO/FI codec has lowest critical path delay when compared to all other codecs. Bit overhead for a flit size of 32 bits transmitted through 4 link for different coding schemes is provided in Table 3. It is seen that for 4-wire link, TBO/FI coding scheme has the least bit overhead and FPC has the highest bit overhead.

It is noted from Table 4, that the area, power, and delay for 8-wire link are the lowest for TBO/FI codec compared to all other schemes although it has higher bit overhead. For 8-wire link, bit overhead of our scheme is higher than that of BI, CBI, ECDBI, and SCS and same as that of COEBI and OEF coding schemes and lesser than FPC scheme as noticed in Table 4. Even here, FPC has the highest bit overhead. Hence, bit overhead of our coding scheme increases with increase in link width and vice versa.

In Fig. 8, power and area overheads of wormhole router without any codec and with different codecs are given for 4- and 8-wire links. Compared to all other similar codecs, TBO/FI codec has less power and area overheads for both 4- and 8-wire links. As the logic complexity of our encoder and decoder is less compared to other codecs, this improvement in power and area is achieved.

4.3 Power Savings Due to the Coding Schemes

Efficiency of TBO/FI coding scheme in minimizing the overall power consumption of the NoC is calculated using 4 × 4 mesh based NoC architecture with routers incorporating codecs at 4 ports (i.e., North, South, East, and West). Packet is transferred through the longest path in the network and its corresponding power consumption

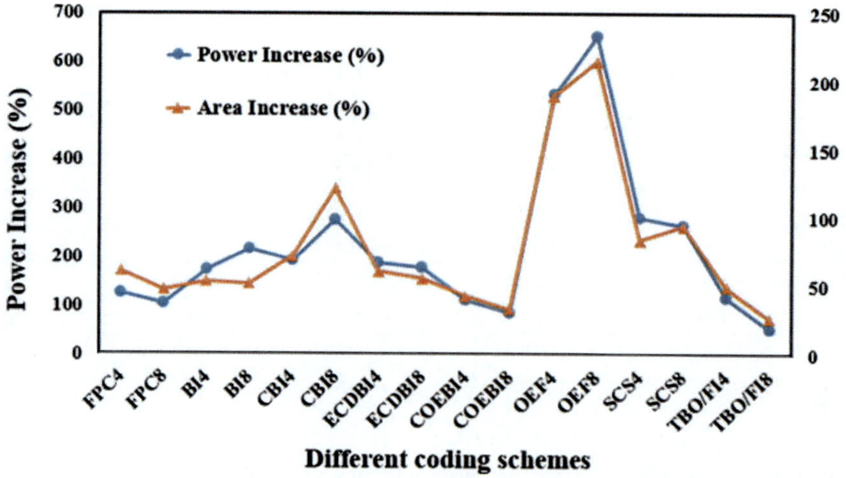

Fig. 8 Percentage increase in power and area of the router due to coding logic implementation for 4- and 8-wire links

is estimated using routers and links in the path. The link power consumption is computed using (1) and the coupling and self switching activities are estimated based on the information acquired from the simulator. Comparing the increase in router power consumption with the decrease in link power consumption due to codec insertion in the router, the percentage of power savings is computed.

Total power savings obtained by different coding schemes at the worst-case, best-case, and typical-case for 4- and 8-wire links are given in Figures 9 and 10 respectively. It is noticed in Figures 9 and 10 that TBO/FI scheme has the highest percentage of power savings of 25.5% and 40.4% for 4- and 8-wire links in worst-case scenario. FPC and COEBI have power saving of 10.5% and 30%, respectively, in worst-case

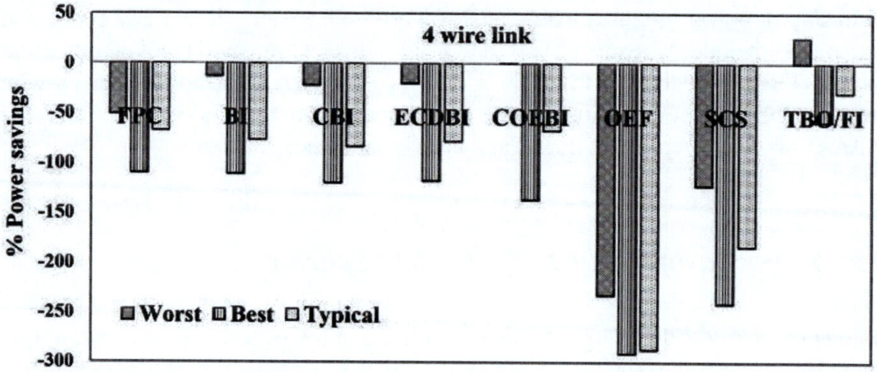

Fig. 9 Total power savings percentage of 4-wire link coding schemes for worst, best, and typical scenarios

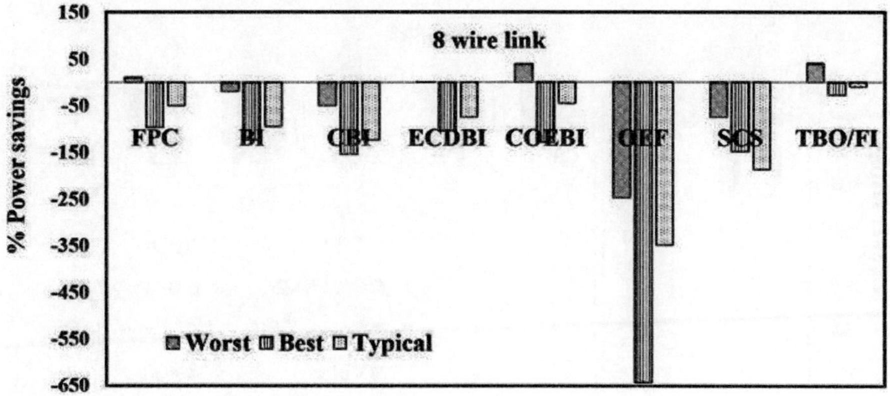

Fig. 10 Total power savings percentage of 8-wire link coding schemes for worst, best, and typical scenarios

scenario for 8-wire link. All other schemes for both the wires result in increase in power. In all the schemes, the flit size is constant but the number of extra bits used and reduction in switching activity varies. In Figs. 9 and 10, it is noticed that for all the schemes in best and typical-case scenarios, power saving is in negative, i.e., there is increase in power consumption. It is seen that the TBO/FI scheme for both the wires and for best and typical cases has the least increase in power. Power consumption increases for best and typical cases in all other schemes as the extra bits generated for a flit are transmitted parallelly through the extra wire to support decoding which adds on to the self and coupling switching activities. In TBO/FI scheme, no extra wire is used but extra bits are used which are transmitted through the same exciting 4- or 8-wire link. Encoder logic used in TBO/FI coding scheme is in such a way that extra bits appended do not lead to the increase in the self and coupling transitions, as the value of the extra single bit is set to 0 or 1 based on the encoded bit to which it is appended. In all other schemes either one or two extra bits are appended for each set of parallel bits transmitted. These bits are set to 0 or 1 in one bit case and 00, 01, 10, and 11 for two bit case. These bits are just appended with encoded bits which result in additional self and coupling transitions in the encoded bits. Therefore, in all the three cases, TBO/FI has better power saving percentage.

Percentage of power savings obtained when different sets of data streams like text, pdf, black, and white BMP picture, color BMP picture, black and white JPEG picture, color JPEG picture, video and audio are used for evaluation is given in Figs. 11 and 12.

It is observed that TBO/FI scheme for 4- and 8-wire links has the highest power saving when compared to all other schemes. All schemes have increase. in power consumption except our proposed TBO/FI and COEBI schemes. In our coding scheme, higher is the activity higher is the amount of power saving. Furthermore, in our scheme at lower activity, probability of increased power consumption is eliminated. Therefore, the percentage of power savings achievable by our coding scheme

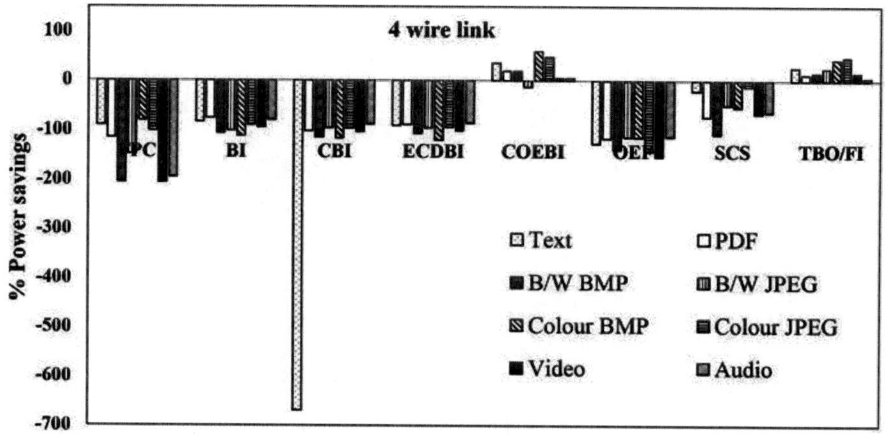

Fig. 11 Total power savings percentage of 4-wire link coding schemes for different data streams

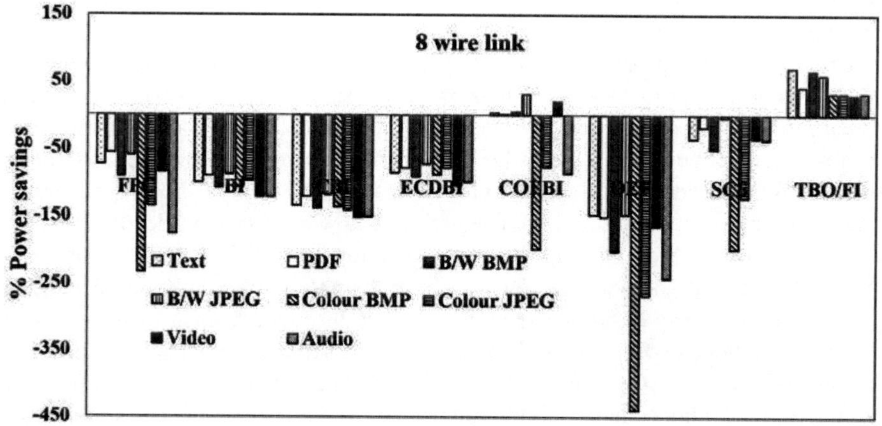

Fig. 12 Total power savings percentage of 8-wire link coding schemes for different data streams

depends on the type and nature of data involved in the computations in an application (i.e., integers, booleans, characters, etc.).

4.4 Performance of the Coding Schemes

Performance of different coding schemes is discussed in terms of average delay and throughput in this section. Average delay and throughput practicable in 4×4 NoC in the presence of codec at switch level are computed. We assumed packet of six flits and 32 bit link width in the simulation. Using synthetic traffic pattern, packets are created

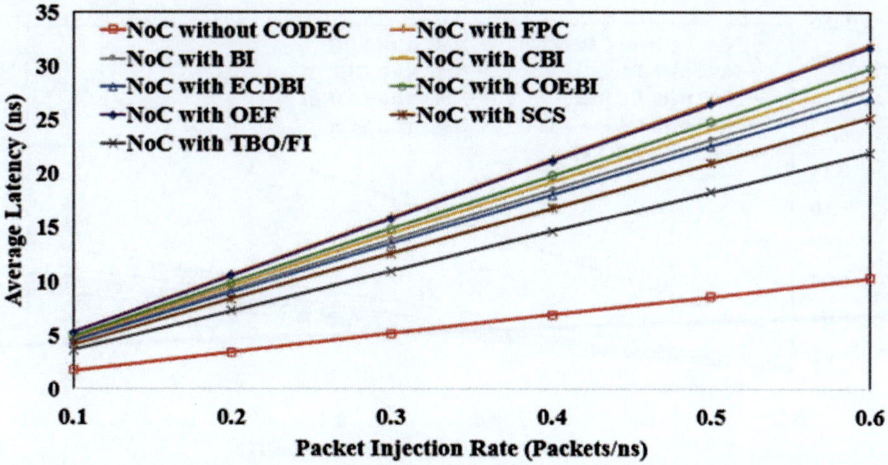

Fig. 13 Average latency versus packet injection rate

randomly and inserted into the network that follows *XY* routing algorithm to transfer the packets to its corresponding destination IP. In the simulation, 1 MB of data traffic is used to investigate the network performance. By varying the packet injection rate, average latency is analyzed in the presence of TBO/FI and other coding schemes and provided in Fig. 13. It is observed in Fig. 13, that with increasing packet injection rate, the average latency in nanoseconds increases for all the coding schemes. NoC without codec has the least average latency. Next least average latency is for the NoC with our coding scheme which is low compared to other schemes. Therefore, TBO/FI coding scheme is at most suitable for applications with low latency and power constraints.

Also for similar values of packet injection rates, the average throughput achievable by different coding schemes is analyzed for 4 wire link and is provided in Fig. 14. For increasing packet injection rate, the average throughput increases for all the coding schemes. Our scheme has the highest average throughput than all other schemes.

5 Conclusion

In this paper, our TBO/FI coding scheme is presented for reducing the crosstalk effects and link power consumption of an NoC. TBO/FI coding scheme has low bit overhead with higher reduction in percentage of self and coupling switching activities as compared to other schemes for 4-wire link. Also, TBO/FI coding scheme has high bit overhead with higher reduction in percentage of self and coupling switching activities as compared to other schemes for 8-wire link. Performance impact in terms of average latency and throughput is assessed for synthetic traffic scenarios for 4-wire link. TBO/FI coding scheme achieved less average latency and high average

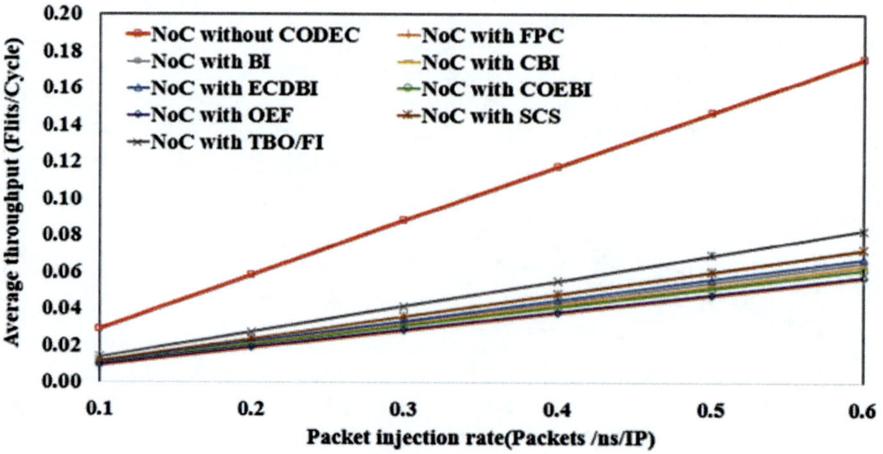

Fig. 14 Average throughput versus packet injection rate

throughput than all other schemes. TBO/FI coding scheme has higher percentage of NoC power savings for the considered cases. The application of the proposed TBO/FI coding scheme allows NoC power savings of up to 25.5% and 40.4% for worst-case scenario and though for the best and typical-case there is increase in power consumption of NoC this increase is lesser than all other similar coding schemes. This is despite the router area and router power overhead of 117% and 50% for 4 wire and 52% and 26% for 8 wire link, respectively. Even the router area and power overhead are lesser compared to other similar schemes. Therefore, this coding scheme is best suitable for applications which prefer low power consumption with signal integrity.

References

1. Benini L, De Micheli G (2002) Networks on chips: a new SoC paradigm. IEEE Comput 35(1):70–78
2. Vinodhini M, Murty NS (2018) Reliable low power NoC interconnect. Micro-Process Microsyst 57:15–22
3. Mohan P, Somasundaram K, Dash SS, Das S, Bhaskar MA (2016) Design and evaluation of 3D NoC routers with quality of service (QoS) mechanism for multi-core systems. Adv Intell Syst Comput 394:429–441
4. Jafarzadeh N, Palesi M, Eskandari S, Hessabi S, Afzali-Kusha A (2015) Low Energy yet reliable data communication scheme for network-on-chip. IEEE Trans Comput-Aided Des Integr Circ Syst 34:12
5. Moreac E, Rossi A, Laurent J, Bomel P (2016) Bit-accurate energy estimation for networks-on-chip. J Syst Architect 77:112–124
6. Khan Z, Arslan T, Erdogan AT (2006) Low power system on chip bus encoding scheme with crosstalk noise reduction capability. In: IEEE proceedings of computer digital technology, vol 153, no 2

7. Fang C, Fan C (2011) Novel low-power bus invert coding methods with crosstalk detector. J Chin Inst Eng 34(1):123–139
8. Jafarzadeh N, Palesi M, Khademzadeh A, Afzali-Kusha A (2014) Data encoding techniques for reducing energy consumption in network on-chip. IEEE Trans Very Large Scale Integr (VLSI) Syst 22(3):675–685
9. Shirmohammadi Z, Miremadi SG (2015) S2AP: an efficient numerical-based crosstalk avoidance code for reliable data transfer of NoCs. In: 10th International symposium on reconfigurable communication-centric systems-on-chip (ReCoSoC)
10. Singha S, Mahanti GK (2014) A novel encoding scheme for cross-talk effect minimization using error detecting and correcting codes. Int J Electron Electr Eng 2(4)
11. Radetzki M, Feng C, Zhao X, Jantsch A (2013) Mechanisms for fault tolerance in network on chip. ACM Comput Survey 44:1–36
12. International Technology Road map for Semiconductors (ITRS) [Online] (2015) https://www.itrs.net
13. Shirmohammadi, Z, Miremadi SG (2016) On designing an efficient numerical-based forbidden pattern free crosstalk avoidance codec for reliable data transfer of NoCs. Microelectron Reliab 63
14. Sotiriadis PP, Chandrakasan A (2001) Reducing bus delay in submicron technology using coding. In: Proceedings of the Asia and South Pacific Design automation conference
15. Palesi M, Ascia G, Fazzino F, Catania V (2011) Data encoding schemes in networks on chip. IEEE Trans Comput-Aided Des Integr Circ Syst 30(5):774–786
16. Stan MR, Burleson WP (1995) Bus-invert coding for low-power I/O. IEEE Trans Very Large Scale Integr (VLSI) Syst 3(1):49–58
17. Parandeh-Afshar FH, Saneei M, Afzali-Kusha A, Pedram M (2007) Fast INC-XOR codec for low-power address buses. IET Comput Digital Tech 1(5):625–626
18. Su CL, Tsui CY, Despain AM (1994) Saving power in the control path of embedded processors. IEEE Des Test Comput 11(4):24–31
19. Benini L, De Micheli G, Macii E, Sciuto D, Silvano C (1997) Asymptotic zero-transition activity encoding for address busses in low power microprocessor-based systems. In: Proceedings of 7th Great Lakes symposium on VLSI, USA, pp 77–82
20. Musoll E, Lang T, Cortadella J (1998) Working-zone encoding for reducing the energy in microprocessor address buses. IEEE Trans Very Large Scale Integr (VLSI) Syst 6(4):568–572
21. Fornaciari W, Polentarutti M, Sciuto D, Silvano C (2000) Power optimization of system-level address buses based on software profiling. In: Proceedings of 8th international workshop on hardware/software codesign (CODES), USA, pp 29–33
22. Sharma O, Saini A, Saini S, Sharma A (2016) A Quadro coding technique to reduce self transitions in VLSI interconnects. In: IEEE international symposium on nanoelectronic and information systems (iNIS), Gwalior
23. Vithyalakshmi N, Rajaram M (2014) A multi coding technique to reduce transition activity in VLSI circuits. J Semicond 35(2)
24. Moulika M, Vinodhini M, Murty NS (2017) Data flipping coding technique to reduce NoC link power. In: IEEE international conference on computational intelligence and computing research
25. Pande PP, Zhu H, Ganguly A, Grecu C (2006) Energy reduction through crosstalk avoidance coding in NoC Paradig. In: Proceedings of the 9th EUROMICRO conference on digital system design
26. Shirmohammadi Z, Miremadi SG (2016) S2AP: an efficient numerical-based crosstalk avoidance code for reliable data transfer of NoCs. In: Proceedings of the international symposium on reconfigurable communication-centric systems on-chip (ReCoSoC), pp 1–6
27. Yan Z, Lach J, Skadron K, Stan MR (1998) Odd/even bus invert with two-phase transfer for buses with coupling. In: Proceedings of international symposium on low power electronic design (ISLPED), Monterey, pp 80–83
28. Fan CP, Fang CH (2011) Efficient RC low-power bus encoding methods for crosstalk reduction. Integr VLSI Journal 44(1):75–86

29. Kim KW, Baek KH, Shanbhag N, Liu CL, Kang, SM (2000) Coupling-Driven signal encoding scheme for low-power interface design. In: IEEE/ACM international conference on computer aided design
30. International Technology Roadmap for Semiconductors [Online]. https://www.itrs.net/Links/2009ITRS/Home2009.html
31. Nanoscale Integration and Modeling (NIMO) Group, Predictive Technology Model (PTM). Accessed 2012.

Feature-Based Model for Landslide Prediction Using Remote Sensing and Digital Elevation Data

Litesh Bopche and Priti P. Rege

Abstract This study aims to generate landslide susceptible maps and landslide hazard zonation maps using the digital elevation model for the prediction of future landslides. The landslide zone is based on the qualitative and quantitative factors combined using the weighted sum of the different features and hydrological parameters. The main aim of the research is to discover the damaged areas with the help of detailed field observation of prior and post landslide events. Landslide hazard zonation is a map classified into six different zones ranging from very low hazard zone to scars hazard zone and to represents the prediction of future landslide occurrence under the area of the study. The result of this study shows that a very high and scars susceptible region depicts a higher chance of landslides.

Keywords Landslides · Landslide hazard zonation · Digital elevation model

1 Introduction

Landslides have always been one of the most destructive natural calamities that affect properties, agriculture land, livestock, and human lives on a large scale. They obstruct the development of infrastructure such as roads, dams, communication lines, bridges, and so on, in that particular region. The various complex geological settings are directly or indirectly responsible for triggering the landslides. There are various factors, affecting the occurrence of the landslide like tectonic activities in the region, presence of thrust and faults, heavy rainfall or snowfall or cloudburst, hill slope, and increased human intervention in the environmental process. Landslides are generally failures in the slopes caused by the various geomorphological activities. According to the statistical survey, hundreds of human lives and properties worth thousand

L. Bopche (✉) · P. P. Rege
College of Engineering, Pune, India
e-mail: litesh.bist02@gmail.com

P. P. Rege
e-mail: ppr.extc@coep.ac.in

© Springer Nature Singapore Pte Ltd. 2021

S. N. Merchant et al. (eds.), *Advances in Signal and Data Processing*,
Lecture Notes in Electrical Engineering 703,
https://doi.org/10.1007/978-981-15-8391-9_22

crores are estimated to be lost almost every year due to the occurrences of the landslides. The landslide event such as Lwara slides and Basukedar landslides (1992), Madhyameshwar and Kaliganga sub-watersheds (1998), Phata cloudburst (2001), Balganga valley landslide (2001), Bhagirathi valley landslide (2003), Ukhimath landslide at Uttarakhand (2012), and many others, have largely affected the valley in terms of human lives and damage to resources [1–3].

In this study, we focus on the prediction map of the landslide detection for risk management. The prediction map is generated through the weighted raster calculation method of the classified slope map, classified aspect map, classified elevation map, hydrological parameters, and relative relief map based on the area under the study. There are various reasons such as high altitude, rugged terrain, shortage of agricultural land, a lesser amount of industrial development, and extreme environmental conditions that limit the economic growth of the mountainous areas like the Uttarakhand state of India. This condition becomes more severe during the monsoon period as the rain causes more instability in the slope surfaces. The interaction between slopes and various other conventional geological factors is different in different regions and hence a susceptibility map depicts this relationship providing a metric to access the safety of the region concerning the landslide occurrences.

A Digital Elevation Model (DEM) is a 3-D representation (X, Y, Z) of the continuous geography of the terrain's surface and generate a terrain elevation model [4, 5]. A DEM is a regularly spaced grid that contains the elevation of a point on the surface that is coincident with the location of the grid cell. The terrain attributes mean not only the elevation values but also includes the slope values, aspect values, PH value of soil, surface values like groundwater-surface. DEM representing a surface is presently used in a large number of geographic environmental applications, such as hydrology, agriculture, geology, cartography, rural and urban planning, disaster risk mitigation, geomorphology, among others [6, 7]. It is one of the essential input rasters for observing the landscape as well as dynamic natural phenomena such as soil erosion, flooding, earthquake monitoring, and landslide detection.

1.1 The Area Under the Study

The area under the study is about 7626 km^2 between 30° 25′ 12″ N, 79° 19′ 48″ E to 30° 42′ 8.66″ N, 79° 33′ 56″ E in the Chamoli district of Uttrakhand (Fig. 1). The district is enclosed by Tibet area on the north, Pauri Garhwal to the south, Rudraprayag to the southwest, and Uttarkashi to the northwest. In the Chamoli district variety of destinations present for tourists interest like Badrinath, Valley of Flowers, and Auli. The altitude of the Chamoli district lies between 603 to 7100 m above mean sea level. In Chamoli 70–80% of the yearly rainfall occurs in the period of July to September. Most of the region is a highly elevated terrain. Alaknanda is the major river in the valley having some other tributaries.

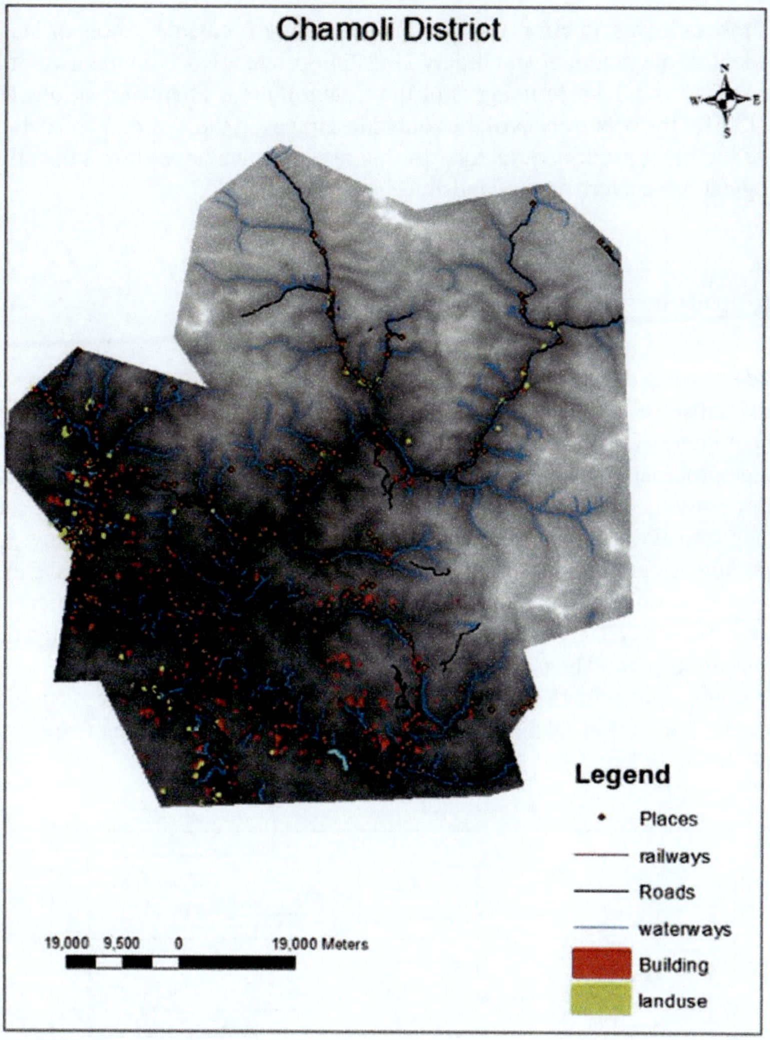

Fig. 1 CARTOSAT-1 DEM image of area under consideration

The data used for the present study is the Cartosat-1 digital elevation model based on the stereoscopic technique. The data is downloaded from the Indian Geo-platform of ISRO, BHUVAN's official website [8]. CartoSat-1 is the first Indian satellite launched in May 2005. Fore-Aft stereo capturing capability and recording stereo images in five days of revisit time. Cartosat-1 has two panchromatic cameras. The panchromatic cameras mounted with a tilt of +26° (Fore) and −5° (Aft) from Nadir point operate in the 0.5–0.85 microns spectral band [9]. The satellite data is processed using Arc-Map and ERDAS Imagine software.

This research was an effort towards the detection of feasible causes of landslide and to identify the potential and highly landslide-prone areas with the help of high-resolution Cartosat-1 DEM imagery of the Chamoli area. There are various factors responsible for the occurrences of the landslide. However, the hydrological parameter is one of the most prominent factors. In this research, we have shown the effect of hydrological parameters on the landslide.

2 Methodology

Landslide susceptible zonation is performed with the help of the weighted sum of the classified slope map, classified aspect map, classified elevation map, and hydrological parameter. It is the combination of the Geographic Information System (GIS) and image processing techniques on the stereoscopic based satellite image [10, 11]. The work consists the selection of features (such as slope map, aspect map, relative relief map, and elevation map, etc.) and hydrological parameter (such as flow stream network, flow accumulation, stream order, and stream feature) from digital elevation model for the monitoring and detection of the landslide in Chamoli district region. There is a necessity for landslide susceptibility mapping for the identification of landslide-prone areas. The monitoring and detection of the landslide are very important for relief management in the landslide-prone area [12, 13]. The methodology involves the generation of digital data and the selection of various features from DEM as shown in Fig. 2.

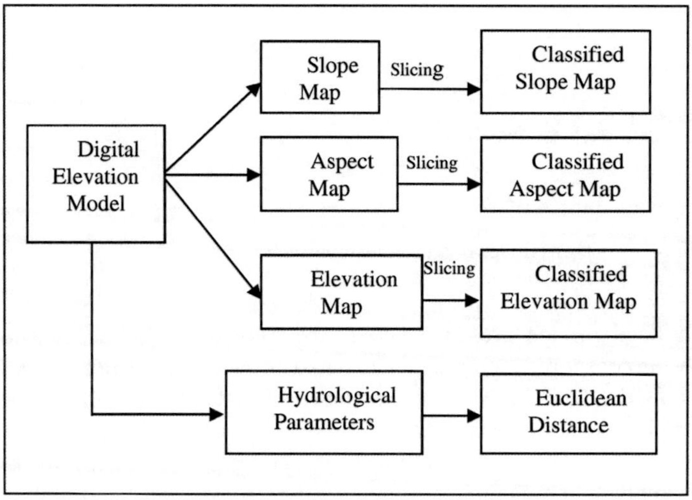

Fig. 2 Block diagram of feature extracted from DEM

2.1 Slope Map

The steep ridges, peaks, valleys, and scarps are the results of the denudation process. The measurement of the slope is a very important factor in the identification of the area, which is more prone to the occurrence of the landslide. The slope map identifies the gradient of altitude from each cell of a raster-scanned surface to its neighbor's cell and categorizes the steepest downwards hill descent from the cell. Cartosat-1 digital elevation model is used to generate a slope map for the area under the study as shown in Fig. 3a. Further, the slope image obtained through the above method is classified into nine classes. Each class consists of uni-variate data. The highly unstable part of landslides lies in the middle slope areas, such as 35°–50°. The slope values are given by Eq. (1):

$$\text{Slope}(\theta) = \arctan\left(\frac{\delta(x)}{\delta(y)}\right) \tag{1}$$

where $\delta(x)$ and $\delta(y)$ are the rates of change in x and y direction respectively. The algorithm for the slope map is based on the 3×3 neighborhood of the elevation values can be stated as below:

- The use of the middle cell values and its neighbors determines the horizontal and vertical deltas as shown in Table 1.
- change in x-direction $\delta(x)$:

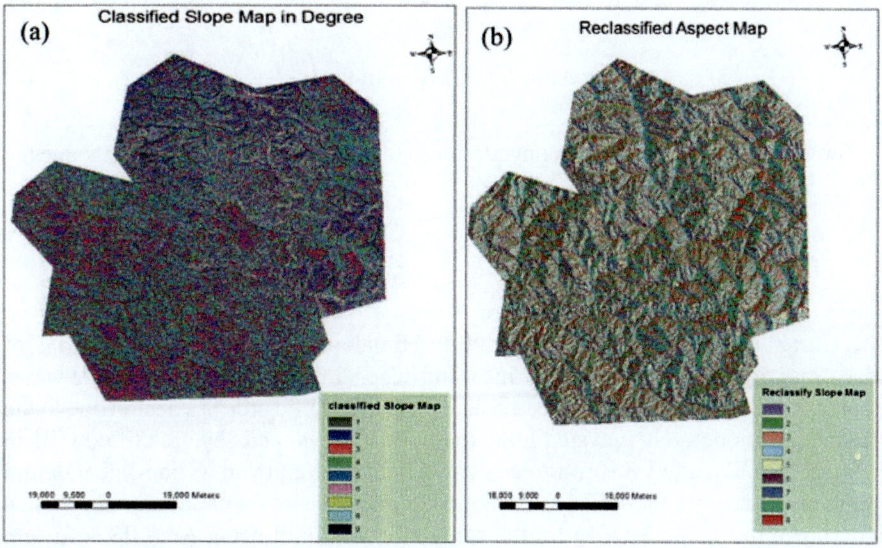

Fig. 3 Feature class map of the study area. **a** DEM classified slope map, **b** DEM classified aspect map

Table 1 3 × 3 block from DEM

a	b	c
d	e	f
g	h	I

$$\delta(x) = \frac{((c + 2f + i) - (a + 2d + g))}{(8 * \text{cellsize})} \tag{2}$$

- Change in *y* direction $\delta(y)$:

$$\delta(y) = \frac{((g + 2h + i) - (a + 2b + c))}{(8 * \text{cellsize})} \tag{3}$$

- The slope is calculated by using Eq. (1) and shown in Table 2.

2.2 *Aspect Map*

Aspect map shows the direction and orientation of the rate of change of the altitude from each cell of the raster surface. Aspect is measured clockwise from the north and expressed in positive degrees from 0 to 359.9. Slope facing in the study area is given in Table 2. There is some unification to the occurrence of landslides with the slope facing and classified aspect map shown in Fig. 3b. The appearance of the Landslides is preferentially on south to east-facing slopes. The aspect value is given by Eq. (4):

$$\text{Aspect}(A) = 57.92578 * \arctan\left(\frac{\delta(x)}{\delta(y)}\right) \tag{4}$$

The value of Eq. (4) is then converted into 0°–360° (compass direction values).

2.3 *Elevation Map*

The Uttrakhand region is characterized by Himalayan geography with a series of criss-cross ridges, weaker rocks, springs, and deep narrow valleys. The altitude varies from 603 to 7100 m above the mean sea level. The specific area is selected due to the presence of moderately elevated hills, disserted valleys, and several clifted hills in the area. Initially, a DEM is re-projected to a linear datum from its non-linear datum and then classifying the DEM such that each of its classes will contain uni-variate data shown in Fig. 4. A DEM for the study area was built using Arc-GIS software. The occurrence of the landslide is more prominent in the high altitude area.

Table 2 Numerical rating scheme

Layer	Factor	Value	Class
1	Slope (°)	1	0°–11°
		2	11°–18°
		3	18°–25°
		4	25°–30°
		8	30°–36°
		9	36°–42°
		7	42°–48°
		6	48°–55°
		5	55°–83°
2	Slope aspect map	1	Flat (–1)
		2	North (337.50–22.50)
		3	Northeast (22.50–67.50)
		8	East (67.50–112.50)
		9	Southeast (112.50–157.50)
		7	South (157.50–202.50)
		4	Southwest (202.50–247.50)
		5	West (247.50–292.50)
		6	Northwest (292.50–337.50)
3	Elevation (m)	1	1091–1142
		2	1142–1193
		3	1193–1244
		4	1244–1295
		5	1295–1346
		6	1346–1397
		7	1397–1448
		8	1448–1499
		9	1499–1550
4	Euclidean distance (m)	3	0–4
		6	4–8
		9	<8

Fig. 4 The elevation map of the area under the study

2.4 *Hydrological Parameter*

The downloaded DEM has some errors during the collection or interpolation techniques that have been used for the generation of an elevation model. Hydrological parameters are used to correct the data set. First, fill sink is used to remove small imperfection in the raster data, and then a flow direction is created from each cell to its nearest steeper downslope. A weight factor applied to generate raster of accumulation flow into each cell as shown in Fig. 5a. After that, a unique value is assigned to each section of a raster between the intersections, and a numeric order is assigned to each branch of the linear network. A stream to feature conversion converts a raster stream network into features representing a linear network. The stream to feature is used for the generation of the Euclidean distance for each cell to the closest sources

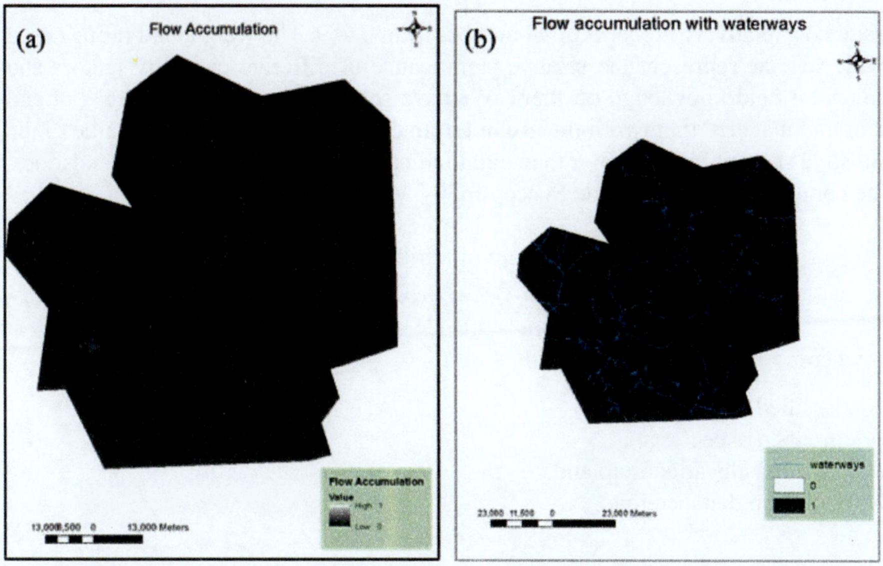

Fig. 5 **a** Flow accumulation map of the study area, **b** flow accumulation map with waterways

and the results show that the waterways and flow accumulation overlap each other as shown in Fig. 5b.

2.5 Rating Scheme Used

The features in the landslide-prone area need to be organized based on their relative importance on the initiation of the landslide. By developing a scheme to assign numerical values to different features the above objective is achieved. In this method, the features are classified into different classes. These classes are formed such that there is unit variance within the class. A numerical weight value from 0 to 9 is assigned to each class in the order of their importance to landslide occurrence, where higher weights indicate greater influence on the landslide. The numerical assignments are given in Table 2.

2.6 Knowledge-Based

Knowledge-Based (KB) is derived from the past landslide occurrences overlapped with the feature set. The KB is used for deciding the suitable weights assigned to each class of the classified features in the numerical assignment step. Table 2 is prepared based on the KB. The feature class that overlaps with the past landslide scars are assigned a relatively higher ordinal number from 0 to 9. The weight and ratings used in the scheme represent the relative significance of different causality factors and the actual field knowledge on them. We have selected the weights using trial and error and matched them with the ground truth obtained from scars. The value of the landslide susceptibility greater than eight indicates higher chances of the landslides. The computation of Landslide Susceptibility Map (LSM) is given by:

$$\begin{aligned} \text{LSM} &= \text{weighted sum of attributes of the data layer} \\ &= [0.5 * \Phi + 0.3 * A + 0.3 * H + 0.3 *]. \end{aligned}$$

where

Φ classified slope angle.
A classified aspect map.
H classified elevation map and.
D Euclidean distance map.

3 Results and Discussions

The combination of all the calculating parameters and different numerical rating scheme, the final landslide susceptible zonation map was prepared and categorized into a low susceptible zone (0) and highly susceptible zone (1) shown in Fig. 6a, b. Figure 6a shows the landslide susceptible zone of all the major known landslides and

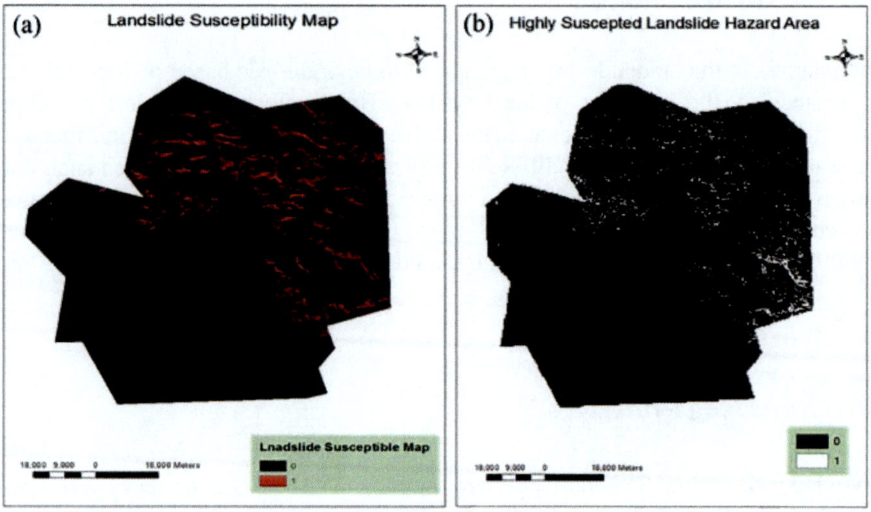

Fig. 6 Landslide susceptible zonation map. **a** The landslide susceptible zone of all the known landslides. **b** Scars and high susceptible landslide hazard area

Fig. 6b shows scars and very high susceptible landslide hazard area, which is very dangerous. Very high and scars susceptible region depicts a higher chance of failure of slopes and it is found to be 8–12% of the study.

Landslide Hazard Zonation (LHZ) map has been prepared using the pre and post landslide scars event shown in Fig. 7. This map shows the level of damage area under the study. The landslide hazard zonation map was finally divided into six hazard levels: very low, low, moderate, high, very high, and scars. The landslide events are also dependent on the climatic condition of the area under study. The allocation of major development and settlements should be avoided in such areas. The present research is an effort to prepare a comprehensive landslide susceptible zonation map of Chamoli district, Uttarakhand. The following conclusion can be drawn on the bases of landslide susceptible zonation map.

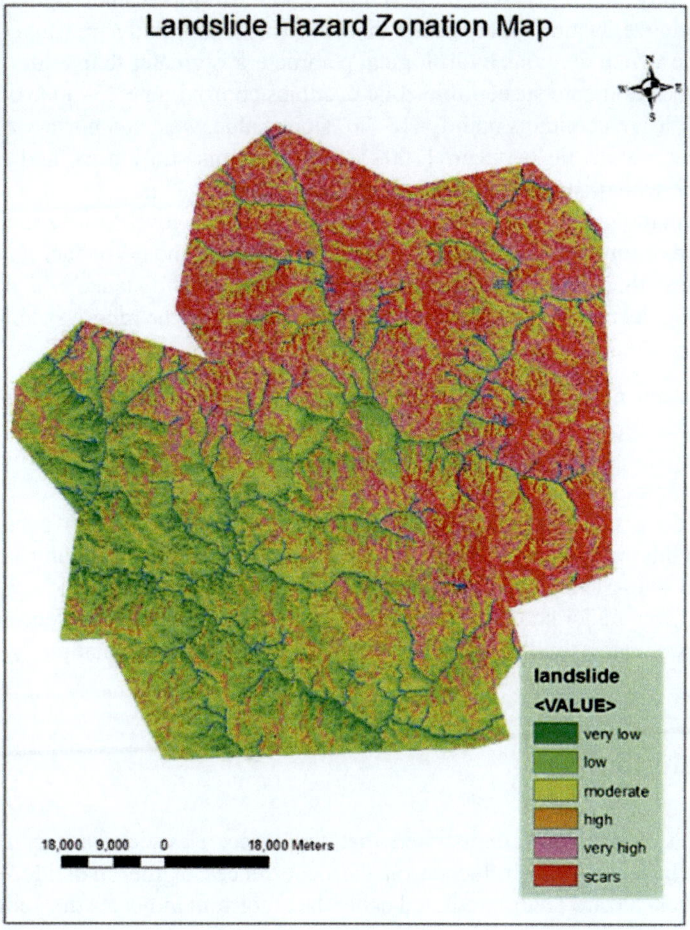

Fig. 7 Landslide Hazard Zonation map derived using a weighted sum of features

Table 3 Range of parameters affecting various landslide

Landslide susceptible zone	Slope (°)	Aspect map	Elevation (m)	Euclidian distance (m)
Scars and high susceptible zone	30–50	South and South-East facing	<1300	<8
Moderate susceptible zone	20–30	West and North-West facing	1200–1300	4–8
Low susceptible zone	>30	Flat and North facing	>1200	>4

1. Scars and very high-susceptibility regions have a very high possibility of slope failure and hydrological parameter effect. This zone occupies as much as 8–12% of the study area. This area contains mainly 30º–50º slope value, south and south-east facing, elevation value greater than 1300 m for this study area, and Euclidian distance which contains hydrological parameters is greater than 8 m.
2. High and Moderate susceptible zone occupies approximate 25–30% of the total area. This area contains mainly 20º–30º slope value, west and north-west facing, elevation values lie between 1200–1300 m for this study area, and Euclidian distance which contains hydrological parameters is 4–8 m.
3. The remaining part of the area occupies a low and very low hazard zonation map. This area contains slope values less than 20º, flat area and north-facing, elevation value less than 1200 for this study area, and Euclidian distance which contains hydrological parameters is less than 4 m. Table 3 gives the ranging of parameters affecting the landslide for the study area.

The contour map has been generated for the study area and is shown in Fig. 8. Landslide susceptible zone can be verified using the contour map of the digital elevation model. It shows that most of the landslide events occur at the high altitude area of the contour map.

Landslide is one of the national disasters in India. It generally occurs in the hilly areas due to the slope failures and results in massive destruction of buildings and human lives. The work depicts an attempt to identify the landslide susceptible map of the area under study in advance using remote sensing. The identification of landslide-prone areas will help in planning and building new structures.

4 Conclusion

The present study has demonstrated that the slope, elevation, aspect ratio, and hydrology have a strong influence on the occurrences of the landslide. The very high and scars hazard zones predicted using the algorithm indicates the geographical unstable area. Slope aspect failures in such areas can be triggered particularly after heavy rainfall. The results show that the slope of 30°–50°, the orientation of southeast

Fig. 8 Contour map showing landslide susceptible area

to east-facing slopes, high altitude region and flow direction play an important role in the occurrences of the landslide. The use of the contour map also validates the prediction of the landslides.

The landslide susceptible and hazard zonation map can be of great help for effective planning in disaster relief events. The occurrence of the landslide is also dependent on seismic activities. It is observed from the map that the population around the landslide susceptible area is meager. The last seismic activity in the study area was observed in the year 1999. The influence of the seismic activities was not considered while processing the information.

References

1. Rawat MS, Dobhal DP, Ray CP, Joshi V, Rawat BS, Bartwal A, Singh D, Aswal A (2015) Study of landslide hazard zonation in Mandakini Valley, Rudraprayag district, Uttarakhand using remote sensing and GIS. Curr Sci 109(1):158–170
2. Chaudhary S, Gupta V, Sundriyal YP (2010) Surface and subsurface characterization of Byung landslide in Mandakini valley, Garhwal Himalaya. Himalayan Geol 31(2):125–132
3. Islam MA, Chattoraj SL, Ray CP (2014) Ukhimath landslide 2012 at Uttarakhand, India: causes and consequences. Int J Geomat Geosci 4(3):544–557
4. Michalis P, Dowman I (2006) Sensor model evaluation and DEM generation for Cartosat-1. Int J Geomat Geosci 36:5–9
5. Ahmed N, Mahtab A, Agrawal R, Jaya-prasad R, Pathan SK, Singh DK, Singh AK (2007) Extraction and validation of Cartosat-1 Dem. J Ind Soci Remote Sens 35(2):121–127
6. Krishna Murthy YVN, Rao SS, Prakasa Rao DS, Jayaraman V (2008) Analysis of DEM generated using Cartosat-1 stereo data over Mausanne Les Alpilles—Cartosat scientific appraisal program (CSAP TS-5). ISPRS 37:1343–1348
7. Giribabu D, Kumar P, Mathew J, Sharma KP, Krishna Murthy YVN (2016) DEM generation using Cartosat-1 stereo data: issues and complexities in Himalayan terrain. Eur J Remote Sens 24:295–302
8. NRSA Data Center (2006) CARTOSAT-1 data user's handbook, National Remote Sensing Agency (NRSA), CARTOSAT-1/NRSA/NDC/HB09/06
9. ISRO's Geoportal Bhuvan Store. https://bhuvan3.nrsc.gov.in/applications/bhuvanstore.php
10. Martha TR, Kumar KV (2013) landslide events in Okhimath, India—An assessment of landslide consequences using very high-resolution satellite data. Landslides 10(4):469–479
11. Teerarungsigul S, Chonglakmani C, Kuehn F (2007) Landslide prediction model using remote sensing, GIS and field geology: a case study of Wang Chin District, Phrase province, Northern Thailand. In: Proceedings of 2007 international conference on geology of thailand: towards sustainable development and sufficiency economy, pp 156–168
12. Guzzetti F, Reichenbach P, Cardinali M, Galli M, Ardizzone F (2005) Probabilistic landslide hazard assessment at the basin scale. Geomorphology 72(4):272–299
13. Ghosh JK, Bhattacharya D, Sharma SK (2012) Fuzzy knowledge-based GIS for zonation of landslide susceptibility. In: Applications of chaos and nonlinear dynamics in science and engineering, vol 2. Springer, Berlin, Heidelberg, pp 21–37

Acoustic Classification of Bird Species

Rashmika Patole and Priti Rege

Abstract The main focus of this paper is identification of bird species or even individual birds on the basis of their sounds. This work compares an audio signal of an unknown bird to a database of known birds. The system has two modes of operation: training mode, and recognition mode. In the training mode, ate a feature model of the available bird sounds in the database is created. The recognition mode will use the information obtained from the training mode to isolate and identify the bird. Mel frequency Cepstral Coefficients and Gammatone frequency Cepstral Coefficients have been employed as feature sets for classification. The classification accuracies are evaluated using Support Vector Machine and Artificial Neural Networks.

Keywords Bird identification · Acoustic classification

1 Introduction

Birds have always been of great interest to people since ages because of their social as well as ecological importance. Bird monitoring has always been of great importance because of many practical reasons. In the context of ecological concern, birds play an important role, since they are one of the classes of living beings that have direct contact with humans. Reasons such as changes in habitat, nest egg loss, mortality during migration human and animal predators, etc. have caused decline in the population of bird species over the last few years [1]. It can be possible to correct the population decline and reduce future risk of extinction by understanding the connection between the bird vocalizations and their behavior patterns. The identification of birds can also aid in the monitoring of migration and population of birds in the ecosystem.

There are numerous engineering applications where identification of birds in real-time is necessary. One such application is used in aircraft monitoring systems where they need to avoid collision between birds and the aircraft. There may be birds in

R. Patole (✉) · P. Rege
College of Engineering, Pune, India
e-mail: rkp.extc@coep.ac.in

© Springer Nature Singapore Pte Ltd. 2021

S. N. Merchant et al. (eds.), *Advances in Signal and Data Processing*,
Lecture Notes in Electrical Engineering 703,
https://doi.org/10.1007/978-981-15-8391-9_23

the neighbourhood of wind turbine generators which may need to be tracked. Also, identification of birds is necessary to understand their seasonal migratory patterns and behaviour, especially at night and when the weather conditions are unfavourable. To study the impact of human development on plants and animals, ornithologists have to identify and count the number of birds in particular site. To identify birds in a particular area, it is easy to rely on their sounds because they are often easier to locate a particular bird by hearing its sound instead of seeing it physically. Hence, it is advantageous to rely on the bird sounds to identify bird species in a particular area. Thus, ornithologists must study the bird sounds and identify the birds in an area by sound alone. To monitor the bird sounds in real time can be a difficult task. Therefore, it can be useful to record unknown sounds so that they can be identified later. Thus, there is a need for automated methods for bird species identification to monitor and also to evaluate the diversity and quantity of birds [2].

Classifying bird species has been a research topic since many years. Different feature sets and classification algorithms for the task of bird classification have been discussed in the literature. In [3], spectrograms are used to represent the bird sound recordings and Dynamic Time Warping (DTW) has been used to measure the difference between the spectrograms [4] uses different feature sets such as Linear Predictive Coding (LPC) coefficients, LPC-derived cepstral coefficients, LPC reflection coefficients, Mel-Frequency cepstral coefficients (MFCCs), log mel-filter bank channel, and linear mel-filter bank channel. DTW and Hidden Markov Model (HMM) are used to form the acoustic models and classify the bird sounds. Neural networks and multivariate statistics have been employed in [5] to identify the bird species [6] gives an overview of previous works in the area of bird classification from vocalizations. A recent study includes recognition of bird species based on a hybrid model including HMM and Deep Neural Networks [7].

In [8], the author uses a decision tree along with support vector machine (SVM) for classification. Some prior work is concerned with classification of bird species from individual syllables [9], while other work is also concerned with identifying species from songs composed of sequences of syllables [8, 10]. The algorithms that have been applied to classifying syllables include nearest-neighbour and distance-based classifiers [8, 11, 12], multi-layer perceptrons [13], and support vector machines [9].

This paper is organized into the following sections: Sect. 2 discusses the sound mechanism in birds. The computation of Mel-Frequency Cepstral Coefficients, implementation of Vector Quantization and K-means algorithm is given in Sect. 3. Section 4 provides the results of the experiment followed by conclusion in Sect. 5.

2 Elementary Concepts and Organization of Bird Sounds

The mechanism through which sound is produced in birds is very similar to the human sound production mechanism. In humans, the vocal chord are responsible for the production of sound. A similar organ is present in birds, which is called Syrinx.

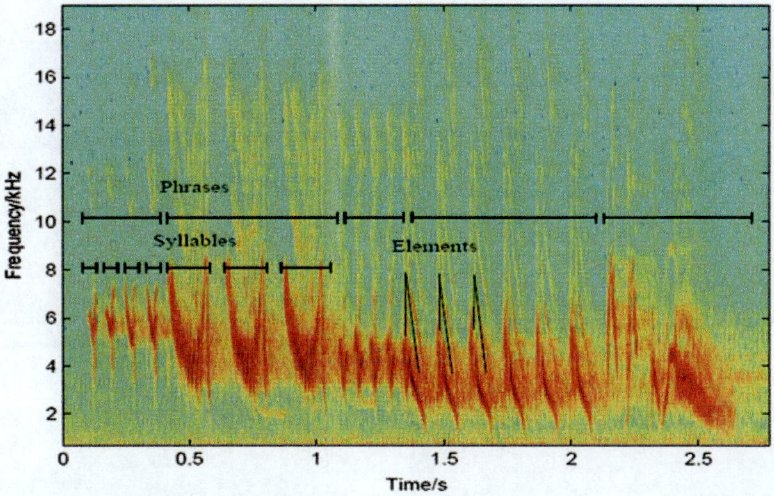

Fig. 1 Hierarchical levels of bird song

Bird sounds can be divided into either songs or calls, which can be further divided into phrases, syllables, and elements or notes as shown in Fig. 1. Similar to human speech, bird sounds can also be divided into voiced sounds and unvoiced sounds. Voiced sounds in birds are similar to the human vowel sounds in structure as well as the way they are produced. Sounds that do not contain any harmonics, e.g. pure tonal or whistled sounds can also be produced by birds. Such sounds are closely related to the unvoiced sounds in human speech. Bird sounds can be also noisy, broadband, or chaotic in structure [14]. Figure 2 shows examples of bird songs and calls from different bird species.

3 System Overview

The problem of Bird Species Classification is similar to other audio or speech classification problems like classification of general audio/speech content, auditory scene recognition, music genre classification, speech recognition, etc. that have been extensively studied during the last few decades. This project involves two modules namely (1) Training module *(feature extraction)* and (2) Testing module *(feature matching)* and classification.

The feature extraction step aims to extract acoustic features from the audio signal waveform. This module converts the audio signal waveform of the bird to some type of parametric representation for further analysis and processing. Feature extraction is about reducing the dimensionality of the input-vector but the discriminating power of the signal is maintained. These features carry the characteristics of the bird sound which are unique to a specific bird. Similar to the human speech signal, the audio

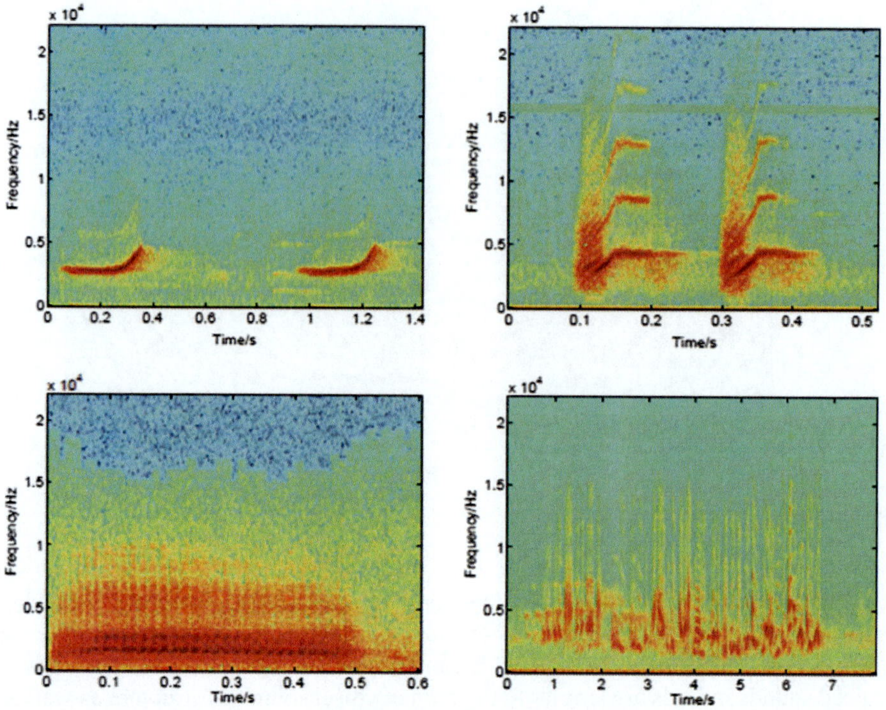

Fig. 2 Bird sounds from Willow Warbler (*Phylloscopus trochilus*) (upper row), Common Chaffinch (*Fringilla coelebs*), Hooded Crow (*Corvus corone cornix*) (second row)

signal of birds is a slowly varying signal. This can be seen in Fig. 3. It is not stationary. Therefore, the signal processing techniques which are commonly used cannot be applied to our signal because of its non-stationary nature. If the audio signal is analyzed over a short period of around 5–50 ms, the characteristics of the signal remain fairly stationary. Therefore, short-time analysis is needed to analyze the audio signal.

Mel-frequency Cepstral Coefficients (MFCCs) and Gammatone filter Cepstral coefficients (GFCCs) are used as features. The pre-processing done and the filter banks for extracting MFCCs and GFCCs have been described below.

3.1 Pre-processing and Filter Banks

The audio recordings of bird sounds available are first framed into short intervals of 25 ms size. The frames have an overlap of 50% and are windowed using a Hamming window. Short Time Fourier Transform (STFT) converts the frames into frequency domain. STFT uses 1024 FFT points. Two filter banks are used in this work, the

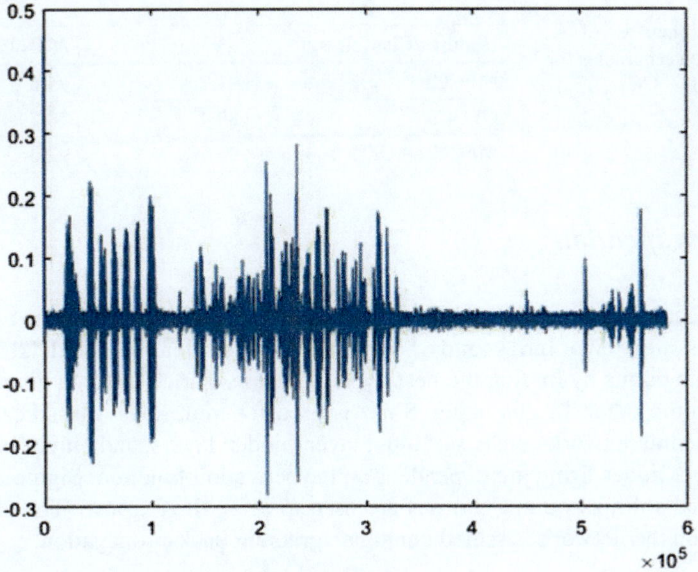

Fig. 3 Audio signal of Greater Racket-tailed Drongo

Mel-bank for obtaining MFCCs and the Gammatone filter bank for obtaining the GFCCs. 32 filters have been used in the Mel-bank. The linearly spaced frequencies are converted to the Mel frequencies, using the formula in Eq. (1).

$$\text{mel}(f) = 2595\log_{10}\left(1 + \frac{f}{700}\right) \tag{1}$$

The first filter is narrower while the filters become broader with increasing frequency and they are triangular in shape.

The Gammatone (GT) filter bank [14] is a biologically inspired bank with ERB (equivalent rectangular bandwidth) especially for effective representation of spectral properties at lower frequencies. The authors have used GT filter bank for another application as given in [15]. The magnitude response of the GT bank is similar to the ReOx function which closely models the human auditory system. Gammatone filter bank has its impulse response similar to Gamma distribution function. 64 filters are used with an ERB scale ranging from frequencies $\frac{\text{fs}}{2}$ to 100 Hz. ERB scale used in this paper is calculated using the Glasberg and Moore parameters of EarQ = 9.26449, minimum B.W. = 24.7 and order = 1. GT filter of fourth-order is implemented using four cascaded filters of order one.

Table 1 Comparison of Classification accuracies for MFCCs and GFCCs

Features/Classifiers	SVM (%)	ANN (%)
MFCCs	91.7	95.6
GFCCs	89.4	90.2
MFCCs + GFCCs	93.5	97.6

3.2 Classification

Support Vector Machine (SVM) and Artificial Neural Network (ANN) are employed for the classification of bird sounds. SVM is a supervised learning model that classifies the data points by finding the best hyperplane to separate the data points of one class from the other. In this paper, SVM is used for multi-class classification. An artificial neural network consists of input layer, hidden layers, and output layer. The hidden layer nodes firing are dependent on the activation function. Sigmoid hidden neurons and softmax output neurons are used to serve the purpose. The algorithm used to train the network is scaled conjugate gradient back-propagation.

4 Dataset and Experimental Results

Our data set consists of bird sounds taken from the Xeno-Canto [16] dataset which consists of bird sounds from all over the world. The data set consists of sounds from 70 different bird classes and 10 recordings from each class. The duration of each audio recording is approximately 3–4 s.

The feature sets consist of MFCCs, GFCCs and MFCC + GFCC features. Table 1 shows the classification accuracies using the above feature sets and classifiers. MFCC features give a good accuracy for SVM as well as ANN. GFCC features when employed alone give accuracy values less than MFCC features. It can be observed that highest classification accuracy of 97.6% is given by ANN when MFCC and GFCC features are used in combination.

5 Conclusion

This paper discusses a methodology for bird species classification based on its sound. In this paper, two sets of classifiers and feature sets have been employed for the classification of bird species. ANN outperforms SVM and gives the highest accuracy with both the feature sets. The accuracy can be further improved by exploiting more feature sets and classifiers. Also, the future work will look into scaling up the database by including more number of bird sounds. Furthermore, the recordings available were free from noise. Real bird recordings will definitely include environment noise. The performance of our system will be assessed in the presence of different noises.

References

1. Dale S (2001) Causes of population decline in Ortolan Bunting in Norway. In: Proceedings in 3rd international Ortolan symposium, pp 33–41 (2001)
2. Brandes TS (2008) Automated sound recording and analysis techniques for bird surveys and conservation. Bird Conserv Int 18:163–173
3. Anderson SE, Dave AS, Margoliash D (1996) Template-based automatic recognition of birdsong syllables from continuous recordings. J Acoust Soc Am 100(2):1209–1219
4. Kogan J, Margoliash D (1998a) Automated recognition of bird song elements from continuous recordings using dynamic time warping and hidden Markov models: a comparative study. J Acoust Soc Amer 103(4):2187–2196
5. McIlraith AL, Card HC (1997a) Birdsong recognition using backpropagation and multivariate statistics. IEEE Trans Signal Process 45(11):2740–2748
6. Rassak S, Nachamai M (2016) Survey study on the methods of bird vocalization classification. In: 2016 IEEE international conference on current trends in advanced computing (ICCTAC). IEEE, pp 1–8
7. Jancovic P, Köküer M (2019) Bird species recognition using unsupervised modeling of individual vocalization elements. IEEE/ACM Trans Audio Speech Lang Process 27(5):932–947
8. Somervuo P, Harma A, Fagerlund S (2006) Parametric representations of bird sounds for automatic species recognition. IEEE Trans Audio Speech Lang Process 14(6):2252–2263
9. Fagerlund S (2007) Bird species recognition using support vector machines. EURASIP J Adv Signal Proces
10. Kogan JA, Margoliash D (1998b) Automated recognition of bird song elements from continuous recordings using dynamic time warping and hidden markov models: A comparative study. J Acoust Soc Am 103(4):2185–2196
11. Lee C-H, Lee Y-K, Huang R-Z (2006) Automatic recognition of bird songs using cepstral coefficients. J Inf Technol Appl 1(1):17–23
12. Chen Z, Maher RC (2006) Semi-automatic classification of bird vocalizations using spectral peak tracks. J Acoust Soc Am 120(5 Pt 1):2974–2984
13. Fletcher NH (2000) A class of chaotic bird calls. J Acoust Soc Am 108(2):821–826
14. Valero X, Alias F (2012) Gammatone cepstral coefficients: Biologically inspired features for non-speech audio classification. IEEE Trans Multimedia 14(6):1684–1689
15. Patil R, Patole R, Rege P (2019) Audio environment identification. In: 2019 10th International conference on computing, communication and networking technologies (ICCCNT). IEEE
16. https://www.xeno-canto.org/
17. Härmä A (2003) Automatic identification of bird species based on sinusoidal modeling of syllables. In: IEEE international conference on acoustics, speech, and signal processing, 2003. Proceedings (ICASSP'03), vol 5, pp V–545–548
18. Härmä A, Somervuo P Classification of the harmonic structure in bird vocalization. In: IEEE international conference on acoustics, speech, and signal processing, 2004. Proceedings. (ICASSP'04), vol 5, pp V–701–704
19. Calvo de Lara JR (2005) A method of automatic speaker recognition using cepstral features and vectorial quantization. In: 10th Iberoamerican conference on pattern recognition, CIARP 2005 proceedings
20. Furui S (1994) An overview of speaker recognition technology. In: ESCA workshop on automatic speaker recognition identification and verification, pp 1–9
21. Song FK, Rosenberg AE, Juang BH (1987) A vector quantisation approach to speaker recognition. AT&T Tech J 66–2:14–26

Analysis of Chronic Joint Pain Using Soft Computing Techniques

Shailaja Suresh Patil and Shubhangi B. Patil

Abstract In recent years, chronic pain in joints is the most common type of musculoskeletal pain among office workers. As the work culture in the office has been changed, it is the common demand that everyone has to work with a computer. In office, the worker has to sit continuously on a personal computer at least for 3–4 h. Most of the time, this leads to chronic pain in joints, generally after the period of few years of repeated exposure to such kind of regular engagement of user-computer interfacing, which is consequently converted into a serious health concern in the considerably large-sized office community. In this paper, an attempt has been made to provide a tool to assess the level of pain in the incumbent worker to have a reliable measure and at the disposal of medical practitioners to know the degree of pain to diagnose the severity of the health concern attached to it and have an optimum analysis which is not available even today. Around 250 samples have been taken and these cases have been categorized based on level of severity of pain, i.e. normal, medium, and severe. The results obtained show a considerable difference between mean and standard deviation values of pain level. Based on the data analysis of rectified EMG data, it has been observed that there is a notable difference in the Power Spectral Density of clean EMG signal w.r.t mean and Standard Deviation. The accuracy of the experimentally determined pain intensity level is a more reliable source for medical practitioners to make their diagnosis more objective one and this can be used as one of the measures to calibrate the pain-intensity.

Keywords Joint pain · Pain-intensity · Feature extraction · Surface electromyogram (SEMG)

S. S. Patil (✉)
Shivaji University, Kolhapur, India
e-mail: shaila.nalawade@ritindia.edu

Rajarambapu Institute of Technology, Rajaramnagar, Maharashtra, India

S. B. Patil
Dr. J.J. Magdum College of Engineering, Jaysingpur, Maharashtra, India

© Springer Nature Singapore Pte Ltd. 2021

S. N. Merchant et al. (eds.), *Advances in Signal and Data Processing*,
Lecture Notes in Electrical Engineering 703,
https://doi.org/10.1007/978-981-15-8391-9_24

1 Introduction

In the era of twenty-first-century globalized and liberalized world, modern digital office culture is adopted in most of the Indian corporate offices of small, medium as well as large scale Industries, Education Societies, Government and Non-government Organizations. Due to work culture at these offices, there is an urgent and genuine need to study, detect and analyze the level of chronic pain in joints and prediction of the consequences on the health of the office staff. Further, depending upon the severity of the pain, different kinds of clinical solutions are required to be suggested for the treatment of these problems.

Methods currently available are focusing on clinical trials and subjective suggestions for the physical setup. Their effectiveness and performance are judged and evaluated by health-care cost data, which is collected through standardized clinical forms, monthly phone interviews, and self-report questionnaire. Figure 1 shows the details of typical joints at knee and ankle.

One of the research method adapted has been found Thalamocortical Dysrhythmia (TCD) like pattern in patients with moderate chronic back pain and compared patients with neuropathic and non-neuropathic pain components. Also assessed the presence of psychopathology and the degree of psychological functioning and examined whether the strength of the TCD-related EEG markers is correlated with psychological symptoms and pain ratings.

In this paper, development of a system which will detect and measure pain intensity in the joints is discussed. The theoretical and experimental analysis of chronic body pain in joints by using soft computing tools and quantifying the major parameters responsible for such chronic pain with an objective to develop the system which will

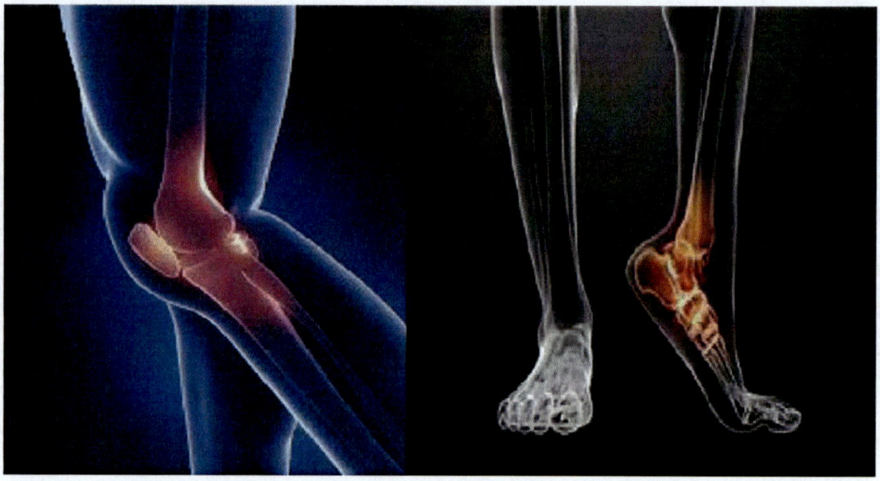

Fig. 1 Details of knee and ankle (courtesy University of Utah Health Care)

provide precise level of severity of chronic joint pain and provide a tool for further prediction of the consequences for clinical decision making and related treatment.

2 Literature Review

In the area of chronic pain analysis, some of the important contributions made are discussed here.

Shah and Patel [1] stated the prevalence of neck pain in computer operators in Ahmadabad. The experiment comprised of 1000 subjects from various institutes where extensive computer-based work. The subjects qualified the criteria of minimum work of 3hrs per day on computer and are considered to be working from the past 6 months and expected to work in the same fashion for next one year. The feedback of only 700 stated the prevalence of neck pain was 47%. Out of the 47, 27% were under 30 yrs, 53% were 30–50 years old and rest were above 50 years old. The parameters taken for consideration were divided into work-related variables and individual variables.

Nimbarte et al. [2] discussed the physical and psychosocial exertions on the neuromuscular fatigue on the neck. Two sessions were carried out comprising of isometric pulling patient from bed to stretcher and the second was a mix of physical and psychosocial activity. The two equipment used for the experiment were Surface Electromyography System and the pulling exertion device. The Surface Electromyography system consisted of 16 telemetry channel and was operated at 1500 Hz. The neuro-muscular fatigue development was processed, analyzed, and quantified using Discrete Wavelet Transform. NASA TLX workload scores were used for fatigue development measurement. The work concludes on a higher note pointing towards stressful considerations that are add-ons to the physical exertions. Healthcare workers had rapid development of fatigue in the neuromuscular region.

Cheng et al. [3] focused on early intervention educators. The strain and fatigue development in the neuro-muscular region for persons in this occupation were investigated. The various factors considered for categorizing the subjects were, work profile as in the Early Intervention institutes for at least three months, direct services to children, no prior injuries other than workplace in the neck, lower back, and shoulders. The sampling method used for this work was of type stratified. Several parameters of educators were also taken into account such as experience, workdays per week, work hours per day, break time, nap habit, etc. The authors came to a point that the lower back, neck, and shoulders were greatly affected and 94% of the subjects assumed to develop musculoskeletal disorders. The study selected the EI institutes with five workdays per week however more workdays cannot be neglected. The authors recommended that co-workers greatly reduces the chances of Musculoskeletal disorders.

Stainmetz et al. [4] undertook musicians under their microscope, especially violinists. The most important part in playing violin is done with the help of neck. It offers firm grip to hold the instruments properly. However, prolong sessions of using violin demand neck to be in the same position for long time which could result in musculoskeletal disorders mostly in the neck area. Fatigue and strain development chances are more as neuro-muscles are very delicate. NDI neck disability Index was used for grading and predicting the neck pain subject. Most of the study includes medical terms such as various kinds of muscles and cervical muscles. The proper guidelines could be formulated for reduction in neck pain for such musicians.

Carmela et al. [5] carried out an extensive study on how input devices such as mouse and touch pads impose restrictions on human body postures. The work included two standardized procedure to take samples of ten subjects those using mouse and touch pads. The study compared the body muscles contractions by investigating the EMG. The subjects included in the study were right-handed and electrodes were used extensively while mouse and touch pad usage. The study revealed that touchpad usage imposes more restrictions on body postures and compels user to maintain a very straight unnatural posture. However, the use of mouse gives user flexibility and needs no specific posture to maintain resulting in less strain and fatigue development in the shoulder and neck region.

Nathan et al. [6] reported work on Muscular demands during common clinical ophthalmologic activities may contribute to the high prevalence of musculoskeletal health outcomes observed among ophthalmologists and other eye care physicians. Characterizations of the muscle activation patterns in the live ophthalmologic environment are lacking. This study was conducted to (1) characterize the frequencies and durations of common activities performed by ophthalmologists during routine clinical examinations, (2) characterize neck and shoulder muscle activation patterns during the whole clinical work day, and (3) explore differences in neck and shoulder muscle activation patterns between common clinical activities.

Chu et al. [7] presented a newly pain intensity measurement method based on multiple physiological signals, including blood volume pulse (BVP), electrocardiogram (ECG), and skin conductance level (SCL), all of which are induced by external electrical stimulation. The proposed pain prediction system consists of signal acquisition and preprocessing, feature extraction, feature selection, and feature reduction, and three types of pattern classifiers. Feature extraction phase is devised to extract pain-related characteristics from short-segment signals.

A hybrid procedure of genetic algorithm-based feature selection and principal component analysis-based feature reduction was established to obtain high-quality features combination with significant discriminatory information. Three types of classification algorithms, linear discriminant analysis, k-nearest neighbor algorithm, and support vector machine were adopted during various scenarios, including multi-signal scenario, multi-subject, and between-subject scenarios, and multi-day scenario. The classifiers gave correct classification ratios much higher than chance probability, with the overall average accuracy of 75% above for four pain intensity. Experimental results demonstrate that the proposed method can provide an objective and quantitative evaluation of pain intensity.

3 Sampling Methodology

Thirty samples of healthy, medium pain, and severe pain (10 each) have been recruited in this analysis. To quantify the pain intensity, the samples of all the categories i.e. normal, medium pain, and severe pain of all ages have been considered. All the samples have been adequately informed about the purpose, procedure, and benefits of the study. Samples have been taken from college students, patients available at medical practitioners and physiotherapists. Algorithm design, classification, and performance validation in this pain-intensity calibration have been done by taking samples by employing surface Electromyogram Electrode.

4 Data Acquisition

Acquisition of high-quality SEMG signal is of prime importance for the pain intensity measurement system. SEMG signal is the primary consideration in this pain measurement system. It is expected that SEMG signal reflects the effect of pain. Quanser myoelectric electrode along with labview software is used to acquire the data. The data is available in the excel form which is processed in the MATLAB.

Figure 2 gives the flowchart of the process employed for experimental analysis of chronic pain by using the SEMG Electrodes.

Figure 3 shows some of the cases of samples of real-time data acquisition of the patients in various categories.

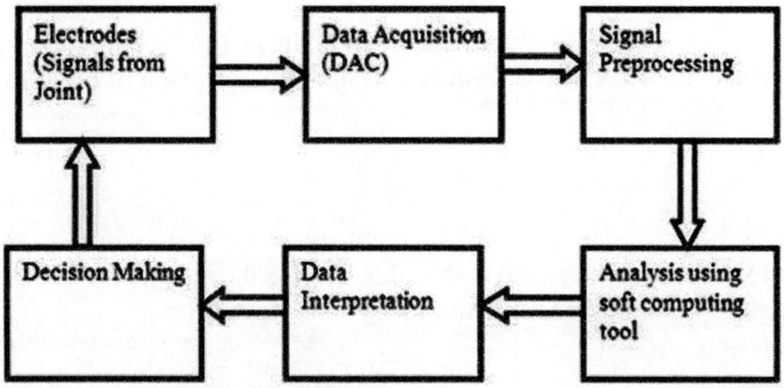

Fig. 2 Experimental analysis flowchart

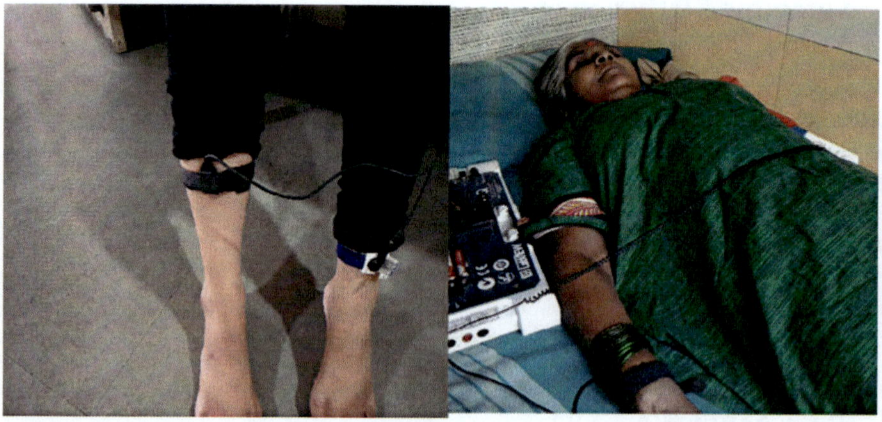

Fig. 3 Data acquisition by SEMG electrode

5 Feature Classification and Analysis

Independent Component Analysis is used for classification. Support Vector Machine is applied for analysis. Classification has been done by deriving parameters such as Mean, Standard deviation, Power Spectral Density of ten samples each, and average is taken for comparison. Based on the values obtained the pain intensity levels have been categorized into three levels as Normal, Medium pain, and Severe pain.

The SEMG data is band-pass filtered and given to the ICA algorithm for feature extraction. For SEMG data classification following features are extracted.

- Mean Absolute value: The signal is contained in χ_0 through $\chi_{N-1,}$ i is an index that runs through these values, where μ is the mean as given in Eq. (1).

$$\mu = \frac{1}{N}\sum_{i=0}^{N-1}\chi_i \tag{1}$$

- Standard Deviation: The signal is stored in χ_i, μ is the mean, N is number of, and σ is the standard deviation, as given in Eq. (2).

$$\sigma^2 = \frac{1}{N-1}\sum_{i=0}^{N-1}(\chi_i - \mu)^2 \tag{2}$$

6 Results and Conclusions

Data of normal, medium pain, and severe pain have been taken and Power spectral density, Signal to Noise Ratio, mean square value, Mean of Raw, and rectified EMG data are calculated. The results of normal, medium pain, and severe pain are shown in Tables 1, 2 and 3, respectively. The samples are tested on the developed code and found 77% accuracy in practice.

Comparison of various parameters of Normal, Medium pain, and Severe pain samples of Raw and Rectified EMG data have been made in order to quantify the intensity of pain. Parameter values of Raw EMG data and rectified EMG data of Normal, Medium pain and Severe pain samples are presented in Tables 4 and 5, respectively.

As per the objectives of the work, data processing is first and important task in analyzing the data. The result of developed algorithm shows the difference between raw and rectified data. The rectified-clean data is used for feature extraction. The difference between raw and rectified of Normal, Medium pain, and Severe pain samples is shown in Fig. 4.

At this early stage of research and based on the data analysis of rectified EMG data, it has been observed that there is a notable difference in the Power Spectral Density of clean EMG signal w.r.t mean and Standard Deviation of Normal Samples, Medium pain samples, Severe pain samples. It can be used as one of the measure to calibrate the pain-intensity. This is conferred in Figs. 5 and 6 respectively.

The samples are tested on the developed code and found 77% accuracy in practice.

The experimental result demonstrates that the projected method is a potential candidature for providing quantitative evaluation of pain- intensity level. The technique might be used to develop a wearable device that is suitable for routine applications at clinics.

Table 1 Parameters of samples of normal EMG data

Subject database file	Power spectral density of raw EMG signal	Power spectral density of clean EMG signal	SNR	Mean square error	Mean value of raw EMG	Mean value of clean EMG
Normal sample 3	33.878704758	2890.75121	53.920946341	1,202,445.4256	30.4505997908	−2.379108544
Normal Sample 4	0.0159110927	1.2648256437	20.023613191	820.29092349	−2.273448657	−0.009411343
Normal sample 5	34.181022834	1136.4655973	45.773275325	369,642.19964	23.9917609476	−1.300510673
Normal sample 6	58.998532837	2955.091618	51.703006883	1,132,234.3888	40.559406376	−0.153485102
Normal sample 7	31.899971262	151.78519569	28.58668538	99,926.420145	35.4555481706	−0.193010312
Normal sample 8	33.928422315	1102.1044735	45.538818652	562,365.57737	20.0296156308	−1.316061429
Mean	29.842120505	1164.460527	38.338208181	511,015.48667	22.4689800045	−0.529908639
Std. dev	20.496230242	1191.1067919	14.350965068	481,125.29525	16.4738222338	1.081482743
Rectified EMG data						
Normal sample 1	0.0003110465	0.2428109879	22.777113799	83.169276575	0.1398353247	−0.000384729
Normal sample 2	29.157427969	1941.0994089	51.113463432	867,594.50876	85.712433574	1.146258494
Normal sample 3	23.002346468	2840.0333045	55.448747253	1,125,092.3685	73.4165338188	−3.041253767
Normal sample 4	9.83E−05	0.1781385484	25.088010358	61.184573275	0.0675744046	0.005108485
Normal sample 5	22.698075519	2879.1479653	55.625389329	954,020.95057	69.7193951136	−1.403796865
Normal sample 6	34.61323624	1302.4527905	46.902817746	625,653.49864	101.71568372	0.263575102
Normal sample 7	22.348925494	114.66465019	27.696024533	71,497.586543	74.1025015306	−0.508403302
Normal sample 8	22.398735818	722.40707811	43.673404776	398,160.68189	67.3948399904	−0.762845575
Mean	19.277394611	1225.0282684	41.040621403	505,270.4936	59.0335996846	−0.53771777
Std. dev	12.653495266	1215.9063848	13.779842221	453,779.21226	37.979365175	1.260983789

Table 2 Parameters of samples of medium pain EMG data

Subject database file	Power spectral density of raw EMG signal	Power spectral density of clean EMG signal	SNR	Mean square error	Mean value of raw EMG	Mean value of clean EMG
Medium pain sample 1	0.0119474423	0.2880066095	8.4152997	153.97992	−1.7680412	0.0007684
Medium pain sample 2	32.6071579806	824.511462849	43.190804	373,311.31	34.060259	−0.7066877
Medium pain sample 3	33.919678757	7231.91476372	61.880549	3,093,064.1	35.235023	1.6701881
Medium pain sample 4	30.5401012445	90.3521185784	24.270062	89,194.156	29.271776	0.2441957
Medium pain sample 5	34.5693639051	660.278848547	41.007634	349,940.38	26.872414	1.3935375
Medium pain sample 6	31.5111640518	149.547010209	28.51091	137,283.34	32.237127	−0.5360175
Medium pain sample 7	0.0083975697	0.8823476582	19.671259	368.76349	−1.328349	0.0125329
Medium pain sample 8	0.0165338979	0.2991370984	7.333653	222.72806	−2.5582754	0.0055303
Medium pain sample 9	72.2400313394	272.169502611	30.10901	186,091.08	−21.405036	0.2022336
Medium pain sample 10	41.4632764055	1109.92280492	44.729219	425,371.71	22.109175	−0.1620618
Medium pain sample 11	7.89E−08	0.000114588	−7.7894092	0.110083	−0.0025013	−0.0009633
Medium pain sample 12	65.3243647354	470.781077546	35.305628	254,865.87	−13.40604	−0.7958573
mean	28.5176681174	900.912266244	28.052885	409,155.63	11.609794	0.1106166
std. dev	24.8275384457	2027.87017652	19.213124	859,560.95	20.316819	0.7466128
Rectified EMG Data						
Medium pain Sample 1	0.0006714396	0.0731596813	9.0153666	35.002702	0.1858806	−0.0017001
Medium pain sample 2	22.6887698877	520.023398812	40.762348	246,575.06	74.053084	−0.0767522

(continued)

Table 2 (continued)

Subject database file	Power spectral density of raw EMG signal	Power spectral density of clean EMG signal	SNR	Mean square error	Mean value of raw EMG	Mean value of clean EMG
Medium pain sample 3	23.7616496275	2186.44294523	53.035997	773,451.89	76.165694	1.0883839
Medium pain sample 4	21.5481582979	80.5601248387	24.788301	71,442.079	69.926249	0.026694
Medium pain sample 5	22.7508174885	1197.87386865	47.998252	528,629.61	71.814043	0.2773953
Medium pain sample 6	21.9424467778	85.3660970946	25.212858	96,436.515	72.416386	−0.4556381
Medium pain sample 7	0.0013037193	0.2648970143	17.3097	126.79828	0.2111236	−0.0104451
Medium pain sample 8	0.0001892958	0.1112090111	18.15139	48.066426	0.0536277	−0.0009861
Medium Pain sample 9	31.5808774529	69.9702412126	21.904025	44,830.428	80.019406	−0.0424457
Medium Pain sample 10	26.1567532113	1485.77339579	49.263213	474,386.43	75.540006	−1.111458
Medium pain sample 11	4.14E−08	3.06E−05	−16.453853	0.0407274	0.0022964	−0.0003361
Medium pain Sample 12	34.6382017053	1456.84870983	47.872736	591,044.98	79.181732	−1.0164971
Mean	17.0891532454	590.275673148	28.238361	235,583.91	49.964127	−0.1103154
Std. dev	12.5844604461	763.81552224	20.538999	269,338.13	37.506167	0.5200391

Table 3 Parameters of samples of severe pain EMG data

Subject database file	Power spectral density of raw EMG signal	Power spectral density of clean EMG signal	SNR	Mean square error	Mean value of raw EMG	Mean value of clean EMG
Severe pain sample 1	0.0124087389	0.8178027379	17.315695	243.22085	−1.7535781	-0.0232255
Severe pain sample 2	0.0200849617	1.4687567005	20.310287	668.07374	−1.5918421	0.0121741
Severe pain sample 3	29.315064237	72.9697783205	22.591952	44,578.876	31.598394	0.0773192
Severe pain sample 4	0.0105866099	0.6424586155	15.909334	249.21495	−1.5830765	−0.0040712
Severe pain sample 5	0.0107663703	1.7357229571	24.468915	657.15891	−1.6009298	0.0096121
Severe pain sample 6	28.348683435	141.967024944	28.518421	89,225.428	29.70029	−0.4554452
Severe pain sample 7	7.89E−08	0.000114588	−7.7894092	0.110083	−0.0025013	−0.0009633
Mean	8.245370633	31.37166555	17.33217	19,374.58	7.823822	−0.05494
Std. dev	14.06602687	55.6719655	11.87236	34,932.87	15.61387	0.179386
Rectified EMG data						
Severe pain sample 1	0.0002656557	0.3201582762	25.864104	85.390364	0.1557027	−0.0066223
Severe pain sample 2	0.0023090715	0.2587401196	14.622902	105.05676	0.4680857	0.0002534
Severe pain sample 3	20.911819094	52.5613238829	21.209408	43,227.994	69.64426	−0.1533798
Severe pain sample 4	0.0004262267	0.1481263024	17.116237	102.02025	0.1611355	−0.006861
Severe pain sample 5	0.0005939908	0.19902367	18.240298	64.954147	0.1773695	0.0120415
Severe pain sample 6	19.940022223	198.179474856	32.943917	159,858	67.878649	−0.4133761
Severe pain sample 7	4.14E−08	3.06E−05	−16.453853	0.0407274	0.0022964	−0.0003361
Mean	5.8364909	35.9524111	16.22043	29,063.35	19.78393	−0.08118
Std. dev	9.970425332	74.15074395	15.66636	59,875.6	33.46219	0.157325

Table 4 Parameter values of raw EMG data of various Pain Samples

Subject database file		Power spectral density of raw (rectified) EMG signal	Power spectral density of clean EMG signal	SNR	Mean square error	Mean value of raw (rectified) EMG	Mean value of clean EMG
Normal	Mean	29.842120505	1164.4605	38.338208	511,015.49	22.46898	−0.5299086
	Std. dev	20.496230242	1191.1068	14.350965	481,125.3	16.473822	1.0814827
Medium pain	Mean						
	Std. dev	28.517668117	900.91227	28.052885	409,155.63	11.609794	0.1106166
Severe pain	Mean	24.827538446	2027.8702	19.213124	859,560.95	20.316819	0.7466128
	Std. dev	8.2453706332	31.371666	17.332171	19,374.583	7.8238224	-0.0549428

Table 5 Parameter values of rectified EMG data of various pain Samples

Subject database file		Power spectral density of raw (rectified) EMG signal	Power spectral density of clean EMG signal	SNR	Mean square error	mean value of raw (rectified) EMG	Mean value of clean EMG
Normal	Mean	19.277394611	1225.0283	41.040621	505,270.49	59.0336	−0.5377178
	Std. dev	12.653495266	1215.9064	13.779842	453,779.21	37.979365	1.2609838
Medium pain	Mean	17.089153245	590.27567	28.238361	235,583.91	49.964127	-0.1103154
	Std. dev	12.584460446	763.81552	20.538999	269,338.13	37.506167	0.5200391
Severe pain	Mean	5.8364909005	35.952411	16.22043	29,063.352	19.783928	-0.0811829
	Std. dev	9.970425332	74.150744	15.666357	59,875.601	33.462188	0.1573248

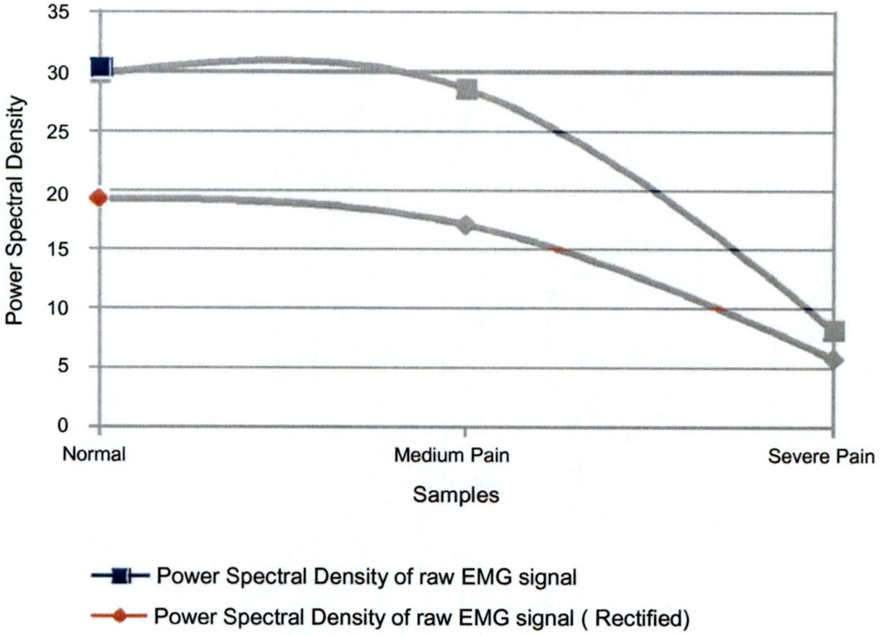

Fig. 4 Difference between raw and rectified normal, medium pain and severe pain samples

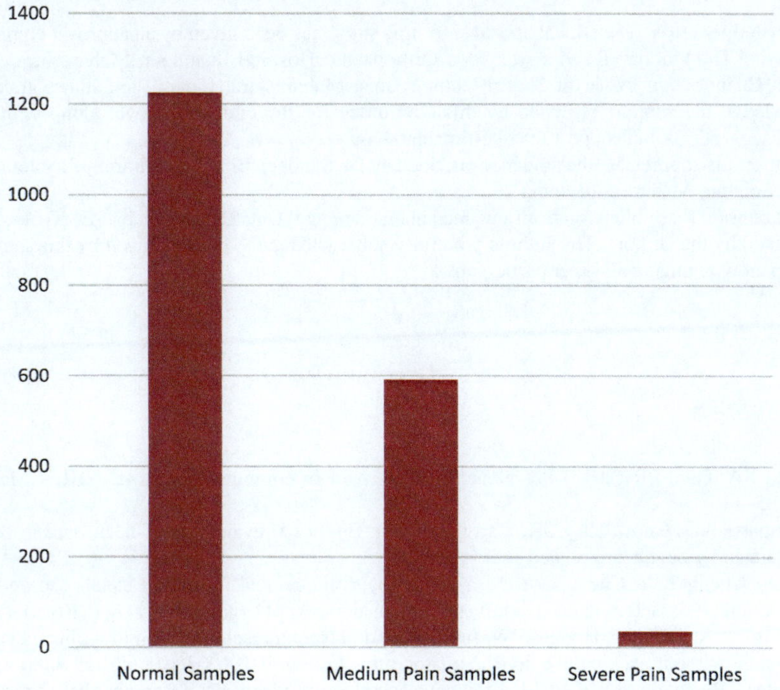

Fig. 5 Statistics showing comparison of power spectral density of clean EMG signal—Mean

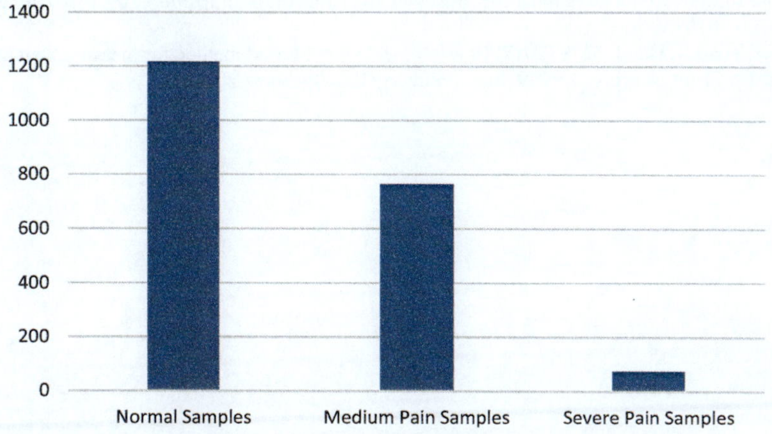

Fig. 6 Statistics showing comparison of power spectral density of clean EMG signal—Standard deviation

Acknowledgements The ethical approval for this study has been given by an approval committee comprising Dr. Vaibhav Raje of Mahadeo Orthopaedic Hospital, Islampur, Maharashtra, India, and Dr. Chandrakant Pawar of Shashi Clinic, Palus, Maharashtra, India. The authors gratefully acknowledge the support provided by this committee for the ethical approval along with their critical suggestions and support for real-time database.

Authors also appreciate the guidance provided by Dr. Sandeep Bhagwat, Niramay Physiotherapy Clinic, Solapur, Maharashtra, India.

The consent for publication from a woman in Fig. 3 image (Data acquisition by SEMG electrode) is obtained by the authors. The authors gratefully acknowledge the consent given by this sampling participant woman and all other participants.

References

1. Shah SA, Patel PR (2015) Prevalence of neck pain in computer operators. NHL J Med Sci 4(1):5–11
2. Nimbarte AD, Chowdhury SK, Cartwright ED. Empirical evaluation of neck muscle fatigue generated by healthcare related exertions
3. Kathy Cheng K-Y, Cheng C-Y, Ju Y-Y (2013) Work-related musculoskeletal disorders and ergonomic risk factors in early intervention educators. Appl Ergon 44:134–141 (Elsevier)
4. Steinmetz A, Claus A, Hodges PW, Jull GA (2015) Neck muscle function in violinists/violists with and without neck pain. Clin Rheumatol. https://doi.org/10.1007/s10067-015-3000-4,July
5. Conte C, Ranavolo A et al (2014) Kinematic and electromyographic differences between mouse and touchpad use on laptop computers. Elsevier, Int J Industrial Ergon 44(2014):413–420
6. Fethke NB, Schall MC Jr et al (2015) Neck and shoulder muscle activity among ophthalmologists during routine clinical examinations. Elsevier Int J Ind Ergon https://doi.org/10.1016/j.ergon.2015.06.0010169-8141
7. Chu Y, Zhao X, Han J, Su Y (2017) Physiological signal-based method for measurement of pain intensity. Front Neurosci 11:279. https://doi.org/10.3389/fnins.2017.00279

Emotion Recognition using Gamma Correction Technique Applied to HOG and LBP Features

Vishal D. Bharate, Devendra S. Chaudhari, and Mayur D. Chaudhari

Abstract Human social interaction, especially facial expressions, is often influenced by non-verbal communication. The surrounding people often watch the face in day-to-day group interaction to understand the inner feelings of a person. Face thus forms an essential source of human emotion recognition that is generally categorized as a surprise, fear, anger, disgust, sad, and happy. Recognition of emotions plays an important role in a variety of fields in behavioral science. In this paper, median filtering is used for pre-processing of an input image. Watershed segmentation is used before extracting features to obtain the necessary image properties. Gamma correction is implemented in this paper, and features are obtained, including Histogram of Oriented Gradient (HOG) and Local Binary Pattern (LBP) techniques. The performance of LBP and HOG is evaluated. kNN and SVM are used as classifiers for comparing the efficiency of recognition. The overall accuracy, along with precision, recall, and f-score has been computed and compared. It is found that all performance parameters with gamma correction give better performance compared to without gamma correction.

Keywords Watershed Segmentation · kNN · SVM · HOG · GFE · LBP · Median filtering

V. D. Bharate (✉)
Government College of Engineering, Amravati, India
e-mail: vishalbharate@gmail.com

D. S. Chaudhari
Government College of Engineering, Jalgaon, India
e-mail: ddsscc@yahoo.com

M. D. Chaudhari
Data Architect, Parkar Labs, Pune, India
e-mail: chaumayu@gmail.com

© Springer Nature Singapore Pte Ltd. 2021

S. N. Merchant et al. (eds.), *Advances in Signal and Data Processing*,
Lecture Notes in Electrical Engineering 703,
https://doi.org/10.1007/978-981-15-8391-9_25

1 Introduction

Face forms mirror of human sentiments. The study of expressions on the face of a person provides a representation of the feelings. Hence, the feelings of a person are interpreted by studying facial expressions that provide the basis for interpersonal communication in the group activity of people. According to a classical review, this proportion is as high as 55% [1, 2]. Active recognition of facial expression can improve various applications in daily life. Such systems include robots, digital equipment, automated machines, security, etc.

Therefore, emotion identification using expressions on the face of a person contributes to the creation of various domestic and industrial applications.

Usually, expressions on the face of a person are categorized into seven types: fear, neutral, anger, happy, disgust, sad, and surprise. Recognition of facial expression in recent decades has seen significant development in the research area. To develop various system automation applications, the researchers are actively exploring the identification of expressions on the face as a field of crucial research investigation.

Usually, different steps for the identification of facial expressions are required. The initial step is to pre-process the image, eliminating the image's latent low-frequency noise. In the next step, the action unit is identified by observing expressions on the face that helps in investigating further. In the next step, the face image is segmented into various logical parts that enable smooth processing in the phase of extracting various features. In the further step, features are extracted to form feature vectors. It represents a critical stage as this step's output offers a practical classification database. The feature vectors are classified into a particular emotion class in the final phase of identification of facial expressions.

In this paper, the facial expression recognition system's characteristic extraction and identification phases are focused. During the extraction phase of the feature, different techniques are used. These include Principal Component Analysis (PCA), Scale Invariant Feature Transform (SIFT), Gabor Wavelets, Speeded-Up Robust Features (SURF), Gray Level Co-occurrence Matrix Moments, Linear Discriminator Analysis (LDA), and so on. Such techniques that add some disadvantages to the established method, although they are common in literature, used in isolation. A combination of extraction techniques for features gives some advantages to the established system [2]. LBP, GFE and HOG techniques are therefore combined in the system described in this paper. In the surveyed literature, various classification algorithms used in various applications are described that are used after the feature extraction process. These include Support Vector Machine (SVM) [3–5], AdaBoost classification [6], Artificial Neural Networks (ANN) [7], Extreme Learning Machine (ELM) [8], Random Forests (RF) [3], Nearest Neighbor (kNN) [9].

Unlike feature extraction methods that are utilized in systems for recognizing expressions on the face of a person, classification algorithms also have individual drawbacks. KNN has limitations such as memory usage, selecting a variety of neighbors. ANN typically requires more training time. In addition to this, the training phase of ANN also requires a large number of samples in various applications. In addition to optimizing the selection of hidden layer nodes, ELM faces the overfitting issue. It is a challenging task to use outliers and noisy data in AdaBoost classifier.

The organization of the paper is as follows. The implemented method is elaborated in Sect. 2. The analysis of performance for the proposed system is presented in Sect. 3. The conclusions from the proposed work are summarized in Sect. 4.

2 Implemented System

In this section, the implemented system to interpret the emotions of a human being using expressions on the face is explained. The flow graph of the implemented system is presented in Fig. 1.

The first phase of implementation is preprocessing. In this step, median filtering is used to smooth out the source image. In this step, segmentation has been used to extract the necessary features from the input image. The Gamma correction technique is used to modify the gamma values of various image segments obtained from the first step. This helps in increasing the overall efficiency of the system. The output image features after gamma adjustment are extracted as part of the system using LBP. The HOG is another technique for feature extraction used in the current system. After applying Gamma correction, the processing of input image is performed to remove the attribute using the HOG technique for this purpose.

Feature vector classification either into HOG or LBP is done separately using SVM and kNN. Classification tests have been compared for system performance analysis.

2.1 Preprocessing

Pre-processing is the first step in a conventional digital image processing system. With the support of pre-processing, image quality is enhanced in terms of visual attributes. It also helps to improve the database for processing. The image's random noise is removed. The image's homogeneity and contrast features are improved. There is also a boundary zone enhancement and low-frequency feature illumination. The operation of noise removal is performed using some kind of filtering technique on an image.

The input image is preprocessed with the aid of the median filter. This is required for removal of impulse noise and to bridge the gaps in between corners and edges.

The purpose of variable size matrix was to cover a set of pixels on which the computation of median was performed. The central pixel is given by ascending the sorting of the pixel value. The value of the pixel at the centre replaces the computed median value. To minimize the noise, each pixel is compared to the neighboring pixel.

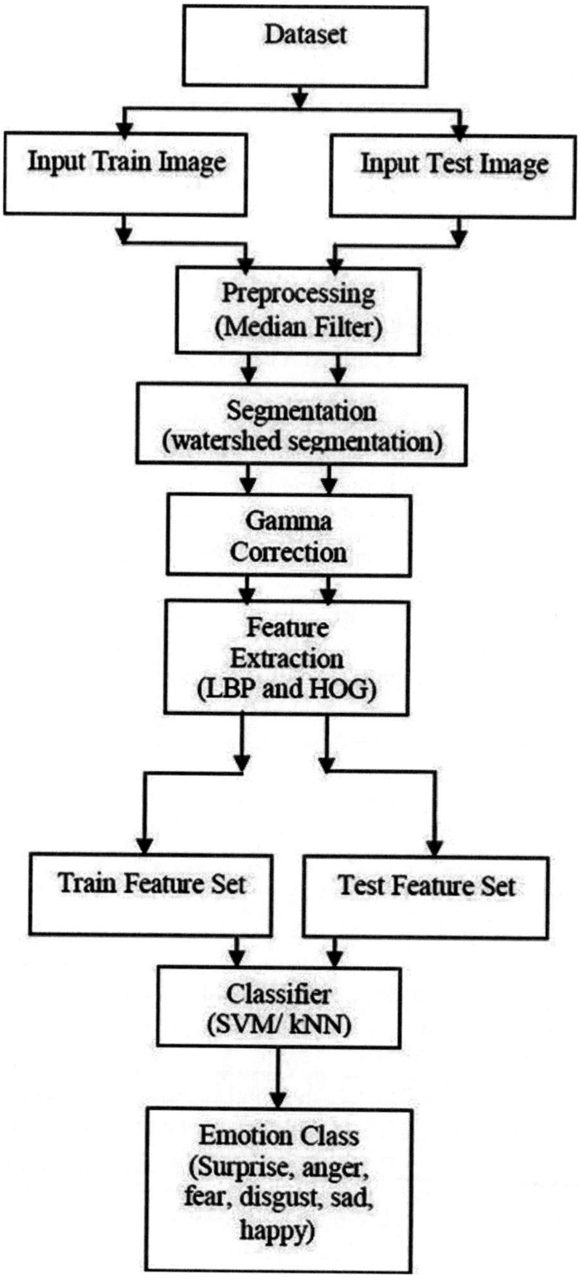

Fig. 1 Block diagram of the proposed system

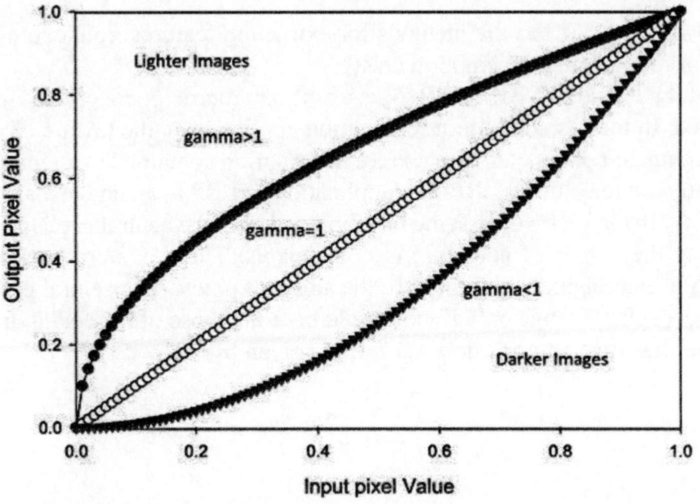

Fig. 2 Gamma curve

Segmentation: The purpose of segmentation is to divide an image into various logical regions. To avoid over-segmentation, which may be inherent in the process of segmentation, the watershed segmentation is used. This process removes the image's background noise. The performance improvement of the system is observed because the process of segmentation removes the noise in the background of an image.
Gamma Correction: Gamma correction is a part of the pre-processing function. Gamma correction I_{γ} replaces the frequency of the grey level I. In this case, γ is the user-controlled parameter. While the bright area is condensed, the darker part is improved using gamma correction.

The phenomenon of gamma correction is as shown in Fig. 2. Its range is between 0.00 to 10.00 that can be controlled by the user. Gamma correction can be seen to change the linear output of the image under consideration. The classification accuracy is improved by this mechanism. The gamma correction, along with LBP is applied in this paper for analyzing the performance of the proposed system.

2.2 *Feature Extraction*

This is the further process after performing preprocessing and segmentation. The significance of feature extraction lies in the fact that the required hidden data for processing can be obtained using it. Another important characteristic of the features is that there is a reduction in the dimensions of the data, ultimately leading to saving the processing time and memory requirements, in case of pattern recognition applications. Storing of the features is essential for forming the feature vectors. This paper presents LBP and HOG as the methods for extracting features from expressions on the face for recognizing the emotion classes.

Local Binary Pattern (LBP): LBP [6] is a non-parametric operator used to retrieve texture data. In the case of pattern recognition applications, the texture of an image forms a significant parameter. The texture information contained in an image can be significantly obtained using LBP. The application of LBP to an image makes the use of 3x3 mask. In the LBP process, the threshold value of the central pixel is computed by utilizing the values of neighboring eight pixels. There is a comparison of the values of the central pixel with those of the adjacent pixels. The central pixel values are in the acceptable range only if they are larger than those of the neighboring pixel values. The resulting vector's decimal form is given by:

$$\mathrm{LBP}(p, q) = \sum\nolimits_{k=0}^{7} 2^k g(m_k - m_{(p,q)}) \tag{1}$$

where

$m_{p,q}$- Central pixel grey value,
m_k- Neighbour pixel grey value,

and

$$g(p) = 1 \text{ if } p \geq 0$$
$$g(p) = 0 \text{ if } p < 0$$

In this case, the transformation of the grayscale affects the LBP operator slightly less. Thus pixel intensities of neighbourhood pixels are maintained.

Histogram of Oriented Gradients (HOG): HOG [9] is one of the robust descriptors of features. The HOG is commonly used in applications of image processing. From the localized area, the gradient orientation frequency is measured. The HOG feature measures the distribution of localized frequency.

The preprocessed image is applied to the extraction block of the HOG feature as an input. This extraction of the HOG feature divides the image into a small number of cells. Every cell's pixels are compiled further using histogram gradient. The compiled histogram gradients are concatenated to form the descriptor of the HOG feature. An important feature of the HOG feature descriptor is its invariance with respect to photometric and geometric transformations. The generation of descriptor blocks is performed after the computation of gradient and orientation binning. To get HOG features, these blocks of descriptors are concatenated and standardized.

$$L_{\mathrm{norm}} \rightarrow f = v_{nm} / \|v_{nm}\| + e \tag{2}$$

where

V_{nn}	is the non-normalized vector and
e	is the error vector.

Classification: One of the important steps following feature extraction is the process of classification. The train image dataset features already stored in the memory are compared with the features of the image being considered in the classification stage. Finding the correct match is essential for allocation of class tag related to feature dataset obtained from a database of training images. Recognition of emotions from facial expression is divided into different classes, namely, disgust, surprise, happy, fear, anger, and sad. So it's a multi-class issue. We use two classifiers, k-Nearest Neighbor (kNN) and Support Vector Machine (SVM), to test the performance of the implemented technique.

K-Nearest Neighbor (kNN) Classification: In pattern recognition, a non-parametric approach for classifying different classes is implemented by kNN [1, 8] classifier. The classifier tests the classes according to the test template. The utilization of the nearest k training pattern sets leads to the formation of the input vector to the classifier. The basis of the classification algorithm is the comparison between a set of an odd number of nearby training patterns that fall in the vicinity with the test pattern. The output of the classifier is computed using the largest number of training patterns class in the vicinity of the test pattern. A class is identified by comparing the feature vector of the test pattern with the odd number of nearby feature vectors from the training patterns. As an example, one nearest neighbor is compared to the feature vector of the test pattern for $K = 1$. Three nearest neighbors are compared to the feature vector of a test pattern for $K = 3$, in order to find the maximum match for identification of a class. This process continues for $K = 5, 7, 9$ and so on. Fig. 3 elaborates the aforesaid classification algorithm.

Figure 3 presents three classes in terms of feature vectors. These three classes are represented by circle, rectangle and triangle. The choice of said patterns leads to a clear understanding of the algorithm. Two shapes, namely the rectangle and triangle, represent the training features. The test feature is represented by a circular shape. The circle as a test feature is compared with the nearby three training features. These three training features contain one rectangle and two triangles. As per the kNN classification algorithm, the circular shape is classified into the triangular shape because the highest numbers of training features in the vicinity of the circular class are the triangles. This explanation can be generalized for all classes found in real-time.

Support Vector Machine (SVM): SVM [2] is a non-linear classifier, one of the supervised learning classifiers. SVM offers widespread quality as it uses the statistical theory of learning. SVM provides optimal performance for a dataset of limited size. The performance optimization of SVM classifier can be achieved by employing

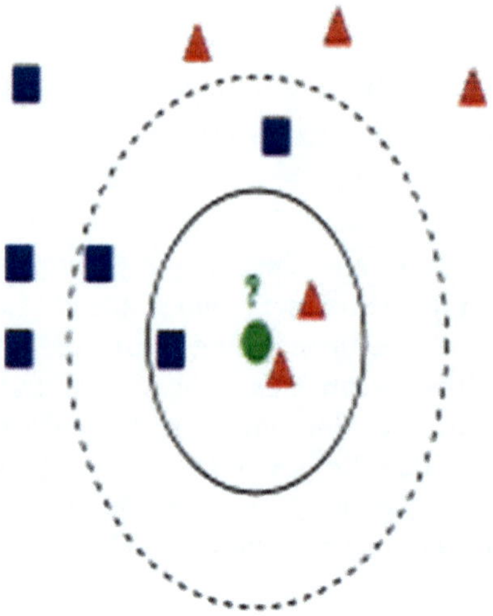

Fig. 3 KNN classification

various kernel functions. This offers a creative and collaborative approach. The selection of the kernel function for high-performance efficiency plays an important role.

In a variety of classification applications, including identity and text recognition as well as DNA microarray data analysis, SVMs are successfully employed in the past.

3 Performance Analysis

This section includes two features of performance analysis, namely HOG and LBP. The initial description relates to the testing database, accompanied by system requirements on which experimentation is performed. The explanation of the criteria used as a performance measure is presented later in the section and is to be considered according to the results presented.

3.1 Setup for Experimentation

The testing phase involves the utilization of 213 images selected from JAFEE [5] database considered as the standard in the field. The file consists of images with dimensions of 256 × 256 pixels in TIFF format and storage size varying from 60 to 70 KB. Using standard database enables reliability, more accurate outcomes and performance comparison The database employed is divided into databases for testing and training The training data set comprises 140 images, while the test data set uses the other 73 images. The setup for experimentation includes a laptop having 3 GB RAM with onboard i3 processor and preinstalled Windows 7 operating system and MatLab 2014a. The dataset is processed by the proposed algorithm using the MatLab program. The simulation results are hence obtained.

3.2 Results

SVM and KNN classifiers are used in this work to train and test data sets. The feature vectors of HOG utilized in the training dataset are determined in the training phase. The testing phase includes measuring the image's HOG features being checked and comparing the entire set of the images in the training database with the HOG features. Basis of the best match is the evaluation feature's nearest distance from the practice feature dataset. By applying the same methodology, computation and application of LBP features is carried out with reference to testing and training datasets.

The results presented in Fig. 4 are obtained by applying the kNN classifiers to HOG features. As exhibited with the aid of Fig. 4, the maximum recognition rate is achieved for Surprise and Happy class with or without the application of Gamma correction prior to feature extraction. The high matching index between the feature vectors leads to indistinguishable characteristic of fear and anger classes, as explored from Fig. 4.

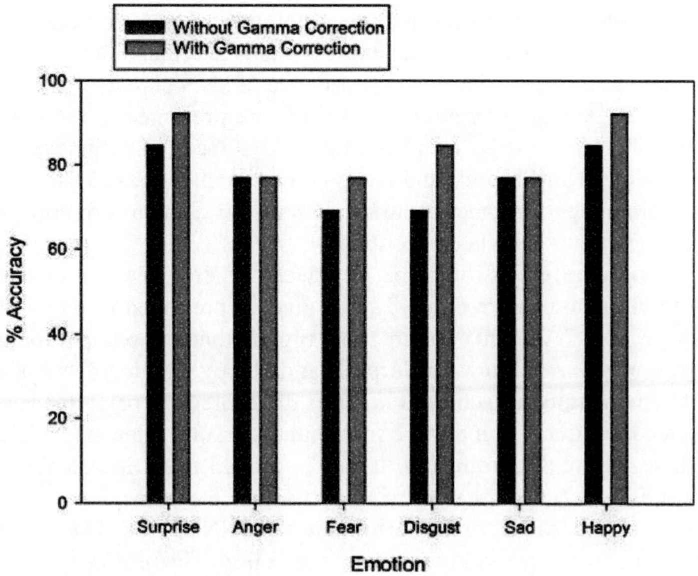

Fig. 4 Performance of kNN with and without Gamma correction for HOG feature

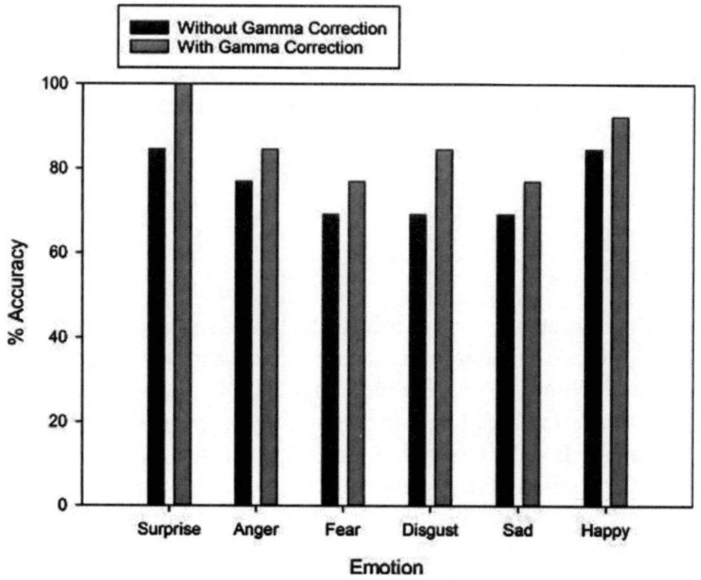

Fig. 5 Performance of SVM with and without Gamma correction for HOG feature

The results depicted in Fig. 5 are obtained by applying the kNN classifiers to LBP features. As exhibited with the aid of Fig. 5, the recognition accuracy with gamma correction has been increased for all emotion classes in comparison with the results obtained without gamma correction. Another finding from Fig. 5. is that the performance accuracy of Fear and Sad expressions is less than 80%.

Figure 6. shows the results for the performance of kNN classifier applied to LBP with and without gamma correction. The results are presented in terms of the bar chart shown in Fig. 6. It can be seen from Fig. 6. that the recognition accuracy with gamma correction is for Surprise and Happy emotion class. Fear, Disgust, and Sad class have minimum performance accuracy. Results for gamma correction are better as compared to without gamma correction.

Figure 7. shows the results for the performance of SVM classifier applied to LBP with and without gamma correction. The results are presented in terms of the bar chart shown in Fig. 7. It can be seen from Fig. 7. that the recognition accuracy with gamma correction is more for Surprise and Happy emotion class. The results exhibit that better results are achieved in terms of accuracy of performance for LBP with Gamma correction. It can also be seen that the results achieved for Happy and Surprise classes are really promising. It can be judged that Disgust, Sad and Fear classes still provide a scope for improving performance accuracy.

Figure8. shows the results for the performance of kNN and SVM classifier applied to LBP and HOG, with and without gamma correction. The results are presented in terms of the bar chart shown in Fig. 8. It can be seen from Fig. 8. that the recognition accuracy with gamma correction is more for Surprise and Happy emotion class. The

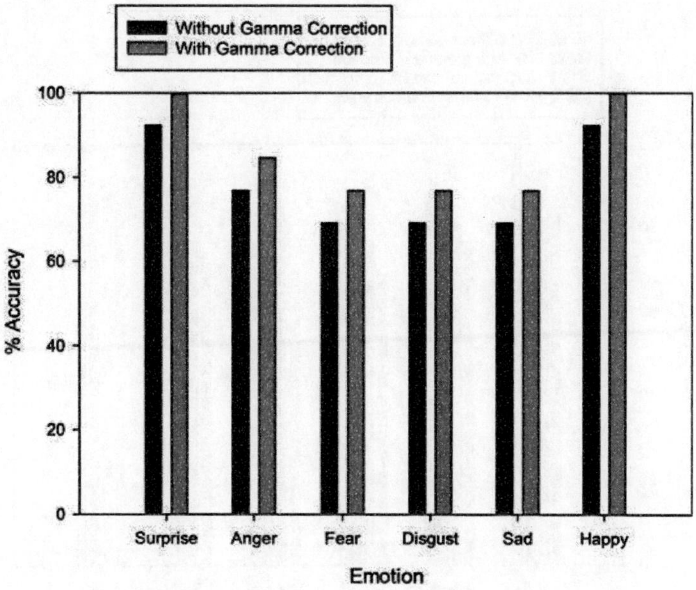

Fig. 6 Performance of kNN with and without Gamma correction for LBP feature

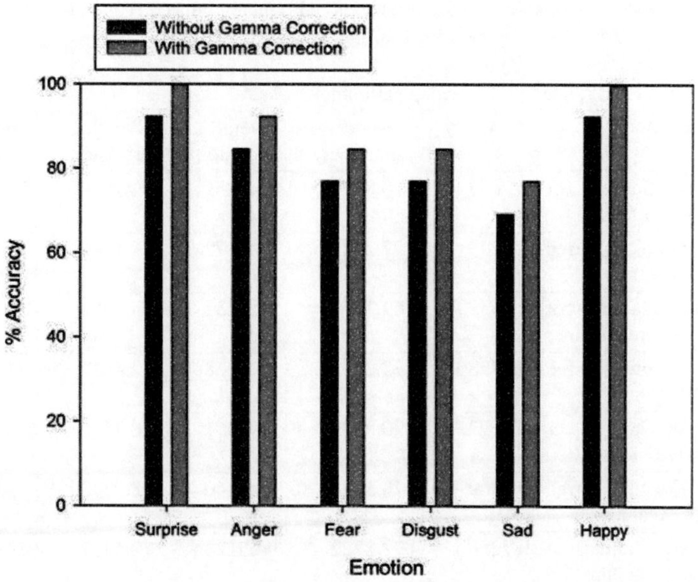

Fig. 7 Performance of SVM with and without Gamma correction for LBP feature

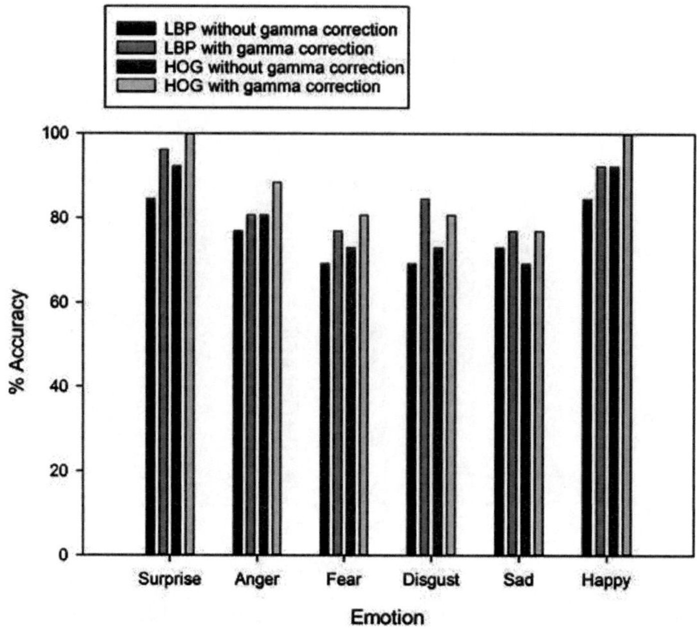

Fig. 8 Performance Rate of kNN and SVM with and without Gamma correction for LBP and HOG feature

Table 1 Methods and related performance parameters

Method parameter	Accuracy	Overall sensitivity	Overall specificity	Overall precision	Overall F score
kNN without Gamma correction for HOG feature	0.7097	70.8333	94.2051	70.7576	70.7419
kNN with Gamma correction for HOG feature	0.7742	77.2222	95.4872	77.7146	77.2487
kNN without Gamma correction for LBP feature	0.6452	64.1667	92.9231	64.5202	64.1466
kNN with Gamma correction for LBP feature	0.7258	72.2222	94.5128	72.4417	72.1429
SVM without Gamma correction for HOG feature	0.7097	70.8333	94.2051	70.7576	70.7419
SVM with Gamma correction for HOG feature	0.7869	78.5185	95.7446	78.9773	78.6151
SVM without Gamma correction for LBP feature	0.7258	72.2222	94.5128	72.4417	72.1429
SVM with Gamma correction for LBP feature	0.7742	77.2222	95.4872	77.7146	77.2487

Table 2 Qualitative results

Input Image	Recognized Output

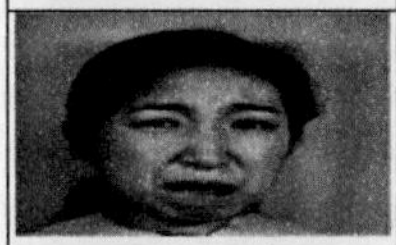	
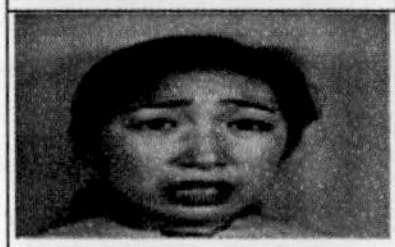	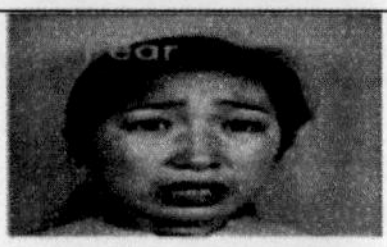

results exhibit that better results are achieved in terms of accuracy of performance for LBP and HoG with Gamma correction. It can also be seen that the results achieved for Happy and Surprise classes are really promising. For all classes except fear, class accuracy is more than 80%.

The methods, along with all performance parameters, are listed in Table 1. From Table 1, it can be observed that gamma correction gives a significant increase in all the parameters.

Table 2 represents the qualitative results obtained for six basic emotions.

4 Conclusion

The implemented framework for facial expression emotion recognition is evaluated based on the reliability of the results. The median filtering technique preprocesses images, and the watershed segmentation algorithm is used for the segmentation of images. The features extracted by LBP and HOG. Two classifiers, namely kNN, and SVM, compare the performance. It is seen from the results that the HOG's performance accuracy using kNN and SVM gives the best value for surprise and happy categories.

On the same platform, another feature, namely LBP, is also compared. By applying two methods, namely with gamma correction and without gamma correction, the output accuracy is compared with this. From the tests, LBP's output reliability with gamma correction can be seen as compared to that of LBP without gamma correction. The only sad class did not give more than 80% accuracy.

Finally, with and without gamma correction, the output accuracy of LBP is compared to the HOG feature without gamma correction. It can be seen from the results that LBP with gamma correction provides better performance for all emotion class compared to LBP without gamma correction and HOG without gamma correction. Happy and Surprise's recognition rate for all implemented methods is better than any other class, whereas the lesser performance accuracy has been provided by the Fear and Disgust class.

References

1. Vaibhavkumar J, Mistry M, Goyani M (2013) A literature survey on facial expression recognition using global features. Int J Eng Adv Technol (IJEAT) 2(4)
2. Pande S, Shinde S (2012) A survey on emotion recognition with respect to database and various recognition techniques. Int J Comput Appl 58(3)
3. Jayalekshmi J, Mathew T (2017) Facial expression recognition and emotion classification system for sentiment analysis. In: 2017 International conference on networks & advances in computational technologies (NetACT), Thiruvanthapuram, pp 1–8
4. Dniz O, Bueno G, Salido J, Torre F (2011) Face recognition using histograms of oriented gradients. Pattern Recogn. Lett. 32(12):1598–1603
5. Kamachi M, Lyons M, Gyoba J (1997) The Japanese female facial expression (JAFEE) database. In: International conference on automatic face and gesture recognition
6. Verma R, Dabbagh MY (2013) Fast facial expression recognition based on local binary patterns. In: 2013 26th IEEE Canadian conference on electrical and computer engineering (CCECE), Regina, SK, pp 1–4
7. Kauser N, Sharma J (2017) Facial expression recognition using LBP template of facial parts and multilayer neural network. In: 2017 International conference on I-SMAC (IoT in social. mobile, analytics and cloud) (I-SMAC), Palladam, pp 445–449
8. Liu Z et al (2017) A facial expression emotion recognition based human-robot interaction system. IEEE/CAA J Autom Sin 4(4):668–676
9. Agrawal S, Yadav S (2018) Approach based on HOG descriptor and K nearest neighbour for facial expressions recognition. Int J Innov Adv Comput Sci 7(5):239–249

Scalable and Rapid Fault Detection of Memories Using MBIST and Signature Analysis

Midhun Sasikumar, Ramesh Bhakthavatchalu, K. N. Sreehari, and Arjun S. Kumar

Abstract A novice scalable MBIT unit with MISR signature analysis is proposed in this project. Enhancement of MBIST architecture with signature analysis significantly improvises resolution of fault detection in memories in comparison to conventional MBIST Architecture. The proposed MBIST Algorithm is optimized by using only 14 states for 7 March algorithms, hence improvising the scalability of the MBIST without area overhead. The proposed architecture has an interface check MISR which features isolated detection of fault in memory interface and memory, hence improving resolution and accuracy. Rapid Fault check is enabled by Memory pre-check using MISR. Benchmarking for this project in terms on the accuracy, feature enhancement, scalability and the ability to detect early faults in less time are done via Random constrained verification of the developed RTL Model....

Keywords MBIST (Memory Built In Self Test) · RTL (Register Transistor Logic) · MISR (Multi Input Signature Register) · DUT (Design Under Test) · SAF (Stuck-at Fault) · HVMT (High Volume Manufacturing Testing) · FPGA (Field Programmable Gate Array) · ASIC (Application Specific Integrated Circuit) · FSM (Finite State Machine)

M. Sasikumar (✉) · R. Bhakthavatchalu · K. N. Sreehari
Department of Electronics and Communication Engineering, Amrita Vishwa Vidyapeetham, Amritapuri, India
e-mail: midhunsasikumar@outlook.in

R. Bhakthavatchalu
e-mail: ramesh.amrita@gmail.com

K. N. Sreehari
e-mail: sreehari.kn86@gmail.com

A. S. Kumar
Intel Technology Pvt Ltd, Bellandur, Bangalore, Karnataka, India
e-mail: arjun.skumar@intel.com

© Springer Nature Singapore Pte Ltd. 2021

S. N. Merchant et al. (eds.), *Advances in Signal and Data Processing*,
Lecture Notes in Electrical Engineering 703,
https://doi.org/10.1007/978-981-15-8391-9_26

1 Introduction

Industrially purposed integrated chips consists of 90% of memory units, memory units are a crucial and integral part of any VLSI hardware. But Memory units are prone to defects while manufacturing. Detection of these faults or defects are very crucial to ensure safety and proper functionality which are done on post-silicon validation. To perform post-silicon validation of memories, it is required to obtain high observability and controllability of memory. Design for Testability ensures that the hardware can be tested against post-silicon errors. For this purpose, design under test (DUT) is pre-equipped with memory built in self test (MBIST) which is a self-checking system that checks whether memory under test contains a faulty cell or not [1].

High Volume Manufacturing Testing (HVMT) of memories consist of scan test and MBIST methodologies to classify the faulty memories and faultless [2]. The memory is classified faulty once it reaches the threshold for faults detected. The MBIST provides a centralized approach and these units were driven by MBIST algorithms that aims at reduced test time.

Conventional MBIST algorithms takes exhaustive time for testing a memories and are less scalable to different algorithms. The conventional MBIST engines driven using algorithms, lacks the capability to differentiate between the faults in the memory interface and the actual memory. This paper proposed a system capable for driving multiple MBIST algorithms, hence extending scalability. Also the proposed system for memory fault checking is able to differentiate between memory interface errors and memory errors, with the help of separate MISR (multi input signature register). Enhanced for rapid fault detection were added to the proposed system by inclusion of a fast pre-check of the memory using a separate MISR. This system can act as a mid layer of enhanced fault check between the scan check and MBIST algorithm based check for HVMT, thus reducing the bucketed size of the memories to be tested against exhaustive MBIST algorithm.

Section 1 gives a brief introduction, Sect. 2 deals with the different type of faults in the memory. Section 3 gives insights on the algorithms available for MBIST. Section 4 glances through the conventional architecture and its disadvantages. In Sect. 5 the proposed new scalable MBIST is explained. Section 6 benchmarks the exhaustive test and validation done on the proposed architecture. FPGA and ASIC implemented results are depicted in Sect. 7. Future scope and conclusion is discussed in Sect. 8.

1.1 Generalized Block Diagram of a MBIST Engine

The Generalized MBIST Engine Block Diagram is shown in Fig. 1 MBIST engine generates the test patterns based on the algorithm used [2]. The MBIST engine comprises of:

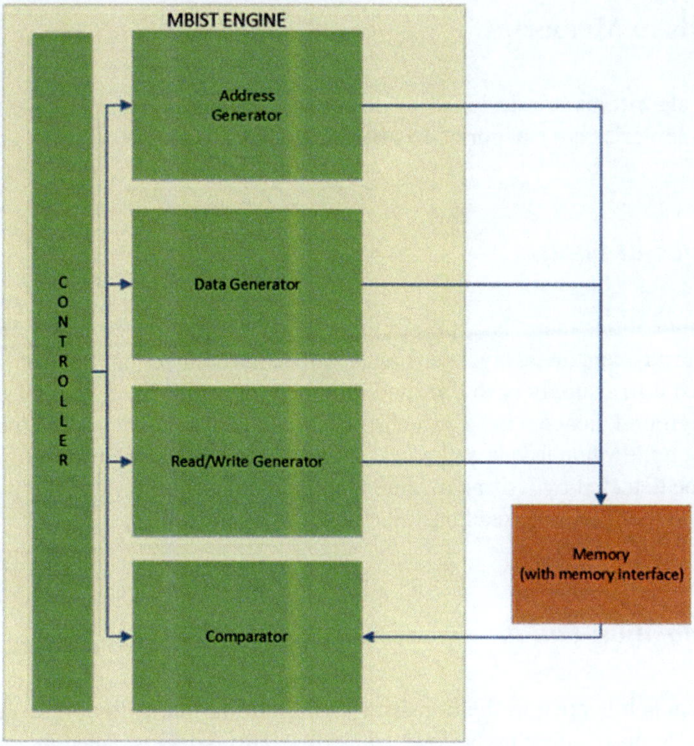

Fig. 1 MBIST Generalized block diagram

- Data Generator: Based on the control signal issued from the MBIST controller, the test pattern to be written to the memory is generated in the Data generator.
- Address Generator: The algorithm issued by the controller generates the address either by incremental fashion or decremental fashion.
- Read/Write Generator: issues the read and write command required for the test pattern to be written to the address in the memory. The read command along with the address are provided to the memory to read the already written test patterns from the memory.
- MBIST controller: as usually FSM based and controls the entire issue and entry of bus data, address and commands from the MBIST engine to the memory.
- Fault Comparator: The data read from the memory is compared against the written test sequence to identify and isolate the faults. The comparator units usually issues the fault detection output and the number of faults observed.
- Memory (DUT): The MBIST engine communicates with the design under test that is the memory. Based on the fault count obtained from the memory, the memory is then classified as faulty or faultless.

2 Faults in Memories

Different algorithms are designed to detect the various faults in the memory [3]. Faults in memories are categorized as below.

2.1 Stuck-at Faults

During fabrication, there is a possibility of a metal layer getting extended and touching an adjacent metal layer causing an unwanted connection. If the new connection formed is to a supply or to a ground rail, then the node will always be shorted to supply or ground, hence a fixed value is observed always in that node. Stuck-at-one faults fix a node value to one and stuck-at-zero fixes a node value to zero. Stuck-at fault can be detected by writing a value complementary to the value the memory is stuck-at and then trying to read back the newly written value.

2.2 Transition Faults

If any mismatch happen to the transistor sizing during fabrication, then cell might not be able to make either up or down transition. This defect is called as a transition fault. If a cell cannot make up transition, then fault can be called as up transition fault and if cell cannot make a down transition, then fault can be called as down transition fault [4]. The transition faults are also called at speed faults, as these are dynamically introduced into the memory on writing. Detection of transition fault needs the debugger to apply both up transition and down transition to the cell under consideration.

2.3 Coupling Faults

In case of a coupling fault, the cell who forces the data value to change in another cell is called the aggressor cell and the cell whose data value changes according to the aggressor cell is called as the victim cell. Detection of a coupling fault is difficult if the coupling happens with a long distance cell. Near Neighborhood coupling fault can be detected by forcing a value to the victim cell, changing the data value in the aggressor cell and then reading back the value from the victim cell.

3 Algorithms for Fault Detection

Fault detection algorithms are defined methodologies of writing data into the cell and reading back these data from the cells in an order. An algorithm has to be efficient in terms of the length (which is a measure of the time taken for the testing) and the fault coverage (types of faults detected by the algorithm) offered by the algorithm. The length of the time taken by the algorithm is measured in xN, where N is the number of bits of the memory. Large memory with long algorithms takes exhaustive time to complete the testing. A classical test algorithm is characterized as the one which is simple but has low fault coverage or the one with higher fault coverage, high complexity and slow. Algorithms of classical nature are not preferred for efficient testing. On the other hand there are march algorithms which are comparatively longer than simplest of classical algorithms but having higher fault coverage and is simpler to implement as well. The march algorithms comprises of March elements, which navigates through the rows and columns while performing required actions. Current MBIST engines popularly incorporate the march algorithms, as they can be easily implemented and has high fault coverage.

All test algorithm follow certain notations, which needs to be discussed to properly understand the algorithm.

$\uparrow$ indicates an action is to be performed in increasing order of address.

$\downarrow$ indicates an actions is to be performed in decreasing order of address.

$\updownarrow$ indicates the operation is to be performed in increasing or decreasing order of the memory address.

R_0 means a zero is to be read from the cell.

R_1 means a one is to be read from the cell.

W_0 means to write a zero into the cell.

W_1 means to write a one into the cell.

Few popular algorithms are depicted in Table 1 First three algorithms in table are

Table 1 March algorithms

Algorithm	Pattern	Complexity	Faults Detected
MATS	$(\updownarrow w_0)(\uparrow r_0, w_1)(\downarrow r_1)$	$4n$	Stuck-at
MATS+	$(\updownarrow w_0)(\uparrow r_0, w_1)(\downarrow r_1, w_0)$	$5n$	Stuck-at
MATS + +	$(\updownarrow w_0)(\uparrow r_0, w1)(\downarrow r1, w0, r0)$	$6n$	Stuck-at, Transition
MARCH C	$(\updownarrow w0)(\uparrow r0, w1)(\uparrow r1, w0)(\downarrow r_0, w_1)(\downarrow r_1, w_0)(\updownarrow r0)$	$4n$	Stuck-at, Transition, Coupling
MARCH X	$(\updownarrow w_0)(\uparrow r_0, w_1)(\downarrow r_1, w0)(\updownarrow r_0)$	$6n$	Stuck-at, Transition, Coupling
MARCH Y	$(\updownarrow w_0)(\uparrow r_0, w_1, r_1)(\downarrow r_1, w_0, r_0)(\updownarrow r_0)$	$8n$	Stuck-at, Transition, Coupling

classified as classical algorithms, while the other are march algorithms.

4 Conventional Architectures and Drawbacks

MBIST architecture are implemented either using FSM based approach or micro-code-based approach [5]. Also a PMBIST (programmable based MBIST) was developed recently. Currently available architectures for MBIST implementation are generalized in this section.

4.1 FSM Based Approach

An FSM based controller in the respective state drives the data generation, address generation and command generation unit to push out the algorithm-based test patterns to the memory interface bus. The controller in the respective state drives the data generation, address generation and command generation unit [6].This methodology of FSM driven march elements are faster compared to the microcode based approach. Although the FSM based approach is quicker, easier to control, such methods lacks scalability and consumes more area.

4.2 Micro-code-based Approach

This approach consists of storing the algorithm generated test patterns in ROMs and LUTs [7]. The pre-stored test patterns are then controlled and applied to the interface bus of the memory. Although this method has an increased scalability to accommodate any algorithm by programming of the ROM/LUTs, the area overhead is large. The micro-code based approach utilizes a storage unit for the test patterns, it is possible that the unit can itself be subject to faults and hence as considered less accurate.

4.3 Programmable MBIST

Recently developed approach PMBIST which combines the effort and advantages of MBIST and Micro-code to produce a scalable MBIST unit [8]. The MBIST controller is issued a control word which selects the different modular FSM for based on the algorithm decrypted from the control word [9]. This hybrid approach/PMBIST has the scalability to choose different algorithms but consumes a lot of area.

5 Proposed New Architecture

The proposed architecture resolves the multi algorithm and enables the ability to detect memory faults and memory interface faults separately. The architecture consists of a novel MBIST Engine coupled with 2 MISRs for the feature enhancement. The high level Architecture of the Proposed Architecture is shown in Fig. 2. The MBIST engine proposed is optimized for the minimum FSM and features 8 algorithms where 7 are march algorithms (mentioned in Table 1) including the most used March C algorithm and an added exclusive algorithm for the pre-checks for MISR. The proposed architecture outputs the fault detected in memory and memory interface along with the fault count.

5.1 Recommended Mode of Operation

The recommended mode of operation of the proposed architecture is conducted in 3 stages as shown in Fig. 3. (1) Memory Interface Check: MBIST configured in MISR algorithm mode, and MISR for interface check activated. This stage solely checks for the interface faults. (2) Rapid Memory Pre-check: MBIST configured in MISR algorithm mode, and MISR for memory pre-check activated. This stage

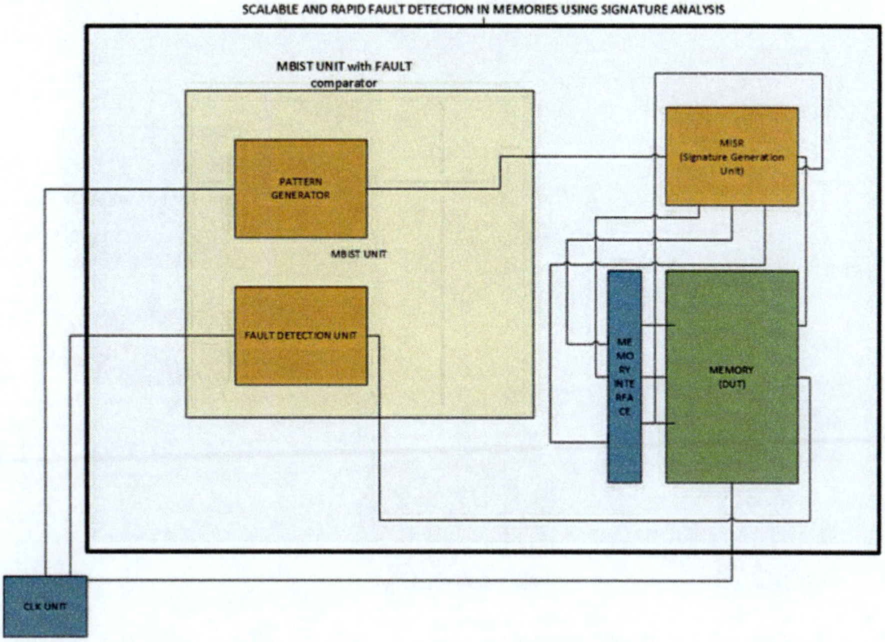

Fig. 2 Proposed Architecture HAS

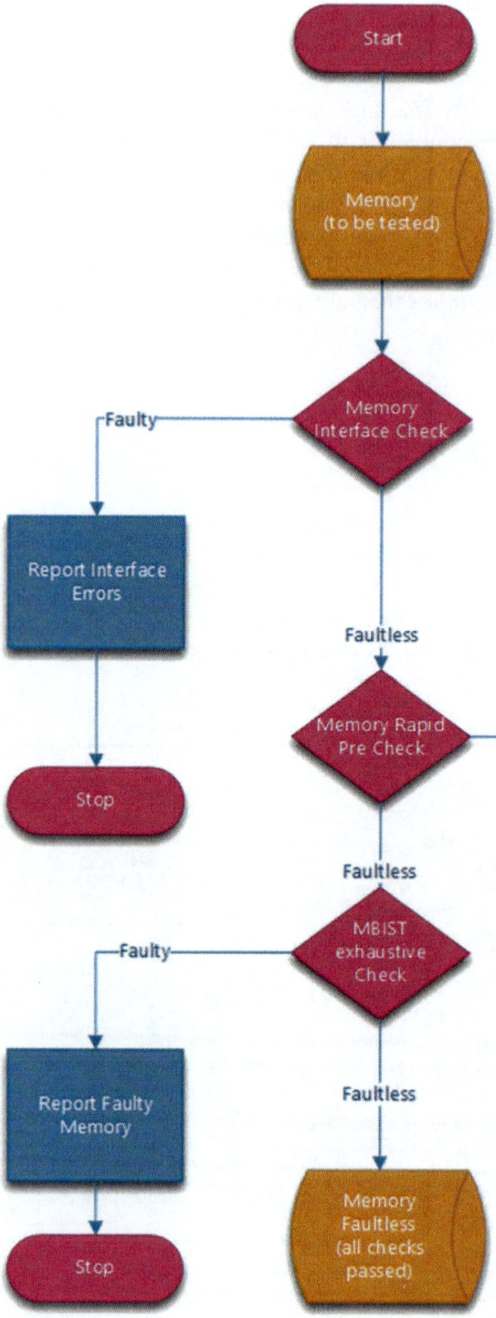

Fig. 3 Recommended Mode of Operation

does a rapid pre-check on the memory with limited accuracy. (3) MBIST exhaustive Check: MBIST configured in any march algorithm mode with both MISRs disabled. This mode activates the MBIST exhaustive testing with the march algorithm thus delivering high accuracy and precision for fault detection.

5.2 MBIST Engine Micro Architecture

The MBIST engine has a multi algorithm controller and a fault detection Unit as shown in Fig. 4. The multi algorithm FSM controller is optimized by implementing the 7 algorithm by using sub level modularity. Modularity is enabled by 14 FSM which interconnects or knits together on the Algorithm selected. The 14 FSM states are of two type, states where the operation proceeds in up or down direction of address -write zero up, write zero down, read zero up, read zero down, write one up, write one down, read one up, read one down, and operation which simple performs an operation—write zero, write one, read one, read zero. Apart from this there is idle state and a Stop state. FSM for March C algorithm [10]using the 14 states are shown in Fig. 5.

Scalable MBIST The scalable MBIST engine consists of the sequential FSM state. Pre-state unit, Algorithm protection Unit, Address Generation Unit, Direction Unit,

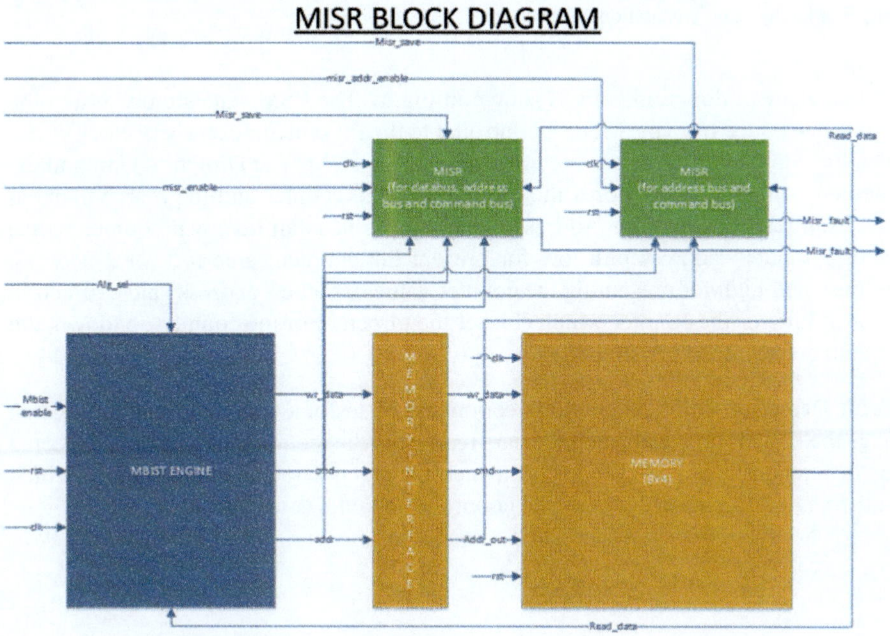

Fig. 4 Proposed MBIST Engine and MISR—MAS

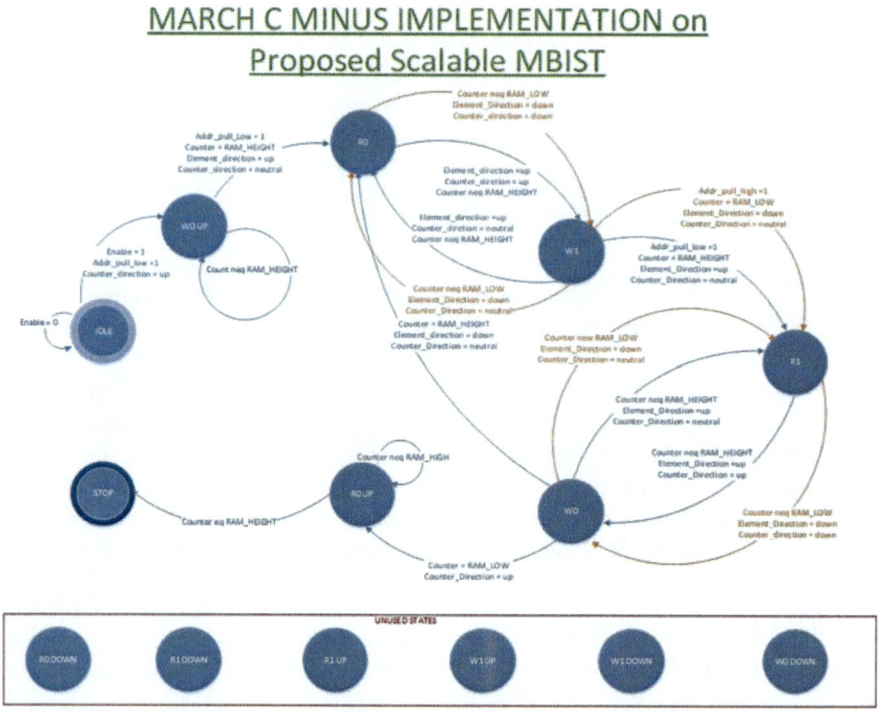

Fig. 5 March C Implementation in FSM

and an element direction Unit as shown in Fig. 6. The FSM state simply assign the Next state issued from the MBIST controller to the Present State on every clock cycle. The Pre-State Unit stores the previous state of the FSM. The Direction Unit and the Element Direction Unit stores the direction of the counter and the march element direction respectively. The Address Generation unit is an up/down counter which takes 3 inputs—address pull low for present the counter value to lowest address, address pull high for presenting the counter value to highest address, and a direction control line for the counter which can set the direction of the counter up, down and in a no count freeze mode neutral.

Fault Detection Unit The fault detection comprises of a combinational unit which flags the fault as detected when the data read from the memory and the data expected are different. The fault flag controls the increment of a sequential fault counter, which at the end of execution contains the count of the faults detected.

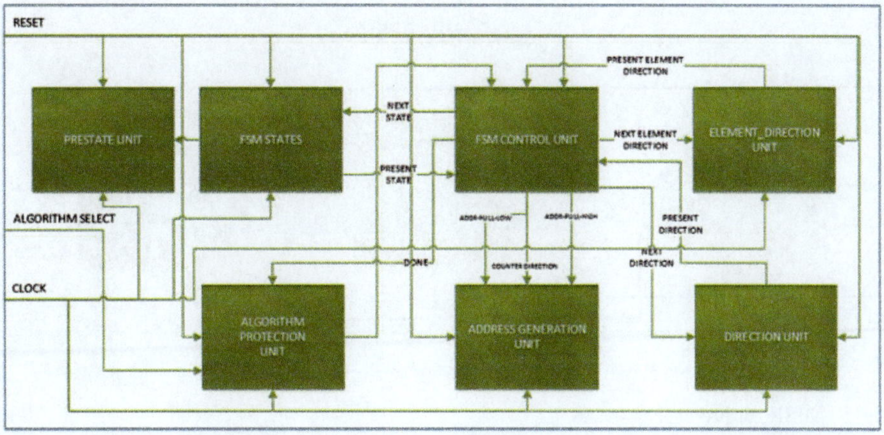

Fig. 6 Scalable MBIST MAS

5.3 *Memory Interface Check MISR*

The memory interface check MISR enabled the functionality to detect the Memory interface for faults. The interface unit Micro level Architecture is shown in Fig. 7 and consists of a PUT function which concatenates the address bus and the control bus, the MIR which has a characteristic polynomial of x4 +x3 +x0, and a MISR comparator which detects the fault. The PUT functions generates the concatenated output and the PUT enable for the MISR, where the MISR computes the signature. The MISR Comparator has 2 mode of operation—Save mode where the generated signature is stored in the comparator as the golden signature and a compare mode where the generated signature is compared with the golden signature for fault detection.

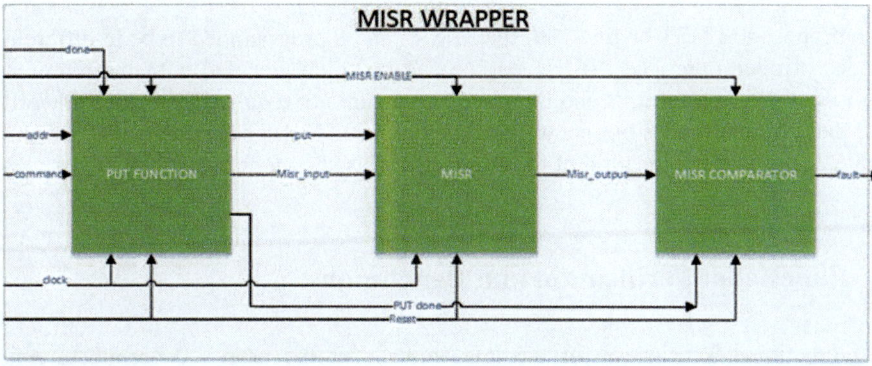

Fig. 7 MISR for Interface check—MAS

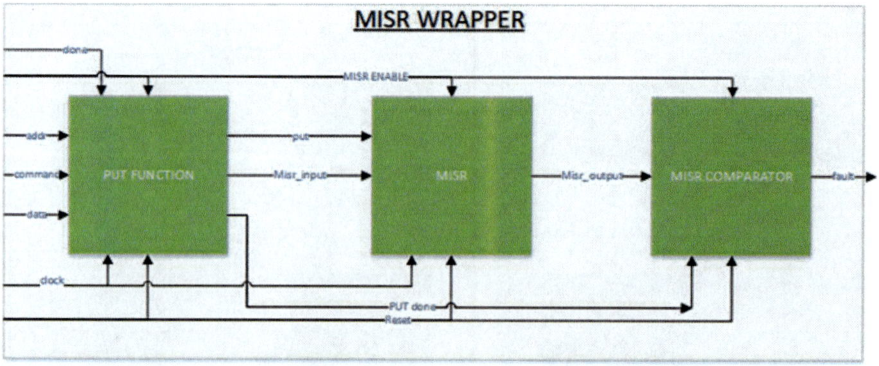

Fig. 8 MISR for pre-check—MAS

5.4 Memory Pre-Check MISR

Similar to the Memory Interface Check MISR, the Memory Pre-Check MISR follows a similar architecture as shown in Fig. 8. The PUT function for the memory pre-check MISR has a delay mode where the address and the control bus is delayed to be in sync with the read data, and a non delay mode where the control bus, address and the data bus. The MISR and the MISR comparator has a higher bus width of the data bus size + address bus size + the control bus.

The characteristic polynomial for the Memory Pre-Check MISR is x14 + x13 + x12 + x2 + 1.

5.5 The Control Word

The proposed MBIST engine with the MISR can be programmed to be in different modes of operation. The MBIST engine can be solely enabled with the required algorithm if the pre-checks and interface checks are not required. The control word and the different modes of operation of the proposed MBIST Engine is given in the Table 2. Bit wise explanation of the control word is shown in Fig. 9.

6 Functional Verification and Benchmark

The functional Verification of the RTL model for the proposed architecture is performed by subjecting the design under test to a faultless test and a faulty test. The Faulty Test injected the below faults into the memory.

Table 2 Control word

Algorithm	Pattern	Control word
MATS (with MISR disabled)	$(\Updownarrow w_0)(\Uparrow r_0, w_1)(\Downarrow r_1)$	8′h86
MATS+ (with MISR disabled)	$(\Updownarrow w_0)(\Uparrow r_0, w_1)(\Downarrow r_1, w_0)$	8′h8e
MATS++ (with MISR disabled)	$(\Updownarrow w_0)(\Uparrow r_0, w_1)(\Downarrow r_1, w_0, r_0)$	8′h96
MARCH C (with MISR disabled)	$(\Updownarrow w_0)(\Uparrow r_0, w_1)(\Uparrow r_1, w_0)(\Downarrow r_0, w_1)(\Downarrow r_1, w_0)(\Updownarrow r_0)$	8′ha6
MARCH X (with MISR disabled)	$(\Updownarrow w_0)(\Uparrow r_0, w_1)(\Downarrow r_1, w_0)(\Updownarrow r_0)$	8′hb6
MARCH Y (with MISR disabled)	$(\Updownarrow w_0)(\Uparrow r_0, w_1, r_1)(\Downarrow r_1, w_0, r_0)(\Updownarrow r_0)$	8′hbe
Rapid Check (with MISR enabled)	$(\Uparrow w_0 w_1)(\Downarrow r_0, r_1)$	8′hbe

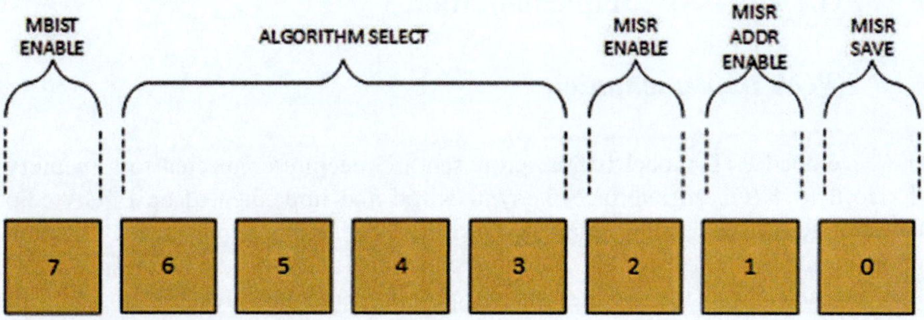

Fig. 9 Control word for the proposed architecture

1. Stuck-at Faults: The memory connected to the proposed architecture is probed randomly for data and both stuck-at-one as well as stuck-at-zero faults are injected in a constrained manner.
2. Transition and Coupling Faults: The memory is probed for transitional changes and both transition as well as coupling faults are injected in a constrained random manner.
3. The Memory interface is injected with faults randomly.

All the random faults injected are logged via a detailed scoreboard test, which records the injected and detected faults. The accuracy obtained for different algorithms are given in Table 3.

Table 3 Accuracy obtained on benchmark test

Algorithm	Pattern	Accuracy (in %)
MATS	$(\Updownarrow w_0)(\uparrow r_0, w_1)(\downarrow r_1)$	55
MATS+	$(\Updownarrow w_0)(\uparrow r_0, w_1)(\downarrow r_1, w_0)$	65
MATS++	$(\Updownarrow w_0)(\uparrow r_0, w_1)(\downarrow r_1, w_0, r_0)$	68
MARCH C	$(\Updownarrow w_0)(\uparrow r_0, w_1)(\uparrow r_1, w_0)(\downarrow r_0, w_1)(\downarrow r_1, w_0)(\Updownarrow r_0)$	93%
MARCH X	$(\Updownarrow w_0)(\uparrow r_0, w_1)(\downarrow r_1, w_0)(\Updownarrow r_0)$	88
MARCH Y	$(\Updownarrow w_0)(\uparrow r_0, w_1, r_1)(\downarrow r_1, w_0, r_0)(\Updownarrow r_0)$	90
Rapid Check	$(\uparrow w_0 w_1)(\downarrow r_0, r_1)$	50

7 FPGA and ASIC Implementation

7.1 *FPGA Implementation*

The developed RTL model for the proposed architecture connected to a memory of size 8 × 4 (via an interface) is synthesized and implemented on FPGA chip "xa7s50ftgb 196-2I" using Xilinx Vivado 2017.4. Timing report and Utilization report for the implemented RTL model is shown in Tables 4 and 5 and respectively.

It can seen from the reports that the proposed architecture uses less area as compared to conventional architectures.

Table 4 Utilization report—FPGA implementation

Resource	Utilization	Available	Utilization%
LUT	220	32,600	0.67
FF	123	65,200	0.19
IO	31	100	310

Table 5 Timing report—FPGA implementation

Setup		Hold	
Worst negative slack	5.750 ns	Worst negative Slack	0.154 ns
Total negative slack	0	Total negative slack	0
Number of failing endpoints	0	Number of failing endpoints	0
Total number of Endpoints	62	Total number of endpoints	62

Table 6 Gate Area report—ASIC

Type	Instances	Area	Area %
Sequential	116	2447.815	53.3
Inverter	52	118.076	2.6
Buffer	1	4.541	0.1
Tristate	37	336.064	7.3
Logic	351	1686.373	36.7
Total	557	4592.869	100.0

7.2 ASIC Implementation

Using GDK90nm Technology the ASIC layout for the proposed architecture connected to a memory size of 8 × 4 via a memory interface is generated using Cadence Virtuoso Digital Implem XL 14.2. The RTL has a timing closure of 7600 ps and can work at the max speed of 131 MHz. The Area report, Power report and Gate area report of the design are shown in Tables 6, 7 and 8 respectively. The total power utilization showed only 299,520.038 nW usage comprising of both static and dynamic dissipation, which is less as compared to conventional FSM based MBIST and PMBIST [11].

Table 7 Area utilization report—ASIC

Instance	Cells	Cell area	Net area	Total area
mbist wrapper top	557	4593	0	4593
mbist fsm instance	301	1786	0	1786
mem instance	76	1086	0	1086
misr wrapper instance	46	522	0	522
misr instance	38	473	0	473
misr compartor instance	8	49	0	49
put instance	56	515	0	515
fault det instance	34	255	0	255
misr wrapper addr instance	22	217	0	217
misr addr instance	18	188	0	188
misr addr compartor instance	4	29	0	29
put addr instance	22	211	0	211

Table 8 Power utilization report—ASIC

Instance	Cells	Leakage power (nW)	Dynamic power (nW)	Total power (nW)
mbist wrapper top	557	25,047.020	274,473.018	299,520.038
mbist fsm instance	301	9009.967	33,892.334	42,902.301
mem instance	76	6254.395	79,194.384	85,448.778
misr wrapper instance	46	3218.685	68,026.461	71,245.145
misr instance	38	3007.073	67,161.902	70,168.975
misr compartor instance	8	211.612	864.558	1076.171
put instance	56	2956.721	28,579.120	31,535.841
fault det instance	34	1282.711	17,001.158	18,283.870
misr wrapp.addr instance	22	1244.149	27,299.848	28,543.997
misr addr instance	18	1100.200	27,010.688	28,110.888
misr addr.rtor instance	4	143.949	289.160	433.109
put addr instance	22	1080.392	8832.674	9913.066

8 Conclusion and Future Scope

The proposed Novel Architecture for the MBIST engine has the highest fault coverage as 93%. It also features enhancements such as ability to separately detect and isolate faults from the memory and the memory interface.

Also the proposed architecture can be scaled in future to accommodate other march algorithm by following the same optimization techniques. Further Enhancements done on the conventional MBIST [13], 14] to add a rapid pre-check stage was able to detect the fault in memory quickly. This new stage added will significantly reduce the bin size of memories to be tested against MBIST exhaustive algorithms in HVMT. Pre-check also acts a second layer of check for memory faults in conjunction with the MBIST march algorithms, thus improvising the accuracy of the proposed MBIST engine. The MISRs are fully scalable except the polynomial which needs to be altered according to the memory size, each MISR features internal methodologies for signature compare and save. Each MISR has the ability to save the latest generated signature, thus allowing the user capability to choose for the algorithm against MISR signature analysis. Also as both the MISRs in the proposed architecture cover the entire bus architecture of the MBIST to Memory Interface and Memory Interface to Memory, the proposed architecture itself is self verifiable against bus faults. Both synthesis and FPGA implementation had ascertained a low power and area. The Power usage can be further reduced by using low power techniques such as UPF oriented modeling and LFSR based low power address generators [15]. The proposed architecture is hence a scalable and accurate, rapid fault detection MBIST engine compared to the conventional architectures.

Reference

1. Ravinder P, Uma Rani. Design and implementation of built-in-self test and repair Int J Eng Res Appl (IJERA) 1(3):778–785
2. Kang W, Lee C, Lim H, Kang S (2016) Optimized built-in self-repair for multiple memories. In: IEEE Trans Very Large Scale Integr (VLSI) Syst 24(6)
3. Bui TQ, Pham LD, Nguyen HM, Nguyen VT, Le TC, Hoang T (2016) An effective architecture of memory built-in self-test for wide range of SRAM, In: International conference on advanced computing and applications 2016
4. Dr. Bhakthavatchalu R, Dr. Nirmala Devi M, Krishnan S (2014) Reconfigurable logic built in self-test technique for SoC applications. In: International conference on communication and computing, ICC 2014, vol 3. Elsevier, Bangalore, India, pp 16–23 (2014)
5. Devika KN, Bhakthavatchalu R (2017) Design of efficient programmable test- per-scan logic BIST modules. In: 2017 International conference on microelectronic devices, circuits and systems (ICMDCS), Vellore, pp 1–6
6. Singh NI, Joshi PV (2018) A brief review for semiconductor memory testing based on BIST techniques. Int J Eng Technol 7(3.1):98–1007
7. Suresh Kumar V, Manimegalai R (2015) Efficient memory built in self test address generator implementation. J Appl Eng Res **10**(7):16797–16813
8. Noor NQM, Saparon A, Yusof Y (2009) An overview of microcode-based and FSM based programmable memory built-in self test (MBIST) controller for coupling fault detection. In: IEEE symposium on industrial electronics and applications (ISIEA 2009), October 4–6, 2009, Kuala Lumpur, Malaysia
9. Lakshmi HR, Varchaswini R, Shirur YJM (2014) Implementation of FSM-MBIST and design of hybrid MBIST for memory cluster in asynchronous SoC. Int J Comput Appl Technol Res 3(4):216–220. ISSN:23198656
10. Wang C-W, Wu C-F, Li J-F, Wu C-W, Teng T, Chiu K, Lin H-P (2002) A built-in self-test scheme with diagnostics support for embedded SRAM. J Electron Testing 18(6):637–647
11. Singh B, Narang SB, Khosla A (2010) Modeling and Simulation of efficient march algorithm for memory testing. In: International conference on contemporary computing IC3 2010: contemporary computing, pp 96–107
12. Dr. Padma Priya (2013) High speed FSM-based programmable memory built-in self-test (MBIST) controller. Int J Comput Sci Mobile Comput IJCSMC 2(2):46–52
13. Manikandan B, Anbuarasan K (2012) Power optimized address generator for MBIST. Int J Adv Sci Res Technol 3(2)
14. Maneshinde N, Hegade P, Mittal R, Palecha N, Suma MS (2016) Programmable FSM based built-in-self-test for Memory. In: IEEE international conference on recent trends in electronics information communication technology, May 20–21, 2016, India
15. Ivanyuk AA, Yarmolik VN (2008) A new approach to the design of built-in internal memory self-testing devices. Autom Control Comput Sci 42(4):169–174
16. Murali Krishna K, Sailaja M (2014) Low power memory built in self test address generator using clock controlled linear feedback shift registers. J Electron Testing 30(1):77–85

Analyzing Vocal Tract Parameters of Speech

Sharada Vikram Chougule

Abstract Speech sounds produced by human depends on movement of various articulators. Dimensions and shape of the various elements of speech production organs also have impact on nature of speech produced. Vocal tract plays a major role like characteristics of linear time invariant system. In this paper, analysis of vocal tract parameters, in terms of resonances of vocal tract, also referred as formants is done. Of the different speech sounds, vowels carry most significant clues of speech. Despite specific properties of different vowels, there is much variability of vowel characteristics among speakers. It gives characteristics of vocal tract related to acoustic resonances. At large, first four formants are useful to categorize vowel sounds. The variation in formants for same speech sounds is a challenge to speech recognition algorithms in which vowel spectral characteristics are assumed to be invariant among speakers. But the same variability of formants among speaker is useful in speaker recognition. The method used for formant estimation is based on designing all-zero filters to track the formants and voicing detection-based formant extraction filters to estimate the first four formants. Vocal tract parameters in terms of formants are analyzed using twelve vowel sounds from different speakers. From the experimental analysis, it is observed that the first formant specifically represents the pertinent characteristics of vowel speech, whereas there is very little consistency of higher-order formants for same speech sounds by different speakers, indicating the impact of physiological nature as well as behavioral aspects of individual on nature of speech produced.

Keywords Resonances · Formants · Speech recognition · Speaker recognition

1 Introduction

Speech signal is a result of variations in articulatory movements and is prone to include variability such as phonetic contents and distressing traits, characteristics

S. V. Chougule (✉)
Finolex Academy of Management and Technology, Ratnagiri, India
e-mail: shardavchougule@gmail.com

© Springer Nature Singapore Pte Ltd. 2021

S. N. Merchant et al. (eds.), *Advances in Signal and Data Processing*,
Lecture Notes in Electrical Engineering 703,
https://doi.org/10.1007/978-981-15-8391-9_27

of individual human speech production structure, and sometimes behavioral state of speaker while speaking [1]. The wider variability in speech signal of individual's speech leads to automatic speech as well as speaker recognition a challenging. The physiology of human speech production system is the fundamental aspect of characterizing speech sounds. From practical perspective, human speech production starts from vocal cords (vocal folds) and end at mouth (lips) or nose.

Practically, the voice production system can be modeled as connected auditory pipe having some peaks (called as formants) as well as valleys generated based on nature of speech. Formants are basically the perception characteristics of vowels, in which concentration of acoustic energy is observed at certain frequencies, which are called as first formant, second formant, and so on. Formants represent characteristics of speech sound, which involves large dynamics because of physiological as well as behavioral aspects of a person while speaking.

Research have been carried out based on use of formant frequencies for automatic recognition of speech. The main approaches adopted in the literature are linear prediction [1–3] investigation, and analysis of speech using Fourier spectrum [4], and using features such as peaks of homomorphically smoothed cepstrum [5]. Vowels which typically have voiced characteristics are the most useful speech sounds in reliable formant estimation [6].

In this paper, analysis of formant estimation model is investigated for vowel sounds. The algorithm proposed is to track the most prominent formants considering the variabilities in speech during speech production. Following section discusses the methodology to track and estimate the formants from different speech sound.

2 Formant Estimation Algorithm

Input analog speech signal is converted in discrete form using a sampling frequency of 8 kHz and framed using 20 ms Hamming window. Pre-emphasis is performed using Butterworth IIR high pass filter. Pre-emphasis helps to reduce spectral tilt and improves spectral flattening providing more gain for high-frequency components. Further, this pre-emphasized signal is passed through Hilbert transform (all-pass filter) to create an analytic signal from a real signal. A set of adaptive FIR (all-zero) bandpass filters with linear phase characteristics is designed and cascaded with formant filter. Before estimating individual formant, speech signal is filtered out using a set of bandpass filters (filter bank). The most recent formant estimates are used to update the magnitude response of filters. This allows tracking of individual formant frequency over time, and in suppression of nearby formants and intrusion of surrounding noise.

The center frequency of these bandpass filters are first formant (F1):0.7 kHz, second formant (F2): 1.5 kHz, third formant (F3): 2.2 kHz, and fourth formant (F4):3 kHz, respectively. These four formants are spectrally separated using the adaptive filter bank. To isolate pitch frequency (F0) from the first formant (F1), additional zero is placed at F1 filter transfer function. The Hilbert transformed signal gives complex-valued filter coefficients, which help in designing filters with normalized gain and zero phase characteristics at the center frequency of each filter.

The kth all zero formant filter transfer function for $k = 2, 3, 4$ is given by [7]:

$$H_{Fk}(z,n) = k_K(n,z) \prod_{l=1, l \neq k}^{4} 1 - r_z e^{-j2\pi Fl(n-1)} z^{-1} \tag{1}$$

Here, $r_z = 0.98$. Above equation of filter transfer function ensures minimum response of formant filters except for the kth formant. The term $k_k(n, z)$ ensures normalized magnitude response and zero phase characteristics of kth estimated frequency component.

$$k_K(n) = \frac{1}{\prod_{l=1, l \neq k}^{4} 1 - r_z e^{-j2\pi Fl(n-1) - Fk(n-1)}} \tag{2}$$

An supplementary zero is added in the transfer function of first formant filter, with zero at pitch frequency at 200 Hz. This zero is to prevent interference of pitch frequency to first formant. Thus, transfer function of first formant frequency filter is given by:

$$H_{Fk}(z,n) = k_1(n) \prod_{l=0, l \neq 1}^{4} 1 - r_z e^{-j2\pi Fl(n-1)} z^{-1} \tag{3}$$

where

$$k_1(n) = \frac{1}{\prod_{l=0, l \neq 1}^{4} 1 - r_z e^{-j2\pi Fl(n-1) - F1(n-1)}} \tag{4}$$

The signal filtered through all-zero FIR filter is further passed through a set of first-order IIR filter. The pole of each of these filters is updated based on formant frequency estimated in previous frame of that filter. The transfer function of kth single-pole IIR filter at time instant n is as below:

$$H(n,z) = \frac{1 - r_p}{1 - r_p e^{j2\pi Fk(n-1)} z^{-1}} \tag{5}$$

Here, $r_p = 0.9$ defines radius of pole, which decides the magnitude/gain at a formant frequency, and estimation of kth formant filter at index $(n - 1)$ is given by $F_K(n - 1)$. Equation (5) gives the design of four formant filters having complex-valued coefficients. Thus, these filters divides the spectrum of Hilbert transformed speech signal into four spectrally separated regions and estimate the formant frequencies based on updated filter coefficients [8].

The analytic speech contains all types of phonetic contents, out of which voiced part is most useful in detecting speech specific formants [9, 10]. A voicing detector is used to distinguish voiced and unvoiced frame, as formants are the characteristics related to voicing properties of the speech, generally of vowels. A simple zero cross-rate detector (ZCR) is used to classify voiced and unvoiced part of the speech. A simple measure of zero cross-rate is the count that the speech signal crosses zero (reference) amplitude. As a general observation, unvoiced or noisy speech is having more ZCR than voiced speech. It is calculated using signum function as:

$$z_n = \sum_{n=-\infty}^{\infty} |\mathrm{sgn}[(s(n)] - \mathrm{sgn}[s(n-1)].w(m-n) \tag{6}$$

nn

Here, sgn() is signum function which is 1 for $n \geq 0$ and -1 for $n < 0$, and $w(n)$ is window function of N samples. Thus, the formant frequencies are estimated over each speech frame based on decision of voicing detector. This reduces the redundant estimation of formants during unvoiced or silence part of speech.

3 Results and Discussion

Figure 1 shows the narrowband spectrogram of two different women speech samples of vowel 'ae' extracted from the formant estimation algorithm discussed in Sect. 2. It is observed that variation of first formant is almost similar over the entire time duration, and the position (frequency) of higher formants is different for the same speech sound. Similar analysis is carried of for 12 vowels sounds from speech samples of five female speakers.

The results in Fig. 2 show the analysis plots of first four formants of 12 different vowel sounds and its variation among five female speakers notes as W1 to W5.

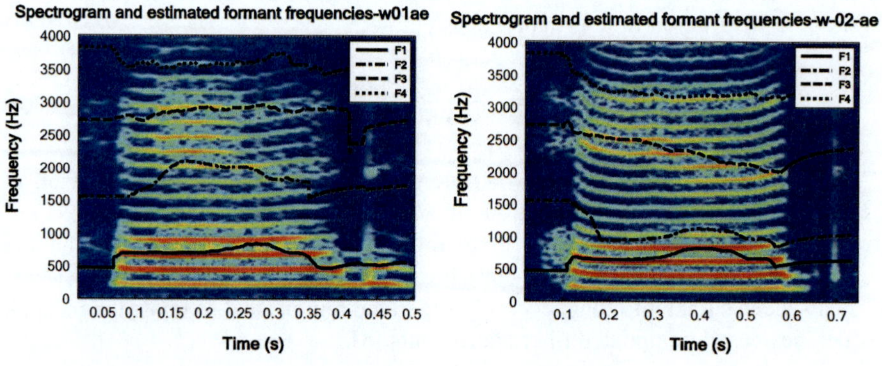

Fig. 1 Narrowband spectrogram of two female speakers for vowel sound 'ae'

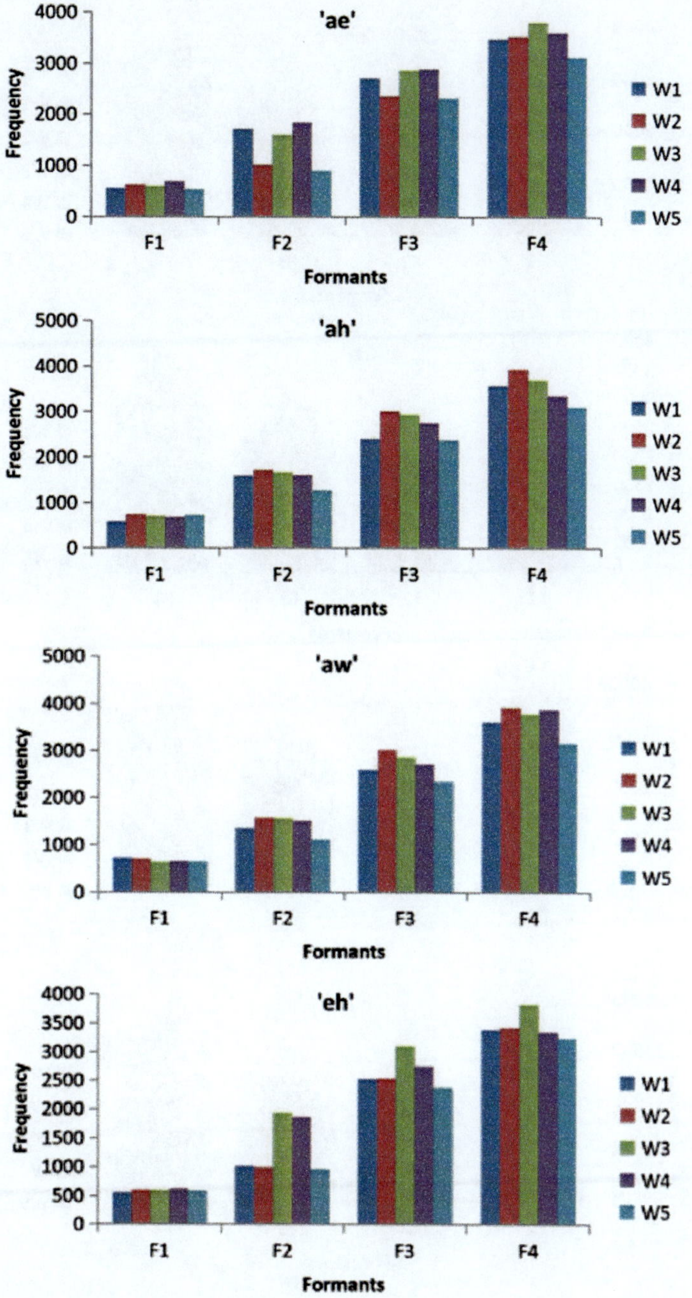

Fig. 2 Formant analysis of vowel sounds of five female speakers

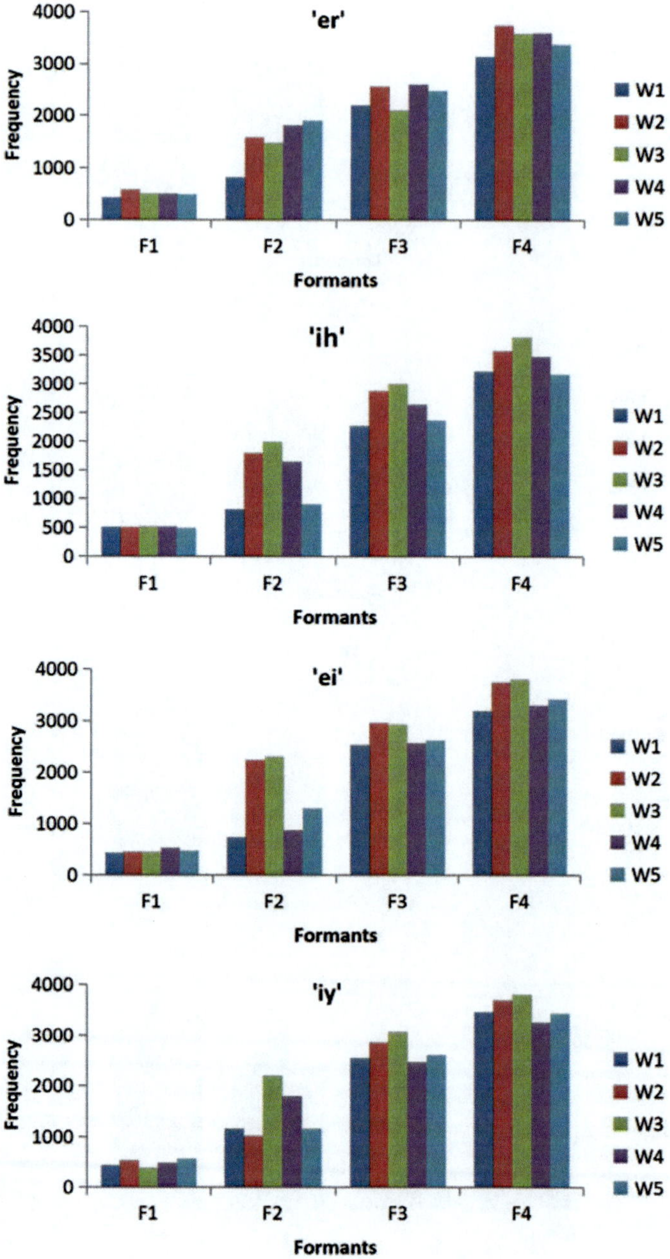

Fig. 2 (continued)

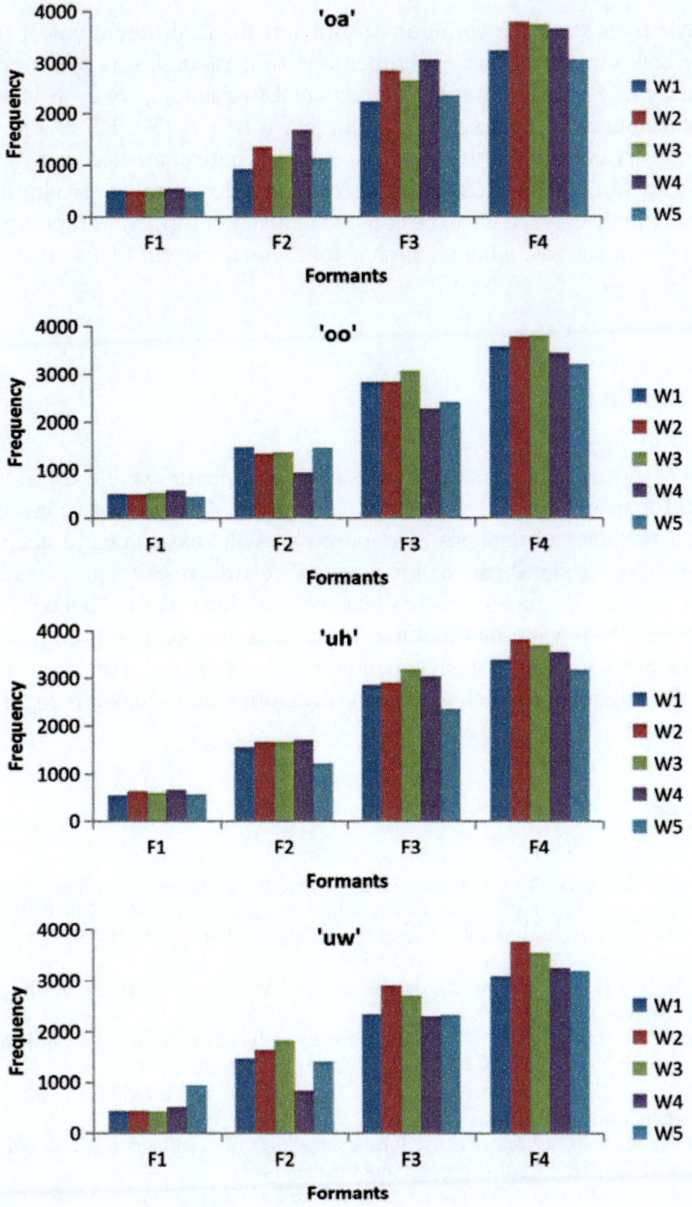

Fig. 2 (continued)

The above plots show the variation of formants for 12 different vowel sounds by five female speakers. From the experimental results plotted, it is observed that the first formant, i.e., F1 (excluding the fundamental frequency), for each vowel sound is almost constant over the speakers. All higher formants (F2, F3, F4) for a vowel single sound vary considerably, over the speakers. These characteristics of formants can be used for formant as a feature for speech as well as speaker recognition. Thus, from experimental analysis, it can be concluded that first formant is most appropriate feature of speech sounds, whereas higher formants represent the characteristics of speaker.

4 Conclusion

In this paper, formant analysis of vowel sounds is done to explore the significance of formants for speech as well as speaker recognition. A set of twelve vowel sounds is used for experimental analysis. The purpose of the experimental analysis is to study and explore the significance of formants in relation to characteristics of speech sounds and speakers. As an initial study, experimentation is carried out on five female speech samples. Using the methodology described, the work will be extended for continuous speech, for a large speaker database and in a variety of real-world dynamic conditions for applications such as speech recognition and speaker recognition.

References

1. Rabiner LR, Juang BH (1993) Fundamentals of speech recognition. Prentice-Hall (1993)
2. Deller J, Hansen J, Proakis J (2000) Discrete-time processing of speech signals. IEEE Press
3. Quatieri TF (2007) Discrete-time speech signal processing: principles and practice, Third impression. Pearson Education
4. Welling L, Ney H (1996) A model for efficient formant estimation. In: Proceedings of IEEE ICASSP, Atlanta, pp 797–800
5. Holmes JN, Holmes WJ (1996) The use of formants as acoustic features for automatic speech recognition. In: Proceedings of IOA, vol 18, part 9, pp 275–282
6. O'Shaughnessy LDD. Speech processing, a dynamic and optimization- oriented approach. Marcel Dekker Inc. New York, NY, USA
7. Mustafa K, Bruce IC (2006) Robust formant tracking for continuous speech with speaker variability. IEEE Trans Audio Speech Lang Process 14(2)
8. Ververidis D, Kotropoulos C (2006) Emotional speech recognition: resources, features, and methods. Speech Commun 48:1162–1181
9. Craciu A, Paulus J, Sevkin G, Backstrom T (2017) Modeling formant dynamics in speech spectral envelopes. In: 25th European signal processing conference (EUSIPCO)
10. Dey S, Alam MA (2018) Formant based bangla vowel perceptual space classification using support vector machine and K-nearest neighbor method. In: 21st International conference of computer and information technology (ICCIT)
11. Hamzenejadi S, Yousef SA, Goki H. Extraction of speech pitch and formant frequencies using discrete wavelet transform. In: 2019 7th Iranian joint congress on fuzzy and intelligent systems (CFIS)

Effect on Quality of 3D Model of Plant with Change in Number and Resolution of Images Used: An Investigation

Abhipray Paturkar, Gourab Sen Gupta, and Donald Bailey

Abstract 3D reconstruction of plants is a challenging task. It is essential in non-destructive plant growth monitoring systems and important to reconstruct plant features precisely. Two parameters which critically affect quality of the 3D model are change in the number of input images and the resolution of the images. In this paper, the effect on 3D models of these parameters is analysed. This analysis enables an appropriate number of input images and resolution of the images to be determined for a precise 3D model. To validate the results, the plant stem height and number of leaves extracted from reconstructed 3D models were compared with ground truth values. We used descriptive statistics for validation and achieved high correlation between extracted and ground truth values.

Keywords 3D reconstruction · Structure-from-motion · Feature extraction · Feature matching · Plant phenotying

1 Introduction

3D models provide information about plant status like plant growth or plant disease for agricultural management [1]. 3D plant models can be used to estimate plant features or parameters to avoid the subjective biases associated with human evaluation [2]. Plant features such as stem height, number of leaves, leaf area and so on can be extracted accurately from the 3D model which can be extremely important in the decision making process. These 3D models can characterize plants with complex architecture, supplying important data to plant breeding plans that is inevitably essential for altering features related to plant stress, shape or agricultural management [3]. 3D modeling makes data, such as plant growth monitoring or treatment, available to farmers or growers. In addition, 3D models can be helpful to farmers in yield estimation, disease detection, weed and crop discrimination, and to describe

A. Paturkar (✉) · G. S. Gupta · D. Bailey
School of Food and Advanced Technology, Massey University, Palmerston North, New Zealand
e-mail: A.Paturkar@massey.ac.nz

© Springer Nature Singapore Pte Ltd. 2021

S. N. Merchant et al. (eds.), *Advances in Signal and Data Processing*,
Lecture Notes in Electrical Engineering 703,
https://doi.org/10.1007/978-981-15-8391-9_28

plant quality. Even though 3D modeling provides the details of plant structure, it still requires technological advancements for capturing images and extracting plant features.

The majority of modeling techniques are based on 2D images, such as hyperspectral and thermal imaging [4]. Reconstruction in 3D is rapidly getting attention and has recently been the focus of much research. Two main classes of 3D measurement are active and passive techniques. LiDAR [5] and structured light [6] are active sensing methods which use their own source of illumination for sensing. In comparison, passive sensing methods which use radiation in the scene (illumination from the sun) include, stereo vision [7] structure-from-motion [8] and many more [4]. Most passive sensing methods use one or two cameras for sensing which makes them economical compared to active sensing methods.

LiDAR measures the distance from the sensor to each part of the object by measuring the time it takes for a laser pulse to return to the sensor. Paulus et al. [9] presented reliability of LiDAR to obtain an accurate 3D point cloud of a plant but the model gave limited information about surface area because of poor LiDAR resolution. Kaminuma et al. [10] successfully reconstructed a plant model in 3D which represented leaves as a polygonal mesh and used this 3D model for plant feature extraction. Even though LiDAR has performed well in reconstructing plants in 3D, it has several disadvantages, such as, high warm-up time, poor resolution, being very costly and needs numerous captures to handle occlusion [4]. Another method to get depth information is to use structured light (one example is the Kinect sensor). This sensor projects a known pattern on the object and deformation of this pattern allows the vision system to infer surface and depth information of the object [11]. Chéné et al. [12] used a Kinect sensor for 3D plant reconstruction and achieved good results but this struggled to perform in an outdoor environment because of poor contrast of the projected pattern due to bright sunlight. In addition, they did not consider change in any parameters for reconstruction process such as, the number of input images or change in image resolution. Baumberg et al. [13] also presented a 3D reconstruction system for a controlled environment. They used a mesh processing approach for generating 3D models of cotton plants and did not consider change in the reconstruction parameters.

In contrast, passive sensing methods use radiation in the scene, which allows this method to work in outdoor conditions efficiently. Stereo vision has several parameters which the user has to keep in mind, such as distance between two cameras, distance between camera and the plant, and the focal length of the cameras [14]. This can make stereo vision systems difficult to set up and use as the user has to adjust these parameters. Takizawa et al. [15] used a stereo vision system to generate a 3D model of plants and extracted plant shape and height data. There's a still scope for experimentation in this system by considering a few more plant features like number of leaves, leaf area, leaf shape. Ivanov et al. [16] extracted plant structure information using a stereo vision system in outdoor conditions. They captured plant images from different angles and views. They generated a 3D model by considering all the camera views for precise reconstruction.

In structure-from-motion (SfM), set of 2D images are used to create a 3D model of an object. Unlike stereo vision, SfM uses a single camera. A 3D model is generated using SfM by capturing a series of images as the camera is moved around the plant. SfM calculates the distance between camera and plant by itself and it does not require any prior camera calibration, this makes SfM easier to use [4]. Jay et al. [17] presented a system to generate a 3D model of plants in a row and then extract the parameters from the 3D model. The camera was translated along the row which gives limited information about the plant, plant height and leaf area information because of occlusion of leaves. Quan et al. [18, 19] used SfM to reconstruct a tree in 3D. It was a semi-automatic system in which user can select which images are used for the reconstruction. However, the authors did not consider the change in number of input images for the generating 3D model. In this research, we will be using structure-from-motion method for reconstruction because of its easier implementation.

1.1 3D Recovery of Plants

Precise reconstruction of an object from multiple images is an on-going problem. In the past few years advancements to these reconstruction techniques have been made. However, these techniques have been applied to simpler objects e.g. human face, vases, buildings or some round objects. Objects with more complex architecture such as, plants, pose more challenges and issues to precisely reconstruct in 3D. Complex architectures are subject to significant occlusion, where a leaf is not visible from current view, and the parallax effect, where plant appearance differs from view to view, making reconstruction of plants more difficult than convex objects. Reconstruction of plants is difficult because of high self-occlusion, presence of many leaves, shiny surfaces, and texture-loss in some camera views making feature matching more difficult. In addition, plants are very sensitive to changes in the environment, right from small changes like foliage reorganisation to life long growth patterns. Consequently, plants have complex architecture which changes over time which makes it complicated to reconstruct especially by fixed, standard camera phenotyping systems. To reduce this complication and to contribute to the solution of this problem we proposed an easy and cost-effective plant 3D reconstruction system in our previous work [1]. However, we determined that there are some more parameters which need attention to reconstruct a precise 3D model such as the number of input images used for reconstruction and resolution of the images. The speed and quality of a reconstructed 3D model largely depends on the number of input images and selection of the views. It is not necessary that every image will contribute equally to overall quality of a 3D model. It is difficult to select the number of images because with a limited number of images it is difficult generate an accurate plant model because fewer images might generate false points in the 3D point cloud. In contrast, a large number of images will result in processing redundant information which will inevitably increase computation time [20, 21]. To get appropriate number of images for precise 3D reconstruction, prior information or manual measurements of plants are required to compare to the corresponding measurements from the 3D model.

Clearly, there is a need to consider the number of input images and their resolution for the reconstruction process. These parameters have not yet been discussed in this area of plant phenotying. In this work, we investigate these two parameters. Our contributions are: (1) To reconstruct plant in 3D using a different number of input images followed by extraction of plant feature, such as stem height and number leaves in 3D. These extracted 3D values are then compared to manual measurements of the plant using descriptive statistics to determine an appropriate number of images for reconstruction. (2) To analyse the effect of image resolution on the quality of the 3D data.

The remainder of the paper is as follows; Sect. 2 discusses the detailed material and methods used in this paper to analyse the effects of the investigation parameters on the quality of the 3D model. The experimental results are discussed in Sect. 3.

2 Materials and Method

For this experiment, we considered a chilli plant (*Capsicum annum L.*) and conducted this experiment on a commercial farm in Palmerston North, New Zealand. The appropriate permissions from the responsible authorities of the commercial farm for conducting this experiment were taken. The chilli plant was selected because of its demand throughout the year and its high value. The stem height of the plant was 15 cm with 11 leaves on it. We aimed at reconstructing a single plant in 3D; hence, the plants were planted in a row having a distance of approximately 90 cm between adjacent plants. Therefore, other plants did not interfere in the 3D model and only one plant is modelled.

2.1 Image Acquisition

The image acquisition process and sample acquired images are shown in Fig. 1. Images are captured while moving the camera around the plant in an approximately circular path. We captured the images in six rounds at different views, angles, distances, and heights. 15 images were acquired in every round [1], with images acquired at approximately 10°–15° intervals. The 90 images from these six rounds had an overall 90% overlap between images. The distance between camera and plant was variable. The experiment was performed in outdoor conditions which gave varying image quality, such as images with shadow, wind, change in sunlight because of cloud movement. Structure-from-motion calculates the camera intrinsic and extrinsic parameters automatically; hence, different positions of the camera do not need any calibration process.

Fig. 1 Image acquisition process and sample acquired images

Fig. 2 3D modeling pipeline

2.2 3D Modeling

Once the images,have been acquired, it is necessary to detect common keypoints and then match these keypoints between other camera view images. Figure 2 shows the 3D modeling pipeline. A scale-invariant feature transform (SIFT) [22] was used to detect the keypoints.

1. **Scale-space extrema detection**: The primary step of SIFT searches over all image locations and scales using a difference-of-Gaussian function to detect promising keypoints that are orientation and scale invariant.
2. **Localisation of keypoints**: Keypoints are filtered to remove those with poor stability. Stability is a measure of the sensitivity of keypoints to changes in position and scale.
3. **Orientation assignment to the keypoints**: Orientations are allocated to every keypoint depending on local image gradient directions. These are calculated based on the detected scale to give scale invariance.
4. **Keypoint descriptor**: Local image gradients are calculated at the chosen scale in the area around every keypoint. These gradients are then transformed into a description which allows for considerable levels of change in illumination and shape distortion.
5. **Keypoint matching**: These keypoints are matched between pairs of images of the chilli plant acquired from various angles and views. Bundle adjustment is used to form a sparse 3D point cloud of the plant and retrieve camera positions and intrinsic, extrinsic parameters simultaneously. The size, position, and orientation of the chilli plant are defined related to the coordinate frame of the reconstructed model.

Once the features are detected and matched from various views, a sparse 3D model is produced. The sparse model is filtered to remove outliers (erroneous peaks form due to keypoint mismatching) and unwanted reconstructed part is then removed manually. Thereafter, the calculated camera intrinsic and extrinsic parameters, positions, and orientations are used to generate a dense 3D point cloud of the plant. Cross-correlation matching method is used to match a pixel in one image with the corresponding pixel in an another image on the epipolar line for an overlapped image pair [23]. The process is repeated for each image pair.

The generated dense 3D point cloud then processed (cleaned, smoother, and managed) using remenshing and filtering tools in *Meshlab* [24] to create the final 3D model.

2.3 Investigation Parameters

This investigation considers two different parameters which have not been discussed yet in the literature.

1. **Change in number of input images for 3D modeling**: The 3D modeling method explained in Sect. 2.2 will be repeated on different subsets of randomly selected images for a particular set of input images. The size of the subset is varied from 25 images through to 78 images. For each subset size, the experiment is repeated five times, selecting a different random subset. The quality of the reconstructed 3D model will be determined by comparing features extracted from the model with ground truth data (manually measured values of the actual plant). Plant features, such as stem height and number of leaves will be extracted. By exploring the correlation of extracted features with ground truth values, the number of images required to give an accurate reconstruction of chilli plant will be determined.
2. **Change in image resolution**: Once the appropriate number of images for reconstruction is determined, the effects of image resolution on the 3D data will be analysed. This will provide information about the best image resolution for reconstruction of the chilli plant.

2.4 Measurement of Plant Features from the Reconstructed 3D Model

Two plant features: stem height and number of leaves are extracted manually from the 3D model. The stem height is calculated by calculating the distance between marked point on the stem tip and bottom in the reconstructed 3D model. These points are selected visually by the user. Figure 3 shows the marked points on the plant and the distance between the marked points. Similarly, the number of leaves can be estimated by zooming and rotating the reconstructed 3D model manually by the user.

Fig. 3 The approximate calculation of stem height from the reconstructed 3D model

3 Results and Discussion

3.1 3D Modeling

Detailed 3D modeling was performed as described in Sect. 2.2. The experiment on five different subsets of randomly selected images from the set of input images.

Using from 5 to 20 input images give poor 3D models. A small number of images do not cover enough plant views to provide sufficient images for an accurate reconstruction. Quality of the 3D model is directly proportional to the number of views and images [21]. Figure 4a shows a reconstructed 3D model using all 90 captured images. This model has replicated the actual plant very well when compared visually. Figure 4b shows the 3D model derived from 78 images. While there are some differences with Fig. 4a, these are minor. With fewer images, some of the details start to get lost, such as leaf stalks Fig. 4c and the stem Fig. 4d.

Consequently, features of the plant derived from the 3D model become less accurate. As leaves become disconnected from the stem, Fig. 4c, the number of leaves decreases, and as the stem becomes more broken (Fig. 4d–f), the measured stem height decreases. With fewer images, the number of points detected on the leaf surfaces decreases, distorting the shape and size of the leaves.

(a) All 90 images (b) 78 images (c) 65 images

(d) 50 images (e) 35 images (f) 25 images

Fig. 4 Example 3D reconstructions of a chilli plant with different numbers of input images

3.2 Investigation Parameters

– **Change in number of input images for 3D modeling**:

For each subset size, the median, range, and interquartile range are represented by a box-whisker- plot. This clearly shows the variations from different random selections (see Fig. 5), and gives a clear idea of the accuracy of the extracted stem height for different numbers of input images. Similarly, Fig. 6 gives an idea about extracted number of leaves from 3D model with change in number of input images.

Using fewer images provide less information about plant features as they do not cover enough plant views. Due to insufficient plant views, the features from one image are not matched efficiently to another image. This causes feature matching error, resulting in unwanted outliers and loss of information. Because of the loss of information, the plant features, such as plant leaves are not reconstructed accurately and missing part of the leaf. Consequently, the leaf becomes disconnected from the stem so fewer leaves are counted on the reconstructed plant compared to the actual plant. Similarly, with fewer input images, the plant stem is not reconstructed accurately. The stem stem gives an inaccurate stem height from the reconstructed model.

Consequently, with more input images, the feature matching error and amount of information loss reduces as the input images cover sufficient plant views. It is important to use sufficient plant views to reconstruct the plant precisely. When input

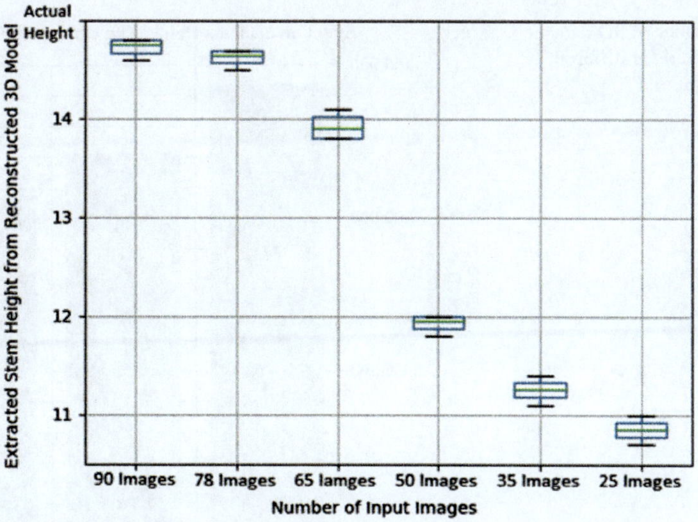

Fig. 5 Extracted stem height from reconstructed 3D model with change in number of input images

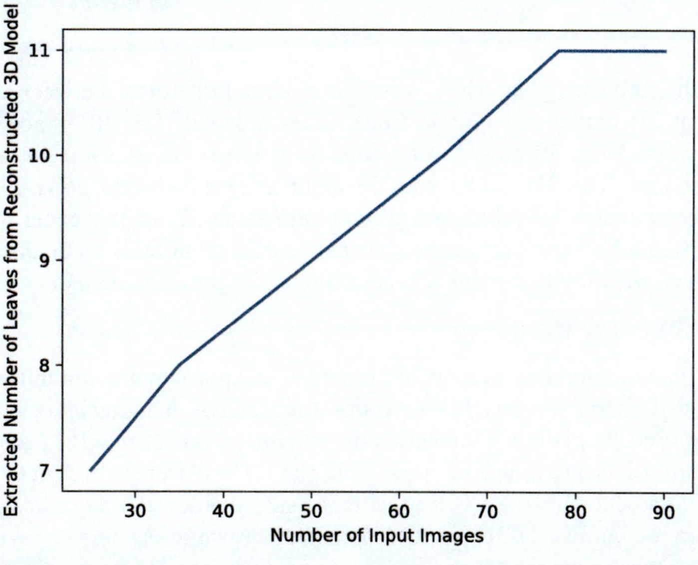

Fig. 6 Extracted number of leaves from reconstructed 3D model with change in number of input images

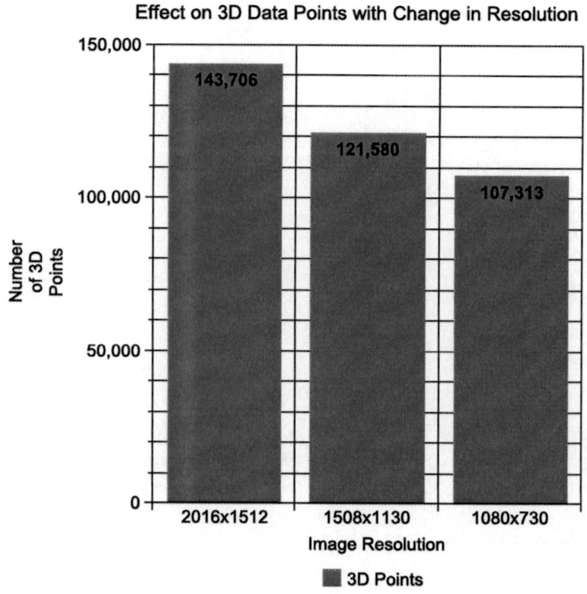

Fig. 7 Number of 3D data points detected for different image resolutions

images with sufficient plant views where considered, the error between extracted values from 3D model and ground truth values reduced. The difference between extracted values from 3D model and ground truth values varies inversely to number of input images. The 3D model with 78 input images provided good correlation between the extracted 3D values and ground truth values. From this experiment, it is clear that, instead of using all images, we can extract plant features precisely using fewer images which will inevitably save computation time and memory.

– **Change in Image Resolution**

The experiment in previous section has provided an approximate minimum number for accurate 3D reconstruction. In this section, the effect of changing image resolution will be explored. The original resolution of each image was 2016×1512 which gave 143,706 data points on average for the sample plant. These were sufficient to generate a precise 3D model. However, when the image size is reduced, there is a significant reduction in the number of 3D points detected, as shown in Fig. 7.

Although these data points are sufficient to reconstruct a 3D model, the quality of the model is poorer with fewer data points. When the resolution is further reduced, the number of data points also reduced to 107,313. The reduction in resolution tends result in inefficient feature detection and matching. Detecting fewer features leads to feature matching errors which are responsible to create unwanted outliers or distortions resulting in poor quality 3D model. Thus, the better the resolution of the images, the more 3D data points are detected.

4 Conclusion and Future Work

The results of experiments showed that 3D reconstruction of chilli plant and extracting plant features using fewer images is possible. However, the number of input images required for precise 3D reconstruction is totally depends on plant architecture. If the plant architecture is complex then more number of input images will be required for 3D reconstruction. Similarly, if plant architecture is simple then fewer images are sufficient for precise 3D reconstruction.

Our contribution to the knowledge is: (1) Investigation of the effect on 3D model when there is a change in the number of input images. (2) Guidelines for the selection of an appropriate number of images for accurate 3D resolution. (3) Successful extraction of plant features non-destructively from a 3D model non-destructively. (4) Investigation of the effect on 3D data when there is change in image resolution. The 3D model reconstructed from fewer images can provide more information about plant like leaf area, leaf length, leaf width, stem diameter and so on. The future work can be extended to implement this method on more plants and analyse the required input images and image resolutions. In addition, it is still needed to analyse, if we carefully select fewer images manually, is it possible to reduce the number of input images and still generate a precise 3D model?

References

1. Paturkar A, Gupta GS, Bailey D (2019) 3D reconstruction of plants under outdoor conditions using image-based computer vision. In: Santosh KC, Hegadi RS (eds) Recent trends in image processing and pattern recognition. Springer, Singapore, pp 284–297
2. Mishra KB, Mishra A, Klem K, Govindjee (2016) Plant phenotyping: a perspective. Ind J Plant Physiol 21(4):514–527 (2016)
3. Li L, Zhang Q, Huang D (2014) A review of imaging techniques for plant phenotyping. Sensors 14(11):20 078–20 111
4. Paturkar A, Gupta GS, Bailey D (2017) Overview of image-based 3D vision systems for agricultural applications. In: International conference on image and vision computing New Zealand (IVCNZ), pp 1–6
5. Guo Q, Wu F, Pang S, Zhao X, Chen L, Liu J, Xue B, Xu G, Li L, Jing H, Chu C (2018) Crop 3D—A LiDAR based platform for 3D high-throughput crop phenotyping. Sci China Life Sci 61(3):328–339
6. Zhang Z (2012) Microsoft kinect sensor and its effect. IEEE MultiMedia 19(2):4–10
7. Scharstein D, Szeliski R (2002) A taxonomy and evaluation of dense two-frame stereo correspondence algorithms. Int J Comput Vis 47(1–3):7–42
8. Jebara T, Azarbayejani A, Pentland A (1999) 3D structure from 2D motion. IEEE Signal Process Mag 16(3):66–84
9. Paulus S, Dupuis J, Riedel S, Kuhlmann H (2014) Automated analysis of barley organs using 3D laser scanning: an approach for high throughput phenotyping. Sensors 14(7):12 670–12 686
10. Kaminuma E, Heida N, Tsumoto Y, Yamamoto N, Goto N, Okamoto N, Konagaya A, Matsui M, Toyoda T (2004) Automatic quantification of morphological traits via three-dimensional measurement of Arabidopsis. Plant J 38(2):358–365

11. Newcombe RA, Izadi S, Hilliges O, Molyneaux D, Kim D, Davison AJ, Kohi P, Shotton J, Hodges S, Fitzgibbon A (2011) Kinectfusion: real-time dense surface mapping and tracking. In: 10th IEEE international symposium on mixed and augmented reality (ISMAR), pp 127–136
12. Chéné Y, Rousseau D, Lucidarme P, Bertheloot J, Caffier V, Morel P, Belin T, Chapeau-Blondeau E (2012) On the use of depth camera for 3D phenotyping of entire plants. Comput Electron Agric 82:122 – 127
13. Baumberg A, Lyons A, Taylor R (2005) 3D S.O.M.A commercial software solution to 3D scanning. Graph Models 67(6):476–495
14. Bergerman M, Van Henten E, Billingsley J, Reid J, Mingcong D (2013) IEEE robotics and automation society technical committee on agricultural robotics and automation. IEEE Robot Autom Mag 20(2):20–23
15. Takizawa H, Yamamoto S, Ezaki N, Mizuno S (2005) Plant recognition by integrating color and range data obtained through stereo vision. J Adv Comput Intell Intell Inform
16. Ivanov N, Boissard P, Chapron M, Andrieu B (1995) Computer stereo plotting for 3-D reconstruction of a maize canopy. Agric Forest Meteorol 75(1):85–102
17. Jay S, Rabatel G, Hadoux X, Moura D, Gorretta N (2015) In-field crop row phenotyping from 3D modeling performed using structure from motion. Comput Electron Agric 110:70–77
18. Quan L, Tan P, Zeng G, Yuan L, Wang J, Kang SB (2006) Image-based plant modeling. ACM Trans Graph 25(3):599–604
19. Tan P, Zeng G, Wang J, Kang SB, Quan L (2007) Image-based tree modeling. ACM Trans Graph 26(3):87
20. Hu Y, Wang L, Xiang L, Wu Q, Jiang H (2018) Automatic non-destructive growth measurement of leafy vegetables based on kinect. Sensors 18:806
21. Liu S-X, An P, Zhang Z-Y, Zhang Q, Shen L-Q, Jiang G-Y (2009) On the relationship between multi-view data capturing and quality of rendered virtual view. Imaging Sci J 57(5):250–259
22. Lowe DG (2004) Distinctive image features from scale-invariant keypoints. Int J Comput Vis 60(2):91–110
23. Remondino F, El-Hakim S (2005) Critical overview of image-based 3D modeling. In: International workshop on recording, modeling and visualization of cultural heritage. CRC Press, p 299
24. Cignoni P, Callieri M, Corsini M, Dellepiane M, Ganovelli E, Ranzuglia G (2008) Meshlab: an open-source mesh processing tool. In: Scarano V, Chiara RD, Erra U (eds) Eurographics Italian chapter conference. The Eurographics Association

Comparative Analysis of Least Squares Method and Extended Kalman Filter for Position Estimation in GPS Receiver

Jyoti S. Kavathekar and Ashwini M. Deshpande

Abstract GPS is a system of obtaining the position of any object on or above the earth surface. Global Positioning System (GPS) has been incorporated into many devices in everyday life. However, GPS receiver design is challenging depending on the user's operating environment. The accuracy of GPS position estimate is affected by various factors like ionospheric delay, tropospheric delay, various multi-path effects, and number of satellites in view and navigational solution employed. Multi-path propagation to low signal strength are examples of some of these challenges. This paper provides a comparative analysis of position estimation techniques in a GPS receiver. These techniques are the Least-Squares (LS) method and Extended Kalman Filter Method (EKF). In this, the data possessed with a dual-frequency GPS receiver is placed at the reference point (X-1687535, Y-5809975, Z-2014102). For this reference point iterative assessments of satellite transmission moment, receiver time, and position are carried out to determine instantaneous estimates of the receiver location. The work explains the design and implementation of a software-defined GPS receiver in real-time. We use five satellites to estimate the position of the receiver. The performance evaluation of position estimation accuracy over the region is carried out based on position coordinates, root mean square error (RMSE), and standard deviation. The experimental evaluation demonstrates that the Extended Kalman Filter provides a more precise and stable estimate than LS method.

Keywords GPS · Extended Kalman filter · Least-squares method · Standard deviation · RMSE

J. S. Kavathekar (✉)
Cummins College of Engineering for Women, Karvennagar, India
e-mail: jyoti.kavathekar@cumminscollege.in

A. M. Deshpande
Savitribai Phule Pune University, Pune, India
e-mail: ashwini.deshpande@cumminscollege.in

© Springer Nature Singapore Pte Ltd. 2021

S. N. Merchant et al. (eds.), *Advances in Signal and Data Processing*,
Lecture Notes in Electrical Engineering 703,
https://doi.org/10.1007/978-981-15-8391-9_29

1 Introduction

The GPS is a satellite-based all-weather, globally, and ongoing convergence system. It is the primary source of data for navigation and timing systems [3]. GPS consists of 24 satellites; those orbit around the earth and provide position coordinates on or above the earth surface. It comprises of three main sections: Space, Control, and User. The space section comprises of GPS satellites and transmits signals. The control segment comprises of a ground station that ensures that the satellites function correctly. The user section consists of receivers [1]. GPS receiver requires data transmitted from the satellites and utilizes the concept of triangulation to calculate the user's position. GPS signal is transmitted from a satellite over 20,200 km from Earth. As a consequence, the measurement precision is impacted. The position estimate is affected by satellite geometry and number of satellites in view, ionospheric and tropospheric delays, various multi-path effects, etc.

Tsui [1] gives in-depth information about the Global Positioning System (GPS), software receiver, and coarse acquisition code generation.

Yang Wang presented technique to estimate position in a GPS software receiver using Extended Kalman Filter (EKF) and Rauch–Tung–Striebel-Smoother (RTS-Smoother) and compared the experimental results [2].

Sarunic [3] described the development of GPS receiver using Kalman filter algorithm; designed for stationary, low dynamics, and high dynamics applications.

Muhammad et al. [4] applied Kalman Filter (KF) to GPS signals with error and reduce the variance to refine the GPS data. There are many sources that are affecting the GPS positioning like atmosphere delay, satellite clock, and orbit error. For overcoming this problem KF is useful.

Jakkaraju Pavan Kumar et al. presented approaches for design and development of a Kalman Filter effective tracking mechanism in a GPS receiver. Their design avoids the need for frequent requisition for weak signal condition [5].

M. R. Mosavi et al. compared the position accuracies for five variants of Least-Squares (LS) methods in high velocities. LS method uses pseudo-range and carrier phase measurements for GPS receiver positioning [6].

David Akopian et al. presented the approach for GPS receiver without navigation data extraction from the received signals. In this approach, there is no need to recover the time of the week from navigation data [7].

In [9], Akash Deep et al. has used Matlab Simulink environment for position estimate of vehicle for noisy position measurements based on Kalman filter. Their results show that Kalman filter gives better position estimate over direct GPS measurements.

Thus, Least-Squares (LS) method and Extended Kalman Filter (EKF) are found useful in position accuracy estimation of GPS receiver. Therefore we have implemented both to compare their performances for a dual-frequency GPS receiver located at a reference point (X-1687535, Y-5809975, Z-2014102).

The paper is organized as follows: Sect. 2 describes working of GPS receiver (Hardware and Software parts). Position estimation techniques are described in Sect. 3. Section 4 presents experimental results and conclusions are given in Sect. 5.

2 Working of GPS Receiver

A GPS receiver setup used in our work comprise of two parts: (a) Hardware section and (b) Software section. A block diagram of the fundamental GPS receiver as shown in Fig. 1 depicts both the sections and their corresponding elements.

2.1 Hardware Section

In hardware section, input is received from active antenna (with range of frequency bands: 1164 to 1254 MHz and 1525 to 1606 MHz and Gain = 32 dB). Signal splitter is used to separate out two signals, i.e., L1 and L5. GPS L1 carrier frequency is 1575.42 MHz and L5 carrier frequency is 1227.60 MHz. For amplification, there is the radio frequency (RF) chain with suitable amplitude and convert the frequency to the required output frequency (discussed in detail in Sect. 4, Experimental Result. Analog-to-digital converter (ADC) is used thereafter to digitize the signal. The digitized signal is passed on to the software section to process further.

2.2 Software Section

In software section, input is received from ADC. This digital signal (with −2, −1, 1, 2 levels) is fed to the acquisition block, where the signal from the visible satellites are acquired. In the process of tracking values of frequency and code phase are extracted. Sub-frames and navigation information is collected by the monitoring. Ephemeris

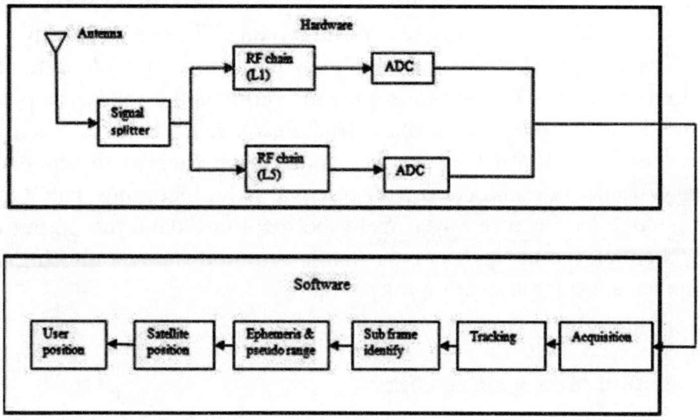

Fig. 1 Block diagram of GPS receiver [1]

data and pseudo ranges are acquired from the navigation information. The ephemeris information gives the satellite location.

In the signal processing part of the GPS receiver removal of C/A code and carrier frequency (to obtain the data message) is carried out. After receiving the information notification, the preamble detection is conducted and the sub-frame start is initiated. The necessary base-band signal processing is splitted into three distinct, sequential tasks: acquisition, monitoring, and decoding of navigational messages.

2.2.1 Acquisition

The acquisition aims to acquire the signals of visible satellites and coarse values of carrier frequency and code phase. It is divided into two parts, first is the down-conversion of the digitized GPS signal to baseband and second is the identification of the acquired C/A code to determine which satellite transmitted the signal. The carrier frequency must be known to down-convert the signal. The real frequency of the received carrier must be estimated accurately, as the Doppler Effect causes a frequency offset on the L1 carrier. Knowing that the receiver is stationary and the Doppler shift must be between ± 5 kHz, a search is conducted across all feasible carrier frequencies.

2.2.2 Tracking

The purpose of tracking is to filter values of frequency and code phase, keep track and demodulate the navigation data from the specific satellite. The monitoring consists of two components: code tracking and frequency/phase tracking of the carrier. Delay Locked Loop (DLL) is used for code monitoring and Phase Lock Loop (PLL) is used for carrier frequency monitoring. Three local replicas are produced for each coarse acquisition (C/A) code to monitor the time alignment. These are replicas *early, late* and *timely*. The version that precisely matches the incoming C/A code alignment is referred to as *prompt*. The monitoring feature requires the following parameters: front-end block of the recorded signal, structure channel, Sine, Cosine, and C/A code sheets. Feedback loops are used in a monitoring channel to produce precise local copies of both code and carrier for removal from incoming data. Once these two components have been removed from the incoming data, the output from the monitoring channel, the navigation bits, can be acquired. The monitoring feature is used to discover the stage transition for navigation data.

2.2.3 Navigation Message Decoding

In the last processing step for the GPS receiver is to identify that the navigation message has been effectively acquired. When the signals are properly tracked, we require only navigation information. So from the tracked signal, we separate out

C/A code and carrier wave. After reading about 30 s of information, the start of a sub-frame must be discovered to identify the moment when the information was transferred from the satellite. When the transmission time is discovered, the satellite's ephemeris information must be decoded. This is subsequently used to calculate the satellite's position at the moment of transmission. At the end, before making the position calculations, pseudo ranges are calculated.

The pseudo rages are calculated based on the time of transmission of signal and time of receiver of signal. The time of receiver is based on the start of the sub-frame. The receiver's final task is to calculate a user position. For position computation Least-Squares (LS) and Extended Kalman Filter techniques are used.

3 Position Estimation Techniques

In this section, Least Squares (LS) method and Extended Kalman Filter (EKF) are described in detail.

3.1 Least-Squares Method

Method of the least-squares offers the approximation to the over-determined system. In this paper, it is used to find out the receiver position from pseudo ranges for five satellites. Considering the clock bias, the basic observation equation for determining the user position can be defined [2].

$$\rho_i = \sqrt{(x_i - x_u)^2 + (y_i - y_u)^2 + (z_i - z_u)^2} + b_u \tag{1}$$

where (x_u, y_u, z_u) and b_u are the unknowns. In the above equation, pseudo range ρ_i and position of the satellites (x_i, y_i, z_i) are known.

How pseudo ranges and position of satellites are calculated is already explained in Sect. 2.2. When pseudo ranges and positions of five different (visible) satellites are substituted in Eq. (1), set of equations are obtained as:

$$\widehat{\rho_i} = \sqrt{(x_1 - x_u)^2 + (y_1 - y_u)^2 + (z_1 - z_u)^2} + b_u \tag{2}$$

$$\widehat{\rho_2} = \sqrt{(x_2 - x_u)^2 + (y_2 - y_u)^2 + (z_2 - z_u)^2} + b_u \tag{3}$$

$$\widehat{\rho_3} = \sqrt{(x_3 - x_u)^2 + (y_3 - y_u)^2 + (z_3 - z_u)^2} + b_u \tag{4}$$

$$\widehat{\rho_4} = \sqrt{(x_4 - x_u)^2 + (y_4 - y_u)^2 + (z_4 - z_u)^2} + b_u \quad (5)$$

$$\widehat{\rho_5} = \sqrt{(x_5 - x_u)^2 + (y_5 - y_u)^2 + (z_5 - z_u)^2} + b_u \quad (6)$$

The receiver position (x_u, y_u, z_u), as well as b_u, can be derived from different approaches to solve the nonlinear equations given in Eq. (2), among which the Least-Squares is most commonly used.

From the analysis, this technique is applicable for fixed GPS receivers and not to dynamic receiver. As the pseudo-range measurements involve more variation, thus take more time to calculate the position.

Therefore, Kalman filter is applied to minimize the variation in the estimation of the receiver position.

3.2 Extended Kalman Filter (EKF)

Extended Kalman filter is used to solve ideal filtering issues that are not linear. The initial consideration doesn't give any information about the rate of change of receiver clock bias. In this, we have taken a stationary receiver model. EKF algorithm is implemented on data in prediction and updating stages.

So the estimation steps followed are given as [3],

1. State prediction:

$$\hat{x}(k+1|k) = F(k)\hat{x}(k|k) \quad (7)$$

where $F(k)$ is state transition matrix and $\hat{x}(k|k) = [xyzr_{t_r}\dot{r}_{t_r}]$ method and after second that can be calculated from previous one). $\dot{r}_{t_r}$ is drift of clock bias.

$$F(k) = \begin{bmatrix} 1\,0\,0\,0\,0 \\ 0\,1\,0\,0\,0 \\ 0\,0\,1\,0\,0 \\ 0\,0\,0\,1\,T \\ 0\,0\,0\,0\,1 \end{bmatrix}$$

Where T is receiver clock bias drift (0.5 s). Here take identity matrix because consider for a stationary receiver.

2. Measurement prediction:

$$\widehat{\rho_i} = \left[X_i(k+1) - \hat{x}(k+1|k)^2 + Y_i(k+1) - \hat{y}(k+1|k)^2 + Z_i(k+1) - \widehat{z}(k+1|k)^2\right]^{1/2} + r_{t_r}(k+1|k) \quad (8)$$

For $i = 1,\ldots, 5$

3. Measurement residual:

$$v(k+1) = \rho(k+1) - \hat{\rho}(k+1|k) \tag{9}$$

where $\rho(k+1)$ is observed pseudo range.

4. Updated state estimate

$$\hat{x}(k+1|k+1) = \hat{x}(k+1|k) + W(k+1)v(k+1) \tag{10}$$

- Evaluation of Jacobians:

$$H(k+1) = \begin{bmatrix} \frac{\partial\rho_1}{\partial x} & \frac{\partial\rho_1}{\partial y} & \frac{\partial\rho_1}{\partial z} & \frac{\partial\rho_1}{\partial r_{t_r}} & 0 \\ \frac{\partial\rho_2}{\partial x} & \frac{\partial\rho_2}{\partial y} & \frac{\partial\rho_2}{\partial z} & \frac{\partial\rho_2}{\partial r_{t_r}} & 0 \\ \frac{\partial\rho_3}{\partial x} & \frac{\partial\rho_3}{\partial y} & \frac{\partial\rho_3}{\partial z} & \frac{\partial\rho_3}{\partial r_{t_r}} & 0 \\ \frac{\partial\rho_4}{\partial x} & \frac{\partial\rho_4}{\partial y} & \frac{\partial\rho_4}{\partial z} & \frac{\partial\rho_4}{\partial r_{t_r}} & 0 \\ \frac{\partial\rho_5}{\partial x} & \frac{\partial\rho_5}{\partial y} & \frac{\partial\rho_5}{\partial z} & \frac{\partial\rho_5}{\partial r_{t_r}} & 0 \end{bmatrix} \tag{11}$$

- State Prediction covariance:

$$P(k+1|k) = F(k)P(k|k)F^T(k) + Q_v(k) \tag{12}$$

where we consider the initial model as stationary so the $Q_{d_t}(k)$ is,

$$Q_{d_t}(k) = \begin{bmatrix} 0\,0\,0 & 0 & 0 \\ 0\,0\,0 & 0 & 0 \\ 0\,0\,0 & 0 & 0 \\ 0\,0\,0 & S_\varphi T + \frac{T^3}{2} S_f & \frac{T^2}{2} S_f \\ 0\,0\,0 & \frac{T^2}{2} S_f & S_f T \end{bmatrix}$$

where S_φ and S_f are the power spectral densities of ω_φ and ω_f. T indicates sampling period and k indicate time index.

- Residual covariance:

$$S(k+1) = H(k+1)P(k+1|k)H(k+1)^T R(k+1) \tag{13}$$

- Filter Gain:

$$W(k+1) = P(k+1|k)H(k+1)^T S(k+1)^{-1} \tag{14}$$

- Updated state covariance:

$$P(k+1|k+1) = P(k+1|k) - W(k+1)S(k+1)W(k+1)^T \tag{15}$$

$x(k+1|k+1)$gives the receiver position [3].

In EKF, prior information of state is estimated from the current estimates of state vector by proper modeling unlike LS method. Due to this, EKF provides more precise and accurate solution.

4 Experimental Results

For the comparative analysis of LS method and EKF technique for receiver position estimation, an experimental setup used in this work comprise of the antenna and Stereo board under hardware section. Software section consists of a Personal Computer (Workstation Z-440) with MATLAB 2016a.

Figure 2 shows the hardware setup for signal tracking. RF front end is connected to Tallysman (7972) antenna. The antenna connected to signal splitter and the USB connector is connected to PC.

In the software section, the important parameters for signal processing selected are: Sampling frequency = 26 MHz, Intermediate Frequency IF = 6.5 MHz, and 2-bit samples. PC equipped with Ubuntu Linux 10.4 is used for collecting the signal data. The collected digital form of data is stored in '.dat' file formats. This data contains $\{-2, -1, 1, 2\}$ digital levels.

The hardware setup is used to collect 8 min of data. The signal characteristics of the data collected is studied and interpreted from its time domain, frequency domain, and histogram plots as shown in Fig. 3.

Figure 3a shows the time domain plot of the segment of the data. This plot shows that data is in random form. Figure 3b shows its frequency-domain plot. This plot shows the increased power around the residual carrier frequency in the center. In this plot, we check presence of intermediate frequency (IF) which proves that we are getting the correct data for processing the position estimation. Figure 3c shows a histogram plot of data. This plot shows the normal distribution of data.

Figure 4 shows the list of observed satellites by the GPS receiver during acquisition phase. This figure shows the Pseudorandom noise (PRN), frequency of the carrier, Doppler offset from 6.49 MHz, and the phase of the spreading code relative to the beginning of the captured data.

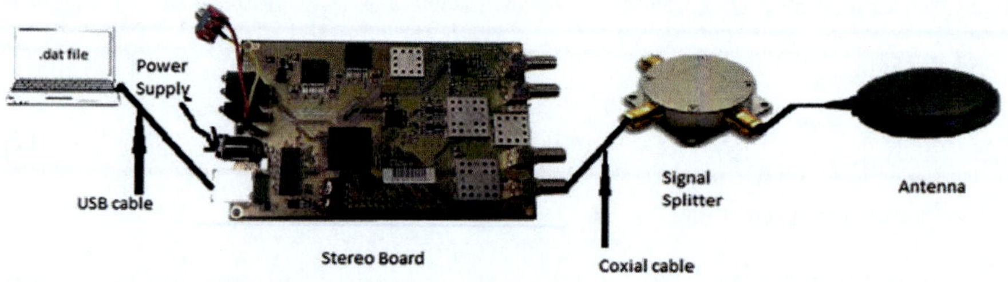

Fig. 2 Hardware Setup for signal tracking

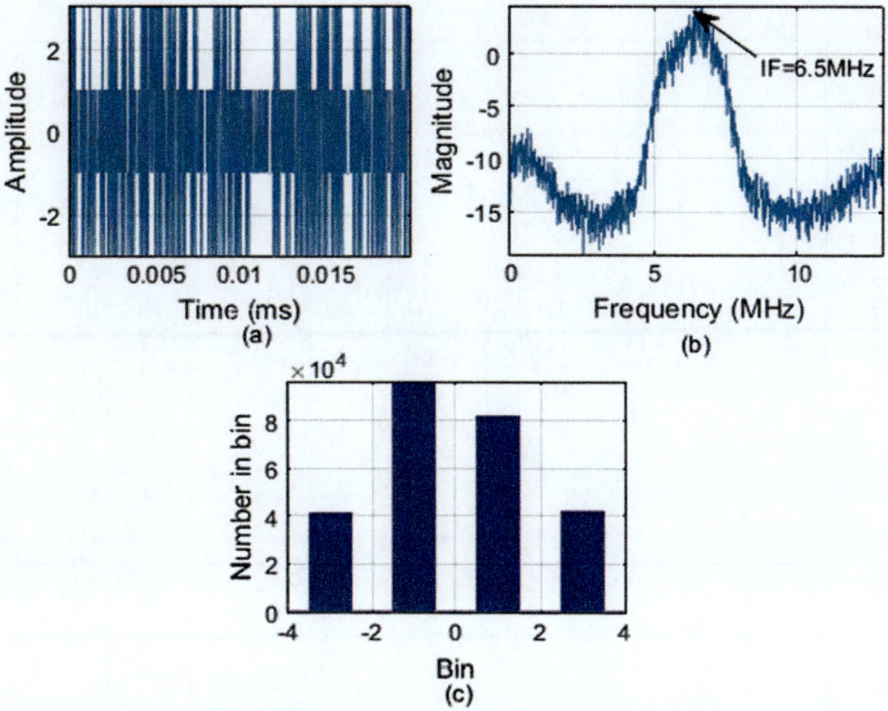

Fig. 3 **a** Time domain plot, **b** Frequency domain plot, **c** Histogram of raw signal

Acquiring satellites...

(. . 03 . . . . . . 09 . . . . 16 . . . . . . 22 23 . . . .)

Channel	PRN	Frequency	Doppler	Code Offset	Status
1	3	6.49826e+06	-1736	15158	T
2	9	6.50316e+06	3161	5146	T
3	23	6.50202e+06	2021	20558	T
4	22	6.49748e+06	-2517	25218	T
5	16	6.49738e+06	-2616	22550	T
6	---	------------	-----	------	Off
7	---	------------	-----	------	Off
8	---	------------	-----	------	Off

Fig. 4 Acquisition satellite and their values

Figure 5 shows position estimation results obtained from the LS method. Figure 5a depicts the coordinate variation in the Universal Transverse Mercator (UTM) system for duration of the 500 ms. In this, *E* indicates Easting, *N* indicates Northing and *U* indicates Upping. From this figure, it is observed that settling point is not reached as

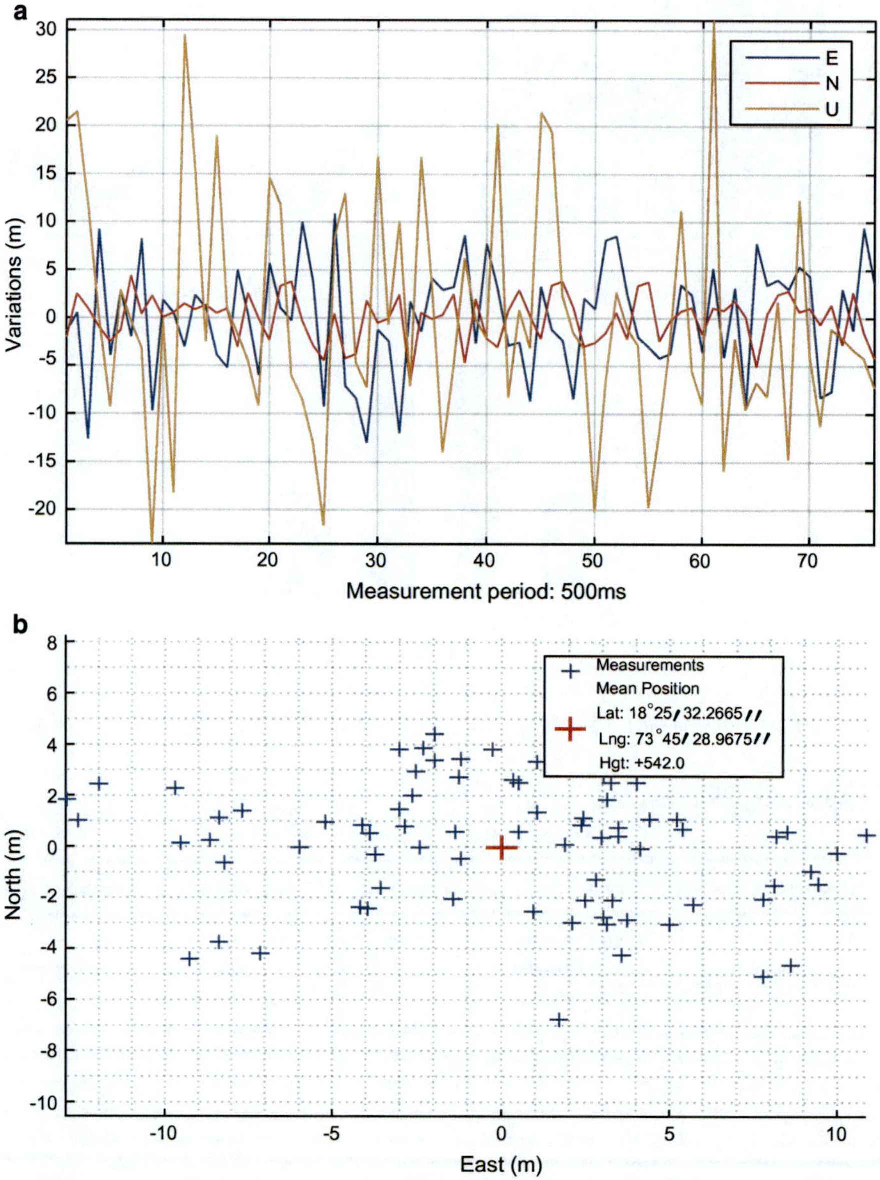

Fig. 5 **a** Coordinate variations in UTM system (LS method). **b** Position in UTM system(3D)(LS method)

there more variations occur with respect to *E*, *N* and *U* in the system. It leads to more errors in position estimation calculation. Figure 5b shows the three-dimensional plot (3D) of the UTM system, where blue markers indicate the approximate calculation for position estimation and a red marker shows exact GPS receiver position in coordinate (Latitude, Longitude, and Height) form. From this plot, it is observed that the estimated values are in the scattered form, as the variation is more, so it is very difficult to find the exact position using LS method.

Figure 6 shows results obtained from EKF method. In Fig. 6a the coordinate variations in the Universal Transverse Mercator (UTM) system for the duration of 500 ms is shown. The settling point is reached very early (0–20 ms) in EKF, as it does not happen in LS method. Thus, in EKF technique minimum variation is observed in *E*, *N*, and *U* as compared to the LS method. So, the accuracy of position estimation calculations is more. Figure 6b shows the UTM (3D) plot. In this plot, blue markers indicate estimated values, which are located close to each other (in the contrary, in case of LS these points are scattered), so the accuracy of a GPS receiver is more. A red marker gives exact GPS receiver position in coordinate (Latitude, Longitude, and Height) form.

Figure 7 shows the skyplot view of five visible satellites. Geometry of satellites depends on Position Dilution of Precision (PDOP) [1].

$$\mathrm{PDOP} = \sigma_{\mathrm{POS}} = \sqrt{\sigma_{\mathrm{e}}^2 + \sigma_{\mathrm{n}}^2 + \sigma_{\mathrm{u}}^2} \tag{16}$$

where σ_{pos} the standard deviation of the receiver position is, σ_{e} is standard deviation of easting, σ_{n} is standard deviation of northing and σ_{u} is standard deviation of upping. The position of accuracy was obtained when the PDOP < 5. In this case PDOP value is 3.6741, which gives the best satellite geometry for calculating the GPS receiver position.

Tables 1, 2 and 3 give a comparative analysis of the LS method and EKF technique. Estimated position coordinates (latitude, longitude, and height) of a GPS receiver in comparison to the reference point are given in Table 1. As shown in this table, parameter estimation along with the estimated height by EKF is more accurate (closer to the reference point) as compared to LS.

Root Mean Square Error (RMSE) position error comparison is given in Table 2. RMSE is calculated over the intervals of 40,000 number of processing bits obtained from ADC. Further, this error is computed for *X*, *Y*, and *Z* position and compared.

Table 3 gives the overall Standard Deviation for two techniques of a GPS receiver position. Standard Deviation for the LS method is 11.6184 is more as compared to the EKF technique. This comparison shows that EKF has better accuracy than the LS method.

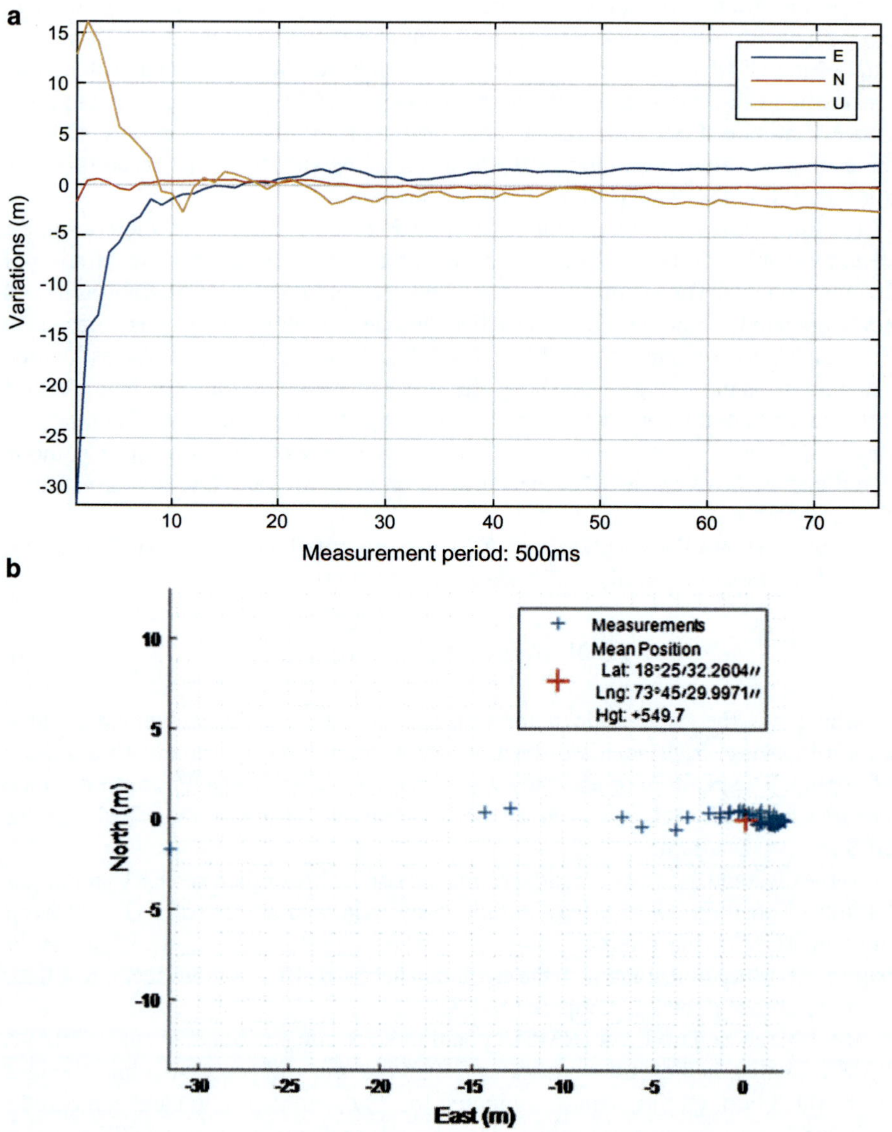

Fig. 6 **a** Coordinate variations in UTM system(EKF). **b** Position in UTM system(3D)

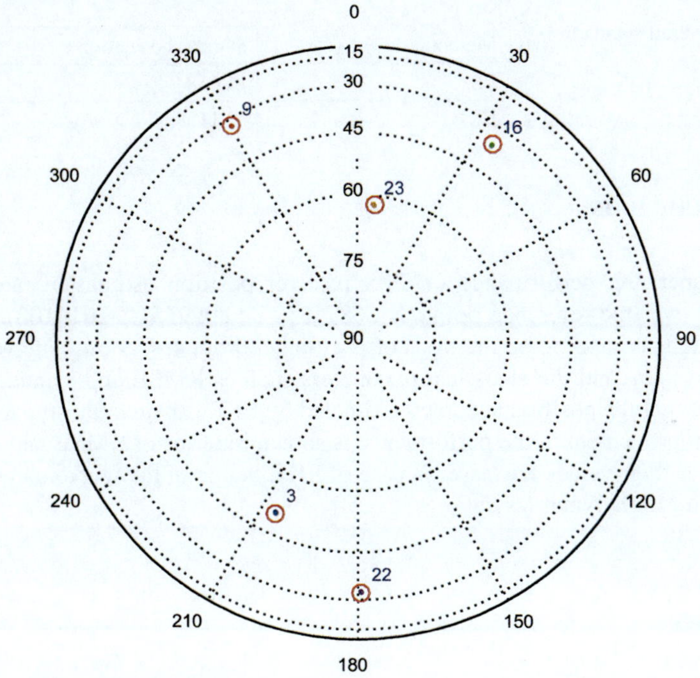

Fig. 7 Skyplot of five visible satellites

Table 1 Position Coordinates

Parameter	Reference point	LS method	EKF
Latitude	18°1′42.4871″	18°25′32.2665″	18°25′32.2604″
Longitude	73°48′13.7103″	73°45′28.9675″	73°45′29.9971″
Height	540.5 (m)	542 (m)	541.7 (m)

Table 2 Comparison of position RMSE error

Bit intervals for processing	LS method X-position error	EKF method X-position error	LS method Y-position error	EKF method Y-position error	LS method Z-position error	EKF method Z-position error
0–40,000	5.8144	5.2650	11.1747	2.1465	5.8779	4.1103
40,000–80,000	4.1697	4.0931	9.0868	6.7152	9.0868	8.1725
80,000–120,000	5.6817	5.0689	15.0973	8.7981	9.8556	7.6428
120,000–160,000	5.9706	5.7214	10.9552	6.5187	8.2415	6.7495
160,000–200,000	5.8187	5.6124	12.1247	7.9142	8.1401	5.6134
Average error	**5.4910**	**5.1522**	**11.6877**	**6.4185**	**8.2404**	**6.4577**

Table 3 Overall standard deviation

Method	Standard deviation
LS	11.6184
EKF	3.1139

5 Conclusions

In this paper, the performance of GPS receiver position estimation methods is compared over the reference point (X-1687535, Y-5809975, Z-2014102). Experimental results compared on the basis of position coordinates, coordinate variations in UTM system, and the statistical parameters such as RMSE and standard deviation, prove that the position accuracy estimated by EKF is better as compared to LS method. Improvement in the performance is shown in terms of RMSE and standard deviation as their values are lesser in case of EKF. Scope of the work can be further extended for high dynamics data.

References

1. Tsui JB-Y (2005) Fundamentals of global positioning system receiver—a software approach, 2nd edn. Wiley (2005)
2. Wang Y (2012) Position estimation using extended kalman filter and RTS-smoother in a GPS receiver. In: 5th International congress on image and signal processin. IEEE, pp 1718–1721
3. Sarunic PW (2016) Development of GPS receiver kalman filter algorithms for stationary, low-dynamics, and high-dynamics applications. Defence Science and Technology Group DST-Group TR3260, pp 1–42
4. Muhammad S, Laghari N-Z, Ali M, Farrukh M (2018) GPS estimation using Kalman Filter. Int J Sci Res Dev 6:692–694
5. Kumar JP, Rarota N, Maheswari U (2015) Design and Implementation of Kalman filter for GPS receiver. Ind J Sci Technol 5
6. Mosavi MR, Azarshahi S, Emamgholipour I, Abedi AA (2014) Least squares techniques for GPS receivers positioning filter using pseudo-range and carrier phase measurements. Iran J Electr Electron Eng 10:9
7. Durga Indira N, Sony K, Nagendram S, Ravi Teja P, Hima Tej G, Prasanna Kumar E (2017). A novel position estimation of GPS receiverutilizing extended Kalman filter. J Adv Res Dyn Control Syst 6
8. Sirish Kumar P, Srilatha Indira Dutt BBS (2019) Navigation solutions for GPS receiver position estimation over the southern region of India. Int J Recent Technol Eng 4
9. Sarras, Gerakios G, Diamantis A, Dounis AI, Syrcos GP (2010) Static single point positioning using the extended Kalman filter. World Acad Sci Eng Technol 6
10. Sarunic, PW (2016) Develoopment of GPS receiver Kalman filter algorithms for stationary,low-dynamics, and high-dynamics applications. Defence Sci Technol Group 55
11. Muhammad S, Laghari N-Z, Ali M, Farrukh M (2018) GPS estimation using Kalman filter. Int J Sci Res Devel 3

12. Syrjarinne Jari DA (2009) A fast positioning method without navigation data decoding for assited GPS receivers. IEEE Trans Veh Technol 58:6
13. Tsui, JB-Y (2005). Fundamentals of global positioning system receiver. Wiley
14. Wang Y (2012) Position estimation using extended Kalman filter and RTS-smoother in a GPS receiver. Int Cong Image Signal Process 4

Fair Scheduling Non-orthogonal Random Access for 5G Networks

Mansi N. Purohit and T. K. Ramesh

Abstract The potential growth of communication medium and its associated devices are rapidly increasing day by day. A most powerful communication support with 5G network features is required nowadays to resolve all communication overheads. The traditional method of embryonic data transfer between base station and the user device follows channel allocation strategy, which is internally used for the maintenance of random entry of data packets resulted in poor scalable nature, poor traffic handling, latency, and so on. The efficiency of network technologies with 5G support suffers from major issues such as the network performance, collision, and delay that are needed to be concerned with. To combat this, the technique of decoding of two or more packets is followed; hence, the concept of NORA was proposed. In analysis of the existing NORA method, Zadoff–Chu sequence was used to obtain the performance of NORA in 5G which gave better performance yet, resulting in collisions. Keeping these issues in consideration, an efficient communication model called fair scheduling non-orthogonal random access (FSNORA) is proposed. In the proposed method, we have scrutinized NORA to improve the fairness of scheduling by using round-robin algorithm so as to make a contentionless system and improve the channel enhancement of 5G networks. Simulation results convey that there is an improvement in the access delay, success probability, throughput, collision probability, and the preamble transmissions in context with contention compared to the existing NORA method.

Keywords 5G networks · Data traffic · Channel allocation · Random access · ORA · NORA · Round-robin algorithm

M. N. Purohit (✉) · T. K. Ramesh
Department of Electronics and Communication Engineering, Amrita School of Engineering, Amrita Vishwa Vidyapeetham, Bengaluru, India
e-mail: mansi.purohit95@gmail.com

T. K. Ramesh
e-mail: tk_ramesh@blr.amrita.edu

© Springer Nature Singapore Pte Ltd. 2021

S. N. Merchant et al. (eds.), *Advances in Signal and Data Processing*, Lecture Notes in Electrical Engineering 703,
https://doi.org/10.1007/978-981-15-8391-9_30

1 Introduction

Today's modern world belongs to communication between one and another entity. A lot of communication providers provide best support to attract their customers and also provide reasonable services to them as well. But the demanding needs of consumers are raised day by day, so that the network and mobile communication industry implements several new technical broadways to improve their network and communication gateways better and better every day. Data traffic is a serious cause need to be concentrated over the network and communication industry, which provides several types of communication frequency levels to its consumers to attain the best services from industry such as broadband services, 3G, 4G, and so on. However, the technical expectation regarding speed and efficient data rate delivery is required to revise, so that a new network communication gateway is introduced to provide better efficiency in case of 5G technologies. Many access schemes have been introduced to boost these technologies to the real-world users. The most common and ultimately used concern called non-orthogonal random access (NORA) is implemented [1, 2]. The concept of NORA is most popular in case of dealing with 5G schemes, to provide most intelligent communication solutions to the consumers in effective way [3]. In this process, the channel allocation method comes into the view where a single communication channel is segmented into separate channels and is accessed by many user devices to carry their own specific challenges [4]. This is the best used and an efficient method where specific frequency is given by which the traffic can also be reduced [5, 6]. But this also has certain limitations. Likewise, consider the first case where there are 10 numbers of users and the respective number of different frequencies. Suppose all the 10 users are using only 6 channels, the leftover 4 channels will be unused by which there will huge power wastage. Considering the second case, the user frequencies may overlap, if all the 10 users are accessing the same channel at a certain time. This will lead to collision in the wireless networks. Hence, the random access channel (RACH) came into the picture where the information is considered in form of packets for communication, and preamble sequences are considered.

Generally, preamble sequences are orthogonal to each other. So, base station can decode these sequences from multiple user devices at the same time, provided each user device is using different preamble sequences. But preamble sequences are selected by user device in random order so it is quite possible that two user devices choose the same preamble sequence and are sent to the base station at the same time. This leads to the possibility of collision [1, 7].

In this paper, we are considering the preamble sequences to be non-orthogonal to each other in other words, NORA along with round-robin algorithm so that decoding of the packets can be done simultaneously and can be scheduled for non-contention-based transmission to achieve better efficiency because of round-robin data packets scheduling nature.

This paper is further divided into sections such as Sect. 2 consists of the previous methods and schema used, Sect. 3 describes about the proposed method and the clear explanation in the implementation of the method, Sect. 4 showcases the implementation results that give the comparison of models 1 and 2 considering of NORA using round-robin algorithm that achieves better fairness performance than the NORA using Zadoff–Chu method, and finally the conclusion of the work is given in Sect. 5.

2 Literature Review

In this section, we discuss the previous methods and technologies used in order to improve the data traffic in the wireless communications.

In [8], they described appliance of random access technique along with mobile and self-arranged networks is complex by means of processing delay over millimeter wave mobile networks along directional beams due to random access preambles. The approaches of proposed two various preambles, such as Random Access Preamble 1 and Random Access Preamble 2, minimized the processing delay in random access position. The main advantage of this proposed approach is achieved guaranteed results via following parameters such as correlation property, false alarm probability, and detection probability of the proposed random access points, so that the results were good in nature but the computational complexity was an issue.

In [9], they described regarding the generation of next network processing called 5G, which is considered as a most dominant and high-speed communication technology. The main concept implemented was the next-generation networks (NGN), which took care of the communication speed and processing power over mobile networks. The major advantage of the approach was it attained good communication abilities over 5G by means of high data rate, real-time data handling, centralized views of the entire network with minimum delay, high security, lesser data losses, and minimized error rate. The major concern regarding the problem was while applying this all in real-time scenario, more number of mobile devices had to be concentrated.

In [10], they described cooperative access class barring with load balancing (CACB-LB) and traffic adaptive radio resource management (TARRM) schemes to improve the performance of mobile-to-mobile communications over LTE-A environments. Concept was used to achieve overlapping avoidance in mobile network scenario as well as for good traffic management constraints. The proposed CACB-LB achieved a better group of barring rates of access class barring than CACB so as to reduce the random access delay experienced by each MTC device or a UE. But this cooperative spectrum coverage ranges were practically complex in implementation as well as the implementation cost was huge, but simulationwise achievements were appreciable.

In [11], they described the communication services by means of wireless mobile networks. They used a new way called access class barring (ACB) with advanced timing data collection process [10], which improved the expected number of MTC devices successfully in each random access slot. The major advantage noticed with the proposed scheme was that to reduce closely half of the random access slots, it is required to serve all MTC devices compared to the existing schemes as well as use timing advance information with ACB only, or cooperative ACB. But the proposed scheme was more efficiently modified according to the LTE norms, and further work is involved in improving efficiency in all real-time scenarios.

The previous generations of cellular technology had unstable backward compatibility. To involve large number of frequency range where lots of mobile devices get connected to mobile network, a robust infrastructure named as 5G network, was introduced because the 4G bands were getting exhausted and could not accommodate lots of smart devices being used now. The performance statistics of chances of collision, chances of success, and mean access delay are calculated in order to bring a model to decrease the traffic congestion [2]. In consideration, the 3GPP standards had to redesign the cellular network so that competent and efficient mobile network service is provided. Motivated by this, the operation of RACH through LTE and LTE-A became crucial in order to ask allowance to the system suffering from network traffic and encumbering in the presence of huge number of devices [10]. The wireless networks provided machine-to-machine interaction. But for the user cells, LTE was delineated. So, the huge number of burst random access requests from machine-type communication devices for the former was feeble to handle. Hence, the RA slots were minimized [3].

To address the RACH traffic, a new design of RA procedure was proposed, wherein it had two important features [14]. First one was the rapid signaling process to reduce the overhead by allowing the user device to transmit data after preamble transmission on a physical channel. The second one was a self-optimization attribute which made access for the cellular system to generate optimal throughput changing resource block (RB) [12, 13]. A new method called SWIPT NOMA was introduced where the usage of simultaneous wireless and information transmission (SWIPT) to non-orthogonal multiple access (NOMA) networks was introduced where the users are located in a random manner [14]. NOMA used successive interference cancellation to receive data where, at the receiver the user device directly decoded its very own signal by treating the other device signal as noise [13]. It offered good quality of service (QoS), lower latency, massive connectivity, and high spectral efficiency, but the main drawback was the worst channel gains of all the remaining users were decoded by every user in the limits of cell. This leads to the receiver being intricate and the energy consumed was higher.

Encouraged by this idea, for 5G networks, non-orthogonal random access mechanism was proposed. Contrary to the orthogonal random access, instead of performing retransferring of data packets, the concurrent transferal of messages of the collided user devices was smoothened by NORA. This avoided the future collisions without rise in demands on PUSCH resources [1, 15]. In this method, we have analyzed the NORA scheme using round-robin algorithm has determined that in the case of probable collisions, more number of user devices are being supported in terms of mean access delay, throughput, and accessible delay.

The existing NORA method used Zadoff–Chu sequence, which is a critical defined processing sequence and it is commonly applied to signal processing strategies to generate new signals with constant or defined amplitude in several period of cycles. In that process, the improvement of 30% w.r.t the traffic model flow between NORA and ORA, the collision probability, and the preamble transmission with access delay is not statistically mentioned [1] and also the performance of NORA in 5G networks limited its concern to the processing accuracy due to the transmission of data packets that did not follow the fair scheduling method. The explanation to the existing NORA technology has been given further in the section.

2.1 Limitations of Zadoff–Chu Involved NORA Method

Here in this model, to decrease the data traffic in 5G networks, NORA technology was introduced to decrease the traffic scenario by improvement of 30%. The technology considered few parameters such as collision probability, average success probability, average access delay, throughput of the successful transmission of packets, and the traffic model flow. By considering these parameters, it assumed two traffic models, Traffic Model 1 and Traffic Model 2 where they involved both conventional orthogonal random access method (ORA) and the existing non-orthogonal random access method (NORA) where each of them was assumed of less number of user requests and more number of user requests, respectively. Across every parameter evaluation, the observation was made that the NORA method had outperformed the ORA method in a way that when the load of UE requests had been rising, NORA managed to reduce down the preamble transmissions by half thereby reducing the collisions for the wireless network but it ended up in minimal reduction of traffic and yet resulted in collisions [1].

Overall the data packet transmission was blindly followed by initiating the random access method without any scheduling algorithm which resulted in collisions when the number of user requests was overloaded. These limitations have been overcome in our proposed model, fair scheduling of NORA method in which the efficient concept of round-robin method is used which will be further discussed in the next section.

3 Fair Scheduling Access for NORA (FSNORA)

Fair scheduling access is an algorithm for any communicating system where the allocation of resources is equally distributed and equally accessed by all the user devices. Therefore, here in the proposed system fair scheduling NORA (FSNORA) follows NORA-based communication model along with round-robin scheduling algorithm used to assign a fixed slot in a periodic cyclic interval. It follows the concept of efficient scheduling in computation period by means of splitting the tasks into slices with equal proportion and in circulation manner the slices are processed without any prioritization over certain defined cycles. Hence, it can be used as a resource algorithm for improving the method of existing NORA by making it contentionless method.

When multiple number of user devices access the channel, fair scheduling disciplines are used to protect from non-responsible flows or non-protected flows. And the allocation of resources is arranged in the increasing order of their demands. Hence, the concept of round-robin algorithm provides the maximum fairness (PM) to user devices where the maximum throughput is grounded on the present highest data rate and grades that each user entity can receive the least data rate, which is considered to be a fundamental method in allocation of resources.

$$\mathrm{PM} = \{R_i(t)\} \tag{1}$$

Though this method gives maximum throughput, it provides poor fairness. Hence, proportional fairness (PF) is the most preferable method that delivers average throughput and high fairness.

$$\mathrm{PF} = \frac{\mathrm{Current}\{R_i(t)\}}{\mathrm{past\ avg}\ \{R_i(t)\}} \tag{2}$$

where $\{R_i(t)\}$ is the data rate of the user device.

Considering this, the system algorithm FSNORA is designed accordingly and the parameter calculations are done which will be discussed ahead.

3.1 System Model of the Proposed FSNORA Method

First for the preamble detection, subframes containing of data requests from the user device are sent for the transmission where the timing offset and initial time are set during the channel modeling. The timing offset can be defined as a one-way time method that permits the system to schedule the burst and install resources before the burst arrival is sent into the network.

$$\text{initial time}(T_i) = \frac{\mathrm{nsf} - 1}{100} \tag{3}$$

where nsf is the combination of subframes. Then to the received wave, noise is added and the implementation delay of the channel modeling is removed. For the success and failure count of devices, we have considered the fading parameters as a part of the random process to measure the efficiency of the fading channels. The transmit power range at the base station is given by

$$P_1 = \mathrm{SNR}(k) - 10\log_{10}(\alpha_1.\beta_1) \tag{4}$$

where k is the window size, α_1 and β_1 are fading parameters of the first and second round, respectively. Transmitting power at the user device which is half of the power at the base station is given by

$$P_2 = \frac{P_1}{2}. \tag{5}$$

At this point of time, when the packets are ready to get transmitted, the scheduling process is excited by the round-robin algorithm. Five processes are considered, and burst time taken is taken as an array of {1, 2, 3, 4, 1, 5}. Quantum time, which is known as the period for which the process is permitted to run and turn-around time which is known as the time interval between the submission and completion of process, is initialized accordingly. The buffer flag is set to zero; if the process has the burst time left after the quantum time is completed the process runs for 1 quantum. If the process gets scheduled, remaining time is set, i.e.,

$$r(t) = r(t) - q. \tag{6}$$

where *r*(*t*) is remaining time and q is the quantum time. If it is not scheduled, the wait time is increased, i.e.,

$$w(t) = w(t) + q \tag{7}$$

where *w*(*t*) is the waiting time. However, if the remaining time is less than the quantum, it will run the process and end it. Otherwise, wait time is increased, i.e.,

$$r(t) = 0;\ \text{if}\ r(t) < q \tag{8}$$

if

$$r(t) > 0, \tag{9}$$

then,

$$w(t) = w(t) + r(t) \tag{10}$$

The packets follow this algorithm to transmit from one end to another end smoothly without any collision. Thereby, the turn-around time is calculated for each process by adding waiting time and the burst time.

$$t_a = w(t) + b(t) \tag{11}$$

This ends the process of data packet transmission. Adding on to this, the probability of collision that is defined by chances at which at least one other station transmits during the desired time slot and equivalent probability of a failed transmission attempt that states a sent frame is received correctly and it does not collide with any other frames are calculated. For the collision resolution between the two stations, back-off time is calculated that decides how much time the station should wait to retransmit the data packet [16]. We further mention 0 few simulation parameters on which the performance of FSNORA is evaluated.

3.2 Performance Evaluation

Throughput of Preamble Transmission The throughput of the header transmission is defined as the simultaneous header transmissions. Let us consider a frame or slot L which contains certain number of preambles P. A user device is said to follow a successful transmission only when all the other $L - 1$ user devices select $P - 1$ preambles otherwise it results in failure transmission.

Throughput of Random Access Process It is termed as the maximum achieved count of retrieved user devices in the RA time interval.

$$S_{\mathrm{RA}} = \max(S_{\mathrm{UD}}) \tag{12}$$

where S_{RA} is the throughput of random access procedure and S_{UD} gives the number of retrieved accessed user devices.

Collision Probability It is coined as the fraction of the maximum count of events when more than two user devices make a random access effort using the exact preamble and the veritable count of the headers in the time slot. It is given by,

$$\mathrm{CP} = \frac{m_{\mathrm{RA}}}{n_{\mathrm{RA}}} \tag{13}$$

where

$$m_{\mathrm{RA}} = I_{\mathrm{RA}} - S_{\mathrm{UD}} \tag{14}$$

Here, m_{RA} is the number of is preambles, n_{RA} is the total number of preambles, and I_{RA} is the number of idle preambles.

Access Success Probability It is coined as the occurrence when a user device achieves the complete cycle of the random access method for the greater count of preamble transferals and is given by,

$$\mathrm{ASP} = \frac{S_{\mathrm{RA}}}{n_{\mathrm{RAP}}} \tag{15}$$

where n_{RAP} is the total number of user devices arrived in the given time interval t_{P}.

Average Count of Preamble Transmissions It is termed as the fraction of the total count of preamble transmissions for all the available retrieved user devices and the veritable achievement count of preamble transferal for the retrieved user devices which is given by,

$$A_{\mathrm{P}} = \frac{\sum(n_{\mathrm{RA}}, S_{\mathrm{UD}})}{nS_{\mathrm{UD}}} \tag{16}$$

where nS_{UD} is the total number of retrieved accessed user devices.

Average Access Delay It is stated as the fraction of the veritable retrieving delay of header transferal for all the retrieving user devices and the achieved veritable count of all the retrieving user devices. This is given by,

$$D_{\mathrm{P}} = \frac{\sum(n_{\mathrm{DA}}, S_{\mathrm{UD}})}{nS_{\mathrm{UD}}} \tag{17}$$

where, n_{DA} is the total delay access of preamble transmissions.

Hence, by evaluating these parameters in order to prove a FSANORA, the results are obtained which will be discussed effectively in the next section.

4 Results and Discussion

In this section, we will discuss the experimental results of FSNORA by considering the parameters discussed in the previous section. The analysis of FSNORA is illustrated by using MATLAB by assuming two traffic models, model 1 and model 2. Model 1 is a traffic scenario that consists of less number of user devices say from (15,000–25,000), and model 2 is also a traffic model that consists of more number of user devices (say >25,000). Considering these two models, both the methods ORA and NORA have been examined. The system parameters considered are shown in Table 1.

In Fig. 1, we are considering model 1 where less number of user devices are considered. Here in the plot using the FSNORA method, the curved peaks indicating NORA method have sharply increased its value from 0 to more than 600 whereas in Zadoff–Chu method since the random method of allocating the sequences is done, the NORA peaks have reached only the limit of value 500 [1]. The unbroken curve gives the maximum throughput achieved comparing to the previous method [1]. This refers to saying that by using FSNORA number of successive and failure, user devices suffer through less collision compared to the existing NORA method.

Figure 2 illustrates the average preamble transmissions, in which the average count of header transferals of both the mechanisms under model 1 and model 2 are plotted. Here, we consider the offered load on the X-axis and the preamble transferals on the Y-axis. By observing the output, we can infer that the efficiency of the header transferal is increasing on par with the heavy traffic under both the models without any collision, whereas in Zadoff–Chu method [1] though the efficiency achieved is

Table 1 System parameters

Notation	Description	Value
N	No. of stations	10
K	Window size	51
b_M	Maximum value of back-off exponent	5
b_m	Minimum value of back-off exponent	3
w_0	Back-off window size	8
α_1, α_2	Alpha fading parameters at first hop	2
β_1, β_2	Beta fading parameters for the second hop	1
R	Information transmit rate	4
L (in bytes)	Preamble length	40
Q	Quantum number	2
N	No. of processes	5
BW (in Hz)	Bandwidth	27×10^6
X	Path loss exponent	4
P_t	Transmitting power	5
d_1	Minimum node distance	1
d_2	Maximum node distance	20

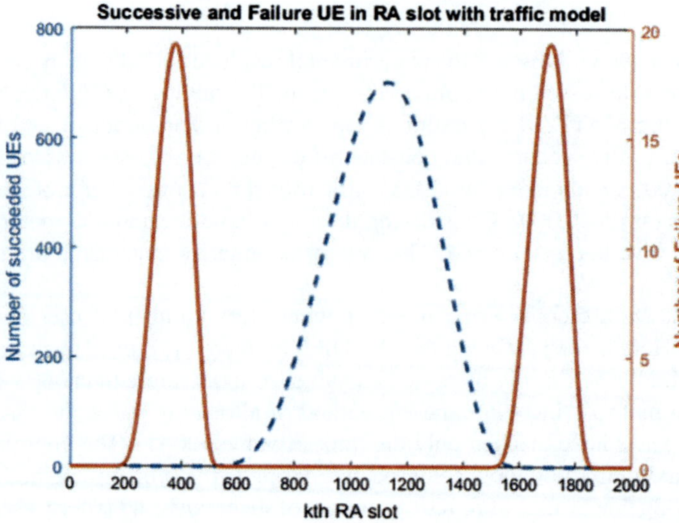

Fig. 1 Throughput of FSANORA in model **1**

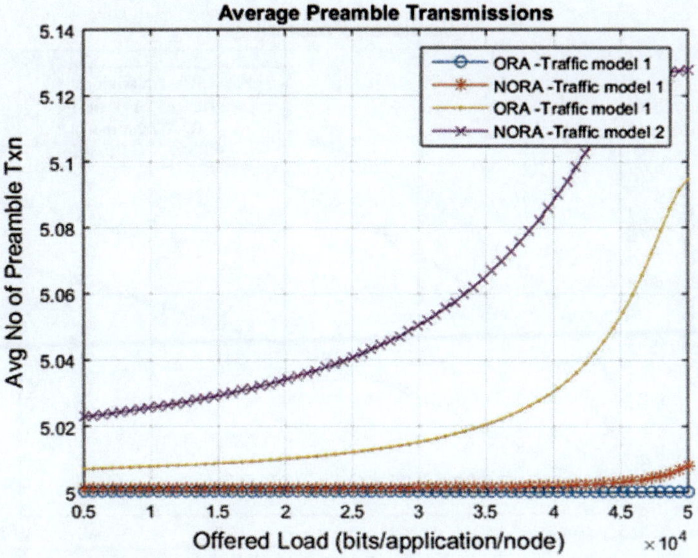

Fig. 2 Average preamble transmissions

more or less than the FSNORA, it results in collisions at extreme points because of the random selection of the user devices without any proper scheduling. Hence, non-contention average preamble transmissions are achieved by using FSNORA.

Figure 3 represents the probability of the collision for both ORA and NORA for model 1 and model 2 which is plotted against total number of user devices. Though the both models consider certain amount of user devices, in case of heavy traffic, it should be supported by enough number of user devices. By using the proposed method, we observe that even if the traffic in the system is too high, it is supported by both the models for a stable time without any collision. Comparing this observation to the Zadoff–Chu method, the collision probability is decreased only up to 30% by not using the scheduling access technique [1]. We can infer that massive number of devices in worst traffic situations will be supported by contentionless method, FSNORA.

Figure 4 represents the average delay for number of arrived user devices plotted against total number of arrived user devices in which the access delay of the orthogonal random access is increasing and that of NORA is reduced thereby proving the efficiency of NORA. Observing the plot, we discover that in huge scenarios of data traffic, the user requests don't tend to collide in the network because of using fair scheduling access. But, in the previous Zadoff-Chu method [1], in terms of huge data traffic, we observe the efficiency of NORA in presence of preamble collisions by not including the scheduling process.

In Fig. 5, the probability graph for the successful access of user devices is plotted against total number of user devices. Observing the graph, we see that from the initial access of maximum number of user devices available, the desired outcome is

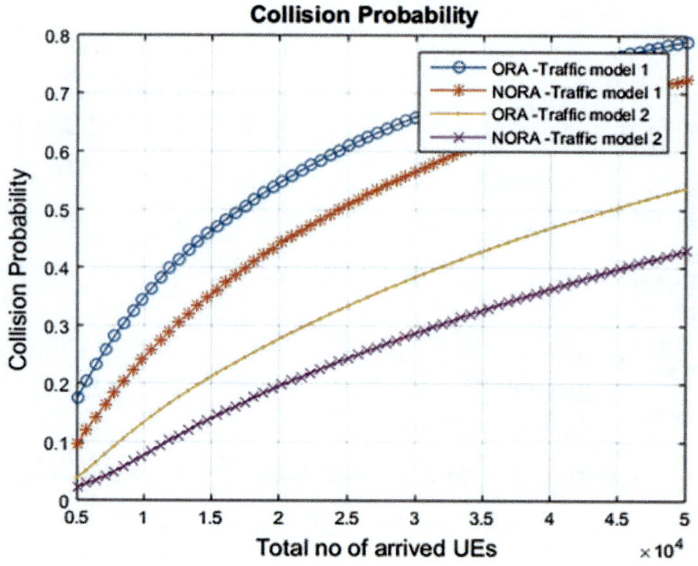

Fig. 3 Collision probability

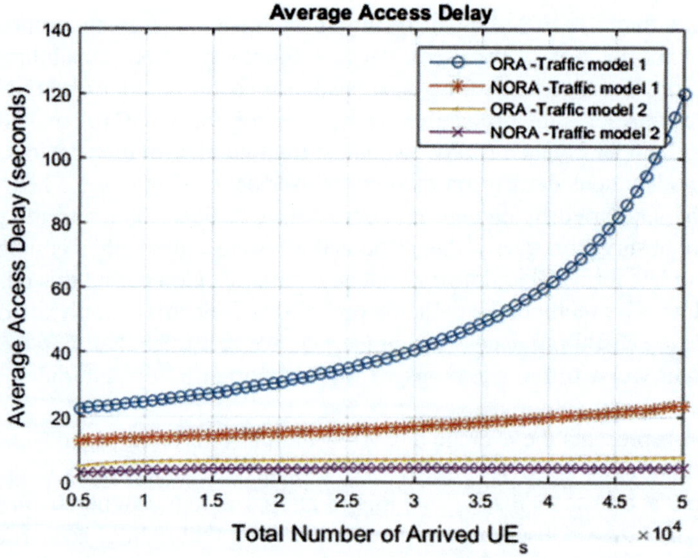

Fig. 4 Average access delay

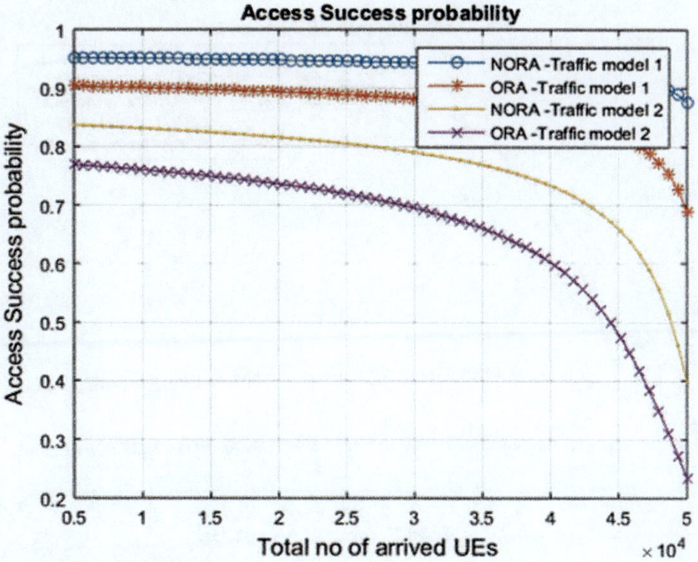

Fig. 5 Access success probability

achieved without any collisions and hence it achieves contentionless-based success probability, whereas in Zadoff–Chu method, at the initial point of access, the collision probability of the user devices is decreased only to 30% because of random preamble transmission [1]. Therefore, at any frequency of traffic, by using FSNORA, the probability of successful access of user devices can be achieved more efficiently.

Figure 6 represents the throughput graph of the arrived user devices in presence of total number of user devices under the models 1 and 2. It is a sensitive approach where we are considering huge amount of user devices in worst traffic scenarios. From the results, we observe that the throughput of ORA and NORA method has gradually increased without any collisions comparing to the previous Zadoff–Chu method of examining NORA [1]. This proves an extra improvement of throughput for non-orthogonal scheme using FSNORA.

Further, Table 2 gives the summary of comparison of NORA using Zadoff–Chu method and the proposed method, FSNORA.

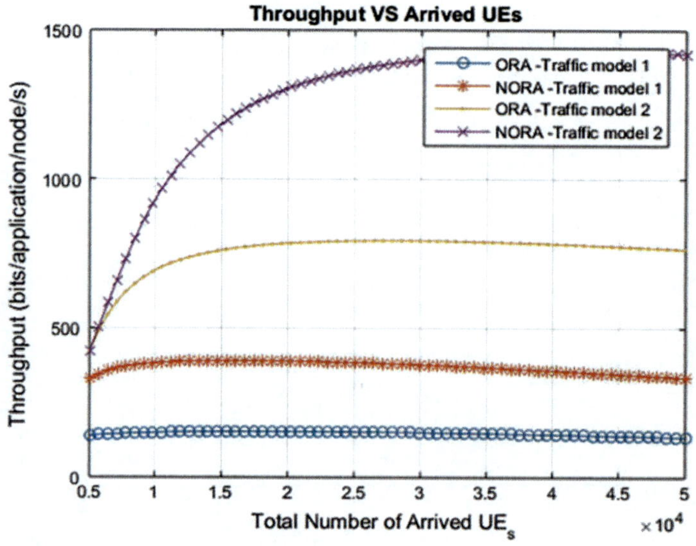

Fig. 6 Throughput versus arrived user devices

5 Conclusion

This paper concludes that by observing the simulation results of the proposed method proves its efficiency by comparing the results with NORA method which is evaluated using Zadoff–Chu sequence. In this proposed approach, the concept of round-robin is included with NORA to achieve better efficiency and failure-free communication abilities to 5G schema, because of round-robin data packets scheduling nature. Thereby, the complete round-robin-based NORA technology showed more effectiveness by supporting enormous amount of user devices. Though many previous works have been implemented to decrease the congestion problems, it still resulted in high consumption of power which is not economically feasible. This is just an embryonic implementation of using NORA in millimeter networks. Therefore, by using the algorithm, we have waved the path for fast communication of NORA as well as for the future technologies in 5G networks.

Table 2 Comparison of NORA and FSNORA

Parameter	NORA	FSNORA	Justification
Throughput of preamble transmissions	500	More than 600	For NORA, by using Z–C sequences which are considered to be random in nature resulted in the throughput but also yielded collisions whereas in, FSNORA more than 600 is achieved since the scheduling algorithm has been used, the throughput has increased without any collisions
Average preamble transmissions	5.2	5.12	The average achieved for preamble transmission for NORA is 5.2 and that of FSNORA is 5.12. Though the average value achieved is more in NORA, it results in collisions at extreme points
Collision probability	0.72	0.7	When there are huge number of user requests for a certain transmission, the collision probability of NORA is 0.72 and that of FSNORA is 0.7. We can infer that massive number of devices in worst traffic situations will be supported by FSNORA
Average access delay	110	22	The average delay of NORA method results to 110 and that of FSNORA results to 22 since in case of huge traffic scenario, the delay caused will be more if no scheduling algorithm is used
Access success probability	1 (max. value)	0.97	The parameter outperforms the NORA method over FSNORA. But the NORA method using the Zadoff–Chu sequences results in many number of collisions as they are random by nature than compared to FSNORA which is a scheduling method
Throughput of arrived user devices	350	400	Throughput achieved for maximum number of user devices in NORA is 350 and that of FSNORA is 400 due to the random sequences involved in the system

References

1. Liang Y, Li X, Zhang J, Ding Z (2017) Non-orthogonal random access for 5G networks. IEEE Trans Wirel Commun
2. Wei CH, Bianchi G, Cheng RG (2015) Modelling and analysis of random access channels with bursty arrivals in OFDMA wireless networks. IEEE Trans Wirel Commun 14(4)
3. Kim T, Jang HS, Sung DK (2015a) An enhanced randomaccess scheme with spatial group based reusable preamble allocation in cellular M2M networks. IEEE Commun Lett 19(10):1714–1717
4. Ramesh TK, Giriraja CV. Study of reassignment strategy in dynamic channel allocation scheme. In: 2016 3rd International conference on signal processing and integrated networks (SPIN)
5. Study on RAN improvements for machine-type communications, 3GPP TR37.868 V11.0.0, Sept 2011
6. Ramesh TK, Konda SK (2016) Energy-efficient resource allocation in WDM networks. In: International conference on communication and signal processing, 6–8 Apr 2016, India
7. Kim T, Jang HS, Sung DK (2015b) An enhanced random access scheme with spatial group based reusable preamble allocation in cellular M2M networks. IEEE Commun Lett 19(10):1714–1717
8. Arana JM, Han JP, Cho YS. Random-access technique for self-organization of 5G millimeter-wave cellular communications. 2016(Article ID 5261089):11 p
9. Le NT, Hossain MA, Islam A, Kim D, Choi Y-J, Jang YM. Survey of promising technologies for 5G networks. 2016(Article ID 2676589):25 p
10. Hsu Y-H, Wang K, Tseng Y-C (2014) Efficient cooperative access class barring with load balancing and traffic adaptive radio resource management for M2M communications over LTE-A. Computer Netw
11. Wang Z, Wong VW (2015) Optimal access class barring for stationary machine type communication devices with timing advance information. IEEE Trans Wirel Commun 14(10):5374–5387
12. Wiriaatmadja DT, Choi KW (2015) Hybrid random access and data transmission protocol for machine-to-machine communications in cellular networks. IEEE Trans Wirel Commun
13. Ding Z, Peng M, Poor HV (2015) Cooperative non-orthogonal multiple access in 5G systems. IEEE Commun Lett 19(8):1462–1465
14. Zhou Y, Wong VWS (2018) Chapter 8-1 non-orthogonal multiple access (NOMA). Springer. Chen J, Lin YT Cheng RG (2015), A delayed random access speed-up scheme for group paging in machine-type communications. In: 2015 IEEE International Conference On Communications (ICC), Jun. 2015, pp 623–627
15. Liang Y, Li X, Zhang J, Liu Y (2017) A novel random access scheme based on successive interference cancellation for 5G networks. In: 2017 IEEE wireless communications and networking conference (WCNC)
16. Ding Z, Yang Z, Fan P, Poor HV (2014) On the performance of non-orthogonal multiple access in 5G systems with randomly deployed users. IEEE Signal Process Lett 21(12):1501–1505

Analysis of Tree-Based Classifiers for Web Attack Detection

Deshmukh Surbhi and Kshirsagar Deepak

Abstract In the last few decades, the increase in the use of Web services has led to an increase in the threats of Web attacks. The severity of such Web attacks is increasing day by day. Intrusion detection systems play a crucial role in identifying Web attacks in proactive manner. There are large numbers of features present in the network traffic. Identification of relevant and irrelevant features is crucial task in machine learning. This paper proposes a Web attack detection system that consists of preprocessing, feature selection, reduced dataset, and tree-based classifiers. The system uses information gain filter method to select relevant features for the classification of Web attack. The system is tested on CIC-IDS-2017 dataset. The experimentation results show that random forest produces high precision of 74.5% for brute force, and J48 produces high precision of 63.8% and 87.5% for cross-side scripting (XSS) and SQL injection (SQLi), respectively, with 65 selected features.

Keywords Web attack · Information gain · Tree-based classifiers · Feature selection

1 Introduction

A Web attack [1] is a malevolent attempt by an attacker to breach a network containing confidential and important data of individuals or organizations. Attackers use many different ways to exploit their victims' networks. Some of the common methods are described here. A brute force attack is the use or trial and error method for guessing an username or password so as to gain access to victim's system. Analysis of previous data breaches helps user to automate the password guessing and easily attack victim's machine. SQL injection [2] is used by attackers to compromise an

D. Surbhi (✉)
Tata Technologies Limited, Pune, India
e-mail: deshmukhsa18.comp@coep.ac.in

K. Deepak
Department of Computer Engineering and IT, College of Engineering, Pune, India
e-mail: ddk.comp@coep.ac.in

© Springer Nature Singapore Pte Ltd. 2021

S. N. Merchant et al. (eds.), *Advances in Signal and Data Processing*,
Lecture Notes in Electrical Engineering 703,
https://doi.org/10.1007/978-981-15-8391-9_31

application's database using malicious SQL code to view or alter the records in the database. An attacker injects malicious code on another user's browser to create vulnerability of XSS. The malicious script can be injected into the current HTTP request, the Web site's database, or from the client side.

Anyone from an individual to large organization may fall prey to such cyber-attacks where strict measures are not in place. The National Vulnerability Database [3] suggests that most of the vulnerabilities are found in the Web applications which include XSS, SQLi, and buffer overflow. The attacker can exploit these vulnerabilities to generates attacks. Hence, there is a need to detect such attacks in an intelligent manner.

An intrusion detection system (IDS) is software which is designed to detect such malicious activities automatically and provide the alert to the administrator. Based on the detection technique, IDS is classified into various types. In misuse detection, the IDS analyze the captured information and compare it to attack signatures or patterns. The anomaly IDS monitors network segments to compare their state to the normal profile and look for anomalies behavior. In a network-based system, the captured packets flowing through a network are analyzed. However, in a host-based system, examine the activity on each individual computer or host which is connected to the network.

Data mining and artificial intelligence techniques using machine learning are used for the development of IDS for the feature reduction and attack detection.

The contribution of this work is as follows:

The proposed system consists of preprocessing feature selection and tree-based algorithms for the effective detection of Web attack.

Conducted experiments on Web attack CIC-IDS-2017 dataset provide consistent performance with reduced features.

The rest of the paper is organized as follows. Section 2 describes the literature review in the domain of IDS. Section 3 presents proposed system with neat sketched diagram. Section 4 describes the implementation and result analysis of the system. Finally, the conclusion is presented in Sect. 5 with future scope.

2 Literature Work

Thaseen et al. [4] analyzed tree-based classifiers for the intrusion detection system. The proposed system selects a feature based on correlation and best-fit search strategy. The system uses tree-based classifiers such as C4.5, REP tree, and random forest for the classification. The system has experimented with NSL-KDD99 dataset. The system produced higher accuracy and low false alarm rate with random forest.

Patil et al. [5] proposed an approach for the detection of DDoS attacks based on feature selection. In this approach, features are classified into primary and secondary based on information gain calculated for each feature. The logistic model tree, random forest, and J48 algorithms are used for DDoS attack detection. The approach is tested on DDoS CIC-IDS-2017 dataset. The experimental result shows that J48 produces higher detection rate of 82.74% with primary features.

Repalle et al. [6] proposed artificial intelligence and machine learning-based intrusion detection systems. The system used support vector machine (SVM)3 and K-nearest neighbors (KNN) algorithms for intrusion detection. The system is tested on the CTU-13 dataset which consists of normal, background, and botnet traffic. KNN produces better results as compared to SVM.

Ahmim et al. [7] proposed a novel IDS which combines rule and tree-based classifiers. The system uses combination of three classifiers namely REP tree, JRip, and forest PA. Comparative study has been done for different classifiers.

Wankhede et al. [8] used neural network and machine learning in intrusion detection for DoS attacks. The system is simulated with the help of WEKA. The system used a multi-layered perceptron (MLP) and random forest for the classification of the DoS attack. The system is evaluated on DoS CIC-IDS-2017 dataset. The system produced higher accuracy of 99.95% for random forest as compared to MLP. Ahmim et al. [7] proposed a novel IDS which combines rule and tree-based classifiers. The system uses combination of three classifiers namely REP tree, JRip, and forest PA. Comparative study has been done for different classifiers.

3 Proposed System

The framework for Web attack detection using tree-based classifiers is as shown in Fig. 1.

$$\text{IG} = \begin{pmatrix} \text{Entropy of distribution} \\ \text{before the split} \end{pmatrix} - \begin{pmatrix} \text{Entropy of distribution} \\ \text{after it} \end{pmatrix} \tag{1}$$

The feature selection-based IG provides relevant features for Web attack detection. Further, relevant features are used to train and test the system. The proposed system uses tree-based classifiers such as random forest, J48, and REP tree for the detection of Web attack with IG based feature selection. The system is evaluated, and the

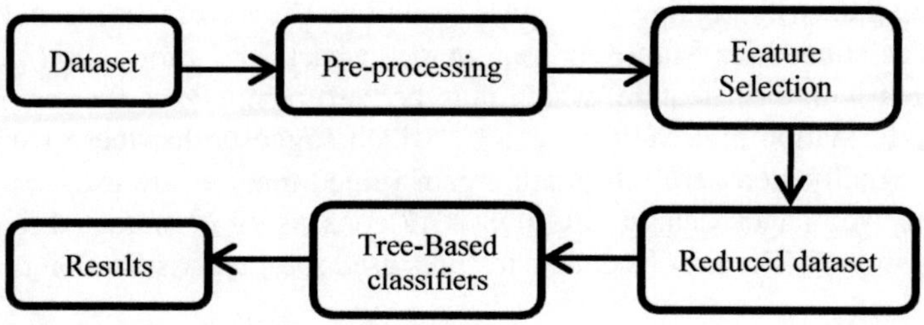

Fig. 1 Overview of the proposed system

performance is measured in terms of precision, recall, false positive rate (FPR) with model built-up time.

4 System Implementation and Result Analysis

WEKA, i.e., Waikato Environment for Knowledge Analysis, tool [9] is used for the system implementation and testing. The proposed system is implemented on the machine having operating system—Windows 10, RAM—32 GB, and CPU—Intel Xeon E3-1271.

4.1 Dataset Description

The Canadian Institute for Cybersecurity (CIC) has provided a benchmark dataset named as CIC-IDS-2017 for the evaluation of IDS. The dataset is generated by capturing the abstract behavior of 25 users with the help of CICFlowMeter based on FTP, HTTP/S, and SSH protocols. The dataset is available in PCAP and CSV formats, and instances are labeled as benign and attack. The CSV format files are available for research in the domain of machine and deep leaning. The dataset consists of benign and Brute Force, DoS, DDoS, Web, Infiltration, and bot. The dataset is created from Monday to Friday for benign and attacks. The flow and machine learning CIC-IDS-2017 datasets consist of 84 and 78 features, respectively. The system is evaluated on a machine learning-based CIC-IDS-2017 dataset consisting of 78 features.

4.2 Implementation

Initially, the captured dataset is preprocessed with handling issues of duplicate features, missing values, and infinite values. Then, this consistent and complete dataset formed with preprocessing is used for feature selection. The system uses information gain with ranking for feature selection. The information gain based on entropy is calculated for features present in the dataset. The information gain value is associated with each feature, which may be zero or non-zero. For some of the features, information gain values are zero, which suggests that these features are irrelevant. Finally, non-zero information gain valued features are used for classification. The Web attack dataset CIC-IDS-2017 consists of 78 extracted features as mentioned in [10]. The list of selected features used for the classification is as given in Table 1.

The above selected 65 features as given in Table 1 are used to train and test the system. The system uses three tree-based classifiers on the dataset viz random forest (RF), J48, and REP tree with tenfold cross-validation. The performance of the system

Table 1 List of the selected relevant feature

Number of selected features	Feature number
65	67, 25, 37, 35, 16, 22, 18, 17, 19, 23, 38, 24, 2, 21, 20, 40, 7, 9, 54, 5, 63, 66, 53, 41, 3, 62, 1, 15, 11, 65, 6, 55, 13, 42, 43, 36, 64, 4, 47, 68, 8, 39, 29, 26, 27, 12, 69, 52, 30, 10, 14, 28, 74, 76, 77, 73, 70, 72, 48, 75, 71, 31, 45, 49, 44

is measured in the terms of precision, recall, and false positive rate (FPR) as shown in Eqs. (2), (3), and (4), respectively.

$$\text{Precision} = \frac{\text{TP}}{(\text{TP} + \text{FP})} \tag{2}$$

$$\text{Recall} = \frac{\text{TP}}{(\text{TP} + \text{FN})} \tag{3}$$

$$\text{FPR} = \frac{\text{FP}}{(\text{FP} + \text{TN})} \tag{4}$$

where TP—True positive, FP—False positive, TN—True negative, and FN—False negative.

The proposed system is evaluated on the Web attack CIC-IDS-2017 dataset with all and selected 65 features. Table 2 shows the analysis of tree-based classifiers with original feature set.

The selected features are used to test the tree-based classifiers, and measured performance is as given in Table 3. Table 3 shows the analysis of tree-based classifiers with selected 65 features from original feature set.

Table 2 Analysis of tree-based classifiers with all features

Classifier	Built-up time (S)	Attack	Precision (%)	Recall (%)	FPR
RF	118.1	Benign	100	100	0.022
		Brute force	74.5	80.9	0.002
		XSS	43.9	32.5	0.002
		SQLi	75	42.9	0
J48	44.72	Benign	100	100	0.011
		Brute force	71.3	97.5	0.004
		XSS	63.8	9.2	0
		SQLi	87.5	66.7	0
REP tree	15.64	Benign	100	100	0.019
		Brute force	72.2	90.6	0.003
		XSS	48.2	18.1	0.001
		SQLi	58.8	47.6	0

Table 3 Analysis of tree-based classifiers with 65 features

Classifier	Built-up time (S)	Attack	Precision (%)	Recall (%)	FPR
RF	127.58	Benign	100	100	0.022
		Brute force	74.5	80.5	0.002
		XSS	43.9	33.1	0.002
		SQL injection	72.7	38.1	0
J48	45.85	Benign	100	100	0.011
		Brute force	71.3	97.5	0.004
		XSS	63.8	9.2	0
		SQL injection	87.5	66.7	0
REP tree	14.79	Benign	100	100	0.019
		Brute force	72.2	90.6	0.003
		XSS	48.2	18.1	0.001
		SQL injection	58.8	47.6	0

Fig. 2 Tree-based classifiers analysis for brute force

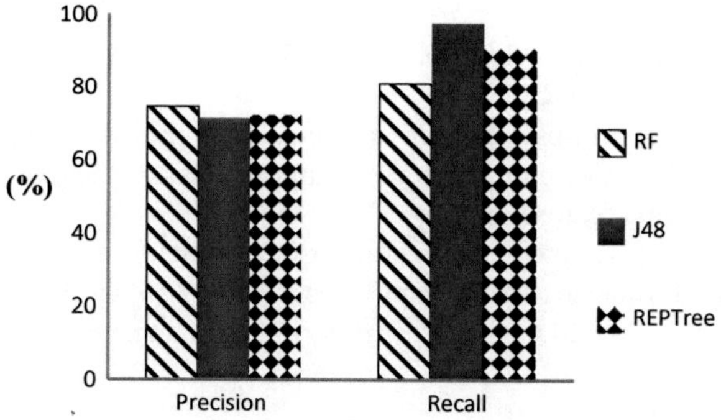

Figures 2, 3, and 4 show the analysis of RF, J48, and REP tree classifiers for brute force, XSS, and SQLi, respectively. Figure 2 shows that RF produces high precision for brute force attack as compared with J48 and REP tree with 65 features. Figures 3 and 4 show that the J48 algorithm produces higher precision for XSS and SQLi as compared to RF and REP tree with 65 features.

5 Conclusion

This paper proposed the Web attack detection system based on information gain filter method with tree-based classifiers. The information gain filter method is used to identify relevant features for Web attacks. The system reduces features from 77 to 65 based on information gain. The system is implemented and tested on the CIC-IDS-2017 dataset. The experimentation result shows that tree-based classifiers produced

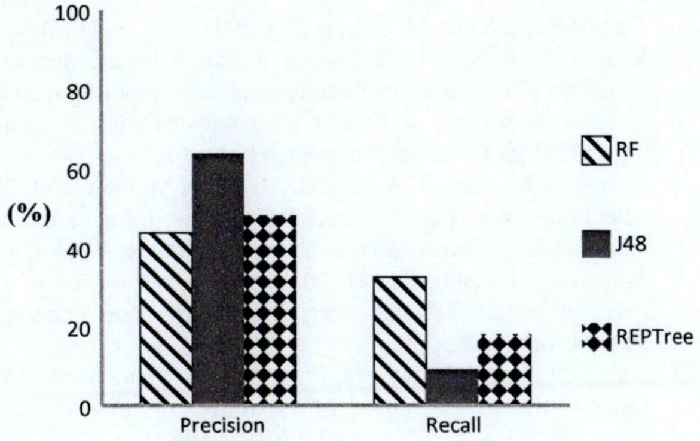

Fig. 3 Tree-based classifiers analysis for XSS

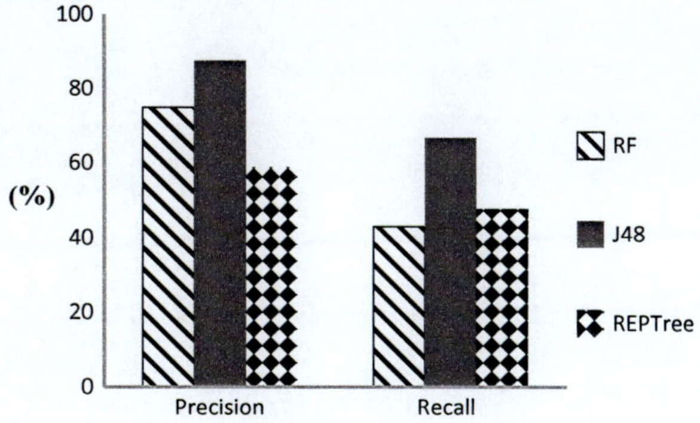

Fig. 4 Tree-based classifier analysis for SQLi

same performance in terms of precision, recall, and FPR with selected 65 features as compared to original feature set.

In the future, other feature selection methods and classifiers in machine learning will be used to improve the performance of Web attack detection.

References

1. Common cyberattacks to look out for, ManageEngine Log360 (2019) https://www.manageengine.com/log-management/cyber-security-attacks/common-types-of-cyber-attacks.html. Accessed 8 Aug 2019
2. Jeff M (2018) In: Top 10 most common types of cyber attacks, Netwrix Corporation. https://blog.netwrix.com/2018/05/15/top-10-most-common-types-of-cyber-attacks/. Accessed 2 Aug 2019
3. Khairkar A, Kshirsagar D, Kumar S (2013) Ontology for detection of web attacks. In: 2013 international conference on communication systems and network technologies. IEEE, pp 612–615
4. Thaseen S, Kumar A (2013) An analysis of supervised tree based classifiers for intrusion detection system. In: 2013 International Conference on Pattern Recognition, Informatics and

Mobile Engineering. IEEE, pp 294–299

5. Patil A, Kshirsagar D (2019) An approach towards feature selection for detection of DDoS attack. In: 4th international conference on computing in engineering and technology
6. Repalle S, Kolluru V (2017) Intrusion detection system using AI and machine learning algorithm. Int Res J Eng Technol 4:1709–1713
7. Ahmed A, Maglaras L, Ferrag MA, Derdour M, Janicke H (2019) A novel hierarchical intrusion detection system based on decision tree and rules-based models. In: 2019 15th international conference on distributed computing in sensor systems (DCOSS). IEEE, pp 228–233
8. Wankhede S, Kshirsagar D (2018) DoS attack detection using machine learning and neural network. In: 2018 fourth international conference on computing communication control and automation. IEEE, pp 1–5
9. Hall M, Frank E, Holmes G, Pfahringer B, Reutemann P, Witten IH (2009) The WEKA data mining software: an update. ACM SIGKDD Explor Newsl 11(1):10–18
10. Panwar S, Raiwani Y, Panwar L (2019) Evaluation of network intrusion detection with features selection and machine learning algorithms on CICIDS-2017 dataset. In: 2019 international conference on advances in engineering science management & technology, pp 1–10

Implementation of Random Pulse Width Modulation Techniques for the Open-End Winding Five-Phase Motor Drives to Reduce Acoustic Noise and Harmonic Distortion

J. Balakrishna, Teegala Bramhananda Reddy, and Marapu Vijaya Kumar

Abstract Multi-phase machine drives nowadays found its applications in electric vehicles, ship propulsion, space craft, drilling machines. High-speed switching of the space vector pulse width modulation techniques (SVPWM) generates high rate of change of currents and voltages in the inverters, which causes serious electromagnetic interference. These generate harmonics in output voltages and currents and high acoustic noise is produced. All these applications produce high acoustic noise and high harmonics distortion which disturbs the human being and some electronics devices that are nearer to the drive. Hence, this paper provides a solution by implementing random pulse width modulation (PWM) techniques for the dual-inverter fed open-end winding (OeW) five-phase induction motor drives to reduce acoustic noise levels that are produced in output effective phase voltages and currents. Randomization in modulating signal; randomization of high-frequency triangular carrier signals; both modulating signal and carrier randomization; variable switching frequency randomization of both modulating signal and high-frequency triangular carrier signals methods are implemented. To show the effectiveness of the proposed random techniques over the conventional continuous PWM techniques, simulation and experimental studies on 1 Hp proto type five-phase motor are presented.

Keywords Acoustic noise · Harmonic distortion · Multi-level multi-phase drives · Open-end winding · Random pulse width modulation · Space vector pulse width modulation

J. Balakrishna (✉) · M. Vijaya Kumar
Department of EEE, Jawaharlal Nehru Technological University, Anantapur, Andhra Pradesh, India
e-mail: balakrishna.caid@gmail.com

T. Bramhananda Reddy
Department of EEE, G Pulla Reddy Engineering College, Kurnool, Andhra Pradesh, India

© Springer Nature Singapore Pte Ltd. 2021
S. N. Merchant et al. (eds.), *Advances in Signal and Data Processing*,
Lecture Notes in Electrical Engineering 703,
https://doi.org/10.1007/978-981-15-8391-9_32

1 Introduction

Field of high-power electrical drives has emerged as one of the most advanced and active areas of research in the last two decades. High voltage and current ratings of the power semiconductor devices have limited the growth of this area. The real-time industrial needs have increased in terms of power-level needs, high production levels, high efficiency, reduced harmonics in outputs, reduced noise, and better power spectrum. In high-power applications, the machines with more than three-phase are able to satisfy the requirements of modern-day industrial drives. The first publication on inverter fed multiphase machine in 1969 by Ward and Harer [1] has drawn steady attention among multi-phase machines. The benefits of multi-phase machines over standard three-phase machines can now be synthesized after evaluating numerous literature's by a great number of analyzes such as reduced rating of power semiconductor switches; improved magneto motive force (MMF) in the machine; reduced DC-link harmonics; improved torque density; reduced torque pulsation; reduced rotor current harmonics; reduced current per phase; increased reliability; improvement in fault tolerance; improvement in power-sharing capability; reduced space harmonic content and higher efficiency; lower acoustic noise. Dual-inverter supply topology is an alternative to multi-level voltage supply and advantages of which is discussed in literatures [2–4]. Multi-phase machines are nowadays using as a replacement to conventional three-phase machines because of its major advantages and are listed in detail in [5]. In conventional space vector PWM techniques, the space vectors are made to switch at high frequencies due to which an increase in harmonic content; torque pulsations; acoustic noise; mechanical vibrations; and electromagnetic interferences. Conventional PWM converters operating under non-sinusoidal power supplies radiate acoustic noise which increases the total harmonic distortion (THD) of the output torque, voltages, and currents [6–8]. The literature work presented in [8–10] will address the increase in acoustic noise and current ripples due to the PWM inverters fed three-phase induction machine drives. Various researchers have proposed different random PWM techniques for the three-phase inverter [11–15] which includes random pulse position PWM (RPP-PWM); random switching frequency PWM (RSF-PWM); random center distribution PWM (RCD-PWM); random zero vector PWM (RZV-PWM); random lag-lead pulse position PWM; and random carrier frequency PWM (RCF-PWM). The authors in paper [16–18] have researched on dual random SVPWM technique that combines both random switching frequency and random zero vector schemes. In applications like electric vehicles (EV) or plugged-in hybrid EV's, the noise emitted by the motor drives should not exceed the prescribed noise levels. Authors in paper [19] have designed compensation circuits for noise reduction methods by spread spectrum. Very few research papers are published upon the application of random PWM techniques for multi-phase machines. A resent work on performance of random space vector PWM techniques for two-level inverter fed five-phase motor drives is studied in [20].

Figure 1 show the simplified diagram of dual-inverter fed OeW five-phase motor drives. Two-level five-phase inverter can generate $2^5 = 32$ voltage space vectors

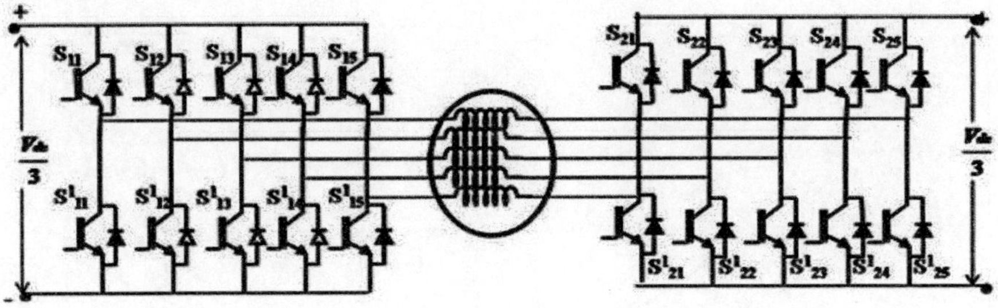

Fig. 1 Dual-inverter five-phase motor configuration

which form 30 active vectors; 2 zero-vectors; and 10 sectors in two-dimensional planes as shown in Fig. 2. Mapping of 32 × 32 switching states results in 1024 switching states to form an open-end winding topology. Figure 2a, b indicates the position of space vectors that are generated by the two inverters. Mapping of space vectors of inverters I and II to form a OeW results in 1024 switching states and 211 sectors which is plotted in Fig. 2c. It is quite hard to evaluate the enormous amount of space vectors, and it is also found from the literature's that the harmonic spectrum concentration at and around the multiples of switching frequency causes space vector modulation methods to produce acoustic noise, vibration, and electromagnetic interferences. The elevated harmonics generated can result in current and voltage distortion; additional energy loss; thermal stress; electromagnetic interference; torque ripples in rotating machines; mechanical vibration; acoustic noise radiation. Hence, a simplified carrier-based approach is followed throughout this paper to implement the random PWM techniques.

Different types of random PWM techniques have been discussed in many literatures which include random switching frequency and random pulse position (constant switching frequency) techniques. By changing rate of switching, RSF-PWM schemes can be obtained. However, constant switching frequency RPWM techniques are simple and easier for implementation. The acoustic noise produced in conventional PWM techniques is due to the concentration of high energy at and around the harmonics of the switching frequency. This needs to be reduced by randomizing the pulse pattern. The two high-frequency carrier signal (V_t) intersects the positive and negative half cycle of reference signal (V_r) to form switching instants. Figure 3a, b indicates the pulse positions in PWM techniques for the conventional and random carrier PWM. It is noted in Fig. 3a pulse width is constant and duty cycle varies. In Fig. 3b for random carrier frequency-based PWM, only the carrier frequency is varied. But for continuously varying the switching frequency, it is necessary to design the filter circuit, which is complicated. Hence, for easier filter design and easier implementation fixed switching frequency random PWM (RPWM) is proposed in this paper. It is observed that spectral line produced in conventional SVPWM coincides with the resonant frequency of the motor parts resulting high noise. By randomly varying the period of the triangular carrier spectrum without energy concentration at descript frequencies can be achieved.

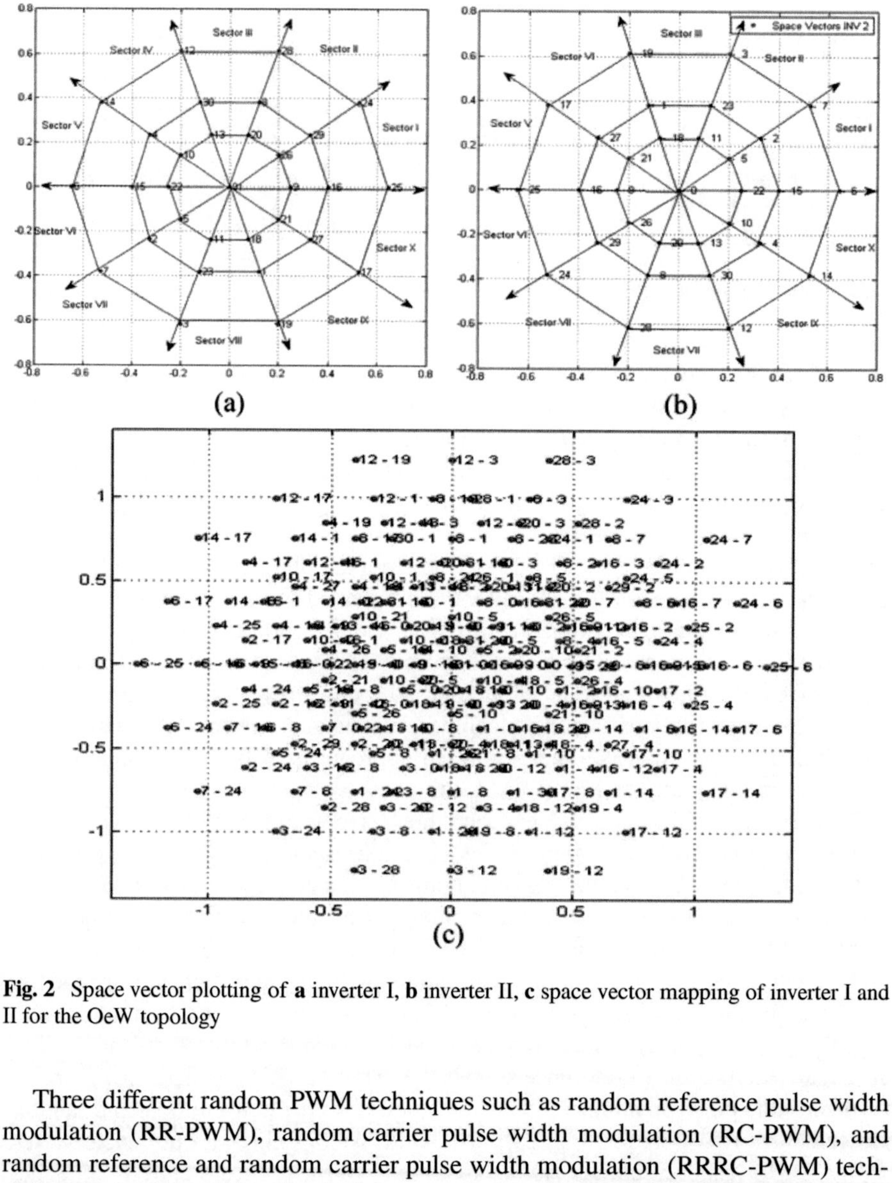

Fig. 2 Space vector plotting of **a** inverter I, **b** inverter II, **c** space vector mapping of inverter I and II for the OeW topology

Three different random PWM techniques such as random reference pulse width modulation (RR-PWM), random carrier pulse width modulation (RC-PWM), and random reference and random carrier pulse width modulation (RRRC-PWM) techniques are proposed for the dual-inverter fed OeW five-phase drive. This paper focuses on a simple scalar-based PWM techniques for four-level dual-inverter configuration fed five-phase induction machine. This PWM technique implementation does not involve complex calibrations like sector and its respective vector identification. Comparative studies between continuous and DPWM3 methods are studied in terms of CMV, number of levels in effective phase voltage and its harmonic spectrum.

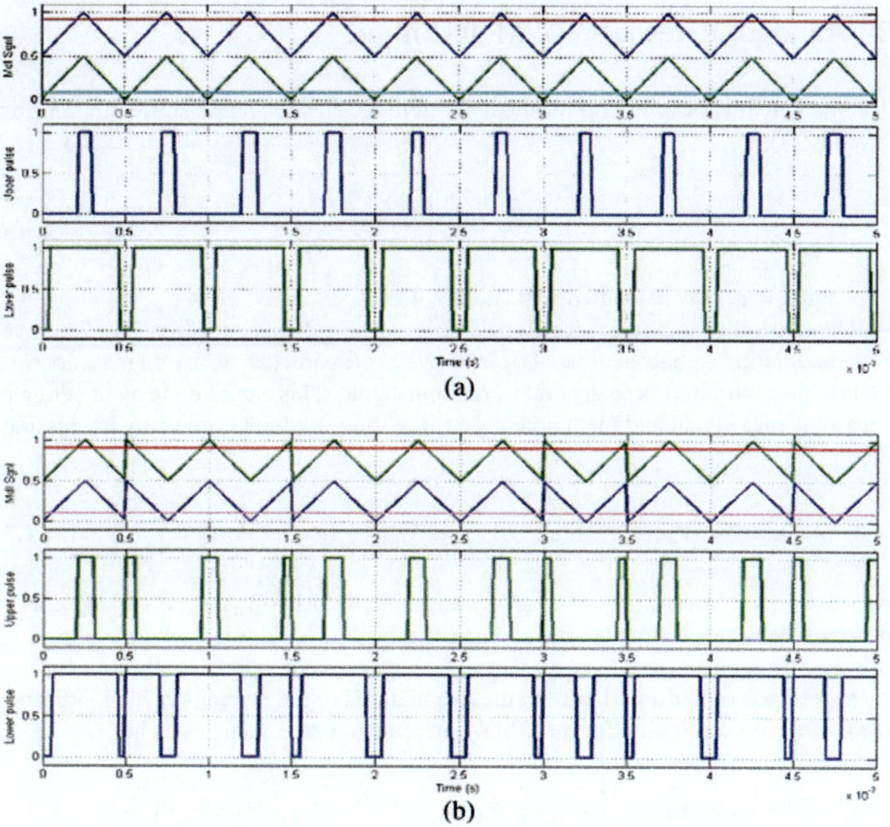

Fig. 3 Pulse generation at $M = 0.8$; $F_s = 2000$ Hz for **a** constant carrier PWM, **b** random carrier PWM

Over chagrining effect is eliminated by implementing an offset addition subtraction method. To validate the results, MATLAB/SIMULINK-based simulation and experimental results are presented.

This paper is subdivided into four sections. Introduction to random PWM techniques for a basic two-level inverter fed five-phase induction motor is discussed, and results for reduction of harmonic spectrum are presented in Sect. 2. Four different random PWM techniques for the five-phase OeW drive analysis is presented in Sect. 3 which include random reference PWM (RR-PWM); random carrier PWM (RC-PWM); random reference and random carrier PWM (RRRC-PWM,); random carrier selection; and random variable switching frequency PWM (RRRC-CFC-PWM) Experimental results are presented in Sect. 4. A comparative result of the conventional PWM and proposed random PWM techniques with respect to the harmonic spectrum analysis of the effective phase voltage is presented in Sect. 5.

2 Random Reference PWM Technique

The mathematical expression for the five reference signals, phase shifted by 2π/5 is given in (1)

$$V_i = V_{\max} \cos\left(\omega_t - \frac{2\pi}{5}\right) \tag{1}$$

where $i = a, b, c, d, e$ and $a = 0, 1, 2, 3, 4$.

The five sinusoidal reference signals have a phase displacement of 72°. The five sine modulating signals generated from above expression can be varied from continuous to discontinuous or to generate a random signal. This can be done by injecting a zero-sequence signal and mathematical expression for which is given by expression (2)

$$V_{zs} = \frac{V_{dc}}{2}(2a_0 - 1) - a_0 V_{\max} + (a_0 - 1) V_{\min} \tag{2}$$

where $V_{\min} = \text{Min}(V_a, V_b, V_c, V_d, V_e)$ and $V_{\max} = \text{Max}(V_a, V_b, V_c, V_d, V_e)$. The new set of reference signals obtained is given by $V_i^* = V_i + V_{ZS}$ where $i = a, b, c, d, e$.

$V_{oo'}$ is the voltage drop between the two neutrals of the inverters, which indicates the CMV. The expression for the CMV derived from pole voltages as in (3)

$$V_{CMV} = V_{OO^1} = \frac{1}{5}((V_{a0} - V_{a'0'}) + (V_{bo} - V_{b'0'}) + (V_{co} - V_{c'0'}) \\ +(V_{do} - V_{d'0'}) + (V_{e0} - V_{e'0'})) \tag{3}$$

The continuous and discontinuous modulating signals can be selected by varying the a_0 value between 0 and 1. For a_0 value of 0.5, continuous modulating signals can be obtained. Injecting a zero-sequence signal V_{zs} to reference modulating signal V_i is shown in Fig. 4. Figure 4a indicates the generation of continuous modulating signal. Addition of a zero-sequence signal V_{zs} to each of the five reference modulating signal V_{ref} employs additional the DC bus voltage utilization by 15.5% more equating with conventional SPWM technique. The positive and negative peaks of resulted signal are little bit flattened making it to form exactly similar to conventional SPWM modulating signal. Randomization of the continuous modulating signals can be obtained by varying the a_0 value from 0 to 1 which has abrupt variations in magnitude is shown in Fig. 4b. In order to produce the control signals for IGBT switches of inverter I and inverter II, the generated modulating signals are compared with carrier triangular signals.

The gating pulses generated for each switch of inverter I (S11, S12, S13, S14, S15) and inverter II (S21, S22, S23, S24, S25) for continuous and random modulating signals are shown in Fig. 5a, b. It is observed that each phase is shifted by 2π/5. It is observed that the pulses produced in random carrier PWM has abrupt variations

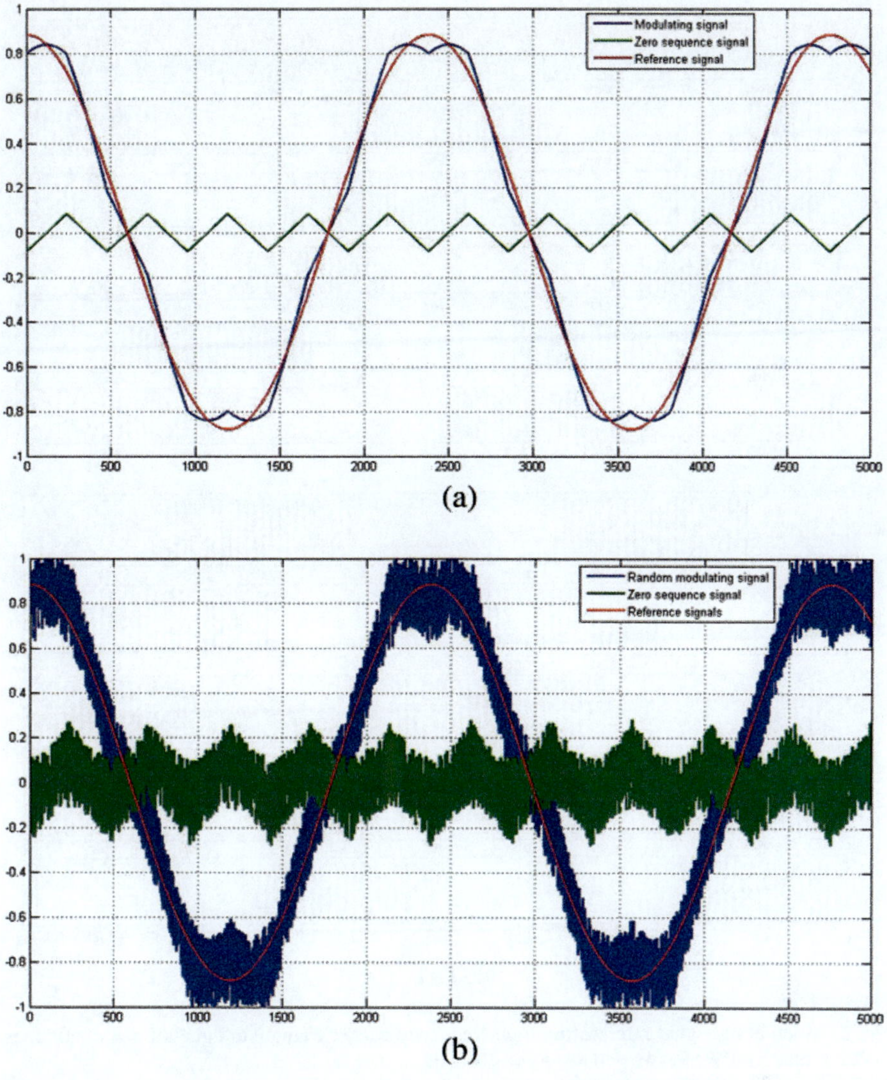

Fig. 4 Generating modulating signals for **a** continuous PWM, **b** RR-PWM at $M = 0.88$; $F_s = 2000$ Hz

in modulating signal which intern reduces the energy concentration at and around harmonics of the switching frequencies.

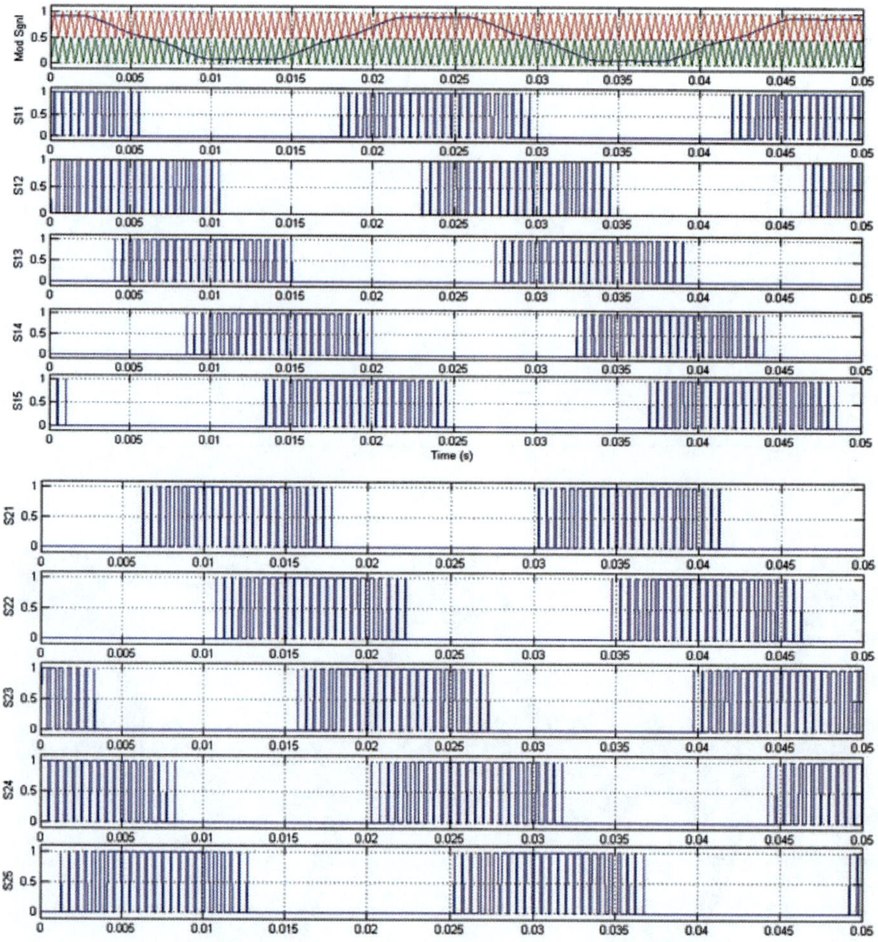

(a)

Fig. 5 Modulating signal intersecting high-frequency carrier triangular signals for **a** continuous PWM, **b** random PWM at $M = 0.88$; $F_s = 2000$ Hz

3 Random Carrier PWM Technique

In this, the level shifting carrier signals are randomly varied by using positive and negative level shifting carriers as shown in Fig. 6. A positive and negative shifting carrier are compared with high-frequency random signal generator as a result of which generates a carrier selector combining the positive and negative level shifting carrier signals. This output carrier selector depends on random generator. The illustration of this scheme is pictured in Fig. 7, at switching frequency of 500 Hz. During the time interval 0–0.001, random generator has an output of 1; hence, positive levels

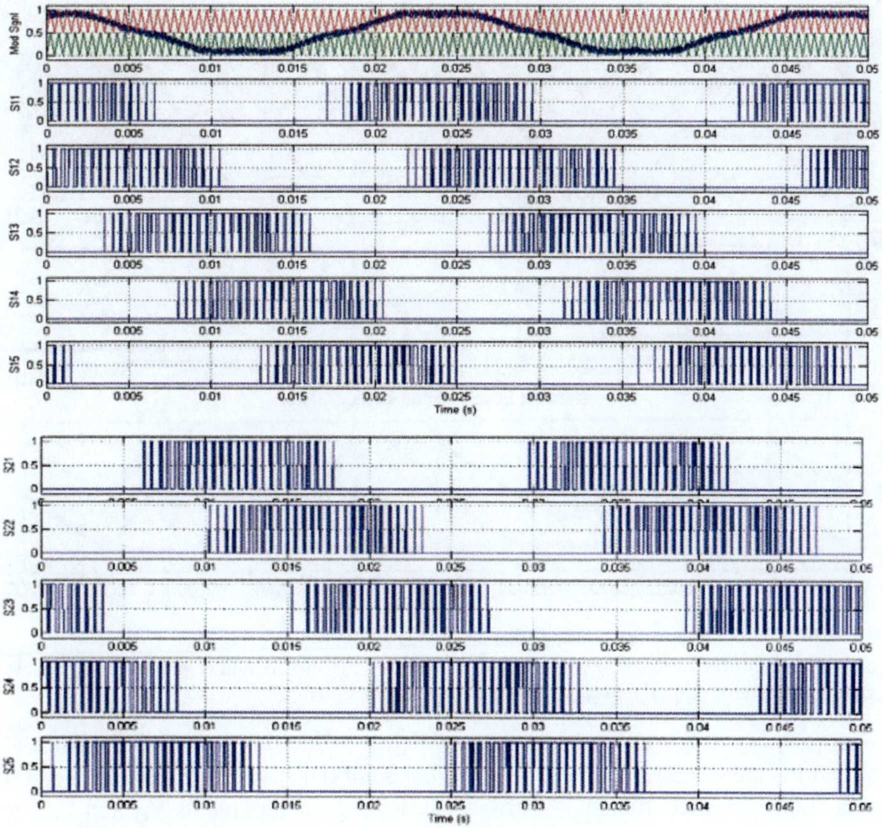

(b)

Fig. 5 (continued)

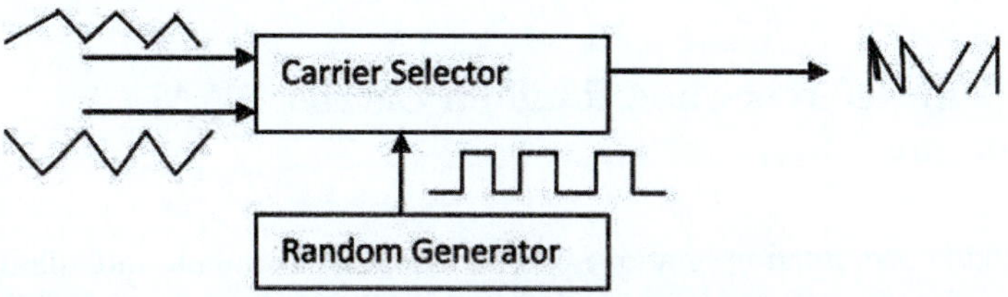

Fig. 6 Illustration of positive and negative level shifting carrier signals comparison with random signal generation

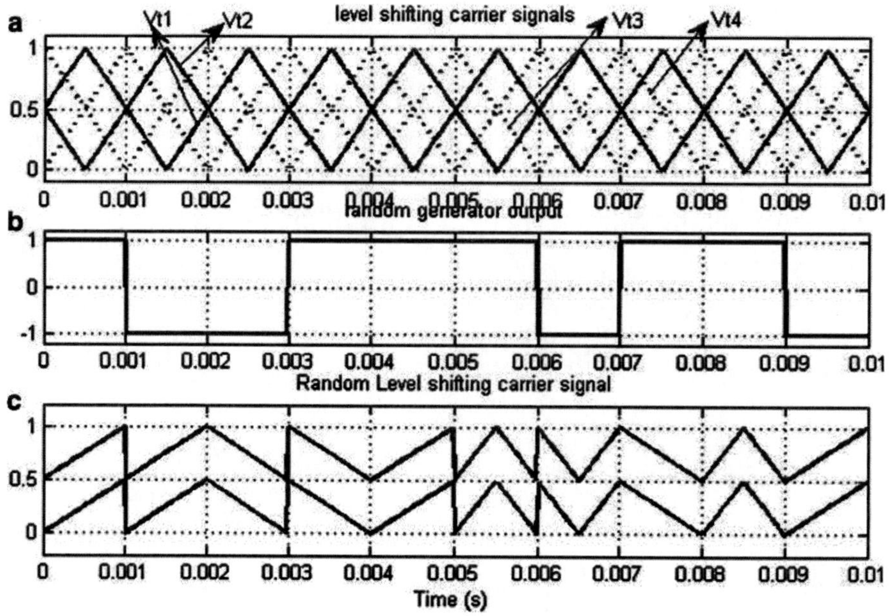

Fig. 7 Illustration of random carrier selection **a** level shifting positive and negative carrier signals, **b** random generator output, **c** random level shifting carrier signals

V_{t1} and V_{t2} are selected. During the time interval of 0.001–0.003, random generator has an output of zero; hence, negative level shifting carrier signals V_{t3} and V_{t4} are selected. This results in a blend of a positive and negative randomly generated level shifting signals. Figure 8 indicates the pulse pattern generation for inverter I and inverter II of the dual-inverter OeW five-phase drive system.

4 Random Reference and Random Carrier PWM Technique

This method is combination of above two methods in which both modulating signal and carrier signals are randomized. An algorithm for implementing this method is shown in Fig. 9. The generated reference signals (continuous or random) are compared with high-frequency triangular carriers. These carriers are randomized using positive and negative level shifting triangular carriers. In this method, magnitudes of harmonics at and around the side bands of switching frequencies can be effectively reduced by randomizing both modulating signals and carrier signals.

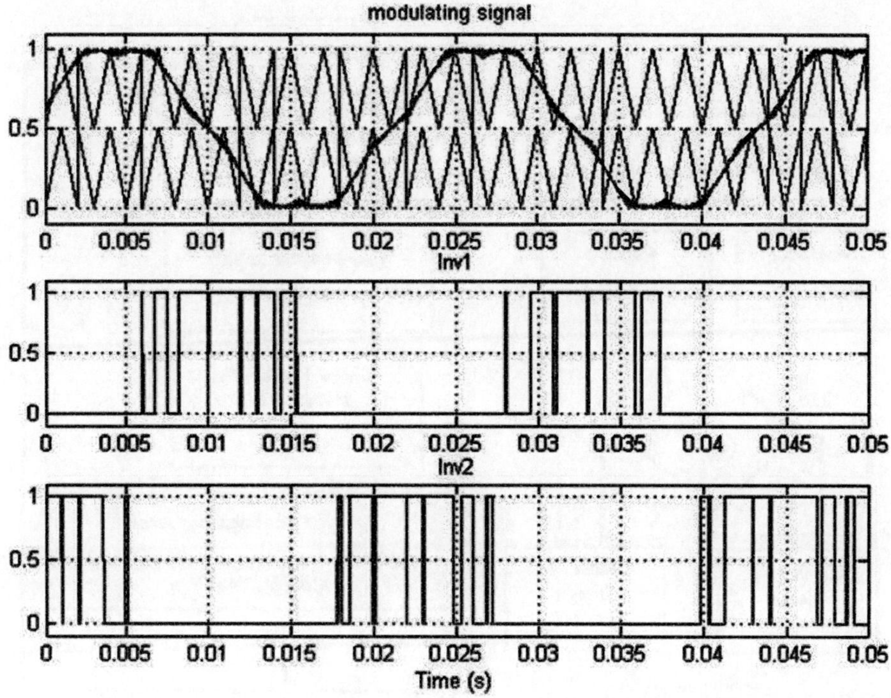

Fig. 8 Illustration of random carrier PWM technique with pulse pattern of inverter 1 and inverter 2

5 Random Carrier Selection and Random Variable Switching Frequency PWM Technique

For this type of RCRVSF-PWM scheme, two sets of random variables switching frequency carrier signals are used. The switching frequencies of positive and negative carrier signals can be varied by using a random frequency generator in a band of ±500 Hz and are shown in Fig. 10a. Figure 10a shows the generation of positive and negative carrier triangles having variable switching frequencies. Figure 10b shows a continuous modulating signal intersecting with variable switching frequency carrier triangles (positive and negative) and the pulse pattern generated for IGBT switches (S11 and S21).

6 Experimental Results and Discussion

The experimental setup shown in Fig. 11 has two two-level VSI inverter setups supplied with a three-phase diode rectifier. The isolation transformer is used between

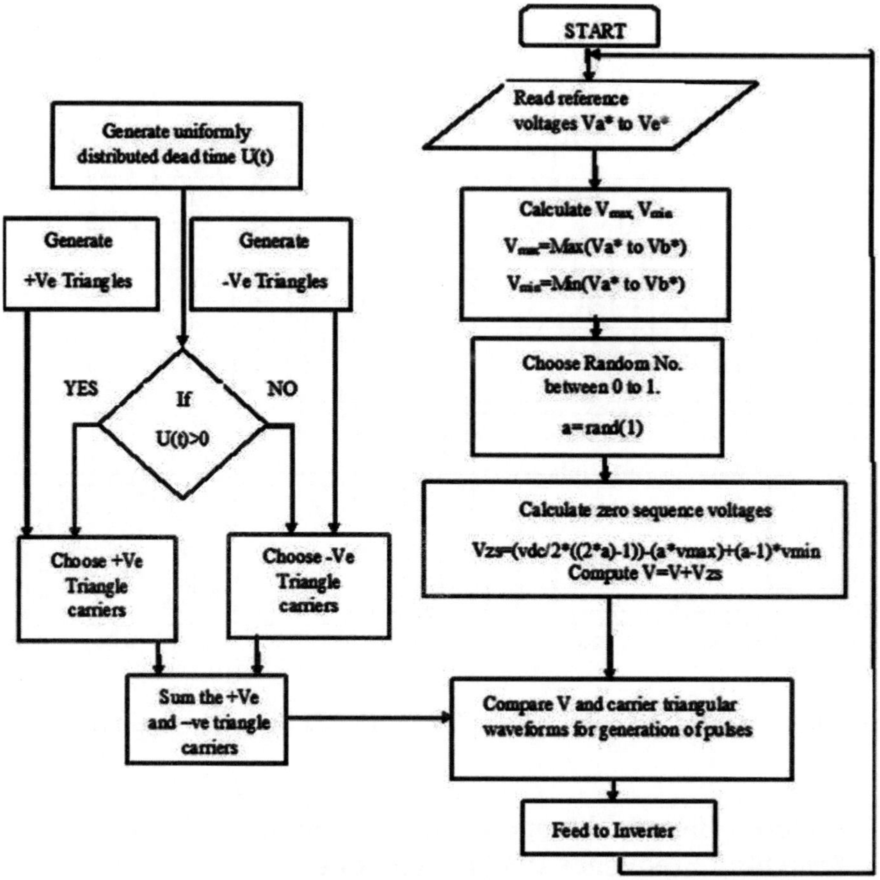

Fig. 9 Algorithm for developing RRRC-PWM modulating signals

the rectifiers (for VS1 and VS2). Two isolation transformers are used to adjust the DC-voltages of the two VSI's. Equal DC voltage of 100 V is applied to VSI I and II. The DC rectifiers are connected to a 3-phase, 415 V, 9.2 kVA, 13 Amp DC-link converters. The inverters are designed externally by parallel connection of five semicron IGBT sets. Each semicron IGBT set consists of one-leg (two semicron IGBTs connected in series) having a rating of 2000 V/100 A. It has provision to provide control signals for the upper and lower gate terminals; DC supply terminals; external triggering terminals. Five-phase motor designed by Ben electrical with ratings of 1 Hp, 200 V, 3.4 A, 1440 rpm, 50 Hz, and parameters of the motor are given in Table 1.

The stator of the five-phase motor is opened, and the terminals are connected to the two inverters on both sides of the motor. dSPACE 1104 is used to generate the control signals with a switching frequency of 1 kHz. LV20-P (500 V to 3.3 V regulator) and LA-55P (50 A to 3.3 V with three turns) are used for voltage and current measurements. Each PWM technique mentioned in this paper is written

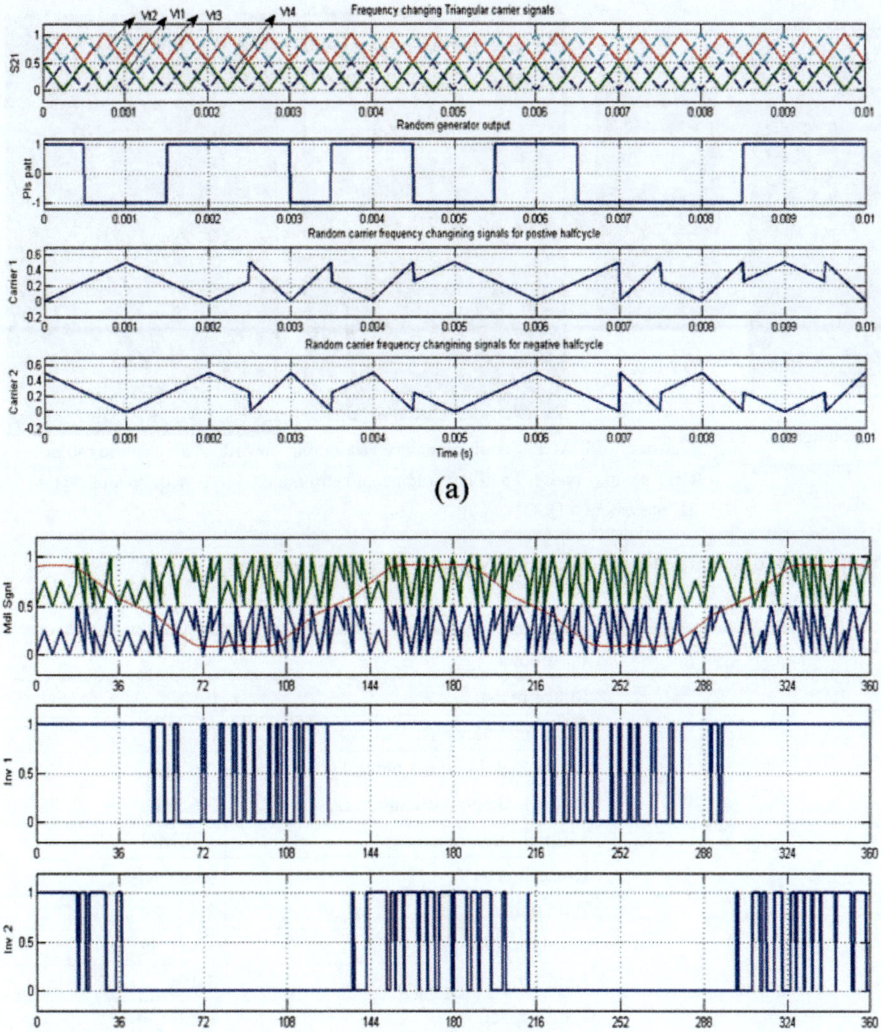

Fig. 10 Illustration of **a** random carrier frequency changing PWM, **b** RCRVSF-PWM modulating signal with pulse pattern

using MATLAB/Simulink environment. The codes written in MATLAB/Simulink are deployed to hardware-in-loop simulation package dSPACE. For this experimental setup, dSPACE 1104 generates 5 control signals for each inverter. Each control signal connected to a complimentary switching board. The function of this board is to generate pulses for the gate terminals of upper and lower IGBTs with a dead time of 20 μs. The generation of gate pulses for CPWM, RR-PWM, RC-PWM, RRRC-PWM, RRRC-CFC-PWM proposed in this paper is shown in Fig. 12.

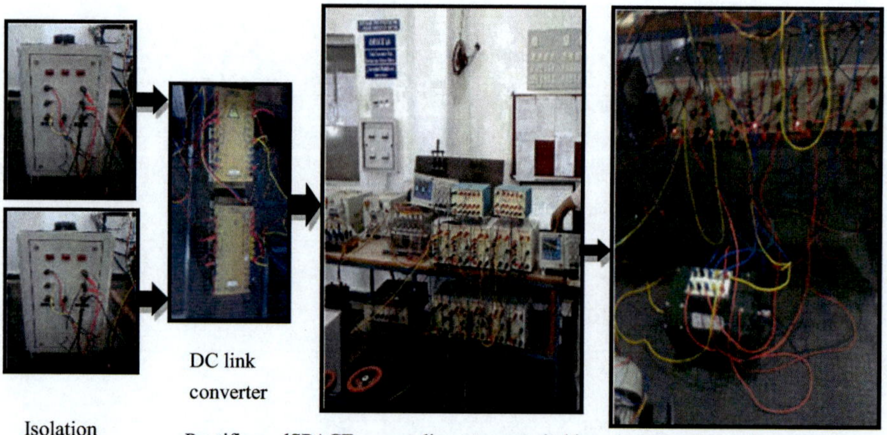

Fig. 11 Experimental setup of random PWM dual-inverter fed open-end

Table 1 Parameters of five-phase motor

Parameter	Value
Stator resistance R_s	0.8 Ω
Rotor resistance R_r	0.6 Ω
Stator leakage inductance L_{ls}	2.6 MH
Rotor leakage inductance L_{lr}	2.6 MH
Mutual inductance L_m	151 MH
Rotor inertia J	0.047 Kg m^2
Pole pair P	2
Switching frequency F_s	1000 Hz
Rated motor current I	3.4 A
Rated speed N_s	1450 rpm

The upper plot showing generated modulating signal intersecting with random carriers and lower plot showing the respective gating pulses of inverter I(S11) and inverter II (S21). The hardware results are taken for conventional SVPWM, RR-PWM, RC-PWM, RRRC-PWM, and RCRVSF-PWM. The effective phase voltage generated in each case is presented. The effective performance in reduction of acoustic noise and harmonic distortion is studied for all PWM techniques presented in this paper. The results are analyzed at two different modulation indices (M) of 0.35; 0.55; 0.88; and 1.0 throughout the experimental analysis switching frequency (F_s) which is maintained constant at 2000 Hz. Harmonic spectrum analysis, periodogram power spectral analysis, eigenvector pseudo-spectrum analysis, and power spectral density estimator are different methods fallowed to study the reduction of

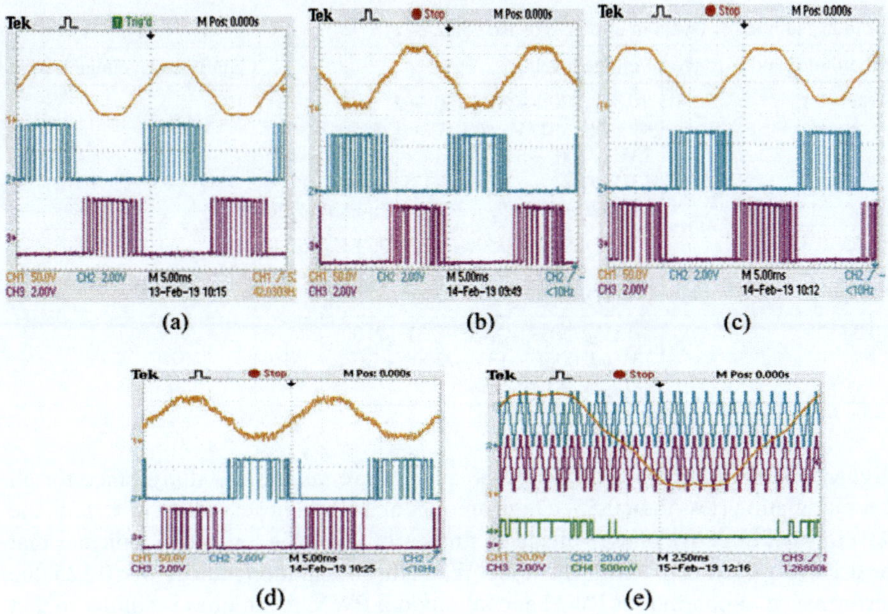

Fig. 12 Modulating signals and pulse patterns of inverter I and II for **a** continuous PWM, **b** RR-PWM, **c** RC-PWM, **d** RRRC-PWM, **e** RCRVSF-PWM at $M = 0.88$

acoustic noise and harmonic distortion. All these methods are explained in detail in fallowing subsection.

6.1 Harmonic Spectrum Analysis

In order to examine the performance of the drive for continuous and different random PWM modulation algorithms, the total harmonic distortion (THD) of the effective phase voltage is calculation at different modulation indices (M) at 0.55; 0.8; 0.96. The THD is calculated using Eq. 4.

$$\text{THD} = \sqrt{\frac{\sum_{n=1,2,3....}^{r} V_n^2 - V_1^2}{V_1^2}} \tag{4}$$

where V_n is the RMS value of the nth harmonic component, V_1 is RMS value of fundamental, n is the order of the harmonic used for calculation, and r is the maximum harmonic order which is taken into account. The THD is calculated up to 5 kHz. Higher modulation indices of $M = 0.88$ and 1.0 are taking into study, since for all low modulation (less than 0.525) a high amount of THD is observed.

Table 2 Number of levels in effective phase voltage

Modulation index range	Voltage levels	Number of voltage levels
0.066–1.15	$+13V_{dc}/20$; $+23V_{dc}/50$; $+2V_{dc}/5$; $+8V_{dc}/24$; $+13V_{dc}/49$; $+2V_{dc}/5$; $+13V_{dc}/100$; $+3V_{dc}/46$; 0; $-3V_{dc}/46$; $-13V_{dc}/100$; $-2V_{dc}/5$; $-13V_{dc}/49$ $-8V_{dc}/2$; $-2V_{dc}/5$; $-23V_{dc}/50$; $-13V_{dc}/20$	17
0.65–0.35	$+4V_{dc}/50$; $+V_{d}/20$; $+3V_{d}/20$; $+V_{dc}/5$; $+V_{dc}/4$; $+7V_{dc}/20$; 0; $-7V_{dc}/20$; $-V_{dc}/4$; $-V_{dc}/5$; $-3V_{dc}/20$; $-V_{dc}/20$	12
0.34–0	$+13V_{dc}/50$; $+V_{dc}/5$; $+13V_{dc}/100$; $+13V_{dc}/200$;0; $-13V_{dc}/200$; $-13V_{dc}/100$; $-V_{dc}/5$; $-13V_{dc}/50$	9

Higher modulation indices of $M = 0.88$ and 1.0 are taking into study, since for all low modulation (less than 0.55) a high amount of THD is observed. At $M < 0.55$, the waveform of effective phase voltage comprises of 9 voltage levels and indicates that the drive is operating in two-level mode. When the modulation indices $M < 0.525$, the performance of continuous PWM and all random PWM techniques is similar to that of a two-level operation and achieve nine voltage levels in effective phase voltage. Since only one inverter is modulated when $M < 0.525$. Due to the poor performance of the drive at $M < 0.525$, the harmonic spectrum analysis of effective phase voltage is studied for $M = 0.88$ and $M = 1.0$. The magnitude of each harmonics present across sidebands around multiples of switching frequency (2 kHz) is changing. In continuous PWM, a high magnitude of harmonics is observed compared to proposed random PWM techniques. The voltage levels formed in effective phase voltage are similar for both continuous and proposed random PWM techniques. These voltage levels are given in Table 2. It is obtained that 17 voltage levels are formed for range $0.65 \leq M \leq 1.15$, and for modulation region, $0.65 \leq M \leq 0.35$ total of 12 voltage levels are observed. Similarly, for low modulation region $M \leq 0.33$ total of 9 voltage levels are observe which is similar to the voltage levels formed under a two-level five-phase VSI.

It is observed from Fig. 13 for continuous PWM high amount of harmonics in sidebands around multiples of the switching frequency (2 kHz). RR-PWM in Fig. 14 shows high harmonics around 2 kHz and slite reduction around 4 kHz. For RC-PWM in Fig. 15, at $M = 0.88$ high harmonics are concentrated at 1 kHz and as the modulation indices increased these harmonics get reduced. When compared with other PWM techniques, RRRC-PWM and RRRC-CFC-PWM shown lesser harmonics at sidebands around multiples of switching frequency (2 kHz) and are clearly shown in Figs. 16 and 17. But in RRRC-PWM, high harmonics are present at high frequencies (of magnitude 2% of fundamental frequency) of around 4 kHz. These are further reduced in RRRC-CFC-PWM. The examined results of harmonic spectra present in effective phase voltage confirms that the developed random PWMs successfully

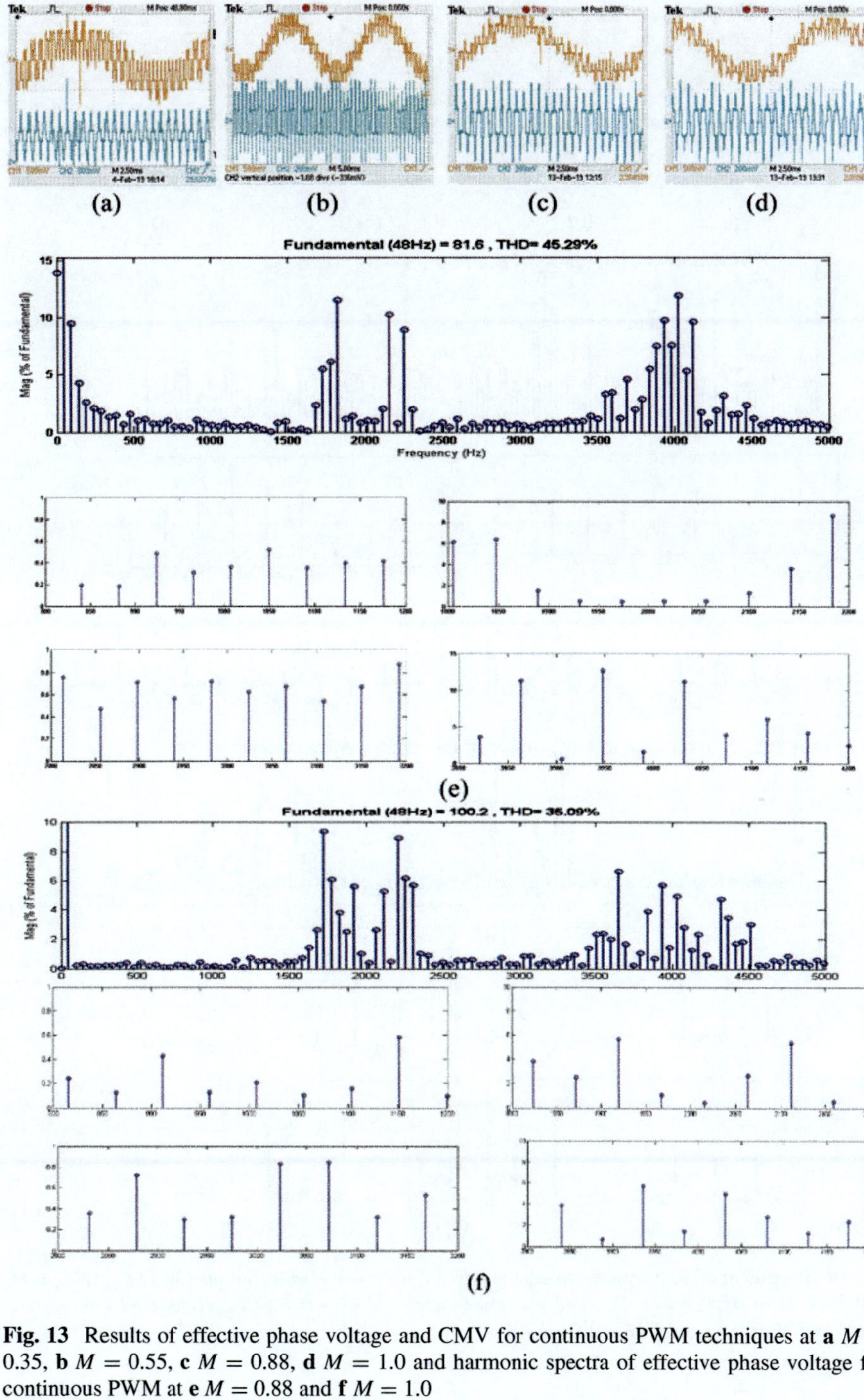

Fig. 13 Results of effective phase voltage and CMV for continuous PWM techniques at **a** $M = 0.35$, **b** $M = 0.55$, **c** $M = 0.88$, **d** $M = 1.0$ and harmonic spectra of effective phase voltage for continuous PWM at **e** $M = 0.88$ and **f** $M = 1.0$

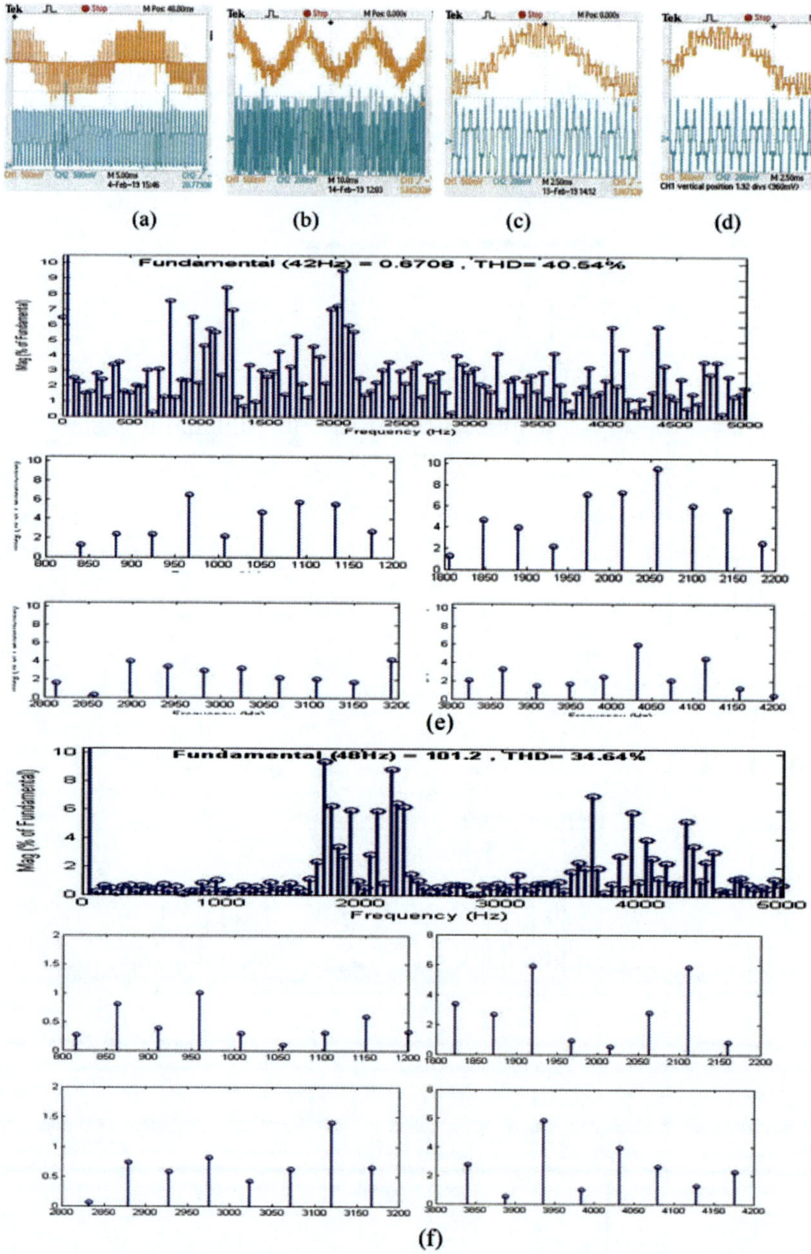

Fig. 14 Results of effective phase voltage and CMV for RR-PWM techniques at **a** $M = 0.35$, **b** $M = 0.55$, **c** $M = 0.88$, **d** $M = 1.0$ and harmonic spectra of effective phase voltage for continuous PWM at **e** $M = 0.88$ and **f** $M = 1.0$

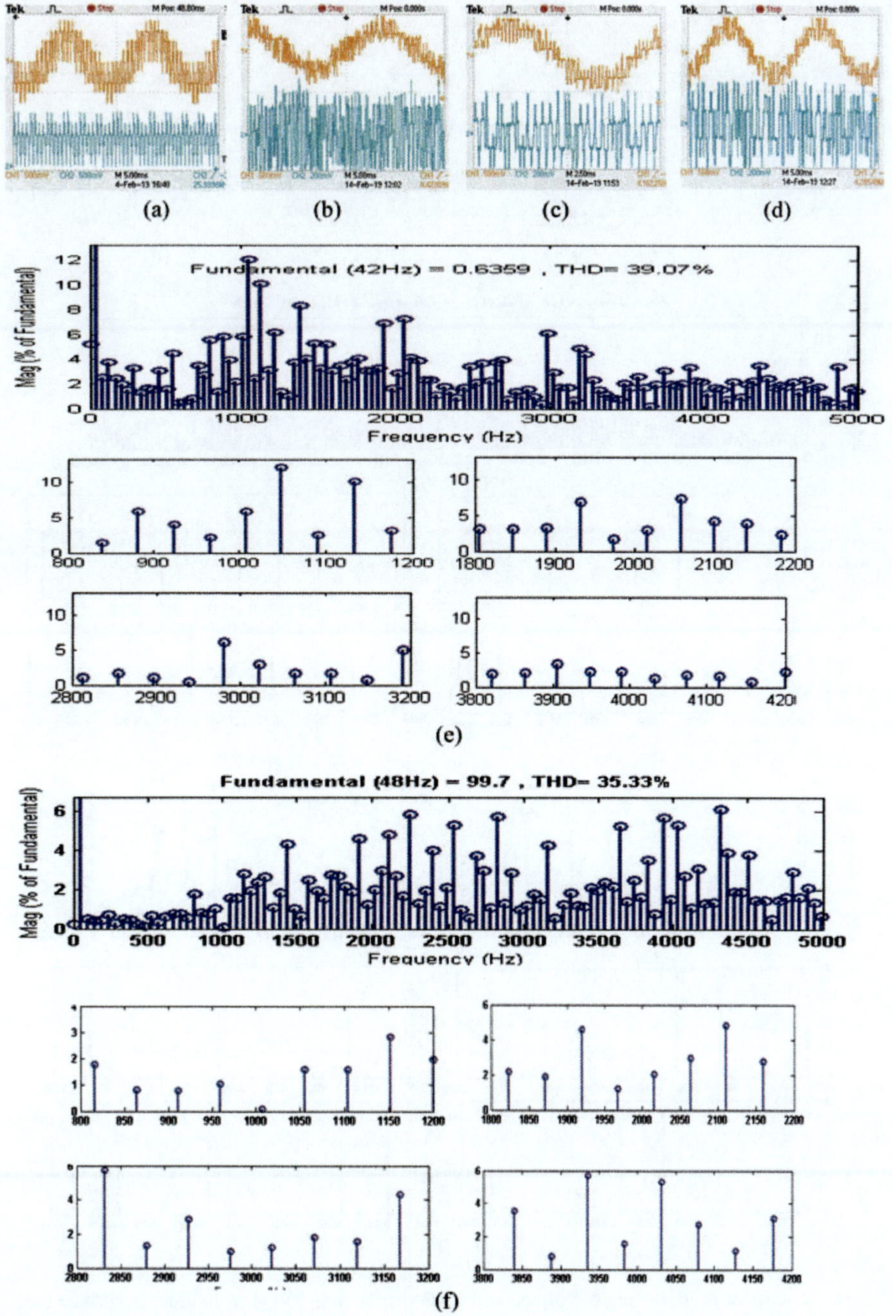

Fig. 15 Results of effective phase voltage and CMV for RC-PWM techniques at **a** $M = 0.35$, **b** $M = 0.55$, **c** $M = 0.88$, **d** $M = 1.0$ and harmonic spectra of effective phase voltage for continuous PWM at **e** $M = 0.88$ and **f** $M = 1.0$

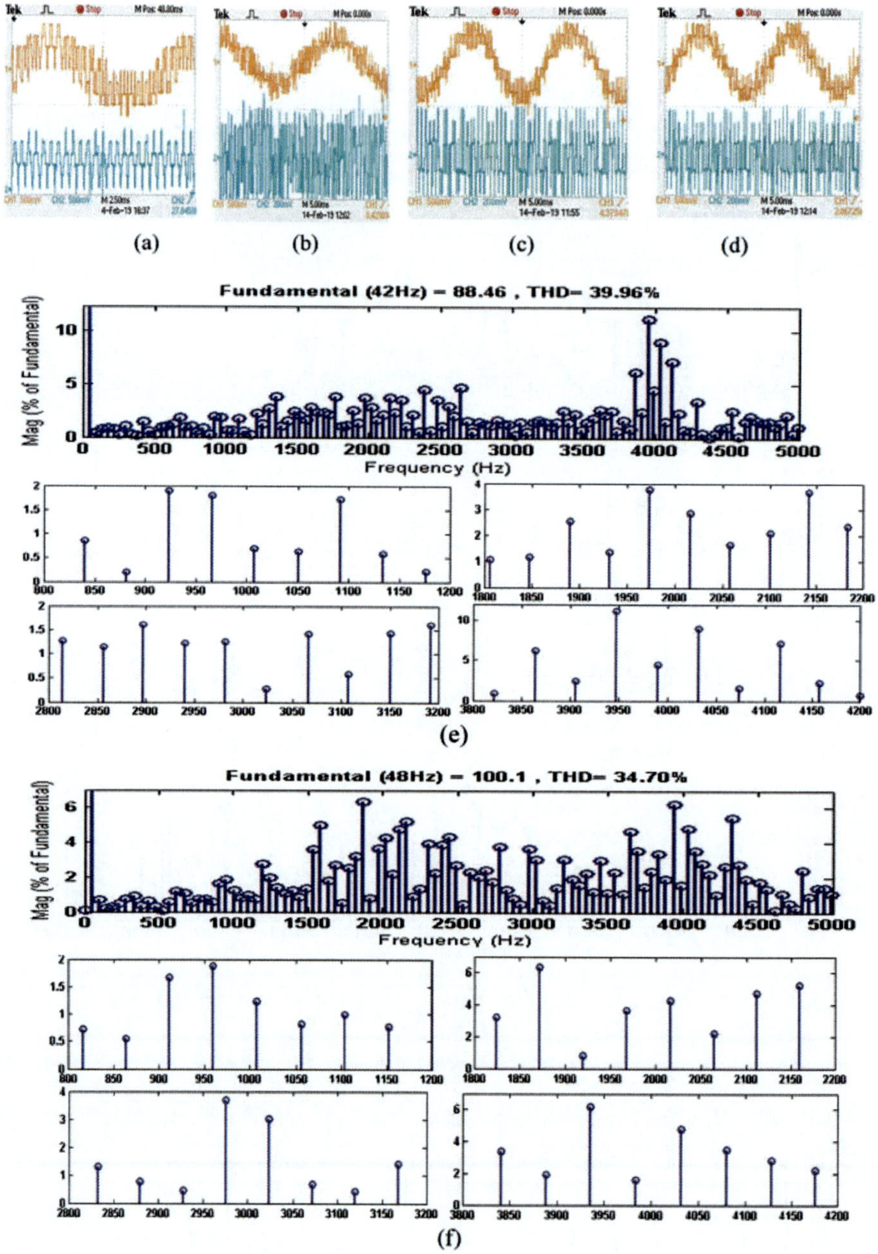

Fig. 16 Results of effective phase voltage and CMV for RRRC-PWM techniques at **a** $M = 0.35$, **b** $M = 0.55$, **c** $M = 0.88$, **d** $M = 1.0$ and harmonic spectra of effective phase voltage for continuous PWM at **e** $M = 0.88$ and **f** $M = 1.0$

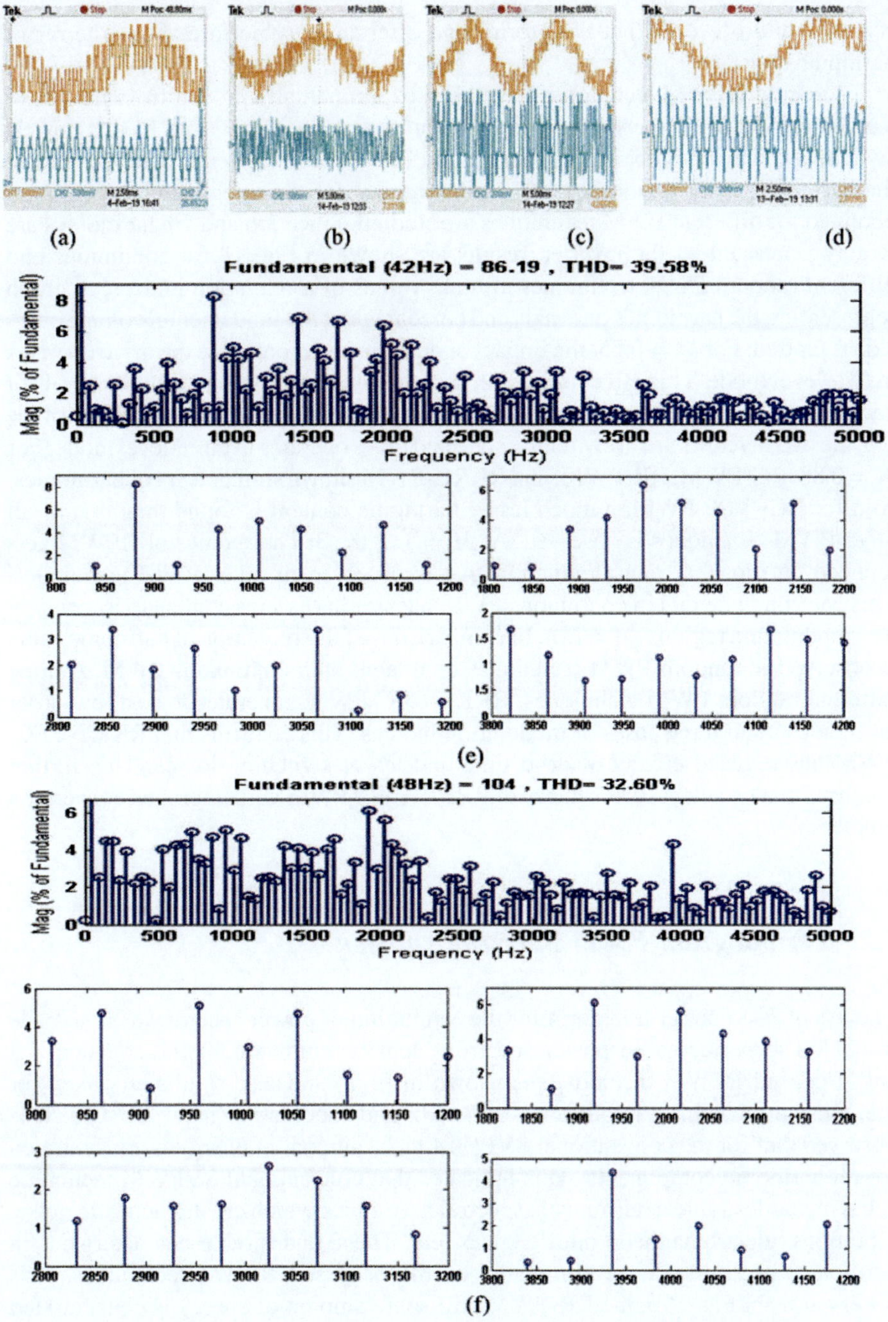

Fig. 17 Results of effective phase voltage and CMV for RCRVSF-PWM techniques at **a** $M = 0.35$, **b** $M = 0.55$, **c** $M = 0.88$, **d** $M = 1.0$ and harmonic spectra of effective phase voltage for continuous PWM at **e** $M = 0.88$ and **f** $M = 1.0$

achieved multi-level output waveforms and offer superior performance in terms of harmonic constant.

The random PWM techniques have been experimentally proven from the above studies to produce lowest harmonic content and at the same time having lowest switching average switching frequency and thus lowest converter loss. To extend the harmonic analysis, the 3rd and 7th harmonics for the effective phase voltages obtained for different PWM techniques are studied. Since 3rd and 7th harmonics are largely generated by the inverter, results are shown in Fig. 18 for continuous and different random PWM techniques at three values of modulation indices. A graph representing the harmonics on x-axis and percentage of the fundamental voltage on Y-axis is plotted. For $M = 0.55$, the impact of dead time is more since the drive operates in two-level mode. This effect is more in case continuous PWM, and similar levels of harmonic content are observed in RR-PWM and RC-PWM. When $M = 0.88$ and $M = 1.0$, the two inverters are in switching state and drive operates in three-level mode. For $M = 0.88$, RC-PWM, RR-PWM, and RRRC-PWM shown similar level of harmonics. And for RCRVSF-PWM, a much lesser harmonic content is found than that of all other PWM techniques is observed. At $M = 1.0$, the 3rd harmonics of CPWM gets reduced but not that much reduction that is observed in other PWM techniques. This can conclude that in continuous PWM a lesser harmonics content exists only at overmodulation regions ($M > 1.0$). In comparison, a 50% reduction of harmonic value is observed in random PWM techniques compared with continuous PWM. Among different random PWM techniques, RRRC-CFC-PWM generates lowest low-order harmonics in all three cases of modulation indices. This confirms that RRRC-CFC-PWM has reduced effects of dead time and lesser switching losses. This further confirms that excellent performance of RRRC-CFC-PWM with improved converter's efficiency.

6.2 Periodogram Power Spectral Analysis

The use of fast Fourier transform for the estimation of power spectra was first made in [21]. The periodogram power spectrum density estimator for the conventional and proposed RPWM techniques is shown in Fig. 19. These results are taken for the constant switching frequencies of 2 kHz and modulation index of 0.85. It is observed that for the conventional SVPWM more amount of energy is concentrated for all harmonic components. This indicates that conventional SVPWM technique is having undesirable deterministic spectral harmonics which creates acoustic noise, vibrations, electromagnetic interferences, etc. These undesirable characteristics in harmonic spectrum can be eliminated by the proposed RPWM techniques. It is observed from Fig. 19b, the RR-PWM still some amount of energy is concentrated odd multiples of harmonic components. Where from Fig. 19c, RC-PWM has less amount of energy is concentrated at lower harmonics (<1500 Hz) and a continuous spectrum is observed for the higher 80/harmonics (>1500 Hz). In Fig. 19d RRRC-PWM, a continuous spectrum with fewer distortions is observed for all harmonic

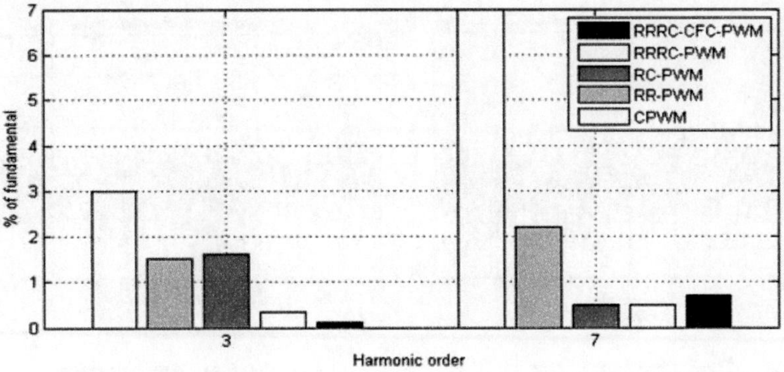

(a)

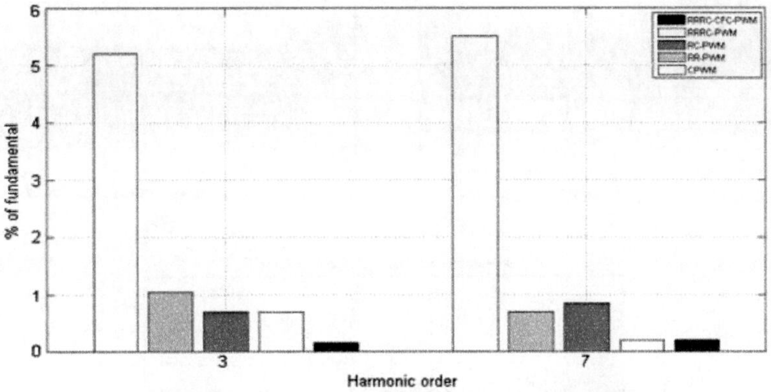

(b)

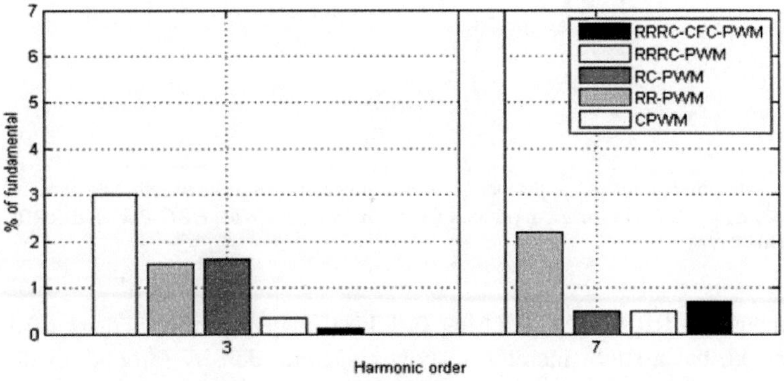

(c)

Fig. 18 Comparison of the 3rd and 7th harmonics for continuous and different random PWM schemes at **a** $M = 0.55$, **b** $M = 0.88$, **c** $M = 1.0$

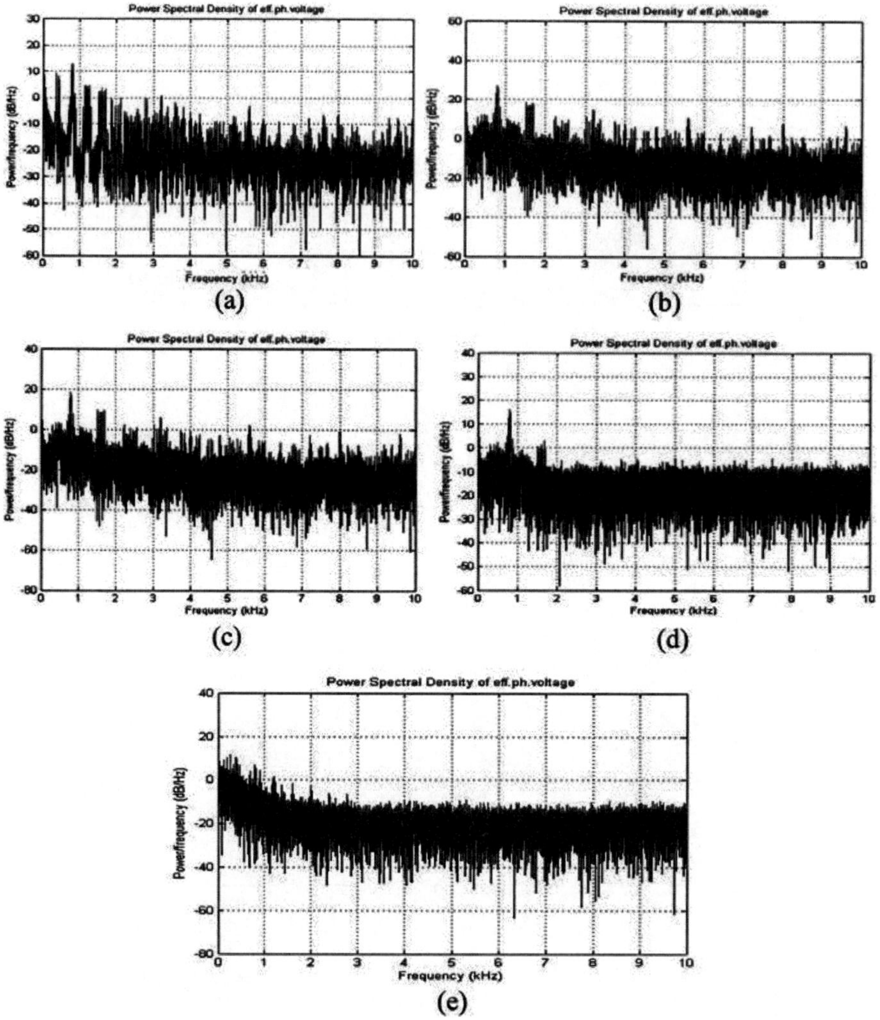

Fig. 19 Experimental results of periodogram power spectral analysis of effective phase voltage at $M = 0.88$; $F_s = 2000$ Hz for **a** continuous PWM, **b** random PWM, **c** RC-PWM, **d** RRRC-PWM, **e** RCRVSF-PWM

components. In RRRC-PWM during 0–1 kHz power spectral density of 0 db/Hz is observed, but a slight increase in power spectral density (around 15 db/Hz) is observed at less than 1 kHz. But in RCRVSF-PWM shown in Fig. 19e has less amount of power spectral density compared to other PWM techniques. It is observed that a less amount of peak value around 4 db/Hz at 500 Hz. This reveals that RRRC-CFC-PWM techniques shown superior performance in terms of elimination of undesirable, deterministic spectral harmonic effects that encounter in using the conventional SVPWM.

6.3 Pseudo-Spectrum Analysis

Every PWM technique generates a periodical switching function, a spectrum of discrete components and these are analyzed to measure energy content using pseudo-spectrums [21]. To further examine the studies, effective phase voltages experimental results are converted into pseudo-spectral waveforms by using mathematical equations. The results obtained are plotted in Fig. 20. All PWM methods have elevated noise of approximately 100 dB at low frequencies. As the increase in frequency CPWM, RR-PWM and RC-PWM radiates some noise and disturbance in the signal are more. But for the RRRC-PWM and RRRC-CFC-PWM disturbances, noise levels

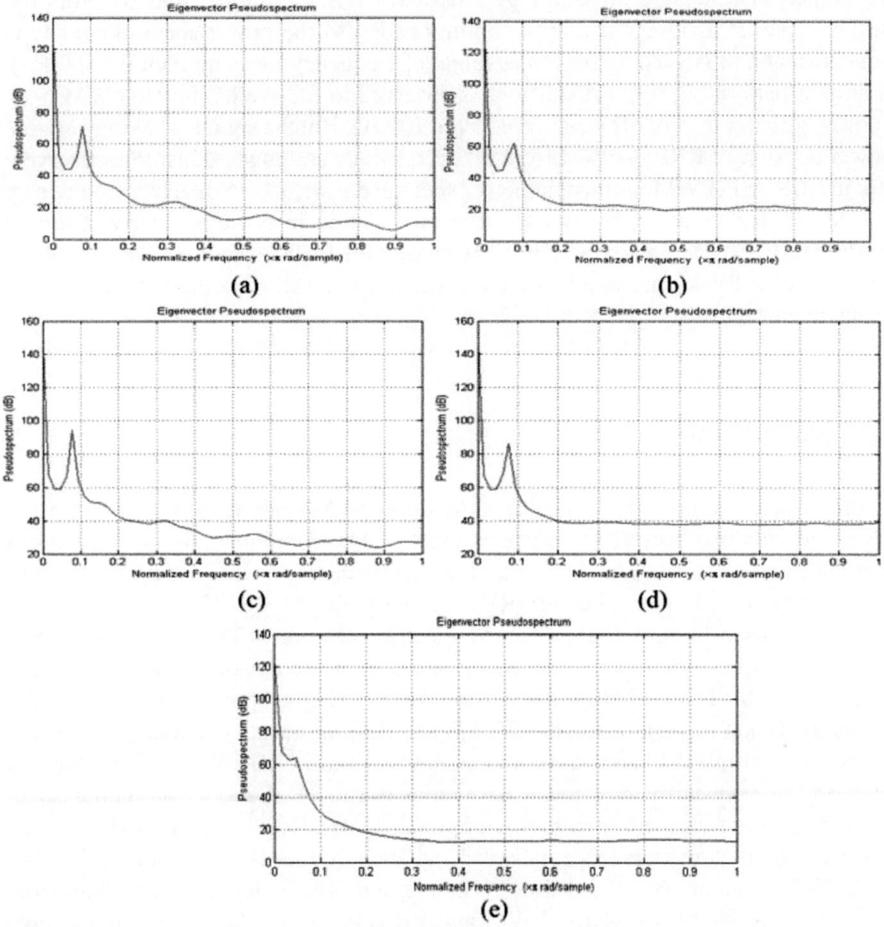

Fig. 20 Pseudo-spectrum diagrams for effective phase voltages obtained at $M = 0.88$; $F_s = 2000$ Hz for **a** continuous PWM, **b** random PWM, **c** RC-PWM, **d** RRRC-PWM, **e** RCRVSF-PWM

are very less. This indicates a superior performance in terms of reduced noise levels in the output phase voltages.

6.4 Power Spectral Density Estimator

Welch proposed a method to measure the power spectral density of a signal proposed in [22–24]. This method is used to determine the strength of a signal. The frequency content of a signal is characterized by the spectral density. The effective phase voltages obtained in experimental results are utilized further to measure the power spectral density. These are plotted in Fig. 21 for all PWM techniques that are proposed in this paper. It is observed that in continuous PWM the power spectral density is not constant and is varying for entire range of frequency measurement (0–500 Hz). A similar pattern but fewer distortions is observed in RC-PWM. In RR-PWM from 100 Hz, a nearer to zero dB is observed from 100 Hz. But the signal is having variable power density. In RRRC-PWM from 100 Hz, a constant power density is observed. But in RCRVSF-PWM, a constant power spectral density of −5 dB/Hz is noted from 100 to 500 Hz. A much better reduction of radiation in noise is clearly observed in RCRVSF-PWM. This concludes that among all PWM schemes, RRRC-PWM, and RCRVSF-PWM measured less power densities and shown efficient reduction in acoustic noise levels.

7 Conclusion

In this paper, a simplified scalar-based random PWM scheme is presented for the open-end winding five-phase induction motor. This technique contains no tedious mathematical computations, as with the standard space vector approach. Four different random PWM schemes (RR, RC, RRRC, and RRRC-CFC-PWM techniques) are used to eliminate unwanted spectral harmonics. The proposed random PWM techniques efficiently reduced the acoustic noise developed in the conventional SVPWM. The results analyzed using the periodogram power spectrum density estimator. Using this approach, the RR-PWM still some amount of energy is concentrated odd multiples of harmonic components, whereas in RC-PWM has less amount of energy is concentrated at lower harmonics (<1500 Hz) and a continuous spectrum is observed for the higher harmonics (>1500 Hz). Results related to harmonics generated by the inverters (especially 3rd and 7th harmonics) it is found that RRRC-CFC-PWM contains a much lesser harmonic and 50% reduction when compared to continuous PWM. In RRRC-PWM and RRRC-CFC-PWM, a continuous spectrum with fewer distortions is observed for all harmonic components. From pseudo-spectral analysis, it is observed that in RRRC-PWM and RRRC-CFC-PWM disturbances and noise levels are very less. The most effective technique for choosing the random PWM schemes in multi-level multi-phase drive applications such as

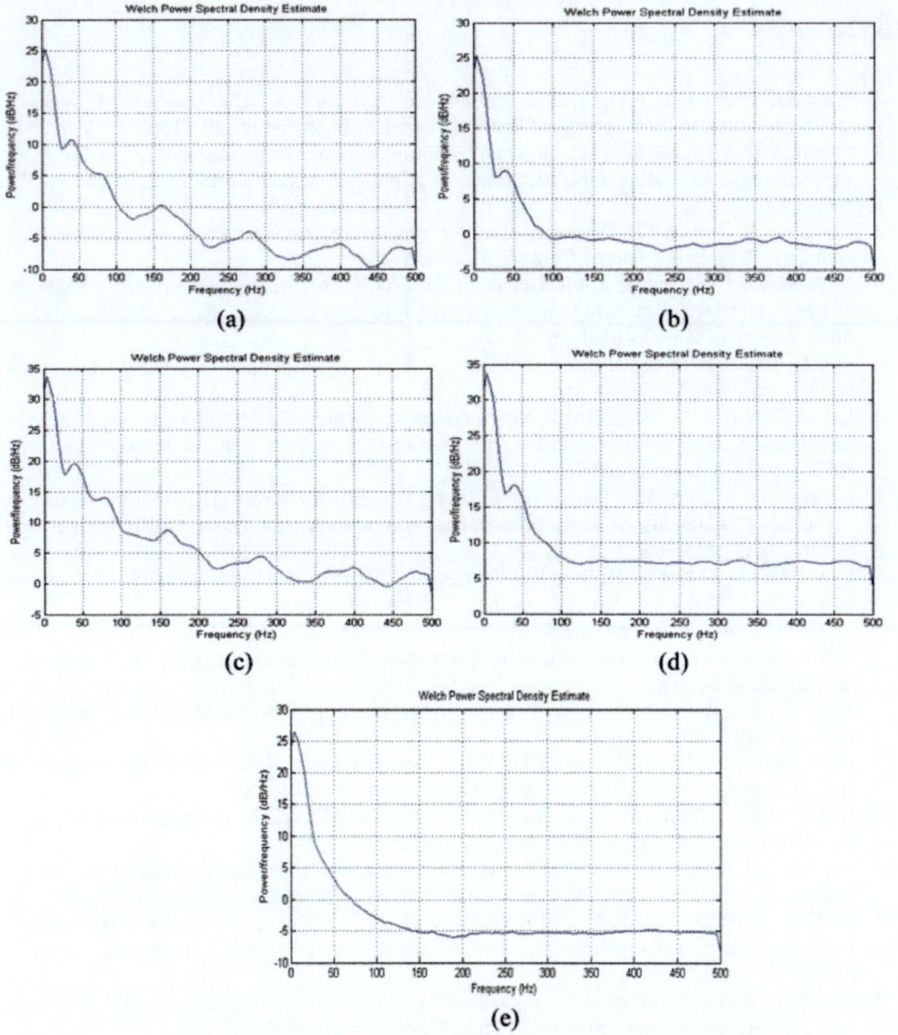

Fig. 21. Power spectral density estimators of effective phase voltage at $M = 0.88$; $F_s = 2000$ Hz for **a** continuous PWM, **b** random PWM, **c** RC-PWM, **d** RRRC-PWM, **e** RCRVSF-PWM

ship propulsion, electric and hybrid vehicles, 'more-electric' aircraft where the drive should operate with a lower acoustic noise levels which has benefits of low-order harmonics, decreased impact of dead time, lower switching losses, and improved overall efficiency.

Acknowledgements The authors would like to thank the management of G. Pulla Reddy Engineering College (Autonomous), Kurnool, Andhra Pradesh, for providing financial support to conduct the experimental work.

References

1. Ward EE, Harer H (1969) Dipl.-Ing, Preliminary investigation of an inverter fed 5—phase induction motor. In: Proceedings of Institute Electrical Engineering, vol 116(6), pp 980–984
2. Stemmler H, Guggenbach P, Configurations of high-power voltage source inverter drives. In: Proceedings of European power electronics and applications conference EPE, Brighton, UK, pp 7–14
3. Corzine KA, Sudhoff SD, Whitcomb CA, Performance characteristics of a cascaded two-level converter. IEEE Trans Energy Convers 14(3): 433–439
4. Mohapatra KK, Somasekhar VT, Gopakumar K (2002b) A harmonic elimination scheme for an open-end winding induction motor drive fed from two inverters using asymmetrical D.C. link voltages. EPE J 12(4): 28–36
5. Levi E (2008) Multiphase electric machines for variable-speed applications. IEEE Trans Indust Electron 55(5):1893–1909
6. Lai Y, Chang Y, Chen B (2013) Novel random-switching PWM technique with constant sampling frequency and constant inductor average current for digitally controlled converter. IEEE Trans Indust Electron 60(8):3126–3135
7. Binojkumar AC, Prasad J, Narayanan G (Feb.) Experimental investigation on the effects of advanced bus-clamping pulse width modulation on motor acoustic noise. IEEE Trans Indust Electron 60(2):433–439
8. Liaw CM, Lin YM, Wu CH, Hwu KI (2000) Analysis, design and implementation of a random frequency PWM inverter. IEEE Trans Indus Electron 15(5):843–854
9. Binojkumar AC, Narayanan G (2014) Variable switching frequency PWM techniques for induction motor drives to spread acoustic noise spectrum with reduced current ripple. In: Proceedings of IEEE PEDES, pp 1–6
10. Kim K-S, Jung Y-G, Lim Y-C (2009) A new hybrid random PWM scheme. IEEE Trans Power Electron 24(1):192–200
11. Liaw CM, Lin YM, Wu CH, Hwu KI (2000) Analysis, design and implementation of a random frequency PWM inverter. IEEE Trans Indust Electron 15(5): 843–854
12. Huang Y, Xu Y, Zhang W, Zou J (2019) Hybrid RPWM technique based on modified SVPWM to reduce the acoustic noise. IEEE Trans Power Electron 34(6)
13. Tse KK, Chung HS, Hui SYR, So HC (Apr.) A comparative investigation on the use of random modulation schemes for DC/DC converters. IEEE Trans Indust Electrons 47(2):253–263
14. Bu F, Pu T, Huang W, Zhu L (2019) Performance and evaluation of five-phase dual random SVPWM strategy with optimized probability density function. IEEE Trans Indus Electron 66(5)
15. Kirlin RL, Kwok S, Legowski S, Trzynadlowski AM (Sep.) Power spectra of a PWM inverter with randomized pulse position. IEEE Trans Power Electron 9(5):463–472
16. Cai JY, Ho YH, Chang YC, Liaw CM (Mar.) On acoustic-noise-reduction control using random switching technique for switch-mode rectifiers in PMSM drive. IEEE Trans Indus Appl 55(3):1295–1309
17. Lee K, Shen GT, Yao WX, Lu ZY (Mar.) Performance characterization of random pulse width modulation algorithms in industrial and commercial adjustable speed drives. IEEE Trans Power Electron 53(2):1078–1087
18. Habetler TG, Divan DM (1991) Acoustic noise reduction in sinusoidal PWM drives using a randomly modulated carrier. IEEE Trans Power Electron 6(3):356363
19. Inoue K, Kusaka K, Itoh JI (2018) Reduction in radiation noise level for inductive power transfer systems using spread spectrum techniques. IEEE Trans Power Electron 33(4): 3076–3085
20. Zhu L, Bu F, Huang W, Pu T (2017) Analysis and performance of two dual random SVPWM for five-phase inverter. In: Proceedings of 20th international conference on electrical machinery and systems (ICEMS), pp 1–4
21. Mathe L, Lungeanu F, Sera D, Rasmussen PO, Pedersen JK (Oct.) Spread spectrum modulation by using asymmetric-carrier random PWM. IEEE Trans Indus Electron 59(10):3710–3718

22. Welch PD (1967) The use of fast fourier transform for the estimation of power spectra: a method based on time-averaging over short, modified periodograms. IEEE Trans Audio Electroacoust AU-15:70–73
23. Heidelberger P, Welch PD (1981) Adaptive spectral methods for simulation output analysis. IBM J Res Dev 25(6):860–876
24. Heidelberger P, Welch PD (1981) A spectral method for confidence interval generation and run length control in simulations. Commun ACM (Special issue on simulation modeling and statistical computing) 24(4):233–245

A Frequency Reconfigurable Antenna for Sub-GHz and TV White Space Applications

Sanjeev Kumar, Rohit Khandekar, and Neela Rayavarapu

Abstract TV White Space (TVWS) band of frequencies are those rendered surplus after conversion of analog TV transmission to digital. The proposed antenna is a meander line antenna (MLA), with a defect ground structure designed for TVWS band viz 470–890 MHz. The antenna has dimensions of 108 × 44 mm^2 printed on FR-4 epoxy substrate with dielectric constant of $\varepsilon_r = 4.4$ achieving size reduction upto 35%. Frequency is switched electronically, using four RF-PIN diodes, mounted in the slots on the ground plane. The simulated resonant frequencies obtained for the different PIN diode biasing combinations are 0.840, 0.660, 0.8560, 0.6460, 0.8700, and 0.828 GHz. The simulated return loss values are well below −10 dB. The measured results of the printed antenna closely follow the simulated ones.

Keywords TVWS band · Meander line antenna · Defect ground and PIN diodes

1 Introduction

TV White Space (TVWS) band of frequencies are broadcasting frequencies rendered surplus after analog-to-digital conversion of TV transmission. TVWS band at sub-GHz frequency range having excellent propagation characteristics and declared unlicensed by the FCC is being widely considered for wireless applications to meet the

S. Kumar
Faculty, Symbiosis Institute of Technology, Symbiosis International (Deemed University), Pune, India
e-mail: sanjeevkumar@sitpune.edu.in

R. Khandekar (✉)
Department of Electronics and Telecommunication, Symbiosis Institute of Technology, Symbiosis International (Deemed University), Pune, India
e-mail: rohit.khandekar@sitpune.edu.in

N. Rayavarapu
HOD of Electronics and Telecommunication, Symbiosis Institute of Technology, Symbiosis International (Deemed University), Pune, India
e-mail: neela.rayavarapu@sitpune.edu.in

© Springer Nature Singapore Pte Ltd. 2021

S. N. Merchant et al. (eds.), *Advances in Signal and Data Processing*,
Lecture Notes in Electrical Engineering 703,
https://doi.org/10.1007/978-981-15-8391-9_33

growing demand for data. The TVWS refers to the frequency range of 470–890 MHz [1–3]. The challenge lies in reducing the size of the patch at the sub-GHz range of frequencies so as to incorporate it into mobile devices.

A meander line antenna is an extension to the basic dipole antenna. The meander line element consists of vertical and horizontal lines so as to form a series of right-angled bends. Combination of horizontaland vertical line forms turns. The vertical elements act as an inductor, and the horizontal elements act as a capacitor. The horizontal lines lie in the short length of the PCB, while the vertical lines are placed along the long length of the PCB [4–6]. The meander line configuration of the monopole allows reducing the occupied space of the antenna element to less than 0.1λ in each dimension. Efficiency increases with the increase in the number of turns present in the meander. If meander spacing is increased, resonant frequency decreases [7–9]. Meander lines increase the path over which the surface current flows and that eventually results in lowering of the resonant frequency than the straight wire antenna of similar dimensions. Radiation efficiency of the meander line antenna is comparable to conventional half and quarter wavelength antennas. The meander line antenna acts as a resonant LC circuit [8–11].

Defect ground structure, DGS, are slots on the ground plane of microstrip circuits. Repetating a single defect with finite spacing or symmetric defects are periodic structure, while non-repetitive or asymmetric are non-periodic structures of DGS. Implementing defects on the ground plane introduces bends/edges/corners disturbing the current distribution on the ground plane. This changes the characteristics of the Tx line by including slot resistance, inductance, and capacitances to the transmission line parameters [10, 11]. The defect etched in the ground plane under the microstrip line changes the effective capacitance and inductance of the microstrip line by adding slot resistance, capacitance, and inductance. Slotted area of the DGS is proportional to the effective inductance and inversely proportional to the effective capacitance. An increment in the slotted DGS area gives rise to effective inductance and thus resulting in a lower cutoff frequency. A decrement in the DGS area reduces the effective capacitance, thereby increasing the resonant frequency [9–13].

Antenna reconfiguration is achieved by changing its frequency, polarization, or radiation characteristics. This change is gained through many techniques that redistribute the antenna currents and thus alter the electromagnetic fields of the antenna's effective aperture. Frequency reconfiguration techniques are based on the use of switches to connect and disconnect antenna parts as well as to redistribute the antenna currents [14–19].

This paper presents a design incorporating the two techniques of meander line along with an optimized defected ground structure to reduce the size of the antenna operating at the TVWS band. PIN diode switches, positioned in the slot of the ground plane, have been used to change current lengths and thereby frequency reconfigure the antenna. The ground plane is partial groung plane, divided in two sections. Section 2 conducts when one of four diodes is ON.

2 Antenna Design and Structure

A meander line monopole of size 108 × 44 mm^2 is designed on the FR-4 epoxy substrate with permittivity $\varepsilon_r = 4.4$, loss tangent $(\delta) = 0.02$, and height of 1.6 mm. The total dimensions of the substrate are 143 × 44 mm^2 as included in Table 1 and depicted in Fig. 1a. The horizontal meander arm is 18 × 2 mm^2 (s × d mm^2), and the total number of turns is 26. An optimized *E* shaped slot, Fig. 1b, is etched in the ground plane to obtain desired results.

Mathematical Design of Meander Patch Antenna:

From [4–6], the design parameter of meander patch is given from follwing equations:

$$d = 0.16\lambda g$$
$$s = 0.42\lambda g \tag{1}$$

where λg is guided wavelength of substrate.

$$\lambda g = \sqrt{\frac{\lambda}{\varepsilon_{\text{reff}}}} \tag{2}$$

where λ is free space wavelength and $\varepsilon_{\text{reff}}$ is the effective dielectric constant given by:

$$\varepsilon_{\text{reff}} = \frac{\varepsilon_r + 1}{2} - \frac{\varepsilon_r - 1}{2}\left[1 + 10\frac{h}{w}\right]^{-1/2}$$

h = height ofsubstrate.
ε_r = dielectric constant of given material.
W = Width of meander line antenna.
$\varepsilon_{\text{reff}}$ = effective dielectric constant.

Calculated values of *d*, *s*, *PL*, and *PW* are rectified using optimisation in HFSS. The correct dimension can be calculated from the below equation:

Table 1 Antenna parameter

Parameter	Values (mm)
Patch length (Pl)	108
Patch width (Pw)	44
Ground plane length (Gl)	72
Ground plane (Gw)	44
Slot width (Yo)	7.5

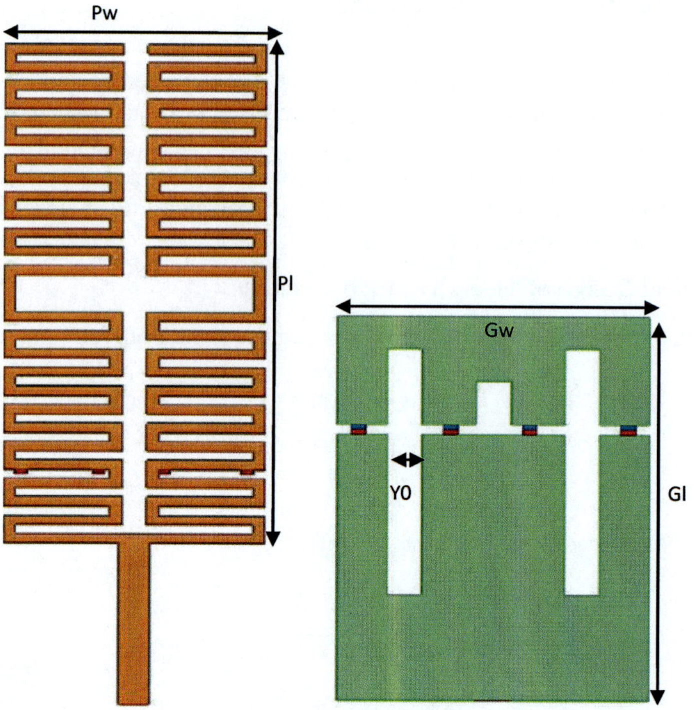

Fig. 1 **a** Meander patch, **b** defected ground plane with diodes

$$\text{sp} = s + \Delta L$$
$$\text{dp} = d + \Delta L$$

where ΔL is length extension
fr = center frequency and c = velocity of light.

The operating frequency for the MLA is 680 MHz having the wavelength (λ) 441.17 mm. Thus, the length of antenna is $\lambda/2 = 220.58$ mm, and designed antenna length is 143 mm, i.e., achieveing 35% size reduction in the length of antenna.

PIN Diode RF switches

Four RF PIN diodes have been positioned in the slots as shown in Fig. 1b. Results are obtained from multiple ON/OFF combinations of PIN diodes state to change the current distribution through the antenna. PIN diodes are placed as switches in the slot in the ground plane. Infineon BAR 63-02V has been selected due to its good isolation value. The equivalent circuit of this PIN diode corresponds to an inductor L, for both states of the switch, in series with a resistance R_s in the ON state and with the parallel connection of a capacitor $C_p$and resistance R_p for the off state as shown in Fig. 2. The data sheet provided by the manufacturer gives these values as $L = 0.6$ nH, $R_s = 1.2\Omega$, $C_p = 0.3$ pF, and $R_p = 15$ kΩ [19].

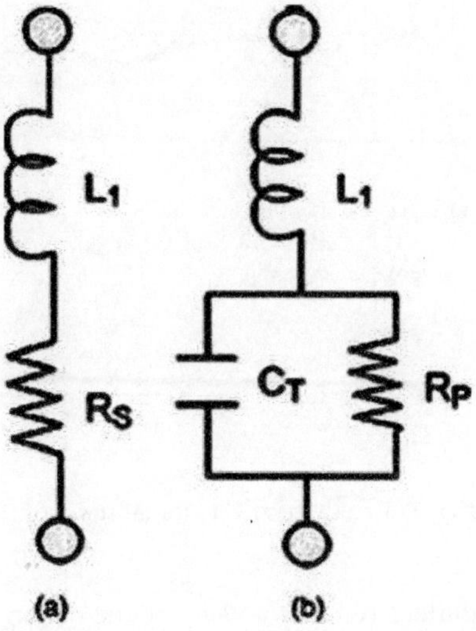

Fig. 2 RLC circuit for RF PIN diode **a** forward bias, **b** reverse bias

3 Simulated Results

The proposed antenna has been designed and simulated using ANSYS HFSS Version.16. Table 2 shows the different ON–OFF states of the four PIN diodes. The combined return loss S_{11} plot for different bias conditions of the switches is shown in Fig. 3. The structure resonates as multiband antenna for different combinations of PIN diode states between 0.646 MHz and 0.8700 GHz for TV White Space band and 1–2 GHz outside the band. Frequencies obtained from TVWS band are 0.840, 0.660, 0.8560, 0.6460, 0.8700, and 0.828 GHz along with the frequencies between 1 and 2 GHz.

The S11 values obtained for the resonant frequencies are satisfied below −10 dB. Radiation pattern plots, Fig. 4, are similar in every state of operation, i.e., the radiation

Table 2 Different states of operations

States	Diodes combinations	Frequencies (GHz)	Return loss S11 (dB)
S1	All OFF	0.8420	−11.89
S2	All ON	1.06	−13.97
S3	OFF, OFF, ON, ON	0.660, 0.8420	−28.63, −22.42
S4	OFF, ON, ON, OFF	0.8560	−15.19
S5	OFF, ON, OFF, ON	0.6460, 0.8700	−24.76, −15.73
S6	ON, OFF, ON, OFF	0.828, 0.660	−29.34, −18.93

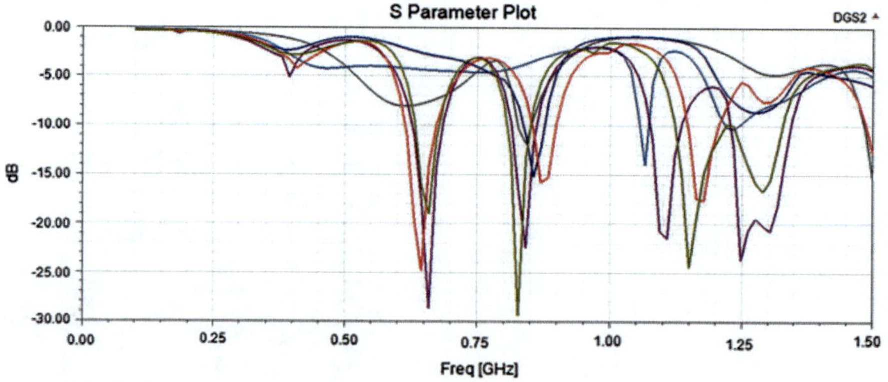

Fig. 3 Return loss (S11) for all states of PIN diode

pattern remains intact for the different resonating TVWS frequencies, meeting the criterion of frequency reconfiguration.

Switching diodes ON and OFF alters the flow of current over surface of the antenna. This change in surface current results into change in resonating frequency or frequencies. Figure 5a–d and Table 3 show the surface current densities for the states S1, S2, S3, and S4, respectively, for the states that are defined in Table 2.

As can be seen from Fig. 5a when the all four diodes are in OFF state, no current flows through the *E* section of the ground plane, and surface current is at the minimal. When all four diodes in Fig. 5b are in ON states, current flows through the section '*E*' resulting into maximum surface current. Switching ON and OFF of the PIN diodes alters the surface current distribution which results into change the resonant frequencies. Figure 5c shows the difference in the surface current distribution when the two diodes under the left arm of the meander line are in the OFF state and two diodes are in ON state under the right arm. Figure 6a, b is ground plane current distribution for state S1 and S2, i.e., when all diodes are in OFF state and ON state, respectively. It was observed that when all diodes are in OFF state, minimal amount of current in the ground plane, i.e., 1.26 A/m, when all diodes are in ON state, Sect. 2 conducts and high amount of current 7.68 A/m was observed in the ground plane.

Figures 7 and 8 show that simulated VSWR for state S4 at frequency 0.6460 GHz is 1.12, and the impedance is 44.54 + 0.03i, i.e., close to 50 Ω.

4 Measured Results

The designed antenna was printed on the FR-4 epoxy substrate, Fig. 9a, b, and tested on Rhode & Schwarz Vector Network Analyzer ZVB 4300 kHz to 4 GHz, 1145.1010.04 as shown in Fig. 9c. The measured results, as shown in Fig. 10a–f show the antenna resonating between 0.470 and 0.890 GHz for TV White Space band and

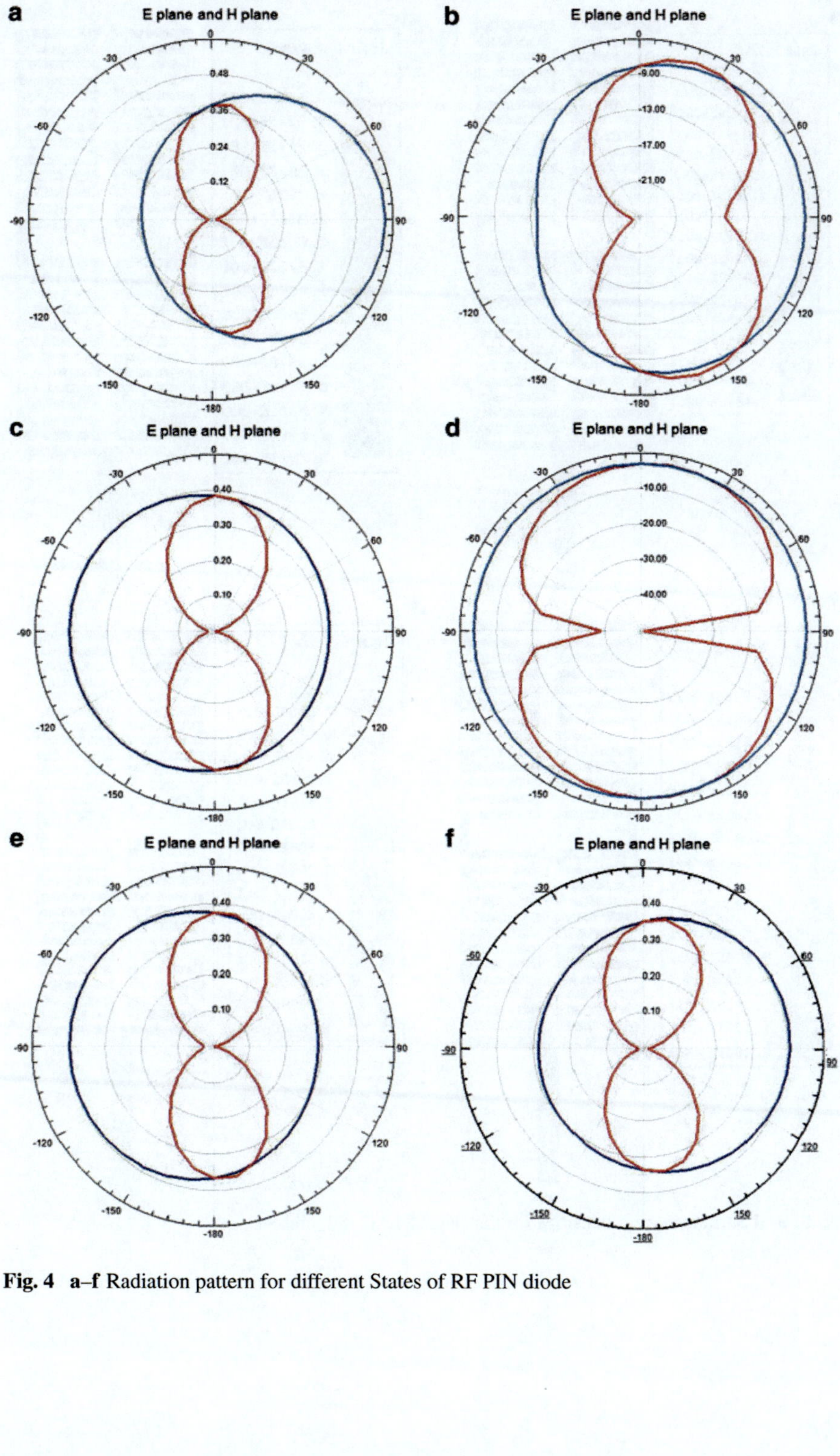

Fig. 4 **a–f** Radiation pattern for different States of RF PIN diode

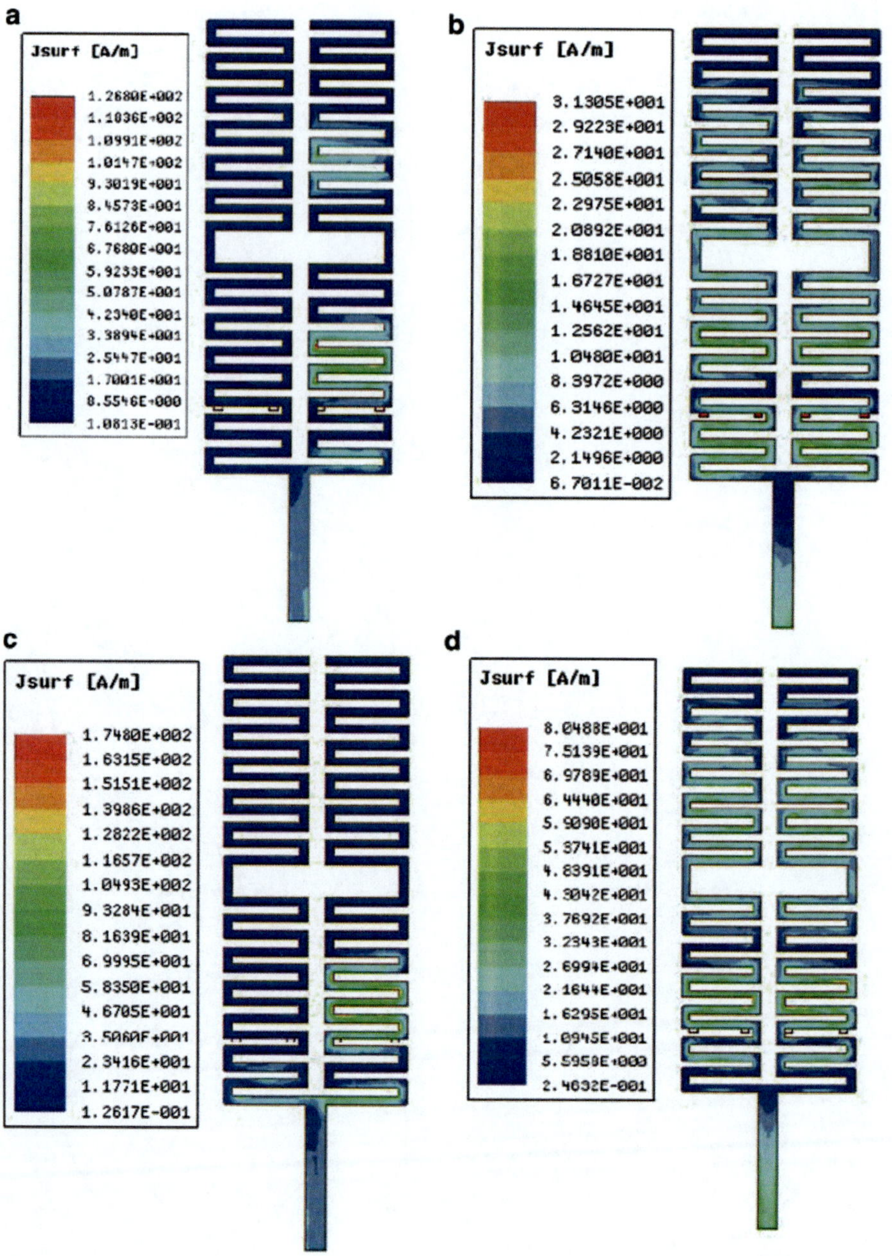

Fig. 5 **a–d** Surface current densities for the states S1, S2, S3, and S4

Table 3 Surface current distribution of MLA

States	Frequencies (GHz)	Jsurf (A/m)
S1	0.8420	1.28
S2	1.06	3.13
S3	0.660, 0.8420	1.74
S4	0.8560	8.04

Table 4 Comparisons of simulated and measured S11

States	Switch position	Simulated		Measured	
		Freq (GHz)	S11 (dB)	Freq (GHz)	S11 (dB)
S1	All OFF	0.8420	−11.89	0.535, 0.835	−16.14, −12.16
S2	All ON	1.06	−13.97	0.423,0.635	−15, −13
S3	OFF, OFF, ON, ON	0.660, 0.8420	−28.63, −22.42	0.547,1.1, 1.18	−11.59 −22.7, −19.1
S4	OFF, ON, ON, OFF	0.8560	−15.19	0.541, 1.1, 1.2	−13.1, −14.27, −19.7
S5	OFF, ON, OFF, ON	0.6040,0.8700	−24.76, −15.73	0.664,1.12, 1.24	−14.5, −18.9, −25
S6	ON, OFF, ON, OFF	0.828, 0.660	−29.34, −18.93	0.535, 1.10,1.20	−12.9, −18, −22.3

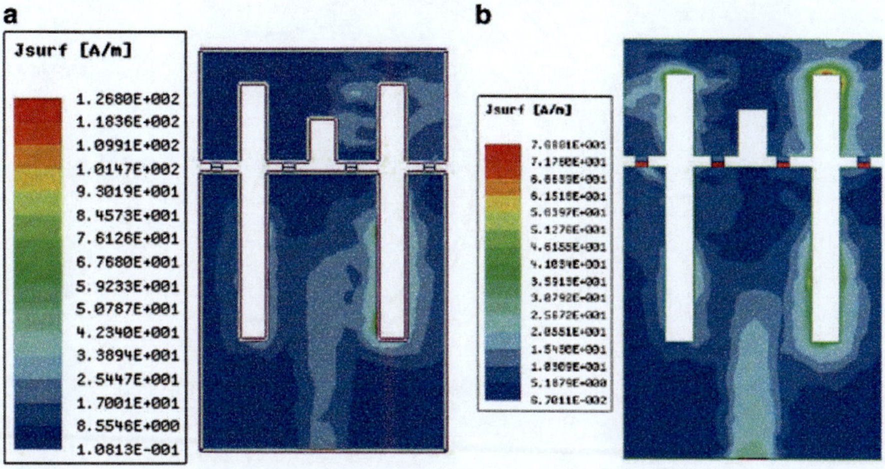

Fig. 6 **a**, **b** Current distribution of ground plane for state S1 and S2

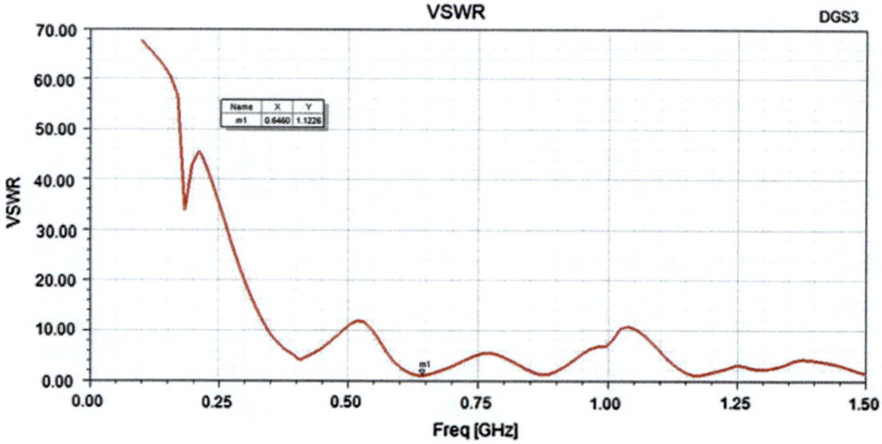

Fig. 7 Simulated VSWR for state S4

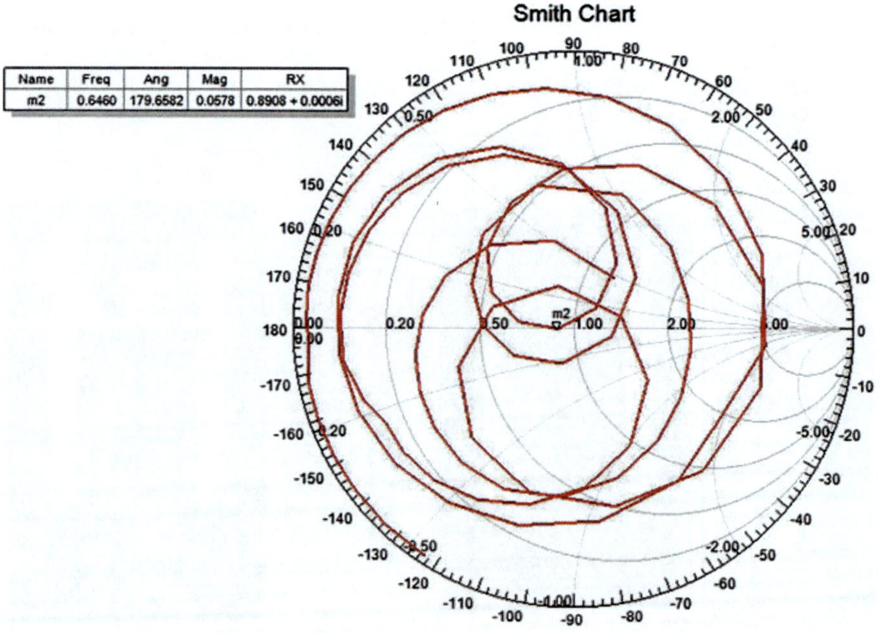

Fig. 8 Simulated Smith chart for state S4

from 1 to 1.5 GHz in the multiband mode. The specific TVWS band frequencies for

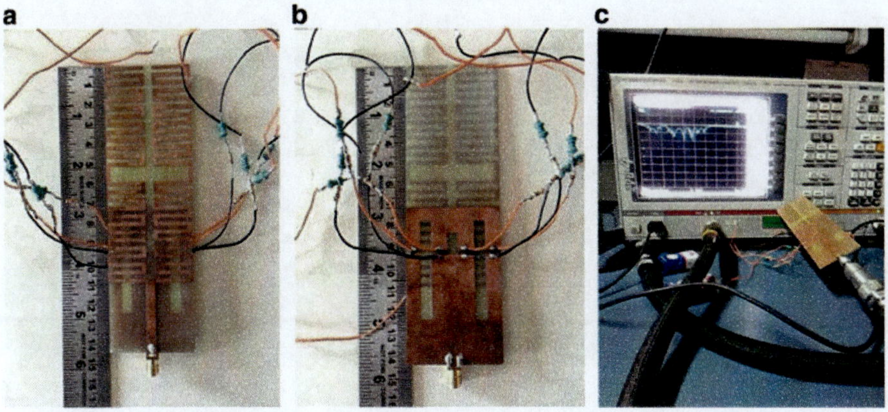

Fig. 9 **a** Top view of prototype, **b** partial DGS incorporated with four PIN diode, **c** setup for the measurement of the prototype

the corresponding PIN diode ON/OFF states are 0.535, 0.835, 0.423,0.635, 0.547, 0.541, 0.664, and 0.535 GHz.

Comparison of results

The simulated and measured results are compared in Table 3. The measured results have been found to be in close approximation of the simulated results. The difference in the results could be attributed to the minor variations in the dimensions of the fabricated antenna as also the interference due to the biasing circuitry.

The VSWR found for the measured results is between 1 and 2 from (4) for all states of simulated and measured results, and the impedance found was close to 50Ω from (6).

$$\mathrm{VSWR} = \frac{1+\Gamma}{1-\Gamma} \tag{3}$$

$$L_r = -20\log_{10}\Gamma \tag{4}$$

where L_r is return loss and Γ is reflection coefficient.

$$Z_L = \frac{1+S11}{1-S11} Z_o \tag{5}$$

where Z_L is load resistance, i.e., antenna, and Z_o is the characteristic impedance with the value of 50 Ω.

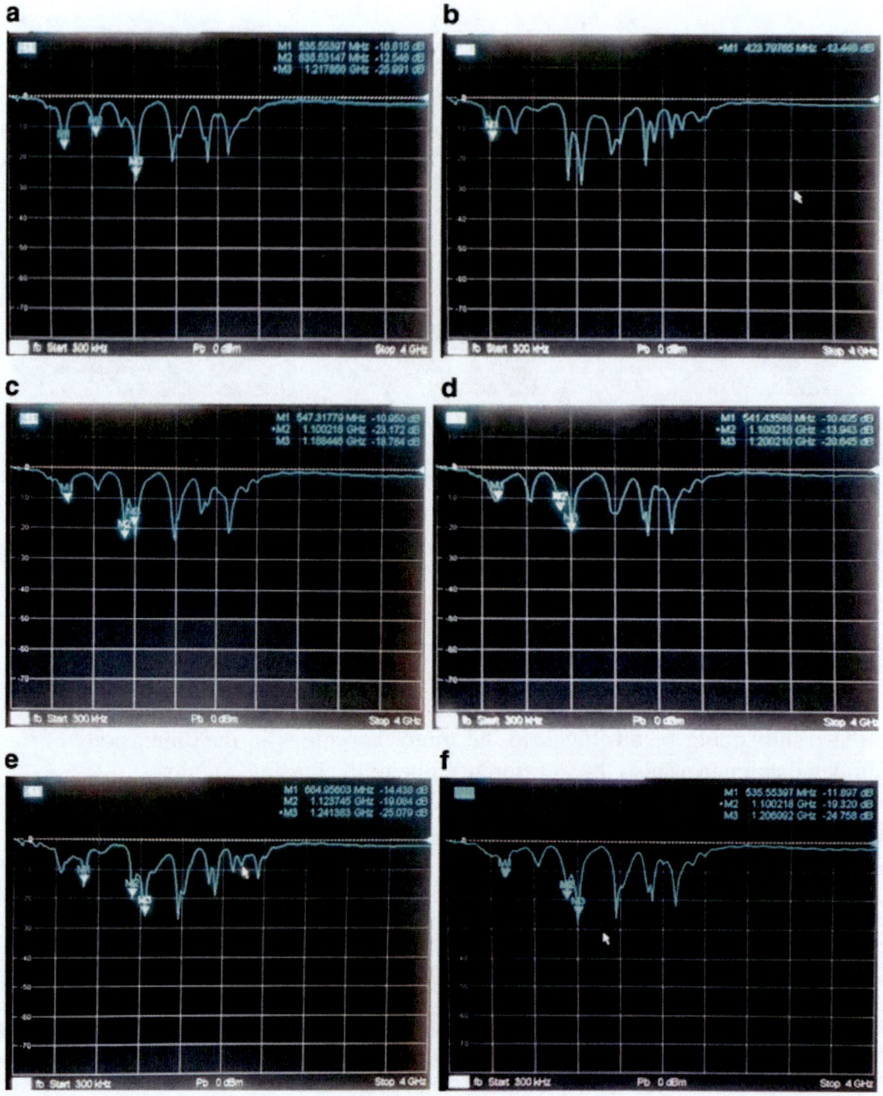

Fig. 10 **a–f** Measured frequencies for different state of operations as per Table 2

5 Conclusion

The proposed meander patch antenna effectively resonates at the sub-GHz range of frequencies and could be suitable for applications in the TV White Space band. The size of the antenna can be further reduced with some more optimization of the defected ground.

The propsed antenna is a frequency reconfigurable antenna whose resonating frequency is changed when the states of the PIN diodes are changed, and the resonating frequency is changed since the current distribution is altered over the surface of antenna.

References

1. Broadband deployment through TV-White Space Radio Division, Department of Telecommunication India
2. Abou-Shahine MY, Al-Husseini M, Ramadan AH, Kabalan KY, Nasser Y (2015) Antenna designs for cognitive radio application in the TV band. Int J Sci Eng Res 6(2): 975–980
3. Ammann MJ, John M (2013) A compact shorted printed monopole antenna for TV white space trials. In: Proceedings of 7th European conference on antennas and propagation, Gothenburg
4. Naik G, Singhal S, Kumar A, Karandikar A (2014) Quantitative assessment of TV white space in India. In: Proceedings of IEEE twentieth national conference on communications, pp 1–6
5. Bong F-L, Lim E-H, Lo F-L (2017) Miniaturized dipolar patch antenna with narrow meandered slotline for UHF tag. IEEE Trans Anten Propag 65(9)
6. Chenand H-D, Tsao Y-H (2010) Low-profile meandered patch antennas for RFID tags mountable on metallic objects. IEEE Anten Wirel Propag Lett 9
7. Warnagiris TJ, Minardo TJ (1998) Performance of a meandered line as an electrically small transmitting antenna. IEEE Trans Anten Propag 46(12)
8. Ogherohwo EP, Mado SD, Eggah MY (2015) Design and analysis of meander microstrip antenna at wireless band. Int J Comput Appl 111(6):975–8887
9. Hongand Bindu J-S, Karyamapudi M (2005) A general circuit model for defected ground structures in planar transmission lines. IEEE Microwave Wirel Compon Lett 15(10)
10. Khandelwal M, Kanaujia B, Kumar S, Defected ground structure: fundamentals, analysis and application in mordern wireless trends
11. Weng LH, Guo YC, Shi XW, Chen XQ (2008) An overview on defected ground structure. Progr Electromagn Res B 7:173–189
12. Liu JX, Yin WY, He SL (2010) A new defected ground structure and its application for miniaturized switchable antenna. Progr Electromagn Res 107:115–128
13. Lim J, Ahn D, Han S-M, Jeong Y, Liu H (2013) A defected ground structure without ground contact problem and application to branch line couplers. Int J Antennas Propag 2013, Article ID 232317
14. Christodoulou CG, Tawk Y, Lane SA, Erwin SR (2012) Reconfigurable antennas for wireless and space application. Proc IEEE 100(7)
15. Constantine J, Tawk Y, Barbin SE, Christodoulou CG (2015) Reconfigurable antennas:design and applications. Proc IEEE 103(3)
16. Debogovic T, Perspectives of reconfigurable antennas in modern wireless communication and sensing system
17. Jay Guo Y, Qin P-Y, Bird TS (2013) Reconfigurable antennas for wireless communications. In: 7th European conference on antennas and propagation (EUCAP 2013), 8–12 April, 2013 Gothenburg, Sweden
18. Yang S, Zhang C, Plan HK, Fathy AE, Nair VK (2009) Frequency reconfigurable antenna for multiradio wireless platform. IEEE Microwave Mag 10(1)
19. Yahya SH, Khraisat, Qubaia AHN (2015) Reconfigurable antenna design. PIERS proceedings, Prague, Czech Republic, July 6–9 2015
20. Data sheet of Infineon RF-PIN Diodes

Human Activity Recognition Using Positioning Sensor and Deep Learning Technique

Aarati Mohite, Priti Rege, and Debashish Chakravarty

Abstract In this paper, Long Short Term Memory (LSTM) deep learning model is used to identify human activities using sequential data obtained from cameras, wearable sensors, or other modalities. The proposed method recognizes human activities by optimizing hyper-parameters for the chosen deep learning model. The proposed approach is validated using public domain UTD MHAD dataset and found to be more accurate outperforming the state-of-the-art in terms of accuracy of activity recognition. Datasets categorized as the depth, skeleton, and inertial modalities have been analyzed for all available features. The dependency of the proposed deep learning model on the hyper-parameters is investigated extensively and discussed in detail. Experimental results depicted in this paper demonstrate the fact that the proposed method is a better choice for indoor activity recognition.

Keywords Indoor human activity recognition (IHAR) · Multimodal datasets · Kinect V1 · LSTM · Hyper-parameters

1 Introduction

In recent times, human activity recognition systems are in demand due to the large requirement of human-aware services. This has kindled the interest of researchers in the area of IHAR. Accurate activity recognition has many applications, and it is the most challenging task due to many problems such as human behavior of performing multitask at a time, intra-class variability, interclass similarity, resource constraints,

A. Mohite (✉) · P. Rege
Department of Electronics and Telecommunication, College of Engineering, Pune, India
e-mail: mohitear17.extc@coep.ac.in

P. Rege
e-mail: ppr.extc@coep.ac.in

D. Chakravarty
Department of Mining Engineering, Indian Institute of Technology, Kharagpur, India
e-mail: dc@mining.iitkgp.ac.in

© Springer Nature Singapore Pte Ltd. 2021

S. N. Merchant et al. (eds.), *Advances in Signal and Data Processing*,
Lecture Notes in Electrical Engineering 703,
https://doi.org/10.1007/978-981-15-8391-9_34

usability, privacy, obtrusiveness, data collection issues, tradeoff, etc. The fundamentals of Human Activity Recognition (HAR) are based on automatically detecting and analyzing human activities. Human actions basically can be categorized into four parts viz. atomic actions, person object interaction, person to person interactions, and group activities [1].

In human's daily life, HAR has many applications, which include home behavior analysis, video surveillance, gait analysis, and gesture recognition. Activity recognition of a human can be achieved in two ways either from videos or from sensors. Video-based HAR uses videos or sequence of images which contains human actions while sensor-based HAR uses motion data obtained from smart sensors viz., accelerometer, gyroscope, magnetometer, sound-based sensors, etc. Different machine learning algorithms like decision trees, random forest, are found to be useful and have made tremendous progress in HAR. HAR from raw video sequences involves pre-processing of raw data, segmentation to remove unwanted data from significant one, and the most crucial part is feature extraction [2]. Initially, HAR based on motion tracking tools is invasive because either the person has to wear wearable sensors or attach them to the body of the person. However, the invention of cameras like Kinect Xbox series introduced by Microsoft, Asus's xition Xpro, markerless body motion tracking is possible. HAR has found useful in surveillance systems that are installed at public places to avoid criminal activities, in the healthcare system—installed at a rehabilitation centre, at hospitals to care for patient's life, in human–computer interaction systems to treat elderly people with neurological disorders.

The paper is arranged in four sections. Section 2 reviews the state-of-the-art in the area of IHAR. Details of dataset and brief introduction to machine learning techniques required for the proposed method, and test environmental setup are discussed in Sect. 3. Section 4 provides the results and discusses the proposed method. We conclude the paper in Sect. 5.

2 Related Work

For clinical purposes, a Kinect has been used to build up virtual reality simulation technology [3]. Kinect is found to be useful in mental health therapy training, clinical skills training [3]. Chang et al. proposed an assessment as a comparison between Kinect usages for motion tracking with an OptiTrack optical system. He found that using Kinect compared to OptiTrack system permeant availability to patients can be achieved so that patients can take rehabilitation treatment in the clinic as well as in-home environments. Lang et al. proposed a system which is an interactive game based rehabilitation tool which is useful as to train adults with the neurological disorder [4]. Continuous action recognition is a challenging issue in machine learning with some obstacles like feature selection from sequential training data, hyper-parameter selection for the design of proper machine learning architecture, finalizing the size of training data. There are several machine learning techniques for HAR. This section gives information about previous work and survey on HAR.

Lara [5] reviewed in their work the state of the art in HAR based on wearable sensors; and explained general architecture and components of HAR along with two-level taxonomy presented in accordance to learning approach and response time. Zubair [6] introduced the state of art sensor modalities in HAR and focused on techniques involved in terms of sensors, activities, data pre-processing, feature learning, and classification. Wang [7] explored the properties of the channel state information (CSI) of Wi-Fi signals and presents a robust indoor daily HAR framework with only one pair of transmission points (TP) and access points (AP). Wang [7] summarized the major techniques in HAR from 3D data with a focus on techniques that use depth data. Vrigkas [1] provided a detailed review of recent and state-of-the-art research advances in the field of human activity classification and proposed a categorization of human activity methodologies and discuss their advantages and limitations. Ramasamy Ramamurthy [8] gives a comprehensive overview of recent machine learning and data mining techniques along with collaborating problems and challenges associated with the existing systems. Hammerlan [9] proposed a deep convolutional neural network (convnet) to perform efficient and effective HAR using smartphone sensors by exploiting the inherent characteristics of activities and 1D time-series signals, at the same time providing a way to automatically and data-adaptively extract robust features from raw data. Yang and Nguyen [10] proposed a generic deep framework for activity recognition based on convolutional & LSTM recurrent units suitable for multimodal wearable sensors. Park et al. [11] proposed a new HAR system via RNN which is one of deep learning algorithms; utilized joint angles from multiple body joints changing in time which are represented a spatiotemporal feature matrix.

3 Dataset and ML Techniques

3.1 UTD MHAD Dataset

In this paper, an LSTM deep learning method is applied to UTD multimodal human action dataset. The dataset includes 27 activities captured using Kinect V1 camera performed by 8 subjects viz., 4 males and 4 females. 4 trials for each action is performed by each subject. The dataset includes 861 data sequences, after removing corrupted files from it. Dataset consists of human actions viz., related to sports action, hand gesture, daily activities, and training exercises. Basically, four modalities are considered in this dataset viz., skeleton joint positions, inertial sensor signals (3-axis acceleration and 3-axis rotation signals), RGB videos, and depth videos. Hereby, using sample code [12, 13] representation of activities is obtained. The 'squat' activity has been chosen here for the representation from available 27 activities of UTD MHAD dataset. In addition to these different modalities, features are extracted and

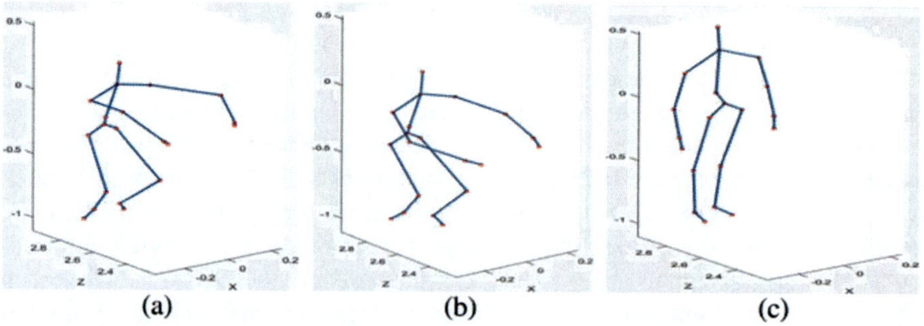

Fig. 1 Skeleton joint frames representation of squat activity **a** arms and legs bent, **b** arms raised and legs bent, **c** arms and legs in straight position

stored in mat files [12]. Inertial signal data contains acceleration and gyroscope data, represented in three axes viz., x, y, z. Thus, it gives six-dimensional data for each activity.

Figure 1, shows the skeleton data representation of one of the 27 activities, viz. squat activity, has shown. Figure 2, gives depth images for squat activity. Figure 3, shows Inertial signal representation for squat activity.

3.2 Machine Learning Techniques for HAR

There are different machine learning algorithms available for HAR. Some of them are decision trees, an adaptive boost for multiclass, random forest, Support vector machine. In addition to this, based on a neural network, there are different classifiers viz. Convolution Neural Network (CNN), Recurrent Neural Network (RNN), LSTM. Benefits of neural network methods like CNN are, they learn features automatically from the sequence of data where features are extracted from raw data viz. spatial and temporal features. Here, in this paper, LSTM based neural network model for activity recognition has been used. Here, in brief, review on neural network methods used for HAR, till the date, are explained.

Convolutional Neural Network

CNN is a type of deep neural network model, and it is used with image data. It has been proven very effective for automatically identifying and localizing the content of an image when they are trained with an ample amount of data images. CNN is comprised of convolution layers and pooling layers. Convolutional layer read an input signal viz 2D image or 1D signal, use filter window which convolves around given input in small segments across the entire input field. It works like a score function. The convolutional filter slides over the special location of the image and results in dot product between their weights and a small region. Pooling layer acts as down

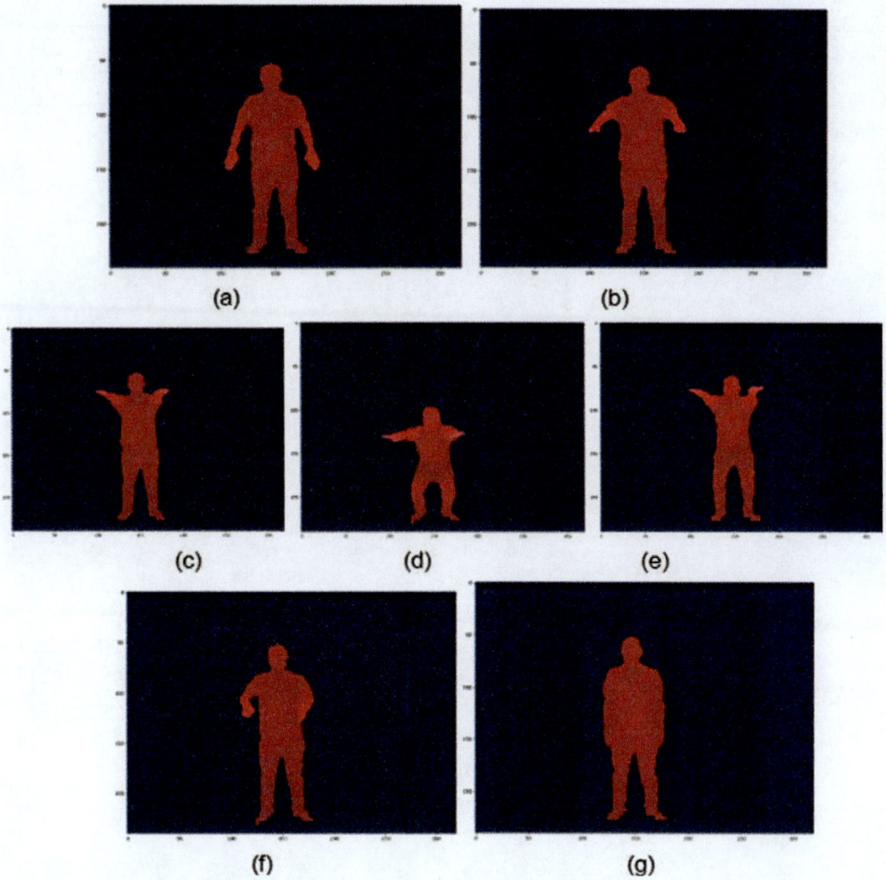

Fig. 2 Depth images of squat activity. Frames **a** straight posture, **b** arms raised, **c** legs bent and arms raised, **d** legs bent and arms raised slightly more than previous posture, **e** legs bent and arms raised less than previous posture, **f** legs bent and arms raised less than previous posture, **g** only arms raised; represents sequential depth motion images for squat activity

sampler. Here, in CNN, it is used to reduce the image size, and hence representation becomes smaller and more manageable.

For accelerometer data, a simple CNN model has been developed by Zeng [14] where separate convolution and pooling layer are used for each axis of accelerometer data. After that concatenated before interpretation by hidden fully connected layers.

Cho and Yoon [15] divided activities into dynamic and static. They have developed a two-stage modeling approach. Instead of classifying activities using single class classifier, divide and conquer approach has been applied where the first binary classifier is used to recognize static and dynamic activities and after that in each class individual activities are recognized using different classifiers. And hence forms two-stage modeling approach.

Fig. 3. 3D Inertial signal representation for squat activity **a** (*X*–*Y*–*Z*) accelerometer signals representation, **b** (*X*–*Y*–*Z*) gyroscope signals representation

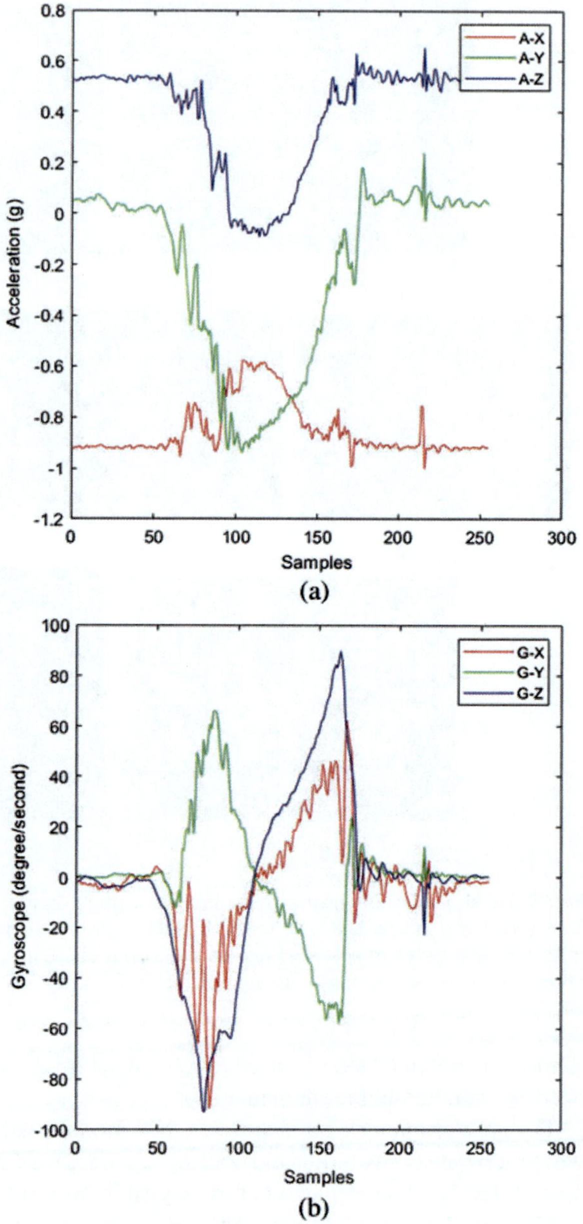

Jiang and Yin [16], combined signals to produce image data rather than using 1D CNN on signals. They have used 2D CNN to process, specifically accelerometer and gyroscope data. They have used 2D DFT to the signal image, and its magnitude is defined as an activity image.

Recurrent Neural Network—LSTM

To learn the sequence of observation, over time or to learn the sequence of words in a sentence, RNN is designed. LSTM network is a special type of RNN which overcomes the problem or difficulties of training RNN on sequential data. LSTM is found to work best for problem statements like handwritten recognition, language modeling, and machine translation. As explained in the previous section, CNN reads across the input sequence, unlike LSTM, which reads observation and predictions in sequential format and makes its own representation of input sequence. It works differently than CNN and is based on backpropagation through time. LSTM can be applied to HAR.

Each window of sensor data is learned as one activity by LSTM. Here, the one-time step consists of one or more variables. In [17], authors claimed that convolutional kernels do not support a wide range of activity recognition because of their limitation to operate on fixed-sized windows of sensor data. Instead, the use of bidirectional LSTM, which works on the sequence of data in both the direction, at a time '*t*', the score generated for that time step predicts activity occurring at that time step. By merging the individual scores into a single prediction, prediction for the entire window is obtained. Liu and Xiao [18] introduced DeepConvLSTM framework. Convolutional and recurrent layers are combined in this architecture. Feature maps are extracted by Convolutional layers from input sensor data, and they provide a compact representation of the data. Recurrent layers modelled the temporal dynamics of the activation of feature maps.

4 Proposed Environment

The system is designed to recognize the activity of human by using dataset, which consists of features extracted from four modalities viz., color data, inertial signals, depth maps, and skeleton data. Figure 5, shows flowchart for the proposed system. Along with it, Fig. 4, shows architecture of designed model. Features from RGBD depth modality, has not been used for this study. The RGBD video dataset is obtained by using the Kinect Sensor [12]. The system is designed by taking the reference from Woo [13]. The proposed environment uses LSTM deep learning model to recognize human activities. And proposed approach, analyses extracted features and apply deep learning model over available modality to recognize activities.

While executing model over fusion of two data modalities viz. skeleton and inertial, both the data-modalities have been merged and then LSTM model is applied, to recognize activities. LSTM model is applied over whole fusion of both skeleton and inertial data forming total 861 samples with 27 actions, hence, 27 number of

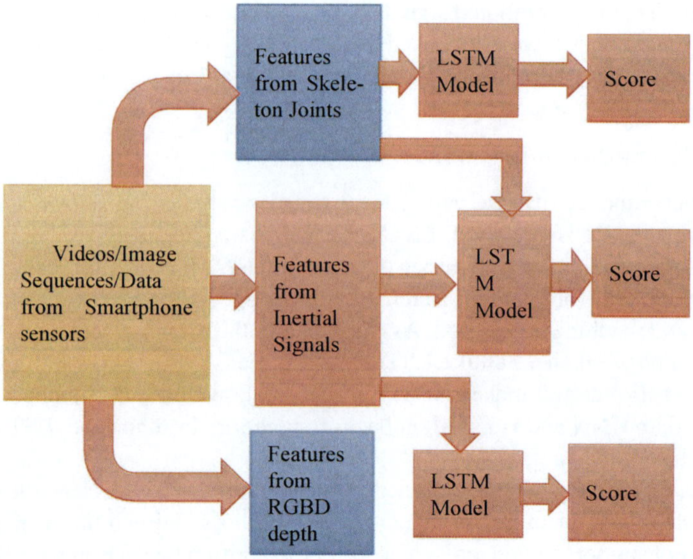

Fig. 4 Architecture for the proposed model

Fig. 5 Proposed methodology for HAR

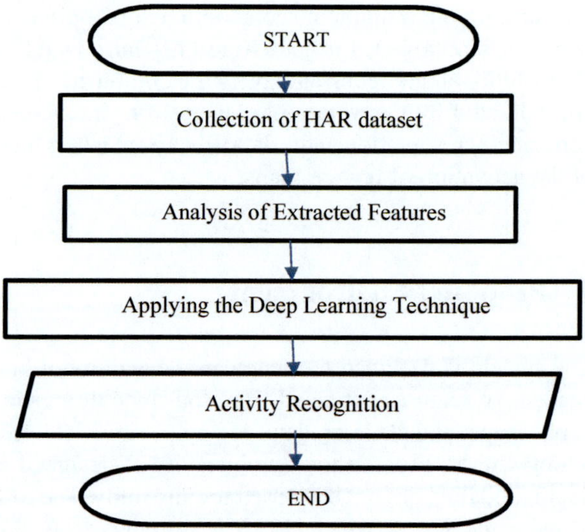

classes, time steps are 41 frames and features count is 78. Similarly, executing with only skeleton data, LSTM model is applied over whole skeleton data forming total 861 samples with 27 number of classes, time steps are 41 frames and features count is 72, for skeleton data.

From the available dataset, while using inertial dataset, three activities have been chosen. Each activity captured using eight subjects and four trials. Thus there are

32 different samples for one activity. For this experiment choosing three activities results in a total of 96 samples. Among that, 72 samples are chosen for training, and 24 samples are reserved for testing. From training data, 8% is used for validation and remaining for training the LSTM model. Nine hyper-parameters are considered for designing the LSTM model. Data is in the form of Samples $*$ timesteps $*$ Features (samples $\times$ timesteps $\times$ Features). Where a sample represents one sequence; a timestep is one point of observation in the sample; a feature is one observation at a time step. Inertial data is chosen for activity recognition. The different activity has different timesteps, thus fixed-sized input shape is obtained by setting maximum length for timesteps. Input training data has a shape of $72 \times 200 \times 6$ (72×200) where 72 are samples, 200 defines timesteps, and 6 denotes features viz., three acceleration features, and three gyroscope features.

The most challenging task in the experiment is to choose hyper-parameters. Chosen hyper-parameters are varied in a particular range to optimize the model's accuracy. Here, on the trial error basis, hyper-parameters are optimized. From the trial error basis, three sets of hyper-parameters chosen in the experiment and is given in Table 1.

5 Results and Discussion

For the proposed system, results are plotted in the form of accuracy, validation accuracy, and the loss occurred while training the model, based on chosen hyper-parameters. The model is designed to recognize activities based on two modalities. Results are obtained over two modalities viz. skeleton data and inertial data. Results of activity recognition are compared over using only skeleton data, only inertial data, and fusion of inertial data and skeleton data. While using inertial data for activity recognition, three activities right-hand wave, two hand push, jogging have been considered. By varying number of times, the learning algorithm applied for the model, i.e., several epochs, and by setting appropriate hyper-parameters, better accuracy has been achieved incrementally.

In Table 2, the performance comparisons of the obtained results of ten reported models [12, 19–28] using UTD MHAD dataset along with the proposed model, are reported. It is observed from this table that the accuracy of the proposed method is acceptable (87.50%).

In Table 3, results compared over different modalities of UTD MHAD dataset, carried out in the present investigation, have been shown. The table shows that better accuracy for activity recognition is obtained using inertial data i.e. data obtained using wearable sensors.

A comparative analyses of the results of accuracy obtained over different modalities of UTD MHAD and the corresponding methods applied on the respective modalities, as obtained from literature review [29], have been shown in Table 4. It may be inferred from this table that our developed technique for this approach worked

Table 1 Sets of hyper-parameters chosen

Hyperparameters	Set-1	Set-2	Set-3
Kernel_regularizer	0.01	0.01	0.009
Recurrent_regularizer	0.01	0.01	0.009
Bias_regularizer	0.01	0.01	0.009
Dropout	0.2	0.3	0.2
Loss	Categorical_crossentropy	Categorical_crossentropy	Categorical_crossentropy
Optimizer	sgd	(lr = 0.01, decay = 1e−6, momentum = 0.9, nesterov = True), sgd	Rmsprop
Epochs	30	130	200
Batch_size	1	1	5
Accuracy (%)	79.10	83.33	87.50

Table 2 Comparison of recognition accuracy (%) with the state-of-the-art

Method	Year	Accuracy (%)
Kinect + inertial [12]	2015	79.10
Cov3DJ [23]	2013	85.58
EIC-KSVD [22]	2014	76.19
DMM [19]	2015	84.2
Optical spectra [26]	2016	86.97
JDM [25]	2017	88.10
3DHOT-MBC [27]	2017	84.40
CNN [20]	2018	87.9
Pose estimation maps [28]	2018	94.51
HDM [21]	2019	92.8
Proposed method- LSTM based on inertial data	2019	87.50

Table 3 Comparison of results of the proposed model over different modalities of UTD MHAD dataset

Modalities used in model from UTD MHAD	Test accuracy (%)
Inertial	87.50
Skeleton	64
Inertial + skeleton	72

Table 4 Activity recognition accuracy—the-state-of-the-art methods compared to the proposed one [29]

Method	Modalities used	Accuracy (%)
DMM-CRC	Depth + inertial	79.10
GF + LF	Depth + skeleton	84.89
SD-SR	Skeleton	86.12
JTM + CNN	Skeleton	87.90
DMM-CT-HOG-LBP-EOH	Depth	88.40
DMM-CRC-LOGP	Depth + skeleton + inertial	91.50
TPM-LLC-BoA	Skeleton	93.02
MDACC	Depth + skeleton + inertial	93.26
VGG-F	RGB + depth + skeleton	94.60
VGG-16	RGB + depth + skeleton	95.11
JD + JMHC	Depth + skeleton	93.33
JD + JMHC + Acc	Depth + skeleton + inertial	94.91
LSTM	Skeleton (proposed)	64
LSTM	Skeleton + inertial(proposed)	72
LSTM	Inertial (proposed)	87.50

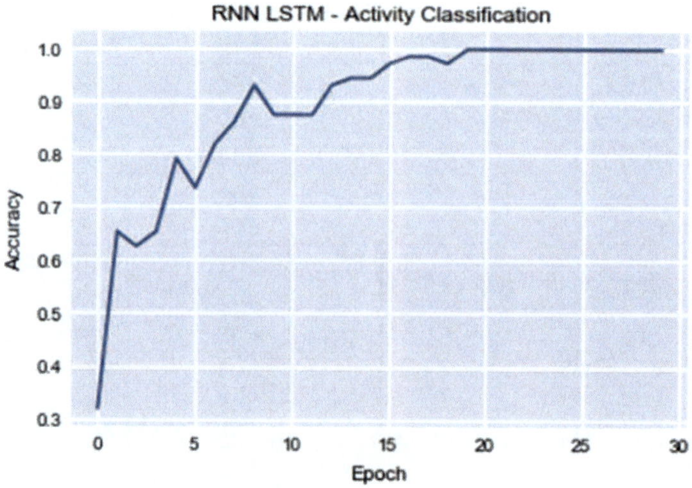

Fig. 6 Accuracy plot for RNN-LSTM model over 30 epochs without using validation model for four parameters

better for inertial modality than for fusion of skeleton and inertial or only for skeleton modality, presented towards the end of the table.

Figure 6, shows accuracy plots for RNN-LSTM model over 30 epochs. Results obtained for simple model without validation and model executed using only basic four hyper-parameters viz loss, optimizer, epochs, and batch_size. Better results are obtained by varying epoch hyper-parameter. As shown in Fig. 6, above 20 epochs, constant accuracy has been obtained for chosen values of hyper-parameters.

Accuracy and loss plots for model executed over individual inertial and skeleton data and on fusion of both the data have been shown in the figures below. Figure 7a shows accuracy for training which keeps on increasing, model executed over fusion of both inertial and skeleton data, but validation accuracy becomes constant after 81 epochs which results in almost above 90% of training accuracy. Test accuracy obtained for the model is observed to be 72%. Figure 7b, shows non-linear curve for training loss which decreases with increasing epochs and the graph for validation loss shows varying changes with an overall decreasing nature. Overall test loss for the model is detected to be 28%.

In Fig. 8a, accuracy for training keeps on increasing. Here, the model is executed on only skeleton data, but validation accuracy becomes constant after 60 epochs which results in 90% of validation accuracy. Test accuracy obtained for this model is 64%. Figure 8b, shows non-linear curve for training loss which decreases up to 60 epochs and the plot for validation loss with respect to epoch show an overall decreasing nature which results in 36% of test loss.

Accuracy and loss plots for set-2 parameters from Table 1, has been shown in Fig. 9. Figure 9a shows that the accuracy for training keeps on increasing, model executed over only for inertial data, but validation accuracy becomes constant after

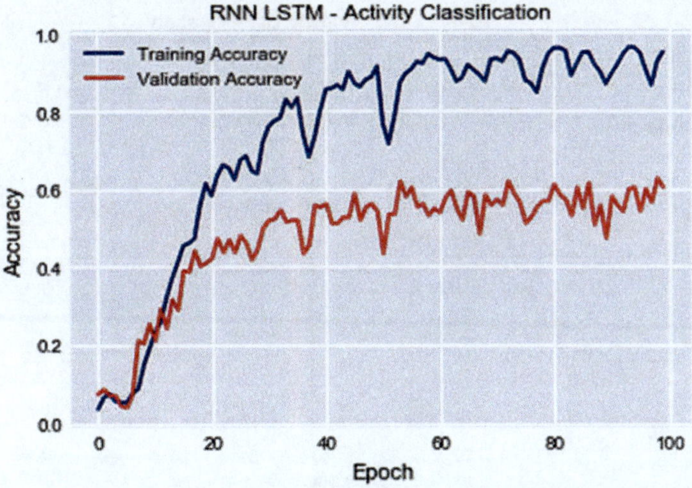

(a)

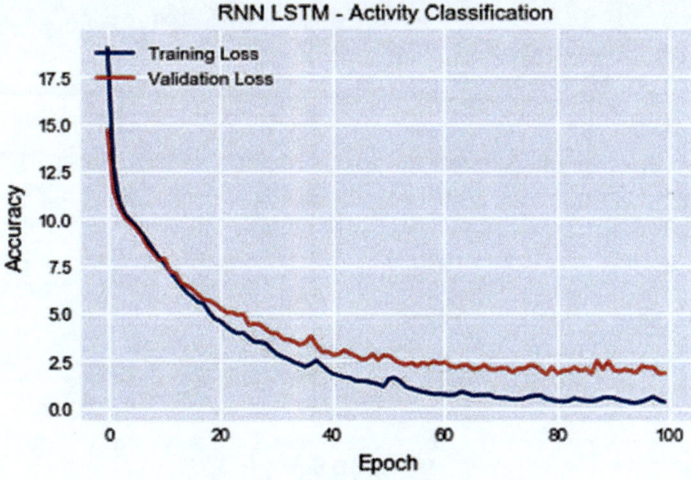

(b)

Fig. 7 **a** Accuracy and **b** loss plots for RNN-LSTM model obtained with validation model using both inertial and skeleton data

115 epochs. Test accuracy obtained for this set is 83.33%. Figure 9b, shows non-linear curve for training loss which decreases up to 108 epochs and graph for validation loss shows varying changes with an overall decreasing trend which results in 16.67% of test loss.

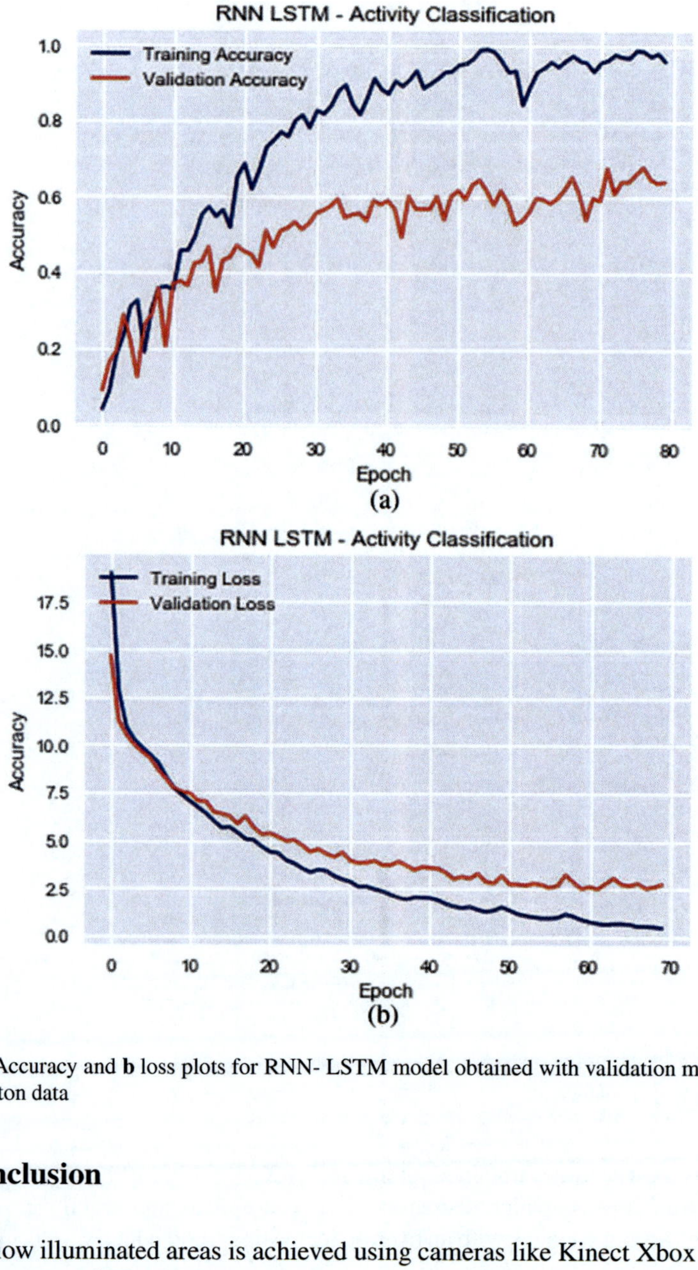

Fig. 8 **a** Accuracy and **b** loss plots for RNN- LSTM model obtained with validation model using only skeleton data

6 Conclusion

HAR in low illuminated areas is achieved using cameras like Kinect Xbox one, due to the presence of IR and depth sensors. Automatic 3D motion tracking is easily achieved using such cameras avoiding previous invasive ways of using markers or inertial sensors. For this proposed system, deep learning technique-LSTM model worked better, in that, crucial task lies in choosing initial hyper-parameters. For the proposed model, better activity recognition is obtained using only inertial data

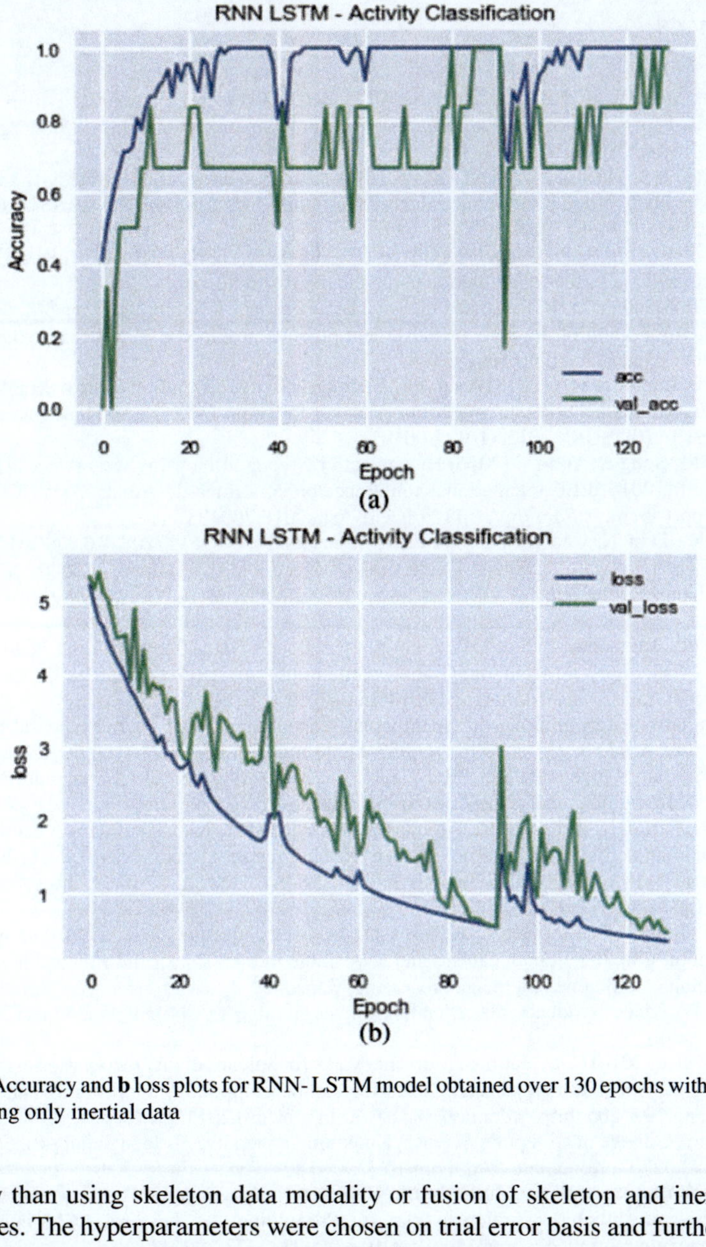

(a)

(b)

Fig. 9 **a** Accuracy and **b** loss plots for RNN- LSTM model obtained over 130 epochs with validation model using only inertial data

modality than using skeleton data modality or fusion of skeleton and inertial data modalities. The hyperparameters were chosen on trial error basis and further results are optimized to achieve better accuracy. In future, parameter selection and tuning of the different hyperparameters may be carried out in detail for better performance of HAR tasks.

References

1. Vrigkas Michalis A, Nikou C, Kakadiaris IA (2015) A review of human activity recognition methods. J Front Robot AI 2:28. https://www.frontiersin.org/article/. https://doi.org/10.3389/frobt.2015.00028
2. Ranasinghe S, Machot FA, Mayr HC (2016) A review on applications of activity recognition systems with regard to performance and evaluation. Int J Distrib Sens Netw. https://doi.org/10.1177/1550147716665520
3. Lange B, Koenig S, McConnell E, Chang C-Y, Juang R, Suma E, Bolas M, Rizzo A (2012) Interactive game-based rehabilitation using the microsoft kinect
4. Lang CE, Macdonald JR, Reisman DS, Boyd L, Jacobson KT, Schindler-Ivens SM, Scheets PL (2009) Observation of amounts of movement practice provided during stroke rehabilitation. Arch Phys Med Rehabil 90:1692–1698
5. Lara OD, Labrador MA (2013) A survey on human activity recognition using wearable sensors. In: IEEE communications surveys & tutorials, vol 15(3), pp 1192–1209, Third Quarte. https://doi.org/10.1109/SURV.2012.110112.00192
6. Zubair M, Song K, Yoon C (2016) Human activity recognition using wearable accelerometer sensors. In: 2016 IEEE international conference on consumer electronics-Asia (ICCE-Asia), Seoul, pp 1–5. https://doi.org/10.1109/ICCE-Asia.2016.7804737
7. Wang Y, Jiang X, Cao R, Wang X (2015) Sensors 15:17195–17208. 10.3390/s150717195 School of Software, Dalian University of Technology, Dalian 116620, China; E-Mails: jiangxl@dlut.edu.cn (X.J.); caocao7066@outlook.com (R.C.); ssdutwxy@foxmail.com (X.W.)
8. Ramasamy Ramamurthy S, https://doi.org/https://doi.org/10.1002/widm.1254 Funding information Alzheimer's Association, Grant/Award Number: AARG-17-533039; Office of Naval Research, Grant/Award Number: N00014-15-1-2229
9. Human activity recognition with smartphone sensors using deep learning neural networks. Expert Syst Appl 59: 235–244. ISSN 0957-4174. https://doi.org/10.1016/j.eswa.2016.04.032
10. Yang J, Nhut Nguyen M, San PP, Li X, Li, Krishnaswamy S, Deep convolutional neural networks on multichannel time series for human activity recognition
11. Park SU, Park JH, Al-masni MA, Al-antari MA, Uddin MdZ, Kim T-S (2016) A depth camera-based human activity recognition via deep learning recurrent neural network for health and social care services. Procedia Comput Sci 100: 78–84. ISSN 1877-0509. https://doi.org/10.1016/j.procs.2016.09.126
12. Chen C, Jafari R, Kehtarnavaz N (2015) UTD-MHAD: a multimodal dataset for human action recognition utilizing a depth camera and a wearable inertial sensor. In: Proceedings of IEEE international conference on image processing, Canada
13. Woo CW, Kinect sequence classification. https://github.com/woo-chia-wei/kinect-sequence-classification
14. Zeng M et al (2014) Convolutional neural networks for human activity recognition using mobile sensors. In: 6th international conference on mobile computing, applications and services, Austin, pp 197–205. https://doi.org/10.4108/icst.mobicase.2014.257786
15. HCI Lab., College of Computer Science, Kookmin University, 77, Jeongneung-ro, Seongbuk-gu, Seoul 02707, Korea. https://doi.org/10.3390/s18041055
16. Jiang, Wenchao and Yin, Zhaozheng, Human activity recognition using wearable sensors by deep convolutional neural networks. In: Proceedings of the 23rd ACM international conference on multimedia. https://doi.org/10.1145/2733373.2806333
17. Deep recurrent neural networks for human activity recognition. Department of Information Communication Engineering, Chosun University, 375 Susuk-dong, Dong-gu, Gwangju 501–759, Korea. aaymurad@chosun.kr
18. Ordóñez FJ, Roggen D, Academic Editors: Yun Liu, Wendong Xiao, Han-Chieh Chao and Pony Chu Received: 30 Nov 2015/Accepted: 12 January 2016/Published: 18 January 2016. Wearable Technologies, Sensor Technology Research Centre, University of Sussex, Brighton BN1 9RH, UK

19. Bulbul MF, Jiang Y, Ma J (2015) Dmms-based multiple features fusion for human action recognition. Int J Multimedia Data Eng Manage
20. Wang P, Li W, Li C, Hou Y (2018) Action recognition based on joint trajectory maps with convolutional neural networks. Knowl-Based Syst
21. Zhao R, Xu W, Su H, JiQ (2019) Bayesian hierarchical dynamic model for human action recognition
22. Zhou L, Li W, Zhang Y, Ogunbona P, Nguyen DT, Zhang H (2014) Discriminative key pose extraction using extended LC-KSVD for action recognition. In: Proceedings of IEEE international conference digital image computing technical application, pp 1–8
23. Hussein ME, Torki M, Gowayyed MA, El-Saban M (2013) Human action recognition using a temporal hierarchy of covariance descriptors on 3D joint locations. In: Proceedings of international joint conference artificial intelligence, pp 2466–2472
24. Wang P, Li Z, Hou Y, Li W (2016) Action recognition based on joint trajectory maps using convolutional neural networks. In: ACM on multimedia conference (ACM MM), pp 102–106
25. Li C, Hou Y, Wang P, Li W (2017) Joint distance maps based action recognition with convolutional neural networks. IEEE Sig Process Lett 24(5):624–628. https://doi.org/10.1109/LSP.2017.2678539
26. Hou Y, Li Z, Wang P, Li W (2016) Skeleton optical spectra based action recognition using convolutional neural networks. IEEE Trans Circ Syst Video Technol
27. Ma S, Sigal L, Sclaroff S (2015) Space-time tree ensemble for action recognition. In: IEEE conference on computer vision and pattern recognition (CVPR), pp 5024–5032
28. Liu M, Yuan J (2018) Recognizing human actions as the evolution of pose estimation maps. In: 2018 IEEE/CVF conference on computer vision and pattern recognition, Salt Lake City, pp 1159–1168. https://doi.org/10.1109/CVPR.2018.00127
29. Malawski F, Kwolek B (2019) Improving multimodal action representation with joint motion history context. J Vis Commun Image Represent 61. 10.1016/j.jvcir.2019.03.026

A Discriminative Model for Multiple People Detection

Smita S. Kulkarni and Sangeeta Jadhav

Abstract Group activity recognition is becoming more important day by day for video surveillance, sports analytics, etc. Monitoring various cameras manually through human resources is a complex job; due to this, computer vision algorithms are being developed to perform lower and higher level tasks. This paper presents multiple people detection using the histograms of oriented gradients (HOG) feature descriptor algorithm through a support vector machine (SVM) based on the different group action class of persons. Multiple people detection for group action identification is a complex problem as accurate detection of individual persons requires extensive computation. To achieve multiple people detection for group activity, HOG feature extraction is proposed. HOG is precise and accurate person detection algorithm in the recent computer vision application. In addition, thresholding algorithm is implemented to collect the HOG feature vectors of definitely detected windows and firms the pathway followed by a person in the video frames. The proposed algorithm is evaluated through different aspects like group action categories and existence of occlusion over Haar and HOG features.

Keywords Multiple person detection · HOG · SVM

1 Introduction

In computer vision, group activity recognition is demanding in many applications. Multiple people carry out group activities. Since this group activity involved multiple people interaction along with varying scene, owing to this, it is hard to differentiate

S. S. Kulkarni (✉)
D. Y. Patil College of Engineering, Akurdi, Pune, India
e-mail: smitak0103@gmail.com; sskulkarni@entc.maepune.ac.in

S. Jadhav
Army Institute of Technology, Dighi, Pune, India
e-mail: djsangeeta@rediffmail.com

S. S. Kulkarni
MIT Academy of Engineering, Alandi(D), Pune 412105, India

© Springer Nature Singapore Pte Ltd. 2021

S. N. Merchant et al. (eds.), *Advances in Signal and Data Processing*,
Lecture Notes in Electrical Engineering 703,
https://doi.org/10.1007/978-981-15-8391-9_35

between people and other object present in the scene. Hence, it is important to detect multiple people as a preprocessing stage in group activity recognition. In cluttered backgrounds under variation in illumination to discriminate multiple people robust features set of persons requires. Detecting multiple people in video frames is a difficult task due to their unpredictable appearance and the variety of poses that they can take on. In computer vision, robust person detection requires visual group activity recognition.

In this paper, multiple people detection module is evaluated on the collective activity dataset [1]. The paper is presented as follows. In Sect. 2 related work is presented for person detection. The feature extraction methods HOG and Haar are discussed in Sects. 3 and 4 respectively. We then discuss the experimental approach in details in Sect. 5. In Sect. 6, we report our experimental quantitative results on collective activity dataset [1], and last, we conclude in Sect. 7. For multiple people detection from video frames, HOG feature descriptor [2] particularly outperform than Haar feature sets [3]. For detection of people from object present in video frames, SVM is used as a trainable discriminative classifier.

Dataset: The proposed technique gives a precise multiple people detection on challenging collectively activity dataset [1] with large collection of poses deviations and background. A collective activity dataset [1] is taken under real-world circumstances. In this dataset, 40 short video clips are involved, in which multiple people are recorded, this represents group activity under the group action walking, queuing, waiting and talking.

2 Related Work

In computer vision, multiple people detection is important and challenging in many applications like surveillance systems and sport analytics where a group of people actively involved. For detecting humans, several researchers have presented many frameworks. Many methods have been developed for feature extraction, which can be applied for SVM [4] to discriminate between object and persons from video frames. The feature extraction should be robust for cluttered background and under change in illumination.

In human detection HOG [2] is widely used. There are several novel methods [5, 6], presented using HOG for human detection. In [7], probabilistic approach developed for distinctive features and AdaBoost for classification of features. Owing to consistency in motion, this performance needs to improve. In [8] depth information, wavelet transform [3] and in [9] local mixtures of parts are used for human detection, but limited due to person's pose for precise detection. Automatic multiple people detection and tracking scheme are presented in [10] which are based on temporal differencing technique and morphology support measures.

In [11] human detection, developed using HOG for people counting whose performance in the presence of occlusion is robust. This framework demonstrates the methods which use classification of HOG features in indicating people counting efficiently and more precisely. In [12] multiple-person recognition has presented in color-with depth sequences. In [13] multiple people are detected in pedestrian zones using surveillance cameras using ACF (Caltech), ACF (INRIA), Viola-Jones, and HOG detectors. In [14] for object detection latent SVM discriminatively multiscale deformable model is presented. Recently using deep model based on recurrent LSTM take image as an input and directly detects people in crowd scenes [15].

3 Histogram of Oriented Gradients-Based Multiple Person Detection

Detecting multiple persons in video frames has confirmed to be a challenging due to the wide inconsistency in the scene. As background, pose and scope of the object, all can be impactful on the appearance of a video frame. For the same robust feature extraction algorithm, specifically, HOG is proposed [2]. The HOG splits each video frame into a grid. Inside the cell of a grid, information is held in the form of orientation of the gradients by concatenation of the histogram. In computer vision for detection of object HOG feature descriptors are used. The main motivation to use HOG is to compute incidents on a dense localized grid of homogeneously spread out cells and usages overlying local contrast regularization for enhanced accuracy. HOG uses spread of intensity gradients to describe the shape of person as edge directions.

The dataset contains labeled multiple person image from diverse video frames. support vector machine (SVM) is used for person detection as a classifier with a HOG feature vector descriptor. In terms of accurateness, HOG features performed better compared to other feature extraction methods. In the paper, HOG performance is tested by an investigation, along with SVM classifier [3] by means of linear kernel and a nonlinear kernel like RBF (radial basis function) for multiple people detection in video frames.

4 Haar Transform Feature Based Multiple People Detection

In multiple people detection, it is important to develop representation of the object class. In a video frame, people illustrate the large uncertainty in the pose, texture and color in clutter background. To avoid these problems, here, Haar features are proposed [3]. Haar features contain several orientations with local intensity differences.

The most visually significant features are computed by Haar transform in the three color channels on specific location and orientation by subtracting average dark and light region pixel. This specific coefficient is used as features of the person class during training of the classifier to discriminate between person and non-person object for multiple people detection. SVM [3] classifier is used on Haar feature set to minimize generalization error for people detection.

5 Experimental Approach

This segment gives details of, the investigations and the results of these experiments. The video is pre-processed, and frame extraction is done, which send through HOG and Haar feature descriptor with SVM to detect multiple people.

The proposed multiple people detection scheme is separated into two elements: a HOG and Haar feature descriptor extractor and an SVM classification method functioning through a sliding windowing system. In first element, video frame transforms into HOG and Haar descriptor. In order to detect all objects present in frames, input vector is composed by the aggregation of all the HOG descriptors associated with each detected window. The SVM computed to classify the presence of multiple people in the window between input vectors of HOG and Haar descriptor using a supervised learning label model. SVM is used as classification in between object and person as supervised learning. SVM is trained as multiclass SVM classifier based on the person descriptors HOG and Haar to detect multiple people from video frame. Due to multiple people, its non-separable optimization problem is shown in Eq. (1).

$$f(w, b) = \|w\|^2 + C \sum \xi_n \tag{1}$$

$$y_n(w^T X_n + b) \geq 1 - \xi_n \tag{2}$$

Multiple people detection is constraint optimization formulation subject to Eq. (2).

Once the multiple people are detected, it is required to estimate the performance of the presented algorithm. The algorithm runs under different group actions and is evaluated as follows.

5.1 Group Activity Categories

In the coverage area of video, each tenth frame consists one of the subsequent group activities: walking, queuing, waiting, crossing and talking in dataset [3]. People detection accurateness needs to check on the different group activity categories. The number of people can alter in different day time and circumstances and the algorithm should perform well for different group activities.

5.2 Occlusion Existence

During group of actions, persons can be occluded by each other or else object within the surroundings. The efficiency of the proposed algorithms during the existence of occlusions offered a clue for appropriateness in real existence situations.

5.3 Performance Parameter

The purpose of the proposed method is to detect the multiple people to facilitate a different group of actions. The following different performance parameters are used for estimating the proposed algorithms.

- **Detection Rate** is defined as proportion of multiple people detected accurately toward the entire group of persons detected into the video scene.

$$\text{Detection Rate} = \text{True Positive}/(\text{True Positive} + \text{False Positive})$$

- **Recall** is defined as the proportion of the number of people accurately detected to the real number of people into the video scene.

$$\text{Recall} = \text{True Positive}/(\text{True Positive} + \text{False Negative})$$

- **Accuracy** is defined as proportion of truth values to the sum of all the possible values.

$$\text{Accuracy} = (\text{True Positive} + \text{True Negatives})/(\text{True Positive} + \text{False Positives} + \text{True Negatives} + \text{False Negative})$$

6 Experimental Results

This section estimates the result of proposed algorithms with respect to the group activity categories and occlusion describes into Sect. 5 in addition to the results.

6.1 Group Activity Categories

In this experimentation, consider group actions with walking queuing, waiting, and talking people. From the result Table 1 understand that Haar transform has less accuracy compared to HOG. Since when more number of people are varying very

Table 1 Effect on group actions on people detection

	Walking	Queuing	Waiting	Talking
	Accuracy	Accuracy	Accuracy	Accuracy
Haar transform	0.7378	0.7323	0.7476	0.7212
HOG transform	0.8154	0.7804	0.8121	0.8237

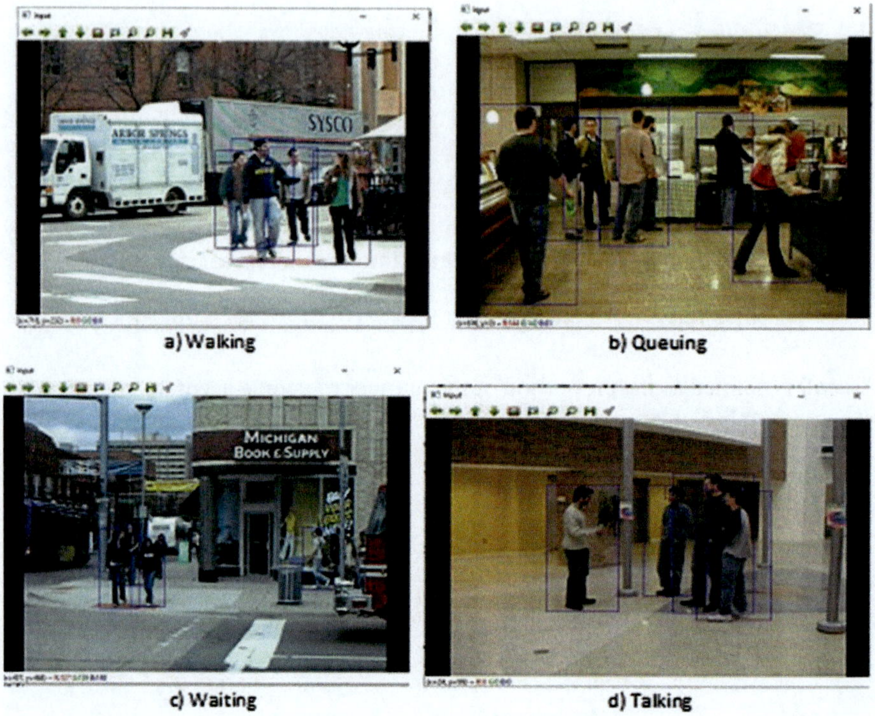

Fig. 1 Multiple people detection

close to everyone, Haar trait them as a single person. HOG implements excellent as it performs without fail irrespective of camera orientation. Also, SVM RBF kernel improves the accuracy to some extent.

In Fig. 1, visualize the multiple people detection results with HOG transform. Consider the example under the group actions (a) walking (b) queuing (c) waiting and (d) talking specified by the area bounded by the blue box.

Table 2 Occlusion effect on people detection

	Occulusion			No occulusion		
	Detection rate	Recall	Accuracy	Detection rate	Recall	Accuracy
Haar transform	0.78	0.67	0.59	0.8213	0.73	0.72
HOG transform	0.83	0.79	0.82	0.86	0.78	0.86

6.2 Occlusion Existence

In this experiment, algorithm is estimated for occluded and non occluded people in the video frame. Table 2 illustrates the results achieved. For the proposed algorithm, it is observed that accuracy is higher for the non-occluded case as measured with to occluded case. During the existences of occlusions, HOG have the maximum accurateness compared the Haar because the HOG person feature descriptors are robust in capturing diverse circumstances. In this case, SVM linear kernel presents enhanced performance.

7 Conclusion

This paper presents multiple people detection from video frames based on HOG and Haar feature in the context of group activity recognition. The performance of the proposed methods is analyzed in context of group actions and the existence of occlusion. The implemented detection experimentations demonstrate the effectiveness of the HOG features along with SVM to discriminate people from another object. In the existence of occlusion, HOG performance is excellent for multiple people detection. Owing to this, in a computer vision domain where a group of person actions involved, such as video surveillance, sports analytics and traffic monitoring, proposed method-based HOG presents excellent performance. Though HOG-based technique with SVM experiences response time less, it needs to be improved. The proposed framework presents the methods to facilitate classification over features in the detection of multiple people efficiently and more accurately.

References

1. Choi W, Shahid K, Savarese S (2009) What are they doing? Collective activity classification using spatio-temporal relationship among people. In: 12th IEEE international conference on computer vision workshops, ICCV workshops. IEEE, pp 1282–1289

2. Dalal N, Triggs B (2005) Histograms of oriented gradients for human detection. In: CVPR, pp. 886–893
3. Mikolajczyk K, Schmid C, Zisserman A (2004) Human detection based on a probabilistic assembly of robust part detectors. In ECCV , pp 69–82
4. Suykens JAK, Vandewalle J (1999) Least squares support vector machine classifiers. Neural Process Lett 3:293–300
5. Zhu Q, Yeh M-C, Cheng K-T, Avidan S (2006) Fast human detection using a cascade of histograms of oriented gradients. In :CVPR , pp 1491–1498
6. Oren M, Papageorgiou C, Sinha P, Osuna E, Poggio T (1997) Pedestrian detection using wavelet templates. CVPR 97:193–199
7. Dalal N, Triggs B, Schmid C (2006) Human detection using oriented histograms of flow and appearance. In: ECCV , pp 428–441 (2006)
8. Yang Y, Ramanan D (2012) Articulated human detection with flexible mixtures of parts. IEEE Trans Pattern Anal Mach Intell 35(12):2878–2890. IEEE
9. Xia L, Chen C-C, Aggarwal JK (2011) Human detection using depth information by kinect. In: CVPRW, pp15–22
10. Barbu T (2014) Pedestrian detection and tracking using temporal differencing and HOG features. Comput Electr Eng 40(4):1072–1079
11. Raghavachari C, Aparna V, Chithira S, Balasubramanian V (2015) A comparative study of vision based human detection techniques in people counting applications. Procedia Comput Sci 58:461–469
12. Muoz-Salinas R (2008) A Bayesian plan-view map based approach for multiple-person detection and tracking. Pattern Recognit 41(12):3665–3676
13. Kurilkin AV, Ivanov SV (2016) A comparison of methods to detect people flow using video processing. Procedia Comput Sci 101:125–134
14. Felzenszwalb PF, McAllester DA, Ramanan D (2008) A discriminatively trained, multiscale, deformable part model. CVPR 2(6):7
15. Stewart R, Andriluka M, Ng AY (2016) End-to-end people detection in crowded scenes. In: Proceedings of the IEEE conference on computer vision and pattern recognition, pp 2325–2333

Jal Sanchay—A Novel Approach for Water Usage Monitoring

Ashish Srivastava, Mandar R. Nalavade, and Debashis Adhikari

Abstract Water is one of the, most important resources on earth. With rapid increase in the world population, water consumption is increasing drastically. People now a days always want something new that can make their life easier. The technological advancements of embedded system as well as articulation of communication by sensing techniques are taking a huge role in recent days. LabVIEW which is a system engineering software for applications requiring test measurement and control with rapid access to hardware and data insights. Design of various virtual instruments (VIs) in LabVIEW provides a strong graphical tool and a platform where automated water usage monitoring system can acquire efficiently as well as with accuracy. In this proposed system GPRS enabled sensors are used to sense water flow in every outlet. The server continuously monitors and collects the data over the internet and tracks usage of water at every outlet via a wireless sensor node. At the point when water is utilized at overabundance it is shown and an alarm is sent to the user. The user can persistently monitor the water usage and the wastage through their mobile devices.

Keywords LabVIEW · Automated · Sensors · Internet · IOT · Mobile devices · Sustainable development

1 Introduction

Very essential and constrained asset required for agriculture, industry and for the presence of living creature on the earth is water. Many people do not even understand that drinking enough water every day is really important. Large amount of water is being wasted by many organizations, industries in an uncontrolled manner. The issue is raised due to poor water allocation, inefficient use, and lack of adequate

A. Srivastava (✉) · M. R. Nalavade · D. Adhikari
Department of Electronics & Telecommunication Engineering,
MIT Academy of Engineering, Alandi, Pune, India
e-mail: aksrivastava@entc.mitaoe.ac.in

© Springer Nature Singapore Pte Ltd. 2021

S. N. Merchant et al. (eds.), *Advances in Signal and Data Processing*,
Lecture Notes in Electrical Engineering 703,
https://doi.org/10.1007/978-981-15-8391-9_36

water management. Efficient use and monitoring of water is therefore a future necessity for the home or office. Engineering applications often require the acquisition of information gauged by different sensors that are positioned at the end or connected to the object of concern. The use and manufacturing of energy is one of those applications that rely on remote monitoring of the sensor equipped manufacturing site.

The physical world is linked to the digital world by capturing and revealing phenomenon of real world and then transforming it into a form that can be processed, stored and implemented. Sensing in a broader way is a technique where some information is gathered about a physical object or process, including the occurrence of event which can be measured and acted upon. Data acquisition refers to the method of evaluating physical world conditions and phenomena like energy, sound, temperature and pressure. This is achieved by using multiple detectors that sample the analog signal and convert them into a digital signal using an analog-to-digital converter. The resulting digital numerical values can then be manipulated directly by a computer so that these data can be analyzed, stored and presented. It includes sensors, to convert physical parameters into electrical signals. The individual sensors connect to controllers and processing stations that houses higher number of sensors directly communicating to a centralized processing station to collect information wirelessly. Many wireless sensor networks collect sensitive information.

The remote and unattended sensor nodes are prone to malicious or unauthorized intrusion and attacks. So proper security measures must be taken to avoid any malicious activity or error. Wireless sensor networks consist of distributed, wireless integrated systems capable of employing number of electronic sensors. In addition to a microcontroller, wireless transceiver, and energy source, each node in a wireless sensor network has one or more sensors. The microcontroller works with the electronic sensors and the transceiver to create an efficient system for relaying small quantities of important data with minimal power consumption. Wireless sensor networks most appealing characteristic is its autonomy.

WaterWiSe platform which is a combination of model-based prediction and data stream analysis is presented in [1] to provide real time monitoring and decision support tools that can be used to reduce water leakage and to improve system management. In [2], an effective way to control the wastage of water at home or industries using Wireless sensor nodes and LabVIEW software is proposed. Continuous monitoring and tracking of water usage is done via wireless sensor nodes with the concept of Internet of Things (IoT).

A system based on wireless sensor networks used for enhancing the quality of drinking water, detection of change in water quality, quantification, evaluation of its consequences and determination of proper action to be taken to limit its effects is designed in [3]. India's economic development is mostly dependent on average rainfall. Due to lower average rainfall, suicides by farmers in India's rural areas have received an immense global and domestic media attention over last ten years. Water management or monitoring system (WMS) is one of the best solution for such kind of situation. Economic distress and suicide in India is reported in [4].

In [5], wireless transmission of measured parameters which are used to monitor quality of water such as pH level, turbidity and temperature to base monitoring station using wireless sensor network is explained. Fundamental design and implementation of WSN highlighting a powerful transmission ZigBee based technology along with IEEE 802.15.4 compatible trans-receiver is used. A water quality monitoring system consisting of data monitoring node, wireless sensing node and remote monitoring node is developed to design a low cost system in [6] to avoid water pollution. ATmega 328, 8 bit Micro-controller, glass electrodes, wireless trans receiver which is compatible with IEEE 802.15.4 are used to design the monitoring system which itself used to measure pH and temperature of water. Online auto monitoring of water temperature and pH value of large scale water environment using wireless sensor networks is explained in [7].

The remainder of this paper is structured as follows. Section 2 presents the system model of the proposed device and software system. Section 3 presents the methodology of the system. Sections 4 and 5 discusses about the result obtained and conclusion.

2 System Architecture

The entire system is divided into four main functional blocks: Sensor array module, a processor module, a wireless sensor module and a display module with an external power source.

The overview of the complete system is shown in Fig. 1. The mechanism in this paper aims at designing a smart home automation application based on Raspberry Pi. The entire system is based on three main modules, (1) Sensor Module (2) Processing Module (3) Display/ Monitoring Module. Wireless sensor module is shown in the Fig. 1.

The schematic diagram of the water management system is given in Fig. 2. Sensor nodes are connected to the sensor module and gives raw data to the sensor module. The raw data is in the analog form, MCP3208 converts this data into digital form. The analog data is given to the respective channel of the MCP3208 ADC. The first input from flow sensor is applied at channel (CHO), and input from sensor two that is from piezoelectric sensor is applied at channel (CH1). Digital data coming out from the MCP3208 ADC is connected to the processing unit that is Raspberry pi. Raspberry pi is the main processing unit, the power supply is given to it and other modules are operated through Raspberry pi. GPRS Module is also connected to the Raspberry pi

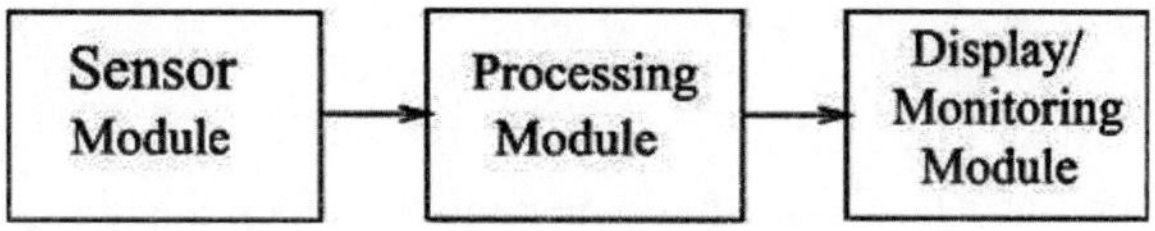

Fig. 1 Overview of wireless sensor based WMS

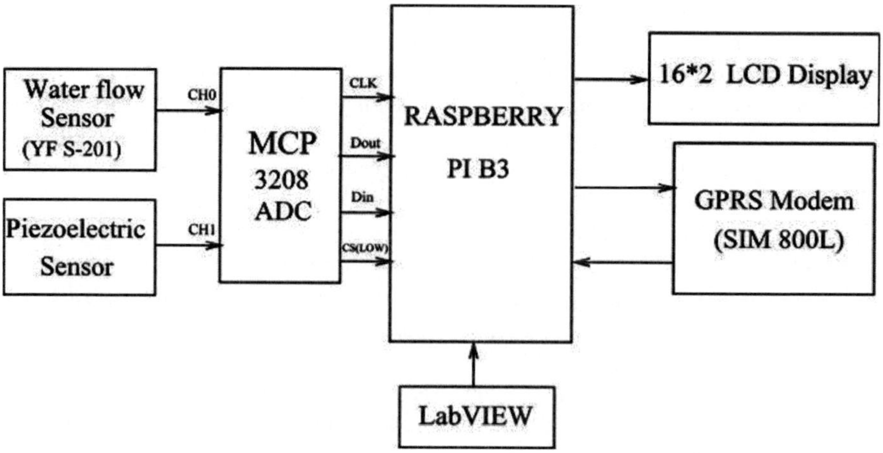

Fig. 2 System overview diagram

which is used for the message alert. GPRS will work as a transceiver device, which is taking data from processing unit sending alert to user. The LabVIEW VIs build using the algorithms, discussed in further sections, are dumped into processing unit.

3 Methodology

The water monitoring system starts the process by checking the status of tap or location of leakage. The system checks weather the tap is ON or OFF. If the tap is ON, message to mobile phone of the user will be sent. If the tap is OFF the system again checks the status of the tap until it is ON for further monitoring and actions.

The flowchart in Fig. 3 explains the functioning of system. Firstly, the raw data from the sensor nodes is given to the sensor module, it converts analog data into digital form. The digital data is then sent from the sensor module using the GPRS Module to the raspberry pi controller. The data is then sent to the LabVIEW interface. The interface checks the values of the flow, pressure and temperature provided to it. The LabVIEW interface then rechecks the previous values which had been provided to it, if there is a difference in the values the user is notified using a message on to his/her mobile number. If there are no changes in value the user is not notified.

4 Results

The LabVIEW VI, front panel and the block diagram, for the water flow monitoring system can be looked as a virtual instrument having function nodes with control and indicators placed within structures. This paper discusses a system implemented

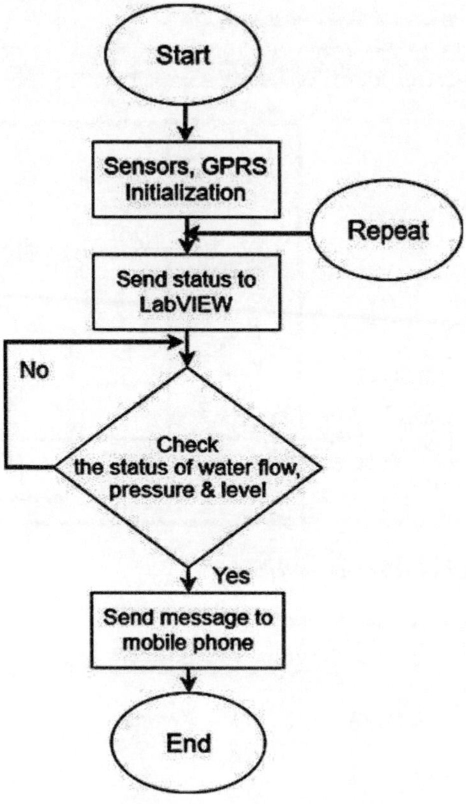

Fig. 3 Flowchart of the water monitoring system

using LabVIEW, a graphical programming language, for programming. For hardware realization of the concept of automated water usage monitoring system Raspberry along with peripherals is used. As LabVIEW is a proprietary product of National Instruments (NI) it is assumed that NI hardwares should be used. For interfacing LabVIEW with Raspberry pi, LINX software is used as an interface requirement of the system. LINX works as a raspberry pi in LabVIEW and it is connected to various sensor node. Analog input from each sensor node is given and specific limit is set at each node.

Figure 4 shows the VI of water flow by analyzing the depth of the water. In this case, maximum and minimum level of water can be set using LabVIEW front panel. Following are the sequence of events happening in the block diagram of flow measurement: The Raspberry Pi controller is initialized, read data from the sensors. We need to specify the pins where the sensors are connected. This function will return the output in the form of centimeters or inches. If the current liquid level is greater than "minimum level" then we give a logic high signal to a motor. The true case of case structure is shown in figure, in the false case we are giving a logic low to that motor which is connected on digital output pin 4. If current liquid level is less

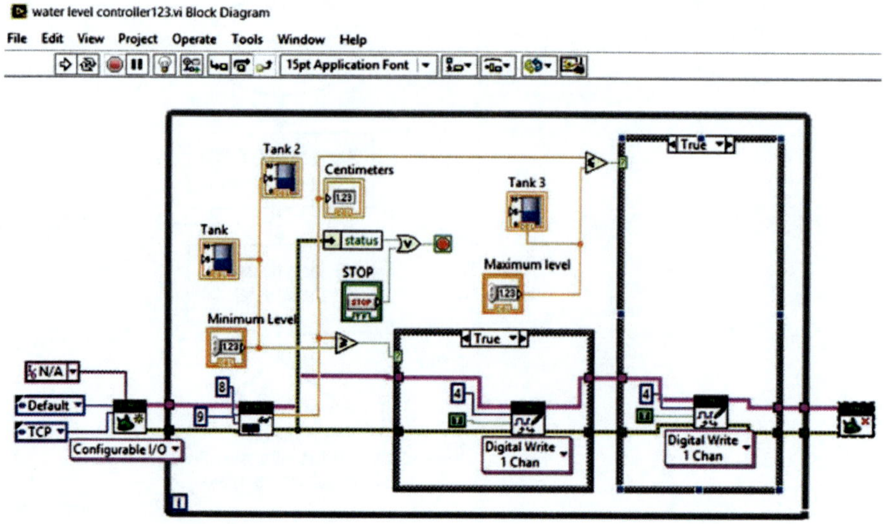

Fig. 4 Flow measurement VI

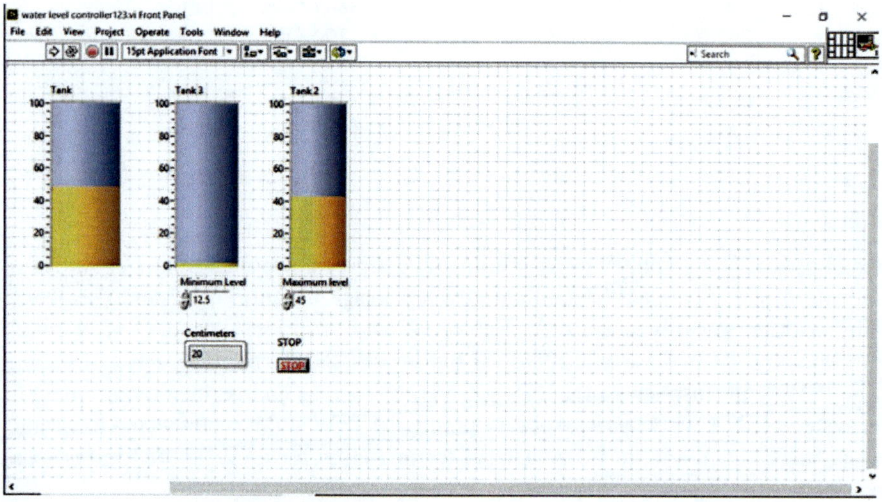

Fig. 5 Front panel of flow measurement VI

than "maximum level" then we give a logic high to motor. The corresponding front panel of flow measurement is shown in Fig. 5. It shows three different level of the tanks low level, moderate level and high level.

Figure 6 shows the block diagram of level measurement VI. The first two events in the sequences happening are same as that of flow measurement. The logical expression is used to display the status of liquid level status that is low, moderate or overflow.

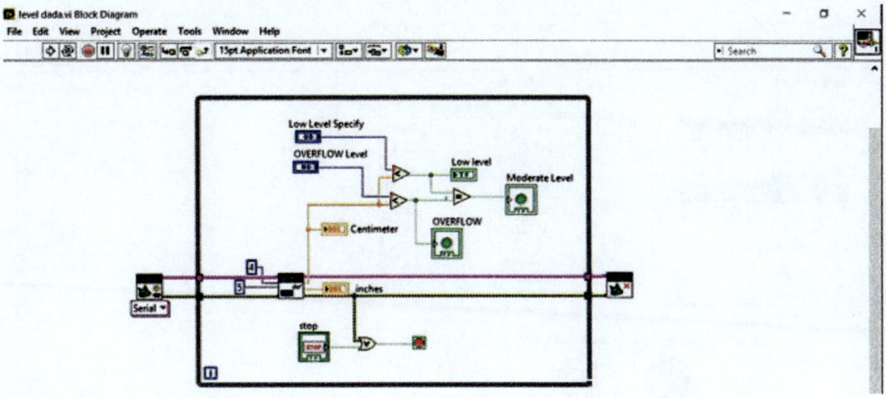

Fig. 6 Level measurement VI

Low level indicator is turns on when current level is below "low level" control, overflow indicator is turned on when current level is above "overflow level" control, and if the current level is between these two limits moderate indicator is turned on. Stop control is used to come out of the loop or stopping the application. Also, in case the controller is not responding due to any error the loop will be terminated. Closing the connection with the hardware/raspberry Pi.

Figure 7 shows the front panel of level measurement VI of the water tank level. It shows three different level of the tank; Low level, moderate level and high level. The output range is from 0 to 300 while input is only 0–10. To increase the limit we can double click on 10 and enter 300 to both the controls. Limits can be changed as per your requirements or it should be changed as per the sensor's output range by referring sensors data sheet. For Liquid level indicators based on current liquid level the respective LED will turn on and other two will be off. For example, in the figure white LED is glowing which indicates tank level is low.

Figure 8 shows the VI which includes pressure, temperature and level of the water at different level and at different conditions. In this VI LINX is used for serial communication. Also, analog read.vi is used. The two outputs from analog read.vi are temperature reading and pressure level. Figure 9 is the corresponding front panel of the final VI. The front panel shows all the parameters under consideration pressure, temperature and flow. Example of low level is considered in the front panel. When all parameters are at low level, the LEDs of low level will glow.

Figure 10 shows deployment of the code onto Raspberry Pi using LabVIEW. The novelty of the idea presented in this paper is a low cost hardware which is implemented using Raspberry Pi and LabVIEW instead of expensive hardware from NI like myDAQ or myRIO. As shown in the figure the deployment status is shown as completed, which means that raspberry pi is successfully connected to LabVIEW. Figure 11 shows the hardware setup for physical verification of the logic after implementation.

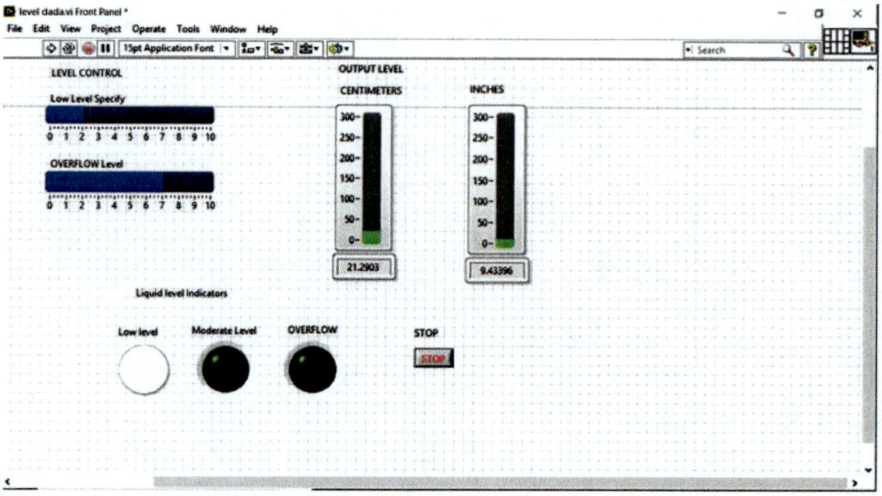

Fig. 7 Front panel of level measurement VI

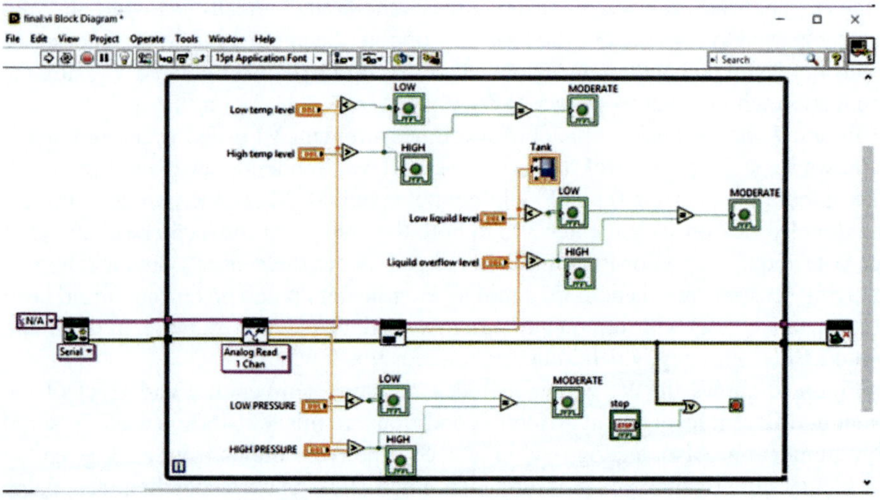

Fig. 8 Final VI

5 Conclusion

The advancements in the field of WSN, IoT and 5G have paved way for better opportunities for building smart systems. The natural resources are being exhausted and obliterated in one or other manner. These resources are important to sustain life, yet with headway in advancement in technology. The advancements should be used to make our life simpler.

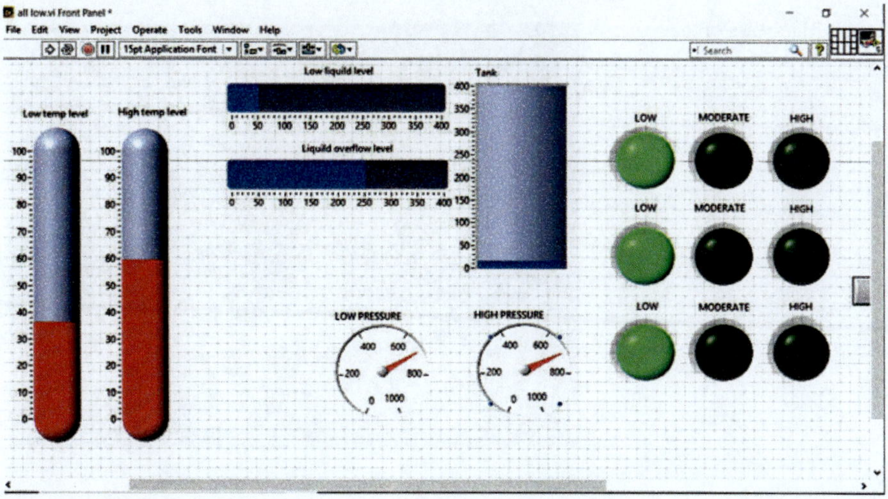

Fig. 9 Front panel final VI

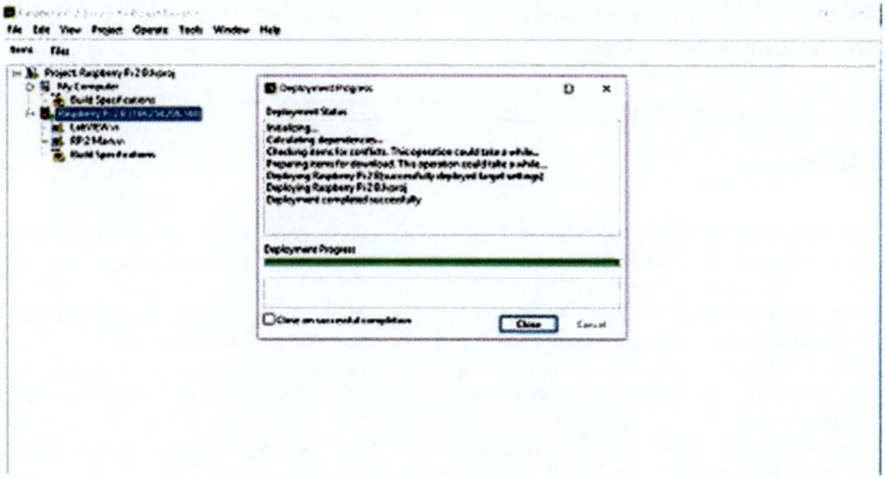

Fig. 10 Deployment of Raspberry pi with LabVIEW

In this paper an idea is presented to keep an eye on the usage or leakage of water assets. A GPRS enabled sensors are used to sense water flow in every outlet. The server continuously monitors and collects the data over internet and tracks water usage at every outlet via a wireless sensor node. The idea can be used to keep equalization with nature and our future generation. Thus, helping in sustainable development.

In future a logic to predict scarcity of water can be developed. The results of the development can be shared with government or agencies so as to take preventive

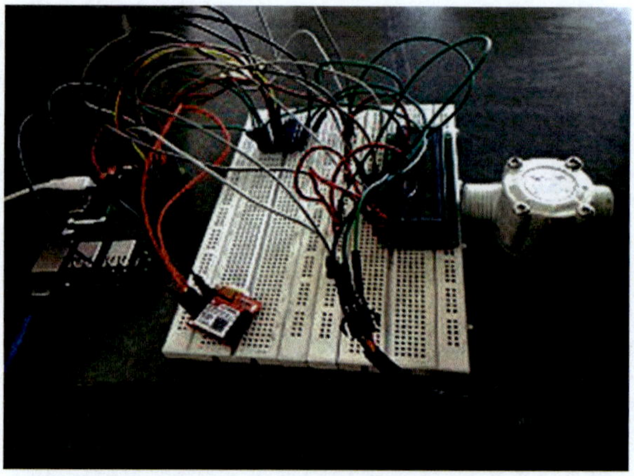

Fig. 11 Hardware of the system

actions. Additionally, the consumption of water by individual house can be predicted and a supervisory data can be generated on usage of water. If the usage of water is over a certain limit an alert message should be sent by the system, leading to an end-to-end automated system.

References

1. Allen M, Peris A, Iqbal M, Whittle AJ (2013) Water distribution system monitoring and decision support using a wireless sensor network. In: 14th IEEE ACIS international conference on software engineering. Honolulu. HI
2. Saseendran S, Nithya V (2016) Automated water usage monitoring system. In: IEEE international conference on communication and signal processing. Melmaruvathur, India
3. Jalal D, Ezzedine T (2017) Towards a water quality monitoring system based on wireless sensor networks. In: IEEE international conference on IOT, embedded system and communication. Gafsa, Tunisia
4. Hebous S, Klonner S (2014) Economic distress and farmer suicides in India : an econometric investigation. University of Heidelberg, Discussion paper series no. 565
5. Sridharan S (2014) Water quality monitoring system using wireless sensor network. Int J Adv Res Electron Commun Eng (IJARECE) 3(4)
6. Aswale P, Patil S, Ahire D, Shelke S, Sonawane M (2014) Water environment monitoring system based on WSN. Int J Adv Res Electron Commun Eng (IJARECE) 4(4)
7. Jiang P, Xia H, He Z, Wang Z (2009) Design of a water environment monitoring system based on wireless sensor networks. Article on sensors
8. Sridharan S (2014) Water quality monitoring system using Zigbee based wireless sensor network. JEST-M 3(2)

An Algorithm for Skew Angle Estimation and It's Application Domain

Unnati Raju Kulkarni, Hemant Goraksh Ghuge, Revati Anand Kulkarni, and Kirti Vasant Thakur

Abstract Angle detection and estimation is important in various fields like line following mobile robots, document analysis, construction sites etc. There are several methods and algorithms proposed earlier to do that. This paper represents a new algorithm to estimate skew angle and its possible applications in different fields. In this algorithm, angle in the preprocessed black-white image is estimated using matrix traversal and slope equation from Euclidean geometry. This algorithm can measure the angle between −90° and +90° efficiently.

Keywords Algorithm · Image processing · Skew angle estimation · Multi-domain application

1 Introduction

Nowadays, digital image processing [1] is flourishing in numerous domains like medical, forensics, defense, robotics, construction works, document processing, space imaging, etc. The basic methods of digital image processing in the medical field are image restoration and image enhancement [2]. Forensic imaging processing uses a variety of computer techniques which involve digital filters that can suppress noise and sharpens the image. However, in satellite interpretations, statistical methods are applied to the digital images and after processing the various discrete surfaces are

U. R. Kulkarni (✉) · H. G. Ghuge (✉) · R. A. Kulkarni (✉) · K. V. Thakur (✉)
Department of Electronics and Telecommunication Engineering, Government College of Engineering and Research, Avasari Khurd, Pune, Maharashtra, India
e-mail: unnatikulkarni46@gmail.com

H. G. Ghuge
e-mail: hemantgghuge@gmail.com

R. A. Kulkarni
e-mail: kulkarnirevati988@gmail.com

K. V. Thakur
e-mail: kirti79@gmail.com

© Springer Nature Singapore Pte Ltd. 2021

S. N. Merchant et al. (eds.), *Advances in Signal and Data Processing*,
Lecture Notes in Electrical Engineering 703,
https://doi.org/10.1007/978-981-15-8391-9_37

identified by analyzing the pixel values [3]. In robotics, image processing is used for navigation, marker detection, angle estimation. Along with this, it can also be used at construction sites for the accurate designing of bridges or for observing the cracks on walls [4]. The essential steps in document analysis are skew detection and correction. Recognition of character begins with data acquisition and ends up with a skew correction method. Currently, there are many skew angle estimation algorithms using image processing which requires connected component analysis and restricted for document analysis. The purpose of this paper is to introduce a new algorithm for skew angle estimation and it's probable application area.

2 Literature Survey

In recent years, many research works were performed for angle estimation using image processing techniques that can be used for line navigation [5], document analysis [6–8], construction sites. However, there are some uncertain issues despite that fact the technology in this field is mature. Hough transform is utilized to identify straight lines in an image. The peak in the Hough space represents the dominant line and its skew. The real downside of this method is that it is computationally expensive and is very challenging to get a peak in the Hough transform when text becomes sparse. The skew angle is the angle made with the normal or perpendicular. Various techniques have recently been proposed for recognizing document image skew angles. Chaudhuri and Chaudhuri [9] propose a precise technique based on the cross-correlation of vertical image slices They exhibit their strategy on both English and Bengali languages, however for huge regions of graphics or pictures the method will fail. The method most frequently utilized is the projection profile method proposed by Baird [10], however, this is only reliable for skew angles within $\pm 15°$. Postl's method [11] is accurate but requires computationally complex integration involving trigonometric functions. It is not clear whether this method is insensitive to graphics in the image. Smith [12] proposed a method of fitting connected components into row bins, updating the vertical shift across the image. This is again precise but has only been tested for angles up to 26.8°. Yu and Jain [13] have developed a very accurate method using the Hough transform of the centroids of connected components.

However, for the applications other than document analysis, skew angle measurement techniques are not yet tested. Wei et al. [14] proposed an algorithm for Lane detection and tracking based on hough transform. In this they controlled the slope of lane lines in two different frames and searched for corner pixel. They claimed for fast operation speed, high accuracy and good robustness but it cannot completely avoid the interference of other lines in the identification.

Thus, applications such as document analysis, navigation, bridges and home construction, posture detection have an emerging area of research and recognizing which is one of the focus of the present work. Generally, real-world situations that involve different angle measurements is in construction work. The roof of a house has to be at least 39° and at max 48° to anticipate downpour water, otherwise the water leaks inside the house. Using skew angle algorithm, angle of the roof can be calculated, also during navigation of a vision based wheeled robot the angle of a line can be detected. Using the proposed algorithm of skew angle estimation the posture of the human body can also be detected. This helps in medical awareness. Thus skew angle estimation and its applications are the area of focus of this paper.

3 Algorithm

The proposed algorithm is related to the skew angle estimation. The algorithm accepts only preprocessed black and white image with extracted region of interest.

Step 1: Original image acquisition.

Step 2: Preprocessing of image(extracting region of interest, converting to BW image). The image preprocessing includes minimizing of noise in the image and extracting the region of interest. User should perform preprocessing according to his/her requirement.

Step 3: There are two methods for vertical and horizontal traversing of matrix of BW image. For convenience they are labelled as method X and method Y. Here, the transition of pixel values from 0 to 1 is observed. In method X, traversing of matrix is row wise from left to right, whereas in method Y, traversing is from top to bottom column wise. The Pixel location of transition is stored in matrix I_{trans}.

$$I_{\text{trans}} = [\,]_{(\text{row}*2)} \ldots \text{method}X \tag{1}$$

$$I_{\text{trans}} = [\,]_{(\text{col}*2)} \ldots \text{method}Y \tag{2}$$

Each row of I_{trans} contains x and y coordinates of transition for that row or column according to the respective method.

Step 4: If no transition is found for a row or column, by default -1 is given for x and y coordinates.

Step 5: Now minimum and maximum positive values of column 1 of I_{trans} are found out and difference is calculated.

$$\text{delta}_x = \max(I_{\text{trans}}(:, 1)) - \min(I_{\text{trans}}(:, 1)) \ldots \text{method}X \tag{3}$$

$$\text{delta}_y = \max(I_{\text{trans}}(:, 2)) - \min(I_{\text{trans}}(:, 2)) \ldots \text{method}Y \tag{4}$$

Step 6: Comparing $delta_x$ and $delta_y$, $y_{\min}$, $y_{\max}$ or $x_{\min}$ and $x_{\max}$ are deducted from I_{trans} matrix of respective method. Points $x_{\min}$, $x_{\max}$, $y_{\min}$ and $y_{\max}$ represent end points of a line.
Step 7: The slope of this line can be calculated as

$$m = \frac{y_{\max} - y_{\min}}{x_{\max} - x_{\min}} \tag{5}$$

Step 8: By using trigonometric formula for tangent of difference between two angles,

$$\tan(\theta_1 - \theta_2) = \frac{\tan(\theta_1) - \tan(\theta_2)}{1 - \tan(\theta_1) * \tan(\theta_2)} \tag{6}$$

slope between two lines is calculated.

Step 9: A virtual straight line is considered as a reference and the angle is calculated w.r.t. to that line. Thus,

$$\tan(\theta_1) = \tan(0) = 0 \tag{7}$$

Required angle,

$$\theta_2 = -\tan^{-1}(m). \tag{8}$$

4 Flowchart

See Fig. 1

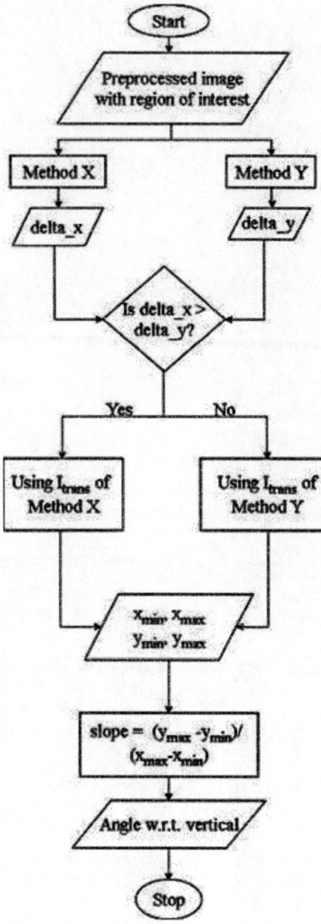

Fig. 1 Flowchart for proposed skew angle estimation algorithm

5 Experimental Analysis

Proposed algorithm is tested for various applications such as line following systems, construction area, document analysis, posture analysis, etc. Figure 2 shows the images for proposed applications on which the skew angle estimation algorithm is tested.

Figure 2a shows the scanned document image as one of the skew angle estimation domain. Nowadays, text recognition using OCR is being more popular. The proposed algorithm can be used for detection and correction of skew angle which ultimately leads to improve the accuracy of OCR. Figure 2b [15] depicts a person who tends to slouch which results in improper posture. It can be harmful to the human body if it happens repeatedly. The proposed algorithm can be used for the detection of bending positions and it is easy to integrate with various software that can notify the person to correct his posture.

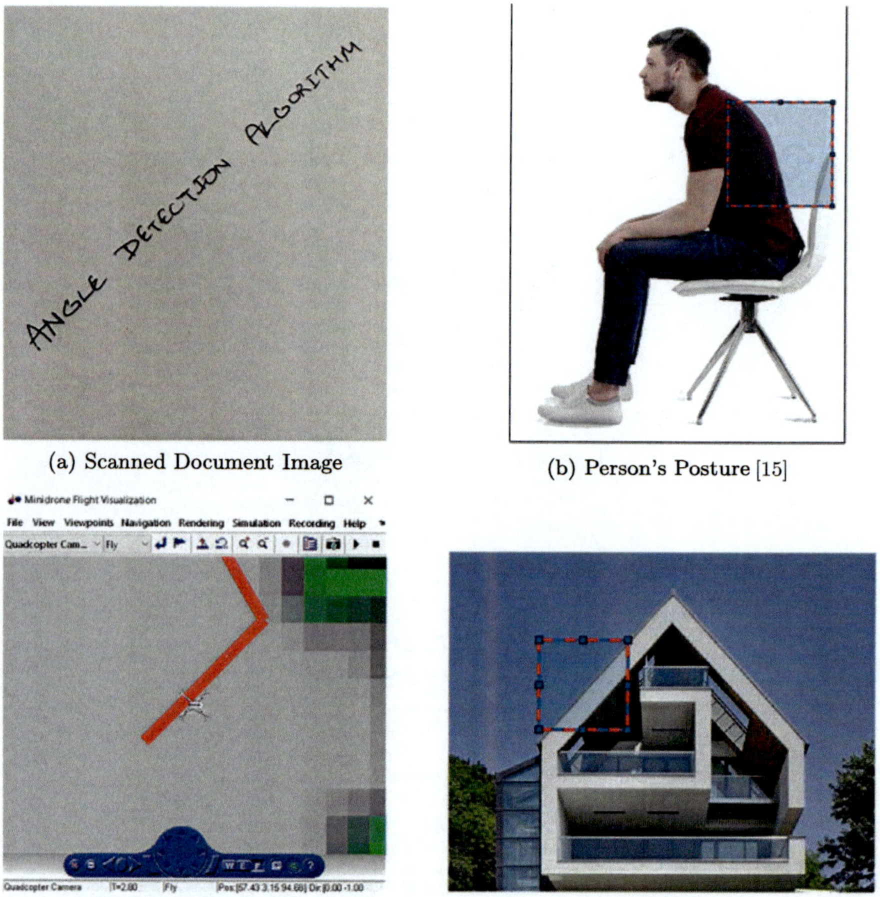

(a) Scanned Document Image

(b) Person's Posture [15]

(c) MiniDrone Environment

(d) House

Fig. 2 Images for proposed applications

Figure 2c portrays minidrone environment mainly consists of the line following including various angles for navigation of the drone from one point to another. The line can be followed by various environments and hardware and their time complexity and accuracy is based on the same. According to the survey, MATLAB has its algorithm for detection of the angle between two intersecting lines but it lacks in single line skew estimation. This algorithm is also can be used for vision-based path navigation for mobile robotics.

Figure 2d shows the house having roof at a particular angle. The proposed algorithm can be used in the construction domain for the roof angle measurement. This is accurate and more time saving than current methods.

Figures 3, 4, 5 and 6 represent how the algorithm deducts respective angle. From three images of each application, the first one is an image containing region of

(a) Resize Image (b) BW + Angle (c) Corrected Image

Fig. 3 Document analysis of Fig. 2a

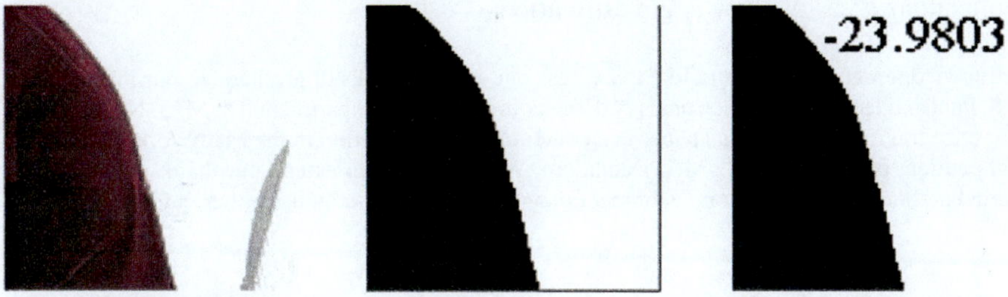

(a) Cropped Image (b) BW Image (c) Angle

Fig. 4 Person's posture of Fig. 2b

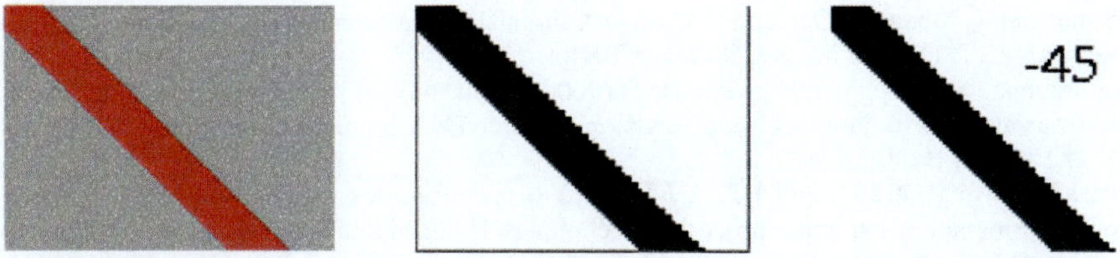

(a) Camera's Image (b) BW Image (c) Angle

Fig. 5 MiniDrone camera's image of Fig. 2c

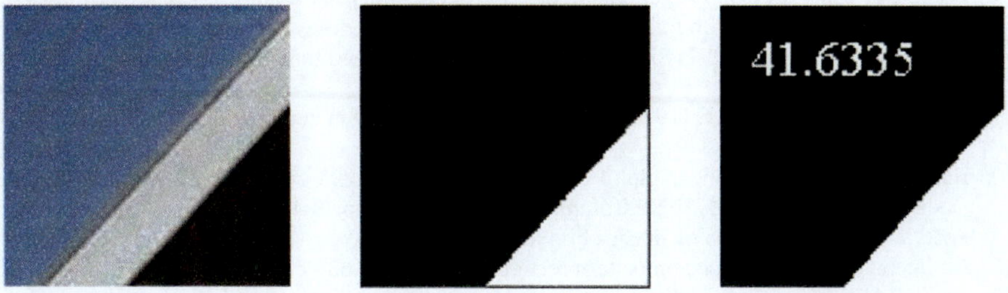

(a) Cropped Image (b) BW (c) Angle

Fig. 6 House image of Fig. 2d

interest, the second one is the image after preprocessing and binarization whereas third image is the output image with required angle. Thus, this algorithm can be successfully used for angle estimation in video processing, posture angle estimation and skew angle estimation.

6 Conclusion

Proposed skew angle measurement algorithm is verified with suggested fields and it satisfied the basic requirement of angle estimation. Earlier, skew angle estimation [6–13] has been restricted to the document analysis. This paper suggest diverse applications areas of skew angle estimation.

Acknowledgements We would like to express our special thanks of gratitude to our Principal Dr A. S. Pant and Head of the Electronics and Telecommunication department, Dr. M. S. Nagmode for their encouraging support. This paper is an outcome of participation in the MathWorks Minidrone Competition, being held at NUMA, Bangalore. We would like to extend our thanks to Maitreyee Mordekar, Student Competition Technical Evangelist, MathWorks India for her guidance.

References

1. Gonzalez RC, Woods RE (2017) Digital image processing, 4th edn. Pearson
2. Simonyan K, Modat M, Ourselin S, Cash D, Criminisi A, Zisserman A (2012) Immediate ROI search for 3-D medical images. Springer, Berlin
3. Sohrabiniaa M, Sadeghianb S, Manavic D (2008) Application of image processing and image analysis methods for large scale map revision. Int Arch Photogramm Remote Sens Spat Inf Sci XXXVII. Part B4. Beijing
4. Park Y-S, Agbayani JA, Lee J-H, Lee J-J (2016) Research article rotational angle measurement of bridge support using image processing techniques. Hindawi Publishing Corporation (J Sens)
5. Elhady WE, Elnemr HA, Selim G (2014) Implementation and evaluation of image processing techniques on a vision navigation line follower robo. J Comput Sci
6. Pal U, Chaudhuri BB (1996) An improved document skew angle estimation technique. Pattern Recognit Lett
7. Peake GS, Tan TN (1997) A general algorithm for document skew angle estimation. In: IEEE international conference on image processing
8. Chaudhuri BB, Pal U (1997) Skew angle detection of digitized Indian script documents. IEEE PAMI 19(2):182–186
9. Chaudhuri A, Chaudhuri S (1997) Robust detection of skew in document images. IEEE Trans Image Process 6(2):344–349
10. Baird HS (1995) The skew angle of printed documents. In: OăĂźGorman L, Kasturi R (eds) Document image analysis. IEEE Computer Society Press, pp. 204–208
11. Postl W (1986) Detection of linear oblique structures and skew scan in digitized documents. In: International conference on pattern recognition, pp 687–689
12. Smith R (1995) A simple and efficient skew detection algorithm via text row accumulation. In: IEEE 3rd international conference on document analysis and recognition, pp 1145–1148
13. Yu B, Jain AK (1996) A robust and fast skew detection algorithm for generic documents. Pattern Recognit 29(10):1599–1629

14. Wei X, Zhang Z, Chai Z, Feng W (2018) Research on lane detection and tracking algorithm based on improved hough transform. In: 2018 IEEE international conference of intelligent robotic and control engineering
15. Belchonock, Sitt A, Sitt S, 123RF, man sitting on chair. https://www.123rf.com/photo_97521727_posture-concept-man-sitting-on-chair-against-white-background.html

Analysis of Accuracy of Supervised Machine Learning Algorithms in Detecting Denial of Service Attacks

Deepa Krishnan

Abstract Intrusion Detection Systems are considered to be one of the primary methods for security attack detection. It is very challenging to design and implement intrusion detection systems that can detect the newer variants of security attacks with greater accuracy. This paper focuses on the detection of the Denial of Service Attacks in particular by DoS attack tools like Goldeneye, Slow HTTP test, Slow Loris, and Hulk. Further, this paper also focuses on the detection of one of the most important Web Application security attack due to the Heartbleed vulnerability. We have used the supervised machine learning algorithms like Support Vector Machine, Decision Tree, K-Nearest Neighbor for analyzing the accuracy of these models in classifying multi-class problems. One of the highlights of this paper is that the CICIDS2017 attack dataset has been used for evaluating the accuracy of various classification models. This research work holds significance as this focuses on the classification of attacks into five categories rather than a binary classification problem which is the focus of majority of the research works.

Keywords Machine learning · Supervised · Denial of service attacks (DoS) · Support vector machine · Decision tree · K-nearest neighbor (KNN)

1 Introduction

1.1 Overview of DoS Attacks and Detection Techniques

Denial of Service attacks (DoS) is one of the major cybersecurity threats that the world is facing today. This category of attacks is very destructive in nature and are capable to bring down major business houses and industries around the world [1]. The authors in [2] have done an extensive classification of the various types of DoS attacks

D. Krishnan (✉)
Mukesh Patel School of Technology Management and Engineering, NMIMS University (Deemed-To-Be), Mumbai, India
e-mail: deepa.krishnan@nmims.edu

© Springer Nature Singapore Pte Ltd. 2021

S. N. Merchant et al. (eds.), *Advances in Signal and Data Processing*, Lecture Notes in Electrical Engineering 703,
https://doi.org/10.1007/978-981-15-8391-9_38

and few among them are described here. DoS attacks can be classified majorly on the basis of the attack traffic dimension, the types of attack packets used and method of scanning techniques employed. There are constant rate attack traffic mechanisms and variable rate traffic mechanism for performing DoS attacks. The attack traffic rate can be of uniformly increasing type or with varying types. The attackers can use TCP SYN packets, ICMP packets, DNS packets, or SMTP packets for crafting attacks. On the basis of scanning methods used, the DoS attacks can be classified into random scanning, hit-list scanning, topological scanning, and permutation scanning. Thus, a wide variety of techniques are available for attackers to design and implement DoS attacks.

In recent times, the attackers are using distributed denial of service attacks (DDoS) with attack traffic of Terabytes per sec and the scale of the resulting damage is huge. Also, there are many DoS attack tools available for use with varying attack intensity. In the research paper by Kaur et al. [3], the authors have given a detailed taxonomy of the various DoS attack tools. The conventional DDoS detection and mitigation mechanisms might prove to be ineffective in this scenario. The techniques for the detection of attacks are mainly of two types: misuse based and anomaly based. The mis-use based techniques or signature-based techniques fail to detect the unseen or Zero-day attacks. However, anomaly-based techniques are effective in detecting such attacks but we need to be very cautious about handling false alarm rates. In the research work by Anna L Buczak et al., they have analyzed the various design parameters while choosing Machine learning algorithms for attack detection [4]. Machine learning algorithms are class of algorithms that can learn from data and use that experience in similar situations. The focus of our work is to analyze the effectiveness of various supervised machine learning algorithms in detecting DoS attacks.

1.2 Dataset

There are various datasets that are available publicly for use by researchers to evaluate the performance of various attack detection mechanisms. KDD 99 is one of the most important datasets that is used for evaluation of anomaly-based Intrusion Detection Systems. However, the KDD 99 dataset is found to have huge number of redundant records resulting in bias by learning algorithms towards frequent records. Also, the comparison of IDS is quite difficult as most of the machine learning algorithms gave a result in the range of 86–100% [5]. These issues in the dataset were addressed in the NSL-KDD dataset. In NSL-KDD there are no redundant records in the training set and no duplicate records in the test set. Further, the dataset also gives the detection rates in a much wider range for most of the machine learning algorithms unlike the KDD dataset. In the research work by Agarwal et al. [6], they have analyzed the NSL KDD dataset with respect to four attribute classes. The study indicates that basic class attributes show high maximum detection rate and traffic class attributes show lower detection rate. However, the NSL-KDD suffers from a major disadvantage that it does not represent the latest categories of DoS attacks, malware attacks, and botnet attacks.

The dataset that we have used in this work is the CICIDS attack dataset of 2017 published by the Canadian Institute of Cyber Security. This dataset is a comprehensive collection of all the recent and relevant attacks. The researchers Sharafaldin et al. [7] have taken care to ensure the dataset is the most real-world representation and it meets all the eleven criteria. The various criteria that the authors have used for evaluating a dataset are complete network configuration, complete traffic, labeled dataset, complete interaction, complete capture, available protocols, attack diversity, heterogeneity, feature set, and metadata. The entire dataset captures the attack and normal activities for five days spanning from Monday to Friday.

In our paper, we have considered the Wednesday morning attack traffic dataset in csv format which is labeled as Wednesday-Working hours pcap_ISCX.csv. This dataset includes benign traffic and prominent categories of DoS attacks like DoS Goldeneye, DoS Hulk, DoS Slowhttptest, DoS Slow Loris, and Heartbleed.

1.3 Statistics of the Dataset

There are 692,703 instances and 79 features in the.csv file that we have used for our study. The various categories of attacks are summarized in Table 1.

2 Related Work

It is highly challenging to differentiate benign traffic from attack traffic considering the extremely heterogenous nature of the traffic patterns. There are several research works that focus on detecting the various categories of Denial of Service Attacks based on anomaly-based techniques. Machine learning (ML) approaches and Artificial Intelligence based approaches have proven to be effective in improving the accuracy of unseen attacks. The efficiency of machine learning and data mining techniques in attack detection are delved deeply by authors Anna L Buczak and

Table 1 Categories of attack traffic

Index	Category	Number of Instances
0	Benign	440,031
1	DoS golden eye	10,293
2	DoS hulk	231,073
3	DoS slowhttptest	5499
4	DoS slowloris	5796
5	Heartbleed	11

et al. in research work 9].In this paper, the authors have described the performance parameters with which machine learning models could be compared for use in cybersecurity. Further, the authors also explains the statistical properties in a dataset that could be leveraged before choosing the model. This research work is indeed very useful for research community to get prior knowledge before using ML approaches in cybersecurity. Some of the significant research contributions done in DoS attack detection is described in this section.

In research work [8], the authors Qian Li and et al. have used machine learning based techniques to detect the DDoS attacks, in particular Recurrent Neural Networks. The authors have investigated the efficiency of using Principal Component Analysis for feature reduction and thereby reducing the complexity of the mechanism. The results by the authors indicate that PCA-RNN can give higher accuracy. However, the authors have used KDD99 dataset for training and evaluation, which is not a realistic capture of the modern DDoS attack traffic. The authors Suresh et al. [9] also have evaluated the use of machine learning algorithms for detecting DDoS attacks. They have compared the effectiveness of machine learning models like Naïve Bayesian, C4.5, SVM, KNN, K-means, and Fuzzy c-means clustering and found that Fuzzy c-means clustering gives better accuracy. The authors have used publicly available dataset CAIDA and traffic based on smart and secure environment network. They have used Chi-Square statistic and Information gain to rank the 23 features in the dataset and to further reduce it to 8 features.

Another significant work [10] in this direction is by Priyadarshini et. al where the authors have used deep learning based approach to improve the accuracy of DDoS attack detection. The authors have used Long Short Term Memory (LSTM) work as it can handle sequential and time-dependent data. They have used modified LSTM by using drop out probability model to prevent over-fitting and mini batch gradient descent algorithm to prevent vanishing gradient problem. The CTU-13 dataset is used for training the machine learning model and further tested using the CTU-13 dataset as well as with the simulated DDoS attack traffic. The researchers obtained promising results for accuracy in both training and testing. In research work [11], the authors have evaluated both supervised and unsupervised machine learning algorithms individually on various instances of virtual machines and also on combination of various instances of virtual machine. They have used joined data from different instances of virtual machines for training and they found the accuracy of all algorithms improved significantly. However noticeable improvement is observed in the K-means, where the accuracy improved by 23%.

The multiple kernel based DDoS attack detection by Cheng et al. [12] is another innovative and relevant contribution in this direction. Despite many promising works in DDoS attack detection, none of the works so far have used a realistic DDoS dataset for training and evaluating the model.

Table 2 Categories of attack traffic after undersampling

Index	Category	Number of instances
0	Benign	11,519
1	DoS golden eye	10,293
2	DoS hulk	8935
3	DoS slowhttptest	5797
4	DoS slowloris	5499
5	Heartbleed	11

Table 3 Categories of attack traffic after SMOTE technique

Index	Category	Number of instances
0	Benign	11,519
1	DoS golden eye	10,293
2	DoS hulk	8935
3	DoS slowhttptest	5797
4	DoS slowloris	5499
5	Heartbleed	11,519

3 Analysis of Results of Various Classification Models

3.1 Preprocessing

Due to the higher computation intensiveness of the dataset described in Table 1, undersampling was performed on the dataset, and the following dataset in Table 2 was created.

The dataset represented in Table 2 is highly imbalanced with respect to the attack category of Heart bleed. Hence, it can result in wrong accuracy for most of the machine learning algorithms. The resulting inefficiency of machine learning models have been addressed by SMOTE- Synthetic Minority Oversampling technique which is available for use by the imblearn.over_sampling.SMOTE. However, in this research work, only the minority class is targeted which in this case is the heart bleed attack category. The following Table 3 shows the statistics of the dataset after SMOTE technique.

3.2 Decision Tree Classifier

Decision trees are used for classification in cases where the relationship between the feature variables are non-linear and where features directly interact with each other. The decision tree algorithm requires a split criterion and there are two most commonly used criteria for splitting are gini index and entropy. Gini index helps us to determine how often a randomly chosen element would be incorrectly classified. This indicates that an attribute with lower gini index should be chosen for the split. On the other hand, entropy is the measure of uncertainty of a random variable and it gives the impurity of a subset of data samples. The higher the entropy the more the information content, and thus we prefer the split on a node with higher information gain. One of the advantages of using decision trees is that it can handle data of high dimension and it follows a typical inductive approach to extract knowledge about classification.

We have used standard metrics such as accuracy, precision, recall, F-score to evaluate the performance of the machine learning models. In Tables 6 and 7 we have summarized the precision, recall, F1-score and support obtained when decision tree classifier is used with criterion gini and entropy. The confusion matrix for illustrating the attained accuracy is also depicted in Tables 4 and 5.

The decision tree classifier gave an accuracy of 59.80% with gini and 70.57% for entropy in the case of training dataset. Decision Tree Classifier with split criterion gini gave an overall accuracy of 48.24% and training time of 0.48 s for test data set. However, decision tree classifier with split criterion of entropy gave an overall accuracy of 61.57%. This timing is obtained while running the model on the Google cloud colab with run time type as Python 3 and Hardware Accelerator as TPU.

Table 4 Confusion matrix: Gini

1448	0	74	0	826	0
0	0	1457	0	5845	0
0	0	1438	0	302	0
0	0	0	0	1108	0
0	0	2	0	1169	0
0	0	0	0	0	3

Table 5 Confusion matrix: entropy

1434	4	11	0	899	0
0	1345	0	0	696	0
0	197	1234	0	309	0
0	0	0	0	1108	0
0	1	0	0	1170	0
0	0	0	0	0	3

Table 6 Decision tree classifier results (Criterion = "gini")

Category	Precision	Recall	F1-score	Support
0	1.00	0.64	0.78	9171
1	0.0	0.00	0.00	8252
2	0.49	0.83	0.62	7195
3	0.00	0.00	0.00	4391
4	0.30	1.00	0.46	4626
5	1.00	1.00	1.00	9171
Accuracy			0.60	42,806
Macro average	0.46	0.58	0.48	42,806
Weighted average	0.54	0.60	0.53	42,806

Table 7 Decision tree classifier results (Criterion = "entropy")

Category	Precision	Recall	F1-score	Support
0	1.00	0.61	0.76	2348
1	0.87	0.66	0.75	2041
2	0.99	0.71	0.83	1740
3	0.00	0.00	0.00	1108
4	0.28	1.00	0.44	1171
5	1.00	1.00	1.00	3
Accuracy			0.62	8411
Macro average	0.69	0.66	0.63	8411
Weighted average	0.73	0.62	0.63	8411

3.3 Support Vector Machine

Support Vector Machines are typically used for binary classification, however, this can be extended to multi-class problems as well. The most common approach followed is to construct a multi-class classifier by combining several binary classifiers [13]. The SVM algorithm is implemented using kernel which can either linear or non-linear. The linear, polynomial and RBF or Gaussian kernel follow different approaches in identifying the hyperplane decision boundary between the classes [14]. The kernel functions help in converting the original dataset into a higher dimensional space in an attempt to linearize the dataset. The dot product is the similarity measure used for linear SVM or a linear kernel because the distance is a linear combination of the input. In the case of polynomial kernels which is used in SVM, it maps the similarity of features in a feature space over polynomials of the original features. One of the advantages of non-linear models is that they consider the similarity of combinations of input features other than considering the input features individually. This helps in capturing the correlations between input features perfectly. However,

Table 8 SVM classifier-one versus rest multiclass classifier

Category	Precision	Recall	F1-score	Support
Benign	0.99	0.96	0.98	2307
DoS golden eye	0.99	0.99	0.99	2100
DoS hulk	0.97	1.00	0.98	1795
DoS slowhttptest	0.97	0.97	0.97	1088
DoS slow Loris	0.96	0.97	0.97	1135
Heart bleed	1.00	1.00	1.00	2288
Accuracy			0.98	10,713
Macro average	0.98	0.98	0.98	10,713
Weighted average	0.98	0.98	0.98	10,713

Table 9 Confusion matrix of SVM classifier using one versus rest classifier

0	2222	23	45	5	12	0
1	14	2070	10	5	5845	0
2	0	0	1795	0	0	0
3	5	2	0	1051	30	0
4	5	0	0	26	1104	0
5	0	0	0	0	0	2288

the most preferred kernel functions used in SVM is the RBF (Radial Basis Function) kernel which has the stationary property that is invariant to translation[15].

In this section, we have compared the accuracy and detection time of the classification model using a linear kernel with OneVsRestClassifier to implement SVC classifier for multi-class classification. Table 8 will summarize the results of SVM with linear kernel with OneVsRestClassifier and decision_ func-tion_shape = ovr.

The confusion matrix for SVM as multi-classifier using One versus Rest Classifier is given in Table 9.

3.4 K-Nearest Neighbour

K-Nearest Neighbour (KNN) is one of the prominent classification algorithms that is basically non-parametric as it doesn't consider the underlying relations in the distribution of data. One of the important parameters that determine the accuracy and prediction time of KNN. The training error rate and the validation error rate are two parameters we need to find on different K-value. We can see from Fig. 1 that the error rate is zero when the value of $k = 1$ for training sample. The optimal value of k is the one in which the test error rate and training error rate are closer. We have identified the value of $k = 9$ where overfitting is minimum and implemented K-NN

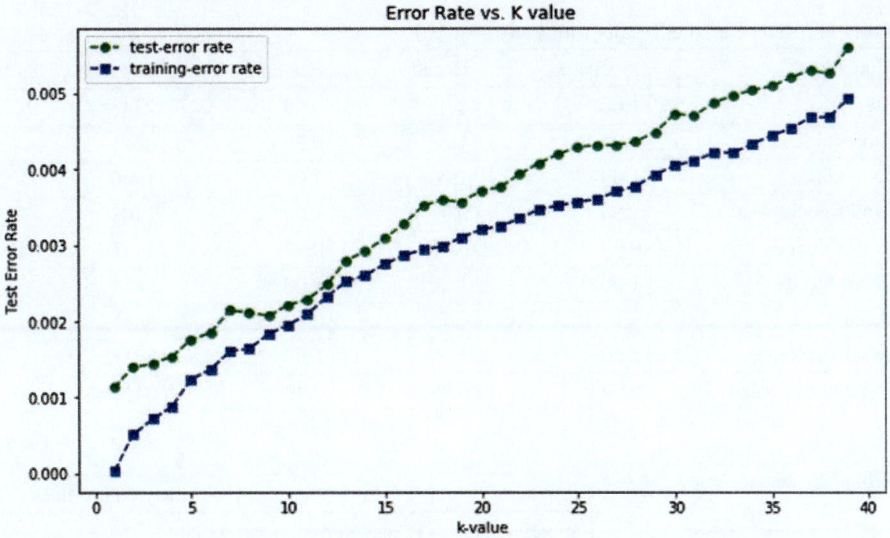

Fig. 1 Elbow method: error rate versus *K* value

Table 10 Confusion matrix of with KNN classifier with $k = 9$

0	9075	52	30	4	10	0
1	2	8243	0	6	1	0
2	10	4	7181	0	0	0
3	13	6	0	4352	20	0
4	14	4	1	14	4593	0
5	0	0	0	0	0	9171

with K value of 9. Tables 10 and 11 shows the confusion matrix and classifier results with $K = 9$.

3.5 Analysis of Accuracy with Decision Tree, Support Vector Machine and KNN

Table 12 shows the test accuracy obtained for Decision Tree, and Support Vector Machine. It is observed that Support Vector Machine and K-NN give higher accuracy of 98.84% and 99.4% respectively. However, Decision Tree is seen to give 47.35% accuracy for Gini Criterion and 77.2% accuracy with Entropy Criterion.

Table 11 KNN classifier results with $k = 9$

Category	Precision	Recall	F1-score	Support
Benign	1.00	0.99	0.99	2348
DoS golden eye	0.99	1.00	0.99	2041
DoS hulk	1.00	1.00	1.00	1740
DoS slowhttptest	1.00	0.99	1.00	1108
DoS slowloris	0.99	0.99	0.99	1171
Heart bleed	1.00	1.00	1.00	3
Accuracy			0.99	8411
Macro average	1.00	1.00	1.00	8411
Weighted average	0.99	0.99	0.99	8411

Table 12 Test accuracy

Classifier		Accuracy on test data
Support vector machine	Linear Kernel	98.84
Decision tree	Gini criterion	47.35
	Entropy criterion	77.2
K-NN		99.4

4 Conclusion and Future Scope

The supervised classification algorithms are proven to be effective in identifying the category of the attacks. The accuracy of Support Vector Machine with linear kernel is higher and training time is lesser compared to lower accuracy and high-er training time of RBF kernel. Linear kernel has given better accuracy in the case of the dataset under consideration as the feature sets have higher dimension. In the case of K-NN the training accuracy and testing accuracy is 99.55 and 99.46%, showing immunity to overfitting.

In future research work, we would investigate techniques to minimize the learning parameters using appropriate feature reduction retaining the accuracy in determining the attacks. We would also be addressing the issue of imbalanced classes without introducing synthetic samples which are overlapping. Nevertheless, supervised machine learning algorithms are effective in determining the multi-class Denial of Service attacks with high accuracy.

References

1. Mahjabin T, Xiao Y, Sun G, Dong W (2017) A survey of distributed denial-of-service attack, prevention, and mitigation techniques. Int J Distrib Sens Netw 13(12)

2. Mirkovic J, Reiher P (2004) A taxonomy of DDoS attack and DDoS defense mechanisms. ACM SIGCOMM Comput Commun Rev 34(2)
3. Kaur H, Behal S, Kumar K (2015) Characterization and comparison of distributed denial of service attack tools. In: International conference on green computing and internet of things
4. Buczak AL, Guven E (2016) A survey of data mining and machine learning methods for cyber security intrusion detection. IEEE Commun Surv Tutor 18(2)
5. Tavallaee M, Bagheri E, Lu W, Ghorbani AA (2009) A detailed analysis of the KDD CUP 99 data set. In: Proceedings of the IEEE symposium on computational intelligence in security and defense applications
6. Aggarwal P, Sharma SK (2015) Analysis of KDD dataset attributes - class wise for intrusion detection. In: 3rd international conference on recent trends in computing. Elsevier
7. Sharafaldin I, Lashkari AH, Ghorbani AA (2018) Toward generating a new intrusion detection dataset and intrusion traffic characterization. In: Proceedings of the 4th international conference on information systems security and privacy (ICISSP 2018), pp 108–116
8. Li Q, Meng L, Zhang Y, Yan J (2019) DDoS attacks detection using machine learning algorithms. In: Zhai G, Zhou J, An P, Yang X (eds) Digital TV and multimedia communication. IFTC 2018
9. Suresh M, Anitha R (2011) Evaluating machine learning algorithms for detecting DDoS attacks. In: Wyld DC, Wozniak M, Chaki N, Meghanathan N, Nagamalai D (eds) Advances in network security and applications. CNSA 2011. Communications in computer and information science, vol 196. Springer, Berlin
10. Priyadarshini R, Barik RK (2019) A deep learning based intelligent framework to mitigate DDoS attack in fog environment. J King Saud Univ Comput Inf Sci
11. He Z, Zhang T, Lee RB (2017) Machine learning based DDoS attack detection from source side in cloud. In: 2017 IEEE 4th international conference on cyber security and cloud computing
12. Cheng J, Zhang C, Tang X, Sheng VS (2018) Adaptive DDoS attack detection method based on multiple-Kernel Learning. Hindawi Security and Communication Networks Volume
13. Hsu CW, Lin CJ (2002) A comparison of methods for multi class support vector machines. IEEE Trans Neural Netw
14. Zhang X (2017) (2017) Support vector machines, Encyclopaedia of machine learning and data mining, 2nd edn. Springer, Boston
15. Van Den Burg GJJ, Groenen PJF, Gen SVM (2016) A generalized multiclass support vector machine. J Mach Learn Res 17: 11–42

An Improved Carrier Frequency Offset Estimation Under Narrowband Interference in OFDM Cognitive Radio

Vibha Patel and Krishna Warhade

Abstract In recent years, overcrowded unlicensed spectrum is devastating spectral efficiency of communications in regional and rural broadband wireless networks. Cognitive radio allows opportunistic use of a licensed spectrum without interfering with primary users (PU) which overcome the scarcity problem of the available spectrum. The occurrence of carrier frequency offset (CFO) degrades the performance of the orthogonal frequency division multiplexing (OFDM). OFDM fulfills the requirements of cognitive radio, and hence OFDM is an appropriate choice for cognitive radio. When OFDM is used for cognitive radio applications, sensitivity to frequency offset remains an issue. This paper surveys various techniques present to estimate carrier frequency offset for OFDM cognitive radio. It covers required parameters to estimate, i.e., training symbols, estimation range, and complexity. This paper also presents simulation results that show extended estimation range of the frequency offset at good performance in the presence of narrowband interference. This method uses correlation among L identical parts of the training symbol at the receiver side to estimate the frequency offset. The estimation range is achieved up to $\pm L/2$.

Keywords Cognitive radio · Carrier frequency offset · Synchronization · OFDM

1 Introduction

Cognitive radio (CR) is a more personalized form of software defined radio (SDR) that allows the use of a licensed spectrum intelligently [1]. Cognitive radio offers the opportunity of spectrum pooling on a rental basis to fulfill the high demand for limited resources for mobile multimedia applications [2]. FCC (Federal Communication Commission)'s Spectrum Task Force published a report and one of the major findings

V. Patel (✉)
Gujarat Technological University, Ahmedabad, India
e-mail: vibha.patel@mitwpu.edu.in

K. Warhade
Dr. Vishwanath Karad MIT World Peace University, Pune, India
e-mail: krishna.warhade@mitwpu.edu.in

© Springer Nature Singapore Pte Ltd. 2021

S. N. Merchant et al. (eds.), *Advances in Signal and Data Processing*,
Lecture Notes in Electrical Engineering 703,
https://doi.org/10.1007/978-981-15-8391-9_39

of the report is "In many bands, spectrum access is a more significant problem than physical scarcity of spectrum, in large part due to legacy command and control regulation that limits the ability of potential spectrum users to obtain such access" [3].

Cognitive radio achieves efficient spectrum utilization by performing three cognitive tasks. And hence, it achieves highly reliable communication [4].

1. Spectrum holes, i.e., unused frequencies of primary users (PUs) are detected by sensing radio environment.
2. Channel capacity is predicted via estimating channel state information (CSI).
3. This estimated CSI will be used to make adaptive operating parameters, i.e., modulation scheme, transmit power, and carrier frequency.

Primary users (PUs) are licensed users (LUs) and secondary users (SUs) are unlicensed users (ULUs).

Orthogonal frequency division multiplexing (OFDM) has the potential to fulfill the requirements of CR. Furthermore, it is also used in current wireless technologies which gives interoperability across several networks. OFDM's strengths are given as below [5].

1. The use of FFT at the transmitter side of OFDM gives ease to spectrum sensing and shaping in the frequency domain. This multicarrier modulation method will give efficient spectrum utilization by just switching off subcarriers used by primary users (PUs.)
2. OFDM gives adaptive CR to interoperate with several recent wireless systems by choosing an appropriate frequency. In addition to this, OFDM changes its parameters like modulation order, number of subcarriers, channel coding, transmit power adaptively to improve performance, and to maximize spectral efficiency.
3. OFDM is commonly used with multiple input multiple output (MIMO) referred to as MIMO OFDM in different recent wireless systems,for example, future High-Definition (HD), Long Term Evolution (LTE) of 3G, IEEE 802.11n, IEEE 802.16e standard, and digital broadcast TV.
4. Cognitive radio should support multiple access techniques to share limited radio resources among secondary users (SUs) efficiently. Orthogonal frequency division multiple access (OFDMA) supports this feature.
5. The use of OFDM gives interoperability to cognitive radio. The reason behind that is the OFDM modulation technique is used by many recent wireless technologies like wireless local area network (WLAN), IEEE 802.11a, IEEE 802.11g, IEEE 802.11n, digital audio and video broadcasting (DAB and DVB), IEEE 802.16 WiMAX and 4G and 5G cellular communication.

Since OFDM has some challenges, i.e., inter-carrier interference (ICI), synchronization, peak-to-average power ratio (PAPR), and also CR has its challenges, i.e., spectrum sensing, cross-layer adaption, and interference avoidance, etc., employing OFDM in CR creates multiple research challenges. Recently, OFDM-based cognitive radio research challenges become major thirst areas in wireless communication.

These challenges are multiband OFDM design, location awareness, signaling the transmission parameters, synchronization, and mutual interference [5].

Synchronization errors like frequency, time, phase offsets, and drifts are present in practice. Upconversion and down conversion at the end of transmitter and receiver, respectively, can generate frequency and phase offset. Frequency offset occurs due to the mismatch between the transmitted carrier frequency and frequency generated by a local oscillator at the receiver side. In addition to this, the moving transmitter/receiver results in a Doppler effect and creates a frequency offset. This may degrade the performance of spectrum sensing algorithms, i.e., cyclostationary feature detector [6].

In this paper, ICI and synchronization are taken as a research problem and a literature survey has been carried out.

2 Literature Survey

2.1 Synchronization in OFDM

The sensitivity of OFDM to timing and frequency offset results in imperfect synchronization, resulting in a loss of throughput. OFDM systems tolerate timing offsets and frequency offsets until and unless the performance of the system does not degrade upto large extent [7]. It is very difficult to maintain small timing and frequency offset in links such as satellite links, mobile digital radio links, hostile propagation environment, i.e., urban areas, terrain areas because of large- and small-scale fading, slow and fast fading, doppler shift and oscillator instabilities [8]. The performance of OFDM is improved by estimating timing offset and CFO and compensating it. For estimating timing offset and CFO in OFDM, blind- and pilot-aided training sequences estimation algorithms are available. For complexity reduction, fractional CFO and integer CFO are separately estimated.

2.2 Synchronization Algorithms in OFDM Cognitive Radio

In [9], under the Rayleigh fading environment, the effect of carrier frequency offset on spectrum sensing is evaluated. The sensing parameters, i.e., the detection probability, false alarm probability, and different OFDM transmission parameters affect the performance of OFDM cognitive radio. It is shown that the average interference to noise ratio degrades the performance of OFDM cognitive radio in the absence of carrier frequency offset. Error floors occured due to inter-carrier interference (ICI) are observed in BER curves for various values of carrier frequency offset. OFDM cognitive radio is vulnerable to ICI as compared to the primary user's interference.

In general, the shape of the whole BER curve is affected by inter-carrier interference (ICI) generated due to carrier frequency offset.

Zsolt and Peter show in [10], OFDM used in cognitive radio is vulnerable to various synchronization errors such as timing offsets, phase noise, frequency offsets, and IQ imbalance. Frequency error of 0.03 subcarrier spacing due to local oscillator mismatch makes OFDM performance worse by 1 dB. It has been proved that improper synchronization in OFDM-based cognitive radio results in performance degradation as well as a decrease in throughput. There are various methods to estimate the carrier frequency offset and then compensate CFO with an estimated value.

Here, a literature survey on various synchronization schemes has been presented.

Morelli and Moretti presented in [11] that conventional synchronization algorithms for OFDM are not enough for OFDM cognitive radio when narrowband interference is present. They propose a scheme that jointly estimates each subcarrier interference power and carrier frequency offset using the maximum likelihood (ML) estimation method. A training sequence which consists of two blocks is used to estimate CFO and subcarrier power. The first block of the training sequence consists of Lth repetitive parts. The acquisition range for FCFO is $\pm L/2$. The estimation of the integer frequency offset is done in the frequency domain using a known PN sequence. The length of cyclic prefix is always kept greater than the length of channel taps to avoid inter-block interference (IBI). Here, Bluetooth interfere is considered as a narrowband interference.

In [12], authors have proposed a synchronization scheme for non-contiguous OFDM which is adaptive. The synchronization requirement is stringent in NC-OFDM as compared to OFDM. The training sequence is modified to make synchronization adaptive. The initial training sequence is constant amplitude zero autocorrelation sequences (CAZAC) of length N. By inserting zeros between adjacent symbols in training sequence, the accuracy of frequency estimation is improved. Furthermore, an increase in pool allocation degrades the accuracy of frequency offset. The correlation method proposed by Schimdl and Cox in [13] is used to estimate the frequency offset.

In [14], a novel ML estimation scheme is shown, which estimate the CFO when the spectrum synchronization information (SSI) is unknown. As this is non-contiguous OFDM, pilot symbols are transmitted only on active subchannels. This training symbol consists of two equal sequences. The conventional CFO estimation schemes cannot be applied because of the lack of SSI information. The identification of active and inactive subchannel can be done by a hard-decision method. Once the active channel is determined, the maximum likelihood algorithm (MLA) is applied to estimate CFO.

The iterative synchronization scheme for OFDM signal for cognitive radio applications is presented in [15]. In this paper, two major synchronization errors, i.e., timing errors and CFO are addressed. Maximum likelihood method is used to estimate the fractional carrier frequency offset. After compensating CFO, the detection of a received signal is done by maximizing the log-likelihood ratio function. Simulations were conducted using the Monte Carlo method. The complexity of this proposed algorithm is proportional to the total iterations which are 500. For complexity reduction and less requirement of memory space, number of iterations should be as small as possible. Furthermore, the proposed algorithm is designed to estimate the fractional carrier frequency offset. So when integer CFO is present, it will not give satisfactory performance.

In [16], author Pawel and Hanna proposed a robust algorithm that estimates time and frequency offset in the presence of in-band, i.e., narrowband interference. The correlation method is used to estimate timing and frequency offset. Here, the assumption is that reference preamble is known at the receiver side. In this, performance is improved but at the same time, complexity is increased.

In [17], the same author proposed a new algorithm that is of low complexity but the estimation range is up to $\pm 1/2$ of subcarrier spacing only. This algorithm uses the Zadoff-chu sequence as a preamble for the estimation of CFO. The specialty of this algorithm is that it gives satisfactory performance in the presence of constant frequency narrowband interference and frequency-modulated interference. It estimates CFO by finding autocorrelation of $N/4$ constant delay of received samples.

For LTE systems, integer frequency offset (IFO) estimation using maximum likelihood is proposed in [18]. In addition to that, it also estimates cell id which helps to initiate the procedure cell search. Here, the assumption is that frequency and timing offsets are already estimated and compensated. This algorithm performed better at the cost of increased complexity. In spite of this, this is a promising candidate for 5G.

Table 1 shows comparison of various synchronization method available for OFDM based cognitive radio.

3 Findings

1. The need for synchronization in OFDM remains the same when OFDM is employed in the cognitive radio system.
2. The traditional synchronization algorithms for OFDM are not performing satisfactorily in OFDM cognitive radio system.
3. Most of the estimation algorithms are having an estimation range of $\pm 1/2$ subcarrier spacing. Some are having low complexity and some are having high complexity.
4. The algorithm presented in [11] gives estimation range $\pm L/2$ of the subcarrier spacing, where L = Number of identical parts in the training sequence. To increase the estimation range, number of identical parts in training symbols should be increased. This algorithm uses the maximum likelihood (ML) to estimate fractional carrier frequency offset.
5. The estimation algorithm presented in [12] shows adaptive synchronization using a modified training sequence for noncontiguous OFDM-based CR. It

Table 1 Comparison of various synchronization scheme in OFDM based cognitive radio

Algorithm	Pilot based/blind	Acquisition range (with respect to subcarrier spacing)	Complexity
Modified maximum likelihood algorithm (MMLE) [11]	Pilot based	$\pm L/2$ where $L =$ No. of identical parts in the training sequence	Low complexity
Zhou and Qiu scheme [12]	Pilot based	±1/2	Not given
A hard-decision-based active subchannel detection assisted ML estimation [14]	Pilot based	±1/2	Not given
Iterative synchronization-assisted OFDM signal detection [15]	Cyclic prefix based	±1/2	Complex multiplication required $3M(2iN_G + N - N_G)$ where $M =$ symbols $N_G =$ length of cyclic prefix $i =$ number of iterations $N =$ Number of subcarriers
Licensed user insensitive synchronization algorithm (LUISA) [16]	Pilot based	±3	Real additions $3N\log_2 N\text{-}N + 12[\upsilon_M] + 5$ Real multiplications $N\log_2 N + N + 18[\upsilon_M] + 6$ where $N =$ number of subcarriers $\upsilon_M =$ CFO
Narrowband interference robust synchronization for NC-OFDM cognitive radio [17]	Pilot based	±1/2	For $N = 256$ Real additions/subtractions 24 real multiplications/divisions 24
Approximated minimum mean square error criterion [18]	Pilot based	± 3	Required flops $291 + 1116N\upsilon + 1494N\upsilon P$

estimates frequency offset using the correlation method. Mean square error depends on pool demand.

6. The authors proposed ML estimation in [14] without the information of spectrum synchronization information. Pilot symbols are transmitted only on to active subchannel and a hard decision is used to identify active subchannel.

7. Iterative synchronization is used in [15], which performs well at low SNR. Again maximum likelihood method is used to estimate CFO. Performance is further improved by the number of iterations which also increase complexity. This algorithm is designed for the fractional CFO.
8. The correlation method is used in [16] to estimate CFO. The proposed algorithm performs better in the presence of a licensed user or any narrowband interference. High complexity for better performance needs to be addressed.
9. Again the correlation method is used in [17] to estimate CFO. It gives estimation range $\pm 1/2$ of subcarrier spacing in the presence of narrowband interference.
10. Integer CFO is estimated using an ML scheme [18] for LTE systems with the assumption that timing and frequency offsets are already estimated and compensated.

4 System Model

The present algorithms SCA [17] and NIRS [17] perform well in the presence of narrowband interference for limited estimation range, i.e. ±½ of subcarrier spacing. When carrier frequency offset is higher than half of the subcarrier spacing, the performance of these algorithms worsened. The objective of this work is to increase the estimation range by using more number identical parts in the training symbol [19]. The performance is measured in the presence of narrowband interference.

4.1 Schimdl and Cox Method

Schimdl and cox presented a synchronization method for OFDM [17]. It exhibits low complexity as it exploits the correlation present in the received samples. For NC-OFDM, a preamble is transmitted only on the available subcarriers. And preamble consists of two identical time-domain sequences. This repeatability will not destroy due to multipath fading and the reason is a cyclic prefix. Narrowband interference can affect fractional frequency offset estimation as correlation-based estimation is implemented in the time domain.

The transmitted sample after IDFT is given by

$$x_n^{(p)} = \frac{1}{\sqrt{N}} \sum_{k=-N/2}^{N/2-1} d_k^{(p)} \mathrm{e}^{j2\pi \frac{nk}{N}}, \quad \text{for} - N_{\text{CP}} < N < N - 1 \tag{1}$$

where $d_k^{(p)}$ is an OFDM symbol transmitted on the kth subcarrier in the pth symbol and N reprsents number of subcarriers. Thus, the transmitted signal consists of NC-OFDM symbols. The cyclic prefix is added in order to mitigate inter-symbol interference (ISI) which occurs due to multipath fading in each NC-OFDM symbol.

The frame occupying all *p*th symbols is represented as:

$$\tilde{x}(n) = \sum_{p=0}^{P-1} x_{n-p(N+N_{CP})}^{(p)} \quad (2)$$

The multipath fading channel, presence of CFO and NBI distort the received signal. The distorted received signal is given as:

$$r(n) = y(n)e^{j2\pi \frac{vn}{N}} + \sqrt{\sigma_i^2} e^{j2\pi \frac{f_c n}{N} + j\emptyset_n} + w(n) \quad (3)$$

where

$$y(n) = \sum_{l=0}^{L-1} \tilde{x}(n-l)h(l) \quad (4)$$

is the received multipath distorted NC-OFDM signal, *h*(*l*) is the channel coefficient at *l*th instant, *L* is the total number of channel paths gain, ν is the normalized CFO with respect to SCs spacing, and *w*(*n*) is a zero mean and variance ${\sigma_w}^2$ complex white Gaussian noise. Here, the narrowband interference (NBI) is characterized by frequency f_c (Normalized with respect to SCs spacing), power of ${\sigma_i}^2$, and phase which is varying slowly with time ϕ_n. This is represented in (3) as the second component. This is valid for narrowband systems, for example, wireless microphones (using frequency modulation) have a phase component of:

$$\emptyset_n = \sum_{i=0}^{\infty} a_i n^i \quad (5)$$

where a_i are the polynomial coefficients.

The autocorrelation between the received signal and its delayed version by *N*/2 is defined as:

$$G(n) = \sum_{m=0}^{N/2-1} r^*(n+m)r\left(n+\frac{N}{2}+m\right) \quad (6)$$

where *G*(*n*) is a metric that defines the autocorrelation, and () $*$ denotes a complex conjugate.

The CFO estimate can be obtained as

$$\hat{v} = \arg\left\{\frac{G(\hat{n})}{\pi}\right\} \quad (7)$$

where $\arg\{\cdot\}$ is the argument of a complex number.

4.2 Narrowband Interference Robust Synchronization (NIRS) Method

A new synchronization method robust against narrowband interference is proposed in [17]. This method exhibits low complexity but limits the estimation range. This assumes that carrier frequency offset (CFO) is less than the subcarrier spacing.

A new function $Q(n)$ is defined, which gives the autocorrelation of a received signal with $N/4$ delay.

$$Q(n) = \frac{1}{2}\sum_{m=0}^{N/4-1}\left[\begin{array}{c} r^*(m+n)r\left(m+n+\frac{N}{4}\right)+2r^*\left(m+n+\frac{N}{4}\right) \\ .r\left(m+n+\frac{N}{2}\right)+r^*\left(m+n+\frac{N}{2}\right)r\left(m+n+\frac{3N}{4}\right) \end{array}\right] \tag{8}$$

$G_{\mathrm{NIRS}}(n)$ is thus:

$$G_{\mathrm{NIRS}}(n) = G(n) - |Q(n)|e^{j2\arg\{Q(n)\}} \tag{9}$$

The CFO can be estimated as

$$\hat{v} = \arg\left\{\frac{G_{\mathrm{NIRS}}(\hat{n})}{\pi}\right\} \tag{10}$$

The synchronization requirement is not limited to 4G, LTE. The need for synchronization in 5G is addressed in [20] for better performance. Frequency synchronization is a key element in 5G because when handover operation is performed, user equipment (UE) must be able to synchronization at the proper frequency at a correct time instant. 5G is targeting a high-speed train mobile environment, which needs to consider the high Doppler Effect.

It is stated in [21], standards organization, i.e., IEEE 1588, 3GPP RAN, IEEE 802.1 TSN, ITU-T Q13/SG15 group focusing on synchronization problem in 5G. In 5G, the size of a mobile network is growing, it is a difficult task to synchronize each terminal to proper frequency and time. The presence of integer CFO (ICFO) creates a circular shift in subcarrier indices. This deteriorates OFDM CR performance. The algorithm presented in [18] estimates IFO for the LTE system. It uses the maximum Likelihood method (ML) which exhibits high computational complexity. In this, for 2 GHz carrier frequency, maximum considered CFO is approximately 2.66 normalized with respect to subcarrier spacing.

4.2.1 M & M Method Proposed in Presence of Narrowband Interference

The SCA method provides an estimation range up to ±1/2 of the subcarrier spacing. The NIRS algorithm gives estimation range up to ±1 normalized to subcarrier spacing. The M & M method [19] suggests an improved frequency offset estimator with an increase in estimation range. The estimation range is further extended by increasing no. of identical parts in the training symbol. The estimation range for L identical parts in the training symbol is $\pm L/2$ of the subcarrier spacing. Here, the assumption is that timing synchronization is already performed using the Schimdl and Cox method [13].

This method uses the correlation of the received signal.

$$R(n) = \frac{1}{N - mM} \sum_{k=mM}^{N-1} r(k) r^*(k - mM) \quad 0 \le m \le H \tag{11}$$

where $M = N/L$, length of the identical segment (in samples) of the training symbol, and H is a design parameter.

The best linear unbiased estimator (BLUE) [22] CFO estimator is given by

$$\hat{v} = \frac{L}{2\pi} \sum_{m=1}^{H} w(m) \emptyset(m) \tag{12}$$

where

$$w(m) = 3\frac{(L - m)(L - m + 1) - H(L - H)}{H\left(4H^2 - 6LH + 3L^2 - 1\right)} \quad 1 \le m \le H \tag{13}$$

$$\emptyset(m) \triangleq \left[\arg\{R(m)\} - \arg\{R(m - 1)\}\right]_{2\pi} \tag{14}$$

Here, $R(m)$ and $R(m - 1)$ represents correlation value at mth and $m - 1$th instant. The argument of any variable will reduce its value to a given interval, i.e., $[a]_{2\pi}$ will reduce a to the interval $[-\pi, \pi)$. We take the design parameter, $H = L/2$ because first $L/2$ samples are enough to find $\emptyset$.

As per discussed in [19], for high SNR and estimation range less than $\pm L/2$, BLUE is Extended Schimdl and Cox Algorithm.

5 Simulation Results

Here, the simulation is carried out using Schimdl & Cox Preamble which is used in the IEEE 802.11 standard. S & C and NIRS method use schimdl & cox Preamble. For

M & M method 4 identical segments are used to estimate CFO. Each OFDM symbol is generated using FFT with $N = 64$. To avoid inter-block interference (IBI), Cyclic Prefix considered is $N/16$. For multipath fading COST 207 6-path urban Rayleigh fading channel is considered. The received signal is distorted by AWGN (Additive White Gaussian Noise). The estimation range achieved is ± 2. Two types of narrowband interference are considered here, i.e., ideal NBI and practical NBI. The unmodulated complex sinusoid of normalized frequency 24.5 is considered as "Ideal" NBI. And the FM modulated signal with a bandwidth of 200 kHz is considered as "Practical NBI" [17].

In Figs. 1 and 2, mean square error (MSE) of S& C and NIRS method for frequency estimate is plotted at −10 dB SIR and + 10 dB SIR, respectively. It shows that NIRS outperforms SCA in the presence of narrowband interference.

The CFO estimation range is limited to ±0.5 of SC spacing for the S & C method and ±1 of SC spacing for the NIRS method. The proposed M & M method has an estimation range ±2 of SC spacing. As shown in Fig. 3, M & M method performs better in the presence of practical narrowband interference with SIR = 10 dB. In

Fig. 1 MSE at SIR = −10 dB

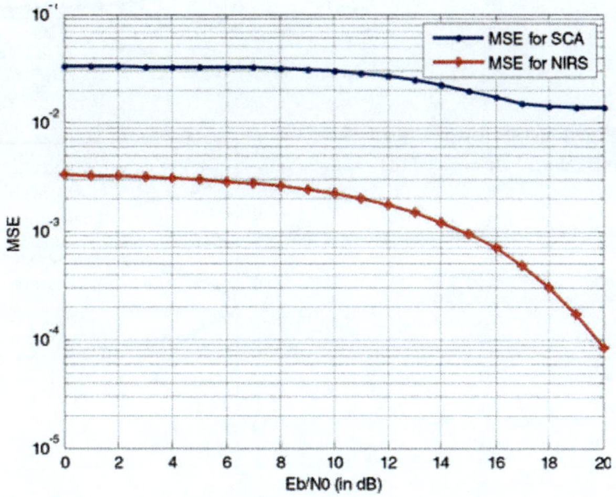

Fig. 2 MSE at SIR = 10 dB

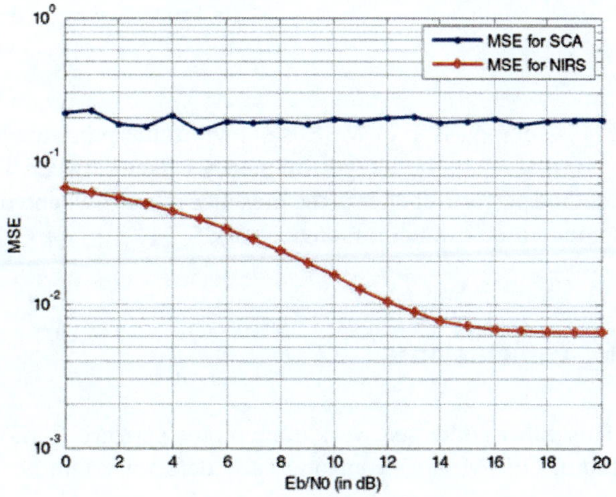

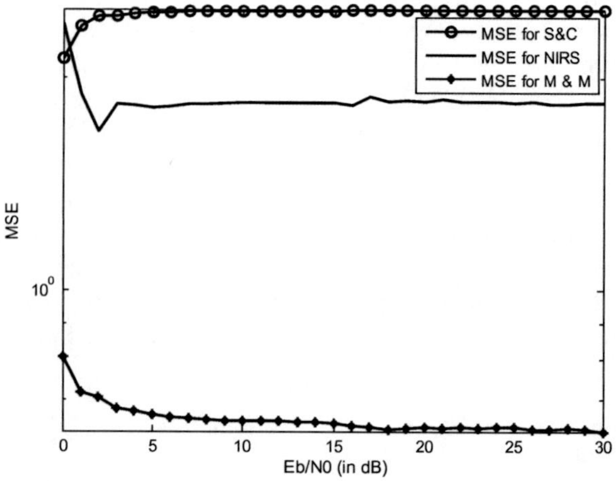

Fig. 3 MSE for practical NBI with SIR = 10 dB

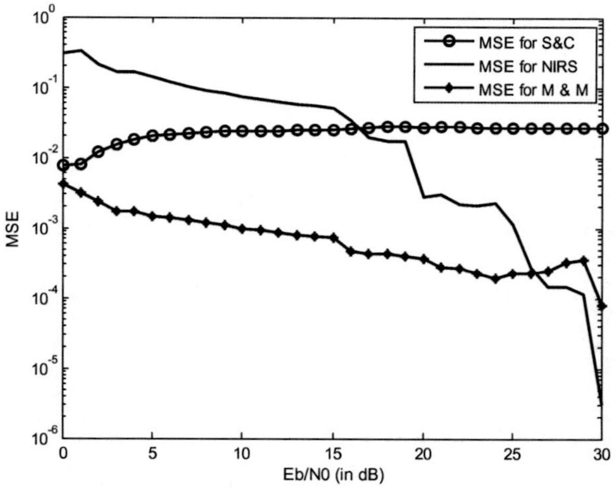

Fig. 4 MSE for practical NBI with SIR = −10 dB

Fig. 4, comparison of MSE is shown in the presence of practical NBI with SIR = −10 dB. Also, the integer frequency offset between 1 and 2 is considered. M&M method performs better in the presence of narrowband interference and also provides an increased estimation range.

6 Conclusion

This paper addressed work done on the synchronization in OFDM-based cognitive radio. OFDM is a promising candidate for cognitive radio due to some inherent features. Comparative analysis of carrier frequency offset (CFO) estimation algorithms is addressed for OFDM-based cognitive radio. The performance of the CFO estimator is improved at the cost of increased complexity. Some algorithms exhibit low complexity but at the same time, estimation range is limited up to fractional CFO. The complexity of these algorithms needs to be addressed for future applications like 5G. The M & M method is proposed in the presence of narrowband interference that achieves increased estimation range.

References

1. Mitola J, Maguire GQ (1999) Cognitive radio: making software radios more personal. IEEE Pers Commun 6(4):13–18
2. Mitola J (2002) Cognitive radio for flexible mobile multimedia communications. In: IEEE international workshop on mobile multimedia communications, 3–10
3. Federal Communications Commission. Spectrum Policy Task Force. Rep. ET Docket no. 02–135(2002)
4. Haykin S (2005) Cognitive radio: brain-empowered wireless communications. IEEE J Sel Areas Commun 23(2):201–220
5. Mahmoud HA, Yücek T, Arslan H (2009) OFDM for cognitive radio: merits and challenges. IEEE Wirel Commun 16(2):6–15
6. Sharma SK, Chatzinotas S, Bao Le L (2015) Cognitive radio techniques under practical imperfections: a survey. IEEE Commun Surv Tutor 17(4):1858–1884
7. Pollet T, Van Bladel M, Moeneclaey M (1995) BER sensitivity of OFDM systems to carrier frequency offset and Weiner phase noise. IEEE Trans Commun 43(2–4. 1):191–193
8. Moose PH (1994) A technique for orthogonal frequency division multiplexing frequency offset correction. IEEE Trans Commun 42(10):2908–2914
9. Chen Y, Tang Z (2012) Effect of spectrum sensing errors on the performance of OFDM based cognitive radio transmission. IEEE Trans Wirel Commun 11(6)
10. Kollar Z, Horvath P (2011) Physical layer considerations for cognitive radio: synchronization point of view. In: IEEE 73rd vehicular technology conference
11. Morelli M, Moretti M (2008) Robust frequency synchronization for OFDM based cognitive radio systems. IEEE Trans Wirel Commun 7(12)
12. Zhou X, Qiu R (2011) An adaptive synchronization algorithm for non-contiguous OFDM cognitive radio systems. In: IET international communication conference on wireless mobile and computing (CCWMC 2011)
13. Schmidl M, Cox DC (1997) Robust frequency and timing synchronization for OFDM. IEEE Trans Commun 45(12):1613–1621
14. Ding J, Huang X, Qu D, Jiang T (2013) Carrier frequency offset estimation for non-contiguous OFDM receiver in cognitive radio systems. In: IEEE global communications conference (GLOBECOM), pp 4192–4197
15. Chin W-L, Kao C-W, Chen H-H, Liao T-L (2014) Iterative synchronization assisted detection of OFDM signals in cognitive radio systems. IEEE Trans Veh Technol 63(4):1633–1644
16. Kryszkiewicz P, Bogucka H (2016a) In band-interference robust synchronization algorithm for an NC-OFDM System. IEEE Trans Commun 64(5):1–13
17. Kryszkiewicz P, Bogucka H (2016b) Low complex, narrowband interference robust synchronization for NC-OFDM cognitive radio. IEEE Trans Commun 64(9):3644–3654
18. Morrelli M, Moretti M (2016) A robust maximum likelihood scheme for PSS detection and integer frequency offset recovery in LTE systems. IEEE Trans Wireless Commun 15(2):1–11
19. Morelli M, Mengali U (1999) An Improved frequency offset estimator for OFDM applications. IEEE Comm Lett 3(3):75–77

20. Lin J-C (2018) Synchronization requirements for 5G: an overview of standards and specifications for cellular networks. IEEE Veh Technol Mag 13(3):91–99
21. Li H, Han L, Duan R, Garner GM (2017) Analysis of the synchronization requirements of 5G and corresponding solutions. IEEE Commun Stand Mag 1(1):52–58
22. Kay SM (1993) Fundamentals of statistical signal processing: estimation theory. Prentice-Hall, Englewood Cliffs

Digital Image Watermarking by Fusion of Wavelet and Curvelet Transform

Jyotsna S. Gaikwad and Usha Verma

Abstract A new hybrid watermarking algorithm using discrete wavelet transform (DWT) and curvelet transform (CT) is implemented in this paper. Curvelet transform is developed to be more efficient than traditional transformations to represent edges along curves. In the proposed work, image watermarking is achieved by applying hybrid transformation and then evaluating the efficacy of the method. Initially, image is decomposed using DWT and then fast curvelet transform is applied on cD (diagonal edges details) sub-band. Secondly, lower coefficients are selected to embed binary-encoded watermark bits. Efficiency of algorithm is tested with different texture images and by varying size of watermark. The effectiveness of proposed algorithm is evaluated by applying attacks like resize, median filtering, and addition of noise. The results shows that, it is sustaining over different texture images even with increase in watermark embedding capacity. The proposed method results in 34.33 dB peak signal-to-noise ratio (PSNR) for high texture image, 47.77 dB for medium texture image, and 48.62 dB for low texture image over maximum size of watermark. It has good robustness against attacks and on large embedding capacity.

Keywords Curvelet transform · DWT · Ridgelet transform · And digital image watermarking

1 Introduction

Security of interactive media content has as of late turned into a significant issue due to inadequate comprehension of licensed innovation. Watermarking is a strategy in PC security by which identifiers of sources or copyright proprietor's implants advanced signs into the particular signs. This makes easy to monitor from where a

J. S. Gaikwad (✉)
VLSI Design and Embedded System, MIT Academy of Engineering, Alandi, Pune, India
e-mail: Jyotsnagaikwad17@gmail.com

U. Verma
School of Electrical Engineering, MIT Academy of Engineering, Alandi, Pune, India
e-mail: uyverma@etx.maepune.ac.in

© Springer Nature Singapore Pte Ltd. 2021

S. N. Merchant et al. (eds.), *Advances in Signal and Data Processing*,
Lecture Notes in Electrical Engineering 703,
https://doi.org/10.1007/978-981-15-8391-9_40

flag originates or who the copyright proprietors are. Watermarking framework must have two attributes, for example, perceptual straightforwardness and power:

- Perceptual straightforwardness: An implanted watermark ought not to present a noteworthy level of mutilation to the spread picture.
- Robustness: A watermark ought to be hard to expel from the watermarked picture. It needs to make due after the picture has experienced some typical flag handling and picture assaults.

In the previous two decades, wavelet is a widely referred tool, used regularly by various researchers. It gives multiresolution portrayal of the spread work. For 1D structure, wavelet gives close ideal portrayal which is piecewise smooth with point singularities. But wavelets are improper for distinguishing, sorting, or to give a minimized portrayal of 2D structures. It requires large number of coefficients to represent line singularities of 2D structure. Around thirty-five years ago, mathematicians and physicists presented curvelet transform trying to beat natural confinements of customary multiscale portrayals. Curvelet change straightforwardly accepts the edge as the fundamental portrayal component; it is anisotropic with solid course, and gives ideally meager portrayals of articles along a general bend with limited arch.

The proposed method maximizes invisibility by using the curvelet transform [1] for embedding a watermark. Curvelet transform can decompose cover image in more than eight different directions but here utilized eight directions, so it became possible to insert a watermark using less amount of energy. Several researchers have considered curvelet domain previously.

Hien et al. [2] implemented watermarking installing framework which offers need of implanting edges of the given picture to fulfill invisibility requirement and obtained robustness against some image processing attacks. Singh et al. [3] proposed medical image watermarking by using DWT-SVD method. In proposed system with increase in watermark size quality of algorithm get degraded. Algorithm based on multiresolution is proposed by Kaushik et al. [4]. In particular, they performed watermarking in wavelet domain and curvelet domain separately. By measuring performance (i.e., PSNR and NCC) of both algorithms shown superiority of curvelet domain over wavelet domain. Thanki et al. [5] proposed a hybrid watermarking technique with compressive sensing (CS) theory. It delivers one additional security layer and provides security to watermark data before embedding.

2 Wavelet Transform

Wavelet transform converts an image from spatial to frequency domain. It is a wave-like oscillations, which are the sines and cosines in the Fourier transform. In general, wavelets are generated with specific properties that make them useful for signal and image representation and used as a base functions. These base functions are achieved after dilation and translation of a mother wavelet $\Psi(x)$ by the amounts of$s$ and τ, respectively, as shown in (1):

$$\Psi_{\tau,s}(x) = \left\{ \Psi\left(\frac{x-\tau}{s}\right), (\tau, s) \in R \times R^{+} \right\} \tag{1}$$

These translation and dilation make possible to wavelet transform to be localized in time and frequency. Apart from sines and cosines, wavelet basis functions can be represented in better way with discontinuities and spikes.

Continuous wavelet transform (CWT) can be defined as in (2):

$$\mathrm{cwt}_{\Psi}(\tau, s) = \frac{1}{\sqrt{|s|}} \int x(t) \Psi^{*}_{\tau,s}(t) \mathrm{d}t \tag{2}$$

where $\Psi^{*}_{\tau,s}$ is the complex conjugate of $\Psi_{\tau,s}$ and $x(t)$ is the input signal defined in time domain. Inverse CWT is obtained as

$$x(t) = \frac{1}{C_{\Psi}^{2}} \int_{s} \int_{\tau} \mathrm{cwt}_{\Psi}(\tau, s) \frac{1}{s^{2}} \Psi_{\tau,s}(t) \mathrm{d}\tau \mathrm{d}s \tag{3}$$

where C_{Ψ} is a constant and depends on the wavelet used.

CWT is discretized using uniform sampling by Nyquist rule as:

$$N_2 = \frac{s_1}{s_2} N_1 \tag{4}$$

Here, N_1, N_2 are number of samples at scales s_1 and s_2, respectively, and $s_2 > s_1$. This rule decreases number of samples at higher scales (lower frequencies). Minimum sampling rate is obtained in order to reconstruct original signal from discrete set of samples. Translation and dilation to the dyadic scale are set by following equations with logarithmic series of base 2 for $\Psi_{k,j}$

$$T = k2^{j} \tag{5}$$

and

$$s = 2^{j} \tag{6}$$

Filters are nothing but above two τ and s coefficients which work as a low pass H and high pass filter G, respectively, and as a quadrature mirror filters.

Cover image is decomposed by using the combination of 1D wavelet transform. Quadrature mirror filters decompose $n \times n$ image I into wavelet coefficients, as shown in Fig. 1. Filters H and G are applied on rows and columns of cover image which gives four sub-images with dimension $n/2 \times n/2$. Sub-images H_cH_rI (cA), G_cH_rI (cH), H_cG_rI (cV), and G_cG_rI (cD) (where subscript r and c denote row and column, respectively) contain the low–low, high–low, low–high, and high–high pass, respectively. Complete DWT decomposition process is shown in Fig. 1.

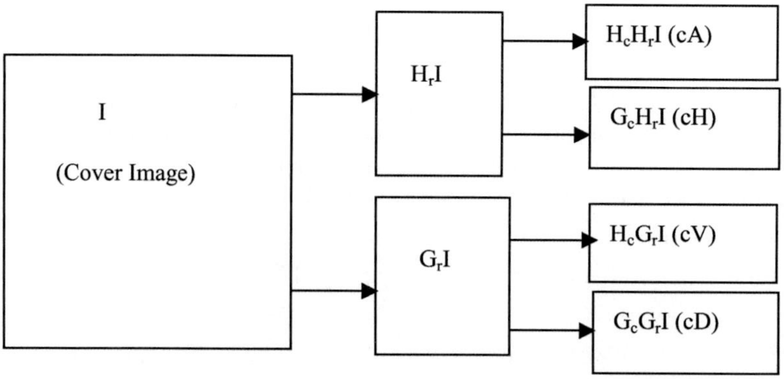

Fig. 1 Wavelet decomposition of an image

3 Curvelet Transform

As described in the introduction, even if DWT has established an impressive reputation as a tool for mathematical analysis and signal processing, still it has the disadvantage of directionality which is shown in Fig. 2. These undetermined its usage in various areas. In 1999, an anisotropic geometric wavelet transform, named ridgelet transform, was proposed by Cande's and Donho [6]. The ridgelet transform is best alternative to represent straight line and curved singularities which is shown in Fig. 2 with the comparison of wavelet strategy. In this, initially, image partition is considered and afterwards, ridgelet transform is applied on the received sub-images. This block ridgelet-based transform is named curvelet transform. Such type of transform was firstly Cande's and Donho in 2000 [7] presented.

As shown in Fig. 2, CT requires less coefficients for a given accuracy of reconstruction. Thus, it has been adequately used as an powerful tool in processing of edge properties, for example, image denoising [8], contrast enhancement [9], texture classification [10], etc. The curvelet decomposition is the sequence of the following steps:

1. **Sub-band decomposition**: It defines a bank of sub-band filters P_0, ($\Delta_{s,}$ $s = 3$). Object cD is filtered into sub-bands:

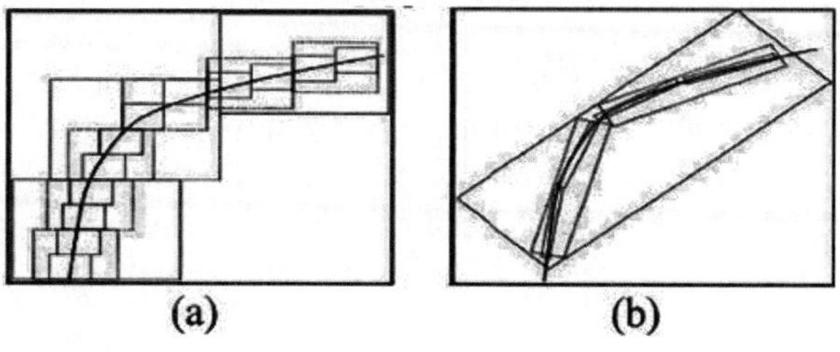

Fig. 2 Comparison between **a** wavelet and **b** curvelet transform

$$\text{cD} \rightarrow (P_0\text{cD}, \Delta_1\text{cD}, \Delta_2\text{cD}, \Delta_3\text{cD}) \tag{7}$$

The three sub-bands Δ_scD contain details about 2^{-2s} wide.

2. **Smooth partitioning**: This states a group of smooth windows $w_Q(x_1, x_2)$ which are localized around dyadic squares.

$$Q = \left[k_1/2^s, (k_1+1)/2^s)\right] \times \left[k_2/2^s, (k_2+1)/2^s)\right] \tag{8}$$

A function with the respective window function w_Q is multiplied to produce outcome situated close Q. Doing this for all Q at a certain scale, i.e., for all $Q = Q(s, k_1, k_2)$ where k_1 and k_2 are varying but s is fixed.

$$\Delta_s f \rightarrow \left(w_Q \Delta_s f\right)_{Q \in Q_s} \tag{9}$$

3. **Renormalization**: For a dyadic square Q, T_Q is the transport operator acting on functions f via

$$(T_Q f)(x_1, x_2) = 2^s f\left(2^s x_1 - k_1, 2^s x_2 - k_2\right) \tag{10}$$

At this point of the operation, each 'square' leading in the earlier phase is renormalized to unit scale

$$g_Q = \left(T_Q\right)^{-1}\left(w_Q \Delta_s f\right), Q \in Q_s \tag{11}$$

4. **Ridgelet analysis**: Each square is analyzed in the orthonormal ridgelet system. This is a system of basis elements ρ_λ making an orthobasis for $L^2(R^2)$:

$$\alpha_\mu = g_Q, \rho_\lambda, \mu = (Q, \lambda) \tag{12}$$

The result of all this is curvelet coefficients with involvement of edge information.

Implementation of fast discrete curvelet transform (FDCT) by using fast Fourier transform (FFT) and wrapping function gives a table of digital curvelet coefficients which are indexed by scale, orientation, and location parameters.

Digital curvelet transform takes a 2D image as input in the form of Cartesian array $f[t_1, t_2]$, where, $0 \le t_1 < T_1, 0 \le t_2 < T_2$ and T_1, T_2 are the dimensions of the array. As illustrated in Eq. (13), the output is a collection of curvelet coefficients:

$$C^D(j, l, k) = \sum_{0 \le t_1 < T_1}^{0 \le t_2 < T_2} f[t_1, t_2] \emptyset^D_{j,l,k}[t_1, t_2] \tag{13}$$

where C is the curvelet coefficient, scale j, an orientation l, and $k = (k_1, k_2)$ are spatial location parameters. An effective parabolic scaling law width $\approx$ length^2

Organization of the curvelet transform:

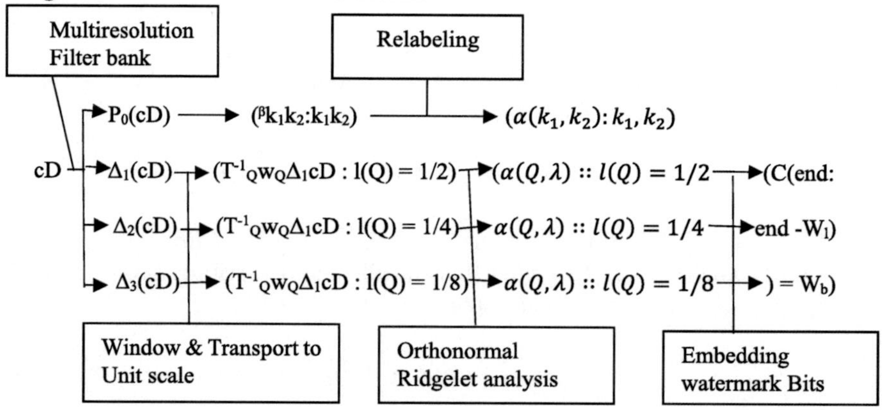

Fig. 3 Organization of the curvelet transform

implements CT on the sub-bands in the frequency domain to catch curved edges of image successfully. The complete organization of curvelet transform after DWT decomposition process is shown in Fig. 3.

where W_l is the watermark length and W_b are watermark bits.

4 Proposed Methodology

The proposed watermarking algorithm incorporates a watermark into selected curvelet coefficients. The CT is applied on DWT coefficients of cD sub-band to get hybrid coefficients of the cover image. Watermark size is varying but cover image in which watermark is embedded has size of 256 × 256. Before embedding, watermark is converted into binary form to simplify the watermarking process.

4.1 Watermark Embedding Process

The proposed block diagram of watermark embedding is shown in Fig. 4. Firstly, image is decomposed by using DWT as shown in Fig. 1. It decomposes image into a set of wavelet bands (i.e., cA, cH, cV, and cD). cD sub-band contains diagonal information of an image so it is chosen for further processing. CT is applied on this sub-band and analyzed through a local ridgelet transform as shown in Fig. 3. The output of a ridgelet analysis is curvelet coefficients of cD image. Out of high and low-frequency curvelet coefficients, low-frequency curvelet coefficients are chosen for embedding process.

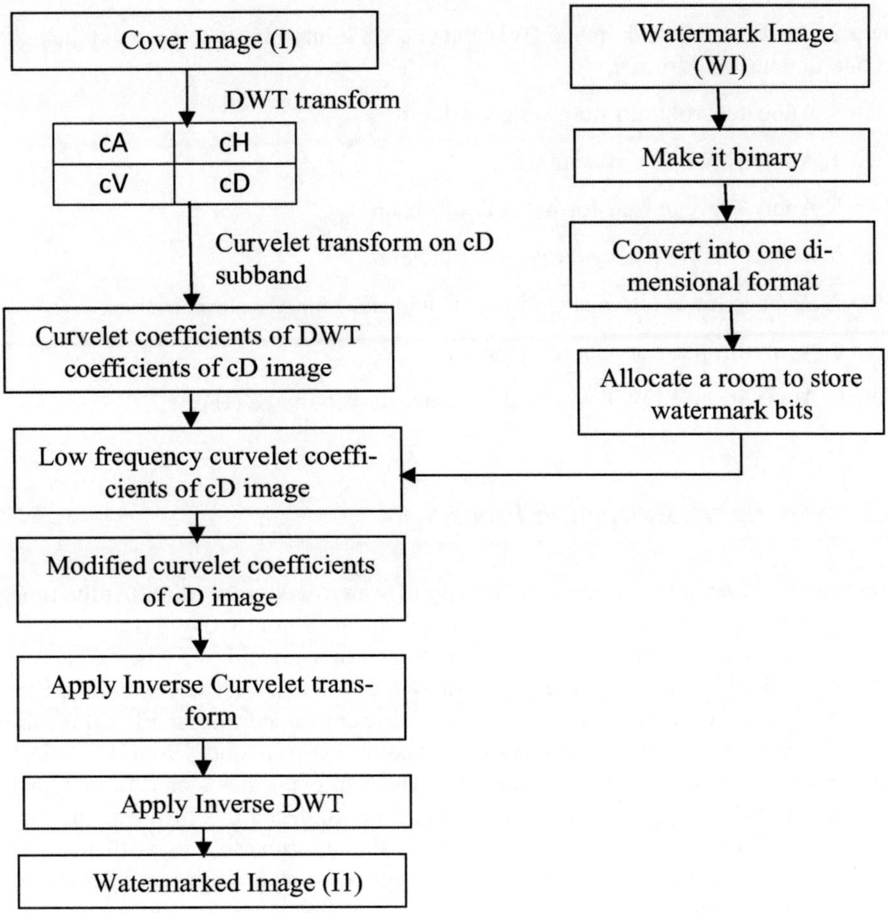

Fig. 4 Watermark embedding

On other side, watermark image (WI) is converted into binary format and rearranged into 1D format. Rearranging of binary watermark bits simplifies the process of embedding. A room is allocated to this binary bits which completes the watermark encoding process. Next, lower curvelet coefficients are replaced with watermark bits. Out of three scales of curvelet which are of size 256 × 256, 128 × 128, and 64 × 64, respectively, lower curvelet coefficients get replaced with watermark bits. After watermark embedding, inverse CT is applied on modified curvelet coefficients of cD image. It is followed by inverse DWT which gives a watermarked image (I1).

Algorithm:

Step 1. Read cover image (I) and resize it into size of 512 × 512.

Step 2. Read a watermark image (WI) and encode it into binary format and make it in one-dimensional format.

Step 3. Allocate a room to store watermark bits.

Step 4. Apply DWT on cover image

Step 5. Apply curvelet transform on cD sub-band.

Step 6. Select low-frequency curvelet coefficients.

Step 7. Replace lower curvelet coefficients with 1D binary watermark bits.

Step 8. Apply inverse curvelet transform

Step 9. Apply inverse DWT which gives watermarked image (I1).

4.2 Watermark Extraction Process

Watermark extraction is an inverted activity of watermark embedding. At the time of extracting a watermark from watermarked image, initially, DWT transform is applied on watermarked image. DWT decomposition gives DWT coefficients of watermarked image which will give four sub-bands cA, cH, cV, and cD. From these, cD sub-band is chosen for further processing, as we embedded watermark in this sub-band. Afterwards, curvelet transform is applied on cD sub-band. From this, select low-frequency curvelet coefficients, arrange them in one-dimensional format, and store in a room. Then, extract watermark bits from replaced bits of lower curvelet coefficients. This extracted bits are in one-dimensional format so reshape them in 2D form which will give extracted watermark image. Complete watermark extraction process is shown in Fig. 5.

Algorithm

Step 1. Read a watermarked image (I1).

Step 2. Apply DWT transform on watermarked image (I1).

Step 3. Apply curvelet transform on cD sub-band.

Step 4. Select low-frequency curvelet coefficients of watermarked image (I1).

Step 5. Arrange them in one column format.

Step 6. Allocate a room to store them.

Step 7. Extract watermark bits from it.

Step 8. Reshape it in 2D form which is extracted watermark image.

Fig. 5 Watermark extraction

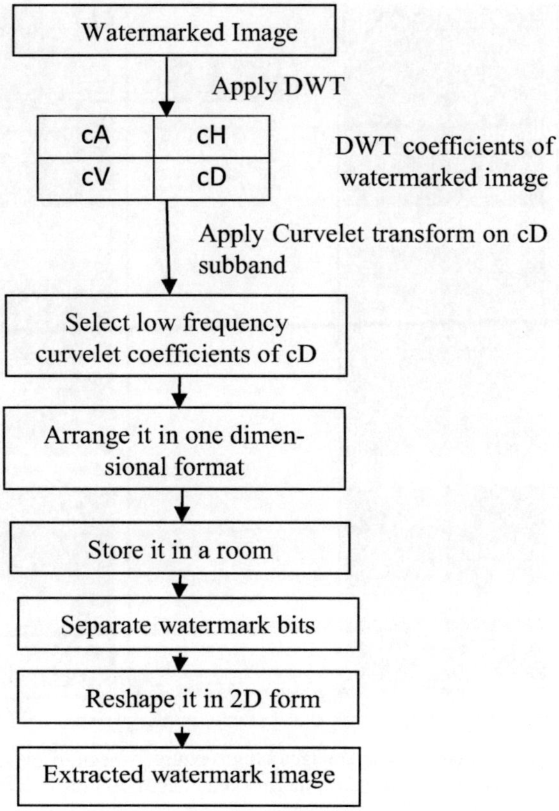

5 Results and Discussion

For experimentation purpose, three different texture images are used as a cover image of size 512 × 512. 'Bamboo' is a high texture image, 'Cameraman' is a medium texture image, and 'Solar' is a low texture image. Binary watermark, 'HELLO' is used as a watermark which varies from minimum size 10 × 10 to maximum size 256 × 256. MATLAB R2015a is used for implementing and testing performance of proposed algorithm. For performance evaluation, parameters used are peak signal-to-noise ratio (PSNR) and normalized correlation coefficient (NCC).

Proposed method is applied over all three different texture cover images. These images are shown in Fig. 6 with a watermark 'HELLO'. Figure 6a is a high texture image, Fig. 6b is a low texture image, and Fig. 6c is a low texture image. Their respective watermarked images are in Fig. 6e–f and extracted watermark images are Fig. 6h–j shown respectively. This results are taken over maximum watermark size 256 × 256 without applying attack.

Figure 6 visualizes that there is no notable dissimilarity between cover image and watermarked image, which illustrates proposed algorithm fulfills invisibility

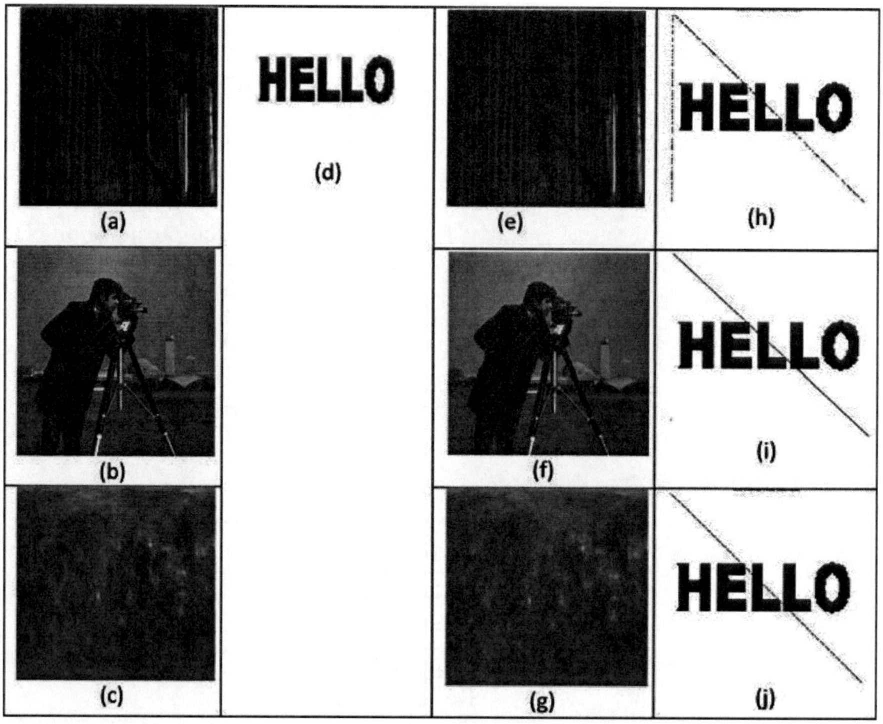

Fig. 6 Different texture images **a** high texture, **b** medium texture, **c** low texture image, **d** watermark image, and **e–g** respective watermarked images, **h–j** respective extracted watermark images

requirement. The imperceptibility of watermarked image is measured using PSNR and for good imperceptibility, it has to be above 30 dB. As proposed algorithm is tested over different textures and various watermark sizes, giving PSNR of 34.33 dB over high texture image, 47.77 dB over medium texture image, and 48.62 dB over low texture image at maximum size of watermark. Similarly, correlation between original watermark and extracted watermark is measured using NCC parameter. From Fig. 6, it can be confirmed that watermarking and extraction of watermark image are done successfully.

Further, robustness of proposed algorithm is tested by applying different attacks such as addition of noise, resizing, and filtering. Comparative analysis of three different textures with variation in watermark size for performance parameter PSNR is shown in Table 1. PSNR is measured between cover image and watermarked image.

PSNR values in Table 1 confirm that proposed algorithm gives best results over medium and low texture images with negligible variation in results for every watermark size. But, somewhat weakens the performance over high texture image but still it is above 30 dB so this also can be acceptable. After applying attacks except resize and filtering of high texture image, small performance degradation is occurring which is nothing but imperceptibility of proposed algorithm.

Table 1 PSNR for imperceptibility of proposed watermarking algorithm

Cover image	Watermark size	PSNR(dB)			
		Attack			
		No attack	Resize	Median filter	Gaussian noise
Bamboo	10 × 10	34.18	27.56	26.24	31.60
	100 × 100	34.21	27.56	26.24	31.63
	200 × 200	34.28	27.56	26.25	31.17
	256 × 256	34.33	27.56	26.25	31.33
Cameraman	10 × 10	45.79	42.88	35.79	37.86
	100 × 100	46.00	42.92	35.79	37.98
	200 × 200	46.94	43.05	35.81	37.73
	256 × 256	47.77	43.15	35.83	37.82
Solar	10 × 10	46.51	46.26	40.11	22.27
	100 × 100	46.74	46.33	40.11	22.29
	200 × 200	47.72	46.58	40.13	22.24
	256 × 256	48.62	46.78	40.17	22.13

Respective graphical analysis of three different texture images for PSNR values over different watermark sizes without and with attacks is shown in Figs. 7, 8, 9 and 10, respectively.

PSNR without applying attack is changing according to texture but it has negligible variation over change in watermark size. In addition, performance slightly increases with increase in watermark size and lowering the texture.

Resizing of image degrades the performance of every texture. Resize attack affects more on high texture image as compare to medium and low texture images. This occurs because high texture image has large high-frequency information and resizing affects a lot on this which cause to degrade the performance.

Sustainability of low texture image is high over median filtering attack as compare to the high and medium texture image. This attack smoothening the image so it affects

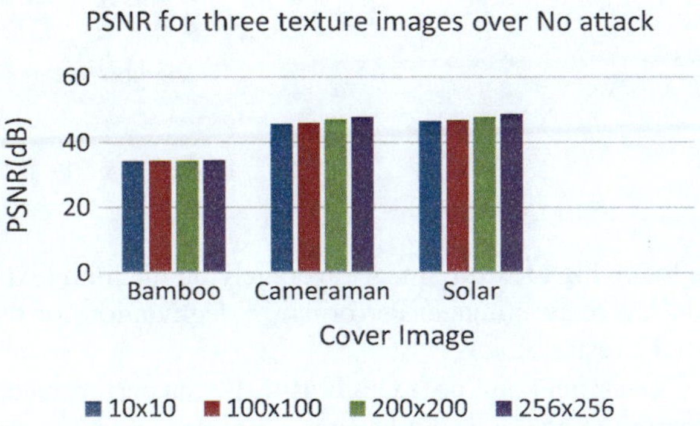

Fig. 7 Performance comparison of three different texture images without applying attack

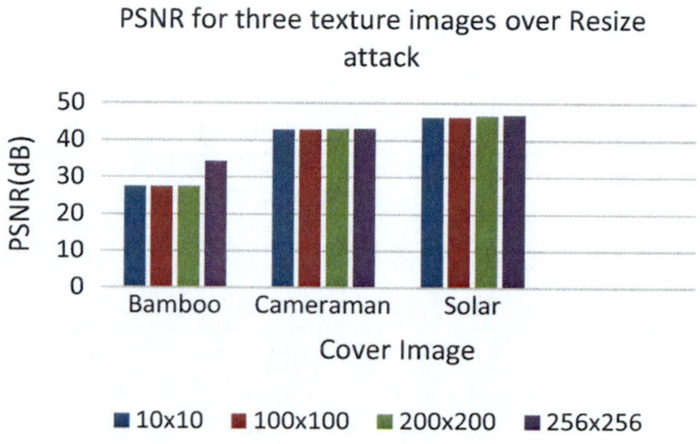

Fig. 8 Performance comparison of three different texture images by applying resize attack

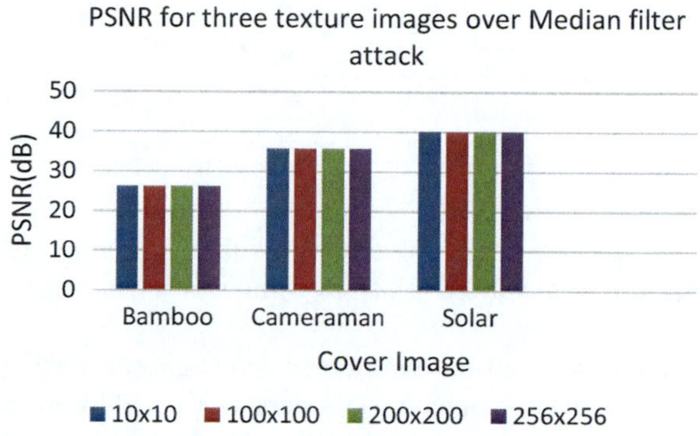

Fig. 9 Performance comparison of three different texture images by applying median filtering attack

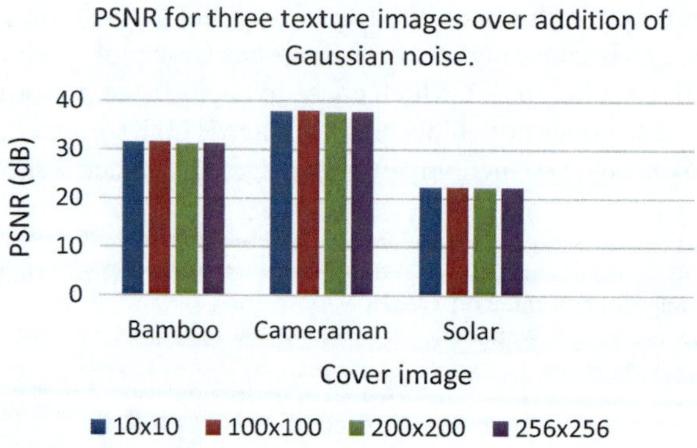

Fig. 10 Performance comparison of three different texture images by adding Gaussian noise

a lot on high texture image, averagely on medium texture image, and has less effect on low texture image. Performance degradation for this attack is more as compare to the resize attack.

Gaussian noise degrades drastically the performance of low texture image, as low texture image does not have any high-frequency content so adding high-frequency

Table 2 NCC to check robustness of watermarking algorithm

Cover image	Watermark size	PSNR(dB)			
		Attack			
		No attack	Resize	Median filter	Gaussian noise
Bamboo	10 × 10	0.010	0.019	0.064	0.051
	100 × 100	0.399	0.392	0.465	0.410
	200 × 200	0.616	0.665	0.695	0.539
	256 × 256	0.672	0.685	0.547	0.691
Cameraman	10 × 10	0.015	0.048	0.007	0.037
	100 × 100	0.400	0.496	0.221	0.361
	200 × 200	0.615	0.625	0.257	0.602
	256 × 256	0.672	0.693	0.262	0.676
Solar	10 × 10	0.014	0.048	0.005	0.011
	100 × 100	0.400	0.423	0.278	0.403
	200 × 200	0.615	0.627	0.283	0.604
	256 × 256	0.672	0.438	0.395	0.672

content in it degrades the performance. Medium and high textures averagely maintain their performance over addition of Gaussian noise.

Similarly, comparative analysis of three different textures with variation in watermark size for NCC is shown in Table 2. NCC is measured between original watermark and extracted watermark image. The range of NCC lies between −1 and 1. In which −1 indicates uncorrelated and 1 indicates totally correlated. In proposed work, NCC is normalized between 0 and 1 so values below 0.05 can be considered as −1 and above it to 1.

Table 2 shows that for low watermark size algorithm is uncorrelated but at high watermark size giving better results. In contrast with increase in watermark size, correlation between original watermark and extracted watermark is increases. So correlation is occurring only at maximum embedding capacity. After application of attacks, there is not so much degradation in NCC, and values are reducing negligibly. This confirms that at low watermark size, proposed method is low robust but showing robustness at high watermark sizes.

Proposed method's results are compared with other existing's works [3, 5]. Table 3 shows that PSNR of proposed method is better than [3]. In both cases, PSNR is measured between original image and watermarked image. From [3], chosen cover image is a 'CT scan' and 'Cameraman' from proposed method for comparison. Somehow correlation results of [3] are little bit better than proposed method. NCC is measured between original watermark and extracted watermark image.

In Table 3 for [3], PSNR value is decreasing with increase in watermark size. It means quality is degrading as payload is increasing. But this is not happening in proposed method; PSNR also increasing with increase in watermark size. In both cases, NCC is gradually increasing with increase in watermark size.

Table 3 Performance comparison with existing work

Method	Watermark size	Performance parameters	
		PSNR	NCC
[3]	16 × 16	48.44	0.582
	32 × 32	45.43	0.931
	64 × 64	42.14	0.954
Proposed method	10 × 10	45.79	0.015
	100 × 100	46.00	0.400
	200 × 200	46.94	0.615
	256 × 256	47.77	0.676

From [5] and proposed method, cover image chosen is 'Cameraman' of size 256 × 256. In [5], to add one extra security layer and to provide security to watermark data before embedding used a complex hybrid watermarking algorithm over watermark and host image. After complex processing on watermark and cover image obtained, PSNR result is 47.19 dB for max watermark size 256 × 256. And proposed method giving PSNR of 47.77 dB for max watermark size 256 × 256. It means proposed hybrid algorithm is giving better results than [5].

6 Conclusion

The proposed hybrid algorithm is tested over three different textures (high, medium, and low textures) and watermark which varies from 10 × 10 (minimum size) to 256 × 256 (maximum size) size. Theoretically, quality of watermarked image should be decreased with increase in watermark size, but proposed algorithm sustaining over various sizes. Curvelet transform reduces the watermarking energy and improves the invisibility and robustness and getting fine results at maximum watermark embedding capacity. Proposed algorithm gives high performance over low texture image and medium texture but somewhat it performs poor for high texture image. After attacks, algorithm performance degrades negligibly except for Gaussian attack on low texture image. It sustains over large embedding capacity and after applying attacks too. Which is nothing but high robustness of proposed algorithm.

References

1. Candes E, Demanet L, Donho D, Ying L (2006) Fast discrete curvelet transforms. SIAM J Multiscale Model Simul 861–899. https://doi.org/10.1137/05064182x
2. Hien TD, Miyara K, Kei I, Ali FF, Chen YW, Nakao Z (2007) Digital watermarking based on curvelet transform. In: Sobh T et al (eds) Innovative algorithms and techniques in automation,

industrial electronics & telecommunications 2007. Springer, pp 305–309
3. Singh HV, Rai A (2019) Medical image watermarking in transform domain. In: Tiwari S, Trivedi M, Mishra K, Mishra A, Kumar K(eds) smart innovations in communication and computational sciences. Advances in intelligent systems and computing, vol 851. Springer, Singapore
4. Kaushik R, Sharma S, Raheja LR (2017) Digital watermarking technique based on multi-resolution curvelet transform. Int J Found Comput Sci Technol (IJFCST), 1–14. https://doi.org/10.5121/ijfcst.2017.7201
5. Thanki R, Dwivedi V, Bvisagar K (2019) A hybrid watermarking scheme with CS theory for security of multimedia data. J King Saud Univ Comput Inf Sci 31:436–451, 2019
6. Candès E, Donoho D (1999) Ridgelets: a key to higher-dimensional intermittency? Philos Trans Math Phys Eng Sci, 2495–2509
7. Candès E, Donoho D (2000) Curvelets—a surprisingly effective nonadaptive representation for objects with edges. Curves and surfaces: Saint–Malo Proceddings, 1–10
8. Starck J-L, Candes EJ, Donoho DL (2002) The Curvelet transform for image denoising. In: IEEE transactions on Image Processing, pp 670–684
9. Starck J-L, Murtagh F, Candes EJ, Donoho DL (2003) Gray and color image contrast enhancement by the curvelet transform. In: IEEE Transactions on image processing, pp 706–717
10. Arivazhagan S, Ganesan L, Kumar TG (2006) Texture classification using curvelet statistical and co-occurrence features. In: 18th international conference on pattern recognition (ICPR'06). IEEE, Hong kong, China. https://doi.org/10.1109/ICPR.2006.1110
11. AL-Nabhani Y, Jalab HA, Wahid A, Noor R (2015) Robust watermarking algorithm for digital images using discrete wavelet and probabilistic neural network. J King Saud Univers Comput Inf Sci 27:393–401

Machine Learning Feature Selection in Archery Performance

Uma Mahajan, Anup Krishnan, Vineet Malhotra, Deep Sharma, and Sharad Gore

Abstract Successful sports performance depends on several physiological and physical fitness components. It is essential to know which fitness features are most important for performance. Sports fitness components are often multicollinear, and the relationship is complex. So there is a need to use more sophisticated methods that can deal with complex multicollinear data. Hence machine learning algorithms are used along with conventional statistical methods for important physiological and physical fitness feature selection in the archery performance of Indian archers. Recursive feature elimination and Boruta algorithm using random forest along with conventional statistical methods are used for feature selection. The root mean square error of stepwise regression was 21.262, recursive feature elimination with 15 features was 19.450 and random forest with 15 features was 8.417 in the training dataset. Further, the root mean square error for random forest with eight confirmed important features was 9.003 and 8.716 for ten non-rejected features in the training dataset. Out of fifteen features, eight features confirmed important are maximum bow hold time, sub-maximal oxygen intake, peak power, average power, core abdominal strength, age, weight, and body fat, while acceleration speed and maximum oxygen intake are tentatively important. Machine learning Boruta algorithm using random forest performs better than traditional statistical and recursive feature elimination method for selecting features as well as predicting performance in unseen data. Thus, eight important features identified through Boruta algorithm are useful to develop battery of test, monitor athletes, and alter the training regimens in real-time and talent selection in the archery.

U. Mahajan (✉) · S. Gore
Shri Jagdishprasad Jhabarmal Tibrewala University, Jhunjhunu, Rajasthan, India
e-mail: umasmahajan@gmail.com

U. Mahajan · A. Krishnan
Army Sports Institute, Pune, India

V. Malhotra · D. Sharma
Armed Forces Medical College, Pune, India

© Springer Nature Singapore Pte Ltd. 2021

S. N. Merchant et al. (eds.), *Advances in Signal and Data Processing*,
Lecture Notes in Electrical Engineering 703,
https://doi.org/10.1007/978-981-15-8391-9_41

Keywords Boruta algorithm · Random forest · Recursive feature elimination · Physiological and physical fitness

1 Introduction

Evaluation of athletes regularly is needed to track the athlete's training progress. Several batteries of tests are conducted to evaluate the physiological and physical fitness of the athletes as a part of routine evaluation. Many studies were conducted in various sports to identify which factors are associated with elite sports performances [1]. Different sports disciplines require different strengths and skills. For example, wrestling, boxing, and weight lifting require anaerobic capacities more, while athletics require more aerobic capacities [2, 3]. Therefore, it is important to identify which physiological/physical fitness components are related to the performance of athletes in that particular sports discipline.

The identification of important variables is known as feature selection. Filter-based, wrapper-based, and embedded/implicit methods are used to select important features. Correlation and linear regression are filter methods, recursive feature elimination is a wrapper method, and the tree-based random forest is the implicit feature selection method [4].

Physiological and physical fitness variables are often highly correlated, and it is difficult to apply conventional regression models to predict sports performance due to multicollinearity [5]. Variance inflation factor (VIF) is used to detect multicollinearity among variables [6]. A variable with VIF $\geq$ 3 indicates that there is high multicollinearity in the data due to this variable. So the traditional approach is to delete the variable with VIF $\geq$ 3 from the linear regression model and then again run the model to get modified VIFs. This process of checking VIFs and eliminating multicollinear variable and re-running the model continues till VIF $<$ 3 for the sub-set of the variables in the model, and there is no multicollinearity. Also, linear/stepwise regression is used to identify important features. A stepwise regression uses forward/backward/both forward and backward directions in each step to select or eliminate the variables based on the a priori rules [7].

In recursive feature elimination (RFE), first, all variables are included in the model. Next, one by one variable is eliminated based on their importance score. Then the model is re-run/re-built. The information gained in each model is compared using the root mean square error (RMSE) and variance explained. This process continues until there is no improvement in the model by eliminating any variable. Finally, the sub-set of the variables in the model which has lowest RMSE are identified as the most important variables/features. Random forest-based RFE algorithm is applied using the randomForest package to arrive at the most important features [8].

Random forest (RF) is a tree-based feature selection machine learning method. It has an in-built mechanism to select variables implicitly based on the optimum splitting criteria in such a way that a branch node is a homogeneous group of variable/s [9]. A Boruta package uses a random forest classifier and trains the algorithm using bootstrap sampling. It creates a shadow variable for each independent variable; if the importance score of the variable is greater than the shadow variable, then that variable is confirmed as an important variable. It separates the variables as important, tentatively important and not at all important and gives decision based on the importance score of the variables [10].

Apart from random forest, other machine learning methods like ridge regression, support vector machines, and neural networks were also used in sports performance prediction, namely swimming time [11], rugby [12], football [13, 14], baseball, basketball, and hockey games [15], predictors of mixed martial arts performance [16], timing of '*tour de france*' [17], talent detection in primary school-going tennis players [18], selection of javelin throwers [19] and cricket team [20], etc.

Recently, a research is done in Malaysia to identify potential elite archers using motor qualities of the archers [21] and other indicators anthropometric, biomechanical, bio-physiological and psychological features [22]. Currently, there is inadequate knowledge about the important physiological and physical fitness factors required for the sports performance in the game of archery in India. So in this paper, machine learning algorithms along with conventional statistical methods are used for the feature selection of physiological and physical fitness variables in Indian archery. The analytical approach is explained in the methods section, outputs are interpreted in the results section, and the findings are discussed and concluded in the remaining sections.

2 Methods

2.1 Sports Fitness and Performance Testing Datasets

Physiological and physical fitness testing of male archers was carried out in an international level sports training institute in India. Various physiological and physical tests were conducted to monitor and evaluate the athlete's progress in sports training. All the archers voluntarily participated in the routine periodic testing and evaluation. Archers who were injured during the evaluation period did not participate in the testing. Also, archers who were under basic training were excluded from the performance testing.

Age (years) at the time of testing was recorded along with height (cm) and weight (kg). In human performance (HP) laboratory, a Wingate cycle ergometry test was conducted to measure lower body peak and average power ($W.kg^{-1}$), and an Astrand cycle test was conducted to measure oxygen consumption VO_2 ($mL.min^{-1}.kg^{-1}$) at sub-maximal level, while bioelectrical impedance analysis (BIA) was done to measure body fat (%) and total body water (%). The push-ups (numbers) and sit-ups (numbers) done in 1 min were measured to assess the upper body and core abdominal strength of the archer, respectively. A standing start 30 m test was conducted to measure acceleration speed, and a shuttle run test was conducted to measure agility.

Time duration to complete these two tests was noted in seconds (sec), and lower time indicates athlete performing better in acceleration speed and the agility. Sit and reach test was conducted to measure the flexibility in centimetres (cm). Canadian stage test was conducted to measure a maximum oxygen intake VO_2max ($mL.min^{-1}.kg^{-1}$), i.e. cardiovascular endurance of the archer. A sport-specific motor fitness test was conducted in which an archer holds the archery bow till his maximum capacity, and timing was recorded in seconds (sec) to measure the handgrip strength of the archer. More time of maximum bow held indicates better handgrip strength of the archer. In all 15 physiological and physical fitness, variables/features were included in the analysis dataset.

The performance of archers was tested periodically at least once in a month as per their training schedule. The archer shot in total 72 arrows which included two rounds of 36 arrows each. Score on the scale of 1 (minimum) to 10 (maximum) was given which depends on the accuracy of the target hit as per Archery Federation rules [23]. Archery performance records of the tests conducted in the same week of physiological and physical testing were considered for the analysis and merged with the fitness dataset.

2.2 *Developing Machine Learning Algorithms for Feature Selection*

All complete cases were included in the dataset. An exploratory data analysis was done to understand the pattern of the data, missing values, and outliers in the data. The last observation was carried forward to replace the missing values in HP lab and physical fitness tests data if the archer could not able to complete any test due to the technical issues or localized pain in the respective body part [24, 25]. So, the final analysis dataset included 270 cases with 15 fitness features and one performance/outcome variable. Summary statistics mean and standard deviation (SD) were used to describe the data. All the analysis was done in *R* 3.5.3. The analytical approach for feature selection is shown in Fig. 1.

The Shapiro–Wilk test was applied to test the normality of all variables. For multicollinearity diagnostics, univariate and partial correlations were calculated using ppcor package [26]. Variance inflation factor (VIF) was calculated using linear regression with archery performance as the target/outcome variable. The equations to calculate univariate and partial correlation coefficient (Corr coeff), variance explained, i.e. coefficient of determination R^2, VIF, and linear regression are as follows

$$\text{Univariate Corr coeff: } R_{XY} = \frac{n\sum XY - (\sum X)(\sum Y)}{\sqrt{n\sum X^2 - (\sum X)^2}\sqrt{n\sum Y^2 - (\sum Y)^2}} \quad (1)$$

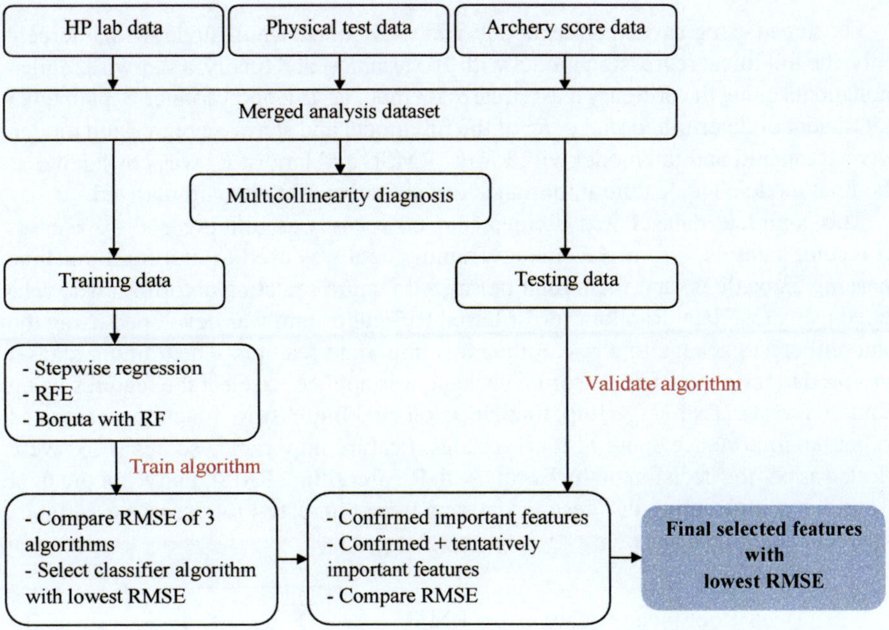

Fig. 1 Analytical approach for feature selection

$$\text{Partial Corr coeff: } R_{XY \cdot Z} = \frac{R_{XY} - R_{XZ} R_{ZY}}{\sqrt{1 - R_{XZ}^2}\sqrt{1 - R_{ZY}^2}} \tag{2}$$

$$\text{Coeff of determination: } R^2 = \frac{\sum \left(\widehat{Y} - \bar{Y}\right)^2}{\sum \left(Y - \bar{Y}\right)^2} \tag{3}$$

$$\text{Variance Inflation Factor: VIF} = \frac{1}{1 - R^2} \tag{4}$$

$$\text{Full regression model: } Y = \alpha + \beta X \tag{5}$$

where

X: data matrix of predictor variables X_1, X_2, …, X_n, i.e. HP lab and physical fitness test variables,

Y: data vector of outcome variable i.e. archery score,

$\bar{Y}$: mean of archery score,

$\widehat{Y}$: predicted values of archery score,

α: intercept of full regression model, and

β: regression coefficient matrix of full regression model.

For stepwise regression, three models were developed: a null model with intercept only, the full linear regression model with 15 variables, and finally, a stepwise regression model using the ordinary least squares method. RMSE and variance explained or coefficient of determination, i.e. R^2 of the full model and stepwise regression model, were compared and the model with lowest RMSE and largest R^2 was considered as the final model. The feature importance of the final model was summarized.

The complete dataset was divided into 60% cases as training and 40% cases as testing datasets with replacement. Training data was used to first train machine learning algorithms, and then, final optimized feature selection algorithm was validated using test data. Random forest-based RFE algorithm was developed using the randomForest package to arrive at the most important features. The training dataset was used to train the RFE. A Boruta package was applied to select the features using random forest classifier. getImp, finalDecision, and ImpHistory functions were used to get the importance score of the variables. Feature importance scores (FIS) were plotted as per the decision of the Boruta with RF algorithm. RMSE and R^2 of the final algorithm with training and testing datasets were compared to assess the variance between the two datasets.

$$\text{Root mean square error: RMSE} = \sqrt{\frac{1}{n}\sum (Y - \hat{Y})^2} \tag{6}$$

3 Results

The mean age of male archers was 23.2 years (SD = 6.12), height was 169.2 cm (SD = 5.79), and weight was 67.3 kg (SD = 8.19). The mean archery score was 653.1 (SD = 25.67). Summary statistics are shown in Table 1.

In univariate correlation analysis age ($p = 0.007$), weight ($p < 0.0001$), fat ($p < 0.0001$), agility ($p = 0.001$), and acceleration speed ($p < 0.0001$) were significantly positively correlated with the archery score while sub-maximal/maximum oxygen intake ($p < 0.0001$), both average and peak powers ($p < 0.0001$), and total body water ($p < 0.0001$) were significantly negatively correlated with the performance. A moderate correlation was observed among physiological and physical fitness variables. Univariate correlations are shown in Fig. 2.

The size of the circles indicated the relationship strength between the two variables. Circles with different shades of blue colour showed positive relationship and the circles in red colour shades represented a negative relationship. Blurred circles indicated statistically non-significant relationship, while dark circles indicated significant relationship.

Sit-ups ($p = 0.046$) were significantly positive and maximum bow hold time was negatively ($p < 0.0001$) partially correlated with the performance. Severe multicollinearity was observed in anthropometric variables fat, weight, and height, while high multicollinearity was observed in power and total body water (Table 1).

Table 1 Summary statistics and multicollinearity diagnostics of fitness and performance

Features	Summary statistics	Correlation of archery score with other features				VIF
	Mean (SD)	Univariate	*p*-value	Partial	*p*-value	
Age (years)	23.2 (6.12)	0.16	**0.007**	−0.10	0.103	**3.055**
Height (cm)	169.2 (5.79)	0.02	0.764	0.01	0.930	**7.917**
Weight (kg)	67.3 (8.19)	0.24	**<0.0001**	0.04	0.511	**22.299**
VO_2 ($mL.min^{-1}.kg^{-1}$)	44.1 (8.76)	−0.29	**<0.0001**	−0.11	0.079	1.891
Peak power ($W.kg^{-1}$)	9.8 (1.08)	−0.25	**<0.0001**	−0.02	0.768	**4.443**
Average power ($W.kg^{-1}$)	7.4 (0.65)	−0.30	**<0.0001**	−0.06	0.363	**5.820**
Body fat (%)	11.3 (2.94)	0.28	**<0.0001**	0.02	0.763	**23.634**
Total body water (%)	60.2 (3.94)	−0.27	**<0.0001**	−0.02	0.733	**3.311**
Agility (sec)	10 (0.41)	0.20	**0.001**	−0.02	0.707	2.459
Standing start 30 m sprint (sec)	4.5 (0.23)	0.27	**<0.0001**	0.10	0.120	2.706
Push-ups in 1 min (number)	55 (9.71)	−0.10	0.109	0.06	0.357	1.417
Sit-ups in 1 min (number)	54.6 (8.69)	−0.05	0.409	0.12	**0.046**	1.513
Flexibility (cm)	17.9 (5.73)	0.01	0.843	0.08	0.194	1.399
Maximum bow hold time (sec)	68 (19.75)	−0.21	**0.001**	−0.27	**< 0.0001**	1.209
VO_2max ($mL.min^{-1}.kg^{-1}$)	51.8 (4.77)	−0.29	**< 0.0001**	−0.10	0.107	2.552
Archery score	653.1 (25.67)					

SD = Standard Deviation, VIF = Variance Inflation Factor
The p-value $\leq$ 0.05 and VIF $\geq$ 3 are indicated in bold

In linear regression full model, only two variables, namely sit-ups ($p = 0.046$) and maximum bow hold time ($p < 0.0001$), were significant predictors of the archery performance. In stepwise regression, weight ($p = 0.005$) and maximum bow hold time ($p < 0.0001$) were significant predictors of the performance, while sub-maximal VO_2 ($p = 0.072$) and average power ($p = 0.056$) were retained in the model (Table 2).

In feature selection analysis, maximum bow hold timing is the most important feature selected by all three feature selection methods viz. stepwise regression, RFE, and Boruta algorithm with random forest using Boruta package. Weight, VO_2, and average power were also common important features in all three algorithms. Boruta algorithm with random forest confirmed eight variables namely maxbowhold, age,

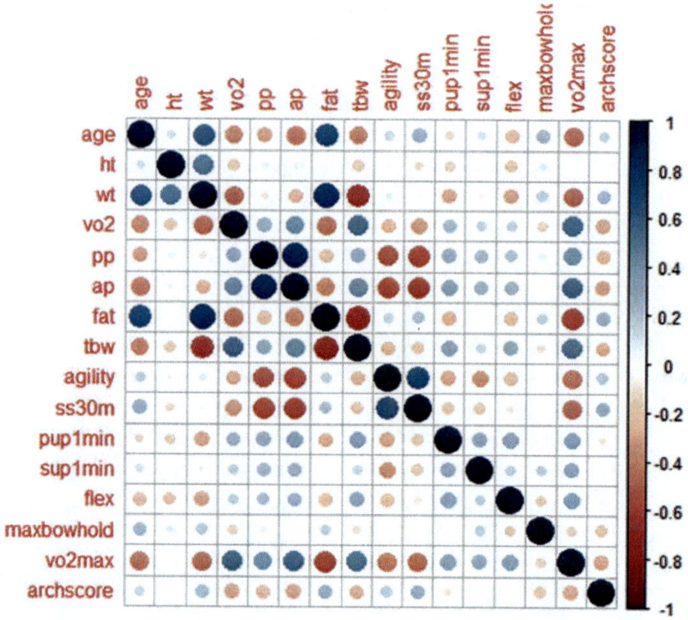

Fig. 2 Correlogram of fitness and the archery performance. ap = average power, flex = flexibility, ht = height, maxbowhold = maximum bow hold time, pp = peak power, pup1min = push-ups in 1 min, ss30m = standing start 30 m, sup1min = sit-ups in 1 min, tbw = total body water, wt = weight

and weight as 100% important, followed by VO_2 (FIS = 96.97%), fat (FIS = 95.96%), sit-ups (FIS = 96.97%), average power (FIS = 87.89%), and peak power (FIS = 80.89%). Two features ss30m and VO_2max were tentatively important with FIS about 45% and rejected the remaining five unimportant features, namely height (FIS = 14.14%), total body water (FIS = 2.02%), agility (FIS = 2.02%), flexibility, and push-ups (FIS = 0%).

A box plot of feature importance as per Boruta algorithm and the decision for each feature is shown in Fig. 3. Green colour boxes indicate that the features are confirmed important by the algorithm, yellow boxes show that features are tentatively important, and the features with red colour boxes rejected, i.e. they are not important for archery performance.

The RMSE for stepwise regression was 21.262 with four variables retained in the model using a training dataset. The RMSE for RFE with 10 and 15 features was 19.59 and 19.45, respectively, in the training dataset. The RMSE in random forest algorithm using the training dataset was 8.417 for all 15 features and 9.003 for 8 confirmed important features. The RMSE was 19.45 for 15 features in RFE, while it was 9.003 for eight features in random forest which was much lower than the RFE. It was observed that the Boruta algorithm with random forest is better than the classical RFE since it reduced the features effectively to eight features out of 15 features.

Table 2 Feature importance by feature selection methods

Features	Stepwise regression	RFE	Boruta with random forest		
	Mean FIS	Mean FIS	Mean FIS	% FIS	Decision
Maximum bow hold time (sec)	4.92	19.63	16.76	**100.00**	**Confirmed**
Age (years)		10.30	9.19	**100.00**	**Confirmed**
Weight (kg)	2.85	9.35	8.59	**100.00**	**Confirmed**
Body fat (%)		7.12	6.82	**95.96**	**Confirmed**
VO_2 ($mL.min^{-1}.kg^{-1}$)	1.81	6.72	6.27	**96.97**	**Confirmed**
Sit-ups in 1 min (Number)		5.85	5.92	**92.93**	**Confirmed**
Average power ($W.kg^{-1}$)	1.93	5.57	5.14	**87.89**	**Confirmed**
Peak power ($W.kg^{-1}$)		4.48	4.57	**80.81**	**Confirmed**
Standing start 30 m sprint (sec)		3.06	2.56	43.43	Tentative
VO_2max ($mL.min^{-1}.kg^{-1}$)		3.13	2.55	44.44	Tentative
Height (cm)		2.69	1.89	14.14	Rejected
Total body water (%)		1.65	1.22	2.02	Rejected
Agility (sec)		1.61	0.93	2.02	Rejected
Flexibility (cm)		2.00	0.63	0.00	Rejected
Push-ups in 1 min (Number)		−1.26	−0.64	0.00	Rejected

% = percentage, FIS = variable importance score, RFE = recursive feature elimination
% FIS and decision columns with bold shows the confirmed important variables

Trained random forest feature selection algorithm was validated on the test data using two sub-sets of important features, first set with confirmed eight features and the second set with 10 features including tentative as well as confirmed features. The RMSE of random forest with eight features was 24.534 and 10 features were 24.002 in testing data. So, the RMSE with 10 features random forest was slightly less than the 8 features.

4 Discussion

Our analysis findings are similar to the research study conducted in Malaysian archers [27]. The authors used PCA method, i.e. principal component analysis to reduce fitness features/variables in two factors. Predicted VO_2max, 1-min sit-ups, BMI and age were important variables and grouped in factor 1 which explained about 43% variability. The second factor explained 21% variability and had 2 fitness variables

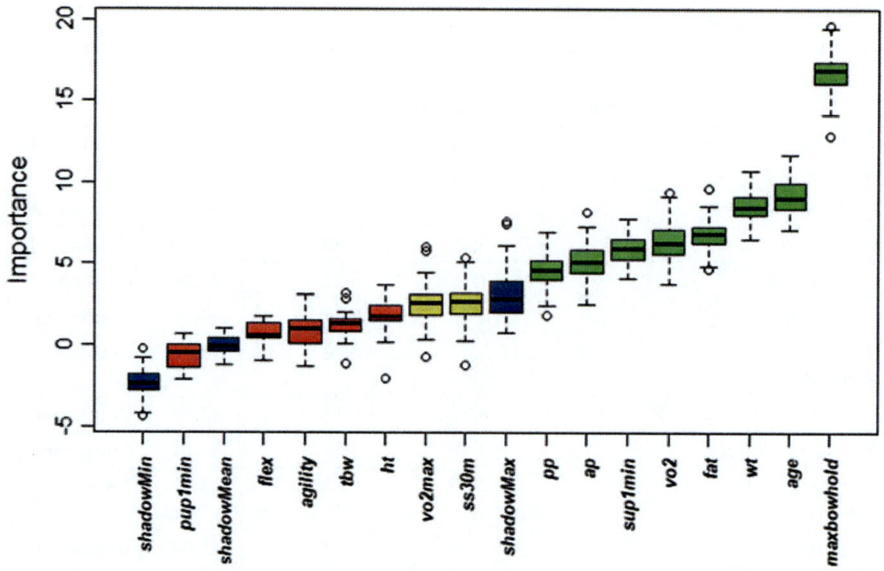

Fig. 3 Feature importance by Boruta with random forest algorithm ap = average power, flex = flexibility, ht = height, maxbowhold = maximum bow hold time, pp = peak power, pup1min = push-ups in 1 min, ss30m = standing start 30 m, sup1min = sit-ups in 1 min, tbw = total body water, wt = weight

with higher factor loadings namely V-sit and reach and max push-ups. However, PCA retained all the variables and just reduced/arrange them into the mutually independent factors, where variables are highly correlated to each other within the factor. Random forest classifier from Boruta package identifies confirmed important variables and thus reduces the variables in the dataset which are not important. This is an advantage of the machine learning algorithm over the PCA feature extraction method.

It is also observed that many fitness variables were weekly correlated with the archery performance, and only four variables were important in stepwise regression with higher prediction error. However, in machine learning algorithms, eight variables were important. The core abdominal strength (Sit-ups), is one of the most important variable by both RFE and random forest, which was not an important variable in correlation/stepwise regression. Core abdominal strength is very important for successful archery performance since an archer needs to stand for a long period during the competition and hold the heavy weighted bow. Hence, traditional filter-based feature selection methods like correlation and linear/stepwise regression have limitations and failed when there is very high multicollinearity and complex relationship present in various physiological and physical fitness variables.

An acceleration speed and maximum oxygen intake, i.e. cardiovascular endurance, are tentatively important variables. These two variables are significantly moderately correlated with the peak power ($r = 0.60, p < 0.0001$) and VO_2 ($r = 0.54$, $p < 0.0001$) respectively. This finding suggests that standing start 30 m and Canadian stage test of VO_2max could be used as an alternative fitness test for Wingate lower body power test and the Astrand test, respectively, in consultation with an Exercise Physiologist and Sports Scientists in resource-limited settings where highly sophisticated human performance laboratory set-up is not available. Also, the difference between the prediction errors of the Boruta algorithm with random forest including eight features and 10 features was very low and proves that eight confirmed important features are sufficient to predict the future performance of archers optimally.

Variable importance score in RFE and Boruta with random forest were similar, but RFE retained all 15 variables in the final trained algorithm, while Boruta with random forest rejected five variables and thus reduced the variable set with the much lower error. Hence, Boruta with random forest algorithm is better than the wrapper-based conventional RFE method.

5 Conclusion

Machine learning Boruta algorithm gives a winning solution to tackle more complex fitness data than the traditional statistical feature selection methods. Maximum bow hold time, i.e. handgrip strength, is the most important feature followed by sub-maximal endurance, peak, and average power, core abdominal strength, fat mass, weight, and age. These important features are helpful to the exercise physiologist, sports scientists, and coaches for monitoring sub-set of fitness components and evaluate the athletes' progress in real time, which will be further helpful to plan effective training plans catering to physiological and physical demands of the individual archer. The analytical approach used in this paper for feature selection in archery could also be applied for feature selection of other individual sports like swimming, diving, weight lifting, skating, track and field events, etc., where the performance indicator is a quantitative variable.

Acknowledgements Brig PS Cheema, VSM, Ex Commandant, Army Sports Institute. Dr. Geetanjali Bhide, Sports Nutritionist, Army Rowing Node, for reviewing the manuscript. All coaches, athletes, and staff at the Army Sports Institute.

References

1. Bangsbo J, Mohr M, Poulsen A, Perez-Gomez J, Krustrup P (2006) Training and testing the elite athletes. J Exerc Sci Fit 4:1–14
2. Bruzas V, Stasiulis A, Cepulenas A, Mockus P, Statkeviciene B, Subacius V (2014) Aerobic capacity is correlated with the ranking of boxers. Percept Mot Skills 119:50–58. https://doi.org/10.2466/30.29.PMS.119c12z9
3. Warnick JE, Warnick K (2007) Specification of variables predictive of victories in the sport of boxing. Percept Mot Skills 105:153–158. https://doi.org/10.2466/PMS.105.1.153-158

4. Saeys Y, Inza I, Larrañaga P (2007) A review of feature selection techniques in bioinformatics. https://doi.org/10.1093/bioinformatics/btm344
5. Slinker BK, Glantz SA (1985) Multiple regression for physiological data analysis: the problem of multicollinearity. https://doi.org/10.1152/ajpregu.1985.249.1.r1
6. Midi H, Sarkar SK, Rana S (2010) Collinearity diagnostics of binary logistic regression model. J Interdiscip Math 13:253–267. https://doi.org/10.1080/09720502.2010.10700699
7. Draper N, Smith H (1998) Applied regression analysis. Wiley
8. Granitto PM, Furlanello C, Biasioli F, Gasperi F (2006) Recursive feature elimination with random forest for PTR-MS analysis of agroindustrial products. Chemom Intell Lab Syst 83:83–90. https://doi.org/10.1016/j.chemolab.2006.01.007
9. Breiman L (2001) Random forests. Mach Learn 45:5–32. https://doi.org/10.1023/A:1010933404324
10. Kursa MB, Rudnicki WR (2010) Feature selection with the boruta package
11. Silva AJ, Costa AM, Oliveira PM, Reis VM, Saavedra J, Perl J, Rouboa A, Marinho DA (2007) The use of neural network technology to model swimming performance. J Sport Sci Med 6:117–125
12. McCabe A, Trevathan J (2008) Artificial intelligence in sports prediction. In: Proceedings—international conference on information technology: new generations, ITNG, pp. 1194–1197. https://doi.org/10.1109/ITNG.2008.203
13. Zhao Y, Yin F, Gunnarsson F, Hultkratz F, Fagerlind J (2016) Gaussian processes for flow modeling and prediction of positioned trajectories evaluated with sports data. In 2016 19th international conference on information fusion (FUSION), IEEE, pp. 1461–1468
14. Constantinou A, Fenton N (2017) Towards smart-data: Improving predictive accuracy in long-term football team performance. Knowl-Based Syst 124:93–104. https://doi.org/10.1016/j.knosys.2017.03.005
15. Dubbs A (2018) Statistics-free sports prediction. Model Assist Stat Appl 13:173–181. https://doi.org/10.3233/MAS-180428
16. James LP, Robertson S, Haff GG, Beckman EM, Kelly VG (2017) Identifing the performance characteristics of a winning outcome in elite mixed martial arts competition. J Sci Med Sport 20:296–301. https://doi.org/10.1016/j.jsams.2016.08.001
17. Kataoka Y, Gray P (2019) Real-time power performance prediction in tour de france. In: Lecture notes in computer science (including subseries lecture notes in artificial intelligence and lecture notes in bioinformatics) (pp. 121–130). Springer. https://doi.org/10.1007/978-3-030-17274-9_10
18. Faber IR, Pion J, Munivrana G, Faber NR, Nijhuis-Van der Sanden MWG (2018) Does a perceptuomotor skills assessment have added value to detect talent for table tennis in primary school children? J Sports Sci 36, 2716–2723. https://doi.org/10.1080/02640414.2017.1316865
19. Maszczyk A, Zajac A, Ryguła I (2011) A neural network model approach to athlete selection. Sport Eng 13:83–93. https://doi.org/10.1007/s12283-010-0055-y
20. Iyer SR, Sharda R (2009) Prediction of athletes performance using neural networks: an application in cricket team selection. Expert Syst Appl 36:5510–5522. https://doi.org/10.1016/j.eswa.2008.06.088
21. Taha Z, Musa RM, Majeed APA, Alim MM, Abdullah MR (2018) The identification of high potential archers based on fitness and motor ability variables: a support vector machine approach. Hum Mov Sci 57, 184–193. https://doi.org/10.1016/j.humov.2017.12.008
22. Musa RM, Taha Z, Majeed APA, Abdullah MR (2019) Machine learning in sports identifying potential archers. Springer
23. Rulebook|World Archery. https://worldarchery.org/rulebook.
24. Gadbury GL, Coffey CS, Allison DB (2003) Modern statistical methods for handling missing repeated measurements in obesity trial data: beyond LOCF. https://doi.org/10.1046/j.1467-789X.2003.00109.x
25. Lane P (2008) Handling drop-out in longitudinal clinical trials: a comparison of the LOCF and MMRM approaches. Pharm Stat 7:93–106. https://doi.org/10.1002/pst.267

26. Kim S (2015) PPCOR: an R package for a fast calculation to semi-partial correlation coefficients. Commun Stat Appl Methods 22:665–674. https://doi.org/10.5351/csam.2015.22.6.665
27. Muazu Musa R, Abdullah MR, Bisyri A, Maliki HM, Kosni NA, Musa RM, Haque M (2016) The application of principal components analysis to recognize essential physical fitness components among youth development archers of Terengganu, Malaysia. Artic Indian J Sci Technol 9. https://doi.org/10.17485/ijst/2016/v9i44/97045.

Skin Lesion Classification Using Deep Learning

Aditya Bhardwaj and Priti P. Rege

Abstract Skin cancer is a common disease and considered to be one of the most prevalent forms of cancer found in humans. Over the years various imaging techniques have shown improvement and reliability in diagnosis process of Skin Cancer. However, quite a few challenges are being faced in generating reliable and well-timed results as adoption of clinical computer aided systems is still limited. With the recent emergence of learning algorithms and its application in computer vision suggests a need for combination of sufficient clinical expertise and systems to achieve better results. Here we attempt to bridge the gap by mining collective knowledge contained in current Deep Learning Techniques to discover underlying principles for designing a neural network for skin disease classification. The solution is based upon merging of top-N performing models used as a feature extractor and a SVM to facilitate classification of diseases. Final model gave 86% accuracy on ISIC 2019 dataset along with high precision and recall values of 0.8 and 0.6, respectively.

Keywords Deep learning · CNN · Neural network

1 Introduction

Melanoma is currently one of the most dangerous types of cancer. The World Health Organization (WHO) indicates that Skin diseases are among the most common of all human health afflictions and affect almost 900 million people in the world at any time. Five common conditions account for over 80% of all skin diseases [1]. Several skin diseases are associated with long-term disfigurement, disability and stigma. The ability to diagnose skin diseases is often made at the late stages of disease development due to late medical attendances and as a result the chances of

A. Bhardwaj (✉)
Department of Technology, Savitribai Phule Pune University, Pune, India
e-mail: adityabhartu@gmail.com

P. P. Rege
College of Engineering, Pune, India
e-mail: ppr.extc@coep.ac.in

© Springer Nature Singapore Pte Ltd. 2021

S. N. Merchant et al. (eds.), *Advances in Signal and Data Processing*,
Lecture Notes in Electrical Engineering 703,
https://doi.org/10.1007/978-981-15-8391-9_42

survival are falling short. The traditional system is based upon manual inspection of dermatoscopic images by experts by using certain imaging tools [2]. The experience of the expert dictates the accuracy of the diagnosis process and also the timeline of detection of the skin lesion.

Dermatoscopy is regarded as a technique in skin cancer screening which provides a higher diagnostic accuracy than the unaided eye. Prior literature infers that Dermascopy when used by expert field doctors improved the diagnostic accuracy in comparison to naïve photography. Without computer-based assistance, the clinical accuracy of skin disease detection especially Melanoma and malignant forms is around 65–75%. Use of dermoscopy helps to increase the accuracy but still the differences between melanoma and benign types are still subtle. One should note that treatment for these diseases require different procedures. If diagnosis is incorrect or delayed it may be fatal. Early detection is imperative and hence gives rise to develop systems that can detect and identify such diseases at an early stage which can then in turn help to cure and save funds as it is a costly process. It is therefore critical to have timely accurate diagnosis. For reasons suggested above a trained intelligent based system can assist physicians to detect and identify skin diseases [3, 4]. In this particular work, we are interested to detect certain kind of malignant diseases, especially melanoma [5].

In this paper, a method for skin lesion classification using deep learning and computer vision is proposed, implemented and successfully test against a public challenge and dataset hosted by ISIC [6]. Besides it also presents the design and evaluation of models and features for objectively detecting diseases by applying deep feature extraction. It uses a CNN (Convolutional Neural Network) which comprises of stacked layers comprising of convolutional layers, pooling layers, rectified linear units, batch normalizers along with a decision layer for final output [7]. The CNN till now is considered the most prevalent architecture in several applications especially in image classification process. The approach involved (1) implementation and analyzing of different CNN based architectures to sort out best performing models, (2) incorporating multiple models to enhance overall classification ability.

2 Literature Survey

Previous literature show that prior efforts to apply machine intelligence to skin lesion detection and classification. Recent years a lot of image processing techniques were used along with Machine Learning Algorithms for extraction of features [8]. These techniques were mainly based upon Gabor, HSV filters [9, 10]. Over the years, significant advancement in GPU computations and hardware has been made along with dataset collection by ImageNet and Kaggle platforms.

Approach proposed by Barata et.al [11] for detection of melanoma is based on global and local features. The global form uses segmentation, histograms and other filters to help extract features such as texture, shape, color. These features are then feed forwarded to a binary classifier. He concludes that color features outperformed texture features. Coella et.al [12] utilizes handcoded feature extraction techniques.

With the emergence of Deep Learning, it has embedded development in medical imaging and proving to be of much better assistance. Kawahara et.al [13] suggests to idea of using pre-trained ConvNets as feature extractor. With the use of filters it then classifies more than three classes of non-dermoscopic images.

Liao [14] attempted to construct universal diseases classifier by applying transfer learning on a deep CNN. Esteva et.al [15] classified around 100,000 skin lesion dataset using GoogleNet. Another work suggests using ResNets in parallel as part of modality fusion.

Adria Rmoero Lopez et.al [16] uses a VGG net along with transfer learning on skin images which achieves an average accuracy of 80%

Jordan yap et.al [17] discusses an idea of Multimodal models which incorporates different resource inputs to make a decision. In this particular paper, dermascopic images as well as macroscopic images were used in parallel and later their results were combined. It achieved an average accuracy of 82% but with greater loss.

Xinyuan Zhang et.al [18] classified four skin diseases by using deep learning algorithms. Dataset was dermatoscopic level images. A hierarchical structure along with domain knowledge. An accuracy of 85% on 1067 images was the result of the experimentation and the hierarchical structure helped to implement a better computer aided support system.

3 Data

The input data comprises of dermoscopic images in jpeg format. These images are part of the ISIC 2019 challenge [6] and all come from HAM1000 dataset [19]. The response data is a csv file consists of a binary classification for each of the 8 disease states. The diagnosis ground truth was established by one of the following methods namely Histopathology, Reflectance confocal microscopy, Lesion did not change during digital dermatoscopic follow up over two years with at least three images. Consensus of at least three expert dermatologists from a single image [6] (Fig. 1 and Table 1).

4 Proposed System

The proposed methods first insist on experimenting with various models and analyze them to find out a set of best-performing ones as we know that Deep Learning models are data specific. This involved studying different Deep Learning Architectures and developing one best suited for AI field. In short, we followed an iterative process (Fig. 2) to gain better insights about each model. Evaluation of every model was done primarily on accuracy and loss curves. Precision, recall and $F1$ score was compared

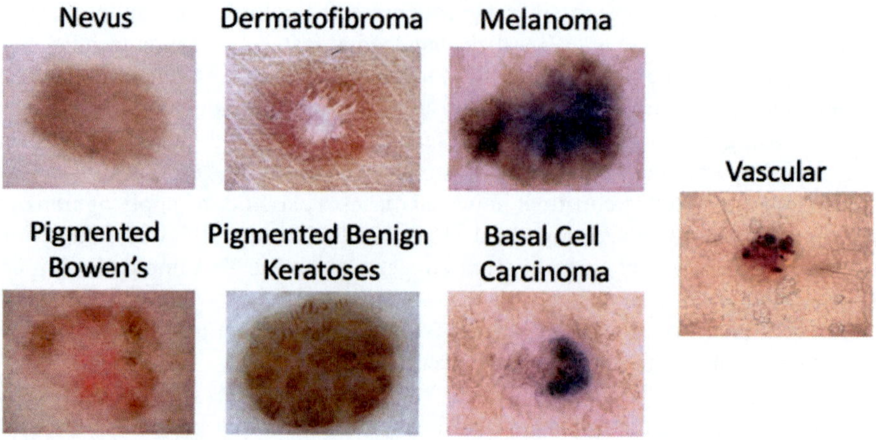

Fig. 1 ISIC 2019 dataset

Table 1 ISIC HAM1000 dataset

Disease	No of images
AK (actinic keratosis)	867
BCC (basal cell carcinoma)	3,323
BKL (benign keratosis)	2,624
DF (dermatofibroma)	239
MEL (melanoma)	4,522
NV (melanocytic nevus)	12,875
SCC (squamous cell carcinoma)	628
VASC (vascular lesion)	253

to get more insights on how good the model is generalizing itself, i.e. performance on an unseen data. Various hyper-parameter tuning methods were applied according to the evaluation parameters [20]. Our approach also consisted of visualizing feature maps and activation maps at each layer to understand and assess how our model processes images at every layer [21]. This also directed the study to venture into Neural Network Design patterns and cater way outside the black box testing [22].

4.1 Model

A final algorithm was developed based on an 'Average of *N*-models' which uses InceptionV along with Deep-CCN and MobileNet. Layers giving insufficient or redundant information were freezed during training process. Multiple networks were trained using additive strategy and corresponding adjustments were made according

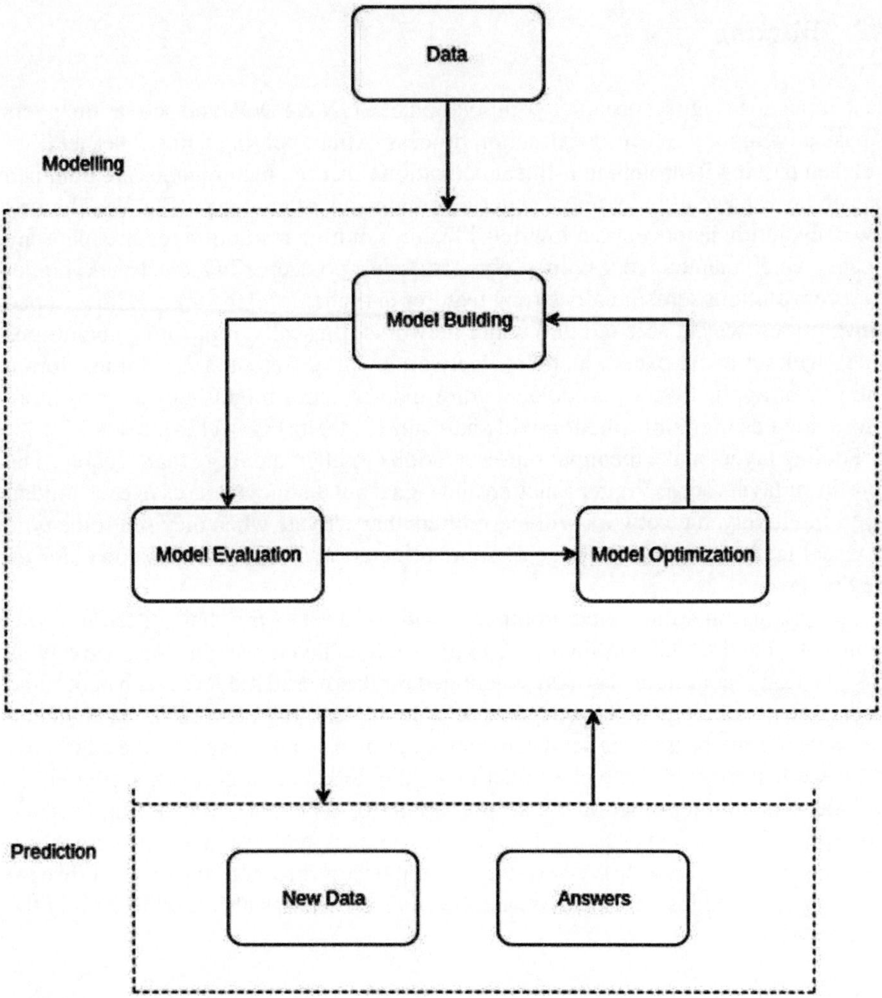

Fig. 2 Basic methodology overview

to the samples. A softmax layer was adjusted according to the number of inputs as used from our datasets. Second, experiments were also conducted by replacing the last softmax function with a Support Vector Machine (SVM) which acts like a classifier by introducing a planar surface between the set of data points in a 2-D scenario [23].

4.2 Working

Each input after it goes through a combination of CONV, POOL and activation layers is basically under the feature extraction process. After applying a linear kernel filter and then passing it through non-linear activation function, feature maps are obtained at each layer basically it extracts generic components of an image (Fig. 3). The first few convolution layers extract low-level features or high resolution feature maps are created which map out edges, lines, corners. As we go deeper in the network, higher level convolutions semantically strong features or higher level features. With grouped convolutions, we are able to build wider networks. Basically, replicate convolutions filters without using excess memory. Networks with different sizes of convolution filter when applied on a particular feature map or input might help us gain more information as more information will share among them(Fig. 4) [24].

Pooling layers make the input representations smaller and more manageable. The non-linear layers act as trigger function and signal out distinct features in each hidden layer. Intuitively, the network will learn filters that activate when they see some type of visual features such as an edge of some orientation or a blotch of some color on the first layer.

The prediction values were compared with actual labels. Their difference was being calculated which was then used to update final layers weights, respectively. A loss function is continuously being calculated for the overall model at each epoch and its derivative is being back propagated to tune the learning process. The calculated parameters from the training set where then examined by applying it on the validation set to see if it fits well. The classification results by a neural network are based on probability summing function where probability of each adds up to 1. On the same intuition, a softmax layer is used which on the back end uses the same math but results is in form of a vector where the 1 in the vector directs to the class of disease it belongs to. Besides, the softmax, and SVM classifier was also used to predict the final decision.

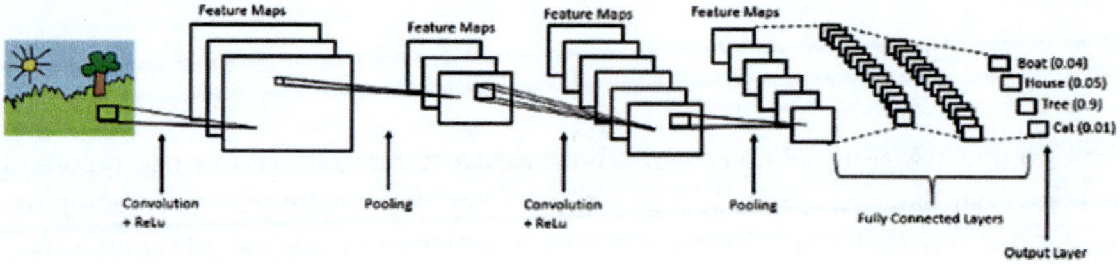

Fig. 3 Feature maps in CNN

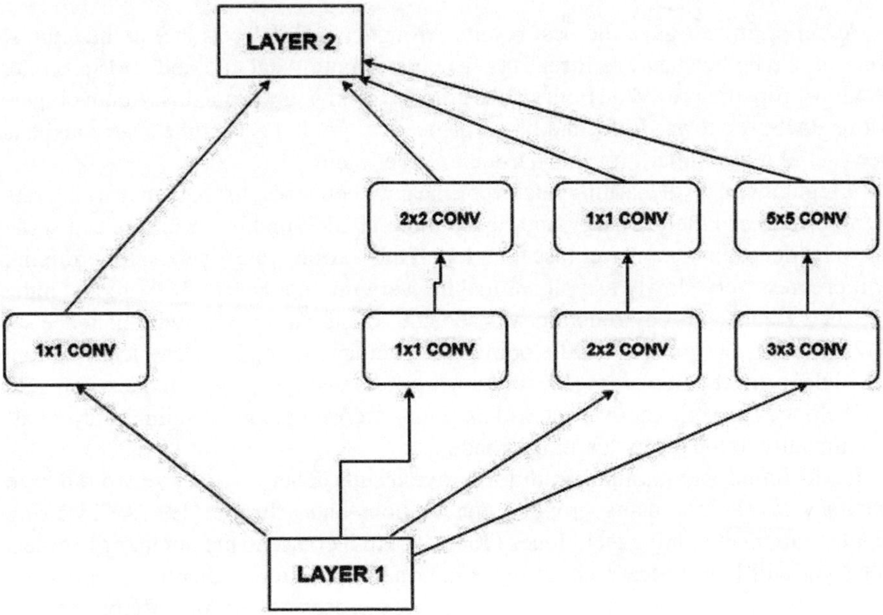

Fig. 4 Grouped convolutions

5 Conduct

Experiments were run using datasets A and B separately. Data set B included a binary set of all the categories while A comprised of 7 categories. Each data set was split into a 80:20 with the larger share towards the training set and lower share for the validation set. Our parallel experiments included one v/s all approach to specifically tackle melanoma classification against all. For the later one, we replace our classification layer to SVM. Average training was done for 50 epochs. Tensorflow and Keras were the frameworks used. Implementation was done on Google Colab, an open-source cloud platform to run ML-based applications.

Parameters:Accuracy, Loss, Loss Function, Optimizing Algorithm, Learning Rate, Batch Size, Hidden Layers.

6 Results and Discussion

Loss function and loss metrics can be considered as the primary parameter along with accuracy to test the models on a given dataset. For binary classification, 'binary-cross entropy' proved to be most effective and 'categorical-cross entropy' for more than 2 classes. A deviation of +3% in accuracy was seen when loss function was switched to binary-cross entropy.

Adam optimizer gave the best results throughout all datasets and architectures. Reason maybe be stated as it realizes benefits of both AdaGrad and RMSprop i.e. tackling properties based on both sparse gradients and noisy problems. Adam outperforms RMSprop towards the end of the problem as gradients become sparse and has been set as a benchmark for this particular experiment.

A regular update in learning rate over time gave better results. A high learning rate signifies higher kinetic energy and hence more chaotic making it difficult to settle down for deeper and narrower loss function. If the learning rate is set too low, training will progress very slowly as you are making very tiny updates to the weights in the network. Hence, a decay function was added to the learning cycle with patience set = 2. However, at times a 'saddle point' was encountered and a decay learning rate scheduler proved unrewarding. A solution suggests that jumping up the learning rate can help the network get over the saddle point inferring towards adding a 'cyclical' functionality to the decay learning scheduler.

It was found that parallelism did improve results. There was a rise from a base accuracy of 71–75% using grouped convolutions and a further 3% rise by using regularization and tuning techniques (Table 2). An increase in the number of kernels per layer will help to learn about more intermediate features, therefore increasing the number of channels in the next layer. This process of using different sets of convolution filter groups on same image is called as grouped convolution (Fig. 4). In short, create a deep network with some number of layers and then replicate it so that there is more than one pathway for convolutions on a single image. This particular idea can be first traced in Alex Net [25]. This concept can further be translated to Data parallelism and Model parallelism.

Among these, Inception Net performed pretty well on both the datasets while VGG was unable to learn anything from this data and generalize it. The concept of grouping and stacking of Convolution layers and subsequently pooling layers didn't work out. On the contrary, we can infer that kernel filters of different sizes in the same network can be further studied especially in medical imaging. One can note in Fig. 5c for Average of N models gave the best visualization and feature maps in comparison to Fig. 5a, b. The feature maps developed in (Fig. 5a, b, c) infers that texture is a more prominent feature than just color and shape. This too suggests that architecture of the Neural Network plays an important role to develop models.

Table 2 shows that the proposed system can produce high accuracy when applied to binary classification. The system was able to classify melanoma v/s others at a

Table 2 Data set A

Model	Accuracy (%)
CNN	70.8
VGG	63
AlexNet	75
AlexNet + fine tuning	**77.3**

Numbers in bold indicate best performance values

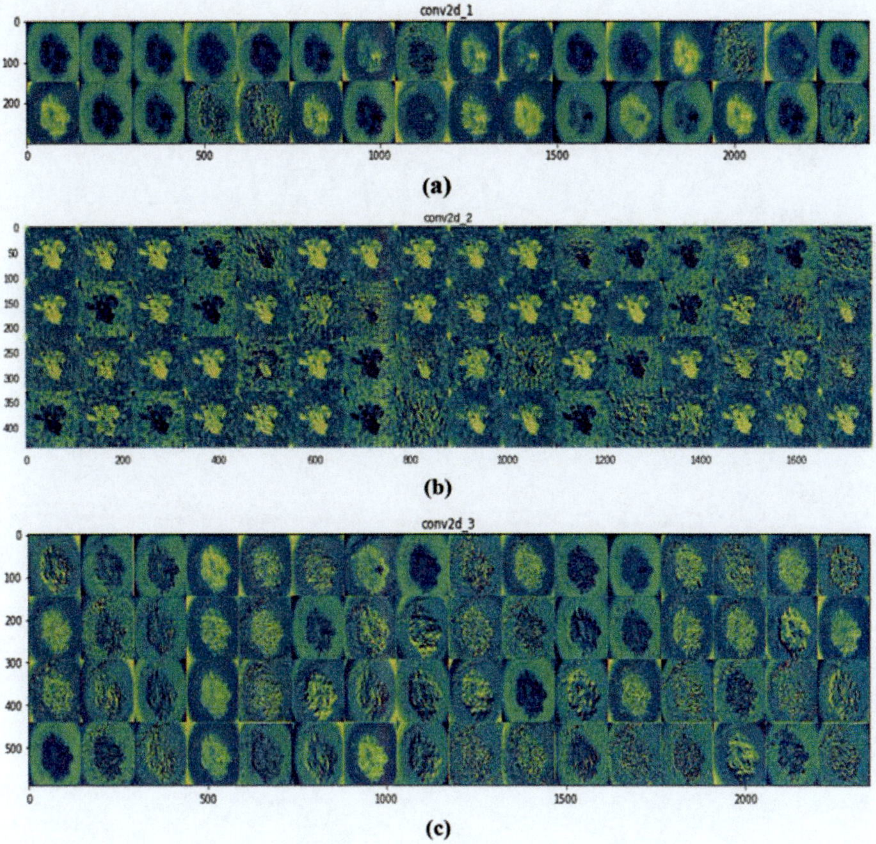

Fig. 5 Visualization of Layers: **a** inception net: **b** AlexNet: **c** Average of '*N* models'

better accuracy and precision, recall, $F1$ score. InceptionNet performed the best Table 2 and Fig. 6c on accuracy metrics but there was still a huge gap on precision and recall parameters. This suggested that model isn't able to generalize the knowledge which then got us to address the issue of imbalanced classes. The dataset was pretty imbalanced with NV having over 12,000 images along with minor representations of AK, DF, SVC ranging from 300–800. A skewed dataset didn't really help us with the precision and recall. A balanced dataset was tested and gave better results in P and R values [26]. Due to hardware constraints, data augmentation was not an efficient choice to address this issue. A comparison between Tables 2 and 3 also proposes that specific models were not able to capture the statistical characteristics of each diseases and failing to apprehend the complexity of dataset. Two class classification performed very well for the same reason.

The SVM classifier instead of the softmax performed way better Fig. 6e and Table 3 as it is a more 'local' based classifier and not as soft as the softmax which calculates probabilities for each class. SVM instantly helped with better precision, recall and $F1$

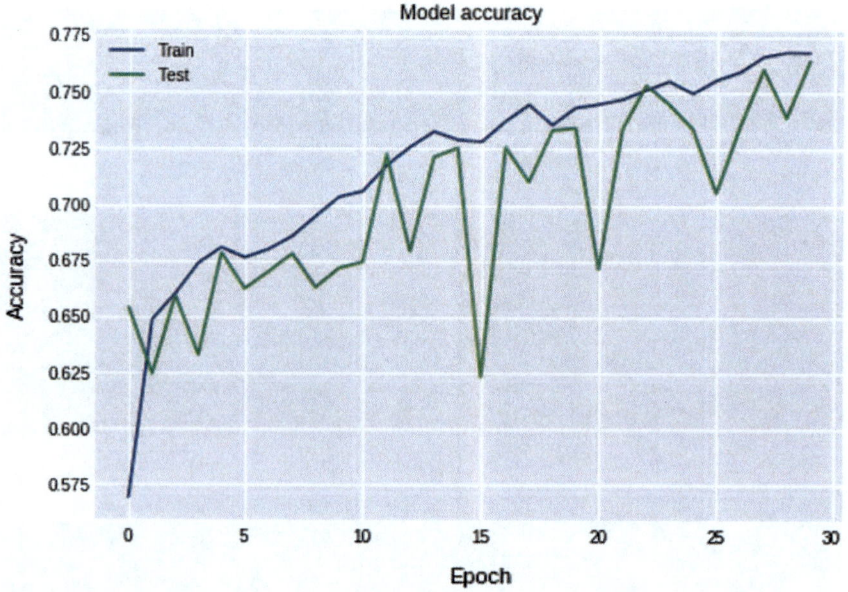

(a)

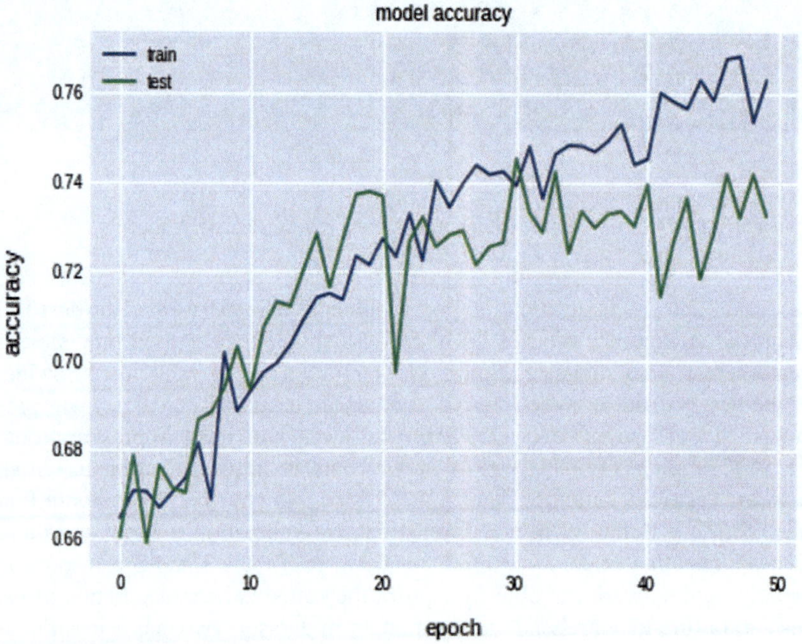

(b)

Fig. 6 Accuracy and loss plots over different models: **a** AlexNet: **b** mobile net: **c** inception net: **d** VGG: **e** average of '*N* models' + SVM: **f** loss graph of SVM

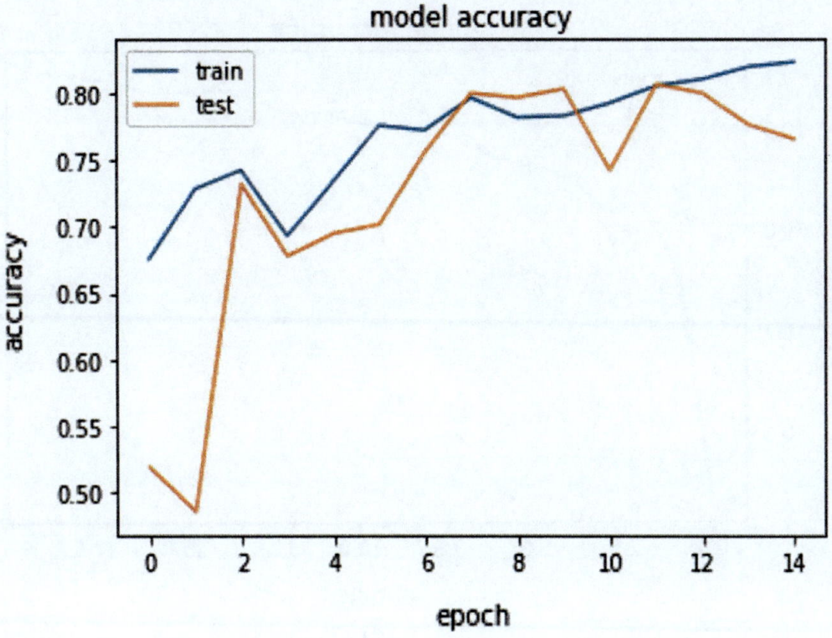

(c)

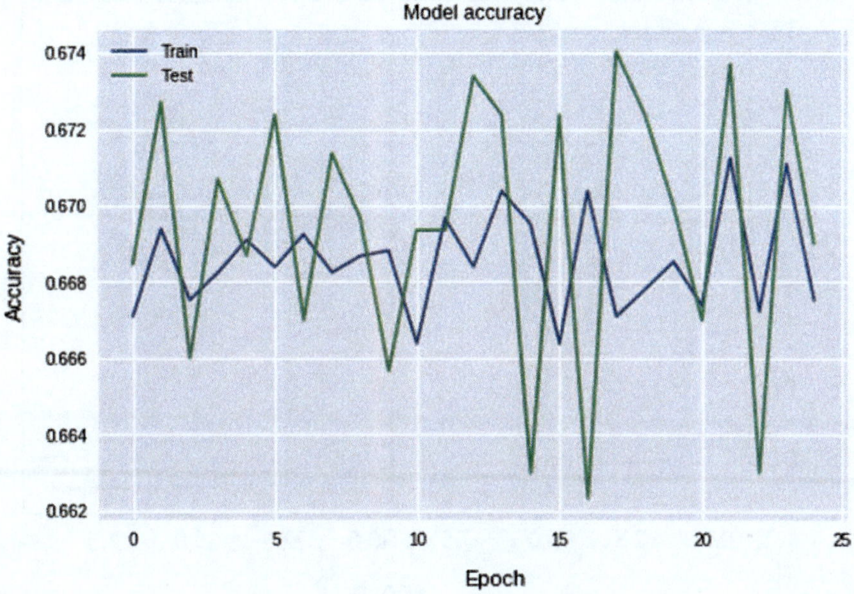

(d)

Fig. 6 (continued)

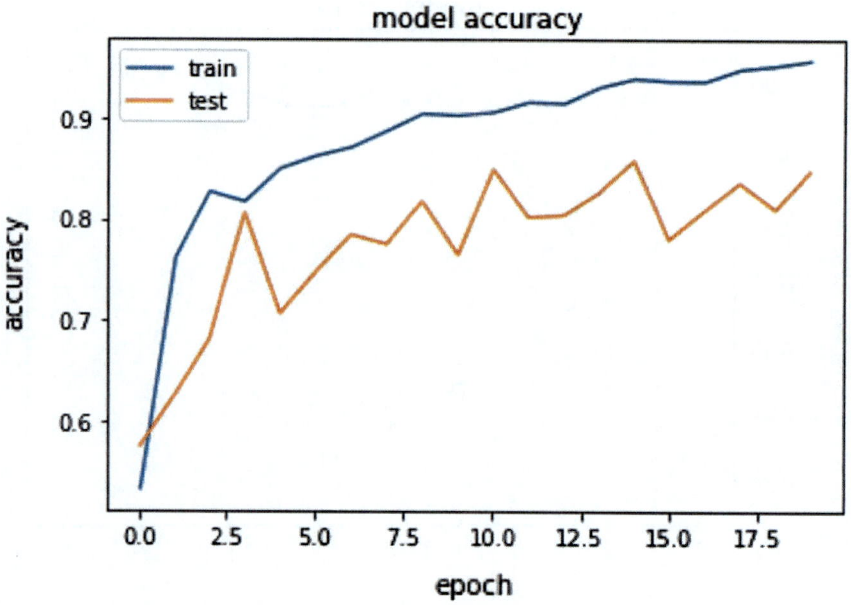

(e)

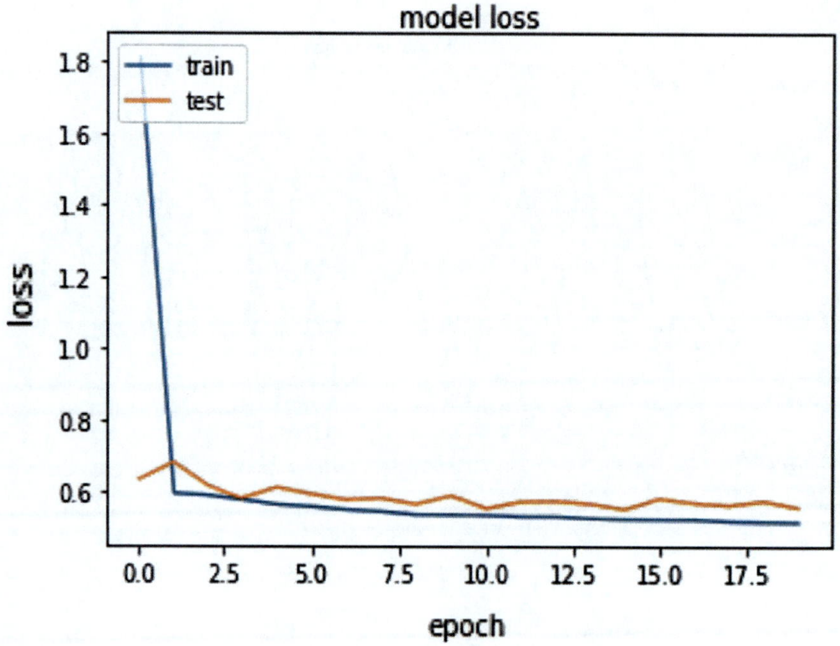

(f)

Fig. 6 (continued)

Table 3 Data set B (melanoma versus others)

Model	Accuracy (%)	Loss	Precision	Recall
MobileNet	77.6	0.55	0.4	0.3
MobileNet + fine tuning	84	0.5	0.5	0.6
Inception + fine tuning	**88**	0.43	0.5	0.4
Average + fine tuning	85.2	0.34	0.5	0.5
Average + SVM	**86.3**	**0.18**	**0.8**	**0.6**

Numbers in bold indicate best performance values

scores. The metrics were double-fold for binary classification. It is also important to note that on a ternary based classification the SVM performed better and gave balanced results on *P* and *R* values as well. We can conclude that SVM along with 'hinge-loss' as a classifier was able to handle data imbalance much better than other classifiers. SVM therefore encourages the correct class to have high scores and is hard with its classification methods on the other hand softmax has un-normalized log probabilities which interpret scores pretty softy and is 'never happy with' low differences in scores of classification.

A predefined class weight improved detection towards melanoma. Besides, it was also found that loading ImageNet Data weights helped the model to train faster [27]. Training from scratch took around 800 sec/epoch while pre-trained weights from ImageNet took around half of the usual time.

7 Conclusion and Future Work

In this paper, an enhanced model based on 'Average of N models' has been proposed while using SVM as the base classifier for decision making. More specifically the system works well on two-class classifier. Dataset B consisting of two classes Melanoma v/s others achieved an accuracy of 86% while showing drastic improvement on the P and R values. Grouped convolutions seem to learn data better as every filter group is learning a unique representation of data. It can be further investigated how Proliferate paths, Incremental Feature Construction, Summation Joining aspects of neural network design can be used to make more robust models for medical imaging applications. SVM as a classifier was able to tackle imbalance of dataset problem and gave better precision and recall values along with accuracy.

Using deep learning models to classify skin diseases would prove useful to doctors in reducing error and work in a capacity of a computer-based assistance. It will help further to carry on diagnosis on remote patients therefore improving accessibility and also reducing costs.

References

1. World Health Organization. https://www.who.int/
2. Hamblin MR, Avci P, Gupta GK (2016) Imaging in dermatology. Academic Press
3. Jerant AF, Johnson JT, Sheridan CD, Caffrey TJ (2000) Early detection and treatment of skin cancer. Am Fam Phys 62:381–382
4. Codella NCF, Gutman D, Celebi E, Helba B, Marchetti AM, Dusza WS, Kalloo A, Liopyris K, Mishra N, Kittler H, et al (2017) Skin lesion analysis toward melanoma detection: a challenge at the 2017 international symposium on biomedical imaging (ISBI), hosted by the international skin imaging collaboration (ISIC). arXiv:1710.05006
5. Sherin DA, Youseff SM, Aildy WM (December, 2016) Computer-aided model for skin diagnosis using deep learning. J Image Graph. 4(2)
6. The International Skin Imaging Collaboration. https://challenge2018.isic-archive.com/
7. He K, Zhang X, Ren S, Sun J (June, 2016) IEEE conference on computer vision and pattern recognition (CVPR), pp. 770–778
8. Chang Y, Stanley RJ, Moss RH, Van Stoecker W (2005) A systematic heuristic approach for feature selection for melanoma discrimination using clinical images. Skin Res Technol 11(3):165–178
9. Litjens G, Kooi T, Bejnordi BE (2018) A survey on deep learning in medical image analysis, Elsevier. EXD13222 (2018)
10. Litjens G, Kooi T, Bejnordi BE, et al (2017) A survey on deep learning in medical image analysis. Elsevier Survey Paper, Med Image Anal 42:60–88
11. Catarina B, et al (2014) Two systems for the detection of melanomas in dermoscopy images using texture and color features. IEEE Syst J 8:965–979
12. Codella N, Cai J, Abedini M, Garnavi R, Halpern A, Smith JR (2015) Deep learning, sparse coding, and svm for melanoma recognition in dermoscopy images. In: International workshop on machine learning in medical imaging, Springer, Cham, Switzerland, pp 118–126
13. Kawahara J, BenTaieb A, Hamarneh G Deep features to classify skin lesions. IEEE international symposium on biomedical imaging (IEEE ISBI), pp 1397–1400
14. Liao Y, Shen L, Yu S (2017) HEp-2 specimen image segmentation and classification using very deep fully convolutional network. IEEE Trans Med Imaging 36:1561–1572
15. Esteva A, Kuprel B, Novoa R, Ko J (2017) Dermatologist-level classification skin cancer with deep neural networks. Nature 542:115–118
16. Lopez AR, Giro-i-Nieto X, Burdick J, Marques O (February, 2017) Skin lesion classification from dermoscopic images using deep learning techniques. https://doi.org/10.2316/P.2017.852-053
17. Yap J, Yolland W, Tschandl P (2018) Multimodal skin lesion classification using deep learning. Exp Dermatol 27:1261–1267. https://doi.org/10.1111/exd.13777
18. Zhang X, Wang S, Liu J, Tao C (2018) Towards improving diagnosis of skin diseases by combining deep neural network and human knowledge. BMC Med Inf Decis Mak 18(2):59
19. Tschandl P, Rosendah C, Kittler H Data descriptor: the HAM10000 dataset, a large collection of multi-source dermatoscopic images of common pigmented skin lesions. www.nature.com/scientificdata
20. Feurer M, Hutter F (2019) Hyperparameter optimization automated machine learning. The springer series on challenges in machine learning. Springer, Cham
21. Zeiler MD, Fergus R (2013) Visualizing and understanding convolutional networks. ArXiv, abs/1311.2901
22. Smith LN, Topin N (2017) Deep convolutional neural network design patterns. ArXiv abs/1611.00847
23. Rifkin R, Klautau A (2003) In defense of one-vs-all classification. J Mach Learn Res 5:101–141
24. Szegedy C, Liu W, Jia Y, Sermanet P, Reed S, Anguelov D, Erhan D, Vanhoucke V, Rabinovich A (2014) Going deeper with convolutions. CoRR abs/1409.4842
25. Krizhevsky A, Sutskever I, Hinton GE (2012) Imagenet classification with deep convolutional neural networks. Adv Neural Inf Process Syst 1097–1105

26. Krawczyk B (2016) Learning from imbalanced data. Prog Artif Intell (Springer Series) 5:221–232. 10.1007/s13748-016-0094-0
27. ImageNet Large Scale Visual Recognition Competition (ILSVRC) (electronic resource). https://www.image-net.org/challenges/LSVRC/. Accessed July 13, 2018

Vehicle-to-Vehicle Driver Safety-Related Data Transmission and Reception Using Li-Fi Technology

Snehal Pacharne and Vinayak Kulkarni

Abstract Light Fidelity Technology is also known as Visible light communication is the form of wireless communication which uses visible light to transfer information such as digital data, Audio and video as well. Light is modulated and amplified to attain desired speed and distance. Vehicle-to-vehicle communication is the technology in which one vehicle transmits and receives data to and from other vehicle so that they share data between each other and will be able to assist each other. Proposed system uses Li-Fi for vehicle-to-vehicle communication system uses Li-Fi module which can be mounted in the Headlamp as well as tale-lamps of the Four-wheeler which will help to transmit real-time information such as speed data, anti-lock braking data, Turn indication, certain emergencies in car, tire related data. System is made for such small applications which can be directly implemented in car with little modifications in Hardware and Software.

Keywords Li-Fi technology · Wireless transmission · Android application · V–V communication

1 Introduction

Nowadays wireless communication use radio waves. But radio waves have a problem of efficiency, availability, security and capacity. Spectrum is very important requirement for wireless communication. With advancement in technology and increase in number of users, existing radio wave spectrum fails to meet the need and hence, the capacity problem. To resolve all the issues, the solution for addressing such issues, we decide to transmit these data using LED wirelessly, that technology referenced as

S. Pacharne (✉)
VLSI and Embedded Systems, MIT Academy of Engineering, Alandi, Pune, India
e-mail: snehal-pacharne@mitaoe.ac.in

V. Kulkarni (✉)
School of Electrical Engineering, MIT Academy of Engineering, Alandi, Pune, India
e-mail: vbkulkarni@etx.maepune.ac.in

© Springer Nature Singapore Pte Ltd. 2021

S. N. Merchant et al. (eds.), *Advances in Signal and Data Processing*,
Lecture Notes in Electrical Engineering 703,
https://doi.org/10.1007/978-981-15-8391-9_43

Li-Fi which is faster, and flexible because of the durability, efficiency and high lifetime characteristics that makes Li-Fi concept a better one. LED lights are nowadays widely used for personal and official purposes for their luminous efficacy improvement. Basic components used for visible light communication are LEDs (having visible range), photodiodes and image sensors.

Being a highly populated country like India and lot of traffic problems, there is always a problem of manual traffic control whenever an ambulance arrives along a particular direction which is not effective. Automatic traffic control and uninterrupted traffic for fire extinguishers and ambulance is highly necessary. The proposed technology Li-Fi by which transmission of data can be enabled between the lights of traffic light system and that of the lights in the siren of ambulance and fire extinguishers on a priority-basis will enable us to have a better traffic control system with un-interrupted services of transport for the ambulances. Vehicle-to-vehicle communications, for instance, is one of the previous trends, which is one of the most effective mechanisms that are implemented in automobiles to provide safety and a protocol of communication.

2 Literature Survey

Usually OEM's are taking efforts to make the driver as well as passenger's experience more comfortable reliable and safer. For safety OEM's are focus are on enhancing active and passive safety features. V2V is inter-vehicle communication made using various wireless communication technology [1]. The purpose to made this type of communication is when in any situation, to make other cars to inform about surroundings vehicle state and action that they are about to take. Since sudden actions or decisions change made by any vehicle may cause adverse situation for other vehicle (e.g., lane departure) and it avoids some problems those are easy to manage if informed prior (e.g., road issues may avoid traffic congestion by informing and changing the routes) [1].

2.1 Overview of V2V Communication

Li-Fi is acronym stands for Light Fidelity was proposed by German Scientist Harald Haas. By this means of technology we will able to send data via light. Li-Fi has an ability to communicate auto-piloted cars using headlights in restricted area. As this technology is based on light communication the speed achieved by this communication is also very fast [2].

Li-Fi is wireless communication technology which is similar to famous Wi-Fi technology. Main feature provided by Li-Fi is fully networked, bidirectional, high speed with secured communication. By typical comparison result shows that Li-Fi is 100 times faster than Wi-Fi communication [3] (Table 1).

Table 1 Comparison of different technologies used in V2V communication till date

Parameters	Technologies used				
	Bluetooth	Zig-Bee	UWB	Wi-Fi	Li-Fi
Data transfer speed	1 Mbps	250 Kbps	100 Mbps	54 Mbps	Greater that 1 Gbps
Bandwidth	2.4 GHz	869/915 MHz	3–10 GHz	2.4–5 GHz	No reserved band
Range of data transfer(meters)	10	10–100	10–100	10–100	Shorter
Power consumption	High	High	High	Medium	Low
Network topology used	Pico net	Star	Pico net	Point-to-point	Point-to-point
Cost	Good	Good	Good	Good	Low
Security	Not	Less	Better	Less	Highly secure
IEEE standard	802.15.1	802.15.4	802.15.3	802.11.a/g/b	802.15.7

In order to achieve Li-Fi communication we need transmitter and receiver.

Transmitter block: contains input, timer and LED bulb. In this input which can be in form text, voices etc. are transmitted via LED bulb in regular interval using timer circuit.

Receiver block: Receiver will collect this transmitted light via photo diode and converts into electrical signal and strengthen them to suitable level via amplifier.

In Li-Fi communication data is transmitted by normal visible light, i.e., LED hence it is falls into category of Visible Light Communication. This communication can take speed up to 224 Gb/sec [4].

2.2 *Advantages of Li-Fi*

- The speed of the Li-Fi communication is very high equals to speed of light.
- In this communication data is transmitted via light without channel hence this communication is more secured. Since hacking is not possible.
- It uses light waves which are harmless.

2.3 *Disadvantages of Li-Fi*

- It requires Line Of Sight communication only which makes communication somewhat difficult.
- This technology may be undergoing in interference caused due to exterior bulb and sunlight.

2.4 Applications Over the Li-Fi Communication

- Traffic Management and Road Safety
- Medical Applications
- Aviation
- Communication Underwater
- Elegant Lighting
- Interior map-reading system for blind people
- In dangerous Environments or In Sensitive regions
- Transportations
- Industrial Areas.

2.5 Use of Li-Fi in V2V communication

Currently V2V communication is based on IEEE 802.11 (or wireless LAN). In radio wave communication when two or more than two vehicles are within range of radio communication then they will connect and form ad hoc network. Since single wireless LAN link range is up to few hundred meters which is quite limited, every vehicle works as router to propagate messages to other vehicles. This routing mechanism depends on position of vehicles, it will capable of handling rapidly changing ad hoc network due to speed of vehicles [1].

Currently are Bluetooth (IEEE802.15.1), Zig-Bee (IEEE 802.15.4), and UWB or Ultra wide band (IEEE 802.15.3a) [5] communication which is also known as Direct short range communication (DSRC).

Li-Fi is line of sight communication and it uses OOK modulation scheme for transmitting data in which supply voltage to the LED's is provided according to the digital data received through photo diode. There are limitations for size of the data that can be transmitted continuously like engine status, speed, braking information and issues in the car driving in front [6]. We need to develop calibration protocol for transmitting long data with the desired speed. The keep safe distance rule can be established using Li-Fi communication [7].

Proposed system we choose the applications which are important for driver assistance and safety in critical situations mapping it to be most relevant with DSRC. Proposed system is intended to be incorporated in the existing vehicles with some modification. So we proposed LED's itself for communication as we will get LED's incorporated in the headlights and tail lamps of the car.

The design proposed here uses Li-Fi for vehicle-to-vehicle communication for following Applications.

1. Emergency situations in car is communicated to the cars behind in case of traffic
2. Breaking Information is transferred
3. Tire Temperature if crossed certain limits will be communicated to vehicle behind
4. Bad road situation is indicated to the vehicle behind

5. Turn indication can be given to car beside with the Li-Fi message.

Distance (We tried for 40 cm) practically possible up to 1 m with the proposed system and data rate—As we are using OOK modulation we practical implementation.

3 System Architecture

The proposed system talks about vehicle security, safety and improve the driving comfort. This is achieved through vehicle-to-vehicle communication made with wireless technology Li-Fi.

To prototype this system transmitter circuit, receiver circuit and android application is been used. Data from transmitting vehicle could be information regarding real-time traffic situation, related to road condition, emergency help related etc. (Fig. 1).

Transmitter in this prototype is one vehicle which is having sensors interfaced with PIC microcontroller for giving input to the transmitter unit. And there 4 motors interfaced to drive the vehicle. When necessary information vehicle wants to transmit, Li-Fi transmitter encrypts that information and sends it over the light.

The different information is provided which senses from sensors are:

Ultrasonic sensor detect path hole and alert latter about same so that latter's driver can make right decision or slowdown vehicle. Proximity sensor is used for Reverse Park and is continuously updated on mobile app so that driver can make safe distance from front vehicle if temperature is high enough to cause fire. Vibration sensor will sense the vibrations and thus alert vehicle if vibration exceeds threshold values.

Motor driver circuit and motor together forms prototype of vehicle (front vehicle). Android app is used for driving front vehicle. Li-Fi (short for light fidelity) is a technology for wireless communication between devices using light to transmit data and position (Fig. 2).

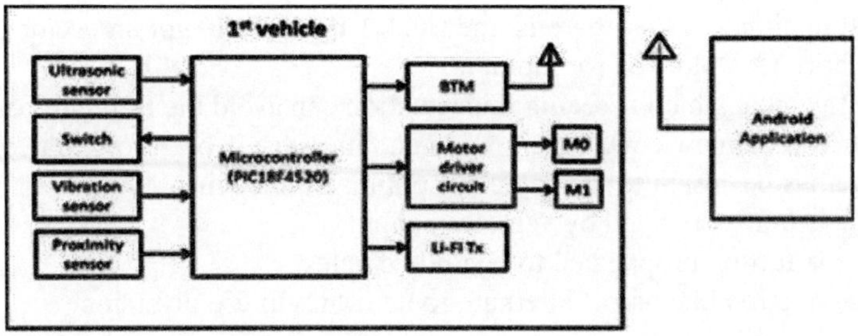

BTM-Bluetooth Module M0,M1-Motores

Fig. 1 Block diagram of the Li-Fi Transmitter

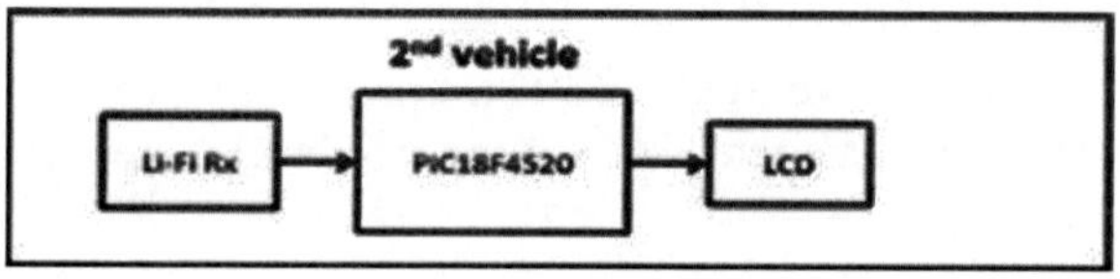

Fig. 2 Block diagram of the Li-Fi receiver

Li-Fi Receiver contains photodiode and Hex encoder IC for converting information received to display it on LCD.

3.1 Applications for Li-Fi in Real Life for Automobile Considered in Project Are

3.1.1 Traffic Application

Generally on highway vehicles are moving with high speed. There are different lanes by which vehicles are moving. When any situation comes due to which traffic congestion is introduced on roads, that lane can be avoided by means on V2V communication using Li-Fi. Consider following scenario:

Vehicle is moving in one lane at constant speed and let us suppose in between there is divergence and this divergence is used by many vehicles coming from behind, then in this case only use of indicator is not enough because some time it makes other vehicles behind to wait. To solve this problem such vehicles will transmit the signal of divergence and is received by behind vehicles then these vehicles take necessary actions [2] (Fig. 3).

In above mentioned and in other un-mentioned scenarios driver in traffic congestion is completely known to the reason behind this traffic congestion queue. By this type of V2V communication from front vehicles reason of congestion can be propagate to behind vehicle's so that they know the reason.

Algorithm for Traffic Situation:

- On road in situation that there is one truck/large vehicle got an accident so that particular road is blocked for moment.
- The first vehicle which is seeing this situation can avoid the traffic to be happen.
- That car will acknowledge the situation and message is been formulated.
- Message has been sent to the vehicles coming from behind.
- Message is being received by vehicles behind.
- Message is further propagated to behind vehicles.
- Necessary action like use of alternate route or it will use diversion.

Fig. 3 V2V communication in traffic situation [8]

3.1.2 Parking Application

In parking if congestion is there, then in that case vehicles are keep on forming queue for parking. Here, vehicle in front of that queue knows how many parking slots are remaining or parking is full such information is shared with other vehicles (using V2V) that are in queue then they can take appropriate decision [9].

Also in parking during vehicles are taking exit, during paring infrastructure (e.g., Pillars) coming vehicle from other end is not visible by driver when it comes directly in front of driver manually it may be dangerous to detect and to apply brake. In such situation V2V communication speed benefits can be used to detect front vehicle and immediately brake can be applied by use if ECU [9] (Fig. 4).

Algorithm for Traffic Situation:

- Vehicle ahead realizing that there is no space left for parking
- Vehicle wishes to take a reverse
- Vehicle applied reverse gear
- Message related to activity is been formulated
- Message will be transmitted to the behind vehicles
- Message has been received by the subsequent vehicles
- Necessary actions by the taken by subsequent vehicles
- Queue is reduced and vacant parking area is found out.

Fig. 4 V2V communication in parking full situation [10]

3.1.3 Drive Application

This feature is comes under the ADAS system. Suppose vehicle is cruising at fix speed and suddenly the pedestrian or any obstacle is comes in front of vehicle, then vehicle will be stopped till obstacle is not pass [11]. But if clear communication is not.

Received then situation may come in which behind vehicles may hit front vehicle due to lack of communication. This can be overcome by making V2V communication using Li-Fi with behind vehicles to slow down or stop the vehicles (Fig. 5).

Figure 6 shows flow of executing the task. Li-Fi communication is wireless and serial type of communication therefore initialization of Serial communication is must. Former vehicle sends information about obstacle ahead in path, any reverse parking or lane departure. Compare Different signals and if value of light intensity is greater than threshold then display that information on LCD of latter vehicle [3].

Algorithm for Drive Application

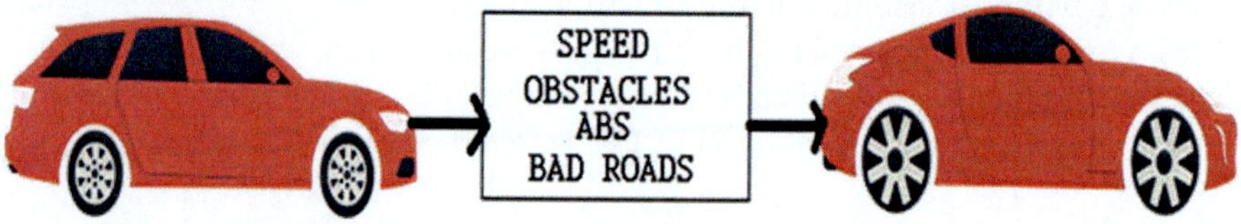

Fig. 5 V2V communication while driving a vehicle [12]

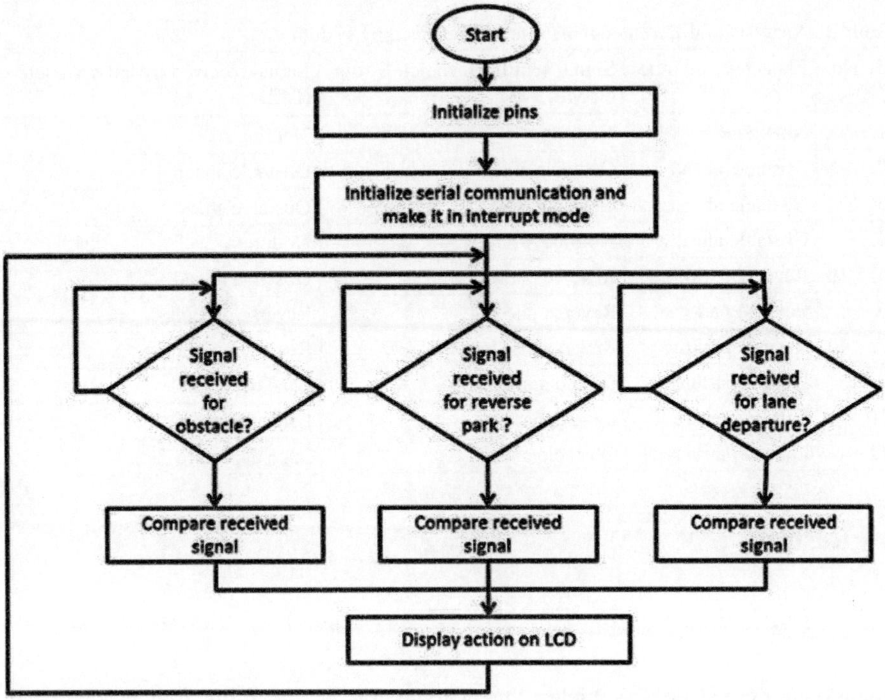

Fig. 6 Flowchart for transmitting data from one vehicle to another

- Vehicle ahead took sudden break or wants to inform vehicle behind about road condition
- Vehicle ahead will formulate the message
- Message is transmitted though encoded LED lights
- Message is received at the receiver end through Photo diode and decoded
- Decoded message is displayed on LCD (Table 2).

$$\text{Accuracy of vehicle1} = \frac{\text{no. of correct output}}{\text{no. of output}} = \frac{9}{11} = 81.81\% \quad (1)$$

$$\text{Accuracy of Li-Fi communication system} = \frac{\text{no. of correct output}}{\text{no. of output}} = \frac{9}{11} = 81.81\% \quad (2)$$

Table 2 Analysis of different signals sent and received by system

S. No.	Expected outcome	Signal sent from vehicle 1 (o/p on MPLAB)	Signal received by latter vehicle (LCD)
1.	No signal	Nothing	Nothing
2.	Obstacle ahead	Obstacle ahead	Obstacle ahead
3.	Obstacle ahead	Obstacle ahead	Obstacle ahead
4.	Obstacle ahead	Nothing	Nothing
5.	Reverse parking	Reverse parking	Obstacle ahead
6.	Reverse parking	Reverse parking	Reverse parking
7.	Reverse parking	Reverse parking	Reverse parking
9.	Lane departure	Lane departure	Lane departure
10.	Lane departure	Lane departure	Lane departure
11.	Lane departure	Obstacle ahead	Lane departure

4 Results

4.1 Simulation Results

The modules consist of 2 Vehicles. One vehicle used as transmitter and another used as receiver vehicle with PIC18F controllers. Different sensors are used to indicate the proximity and sudden Break, turning of vehicle in front to left or right (Figs. 7, 8, and 9).

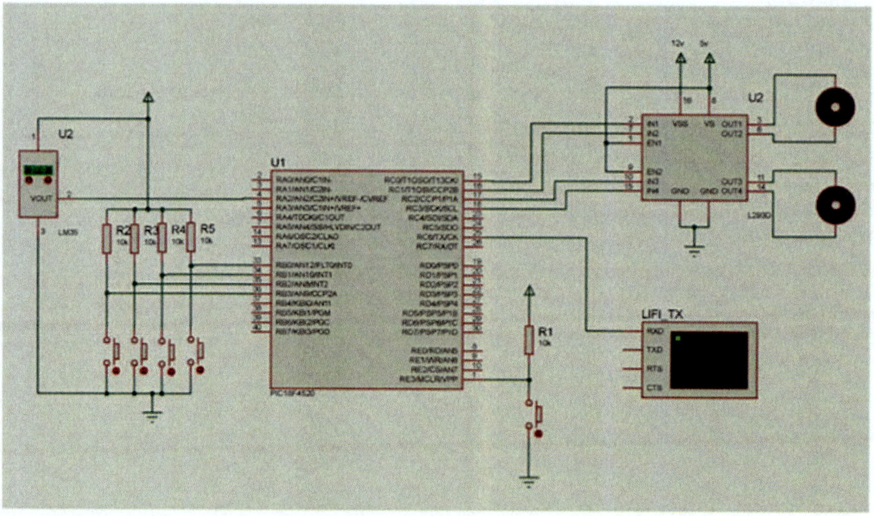

Fig. 7 Simulation results for TX module

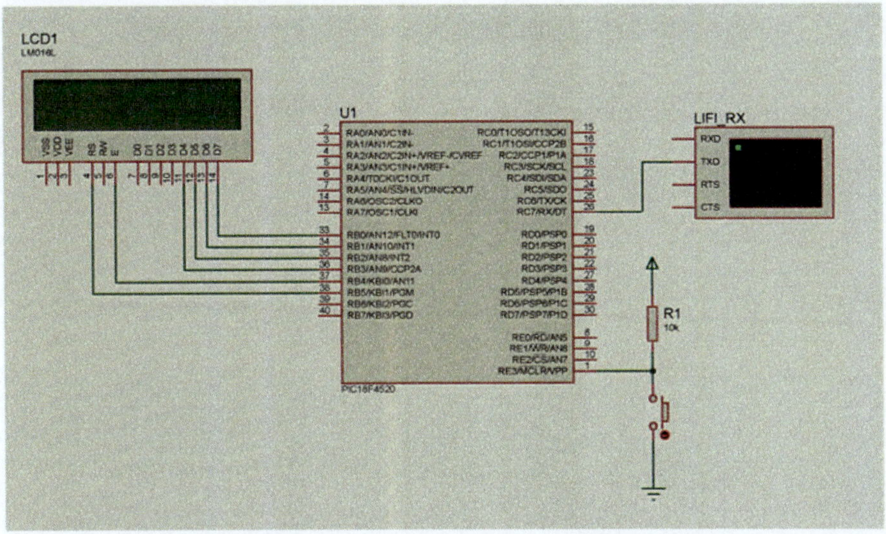

Fig. 8 Simulation for RX module

Fig. 9 Developed Li-Fi communication demo module

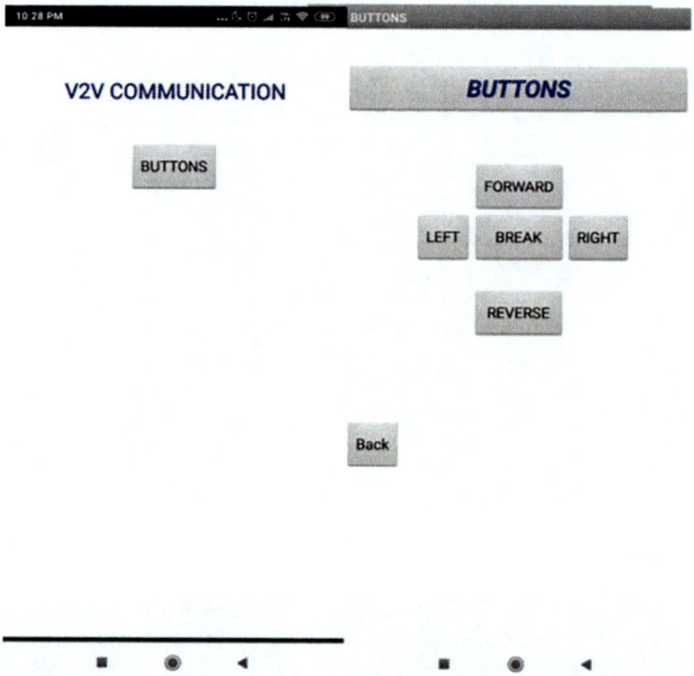

Fig. 10 Andriod application developed to control the Front vehicle remotely

Figure 10 shoes the buttons and GUI developed for controlling the vehicle in front remotely as it was difficult to press and operate the buttons on demo vehicle while running.

4.2 Results Parking Application

Transmitter Vehicle is in driving mode, it decided to take a right turn. So it will turn ON the right indicator. When right indicator is turned ON simultaneously this message information is been transmitted through Li-Fi transmitter module. This information is received by vehicle which is LOS. The received message is after decryption displayed on LCD unit (Fig. 11).

4.3 Results Traffic Situation

Transmitter Vehicle is in driving mode, it decided to take a right turn. So it will turn ON the right indicator. When right indicator is turned ON simultaneously this

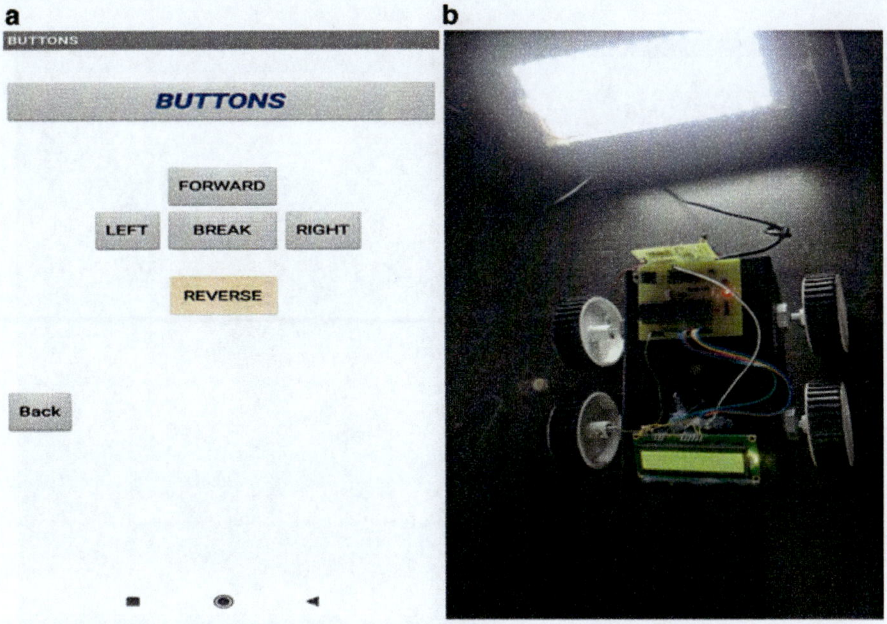

Fig. 11 **a** Instruct from android application for taking reverse. **b** Reveres indication on LCD

message information is been transmitted through Li-Fi transmitter module. This information is received by vehicle which is LOS. The received message is after decryption displayed on LCD unit (Fig. 12).

4.4 Results Drive Application

Transmitter Vehicle is in driving mode, it decided to take a right turn. So it will turn ON the right indicator. When right indicator is turned ON simultaneously this message information is been transmitted through Li-Fi transmitter module. This information is received by vehicle which is line of sight. The received message is after decryption displayed on LCD unit (Fig. 13).

5 Conclusion

Vehicle-to-Vehicle Communication is easy and simple using Li-Fi. The proposed system is used to send the warning messages at faster a rate that helps drivers to make strategic decision at faster rate which is vital in making strategic decision avoiding accidents and congestion. Though different modulation techniques are used

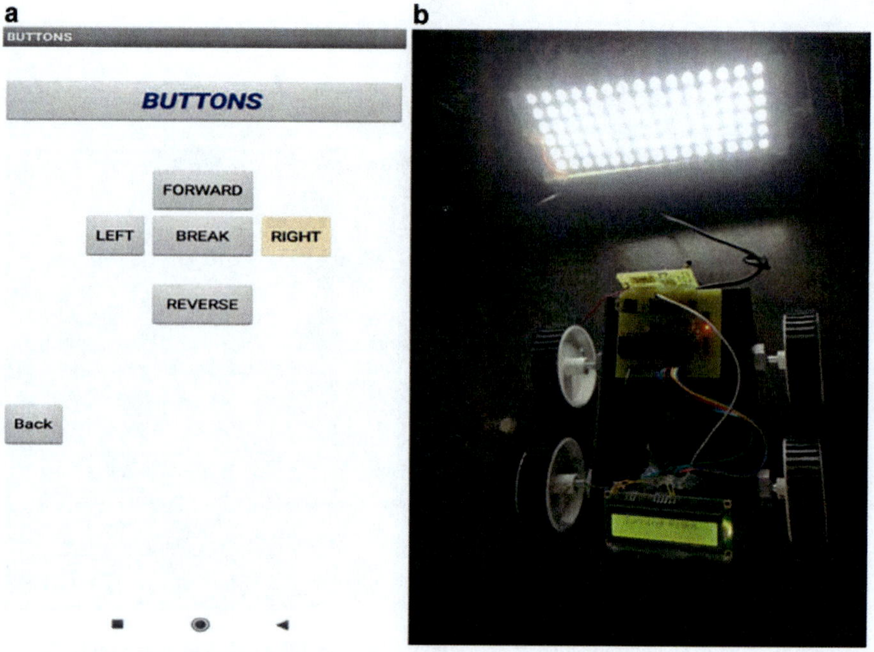

Fig. 12 **a** Input from android application in traffic situation. **b** Output that can be seen ON LCD screen of other CAR

to increase the strength of signal, the important research work is still going on to increase the communication range of Li-Fi communication. Thought is also given for use of lasers instead of LED's.

As this is line of sight communication currently proposed systems can fully form the cluster for communication between cars, means the placement of the system in car is important.

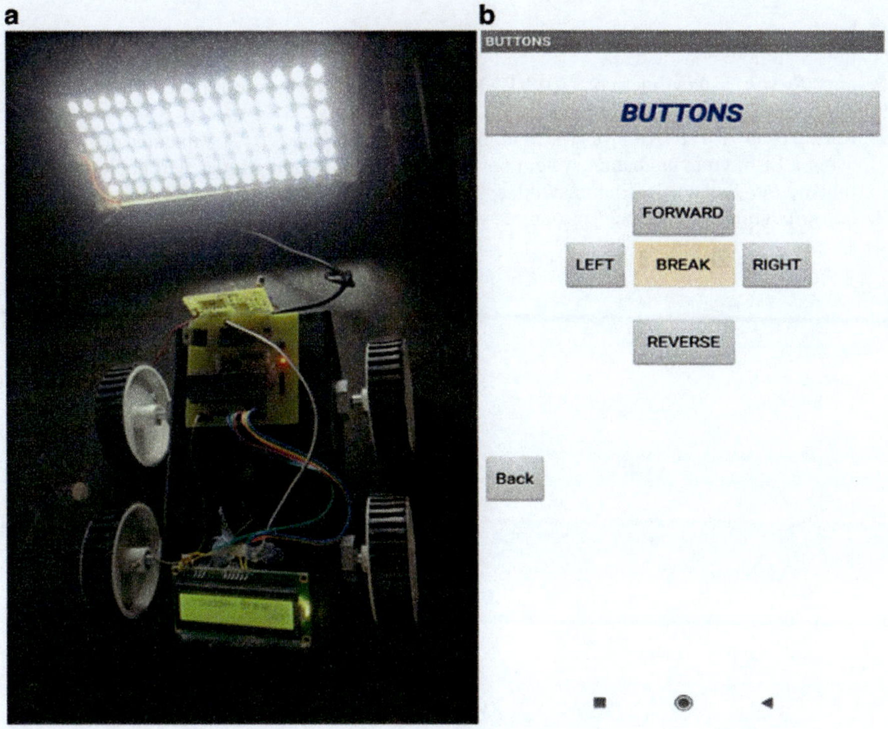

Fig. 13 **a** Break applied on front vehicle through android application **b** sudden break application shown on receiver LCD

References

1. Mali1 S, Patil S, Pathak M, Chokwala H, Patil S (June, 2017) Vehicle to vehicle communication system using Li-Fi technology. IJSART 3(6):34
2. Kalaiselvi VKG, Sangavi A, Eugureyg D (2017) Li-Fi technology in traffic light, vol 404. 978–1–5090–6221–8/17/$31.00c_2017 IEEE
3. Dhayanithi G, Pooja B, Ratul M, Balaji S (2016) Smart vehicular communication system using Li-Fi technology. In: International conference on computation of power, energy information and communication (ICCPEIC)
4. Vidhya Krishn AG, Nagarajan R, Durka T, Kalaiselvi M, Pushpa M, Shanmugapriya S (March, 2017) Vehicle communication system using Li-Fi technology. Int J Eng Comput Sci 6(3):20651–20657
5. Thakur R, Balkawade S, More V, Patil B (January, 2018) Inter vehicular communication using Li-Fi. Int J Res Appl Sci Eng Technol (IJRASET) 6(1)
6. Tali MM, Patil A, Chavan S, Jadhav S (March, 2017) Inter-vehicle communication using Li-Fi technology. Int J Adv Res Comput Commun Eng (IJARCCE) 6(3):5940
7. Al Abdulsalam N, Al Hajri R, Al Abri Z, Al Lawati Z, Bait-Suwailam MM (2015) Design and implementation of a vehicle to vehicle communication system using Li-Fi technology. In: 2015 international conference on information and communication technology research (ICTRC, 2015)
8. https://www.parknsecure.com/automatic-traffic-management

9. https://www.extremetech.com/extreme/176093-v2v-what-are-vehicle-to-vehicle-communications-and-how-does-it-work
10. https://www.eastbaytimes.com/2013/12/31/barts-full-parking-lots-top-mr-roadshows-dirty-dozen-list/
11. Karthikeyan A, Iyer AK, Kar M, Manimegalai CT (February, 2017) Vehicular management using a Li-Fi communication system powered by BIPV (building integrated photo-voltaics). Indian J Sci Technol 10(5). https://doi.org/10.17485/ijst/2017/v10i5/111137
12. https://pngtree.com/freepng/cartoon-red-car_4183386.html

A Novel Approach for CBIR Using Four-Layered Learning

Shweta Salunkhe, S. P. Gaikwad, and S. R. Gengaje

Abstract Content-based image retrieval (CBIR) comprises recovering the most outwardly comparative images to a given question image from a database of images. CBIR from therapeutic image databases does not plan to supplant the doctor by anticipating the sickness of a specific case however to help him/her in analysis. The visual attributes of an ailment convey analytic data, and periodically outwardly comparative images relate to a similar infection class. By counseling the yield of a CBIR framework, the doctor can acquire trust in his/her choice or considerably think about different potential outcomes. With high-dimensional information in which every point of view on information is of high spatiality, determination of highlights is imperative to further build the aftereffects of bunching and characterization. To ease the enthusiastic miscellany in the precision of image retrieval, we developed another graph-based learning strategy technique to successfully recover images from remote detecting. The proposed strategy utilizes a four-layered framework that joins the feature level fusion of Gabor and ripplet Transform of selected query along with SVR. In the first layer two image sets are retrieved utilizing the Gabor and Ripplet-based wavlet Decomposition separately, and the besides, the top ranked retrieved images from both the top up are further used to find their queries. Using each individual part, the chart grapples recoup six image sets from the image database as an augmentation request in the subsequent layer. The photos in the six image sets are evaluated for positive and negative data age in the third layer, and Simple MKL is associated with gain proficiency with the proper inquiry subordinate combination loads to accomplish the last consequence of image recuperation. This research is based on building fully-automatic four layer systems capable of performing large-scale image search based on texture information. An effective four layer architecture

S. Salunkhe (✉) · S. P. Gaikwad · S. R. Gengaje
Bharati Vidyapeeth (Deemed to Be University) College of Engineering, Pune 411046, India
e-mail: shwetasalunkhe16@gmail.com

S. P. Gaikwad
e-mail: spgaikwad@bvucoep.edu.in

S. R. Gengaje
e-mail: gosachin22@gmail.com

© Springer Nature Singapore Pte Ltd. 2021

S. N. Merchant et al. (eds.), *Advances in Signal and Data Processing*,
Lecture Notes in Electrical Engineering 703,
https://doi.org/10.1007/978-981-15-8391-9_44

with the application of SVR was proposed in this study for the purpose of retrieving images from CIFAR dataset.

Keywords Graph-based learning · Content-based image retrieval · Color co-occurrence feature · Ordered dither block truncation coding · Rotation invariant

1 Introduction

Because of its capacity to search and index multimedia images, the ever-increasing amount of digital images on the internet makes the image retrieval based on content shine golden. Content-based picture recovery methods are inherent requirements in multimedia apps through Internet use. Most probably, CBIR systems use color, texture, shape or any other data that is automatically obtained from query and database pictures. The retrieved pictures are refined by ranking in perspective of the proximity between the initial and database pictures. Inefficiency of CBIR systems is anticipated in the gap between highlights of small dimensions and semantic characteristics. Numerous papers are acquainted which demonstrate the assortment of client communication with the question image, and the assortment of inquiry and database image highlights preparing plans. It is conceivable to isolate the CBIR plans into couple of significant classes. Anybody with information of these classifications can receive their advantageous CBIR framework with the sorts of system, gadget and image.

Ripplet and Gabor highlights are two sorts of image portrayal descriptors. Ripplet highlights are particularly fit for taking care of nearby picture examples or surfaces, while Gabor highlights depict a picture's general format. One drawback of considering two features is that the image regain consistently take after the other the same yet may be irrelevant to the request since image classification regularly address large texture of natural scenes containing inexhaustible and complex visual objects. The extreme noise from irrelevant objects are more often than region of interest. Unmistakably, recovery precision can be tremendously improved by coordinating their qualities. Since the element and algorithmic methods are essentially one of a kind, it is definitely not a brilliant idea to straightforwardly join particular segment vectors into one vector to improve image recuperation precision. In spite of the fact that inquiry development can accomplish precise recovery results, because of false positive recorded records, the show of inquiry augmentation will all in all degrade.

Spurred by the characteristics of the inquiry expansion and the particular technique for combination of positions, we propose a four-layered diagram-based learning approach for remote detecting images. We separate the Gabor include and the ripplet highlight in this methodology. Next, there are four layers of the image recuperation process. Utilizing the Gabor learning calculation in the primary layer, a image rundown is recovered from the remote detecting picture database in which each image is like the question. As needs be, we acquire another image list by utilizing the ripplet highlight learning calculation.

We first re-rank the pictures in the over two records in the subsequent layer and afterward get three kinds of chart grapples: PH, PL, and PC. PH and PL are the two records 'top-positioned pictures.' PC contains comparable normal pictures of the two records. Utilizing the Gabor include or the ripplet highlight learning calculation, we take PH, PL, and PC as the inquiries for recovering information base pictures. Along these lines, it is possible to get six records containing recovered pictures. In the third level, by assessing the pictures in the six records, positive and negative sets are chosen. The parameters in Simple MKL are prepared to meld the eventual outcomes of the recovery, and we get the last pictures that have been recovered.

The fundamental commitments of this paper are abridged as pursues.

1. A tale, four-layered diagram-based learning approach is being created to recover remote detecting images. The methodology refines the first inquiry contribution by uniting it with the recovered images which got top-positioned. The nature of the outcomes got using ripplet or Gabor highlight strategies is checked, and the arranged picture sets are intertwined to produce the last consequence of the recovery. The exactness of the recovery is fundamentally improved without giving up the adaptability of the technique proposed.
2. Another strategy for extension inquiry is acquainted that is strong with concentrate insecure highlights. Its primary bit of leeway is that numerous significant images are mined by a solitary info images as opposed to requiring multi-pertinent pictures to be contribution by clients. An increasingly extreme expansion request images set is formed as before learning for the accompanying recuperation by getting the recouped images together with the main inquiry. The exactness of images recuperation can be improved through the set. The proposed methodology can dispense with the inadequacy of the single picture-based recovery.
3. To engage accurate assessment of the nature of each arrival, a novel methodology is exhibited to combine diverse image recovery results. For Gabor and ripplet highlights, Simple MKL is connected to learn appropriate inquiry subordinate combination loads. Various consequences of the image recuperation are intertwined to improve the precision of the recuperation.

In CBIR tasks, various feature extraction methods are documented. Sparse based methods are most popular due to their compact representation of image pixels. DWT transforms are good example of sparse methods, but the failed to represent the texture when edges ames in picture. Gabor features are well known for its predictability. It extract the texture in various scales and directions [1]. Ridglet transforms are mostly used for curvilinear texture [2, 3] but it needs more coefficient to represent the image and also it has more complex representation [4]. To overcome those limitation curvelet features are used [5]. Thus the documented method failed to extract fully detail textures in the image such as curves. To overcome the problem Cuve features are extracted which is capable to extract the all the ridges in an image [6]. Ripplet transform is used for represent the image with this non linear singularities at different scales and orientation. This issue in giving accurate recovery images and minimizing the assistance to the analyst to utilize (CBIR) framework.

The strategy in [7] demonstrated the all-encompassing portrayal of the spatial envelope with extremely low dimensionality to make the picture of the occurrence. This methodology introduced an outstanding outcome in the classification of the scene. To achieve a perfect result, the system in [8] proposed a propelled approach to manage picture portrayal with open field structure and the possibility of over-satisfaction technique as revealed in [8], this methodology achieved the best execution in characterization with much lower spatiality contrasted with the past plans in the undertaking of arrangement of pictures. Tiwari et al. developed a US-based patent database CBIR framework [PATSEEK] as a patent comprising of a picture and literary data. The client must enter catchphrases alongside the question picture that may show up in the patent content for closeness search [9]. Krishnan et al. made concrete setup with CBIR based on the edge, texture and color histogram in the image, which just gives the image's semantics. By using this along with edge orientation greatly reduce time. By using frontal territory challenges alone, overpowering concealing distinctive confirmation can recoup number of similar pictures paying little regard to gauge, considering the bleeding edge concealing. Higher typical precision and survey rates were cultivated adequately appeared differently in relation to the standard dominant color strategy [10]. The image is addressed in another structure by a fuzzy attributed relational graph (FARG) depicting everything in the image, its properties and spatial relationship. The qualities of surface and concealing are resolved to demonstrate the human vision system (HSV) [11]. Using Gabor wavelets, the semantics of surface is recouped. Use gradient vector flow fields to isolate the shape incorporate. It shows the makers in [12] a precision of 60.7 percent; anyway, the inconvenience is that it has incredibly low accuracy. In [13], the makers propose a system that uses an image's concealing characteristics to outline a vector of features. Computer-based intelligence classifiers by then use these features to arrange the photos, yet texture and shape features are not considered.

2 Proposed System

2.1 Holistic Descriptor (HD)

Over the last few years, different approaches have been proposed to improve holistic methods for feature extraction. One of the most successful strategies has shown to be the use of Gabor representation of the images. HD features are a extension of holistic feature descriptor. The Gabor channel (Gabor wavelet) speaks to a band-pass direct channel whose motivation reaction is characterized by a symphonious capacity increased by a Gaussian capacity. Along these lines, a bidimensional Gabor channel establishes a complex sinusoidal plane of specific recurrence and direction adjusted by a Gaussian envelope [1]. Gabor features have been known to be effective for representation. But, only a few approaches utilize phase feature, and they usually perform worse than those using magnitude feature. For this reason, only the magnitudes of the Gabor coefficients are thought of as being useful for feature extraction. It accomplishes an ideal goals in both spatial and recurrence areas.

Our methodology structures 2D odd-symmetric Gabor filter, having the accompanying structure:

$$\mathrm{HD}(F)_{\theta_k, f_i, \sigma_x, \sigma_y}(x, y) = \exp\left(-\left[\frac{x_{\theta_k}^2}{\sigma_x^2} + \frac{y_{\theta_k}^2}{\sigma_x^2}\right]\right) \cdot \cos(2\pi f_i x_{\theta_k} + \varphi) \quad (1)$$

2.2 *Nonlinear Approximated Ripplet Transform (NART)*

To defeat the confinement of wavelet, ridgelet change [2, 3] was presented. Ridgelet change can resolve 1D singularities along a self-assertive heading (counting level and vertical bearing). Ridgelet change gives data about direction of straight edges in pictures since it depends on radon change [4], which is fit for extricating lines of subjective direction. Since ridgelet change cannot resolve 2D singularities, Candes and Donoho proposed the original curvelet change dependent on multi-scale ridgelet [5, 6]. Afterward, they proposed the second era curvelet change [7, 8]. Curvelet change can resolve 2D singularities along smooth bends. Curvelet change utilizes an illustrative scaling law to accomplish anisotropic directionality. From the point of view of microlocal investigation, the anisotropic property of curvelet change ensures settling 2D singularities along C2 bends [9, 7, 8, 10]. Like curvelet, contourlet [11, 12] and bandlet [13] were proposed to determine 2D singularities.

Be that as it may, it is not clear why allegorical scaling was picked for curvelet to accomplish anisotropic directionality. With respect to, we have two inquiries: Is the explanatory scaling law ideal for a wide range of limits? If not, what scaling law will be ideal? To address these two inquiries, we plan to sum up the scaling law, which results in another change called ripplet change Type I. Ripplet change Type I sums up curvelet change by including two parameters, i.e., bolster c and degree d; thus, curvelet change is only an extraordinary instance of ripplet change Type I with $c = 1$ and $d = 2$. The new parameters, i.e., bolster c and degree d, give ripplet change anisotropy ability of speaking to singularities along self-assertively formed bends.

Substitute with discrete parameters

$$\begin{aligned} a_j &= 2^{-j} \\ \overrightarrow{b_k} &= [c \cdot 2^{-j} \cdot k_1, 2^{-\frac{j}{d}} \cdot k_2]^{\mathrm{T}} \\ \theta_l &= \frac{2\pi}{c} \cdot 2^{-|j(1-1/d)|} \cdot l \quad j, k_1, k_2, l \in \mathbb{Z} \end{aligned} \quad (2)$$

Forward transform

$$R\left(j, \vec{k}, l\right) = \sum_{n_1=0}^{M-1} \sum_{n_2=0}^{N-1} f(n_1, n_2)\overline{\rho_{j,\vec{k},l(n_1,n_2)}} \quad (3)$$

Nonlinear approximation (NLA).
Sort coefficient in descending order

$$|c_0| \geq |c_1| \geq |c_2| \geq \ldots |c_0| \geq |c_{n-1}| \geq |c_n| \geq \ldots$$

Approximate signal by n-largest coefficients

$$g \approx \hat{g} = \sum_{i=0}^{n-1} c_i \phi_i \quad (4)$$

The ripplet change has the accompanying capacities:

Multi-goals: Ripplet change gives a progressive portrayal of pictures. It can progressively rough pictures from coarse to fine goals.

- Good restriction: Ripplet capacities have minimized help in recurrence area and rot exceptionally quick in spatial space. So ripplet capacities are all around restricted in both spatial and recurrence areas.
- High directionality: Ripplet capacities situate at different headings. With the expanding of goals, ripplet capacities can acquire more headings.
- General scaling and backing: Ripplet capacities can speak to scaling with subjective degree and backing.
- Anisotropy: The general scaling and bolster result in anisotropy of ripplet capacities, which certifications to catch singularities along different bends.
- Fast coefficient rot: The sizes of ripplet change coefficients rot quicker than those of different changes, which means higher vitality focus capacity.

2.2.1 First Layer

We only consider the query image as the labeled image, and the remaining images are considered unlabeled.

For each query image, a weighted cum directed chart is constructed from each individual feature-based retrieval method, where the retrieval quality or the relevance is modeled by the weights on the edges, where

$$C = F(V, E, W) \quad (5)$$

v is a lot of vertices,
e is a lot of edges and
w is a lot of edge weights.

Each database image corresponds to a vertex in *C*. For each image, we identify its k-nearest neighbors and connect the corresponding vertices in *C* with edges that are associated with the distance between the two vertices.

Given an image dataset $Y = \{y1, ..., yl, yl + 1, ..., yn\}$.For each image, we identify its *k*-nearest neighbors and connect the corresponding vertices in *C* with edges (*E*) that are associated with the distance between the two vertices (*V*).*E* corresponds to the similarity among vertices with the weight *W* defined by,

$$W_{ij} = \exp\left(-\frac{\mathrm{d}^2(Q, A_i)}{\sigma \mathrm{y}^2}\right) \tag{6}$$

where $d(V_i, V_j)$ denotes the feature distance between the vertices V_i and V_j, and W_{ij} is the edge weight of E_{ij}. σ is a constant that controls the strength of the weight, and it is set as the median distance among all images. After that, pairwise relevance is obtained through maximum correlation algorithm.

We have obtained two retrieval image lists by using the graph-based holistic and local feature learning algorithms. The two lists represent different retrieval results and have different image arrangements.

2.2.2 Second Layer Graph

We select the top-positioned pictures from the lists as the new input to further refine the retrieval results. To achieve this, we utilize the re-ranking method to obtain the top-positioned pictures as graph anchors. Then, these anchors are taken as new queries to further improve the recovery exactness.

To generate accurate graph anchors for learning the joint relevance of the holistic and local features, it is necessary to precisely measure the similarity among the images. We define the similarity degree of the two images as the relative similarity score (RSS). RSS can be computed by using the following re-ranking method.

$$\begin{pmatrix} L_Q \\ L_{A1} \\ L_{A2} \\ \dots \\ L_{An} \end{pmatrix} = \begin{pmatrix} A_1 & A_2 & \dots & A_m \\ N_{11} & N_{12} & \dots & N_{1m} \\ N_{21} & N_{22} & \dots & N_{2m} \\ \dots & \dots & \dots & \dots \\ N_{m1} & N_{m2} & \dots & N_{mm} \end{pmatrix} \tag{7}$$

where

Q is the question picture.

N is used to represent the retrieved image matrix.

$A_1, A_2, ..., A_m$ are the top-positioned pictures in the retrieval results in the first layer, which are stored in a list LQ.

The images in L_Q are further used as queries to perform the search. For example, $N_{11}, N_{12}, ..., N_{1m}$ are the query results of A_1. Similarly, each of the other images in LQ retrieves the m best matching database images. Finally, $m + 1$ lists are generated.

We compute the normalized similarity score (NSS) between two images. The SS between the retrieved images N_{ij} and Q is derived by,

$$\mathrm{NSS}(Q, N_{ij}) = \mathrm{norm}\|\mathrm{NSS}(Q, A_i) \cdot NSS(A_i, N_{ij})\| \tag{8}$$

When NSS (Q, N_{ij}) is computed, we first consider the relationship between $_Q$ and A_i. If they are very similar, many common images exist in the lists L_Q and L_{Ai}, and the spatial distributions of the image features are similar between the two lists.

After the images in H_LQ and L_LQ are re-ranked, the top retrieved images are taken as the prior knowledge for further image retrieval. In the following, we obtain three types of graph anchors: PH, PL and PC. PH and PL are the top re-ranked images of H_LQ and L_LQ, respectively, whereas PC contains the common imagesin both of the top re-ranked lists (H_LQ and L_LQ). If PC is no smaller than the given similarity threshold, the corresponding common image in the lists is taken as a graph anchor in PC. The aforementioned process is continued until all of the common images in H_LQ and L_LQ are traversed.

We take PH, PL and PC as query images to retrieve images from the image database using the learning algorithm of holistic and local features introduced in the first layer. We label these queries as the prior knowledge, where the values of the graph anchors are all equal to 1.

Thus, we obtain six retrieval lists through the two graphs GH and GL. That is, through the graph GH, we separately derive three retrieved results (LHH, LHL, and LHC) corresponding to the queries PH, PL and PC. Accordingly, we also obtain other three retrieved results (LLH, LLL and LLC) by GL. LHH and LLL are the results of reinforcement learning. LHL, LLH, LHC and LLC are the results of the preliminary fusion. We know that LHL and LHC are the retrieval results corresponding to the graph GH of the holistic feature, but the queries are the graph anchors PL and PC which are not generated from GH. Similarly, LLH and LLC are the retrieval results corresponding to GL, but the queries are the graph anchors PL and PC which are not generated from GL. LHC and LLC are also the results of the graph-based learning algorithm of holistic and local features, which use the common graph anchors as the queries. This approach can reduce the influence of some inappropriate graph anchors on the retrieval result to a certain extent.

2.2.3 Third Layer

To generate the fusion retrieval result of different features, we assess the exhibition of the retrieved images in the aforementioned six lists and find similar and dissimilar images. Then, we gain proficiency with the loads of holistic and local features and related parameters of the fusion. For the fusion process, positive data and negative data are needed. Thus, we introduce an image list (LG) that contains similar images (positive data), as well as a second one (LD) that contains the dissimilar images (negative data).

To create LD, we randomly selected a certain number of images from the bottom of LHH, LHL, LHC, LLH, LLL and LLC separately and stored them in LD.

The images in LHH, LHL, LHC, LLH, LLL and LLC are re-ranked separately. The top-positioned pictures are usually very similar to the query image and have a place with a similar class. We apply the retrieval consistency to evaluate the aforementioned re-ranked results. For example, we select the *cn* top-ranked images from re-ranked LHH as the graph anchors for an expansion query. From the retrieved result LHH, we can obtain a specific number of the top-ranked images by using the retrieval result evaluation. These images are stored in the aforementioned image list LG.

The evaluation processes of other retrieval lists are the same as the process of LHH. To the retrieved results LHL, LHC, LLH, LLL and LLC, we compute their consistency degree and choose the good retrieval results for the fusion of the final result. These images are also stored in LG.

After the evaluation, we need to address the problem of obtaining the best fusion weights. In the image lists LG, the images are taken as the positive data, and the images in LD with low similarity are taken as negative data. By using distance metric, we rearrange LG & LD according to query.

2.2.4 Fourth Layer

We train the parameters of ISVR through these data and the two features. ISVR determines the combination of the feature weights by solving a standard SVM advancement issue dependent on a gradient descent method.

2.3 Computational Complexity Analysis

The regular method to express the multifaceted nature of one calculation is utilizing enormous o documentation. Assume that we have n database images. The computational cost of the proposed method mainly lies in the four layers. In the first layer, the image retrieval is computed with the cost of $O(n3 + n2)$. In the second layer, since the number of re-ranking is far not exactly n, the complexity of the re-ranking process can be negligible. The cost for the second layer is $O(n3 + n2)$. In the third and fourth layer, the main computational cost is for optimizing the parameters in distance metric and ISVR with the time complexity of $O(N3 + lN2 + dlN)$, where d is the feature dimension, l is the number of training samples and N is the quantity of help vectors. Since $d << n$ and $l << n$, in the third layer, different features can be fused quickly.

2.3.1 Incremental Support Vector Regression:

SVM can be utilized as a relapse show, keeping up all the primary highlights that contribute to maximal edge. ISVR utilizes indistinguishable standards from the SVM for grouping, with just a couple of minor changes. As a matter of first importance, since yield is a genuine number, it ends up being exceptionally hard to anticipate the current data, which has limitless conceivable outcomes. On account of relapse, an edge of resistance (epsilon) is set in estimation to the SVM which would have effectively asked for from the issue. Yet, other than this reality, there is likewise a more confused reason, and the calculation is more convoluted in this manner to be taken in thought. Notwithstanding, the principle thought is dependably the same: to limit mistake, individualizing the hyperplane which augments the edge, remembering that piece of the blunder is endured.

Kernel functions:
For Polynomial

$$\mathrm{RBF}\left(f_i, f_j\right) = \exp\left(-\frac{\left|f_i - f_j\right|^2}{2\sigma^2}\right)\text{For Gaussian RBF} \tag{9}$$

3 Experiments and Results

The CIFAR-100 dataset is a collection of images with 32 × 32 size and that are commonly used for classification purpose. It is one of the most widely used datasets for image retrieval method validation. The CIFAR-100 dataset contains 60,000 (32 × 32) color images in 10 different classes. In our case, we test algorithm for 500 images and calculated accuracy as number of images retrieved from a class.

3.1 Retrieval Accuracy:

Retrieval accuracy is delineated as the competency to distinguish the relevant as well as irrelevant images. The accuracy is a gauge of the extent of the closeness of a calculated or measured value to its original value. Accuracy also stands as the degree to which the outcome of a calculation, specification or measurement matches to the standard (correct) value. The accuracy of retrieval is ascertained utilizing the equation.

Retrieval Accuracy = [Correctly Retrieved Images]/[Total No of Retrieved Images].

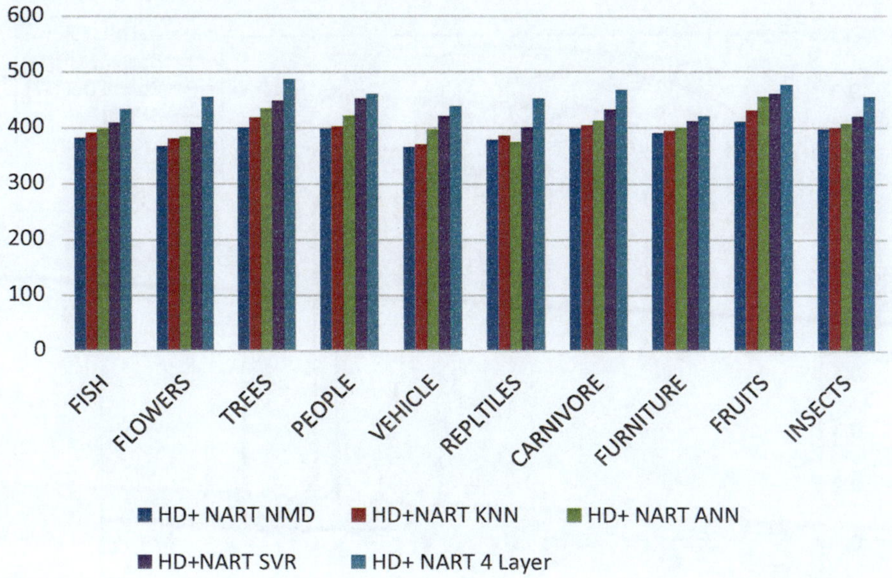

Fig. 1 Class-wise accuracy for different methods

Retrieval Time = Feature Extraction Generation Time + Similarity Measurement Time.

The following table shows class-wise accuracy (Figs. 1–2 and Tables 1– 2).

The above table shows accuracy of the proposed method with other method; from the figure, it is clear that our method gets 94% accuracy, which is very far from the traditional method (Table 3).

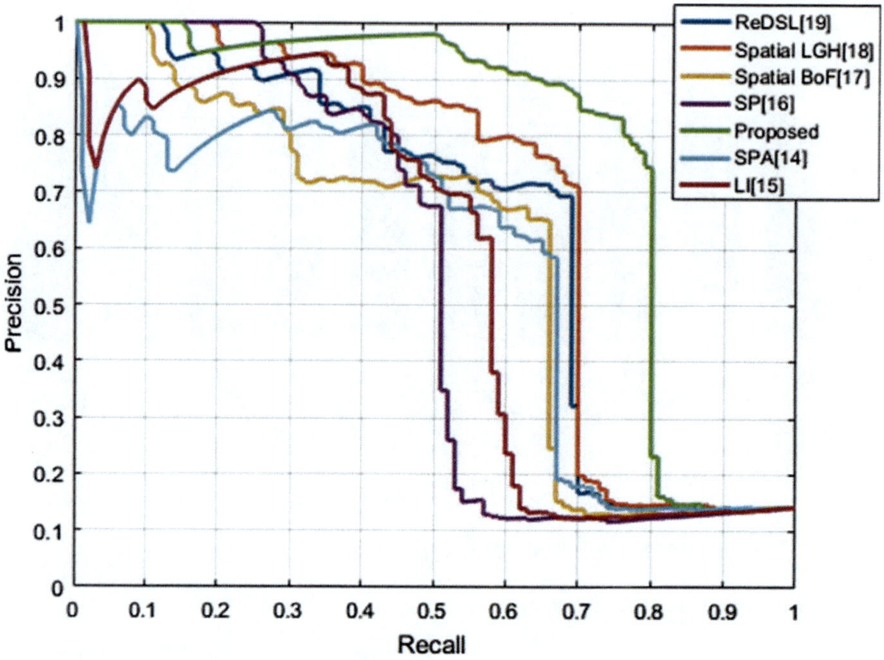

Fig. 2 Comparison of several methods for precision versus recall

Table 1 Class-wise accuracy

Parameter/dataset	Image retrieval precision				
Query image category	HD + NART NMD	HD + NART KNN	HD + NARTANN	HD + NART SVR	HD + NART 4 Layer
Fish	383	392	399	410	434
Flowers	368	381	385	401	456
Trees	401	419	435	449	487
People	399	403	422	452	461
Vehicle	366	371	397	421	439
Repltiles	379	386	375	401	452
Carnivore	399	405	413	433	468
Furniture	391	395	400	412	421
Fruits	411	431	455	461	477
Insects	398	400	408	420	455

Table 2 Comparison with the existing methods

Method	Accuracy
ReDSL	0.84
Spatial LGH	0.90
Spatial BoF	0.73
SP	0.83
SPA	0.81
LI	0.89
Proposed	0.94

Table 3 Confusion matrix

	A	B	C	D	E	F	G	H	I	J
A. Fish	471	0	17	0	0	8	0	0	0	4
B. Flowers	0	453	0	9	13	0	9	0	5	11
C. Trees	8	20	467	0	0	0	0	3	0	2
D. Peoples	0	0	7	478	9	3	0	0	3	0
E. Vehicles	0	10	0	4	473	5	7	1	0	0
F. Repltiles	1	0	0	0	4	458	0	19	5	0
G. Carnivo	0	0	3	0	0	0	483	0	0	14
H. Furniture	5	8	0	0	1	9	0	476	0	1
I. Fruits	0	9	0	7	0	0	1	0	483	0
J. Insects	2	0	6	2	0	17	0	1	4	468

In the confusion matrix, table shows 10 classes used for testing reability of our project. We take 10 classes from CIFAR-100 database; out of that each class, we took 500 images for testing. The above table shows the correctly retrieved images out of 500 proposed method retrieve 471 images correctly for fish, 453 images for flowers, etc.

4 Conclusion

During the previous decade, exceptional advancement has been made in both hypothetical research and framework improvement. The driving force behind substance-based image recovery is given by the wide accessibility of computerized sensors, the Internet and the falling cost of capacity gadgets. Given the extent of these main impetuses, it is to us that content-based recovery will keep on developing toward each path: new crowds, new purposes and new styles of utilization, new methods of association, bigger informational collections and new strategies to take care of the issues. A wide investigation has been made on image recovery. Each work has its very own methods, commitment and impediments. This paper is generalizable to other categorization tasks, and is applicable to any Dataset. Experimental results endeavour that proposed method significantly outperforms the baselines including deep learning models based on coarse and fine categories.

We proposed a four-layered remote detecting image recuperation technique dependent on ripplet and Gabor qualities in this paper. In contrast to the past techniques for recovering pictures, which frequently link a wide range of highlights into one vector to recover image, we have broadened the inquiry and connected another four-layered way to deal with recovering images to meld various highlights. The proposed methodology refines the first information inquiry to broaden the question by joining it with the top-positioned recovery results acquired utilizing strategies dependent on the Gabor and ripplet highlights. For further recovery of six image sets, the extension inquiry images are taken as chart grapples.

To create positive information and negative information, the images in each set are assessed. For the Gabor and ripplet highlights, Simple MKL is associated with learn suitable request subordinate blend loads. Assessments were coordinated in each layer, demonstrating that the exactness in the present layer surpasses those in the past layers. The created system for remote distinguishing picture recuperation is in this manner sensible and adaptable. Examinations of our technique with different strategies further demonstrated that our strategy produces extremely aggressive recuperation execution with a solitary information question image.

References

1. Movellan JR Tutorial on Gabor filters. https://mplab.ucsd.edu/tutorials/gabor.pdf.
2. Candes EJ, Donoho DL (1999) Ridgelets: a key to higher-dimensional intermittency? Philos Trans Math Phy Eng Sci 357(1760):2495–2509
3. Do M, Vetterli M (2003) The finite ridgelet transform for image representation. IEEE Trans Image Process 12(1):16–28
4. Deans SR (1983) The radon transform and some of its applications. Wiley, New York
5. Starck JL, Candes EJ, Donoho DL (2002) The curvelet transform for image denoising. IEEE Trans Image Process 11:670–684
6. Donoho DL, Duncan MR (2000) Digital curvelet transform: strategy, implementation and experiments. In: Proceedings of the aerosense 2000, Wavelet Applications VII. SPIE, vol 4056, pp 12–29.
7. Candes E, Donoho D (2005a) Continuous curvelet transform: I. Resolution of the wave front set. Appl Comput Harmon Anal 19(2):162–197
8. Candes E, Donoho D (2005b) Continuous curvelet transform: II. Discretization and frames. Appl Comput Harmon Anal 19:198–222
9. Hormander L (2003) The analysis of linear partial differential operators. Springer, Berlin
10. Candes EJ, Donoho DL (2003) New tight frames of curvelets and optimal representations of objects with piecewise c2 singularities. Commun Pure Appl Math 57(2):219–266
11. Do MN, Vetterli M (2005) The contourlet transform: an efficient directional multire solution image representation. IEEE Trans Image Process 14(12):2091–2106
12. Do MN, Vetterli M (2003) Contourlets. In: Welland GV (ed) Beyond Wavelets. Academic Press, New York

13. Le Pennec E, Mallat S (2005) Sparse geometric image representations with bandelets. IEEE Trans Image Process 14(4):423–438
14. Bosch SPAA, Zisserman A, Munoz X (2007) Representing shape with a spatial pyramid kernel. In: Proceedings of international conference on image video retrieval. New York, NY, USA, pp 401–408
15. Xia S, Hancock E (2008) 3-D object recognition using hyper-graphs and ranked local invariant features. In: Structural S, Recognition StatisticalPattern (eds) New York. Springer-Verlag, NY, USA, pp 117–126
16. Lazebnik S, Schmid C, Ponce J (2006) Beyond bags of features: Spatial pyramid matching for recognizing natural scene categories. In: Proceedings of IEEE conference on computer vision pattern recognition, pp 2169–2178
17. Kavitha S, Varuna S, Ramya R (2016) A comparative analysis on linear regression and support vector regression. In: Online international conference on green engineering and technologies (IC-GET), IEEE
18. Nhat HTM, Hoang VT (2019) Feature fusion by using LBP, HOG, GIST descriptors and canonical correlation analysis for face recognition. In: International conference on telecommunications (ICT)
19. Ali Z, Christofides N, Polycarpou A (2017) Performance enhanced RES current controller with reduced computational complexity. In: International conference on modern power systems (MPS 2017)
20. Wang C, Zhang B, Qin Z, Xiong J (2013) Spatial weighting for bag-of-features based image retrieval. In: Integrated uncertainty in knowledge modelling and decision making. Springer, pp 91–100
21. Mehmood Z, Anwar SM, Ali N, Habib HA, Rashid M (2016) A novel image retrieval based on a combination of local and global histograms of visual words
22. Wu P, Hoi SCH, Xia H, Zhao P, Wang D, Miao C (2013) Online multimodal deep similarity learning with application to image retrieval. ACM Multimedia 153–162
23. Vadivel SP, Yuvaraj D, Krishnan SN, Mathusudhanan SR (2019) An efficient CBIR system based on color histogram, edge, and texture features. Concurrency Comput: Pract Experience 31(12)

Design of a Power Efficient Multiband Patch Antenna

Punam Deotare and Debashis Adhikari

Abstract A psi-shaped antenna design is proposed using Ansoft's HFSS software in this paper. In this structure, there is a rectangular slot above the monopole antenna and two metal strips besides that structure on the substrate with a slotted rectangular defected ground structure for wireless application. This monopole antenna can be used for WLAN, WiMAX applications. The structure is designed and optimized to operate at 2.5, 3.4, and 5.6 GHz frequencies. The size of the antenna is $34 \times 18 \times 1.6\ \text{mm}^3$. The presented multiband antenna has been designed, simulated by using HFSS software. The antenna has isolation more than 20 dB and peak gain is 3.81dBi. The antenna utilizes microstrip feed. The tri-band good resonance is obtained by rectangular strips. The performance parameters are satisfying the requirements. The gain, return loss, radiation pattern, efficiency, VSWR, 3D polar plot results have been studied through HFSS software. The simulation results satisfy general requirements for commercial use.

Keywords Monopole antenna · HFSS · Multiband antenna · WLAN · WiMaX

1 Introduction

Wireless communication has become the backbone of connectivity in the present-day scenario. To make wireless communication possible, one of the most important devices in the loop is the antenna. For hand-held devices, the most used antenna in the present-day communication devices is the microstrip antenna (MSA). Because of its already known advantages, for example, low weight, preferences of slight profile, size, simplicity of creation, and similarity with integrated circuitry, the MSA has acquired an importance in numerous applications where it is now. MSA is a device

P. Deotare (✉) · D. Adhikari
School of Electrical Engineering, MIT Academy of Engineering, Pune, India
e-mail: punamdeotare7@gmail.com

D. Adhikari
e-mail: dadhikari@entc.mitaoe.ac.in

© Springer Nature Singapore Pte Ltd. 2021
S. N. Merchant et al. (eds.), *Advances in Signal and Data Processing*,
Lecture Notes in Electrical Engineering 703,
https://doi.org/10.1007/978-981-15-8391-9_45

which fulfills all prerequisites of wireless devices with features like less weighing, affordable for fabrication, ability of producing dual- and triple-band operations, scattering cross section is low, etc. An antenna can be a wideband antenna manufactured by designing antenna component on metal tracks attached to an insulating substrate. A PCB is reloaded with a metal layer on one side and on another side there is a ground plane. Multiband antenna is used for operating on multiple frequencies.

In [1], the authors provide the information about proximity coupled feed antenna for new multiband application, in which the patch is of V shape and a rectangular strip is in center of it. The ground plane is not a planar. The defected ground plane structure is obtained by inserting slots. By adding rectangular strip in center of the patch, WLAN and WiMAX applications are achieved. This design was found to operate on various applications like Bluetooth, WLAN, WiMAX, ISM bands. The authors in [2] proposed a slotted square patch microstrip antenna by introducing slots in a square patch. Here, the slotted design uses a single structure for multiband operation.

In [3], an antenna on PIFA configuration with multiband features is presented. This is mostly used in mobile devices. In [4], the authors have proposed a new scheme of multiband operations where the antenna is resonant at 6.18, 4.02, and 11.13 GHz. In the paper, a T-shaped slot is attached on radiating conducting surface and using that slot the required frequency band is getting improved gain and return loss. The customized antenna structure has different frequency bands operating at 5.82, 3.92, 11.35, and 7.88 GHz frequencies. The simulation is done in the electromagnetic (EM) simulation software of this antenna. In [5], the structure of antenna patch is traditional. The geometry consists of two ring-shaped and semicircular slots at the edges. TD/Duroid 6002 substrate is used for this structure. The 4.82 GHz gain is achieved with this customized antenna.

The paper claims to achieve more efficiency and wide bandwidth for the application in a wireless communication system. In [6], the fork-shaped antenna structure is proposed with a rectangular cavity above the monopole antenna and a small rectangular part removed from the ground plane. The design gives us triple band of resonant frequency which will operate on 2.5, 5.5, and 3.4 GHz of frequency. In [7], authors proposed a tuning and reconfigurable antenna used in various wireless applications. The major issues are the size and complexity of the device. The design uses a U-shaped slot antenna with rectangular patch and pin diodes, to make the antenna reconfigurable. Depending upon the application, we can select that frequency band. This antenna is simulated using Advance Design System Momentum (ADS) software. In [8], authors propose broadband and multiband antennas. Here, the fractal antenna configuration is used leading to a star-shaped microstrip antenna and simulated in HFSS software.

In [9], authors proposed a fractal antenna based on rectangular microstrip patch for multiband application. The excitation is provided by the microstrip feed line to the antenna. The antenna is working in between 1 and 10 GHz frequency. The antenna is designed and simulated in Ansoft HFSS V13 software. MIMO techniques are discussed in [10]. This method has very high data transmission rate than SISO techniques. The MIMO antenna is used in receiver section as well as transmitter section.

There are two radiating surface on the substrate which gives different frequencies for different applications. The authors have used Agilent N5230A Network Analyzer for return loss measurement. In [11], a multiband and array design of antenna is proposed for 5G smartphone application working on LTE band for high-speed operations. In [12], the paper discusses the application of ultra-wide band antenna.

The simulation and design is done on HFSS software. An inverted F-antenna mounted on the top of the solar panel for dual functionality is proposed in [13]. The RF energy is collected and converted into DC power, and hence in the absence of solar energy, it will act as a supplement. A metamaterial is used in this antenna presented in [14] to enhance impedance bandwidth and reduce the return loss of –29.263176 dB for WiMAX application.

Flow of the paper after introduction is: In Sect. 2, general design of patch antenna is explained. In Sect. 3, design process of the proposed multiband antenna is discussed. In Sect. 4, paper presents performance analysis of the designed multiband antenna. Finally, conclusion is in Sect. 5.

2 General Design Process of a Patch Antenna

In [15], from this book, the dimensions of an antenna are calculated with the help of given equation.

Equation (1) gives the formula of width calculation.

$$W = \frac{1}{2f\sqrt{\mu_a \varepsilon_a}} \frac{\sqrt{2}v_a}{\varepsilon_r + 1} \frac{\sqrt{2}}{2f_r \varepsilon_r + 1} \tag{1}$$

Effective dielectric constant of substrate is given in Eq. (2)

$$\varepsilon_{\text{reff}} = \frac{\varepsilon_r + 1}{2} + \frac{\varepsilon_r - 1}{2}\left[1 + 12\frac{h}{w}\right]^{\frac{-1}{2}} \tag{2}$$

Due to fringing effect, electrically the patch of the microstrip antenna looks greater than its physical dimensions. For the principal *E*-plane (*xy*-plane), the patch dimensions of the patch along its length have been extended by ΔL. The formula is mentioned in the Eq. (3)

$$\Delta L = 0.412h \frac{(\varepsilon_{\text{reff}} + 0.3)\left(\frac{W}{h} + 0.264\right)}{(\varepsilon_{\text{reff}} - 0.258)\left(\frac{W}{h} + 0.8\right)} \tag{3}$$

As length on each side is increased by ΔL, the formula of effective length of the patch is in Eq. (4)

$$L_{\text{eff}} = L + 2\Delta L \tag{4}$$

The original patch length is given by Eq. (5)

$$L = \frac{1}{2fr\sqrt{\varepsilon_{\text{reff}}}\sqrt{\mu_0\varepsilon_0}} - 2\Delta L \tag{5}$$

Using above formulas, the width and length of the proposed geometry are calculated. So, the width of the outer side of the rectangular microstrip antenna is 15.5 mm and length is 17 mm. The width of the inner rectangular microstrip antenna is 13.5 mm and length is 15 mm. The psi-shaped structures are also calculated using same formulas as mentioned in Table 1. After calculating all the required values, we use HFSS software for designing the required structure.

Table 1 Key dimensions of multiband antenna design

Parameter	Size (mm)
A	18
B	34
C	13.5
D	6.4
E	1.4
F	7
G	1
H	3
I	1
J	3
K	13.4
L	3
M	18
N	8.2
O	1.2
P	8.7
Q	4
R	4.4
S	0.25
T	1
U	15
V	2.6

3 Design Process of Proposed Multiband Antenna

Figure 1 is showing the antenna structure of multiband antenna. The 50Ω microstrip line is excited through the feed. FR4 substrate is selected in the HFSS software during the design and its thickness is 1.6 mm and 4.4 is relative permittivity. The structure is like monopole antennas are in one rectangular structure and this structure is on rectangular microstrip feed line. Two rectangular strips are designed besides the fork-like shape structure to improve return loss. The antenna ground structure is a defected ground structure and there is one slot in this antenna structure that is designed in the ground surface. All calculated results are obtained below −19 dB range at resonant frequency in S11 parameter. The parameters of the structure are simulated in the HFSS software and results are optimized. The parameters are demonstrated in Table 1.

Figure 2 shows the stages of design process of antenna system. The Antenna1 is first developed antenna and its S11 pattern is given in Fig. 3. So, after modifying the structure, we get the structure of Antenna2, which is giving multiband resonant frequency. In this modified structure, above the feed line, one positive sign-shaped structure is plotted and its S11 pattern is shown in Fig. 4. For better results, we add two metal strips in the rectangular cavity, after that modification, antenna is giving multiband frequency with less input impedance loss in Antenna3. The final structure is represented after modification in Antenna3 and Fig. 1. This structure provides best

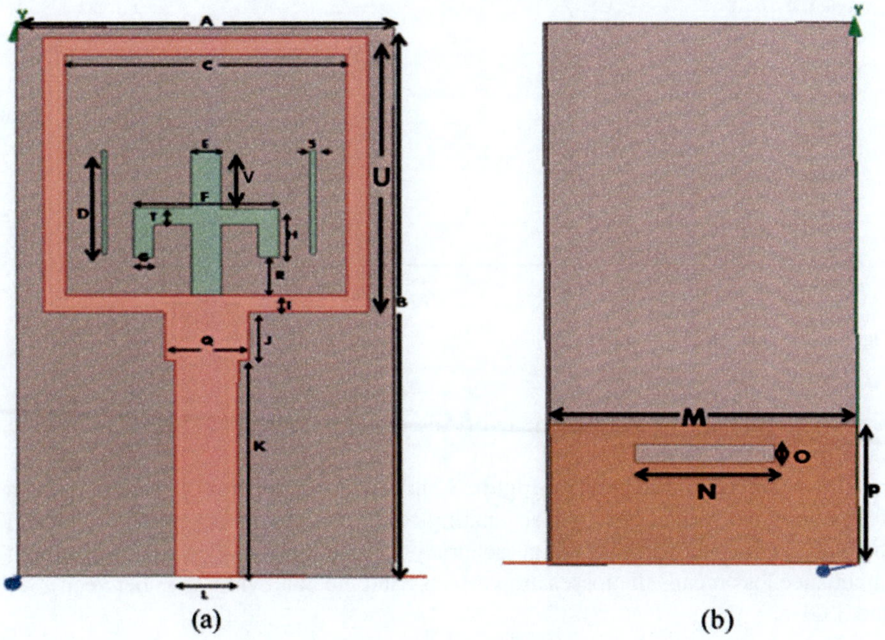

Fig. 1 **a** Multiband antenna design (front view), **b** ground plane

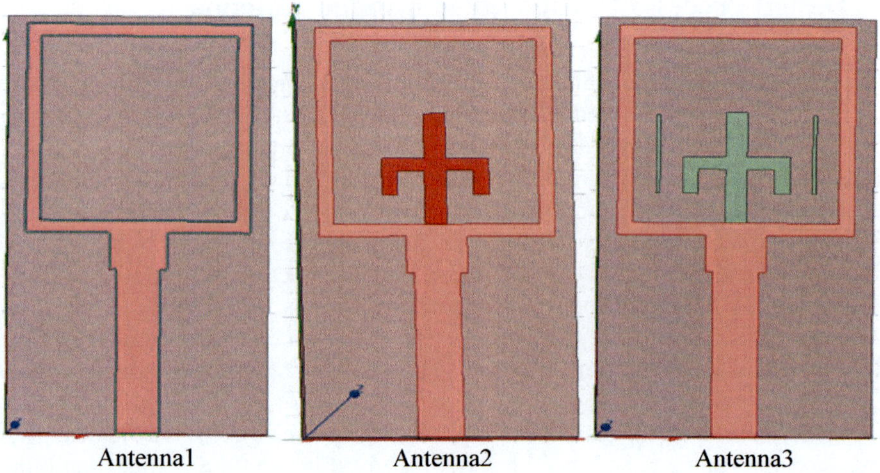

Fig. 2 Design process of the proposed multiband antenna

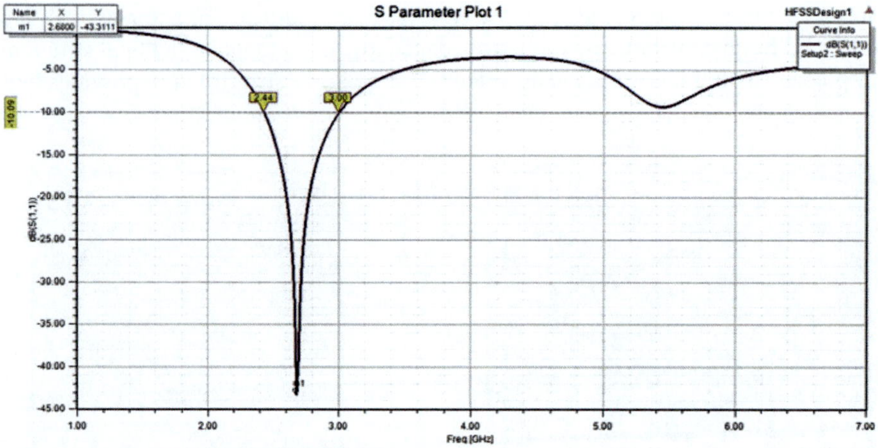

Fig. 3 S11 parameter of Antenna1

results compared to the S11 parameter of Antenna1 and Antenna2 structures and S11 pattern of work is given in Fig. 5.

S11 parameter of Antenna1—Figure 3 shows S11 parameter of Antenna1 from this pattern and we get only one resonating band. It is resonating at 2.68 GHz and the S11 graph is showing very high accuracy near about (−43.311 dB) down input impedance loss means the losses are very less and the bandwidth is in between 2.44 and 3 GHz.

S11 parameter of Antenna2—The structure is resonating in three different frequency bands. First band is resonating at 2.58 GHz frequency (S11 = −27.38 dB),

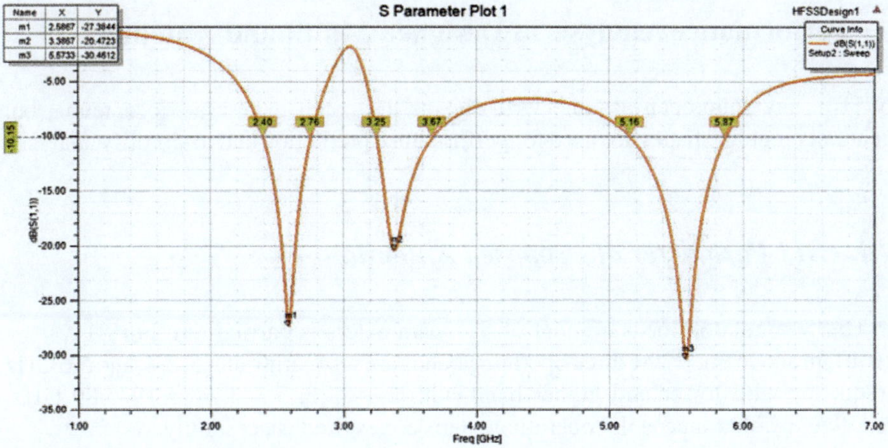

Fig. 4 S11 parameter of Antenna2

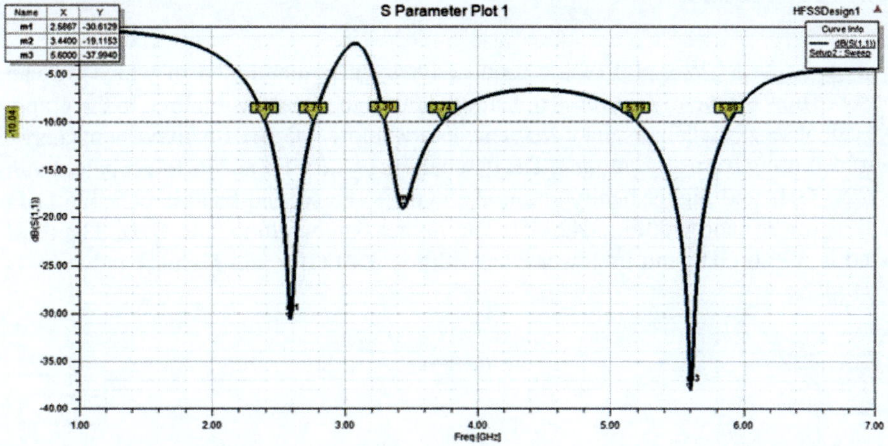

Fig. 5 S11 parameter of proposed antenna

its lower bandwidth is 2.4 GHz, and higher bandwidth is 2.76 GHz frequency. Second band is resonating at 3.4 GHz frequency (S11 = −20.5 dB) and its lower bandwidth is 3.25 GHz and higher bandwidth is 3.67 GHz frequency. Last band is resonating at 5.6 GHz frequency (S11 = −30.5 dB) and its lower bandwidth is 5.16 GHz and higher bandwidth is 2.87 GHz frequency.

4 Performance Analysis of Designed Multiband Antenna

In [16], any shape can assume virtually for the patch in microstrip antenna, but generally regular shapes are used for performance prediction and to simplify analysis.

4.1 S11 Parameter of Proposed Antenna

In HFSS simulator, measurement of S11 parameters is carried out. For S11 ≤ -10 dB, in above Fig. 5, the three operating bands are resonating at 2.5, 3.4, and 5.6 GHz frequency with lower and higher bandwidth 2.4–2.76, 3.3–3.74, 5.19–5.89 GHz, respectively, that means the operating bands are excited successfully.

4.2 VSWR Plot

In Fig. 6, the VSWR plot of the proposed system is mentioned in which we can observe how efficiently power is transmitted through transmission line. In the graph, we can observe that all resonant frequencies are below − 2 dBi. So, maximum energy is transmitting from this antenna. The first band is of 2.6 GHz and its lower bandwidth is 2.53 GHz and higher bandwidth is 2.65 GHz. The second band is of 3.426 GHz and its lower bandwidth is 3.42 GHz and higher bandwidth is 3.46 GHz. The third band is of 5.6 GHz and its lower bandwidth is 5.49 GHz and higher bandwidth is 5.71 GHz.

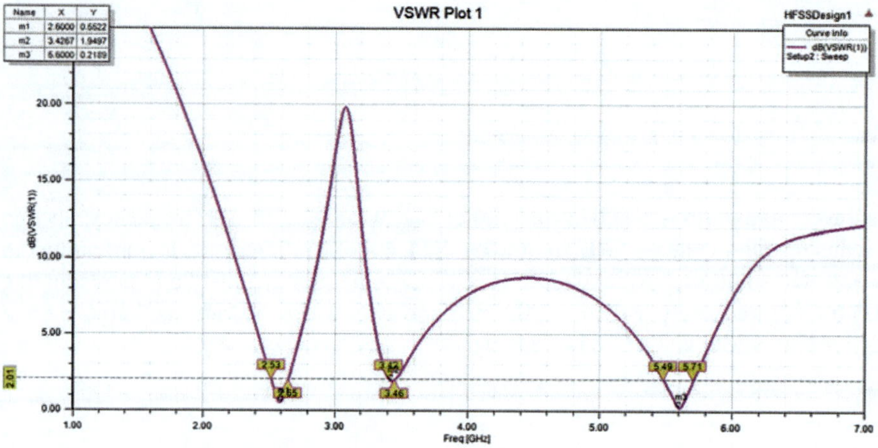

Fig. 6 VSWR plot of proposed antenna

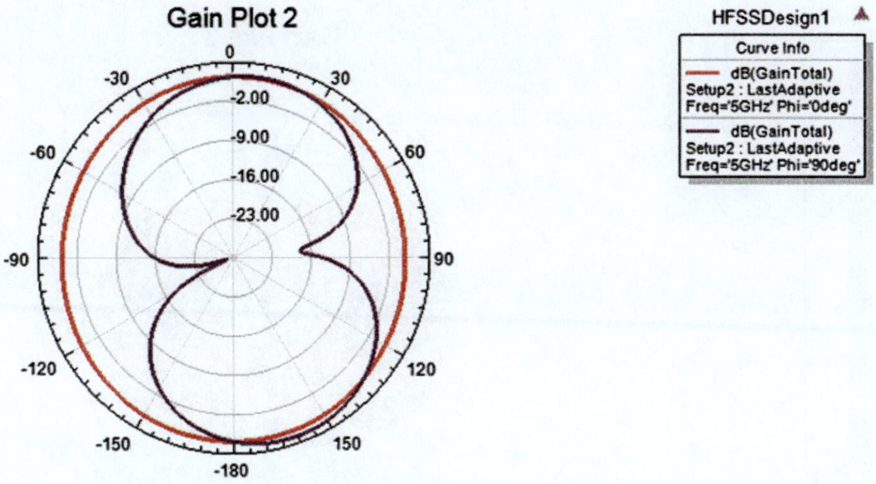

Fig. 7 Radiation pattern of proposed antenna

4.3 Radiation Pattern

In Fig. 7, radiation pattern of the antenna is presented, and from this graph, we are getting omnidirectional radiation pattern, means it radiates equal in all directions.

4.3.1 3D Polar Plot

Figure 8 is showing 3D polar plot of the antenna design, from this apple-shaped structure, we are getting minimum −31.50 dBi minimum gain and 3.84 dBi maximum gain.

4.3.2 Surface Current

Figure 9 shows the surface current distribution of the central frequencies which we are getting from the designed antenna structure. As shown in figure, the current flows mainly along the edges of the rectangular slot. By adding rectangular strips on the substrate, the return loss is improved. Adjusting the length and width of the strip makes changes in the domain of current distribution.

4.3.3 Smith Chart

In Fig. 10, Smith chart of the proposed system is presented, in which we can observe that the line is nearer to 1 in the graph. Means, it is in the center of the Smith chart

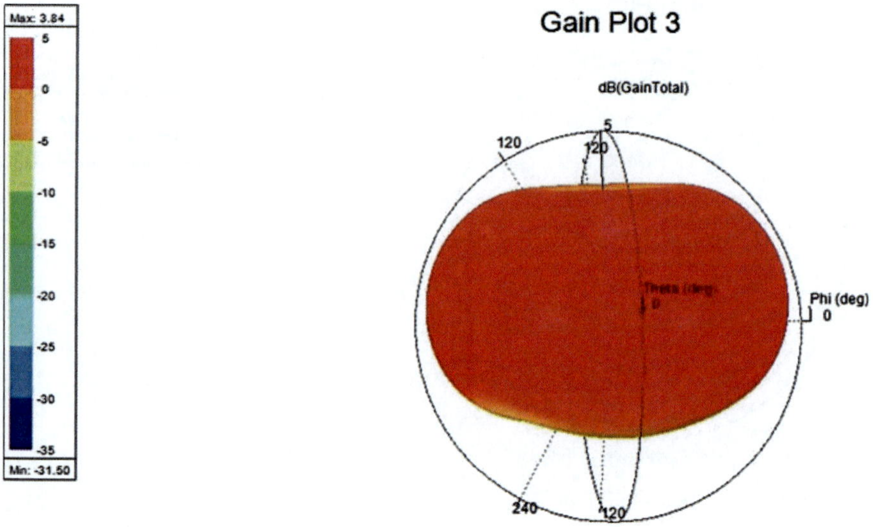

Fig. 8. 3D polar plot of proposed antenna

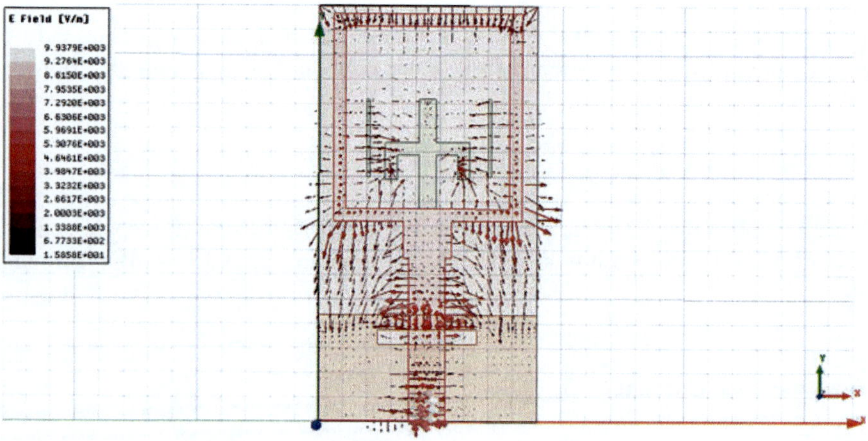

Fig. 9 Simulated surface current distribution of the proposed antenna

which provides 50 Ω microstrip line through the feed from which achieved perfect impedance matching from this antenna.

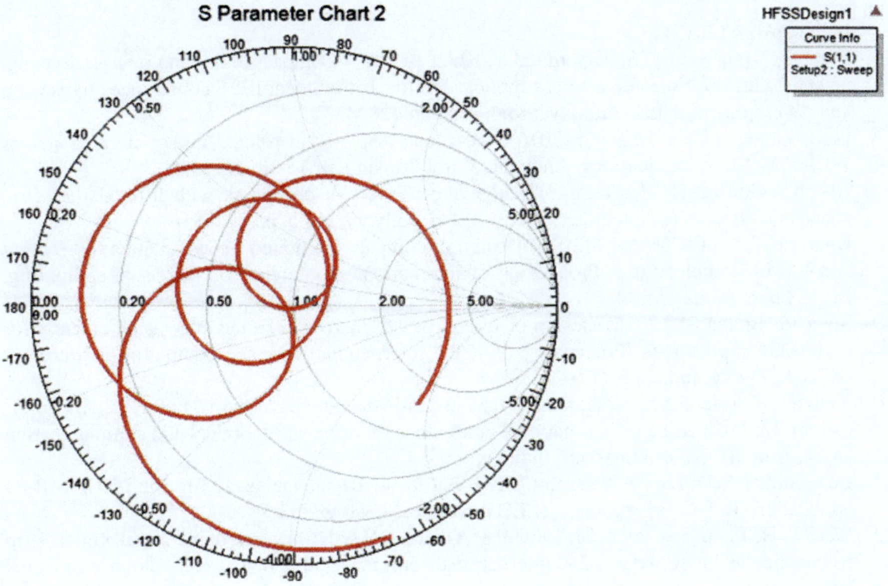

Fig. 10 Smith chart of the proposed antenna

5 Conclusion

A multiband antenna structure for wireless application is presented. Rectangular slot is etched from the ground plane for the proposed work. The antenna can generate three resonant bands of 2.5, 3.4, and 5.6 GHz. Worldwide interoperation for microwave access operates at: 2.5–2.6, 3.4–3.69, and 5.25–5.85 GHz frequency. So, this compact multiband monopole antenna will work for the WLAN/WiMAX applications. This paper limits to software only and all results are checked and simulated in HFSS software and presented in the paper. From this designed antenna, omnidirectional radiation pattern is achieved. Since the return loss of this designed antenna is very low, we can infer that the designed antenna is a power efficient antenna operating at multiband frequencies.

References

1. Bakariya PS, Dwari S, Sarkar M, Mandal MK (2013) Proximity-coupled microstrip antenna for bluetooth, WiMAX, and WLAN applications. IEEE Antennas Wirel Propag Lett 14:755–758
2. Ghosal A, Majumdar A, Das SK, Das A (2018) A multiband microstrip antenna for communication system. Emerg Trends Electron Devices Computat Tech, 1–4
3. Elfergani ITE, Hussaini AS, Rodriguez J, Tong DLH, Abd-Alhameed RA (2017) Practical multi-band antenna for 3G and 4G mobile services. sIEEE. Wrexham, UK, pp 309–314
4. Mahesh A, Shushrutha KS, Shukla RK (2017) Design of multi-band antennas for wireless communication applications. In: Applied electromagnetics conference. IEEE, Aurangabad,

India, pp 1–6 (2017)
5. Nataraj C, Ismael AA, Selvaperumal S, Khan S (2017) Compact multiband microstrip patch antenna with slot-rings for wireless applications. In: 15th student IEEE conference on research and development. IEEE, Putrajaya, Malaysia, pp 428–433
6. Li L, Zhang X, Yin X, Le Z (2016) A compact triple-band printed monopole antenna for WLAN/WiMAX applications. Antennas Wirel Propag Lett 15:1853–1855
7. Ghosh S, Ghosh A, Sarkar I (2017) Design of probe feed patch antenna with different dielectric constants. Devices for integrated circuit. IEEE, Kalyani, India, pp 813–816
8. Kharat K, Dhoot S, Vajpai J (2015) Design of compact multiband fractal antenna for WLAN and WiMAX applications. Proceeding of international conference on pervasive computing. IEEE, Pune, India, pp 1–4
9. Sivia JS, Bhatia SS (2015) Design of fractal based microstrip rectangular patch antenna for multiband applications. Proceeding of IEEE international advance computing conference. IEEE, Banglore, India, pp 712–715
10. Anuvind R, Joseph SD, Kothari A (2015) 2x2 MIMO antenna at 2.4 GHz for WLAN applications. In: Proceeding of international conference on microwave, optical and communication engineering. IEEE, Bhubneswar, India, pp 80–83
11. Li Y, Sim C-Y-D, Luo Y, Yang G (2018) Multiband 10-antenna array for Sub6 GHz MIMO applications in 5-G smartphones. IEEE Access 6:28041–28053
12. Werfelli H, Tayari K, Chaoui M, Lahiani M, Ghariani H (2016) Design of rectangular microstrip patch antenna. Proceeding of 2nd international conference on advanced technologies for signal and image processing. IEEE, Monastir, Tunisia, pp 798–803
13. Tawk Y, Costantine J, Ayoub F, Christodoulou CG (2018) A communicating antenna array with a dual-energy harvesting functionality [wireless corner]. Antennas Propag. Maga. 60(2):132–144
14. Bagri RK, Meena S (2015) Design and analysis of rectangular microstrip patch antenna using metamaterial for WIMAX application at 2.5 GHz. In: International conference on applied and theoretical computing and communication technology. IEEE, Davangere, India, pp 697–700
15. Balanis A (2005) Antenna theory analysis and design, 3rd edn. Wiley, Hoboken, New Jersey
16. Garg R (2001) Microstrip antenna design handbook. Artey House, UK

Brain Activity Analysis for Stress Recognition

Aishwarya Wakale and Usha Verma

Abstract The stress is the major problem that occurs in daily life, which affects on the physical and mental health. There are various methods for detection of stress. In this paper, EEG signal analysis is used for stress recognition. The output of a neurosky mindwave mobile 2 sensor is waves like alpha, beta, and gamma in a specific range. By analyzing these values and keeping a threshold, the dataset formation occurs, and further to train the data, artificial neural network technique (RBFN algorithm) is used. The system learns and is trained using RBFN. The states (stress) are detected. The work is tested on hundred cases and found 80% accurate.

Keywords Brain–computer interface (BCI) · Neurosky mindwave mobile 2 headset · Bluetooth (BlueSMiRF) module · Arduino Uno

1 Introduction

Human stress is the response of the body that has to be adjusted. Human body is intended to experience and respond to stress. A stress that holds the individual informed and prevents risk can be positive. If one can think emotional, mental, or physical imbalances, stress is harmful. There are various methodologies to reduce stress. Basically there are three types of a stress: acute stress, episodic stress, and chronic stress. Acute stress is of very short time period, when there is some excitement or big thrill, for example, roller coaster ride, watching a horror movie, episodic stress is of longer duration of time in which people makes self-harm or having unreasonable demands or worrying, chronic stress is long term which results in unhealthy and dangerous for human health. There are various methods for identification of stress which includes the EEG, ECG, EMG, etc. To distress the person, there are also

A. Wakale (✉) · U. Verma
School of Electrical Engineering, MIT Academy of Engineering, Pune, India
e-mail: aishwaryawakle@gmail.com

U. Verma
e-mail: uyverma@etx.maepune.ac.in

© Springer Nature Singapore Pte Ltd. 2021

S. N. Merchant et al. (eds.), *Advances in Signal and Data Processing*, Lecture Notes in Electrical Engineering 703,
https://doi.org/10.1007/978-981-15-8391-9_46

various methods or techniques including yoga, relaxation therapy, drugs, but all these methods are not practical, and music therapy is one of the easiest techniques to distress the person. This can be proved with the help of EEG system and brain–computer interfacing (BCI).

The EEG is the electroencephalogram which collects the electrical data (signals) of the brain, which applies to the scalp of the brain. The electrical activity is measured in microvolt (μv). Main frequency of the human EEG waves is in the range of 1–50 Hz. Delta wave is in the range of 1–4 Hz (during sleep, awake), theta wave is 4–7 Hz, alpha wave is 8–15 Hz, beta wave is 16–31 Hz, and gamma wave is >32 Hz. Until now, several researchers worked on stress detection.

Ankita Tiwari et al. describe that the EEG helps track electrical activity in the human brain by using metal electrodes [1]. Signals are collected from the brain by applying the electroencephalogram. The system is good at detecting the human eye's focus, concentration, and blinking. This is in the 0.5–50 Hz frequency range. This system is good in sensing attention, meditation, and blinking of the human eye. It is in the frequency range of 0.5–50 Hz. When the person is considered in stressed stage, then it will send a message which is displayed on screen that person is in stress. Similarly, Jun et al. used the detection of patterns in brain waves, whereas Seyyed Abed Hosseini concluded that the emotions are classified into two areas by using physiological signals and used a 2-s rectangular window without overlap for data segmentation [4]. This can be done by using EEG signals. In the same way, Deepika R. Chavan et al. also used EEG [8]. The brainwave becomes beta when person is stressed. The stressed person gains alpha wave to overcome it by effective music of his choice the result showing the change in alpha and beta wave.

Akar Jain et al. describe by using a neurosky sensor. Readings will be inferred from processing beta and alpha waveform activity. Two 100-scale outputs described by the neurosky mind wave mobile chip as "Attention" and "Meditation" are given. Gayathri. P et al. also used the neurosky mind wave mobile 2 sensor for stress detection. By same manner, Tong Chen et al. present a non-contact approach for detecting psychological stress using (HSI) camera to obtain data on tissue oxygen saturation (StO2) to detect human stress [11].

Foteini Agrafioti et al. used ECG [2] which is based on the empirical mode decomposition. The proposed framework is comprised in three steps ECG synthesis, signal decomposition, and feature extraction. Emotion is detected easily.

From the referred work, most of the researchers used the EEG and ECG techniques for recognition of stress, few of them used the neurosky mindwave mobile 2 sensor based on EEG analysis. In this paper, the neurosky mindwave mobile 2 sensor is used for stress detection. The paper is structured in various sections. Section 1 contains the introduction of the proposed work. Section 2 describes the methodology, while Sect. 3 contains the results and discussion, and Sect. 4 is the conclusion part.

2 Methodology

The technology that will be used is electroencephalogram. In this project, the neurosky mindwave mobile 2 sensor is used for measuring brain waves/signals. The brainwaves change according to persons feeling and emotions. Due to slower brainwaves, we feel tired, sleepy, and drowsy. The higher frequencies make alert, active, etc.

EEG measures fluctuation of voltage in the brain. The brain contains several neurons which are building blocks of the brain that communicate with one another by the change in electrical activity. The device consists of a headset and sensor arm. Mindwave mobile 2 sensor used for measuring brain waves/signals in Fig. 1 shows the neurosky mindwave mobile 2 sensor.

To propose a system first, the data is collected from the brain by using a sensor. With the help of Arduino Uno and Bluetooth module, the mind wave sensor is interfaced to the PC. The port which is selected for the use is com66. Further, the EEG raw data in the form of alpha, beta, gamma, theta, etc., is displayed on the screen, and this data is signed 16-bit integer in the range of −32,768 to + 32,767. The scaling or reconstruction of raw data is completed, after that thresholding of data is done and database formation completed. By using the RBFN algorithm, training and learning of data are completed, and the stress recognition is possible.

Fig. 1 Neurosky mindwave mobile 2 sensor

When the person is in stress, the theta waves' range is in between 20–40. In the meditation case, the gamma wave ranges in between 60–80; when the person in a neutral state, the range is 40–60, and for the high level of E-sense or meditation/concentration, the gamma wave range is 80–00. The alpha and beta values between 1 and 20 show the lower level of e-sense that means sleepy, angry, and sad moods. By this way, the stress can be detected. Further, those values can plot using of MATLAB software. Brain wave signals are in the form of this meaningful waveform. The samples of the brainwaves are recorded. The proposed work is mentioned in Fig. 2.

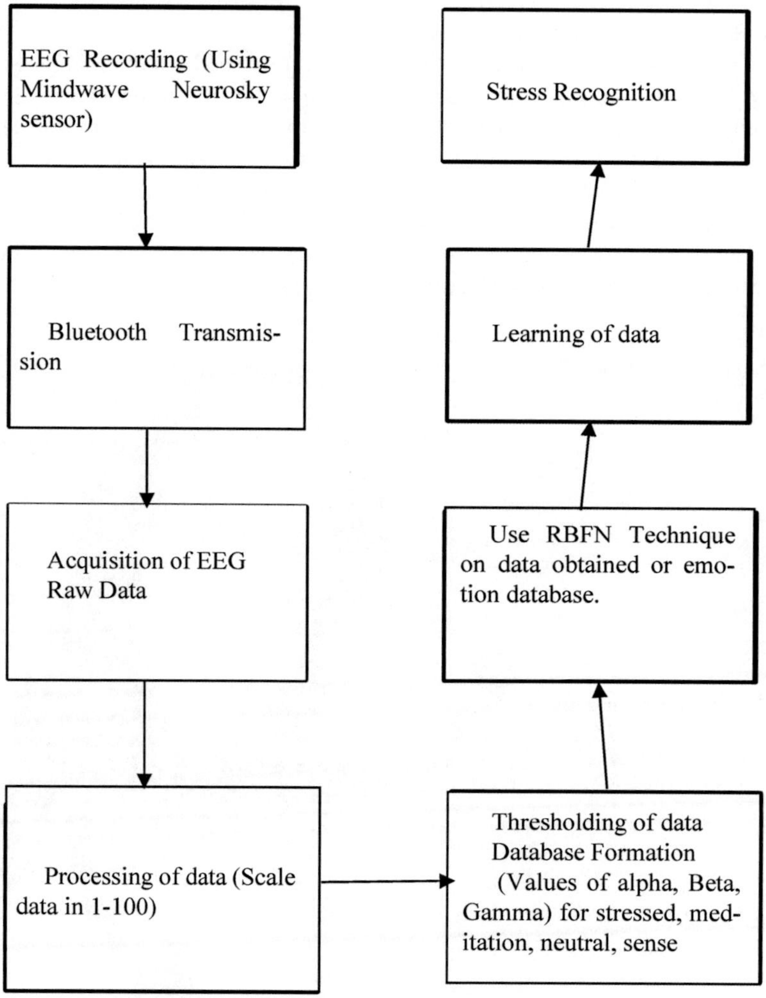

Fig. 2 Block diagram of stress detection using neurosky mindwave mobile 2 sensor

The technique used for stress detection is ANN. An artificial neuron network (ANN) is a computational model based on the structure and functions of biological neural networks. Information comes through the network aspects and the structure of the ANN is because a neural network changes or learns, in a sense based on that input and output. Artificial neural networks are built like the human brain, with neuron nodes interconnected like a Web. The human brain has hundreds of billions of cells called neurons. Each neuron is made up of a cell body that is responsible for processing information by carrying information toward (inputs) and away (outputs) from the brain.

ANN has hundreds or thousands of artificial neurons called processing units, which are interconnected by nodes. These processing units are made up of input and output units, the input units receive various forms and structures of information based on an internal weighting system, and the neural network attempts to learn about the information presented to produce one output report. Just like humans need rules and guidelines to come up with a result or output, ANNs also use a set of learning rules called backpropagation, an abbreviation for backward propagation of error, to perfect their output results.

An ANN initially goes through a training phase where it learns to recognize patterns in data, whether visually, aurally, or textually. During this supervised phase, the network compares its actual output produced with what it was meant to produce, i.e., the desired output. The difference between both outcomes is adjusted using back propagation. This means that the network works backward going from the output unit to the input units to adjust the weight of its connections between the units until the difference between the actual and desired outcome produces the lowest possible error. During the training and supervisory stage, the ANN is taught what to look for and what its output should be, using Yes/No question types with binary numbers.

The RBFN technique is used for the data processing and stress recognition. Radial basis function (RBF) networks typically have three layers: an input layer, a hidden layer with a nonlinear RBF activation function, and a linear output layer. RBFN neural network: The RBFN is a "three-layer network" in which first layer is "input layer," second layer is "hidden layer" which contains RBF neurons, third layer is "output layer." The prototype vectors used by the RBFN neurons are sometimes referred to as the "input weights." Generally, the weights as being coefficients, meaning that the weights will be multiplied against an input value. Here, though, in this project, computing the distance between the input vector and the "input weights" (the prototype vector), this can be done using radial basis function algorithm.

There are several steps in RBFN, the first is RBF mapping, then network architecture of RBF, computational power of RBF, training of RBF network, unsupervised optimization of basis function, computing on output weights, and supervised RBF network training. This algorithm can be proved with the help of mathematical expression. Every RBF neuron is a measure of the similarity of the input to its vector model (taken from the training set). The prototype like input vectors returns a value closer to 1. Figure 3 shows the structure of the RBFN.

There are various possible similarity function options, but the most famous one is based on the Gaussian. Below is a one-dimensional input equation for a Gaussian.

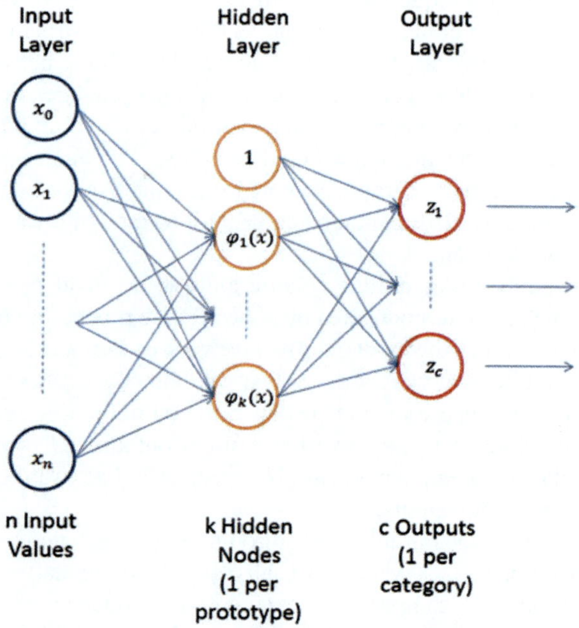

Fig. 3 RBFN architecture

$$f(x) = \frac{1}{\sigma\sqrt{2\pi}} e^{-\frac{(x-\mu)^2}{2\sigma^2}} \tag{1}$$

x is the input, mu is the mean, and sigma is a standard deviation.

The function used for activation of neuron function is

$$\varphi(x) = e^{-\beta\|x-\mu\|^2} \tag{2}$$

The training process for an RBFN consists of selecting three sets of parameters: the prototypes (mu) and beta coefficient for each of the RBF neurons, and the matrix of output weights between the RBF neurons and the output nodes.

Making an intelligent selection of prototypes is to perform *K*-means clustering which is conducted on the training set, and cluster centers are used as models.

The cluster centers are determined as the sum of all the cluster points.

$$\sigma = \frac{1}{m} \sum_{i=1}^{m} \|x_i - \mu\| \tag{3}$$

The *k*-means clustering is used to pick models, then a simple method to determine beta coefficients is to set sigma equal to the average distance between all cluster and cluster center points.

Here, mu is the centroid cluster, m is this cluster's number of training samples, and xi is the cluster's *i*th training sample. Once the cluster has the sigma value, beta can be measured as:

$$\beta = \frac{1}{2\sigma^2} \tag{4}$$

The final set of train parameters is the weights of the output. Using gradient descent (also known as least mean squares), these can be equipped. Next, measure the activation values of the RBF neurons for each data point in the training set. Such values of activation become the outputs for learning for gradient descent.

The linear equation needs a bias expression, so always add a fixed value of "1" to the activation values start of the vector. For each output node, gradient descent must be performed separately (i.e., for each group in your dataset).

For the output labels, use the value "1" for samples that belong to the same category as the output node, and "0" for all other samples. For example, if our dataset has three classes, and we are learning the weights for output node 3, then all category 3 examples should be labeled as "1" and all category 1 and 2 examples should be labeled as 0.

The algorithm for the complete system are as follows.

Step 1: Start (Flag = 1).

Step 2: Create serial network (by selecting port which is com66 and interfacing serially).

Step 3: Receiving of data.

Step 4: Parsing of data (conversion of data from string to num).

Step 5: Plotting received data (data will be plotted on MATLAB).

Step 6: Categorizing values in different modes.

Step 7: Taking a count from receiving values (taking a count of each wave for defined specific range).

Step 8: Comparing the count (compare count from each wave).

Step 9: Decision taking by using RBFN Algorithm.

Step 11: Repeating the process (till flag becomes 0).

Step 12: Stop (Flag = 0).

3 Results and Discussion

In proposed work, the sensor captures the brain signals and that further send with the help of microcontroller to the PC. Further, the obtained raw data is scaled, and by thresholding the values, the database is formed and that is trained by using the RBFN algorithm. By using the sensor and taking readings of 100 different person, the analysis can be done. The actual state and after experiment, detected state can be compared. From these data accuracy of the sensor is calculated. Table 1 shows threshold values for various states of brain.

Table. 1. Threshold values for various states of brain

Waves	Thresholding range	State of brain
Gamma	60–80	Concentrated/mediation
Alpha, beta	1–20	Low e-sense (sleepy, angry)
Theta	20–40	Stressed
Gamma	40–60	Neutral
Gamma	80–100	Highly concentrated

Fig. 4 Confusion matrix

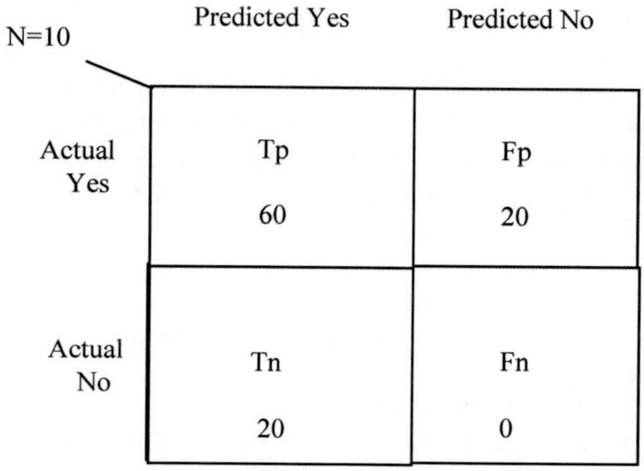

The confusion matrix in Fig. 4 shows the comparison between the predicted and actual results. Here, *N* is the number of samples there are four cases T_p, F_p, T_n, F_n etc. taken for analysis.

From these, T_p is the case in which the predicted and actual results are same or true. F_p is the case in which the prediction is positive but actually not, (person is in stress but he is actually not). T_n is the case in which the predicted and actual both are false. F_n is the case in which the prediction is negative and actual is positive. The analysis can be done by taking number of samples.

Accuracy can be calculated as:

$$\text{Accuracy} = \frac{T_p + T_n}{T_p + T_n + F_p + F_n} \times 100 \quad (5)$$

$$\text{Accuracy} = 80\%$$

4 Conclusion

These techniques play an important role to analyze brain activities for recognition of stress using effective use of system controls, sensor, and RBFN. The proposed system is helpful in detecting the stress, which is very common nowadays. Once stress is detected it can be cured at earliest stage. The advantage of using the proposed RBFN method is easy and fast to determine. The accuracy of the work for recognition of stress is found 80% by using the ANN Technique.

Reference

1. Tiwari A, Tiwari R (2017) Design and implementation of a brain computer interface for stress management using labVIEW. In: International conference on computer, communications and electronics
2. Agrafioti F, Hatzinakos D, Anderson AK (2012) ECG pattern analysis for emotion detection. IEEE
3. Jun G, Smitha KG (2016) EEG based stress level identification. IEEE
4. Hosseini SA, Emotional stress recognition system using EEG and psychophysiological signals
5. Jain A, Petiwala D, Narayan P, Shobhit S (2017) Automated wheel-chair on eyebrow movements using brain wave sensor by EEG. IJMR
6. Prabhu NG, Singh NS, Singh SV, Patil N (2016) Affective e-learning using emotion detection. IJTR
7. Gayathri PA, New step in brain computer interaction towards emotion recognition and prediction
8. Chavan DR, Kumbhar MS, Chavan RR (2016) The human stress recognition of brain using music therapy. ICCPEIC
9. Lim CK, Chia WC. Analysis of single-electrode EEG rhythms using MATLAB to elicit correlation with cognitive stress. Int J Comput Theor Eng
10. Xu Q, Nwe TL, Guan C (2015) Cluster-based analysis for personalized stress evaluation using physiological signals. IEEE J Biomed Health Inf 19(1). doi: https://doi.org/10.1109/JBHI.2014.2311044
11. Chen T, Yuen P, Richardson M, Liu G, She Z (2014) Detection of psychological stress using a hyperspectral imaging technique. IEEE Trans Affect Comput 5(4):391–405
12. Healey JA, Picard RW 2(005) Detecting stress during real-world driving tasks using physiological sensors. IEEE Trans Intell Transp Syst 6(2):156–166
13. NeuroSky -Body and Mind. Quantied. https://store.neurosky.com/products/mindwave mobile
14. Electroencephalography:-https://en.wikipedia.org/wiki/Electroencephalography

Deep Learning-Based Paperless Attendance Monitoring System

Pallavi Derkar, Jitesh Jha, Mayuresh Mohite, and Rushikesh Borse

Abstract This paper concerns about the paperless attendance monitoring system. Major focus of this system is on the concept of face recognition. The goal is to mark the attendance without disturbing the class by using face recognition technique. Marked attendance using this monitoring system can be directly sent to the mentor/coordinator through email. Face recognition uses biometric features, and it extracts person's facial features and stores data. Faces will be recognized based on deep learning algorithm to train the system and to compare or test to identify the person. Time is an important factor in recognition. This factor has been brought up to get the attendance in hand of the respected person instantly. To implement this efficiently, we have gone for deep learning model based on CNN. One important aspect is to deal with the results of the live detection and the immediate prediction of the person. The proposed system can identify an individual from various angles and positions. One of the important parameters that decides the accuracy is the quality of image, better the quality better the accuracy. Quality of the captured image defines the accuracy of the system. After implementing CNN on created database, we achieved an accuracy of 80.5%. Accuracy value can change depending on the testing database.

Keywords Machine learning · Deep learning · Convolutional neural network

P. Derkar · J. Jha (✉) · M. Mohite · R. Borse
Department of Electronics and Telecommunication, School of Electrical Engineering, MIT Academy of Engineering, Alandi Pune, India
e-mail: jsjha@mitaoe.ac.in

P. Derkar
e-mail: prderkar@mitaoe.ac.in

M. Mohite
e-mail: mymohite@mitaoe.ac.in

R. Borse
e-mail: rpborse@gmail.com

© Springer Nature Singapore Pte Ltd. 2021

S. N. Merchant et al. (eds.), *Advances in Signal and Data Processing*,
Lecture Notes in Electrical Engineering 703,
https://doi.org/10.1007/978-981-15-8391-9_47

1 Introduction

Monitoring the attendance has always been a major concern for various institutes and organization, and this is done in order to get defaulter's and salary record of students and employees, respectively. Following are the traditional techniques of attendance monitoring:

(i) In various institutes and organizations, marking and monitoring attendance is still done using traditional method, i.e., using pen and paper.
(ii) Institutes and organizations located in urban areas have adopted recent technologies, i.e., biometric units have been installed which use thumb impressions.
(iii) There are certain biometric units which uses both thumb impression and face recognition by standing in front of the unit installed.

1.1 Past Works

Out of many techniques of attendance monitoring, face recognition has received good remarks. In order to build algorithm for face recognition, human face has a complicated structure and model. Hence, the vertical symmetry as well as multidimensional visualization of human faces plays an important role. Working in the field of deep learning brings us most popular method in case of the feature extraction. According to Laurene Fausett [1], neural network has become of more interest to any of the research field whether be it electrical or computer. In computer, it has main role in recognition, detection, etc.

Feature extraction from the images is an effective process using convolution neural networks (CNN) in deep learning. Deep learning is a practical approach which consumes less time in training process. Dr. Priya Gupta et al. [2] state convolutional neural network to be the powerful tool in extraction of the features from an image and it deliberately focuses on the Har cascade for easy extraction of the features. Digital image processing (DIP) is the process which does various needful operations on image such as image preprocessing, edge detection, image compression, and segmentation for the ease of further process. Working on the large data samples Liu Xia et al. [4] has given the conclusion on the face detection using the AdaBoost algorithm and based on the eyes detection and based on the number of pixels. Apart from this, in order to bring the accuracy further, experimentation on the Euclidean distances and vector cosine is also proposed. Gosavi et al. [8] compute the feature extraction method based on geometric-based feature extraction and also states CNN to use for the feature extraction from various angles as well as proposed various neural network techniques and the feature extraction techniques. Bishop [6] in his book has thoroughly explained that CNN is the approach of creating models that inverts the transformations of the inputs which further builds the properties to the fully connected layer of CNN [7, 5]. Sufficiently larger training dataset is being given and the invariances are extracted by the examples meanwhile states to be the learning approach. According to the book, when we extract the feature from any of image from CNN approach, we tend to do it based on the pixels which are likely to correlate the nearby pixels not the distant ones. So, they are partly done and the local features are extracted from each of the part then they are merged to get the higher-order features which are ultimately the feature of the whole image.

1.2 Our Proposed System

Deep learning-based attendance monitoring system will be an unit which will be installed wherever we need to monitor attendance of students and employees, for example, classrooms and conference hall, respectively.

What it expects is just the physical presence of the person.

The aim of the proposed method is to detect the live captured face in the form of an image and identify the person using a standard image database with a better efficiency.

This paper is organized as follows. Section 2 presents the proposed algorithm for the entire process of face detection and recognition. Section 3 presents implementation and the results, and Sect. 4 concludes this paper.

2 Proposed Methodology

The process of the proposed system is as shown in the flow of Fig. 1.

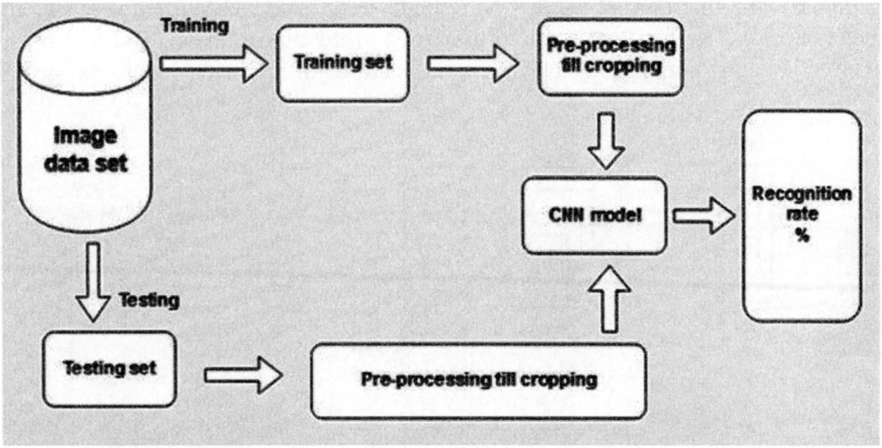

Fig. 1 Proposed methodology for face recognition

The method as shown in Fig. 1 is used for the face detection and the recognition being independent of the variations such as hairstyle, facial expressions as well as color, purposefully working on the CNN model for easier extraction of the features from image.

Processing starts with the standard available image dataset which is not dependent of the color, i.e., it can be a grayscale image or the RGB image. In the image dataset itself, the 70% data will be for training while the remaining 30% will be for testing dataset. The image when given as the input, it is being cropped for the further process of CNN.

CNN model is the heart of the system as it works inherently on extracting the image features which is an important step in detection as well as the recognition. CNN through its meaningful layers brings out the unique feature of each differing inputs. The last stage which is the recognition of the name depends on the name given to the image of that person in the training dataset, otherwise reverts unknown if not in the dataset values. The recognition rate simply gives out the percentile rate of the recognition of the image.

2.1 CNN Algorithm

In the neural networks, CNN states to be the useful approach for the feature extraction, image detection and recognition, classification as well as object detection. CNN stands for convolutional neural network. It works from low-level featuring to the high level (Fig. 2).

CNN model basically consists of layers such as convolutional layer along with the ReLU function, pooling layer, Flatten, and the fully connected layer.

Image and a filter/kernel are the main inputs to this model. At the first stage, an image is passed as the input through the series of the above layers. The first most layer called convolutional layer extracts the features of the given input, i.e., given

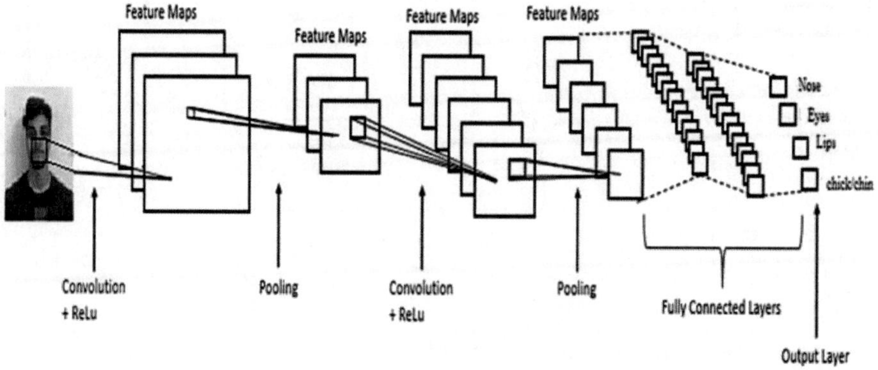

Fig. 2 CNN block diagram

Fig. 3 Max pooling

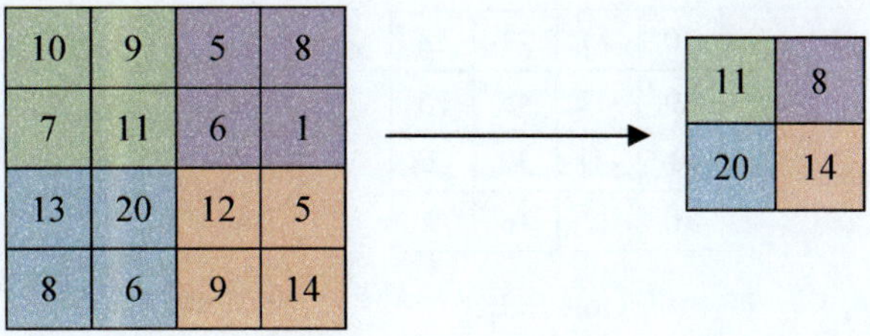

image. From small squares of an input image, the relationship between pixels is preserved by the convolution. Hence, the image forms the matrix of volume having three dimensions.

So, the image matrix dimension can be given as $H \times D \times W$.

A filter/kernel is given by $f_H \times f_W \times D$.

The output which is obtained is of dimension $(H - f_H + 1) \times (W - f_W + 1) \times 1$.

Hence, from the convolution of an image with the applied filters, various outputs can be achieved such as sharpening, making blur, edge detection. Stride gives the number by which the pixels shift over the input image matrix. For example, we move the filters that are moved to one pixel at a time when the stride is one and so on.

$$y[m, n] = h[m, n] * x[m, n] \tag{1}$$

$$y[m, n] = \sum_{j=-\infty}^{\infty} \sum_{i=-\infty}^{\infty} h[i, j] \cdot x[m - i, n - j] \tag{2}$$

After feature extraction, rectified linear unit (ReLU) function is a nonlinear operation and is used to bring in the nonlinearity in the Conv net. Passing this further through the pooling helps in reduction of the spatial size of the convolved features as well as decreases the computational power needed for the processing of the data. Pooling again brings in the types such as 1. Max pooling used in order to reduce the noise, i.e., filtering. 2. Average pooling in order to reduce the dimensions. 3. Sum pooling. Below figure is the example of the max pooling with the stride value 2 (Fig. 3).

Going through the number of epochs in this, we come up with the salient features of the given input image. The given features are then flattened in a network and then processed to the fully connected layer with the features in the form of column.

3 ReLU

The most commonly used activation function among others is ReLU, the rectified linear unit. It is used in various deep as well as machine learning models. Does not

10	15	-9	14
20	-11	45	23
40	-41	30	-26
10	32	16	9

Before ReLU

→

10	15	0	14
20	0	45	23
40	0	30	0
10	32	16	9

After ReLU

Fig. 4 Applying ReLU function

react to the negative values but if positive then gives out the values. So, it can be written as:

$$f(x) = \max(0, x) \tag{3}$$

ReLU is mainly used for two reasons, for modeling the 1. iteration effect 2. nonlinear effect (Fig. 4).

Which results in the following linear graph (Fig. 5).

It is strange to see the linear parts in the ReLU but it works better than any other functions in case of the iteration and the nonlinear effect.

Explaining in simple way what ReLU does to an image is it removes the negative parts, i.e., the black pixels convert them to the non-negative values, i.e., black to gray else gray to white. As the images are nonlinear, we again tend to make it nonlinear with the help of ReLU in order to remove the linearity and make it follow the CNN process easily.

As shown in Fig. 3, the CNN has the various layers described above along those layers when we are done with the pooling layer then it splits and follows the face detection path which is a probabilistic approach. The face gets with its encodings (i.e., numbers); this is in the Boolean format which is true/false or 1/0. As the Bayesian law gives out the result to be.

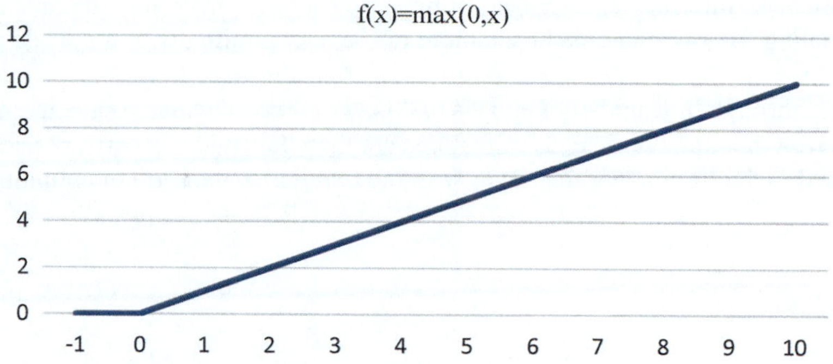

Fig. 5 Graphical representation of ReLU

$$P(\text{face detected}|\text{face in training set}) = \frac{P(\text{detected faces}) \times P(\text{face from training set}|\text{unknown face})}{P(\text{face from training set})} \quad (4)$$

Hence, the face would get detected if and only if it is available in the training set else it is given to be not recognized one.

4 Implementation and Results

4.1 Implementation Details

Our live attendance monitoring system is implemented using Python code. We have produced the Python code using various libraries available such as shape_detector, OpenCV, dlib, numpy, PIL which are needed for the image processing, feature extraction, and the camera interfacing. OpenCV is used to detect the faces where we are providing the orientations from –90° to 90° incrementing with 10°. The results represent the orientation with the increment of 30° as the images more than the 90° will not be detected. The features from the faces are localized in a proper path with the code. In order to increase the performance, changes are being made as required. The image dataset is prepared on the two basis one for the training purpose and the other one for testing. The training dataset is the captured images from us as shown below with varied angles for detection.

4.1.1 Input Database

4.1.2 Execution Steps for Implementing Face Detective Algorithm

Here, we have the system available with the webcamera for the live detection.

1. Initialize and Setup
   We have imported the library which is required for the face recognition and also the parameters needed.
2. Get the encodings of the face through the images
   Here, we are writing the function that will take an image filename and give us the face encodings for that image.
3. Compare faces
   Here, we are writing the function that will compare a given face encoding with a list of known face encodings. It will return an array of Boolean (true/false) values that indicate whether or not there was a match.
4. Find match

Here, we are writing the function that will take a list of known face encodings, list of names of people (corresponding to the list of known encodings of the face) and a face to find a match for. It will call the function in 4(a) and return the name of the person with whom the given face matches.

5. Getting face encodings for all faces
   Numbers for the proposed system which will be useful for multiple comparisons.
6. Matching each image with the known faces (one by one)

4.1.3 Below is the Algorithm for the Training Dataset

Algorithm 1 Face Detection

Initialize and **setup** with the libraries needed for the face detection and recognition;
Gather with the training dataset;
for Each image/face in training dataset **do:**
  if the face is detected based on the training dataset then:
    output = "person's name" Found;
  else
    output= Not found;
  end if
end for
Move back to the faces stating not found and check with the cons in order to retrain the dataset.

4.2 Discussions

Our algorithm of deep learning based on the CNN model detects a face with various angles and alignment as well. The training database is enough for the system to train the system and the testing is done with the live video. As shown in Fig. 6a, the labeled image is given which is the main image to be given in training set which detects the faces at any of the angle from left to right side (-90° to + 90° as shown in Fig. 6b. Also, the images with varied orientations are detected (Fig. 6c). The participant in Fig. 6a–c, Fig. 7a, b and also in Fig. 8 is the author Jitesh Jha of this paper. This is in order to show the real results.

As explained in the above paragraph, the faces at −90° to +90° are detectable because the facial attributes are extractable while in case of the angle greater than 90° are not being detected due to non-facial attributes as shown in Fig. 7a, b. In case of Fig. 7b, it seems to be the case of face angle out of the defined range. The reason behind this is the CNN model which is not able to extract the features from non-facial attributes. These are the sample images which are tested by us (Table 1).

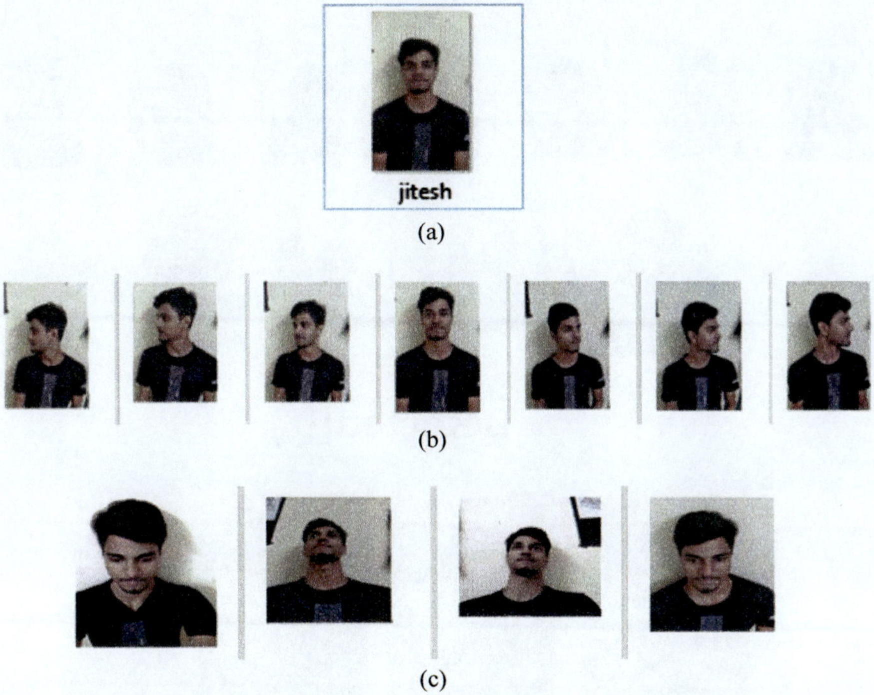

Fig. 6 **a**. Labeled input image. **b**. Training dataset from varied angles for detection. **c**. Few more orientations which are detectable

5 Confusion Matrix

Formula for calculating precision, recall, and F1 score for individual person (Table 2):

$$\text{Precision} = \frac{\text{Truly detected}}{\text{Truely detected} + \text{Positive Misclassification}} \tag{5}$$

$$\text{Recall} = \frac{\text{Truly detected}}{\text{Truely detected} + \text{Negative Misclassification}} \tag{6}$$

$$\text{F1Score} = \frac{2 \times \text{Recall} \times \text{Precision}}{\text{Recall} \pm \text{Precision}} \tag{7}$$

Above matrix is the standard confusion matrix for 200 sample images. None of the images were recognized as other's image. Instead the result found as misclassification with various classes. Some images were classified as zero faces, some images as more than one face, and in some images with no faces.

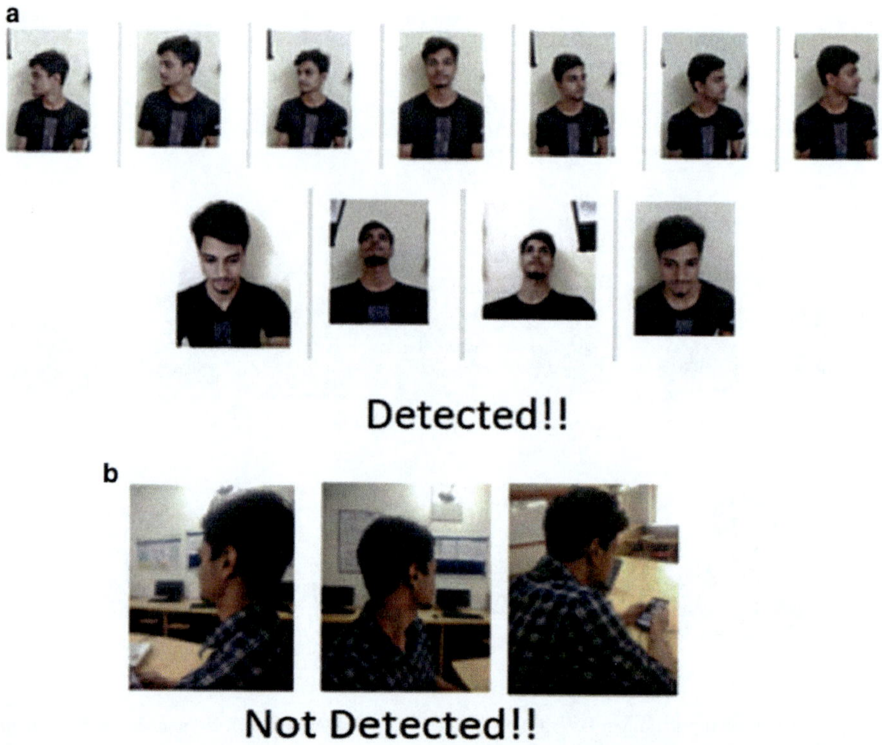

Fig. 7 **a**. Detected sample images on angular basis and elevation angles. **b** Non-detected images

Fig. 8 Farther and blur image

Table 1 Result based on some variable parameters

Parameter	Discussion/inference
Tolerance: It is for face comparison	The lower the tolerance number, stricter the comparison Lower value is used to avoid false matches Higher value is used to avoid false negatives
Face orientation	Between −90° to + 90° face position is detected
Stride: used in convolution	Larger the stride smaller is the output feature map
Accuracy	Good accuracy for images with clearly visible face

Table 2 Standard confusion matrix

Samples 200	**Varad**	**Mayuresh**	**Jitesh**	**Pallavi**	
Varad	**47**	**0**	**0**	**0**	**47**
Mayuresh	**0**	**41**	**0**	**0**	**41**
Jitesh	**0**	**0**	**33**	**0**	**33**
Pallavi	**0**	**0**	**0**	**40**	**40**
	47	**41**	**33**	**40**	↘

Truly Detected

These results made to modify the confusion matrix. Below is the modified confusion matrix (Table 3) with additional classes according to the result found while testing the sample images (Table 4).

Total of the truly detected images are 161/200.

Hence, the accuracy (Table 5),

$$\text{Accuracy} = 161/200 = 0.805$$
$$\text{Error Rate} = 1 - \text{Accuracy} = 0.195$$

Total of the truly detected images is 185/200.

Hence, the accuracy (Table 6),

Table 3 Modified confusion matrix

Samples 200	Varad	Mayuresh	Jitesh	Pallavi	More faces	Error	
Varad	47	0	0	0	2	0	49
Mayuresh	0	41	0	0	2	0	43
Jitesh	0	0	33	0	3	1	37
Pallavi	0	0	0	40	0	0	40
Zero faces	1	3	13	7	0	0	24
No faces	0	4	0	3	0	0	7
	48	48	46	50	7	1	

Truly Detected

Table 4 Precision, recall, F1 score

Name	Precision (%)	Recall (%)	F1 score (%)
Varad	97.91	95.91	96.9
Mayuresh	85.41	95.34	90.1
Jitesh	71.74	89.18	79.51
Pallavi	80	100	88.88

$$\text{Accuracy} = 185/200 = 0.925$$

$$\text{Error Rate} = 1 - \text{Accuracy} = 0.075$$

5.1 Result

As shown in Table 7, we have used the standard database and this shows the following results based on accuracy.

Accuracy of the standard database is way better than our database. There are some reasons by which accuracy value is better in standard database.

In standard database, images are already preprocessed before testing on algorithm. Images are with good quality, well cropped, and with good pixel which helps in easy recognition. Image quality in our database are not much good which causes unknown

Table 5 Confusion matrix for the standard database from Kaggle

Samples 200	**Alexandra**	**Debnam**	**Blake**	**Cameron**	**More faces**	**Error**	
Alexandra	**49**	**0**	**0**	**0**	**0**	**1**	**50**
Debnam	**0**	**46**	**0**	**0**	**0**	**0**	**46**
blake	**0**	**0**	**43**	**0**	**0**	**3**	**46**
Cameron	**0**	**0**	**0**	**47**	**0**	**1**	**48**
Zero faces	**0**	**4**	**4**	**2**	**0**	**0**	**10**
No faces	**0**	**0**	**0**	**0**	**0**	**0**	**0**
	49	**50**	**47**	**49**	**0**	**5**	

Table 6 Precision, recall, F1 score

Name	Precision (%)	Recall (%)	F1 score (%)
Alexandra	100	98	96.9
Debnam	92	100	90.1
Blake	91.48	89.18	93.47
Cameron	95.91	100	97.91

Table 7 Accuracy using the standard databases

Sr no.	Databases	Total number of samples	True recognized images	Accuracy
1	PINS	200	185	0.925
2	14 celebrity	96	92	0.95
3	Our database	200	161	0.805

detection for some images and thus the accuracy value is less than the for the standard image database.

If the images from our database are preprocessed before testing, the accuracy value might increase than the current value as given in Fig. 8. Though the image is blur, in the small pixels, it was detected. If the preprocessing is done, then we will be able to bring better accuracy.

6 Conclusion

Face recognition is a technology to identify or verify a person from the captured image or video using CNN model. The paperless/live attendance monitoring system mainly focuses on the face recognition and the prediction of that respective person. Age, expressions, and color are not main constrains in the attendance monitoring system irrespective of other environmental parameters. Deep learning model CNN is the dominant factor of the proposed system. The proposed system is able to recognize the various test images of range $-90°$ to $+ 90°$ with good accuracy. With this, it is observed that the study of computer vision problems needs the large amount of data availability as it brings a major impact on training the machine properly with less time-related issues.

References

1. Fausett LV (2004) Fundamental of neural networks architectures, algorithms and applications, 1st edn. Pearson Education India,
2. Gupta P, Saxena N, Sharma M, Tripathi J (2018) Deep neural network for human face recognition. Int J Eng Manuf (IJEM) 8(1):63–71
3. Deshpande NT, Face detection and recognition using viola-jones algorithm and fusion of PCA and ANN. Proc Adv Comput Sci Technol, 1173–1189
4. Xia L, Tingjun L, Wei J, Qingchang G (2008) The study of face recognition based on the large samples. In: Proceedings of international conference on computer science and software engineering
5. Murphy KP (2019) Machine learning: a probabilistic perspective. The MIT Press, Cambridge
6. Bishop CM (2006) Pattern recognition and machine learning. Springer, New York
7. Alpaydın E (2010) Introduction to machine learning, 2nd edn. The MIT Press, Cambridge
8. Gosavi VR, Sable GS, Deshmane AK (2018) Evaluation of feature extraction techniques using neural network as a classifier: a comparative review for face recognition. Int J Sci Res Sci Technol 4(2):1082–91. (IEEE Student Member, G. S. Mandal's, M.I.T, Aurangabad Maharashtra, India 2018 IJSRST)
9. Han X (2018) Research on face recognition based on deep learning. In: Proceedings of 2018 6th international conference on digital information, networking, and wireless communications (DINWC), IEEE
10. Yuan L, Qu Z, Zhao Y, Zhang H, Nian Q (2017) A convolutional neural network based on tensor flow for face recognition. In: Proceedings of 2nd advanced information technology, electronic and automation control conference (IAEAC). IEEE
11. Zhi-Peng FU et al (2014) Survey of deep learning in face recognition. In: Proceedings of international conference on orange technologies. IEEE
12. Eidinger E, Enbar R, Hassner T (2014) Age and gender estimation of unfiltered faces. IEEE Trans Inf Forensics Secur 9(12):2170–2179
13. Sandıkcı EN, Erdem ÇE, Ulukaya S (2018) A comparison of facial landmark detection methods. In: Proceedings of 26th signal processing and communications applications conference (SIU). IEEE

Image Analytics to Detect Cigarette in an Image Using Deep Learning

Abhijeet Kharade, Kumar Abhishek, Debaraj Dwibedi, Siddharth Mehta, Hemanth Meruga, Pratap Gangula, D. Narayana, and Rushikesh Borse

Abstract Significant number of modern films depict some form of tobacco use, but rarely depict its real-life consequences such as addiction, illness and death. As per [1], anti-tobacco health warnings are mandatory for scenes depicting smoking scenes. In this paper, an automated recognition system is proposed to identify images with smoking activities and tag them accordingly. The proposed approach implements the technique of object detection based on deep learning. Convolutional neural network is used to generate feature maps from the images. These machine-learnt features are used to classify the images. The system can detect the smoking events of uncertain actions with various cigarette sizes, colors and shapes. We have experimented our work by applying the proposed approach to two real-world datasets and that have demonstrated the effectiveness of our solution with a decent model accuracy.

Keywords Cigarette detection · Deep learning · Convolution neural network

1 Introduction

Tobacco and e-cigarette companies spend billions of dollars every year promoting their products at stores, in magazines and online. Significant number of modern films depict some form of tobacco use, with potentially detrimental effects on the behavior of young people. As such, an automated recognition system for this action may be useful for the purposes of film classification. As per [1], anti-tobacco health warning is to be displayed at the bottom of the screen as a prominent static message when

A. Kharade · K. Abhishek · D. Dwibedi (✉) · S. Mehta · H. Meruga · P. Gangula · D. Narayana
Great Lakes Institute of Management, Manamai, Tamil Nadu, India
e-mail: dwibedidebaraj@gmail.com

D. Narayana
e-mail: narayana@greatlearning.in

R. Borse
E&TC Engineering, MIT Academy of Engineering, Alandi, Pune, India
e-mail: rpborse@mitaoe.ac.in

© Springer Nature Singapore Pte Ltd. 2021

S. N. Merchant et al. (eds.), *Advances in Signal and Data Processing*,
Lecture Notes in Electrical Engineering 703,
https://doi.org/10.1007/978-981-15-8391-9_48

tobacco products are shown in a film or a television program screened on Indian televisions. A task to blur the cigarettes/cigars and other smoking apparatus may be applicable as well. A manual process requires a person to watch entire video and identify the frames that require statutory warning. For the smoking event analysis, the uncertainties in cigarette size, orientation and colors make it very challenging. Automation of this tedious and repetitive audit processes would significantly enhance audit effectiveness and efficiency.

Existing projects in the area of smoking detection typically rely on a combination of face detection, object recognition and smoke detection. The majority of techniques deployed for the task in hand are based on regular data processing techniques with traditional features. In the recent times, with the availability of huge annotated image dataset like ImageNet [2] and Microsoft COCO [3], and with the incremental development of GPUs, it is possible to train large and complex CNN models for accurate recognition. This surpasses most of the methods adopted using hand-engineered features [4]. An attempt is made in this paper, where we have employed deep learning-based method (CNN), rather than the hand-engineered features, to achieve a much higher accuracy.

This paper proposes a solution built using [5] that will allow the auditors to automate many tasks that have traditionally been conducted manually. The proposed approach is based on convolutional neural network [6] (CNN) with few optimal changes. The experimental results of applying the proposed approach to two real-world datasets have demonstrated the effectiveness of our solution.

We will discuss in detail the proposed approach and algorithm for object detection using deep learning in Sect. 3, while Sect. 4 gives about the upper bounds on the object detection results. We are providing the acknowledgement and references in Sects. 5 and 6, respectively.

1.1 Motivation of the Work

Roughly, half of the highest-grossing films in the USA from 2010–2016 (across all age ratings) depicted some form of tobacco use [7]. Studies [8, 9] have found a link between young people being exposed to on-screen tobacco use and taking up the habit themselves. In their 2015 report on tobacco use in films, the World Health Organization went so far as to recommend that films depicting smoking be given an adult rating [9]. This suggests a potential need for film classification to take depictions of smoking into account.

2 Related Work

Éamon Dunne[11] proposed a system that classifies a smoking event by searching for smoke near a person's hand or face (Fig. 1).

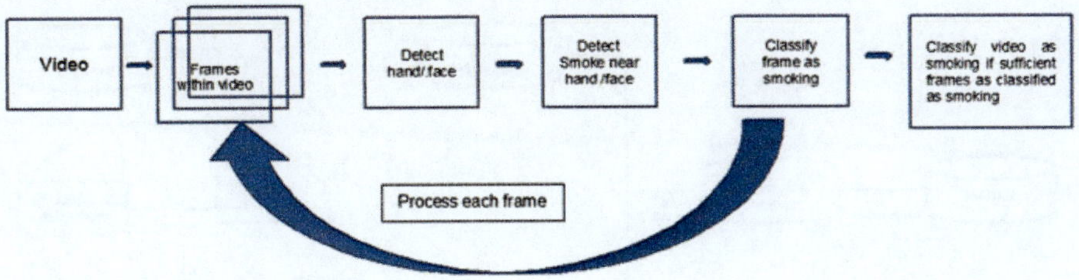

Fig. 1 Smoke fumes detection

For each frame, the system first identifies all faces present using Haar-based cascade classifiers [12]. For each face obtained from the previous stage, the system extracts a small section from the center as a sample of skin color. The system then verifies that all these areas are skin colored using a color analysis technique put forward by Kolkur [13]. Having defined face and hand areas, the system searches for smoke in these regions using a Gaussian mixture model, GMM [14]. First, the system identifies areas of movement using [15]. The system identifies smoke-colored pixels, using a color analysis algorithm by Yuan [15]. The system verifies that each moving pixel's RGB values fall within certain ranges, giving them a grayish color. Pixels that fail this stage of analysis are discarded, giving a binary image. To conclude whether these smoke-colored pixels represent actual smoke, a disorder analysis algorithm used by Chen et al. [16] is employed. If smoke is found in these areas, that frame of the video is classified as a potential smoking frame.

A major limitation of this system is that it requires the test footage to use a stationary camera. The smoking detection step uses GMM to identify moving objects, and GMM is unable to properly handle moving backgrounds. If the camera moves, GMM treats everything in the frame as moving, and the motion detection fails. Additionally, the system assumes that in any depiction of smoking, the smoker's face will be clearly visible. This is not always true: in some cases, the smoker's face may be obscured, turned away or simply not visible in the frame. The assumption that the smoker's hand will always be present is similarly problematic. Ideally, if the hand is not present in frame, the largest skin-colored contour outside of the face region should be very small. Hence, the 8 extra area chosen by the system to search for smoke should be negligible.

3 Proposed Approach

In this section, we discuss some techniques related to object detection based on deep learning. This section also introduces the concept of transfer learning used in this paper. These algorithms are very important in the object detection tasks of natural images. Some algorithms tend to have high precision, while others win in terms of speed (Fig. 2).

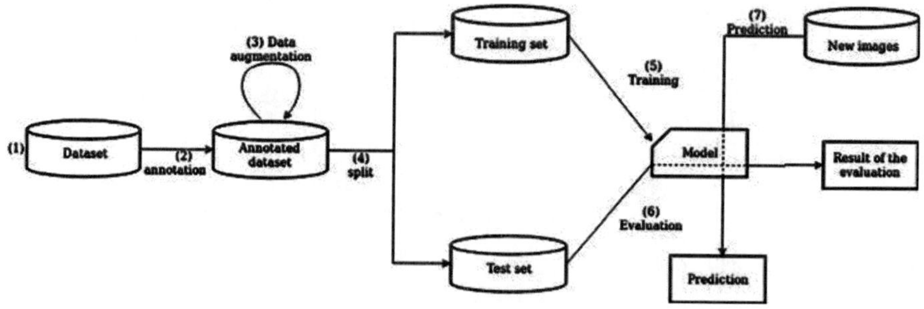

Fig. 2 Block diagram

3.1 Explanation of Block Diagram

Above is an abstraction of the tasks to be performed in the project.

3.1.1 Dataset

The dataset of intent contains two types of data, namely for positive and negative test cases. Positive test cases entail images which explicitly display the object of desire (as stated above) and are labeled accordingly, whereas negative test cases are divided into two subgroups (for the purpose of increased accuracy on test data)—images which do not contain the object of desire in its entirety, i.e., no cigarette and/or smoke vapor present in image and the second being images which contain merely smoke but no cigarette. This would prove useful for model training purposes as it will better assess test data in positive test cases. Much data labeling work for training and testing has been done. The annotated sample is input directly to the convolutional neural networks for training to realize end-to-end training.

Data was formatted into appropriately preprocessed floating-point tensors before being fed into the network. The data sits on a drive as JPEG files, so the steps that were followed for getting it into the network are as follows:

- Read the image files.
- Decoded the JPEG content to RGB grids of pixels.
- Converted these into floating-point tensors.
- Resized the images.
- Rescaled the pixel values (between 0 and 255) to the [0, 1] interval.
- Annotated resized images with labelImg
- Created the train and test folders for images.

3.1.2 Model

The network framework is based on the principle of SSD detection model. SSD model integrates faster R-CNN's anchor with YOLO's single convolutional neural network. The primary difference is that the SSD sets anchors on multiple feature maps (called the default box in SSD), which locates the objects more precisely. At the same time, SSD output is a series of fixed-size bounding boxes which are defined well in advance. This method is superior to the method of extracting region proposals in terms of speed.

3.1.3 Training

Data augmentation has been used for the purpose of improving the accuracy of the trained model, as it would allow us the facility of manipulating the data (flipping, scaling, rotation, light condition tuning, salt and pepper noise, etc.) so as to replicate the production environment in addition to increasing the availability of heterogenous training data, which would eventually improve the accuracy of the model on test or 'live' data.

Step 1: Annotated the training images

- The resize_images.py script is used to resize the images.
- Those files are splitted into two directories,./data/images/train and./data/images/test.
- Annotate resized images with labelImg.

Step 2: Prepared tfrecord files

After running the above step, two files train.record and test.record have been generated, both are binary files with each one containing the encoded jpg and bounding box annotation information for the corresponding train/test set. The tfrecord file format is easier to use and faster to load during the training phase compared to storing each image and annotation separately.

There are two steps in doing so:

Converting the individual *.xml files to a unified *.csv file for each set(train/test).

Converting the annotation *.csv and image files of each set(train/test) to *.record files (TFRecord format).

Scripts have been used to generate the tfrecord files as well as the label_map.pbtxt file which maps every object class name to an integer.

Step 3: Configured a training pipeline

Instead of training the model from scratch, transfer learning technique from a model pre-trained to detect the toxic objects has been used.

To do the transfer learning training, first the pre-trained model weights/checkpoints and then configured the corresponding pipeline config file were downloaded, to tell the trainer about the following informations.

- the pre-trained model checkpoint path(fine_tune_checkpoint),
- the path to those two tfrecord files,
- path to the label_map.pbtxt file(label_map_path),
- training batch size(batch_size)
- number of training steps(num_steps)
- number of classes of unique objects(num_classes).

Step 5: *Exporting and download a trained model*
Once the training job was completed, the newly trained model was extracted, as an inference graph, which we used in later stage to perform the object detection.

Step 6: *Inference test run*
Test with images in repository object_detection_demo/test directory.

3.1.4 Results of Evaluation

Intersection over union (IOU)

Intersection over union (IOU) is measure based on Jaccard Index that evaluates the overlap between two bounding boxes. It requires a ground truth bounding box and a predicted bounding box. By applying the IOU, we can tell if a detection is valid (true positive) or not (false positive). IOU is given by the overlapping area between the predicted bounding box and the ground truth bounding box divided by the area of union between them:

Although many levels can be taken, we have set six levels of aspect ratios (1, 2/1, 3/1, 1/2, 1/3) for the default box to make the object scalable. In the process of training, the default box is matched with the ground truth before selecting + ve and −ve samples.

The image below illustrates the IOU between a ground truth bounding box (in green) and a detected bounding box (in red) (Fig. 3).

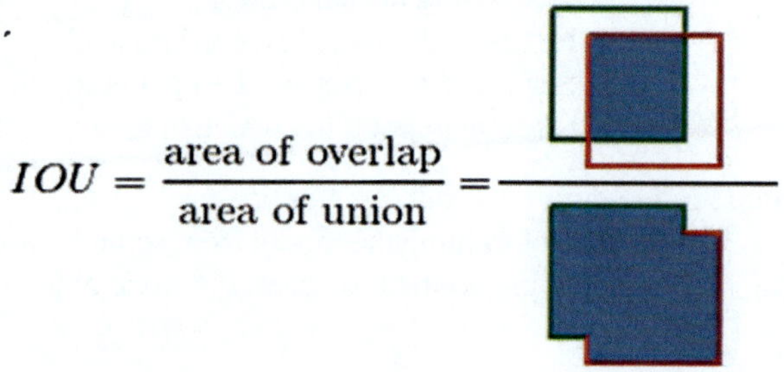

Fig. 3 Intersection over union (IOU)

Some basic concepts used by the metrics:

- True positive (TP)
- Correct detection. Detection with IOU $\geq$ threshold.
- False positive (FP)
- Wrong detection. Detection with IOU $<$ threshold.
- False negative (FN) A ground truth not detected
- True negative (TN):

Does not apply. It would represent a corrected misdetection. In the object detection task, there are many possible bounding boxes that should not be detected within an image. Thus, TN would be all possible bounding boxes that were correctly not detected (so many possible boxes within an image). That is why it is not used by the metrics.

Precision

Precision is the ability of a model to identify only the relevant objects. It is the percentage of correct positive predictions and is given by:

$$\text{Precision} = \text{TP}/(\text{TP} + \text{FP}) \tag{1}$$

Recall

Recall is the ability of a model to find all the relevant cases (all ground truth bounding boxes). It is the percentage of true positive detected among all relevant ground truths and is given by:

$$\text{Recall} = \text{TP}/(\text{TP} + \text{FN}) \tag{2}$$

F1 Score

F1-score is the harmonic mean of recall and precision. F1-score is computed by the following equation:

$$\text{F1} = 2 * (\text{precisionrecall})/(\text{precision} + \text{recall}) \tag{3}$$

In this particular case, the key metrics of desire are precision and recall. Due to the sensitivity of the task in that, it requires high accuracy in the case of correctly identifying the positive test cases and avoid rejecting ground truth cases or false negatives (where the object of desire is incorrectly labeled to be not present in the image, whereas in reality is present), as stated earlier.

Further, a high F1 score, in addition to mentioned classification metrics, would bolster the case for a welltrained model, which has both high precision and recall.

To ultimately arrive at the classification metrics, the intersection over union (IOU) metric would be utilized in order to derive a threshold which would qualify a prediction for accurate binary classification.

Fig. 4 Histogram ratio

$$r_t^+(i) = \frac{h_t(i)}{1+h_{t-1}(i)},$$

Cigarette Detection Using Ratio Histograms

Let f be the detected face region. In addition, assume that () $t\ h\ i$ and 1() $t\ h\ i$—enote the color histograms of f at time t and $t-1$, respectively. Then, we define the ratio histogram of forward type as (Fig. 4).

where (.) $t\ h$ is the color histogram of f, i.e., () (()) $t\ i\ t\ h\ I = \Sigma\delta\ f\ p - i$. Furthermore, the 1 L -form norm of ratio histogram $t\ r$ + is defined as:

$$\|(.)\|\|()\|t\ i\ t\ r+ = \Sigma\ r + i,$$

which measures the difference between two adjacent color histograms 1(.) $t\ h$ − and (.) $t\ h$. Assume that $rt\ T$ + is the average value of all bins in $t\ r$ + . For a color bin k, if rt + $(i) > 1.5\ rt\ T$ + , the kth color will be an important color for highlighting the cigarette region. Figure 2 shows the result of cigarette detection. (b) is the re-projection result of colors found from rt + (i). Clearly, most of important cigarette colors locate centrally around the cigareete area. Then, a connected component analysis can be further used to find the desired cigareete location. Then, we can threshold the magnitude || (.) || $t\ r$ + to highlight the intrusion cue in the analyzed face zoen, i.e., || || tr + $> T$. Like Fig. 5, an object newly comes by, and then the pulse signal also denotes an initiation cue of event occurrence.

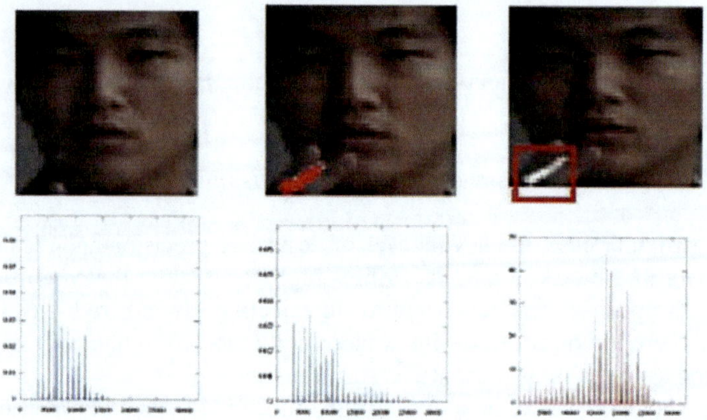

(a) (b) (c)

Fig. 5 Initiation cue of event occurrence

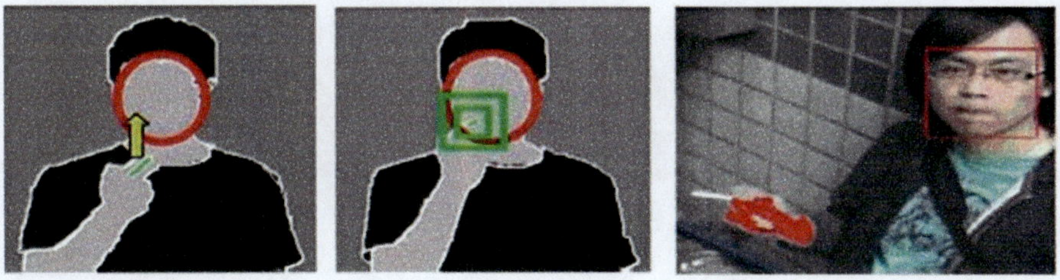

Fig. 6 Conceptual representation for smoking event analysis (left and middle), and the result of object tracking (right)

Figure 5 Result of cigarette detection. (a) and (b): the raw images and color histograms of detected face, taken at time $t - 1$ and t, respectively. (c) The ratio histogram of (a) and (b) and its back-projection regions is shown with red color in (b) (Fig. 6).

3.1.5 Algorithm

Transfer learning from a pre-trained Model—SSD with Mobilenet, i.e., single shot detection > ssd_mobilenet_v2.

The Single Shot Detector Mobilenet architecture model is derived from depth wise separable convolutions, defined as a form of factorized convolutions which factor a standard convolution into a depth wise convolution and a 1X1 convolution known as a pointwise convolution. The Mobilenet model applies a default single filter to each neural input channel to begin feature extraction. Following a depth wise convolution, a 1X1 pointwise convolution follows to combine the outputs of the depth wise convolution. The depth wise convolution will eventually split the resulting outputs into two layers, a separate layer for filtration and a layer for combination. The mixture of both output filtration and combination minimizes the model size reducing computational power demands (Howard et al. 2017). Moreover, depth wise convolution maximizes model efficiency by preventing GPU overconsumption on less demanding devices (e.g., mobile devices and laptops). However, a reduction in GPU consumption also creates a lack of usage equilibrium, which hinders the training model causing slow progress and intervals. Figure below exhibits the consecutive and systematic pooling layers of the SSD Mobilenet model and the compression/condensing of pointwise and depth wise outputs which provides the small-scale nature of the Mobilenet architecture (Fig. 7).

The networking integrity of the Mobilenet system uses default mechanisms such as a batch normalization and ReLU nonlinearity operation (excluding the final fully connected layer possessing no nonlinearity and utilizes a softmax layer for classification) for dataset processing. The compact nature of the SSD Mobilenet framework is associated with a reduction in both parameters and Mult-adds. Although the neural network structure of SSD Mobilenet is notorious for minimal accuracy in relation

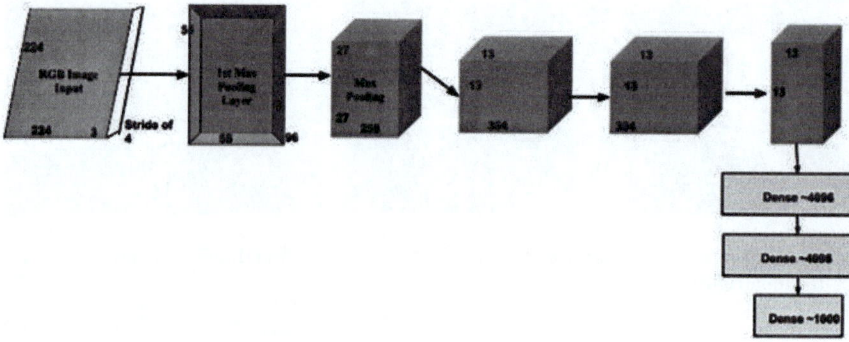

Fig. 7 Mobilenet architecture

to the faster R-CNN inception models, an intensively trained Mobilenet model can equate accuracy and precision on output object detection. Nevertheless, preliminary average precision scores (mAP) on small background objects reveal reduced values in relation to the inception models.

3.1.6 Flowchart

Step 1: *Annotated the training images*

- The resize_images.py script has been used to resize the images
- Those files have been splitted into two directories,./data/images/train and./data/images/test
- Annotate resized images with labelImg (Fig. 8).

Step 2: *Prepared tfrecord files*

After running the above step, two files train.record and test.record were generated, both are binary files with each one containing the encoded jpg and bounding box annotation information for the corresponding train/test set. The tfrecord file format is easier to use and faster to load during the training phase compared to storing each image and annotation separately.

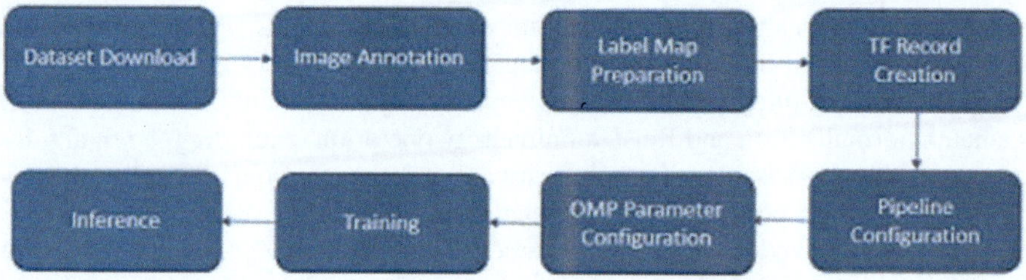

Fig. 8 Object detection flowchart

There are two steps in doing so:
Converting the individual *.xml files to a unified *.csv file for each set(train/test).
Converting the annotation *.csv and image files of each set(train/test) to *.record files (TFRecord format).
Scripts have been used, to generate the tfrecord files as well as the label_map.pbtxt file which maps every object class name to an integer.

```
%cd {repo_dir_path}

# Convert train folder annotation xml files to a single csv file,
# generate the `label_map.pbtxt` file to `data/` directory as well.
!python xml_to_csv.py -i data/images/train -o data/annotations/train_labels.csv -l data/annotations

# Convert test folder annotation xml files to a single csv.
!python xml_to_csv.py -i data/images/test -o data/annotations/test_labels.csv

# Generate `train.record`
!python generate_tfrecord.py --csv_input=data/annotations/train_labels.csv --output_path=data/annotations/train.record --img_path=data/images/train --label_map

# Generate `test.record`
!python generate_tfrecord.py --csv_input=data/annotations/test_labels.csv --output_path=data/annotations/test.record --img_path=data/images/test --label_map da
```

Step 3: *Configured a Training Pipeline*
Instead of training the model from scratch, transfer learning technique from a model pre-trained to detect the toxic objects has been used.
To do the transfer learning training, first downloaded the pre-trained model weights/checkpoints and then configured the corresponding pipeline config file to tell the trainer about the following informations.

- the pre-trained model checkpoint path(fine_tune_checkpoint),
- the path to those two tfrecord files,
- path to the label_map.pbtxt file(label_map_path),
- training batch size(batch_size)
- number of training steps(num_steps)
- number of classes of unique objects(num_classes).

Step 4: *Train the Model*
The model is trained, where the model_dir was the path of a new directory to store our output model.

```
model_dir = 'training/'
# Optionally remove content in output model directory to fresh start.
!rm -rf {model_dir}
os.makedirs(model_dir, exist_ok=True)


!python /content/models/research/object_detection/model_main.py \
    --pipeline_config_path={pipeline_fname} \
    --model_dir={model_dir} \
    --alsologtostderr \
    --num_train_steps={num_steps} \
    --num_eval_steps={num_eval_steps}
```

Step 5: *Exporting and Download a Trained Model*

Once the training job was completed, the newly trained model is being extracted, as an inference graph, which is used in later stage to perform the object detection. The conversion can be done as follows:

```
import re
import numpy as np

output_directory = './fine_tuned_model'

lst = os.listdir(model_dir)
lst = [l for l in lst if 'model.ckpt-' in l and '.meta' in l]
steps=np.array([int(re.findall('\d+', l)[0]) for l in lst])
last_model = lst[steps.argmax()].replace('.meta', '')

last_model_path = os.path.join(model_dir, last_model)
print(last_model_path)
!python /content/models/research/object_detection/export_inference_graph.py \
    --input_type=image_tensor \
    --pipeline_config_path={pipeline_fname} \
    --output_directory={output_directory} \
    --trained_checkpoint_prefix={last_model_path}
```

Step 6: *Inference test run*

Test with images in repository object_detection_demo/test directory.

```
import os
import glob

# Path to frozen detection graph. This is the actual model that is used for the object detection.
PATH_TO_CKPT = project_path + 'fine_tuned_model/frozen_inference_graph.pb'

# List of the strings that is used to add correct label for each box.
PATH_TO_LABELS = project_path + 'fine_tuned_model/label_map.pbtxt'

# If you want to test the code with your images, just add images files to the PATH_TO_TEST_IMAGES_DIR.
PATH_TO_TEST_IMAGES_DIR =  os.path.join(repo_dir_path, "test")

assert os.path.isfile(PATH_TO_CKPT)
assert os.path.isfile(PATH_TO_LABELS)
TEST_IMAGE_PATHS = glob.glob(os.path.join(PATH_TO_TEST_IMAGES_DIR, "*.*"))
assert len(TEST_IMAGE_PATHS) > 0, 'No image found in `{}`.'.format(PATH_TO_TEST_IMAGES_DIR)
print(TEST_IMAGE_PATHS)
```

4 Results and Discussion

4.1 Database Description (Table: Dataset, Images, Size, Classes)

The model, i.e., single shot detector (SSD), trained on the Common Objects in Context (COCO) dataset is used in this paper to detect the toxic object, i.e., cigarette/cigar. Normally, it is recommended to use TensorFlow for the abstraction layer of implementing machine learning algorithms and Python is being used to deal with nodes and tensors. For instance, photos need to be taken from all sides of the entity of interest. The data for our purpose will consist of:

Image category	Quantity	Classes
Training	1500	Cigarette/no cigarette
Testing	270	Cigarette/no cigarette

The dataset of intent contains two types of data, namely for positive and negative test cases. These datasets help the algorithm to learn different types of dataset to help for better predictions. Positive test cases entail images which explicitly display the object of desire (as stated above) and will be labeled accordingly, whereas negative test cases will be divided into two subgroups (for the purpose of increased accuracy on test data)—images which do not contain the object of desire in its entirety, i.e., no cigarette and/or smoke vapor present in image and the second being images which contain merely smoke but no cigarette (Fig. 9).

Data was formatted into appropriately preprocessed floating-point tensors before being fed into the network. The data sits on a drive as JPEG files, so the steps that were followed for getting it into the network are as follows:

- Read the image files.
- Decoded the JPEG content to RGB grids of pixels.
- Converted these into floating-point tensors.
- Resized the images.
- Rescaled the pixel values (between 0 and 255) to the [0, 1] interval.
- Annotated resized images with proper labeling.
- Created the train and test folders for images.

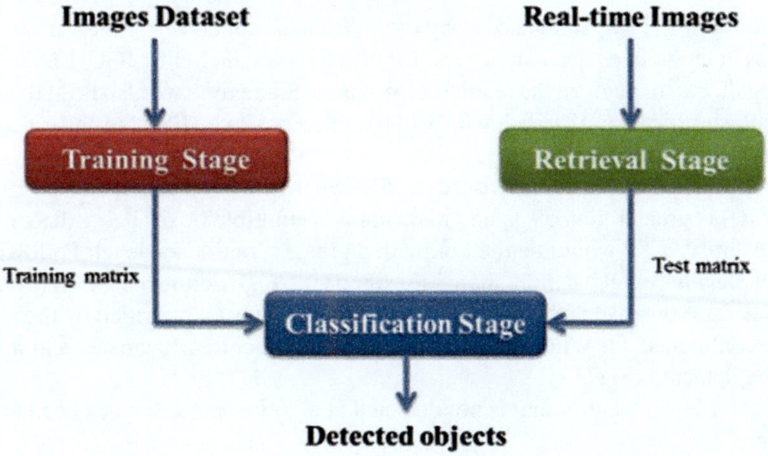

Fig. 9 Data Preprocessing

4.2 Generate the TF File

Test and training data are separated and TF-records can be generated from these files. A TF file is a binary file format that ables to combine multiple datasets and provides performance benefit compared to individual images, since it is more lightweight, thus using less disk space and can be read faster and process faster ways. The resulting files would be in a similar structure as shown below, allowing us to generate the TF file. The script used to generate these records is provided by the object detection dataset tools. The one that is applicable for the COCO dataset as base cane be found in:

--data/
--test_labels.csv
--test.record
--train_labels.csv
--train.record
--object-detection.pbtxt -images/
--test/
---[image.jpg]
--train/
---[image.jpg] -training
--ssd_mobilenet_v2.config

4.3 Primary Results (Correct and Incorrect Results)

It is important to analyze the accuracy of the trained model. There are various such methods or techniques that can be used for the analyses, including IOU, MAP, precision, recall, etc. to analyze the result performance. Same time we also need to understand the significance of true positive (TP), false positive (FP) and false negative (FN) in our mentioned case:

TP: A true positive is defined as a detection where the intersection area over union area (IOU) is greater than 50%, and in the case of multiple boxes being detected for a single entity, a TP would be the box with a greater confidence level. Following is an example where IOU is more than 50% signifying high confidence level (Fig. 10).

FP: A false positive is the predicted boxes that have been preceded by those with greater confidence. Or When an entity is completely incorrectly labeled and it is the only one detected (Fig. 11).

FN: An FN is an entity that is not detected at all, even though it has been trained for (Fig. 12).

Fig. 10 True positive case

4.4 Discussion of Results (Inferences)

An increase in precision generally means that the recall decreases, or the other-way around, i.e., with a low threshold value, lots of false positives are detected, whereas with a high threshold value, not as much false positives are detected, but typically results in certain entities not being detected at all. Test with images in repository object_detection_demo/test directory.

```
import os
import glob

# Path to frozen detection graph. This is the actual model that is used for the object detection.
PATH_TO_CKPT = project_path + 'fine_tuned_model/frozen_inference_graph.pb'

# List of the strings that is used to add correct label for each box.
PATH_TO_LABELS = project_path + 'fine_tuned_model/label_map.pbtxt'

# If you want to test the code with your images, just add images files to the PATH_TO_TEST_IMAGES_DIR.
PATH_TO_TEST_IMAGES_DIR =  os.path.join(repo_dir_path, "test")

assert os.path.isfile(PATH_TO_CKPT)
assert os.path.isfile(PATH_TO_LABELS)
TEST_IMAGE_PATHS = glob.glob(os.path.join(PATH_TO_TEST_IMAGES_DIR, "*.*"))
assert len(TEST_IMAGE_PATHS) > 0, 'No image found in `{}`.'.format(PATH_TO_TEST_IMAGES_DIR)
print(TEST_IMAGE_PATHS)
```

The lower the loss value, the better trained the model is.

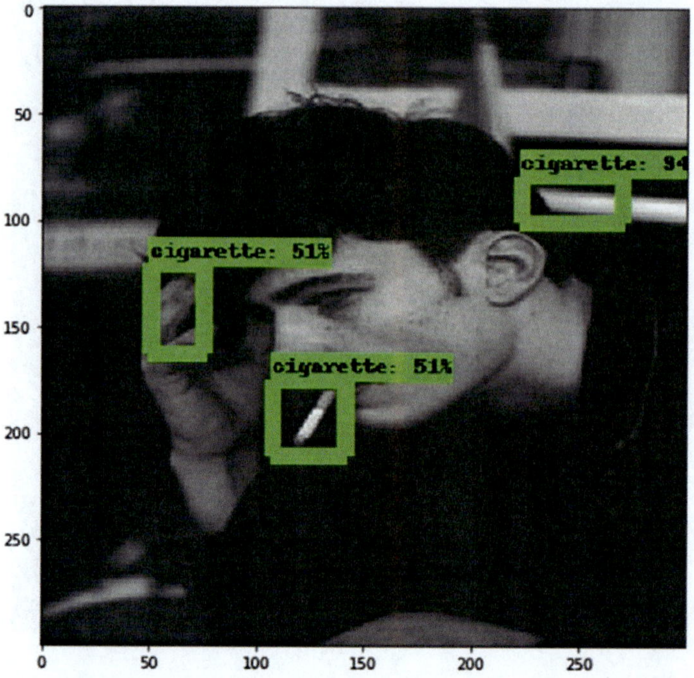

Fig. 11 False positive case

4.5 Result Analysis (Graphs and Table) Confusion Matrix

Below shows the training result of the total loss function. This is an indication of the total number of losses, and the goal would typically be for the training algorithms to reach a loss of about 1%. It may, however, end up that a few hundred thousand more steps do not necessarily result in a lower loss percentage (Tables 1 and 2; Figs. 13 and 14).

5 Conclusion

After having satisfactory training results, it is possible to export a Tensorflow graph proto file from a certain checkpoint, which can finally be used for object detection. Each checkpoint reflects different losses, etc. Model training was analyzed by both fine tuning and learning models from scratch. Fine tuning allowed objects to be trained relatively quickly (a few hours on a GPU) and detected objects with great confidence. The problem, however, is that it would do lots of false positive detections, i.e., random other types of objects were detected (i.e., the two classes that have been trained), along with the actual object. In an attempt to solve this, training images

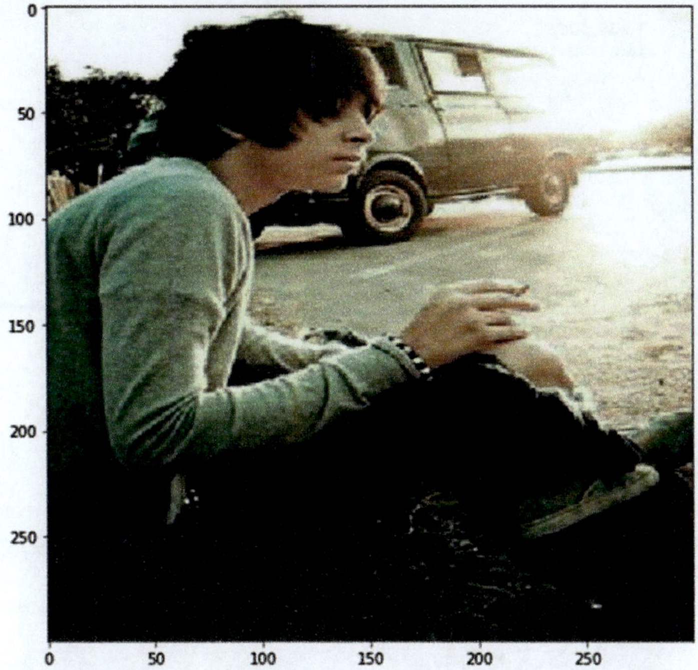

Fig. 12 False negative case

Table 1 Confusion matrix

Class	Cigarette	No cigarette
Cigarete	**792**	330
No cigarette	98	**280**

Table 2 Evaluations: cigarette detection

Evaluations	Percentage
Model accuracy	71.45
Misclassification rate	28.54
True positive rate	52.80
True negative rate	18.66

were presented with completely different backgrounds and image augmentation was implemented in order to learn the model to distinguish an object from different backgrounds (i.e., negative training). This showed better results, but some difficulty was still experienced with a couple of false positive detections.

Training the model from scratch with various backgrounds was then attempted—this showed even better results, although there was still some difficulty with lots of false positive detections for other types of objects, little false positives were detected.

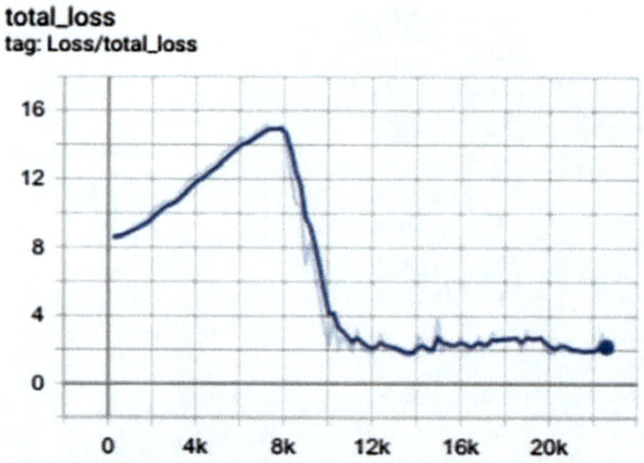

Fig. 13 Total loss versus steps

Fig. 14 Object detected with high accuracy

An automated recognition system is used to identify images with smoking activities and tag them accordingly. The used approach in certain cases increases the accuracy to a good level, with more images as well as the training steps.

Acknowledgements We would like to thank *Ms Gurvani Bassi* for giving us the opportunity to undertake this ambitious project as well as for facilitating interactions with our mentors in between their busy schedule.

References

1. The cigarettes and other tobacco products (Prohibition of advertisement and regulation of trade and commerce, production, supply and distribution) Rules, 2004
2. J. Deng, W. Dong, R. Socher, L.-J. Li, K. Li, and L. Fei-Fei, ImageNet: a largescale hierarchical image database. In: Proceedings of IEEE computer vision and pattern recognition (CVPR), Miami, Florida, USA, 2009
3. T.-Y. Lin, M. Maire, S. Belongie, J. Hays, P. Perona, D. Ramanan, et al (2014) Microsoft coco: Common objects in context. In: Computer vision–ECCV 2014, ed, Springer, pp 740–755
4. Krizhevsky A, Sutskever I, Hinton GE (2012) ImageNet classification with deep convolutional neural networks.In NIPS, p 4
5. Deep learning in neural networks: an overview. Technical report IDSIA-03–14/arXiv:1404.7828 v4 [cs.NE] (88 pages, 888 references)
6. ImageNet classification with deep convolutional network. https://yann.lecun.com/exdb/publis/pdf/lecun-01a.pdf
7. Tynan MA et al (2017) Tobacco use in top-grossing movies—United States, 2010–2016. In: 66, pp. 681–686.
8. Surgeon General (2012) U.S. Department of Health and Human Services. Preventing Tobacco Use Among Youth and Young Adults: A Report of the Surgeon General. Tech. rep. U.S. Department of Health et al., 2012
9. World Health Organization (2015) Smoke-free movies: from evidence to action, 3rd edn. Techchnical report 2015. https://www.who.int/iris/handle/10665/190165
10. MobileNetV2: Inverted Residuals and Linear Bottlenecks. arXiv:1801.04381v4 [cs.CV] 21 Mar 2019
11. Smoking Detection in Video Footage. https://scss.tcd.ie/publications/theses/diss/2018/TCDSCSS-DISSERTATION-2018-009.pdf.
12. Viola P, Jones M (2001) Robust Real-time Object Detection. Int J Comput Vision
13. Kolkur S et al (2017) Human skin detection using RGB, HSV and YCbCr Color Models. In: ArXiv e-prints. arXiv: 1708.02694 [cs.CV].
14. Stauffer C, Grimson WEL (1999) Adaptive background mixture models for real-time tracking. In: Proceedings of 1999 IEEE computer society conference on computer vision and pattern recognition (Cat. No PR00149). vol 2. p 252. vol 2. doi: https://doi.org/10.1109/CVPR.1999.784637
15. Yuan F (2008) A fast accumulative motion orientation model based on integral image for video smoke detection. Pattern Recog. Lett. 29(7):925–932
16. Chen TH, Wu PH, Chiou YC (2004) An early fire-detection method based on image processing. In: Image Processing, 2004. ICIP '04. 2004 International Conference on. vol 3. Oct 2004, 1707–1710 vol 3. doi: https://doi.org/10.1109/ICIP.2004.1421401
17. Xu G, Zhang Y, Zhang Q, Lin G, Wang J, Deep domain adaptation based video smoke detection using synthetic smoke images. State Key Laboratory of Fire Science, University of Science and Technology of China, Hefei 230026, China
18. Dunne É, Dawson-Howe K (2018) Smoking detection in video footage. A dissertation submitted in partial fulfilment of the requirements for the degree of MAI (Computer Engineering), School of Computer Science and Statistics, University of Dublin, Trinity College, May, 2018

19. Lessons learned building a smoking detector for video with machine learning. https://towardsdatascience.com/lessons-learned-building-a-smoking-detector-for-video-with-machine-learning-5a0e93dfa8ec
20. Wu P, Hsieh JW, Cheng JC, Cheng SC, Tseng SY (2010) Human smoking event detection using visual interaction clues. In: 2010 international conference on pattern recognition